Physiological Mapping of the Cerebral Cortex in Neuropathology

Avraham Mayevsky · Amir Livnat · Revital Etziony ·
Avivit Mendelman

Physiological Mapping of the Cerebral Cortex in Neuropathology

Mitochondrial Function, CBF, Ionic Homeostasis and Electrical Activities

Springer

Avraham Mayevsky
Faculty of Life Sciences and the
Multidisciplinary Brain Research Center
Bar-Ilan University
Ramat-Gan, Israel

Amir Livnat
Faculty of Life Sciences and the
Multidisciplinary Brain Research Center
Bar-Ilan University
Ramat-Gan, Israel

Revital Etziony
Faculty of Life Sciences and the
Multidisciplinary Brain Research Center
Bar-Ilan University
Ramat-Gan, Israel

Avivit Mendelman
Faculty of Life Sciences and the
Multidisciplinary Brain Research Center
Bar-Ilan University
Ramat-Gan, Israel

ISBN 978-3-032-13018-1 ISBN 978-3-032-13019-8 (eBook)
https://doi.org/10.1007/978-3-032-13019-8

This Springer imprint is published by the registered company Springer Nature Switzerland AG
The registered company address is: Gewerbestrasse 11, 6330 Cham, Switzerland

Preface by Avraham Mayevsky

The title of the current book reflects the main scientific concept developed over more than 50 years of research activity, which started in 1972 in Philadelphia. I started my postdoctoral training with Prof. Britton Chance at the University of Pennsylvania, Philadelphia.

The first aim of my studies was to develop the technology of fiber optic-based NADH fluorometry that provides real-time information on mitochondrial activity in vivo. The initial data indicated that monitoring a single parameter in brain tissue in vivo is very important but also very limited in its capacity to explain brain function.

The next step in my work was a shift from a single parameter monitoring approach to multiparametric monitoring systems. This process was gradual and after few years a new experimental concept was developed and named Physiological Mapping of the Cerebral Cortex.

The idea was to simultaneously monitor various physiological parameters representing hemodynamics, mitochondrial function, ionic homeostasis, and electrical activities of the brain.

The idea is to compare and integrate various physiological activities in the minimal tissue volume monitored and not to map the brain and compare the physiological activities to other parts of the cerebral cortex, as done in anatomical mapping of one to two parameters in various parts of the cortex or brain.

Chapter 1 provides an historical background and explains the concept of physiological mapping. It also covers homeostasis of oxygen in the brain as well as the discovery of oxygen. The discovery of the mitochondria as well as its in vivo monitoring systems are presented.

Chapter 2 describes the biochemical and physiological activities of the brain that are presented in the other chapters of the book. Chapter 3 summarizes the results obtained using various models of brain ischemia. Chapter 4 is unique and describes oscillations recorded in parameters related to brain hemodynamics and energy balance under brain ischemia.

Chapter 5 provides detailed information on the pathophysiology of various models of brain ischemia.

Chapter 6 describes the involvement of nitric oxide (NO) in the regulation of brain physiological activities such as hypercapnia and hypoxia.

Chapter 7 describes the responses of the normal brain as recorded in the epileptic brain. It also discusses the interaction between epileptic activity and ischemia, hyperoxia, and anesthesia.

In Chap. 8, the effects of aging and traumatic brain injury on brain physiological function are discussed.

Acknowledgments

This book is dedicated to my dearest in-laws Haim Yaakov Shapira and Miriam (Kaplan) Shapira for their help, patience, and support for more than five decades. Without their encouragement, my academic achievements and writing of this book were impossible to be accomplished.

Also, I dedicate this book to my dearest wife, Zipora, and our four children, Amotz, Sarit, Hagit, and Shalev, who understood my devotion to the unusual career and enabled me to fulfill all of my obligations and academic activities.

My deepest appreciation goes to my parents, Shabtai and Rachel Mayevsky, for their devotion and support throughout my life span.

My appreciation also goes to the late Prof. Britton Chance, an outstanding scientist, teacher, and colleague for almost four decades of fruitful collaboration.

During the 50 years of research activities at Bar-Ilan University I trained many M.Sc. and Ph.D. students. I would like to thank all of them for their confidence and for the efforts they made during the experimental work, as well as in the preparation and publication of the manuscripts.

Special appreciation goes to a few of my graduate students who joined my research team after their graduation. Their assistance in training new undergraduate and graduate students is appreciated very much. Nili Zarchin, M.Sc., joined my laboratory in the early 1980. Judith Sonn, Ph.D., started her activities in my laboratory in the early 1990. Efrat Barbiro-Michaeli Ph.D. started her activities in the mid-1990 and collaborated with me until my nomination as Professor Emeritus at Bar-Ilan University more than 15 years ago. Over the past few years, she took part in the preparation of this book, and her assistance is sincerely appreciated.

Prof. Gennady Rogatsky immigrated to Israel in the early 1990 and was active in my laboratory until his retirement. He was involved in the research activities and helped me to establish the literature classification for this book. Special appreciation is extended for his contributions.

Special thanks and deep appreciation are extended to Eli Zimerman from Bar Ian University for his long-term support in the development and maintenance of various instruments in my laboratory, which enabled us to perform our studies.

I want to thank Mrs. Avrille Goldreich from the Faculty of Life Sciences at Bar Ian University for her many years of help and support in the preparation of my manuscripts, and especially this book.

Ramat-Gan, Israel Avraham Mayevsky

Preface by Amir Livnat

My academic journey began at Bar-Ilan University, Ramat Gan, Israel, where I completed my B.Sc. in multidisciplinary brain sciences, followed by M.Sc. and Ph.D. at the Faculty of Life Sciences and the Brain Research Center. During these years, I gained a strong foundation in physiology and neuroscience, combining theoretical knowledge with hands-on experimental research in brain science.

Alongside my academic studies, I served as a tutor and teaching assistant. This period provided me with important tools, sharpening my ability to communicate complex ideas clearly, to guide and inspire students, and to develop a disciplined and structured scientific mindset.

After completing my doctoral studies, I transitioned into the field of clinical research, where I accumulated broad experience across multiple therapeutic areas and trial phases. I worked in various positions of increasing responsibility in a variety of pharmaceutical companies.

In my current position as a Clinical Research Lead at Eli Lilly and Company, I bring together my academic background and industry expertise, contributing to clinical development programs. I see my professional path as a bridge between fundamental science and practical application in patient care. The skills instilled during my academic years—critical thinking, scientific rigor, and collaboration—continue to guide me in my daily work.

It is a privilege to contribute to this volume, which reflects the collective efforts of scientists dedicated to deepening our understanding of human physiology and disease. I hope the insights presented herein will serve as a valuable resource for future research.

Acknowledgments

I extend my heartfelt gratitude to my mentor, Prof. Avraham Mayevsky, for his invaluable guidance and mentoring during my M.Sc. and Ph.D. studies. His scientific insight and dedication have profoundly shaped my academic and professional path. I would also like to thank my colleagues and collaborators over the years, whose companionship and shared commitment to science have inspired me and enriched my journey.

To my beloved parents, Michael and Aliza Livnat, whose love and support have accompanied me throughout my life and shaped me into who I am today. I would also like to express my heartfelt gratitude to my brother, Ofir, for being a significant part of my life.

My deepest love and appreciation to my wife, Mira Livnat (Mandelbaum), for being my life partner, for her endless patience, encouragement, and faith in me. Mira and I shared an academic path at university; she continues to be my scientific collaborator and inspiration till this day.

To our children, Uri, Yonatan, and Shira, I thank you for filling our lives with joy and meaning. Your curiosity and love serve as daily inspiration.

Petah Tikva, Israel Amir Livnat

Preface by Avivit Mendelman

My academic journey toward advanced degrees began in the third year of my undergraduate studies, when I carried out my seminar project in Professor Avraham Mayevsky's laboratory. At that moment, I was drawn to the complexity of the human brain and excited to explore its activity in a deep and research-driven way.

Over the past two decades, I have been a researcher and lecturer in the fields of anatomy, physiology, and pathology, with extensive academic and practical experience. I hold a Ph.D. from Bar-Ilan University, where I investigated the effects of epilepsy on brain activity using real-time in vivo models, under the supervision of Professor Avraham Mayevsky. My post-doctorate research was conducted in collaboration with Bar-Ilan University and Tel Hashomer Hospital, during which I developed a novel screening model for PTSD medications.

Alongside my research work, I lecture at several academic institutions. In addition, I served as a project manager at Harlan Biotech, within a pharmacological research consortium funded by the Chief Scientist's Office. My areas of expertise include the development of preclinical models, physiological and metabolic analysis of the brain, research on neurological disorders such as epilepsy and Parkinson's disease, and implementation of international standards (GLP, OECD) in animal studies.

Throughout my career, I have been awarded several scholarships and research grants, including the pharmacological consortium framework.

Acknowledgments

I dedicate this work to my dear parents—my late father, Tzadok Hacham, and my mother, Dina Hacham, may she live a long life—who believed in me, supported me throughout my journey, and helped raise my children while I was studying.

To my husband, who has believed in me and supported me every step of the way, and to my beloved children—Harel, Idan, Chen, and Ravid—who are a source of my pride.

To Prof. Avraham Mayevsky—for his guidance, expertise, and above all, his kindness and humanity throughout the entire journey. There are very few good people in life like him who can influence us and shape the course of our lives.

I would also like to thank my colleagues in the laboratory for their collaboration and support—Mrs. Nili Zarchin, Dr. Esther Guggenheimer-Forman, Dr. Sigal Maylin, and Prof. Gennady Rogatsky.

Ramat-Gan, Israel Avivit Mendelman

Preface by Revital Etziony

Since childhood, I have been deeply curios about how living systems work. Biology, in all its forms, has always fascinated me.

This curiosity eventually led me to study life sciences at Bar-Ilan University. I earned my M.Sc. in life sciences under the supervision of Professor Avraham Mayevsky. My thesis focused on the involvement of nitric oxide in the regulation of blood flow and brain activity in vivo—a topic that inspired me from the beginning.

Working in the lab was both challenging and inspiring. Using a multiparametric measuring system developed there, I was able to follow several physiological processes by simultaneously monitoring different parameters in order to examine the metabolic, hemodynamic, ionic, and electrical changes occurring in the brain in real time. Seeing these interactions come to life and the way that one signal influenced another gave me a deep sense of wonder about the brain's complexity and its remarkable ability to stay balanced under various changes.

Working with Professor Avraham Mayevsky was a unique privilege. His intellectual rigor, openness to ideas, commitment to excellence, and generous guidance have deeply influenced my thinking and research approach

For me, contribution to this book fulfills a lifelong curiosity and reflects years of collaboration, learning, and shared exploration—and a belief that studying brain physiology is a way to understand life itself and understanding how living systems maintain balance and resilience under stress.

Acknowledgments

I am grateful to all those who contributed to my research: Professor Avraham Mayevsky for his mentorship, inspiration and support throughout this project and beyond; and Judith Sonn, Ph.D., for her instructions and guidance of my thesis work.

I dedicate this work to my beloved family: my parents, Georgette and Eliahu Kadussi, whose love and support have shaped me into the person I am today and encouraged me along the way; my husband Amir Etziony for his patience, belief in me and unwavering support through all the years; and my three children: Lior, Shachar, and Tomer, who fill me with endless hope and pride.

Kfar Saba, Israel Revital Etziony

Contents

Chapter 1
Historical Background and Introduction

Abstract In the first chapter, we will describe the introduction to the new book. The five main subjects that will be described are as follows:

1. Cerebral Cortex Physiological Mapping—CCPM—The Concept
2. Homeostasis of Oxygen in the Brain
3. The Discovery of Oxygen
4. The Discovery of the Mitochondrion
5. History of Mitochondrial Function and its Monitoring Systems

The first item in the list describes the concept of our monitoring approach. The second item on the list describes the concept of oxygen homeostasis in the body. The third and fourth items describe the discovery of oxygen and the mitochondrion. The last item describes the foundation of our knowledge regarding the activity of the mitochondria in vivo and the contribution of the late Prof. Britton Chance to our current understanding of the function of the mitochondria in vitro and, more importantly, under in vivo conditions.

Keywords Brain metabolism · NADH redox state · Tissue energy balance · NADH fluorescence · In vivo monitoring of mitochondrial NADH

1.1 Cerebral Cortex Physiological Mapping—CCPM—The Concept

Normal brain cortical activity is an integration of many biochemical and physiological processes, including hemodynamic, metabolic, ionic homeostasis and electrical activities. In order to evaluate the functional state of the brain, it is necessary to

A. Mayevsky et al., *Physiological Mapping of the Cerebral Cortex in Neuropathology*,
https://doi.org/10.1007/978-3-032-13019-8_1

monitor in real time as many parameters as possible. We developed the concept of **"Cerebral Cortex Physiological Mapping"** that describes the interrelations between the various parameters measured by the multiparametric monitoring system (MPA). We used these systems in experimental animal models as well as in neurosurgical patients.

As can be seen in Fig. 1.1, our approach aims to monitor, in real-time, a small volume of the cerebral cortex containing all the tissue elements that are parts of a typical functioning brain. We are interested in the microenvironment of the brain containing neurons, glia, synapses and the microcirculatory elements (small arterioles and capillaries). The various parameters and the technology developed are presented in Fig. 1.1. During the development process, we pursued the goal of being minimally invasive in terms of the penetration to the cortical tissue itself. It was obvious that the various probes could not monitor the same volume of tissue due to the size of each probe used. Therefore, we attempted to minimize the diameter of the various probes located in the MPA that had a 5–6 mm contact area with the cerebral cortex. In most of the perturbations used, such as global ischemia, anoxia, hypoxia or hemorrhage, most of the areas in the cortex will respond in the same way.

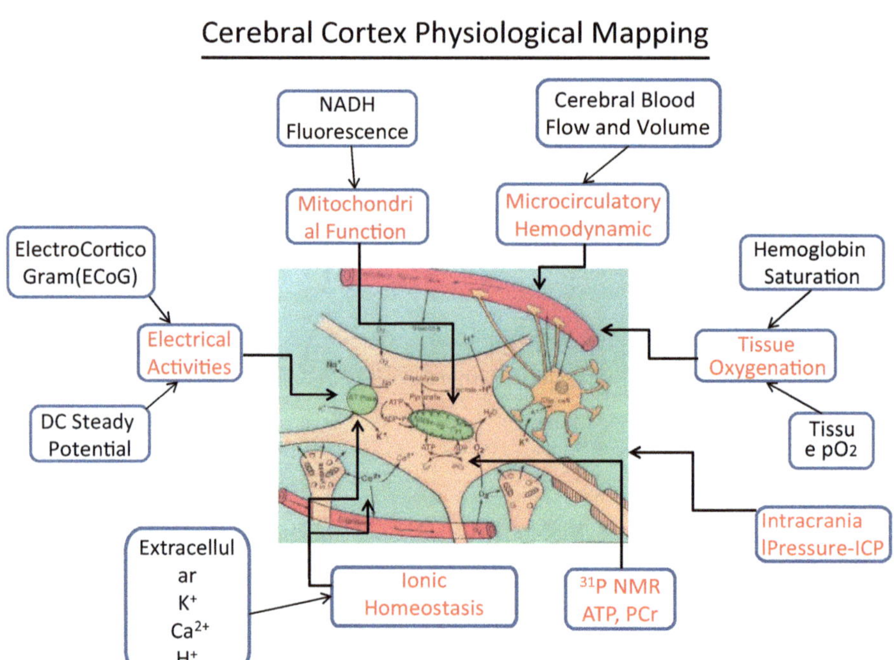

Fig. 1.1 Schematic presentation of the concept "Cerebral Cortex Physiological Mapping" used in our studies to be discussed in the current book. The available techniques for real-time monitoring of Cerebral blood flow, mitochondrial function, ionic homeostasis and electrical activities in the intact brain are presented (Mayevsky 2015). (Modified)

1.2 Homeostasis of Oxygen in the Brain

According to the Royal Society of Chemistry definition "Oxygen Homeostasis" could be described as follows: "A homeostatic process involved in the maintenance of an internal steady state of oxygen within an organism or cell."

The two physiologists presented in Fig. 1.2, namely, Claude Bernard and Walter Bradford Cannon, were the first scientists who described the concept of Homeostasis. Milieu intérieur was the key concept with which Bernard was associated. He wrote, "The stability of the internal environment [the *milieu intérieur*] is the condition for the free and independent life." This is the underlying principle of what would later be called homeostasis, a term coined by Walter Cannon.

Figure 1.3 demonstrates the relationship between oxygen supply and oxygen demand at the tissue level that could be evaluated by monitoring the redox state of the mitochondria. It is important to note that the mechanisms that regulate the oxygen supply to any tissue are common to all organs and tissues in the body. The process of using the oxygen or the energy in the various organs is different in the various organs and tissues as seen in the right side of Fig. 1.3.

Figure 1.4 demonstrates the components of the brain oxygen balance typical to the brain. As seen, the supply of oxygen is dependent on the microcirculatory blood flow and volume, oxygenation of hemoglobin and partial pressure of oxygen. The oxygen demand in the brain is shown on the right side of the figure and is used mainly for the electrical activities in the neurons and glia cells as well as for synaptic activities and ionic homeostasis.

This concept—oxygen homeostasis in the brain was investigated at the end of the nineteenth century and was published by Roy and Sherrington (1890). They studied

Fig. 1.2 The two physiologists (Claude Bernard and Walter Bradford Cannon) were the first scientists to describe the concept of Homeostasis

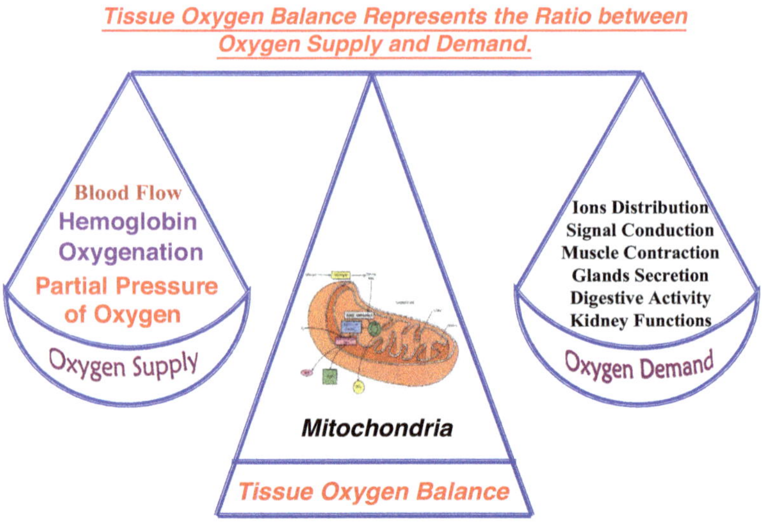

Fig. 1.3 Schematic presentation of tissue energy balance evaluated by energy supply and demand. Energy supply is measured by tissue blood flow (TBF), hemoglobin Oxygenation and partial pressure of Oxygen which is similar in all tissues. Energy demand varies between the different tissues and may include: Ionic Homeostasis, Signal Conduction, Muscle contraction, Glandular Secretion and G-I tract and kidney function. Mitochondrial NADH redox state serve as an indicator for tissue energy or oxygen balance (Mayevsky 2015)

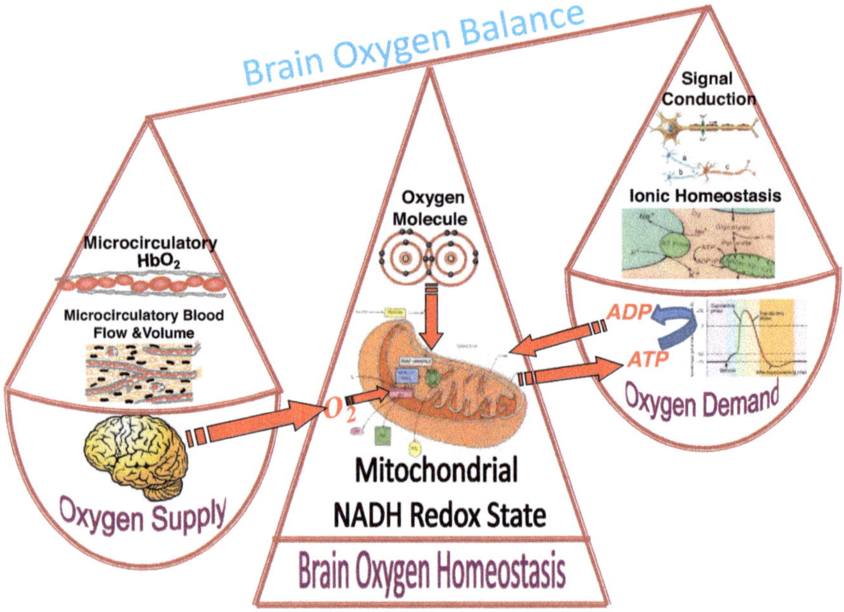

Fig. 1.4 Presentation of Brain Oxygen Balance or Brain Oxygen Homeostasis evaluated by the ratio between brain oxygen supply and brain oxygen demand (Mayevsky 2013)

the effect of various factors on the regulation of blood flow to the brain and came to the following conclusion:

> "We conclude then, that the chemical products of cerebral metabolism contained in the lymph which bathes the walls of the arterioles of the brain can cause variations of the caliber of the cerebral vessels: that in this reaction the brain possesses an intrinsic mechanism by which its vascular supply can be varied locally in correspondence with local variations of functional activity."

The interpretation of their conclusion is that, in the brain, changes in oxygen consumption due to changes in functional activity are due to variations in blood vessels diameter. The next step in the development of the concept Oxygen Homeostasis was made by the leading English physiologist Joseph Barcroft (Fig. 1.5) that published

THE RESPIRATORY FUNCTION
OF THE BLOOD

BY

JOSEPH BARCROFT, M.A., B.Sc., F.R.S.
FELLOW OF KING'S COLLEGE, CAMBRIDGE

Sir Joseph Barcroft 1872-1947.

Cambridge:
at the University Press
1914

Fig. 1.5 The title page of the book published by J. Barcroft (1914)

CAMBRIDGE UNIVERSITY PRESS
London: FETTER LANE, E.C.
C. F. CLAY, Manager

Edinburgh: 100, PRINCES STREET
London: H. K. LEWIS, 136, GOWER STREET, W.C.
WILLIAM WESLEY & SON, 28, ESSEX STREET, STRAND
Berlin: A. ASHER AND CO.
Leipzig: F. A. BROCKHAUS
New York: G. P. PUTNAM'S SONS
Bombay and Calcutta: MACMILLAN AND CO., Ltd.
Toronto: J. M. DENT AND SONS, Ltd.
Tokyo: THE MARUZEN-KABUSHIKI-KAISHA

Fig. 1.6 The publisher of the book by J. Barcroft (1914)

the book "The Respiratory Function of Blood" in (1914) seen in Fig. 1.6. In Chapter VI of the book, seen in Fig. 1.7 he discussed the subject of "The call for oxygen by the tissues." He proved that "There is no instance in which it can be proved that an organ increases its activity, under physiological conditions, without also increasing its demand for oxygen." Barcroft came to the second conclusion (seen in Fig. 1.8) that "in no organ excited by any form of stimulus can it be shown that positive work is done without the blood supply having to respond to a call for oxygen." Longo (2016) published a review entitled "Sir Joseph Barcroft: one Victorian physiologist's contributions to a half century of discovery." "Barcroft's scientific career may be divided into four major phases of inquiry: that of oxygen (O_2) consumption in tissues; the definition of the oxyhemoglobin saturation [HbO_2] curve; establishing the limits of human tolerance to high altitude; and pioneering the exploration of various aspects of oxygenation of the fetus in utero."

It is important to remember that Oxygen is a drug that may need to be optimized in the level given to patients thus avoiding under- as well as overdosing (Martin and Grocott 2013; Branson and Robinson 2011).

As seen in Figs. 1.3 and 1.4, the mitochondrial NADH redox state represents the tissue oxygen balance. The supply of oxygen is dependent upon microcirculatory blood flow (TBF), blood volume (TBV) and the level of oxygen bound to the

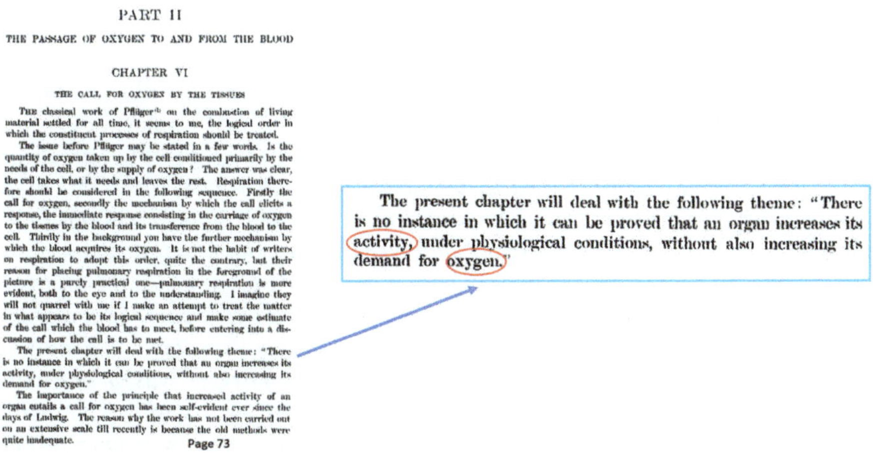

PART II

THE PASSAGE OF OXYGEN TO AND FROM THE BLOOD

CHAPTER VI

THE CALL FOR OXYGEN BY THE TISSUES

THE classical work of Pflüger[1] on the combustion of living material settled for all time, it seems to me, the logical order in which the constituent processes of respiration should be treated.

The issue before Pflüger may be stated in a few words. Is the quantity of oxygen taken up by the cell conditioned primarily by the needs of the cell, or by the supply of oxygen? The answer was clear, the cell takes what it needs and leaves the rest. Respiration therefore should be considered in the following sequence. Firstly the call for oxygen, secondly the mechanism by which the call elicits a response, the immediate response consisting in the carriage of oxygen to the tissues by the blood and its transference from the blood to the cell. Thirdly in the background you have the further mechanism by which the blood acquires its oxygen. It is not the habit of writers on respiration to adopt this order, quite the contrary, but their reason for placing pulmonary respiration in the foreground of the picture is a purely practical one—pulmonary respiration is more evident, both to the eye and to the understanding. I imagine they will not quarrel with me if I make an attempt to treat the matter in what appears to be its logical sequence and make some estimate of the call which the blood has to meet, before entering into a discussion of how the call is to be met.

The present chapter will deal with the following theme: "There is no instance in which it can be proved that an organ increases its activity, under physiological conditions, without also increasing its demand for oxygen."

The importance of the principle that increased activity of an organ entails a call for oxygen has been self-evident ever since the days of Ludwig. The reason why the work has not been carried out on an extensive scale till recently is because the old methods were quite inadequate.

Page 73

Fig. 1.7 Citation from the book published in 1914 by Barcroft regarding the call for oxygen by tissues at rest or during activation

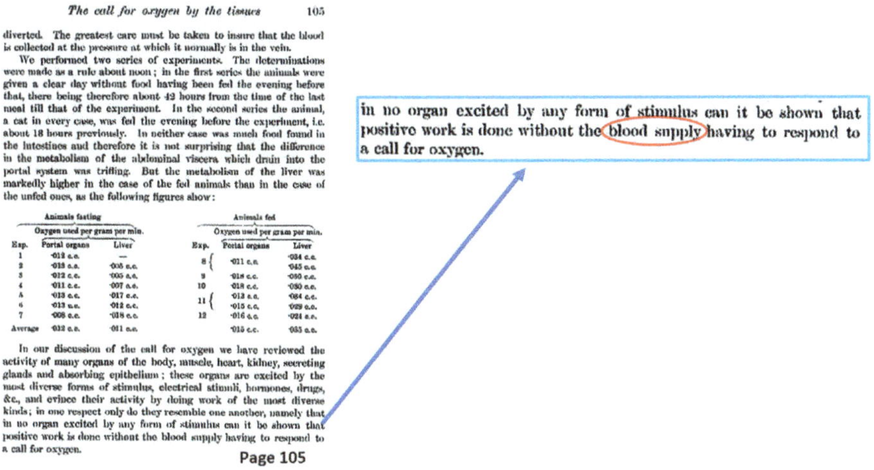

The call for oxygen by the tissues 105

diverted. The greatest care must be taken to insure that the blood is collected at the pressure at which it normally is in the vein.

We performed two series of experiments. The determinations were made as a rule about noon; in the first series the animals were given a clear day without food having been fed the evening before that, there being therefore about 42 hours from the time of the last meal till that of the experiment. In the second series the animal, a cat in every case, was fed the evening before the experiment, i.e. about 18 hours previously. In neither case was much food found in the intestines and therefore it is not surprising that the difference in the metabolism of the abdominal viscera which drain into the portal system was trifling. But the metabolism of the liver was markedly higher in the case of the fed animals than in the case of the unfed ones, as the following figures show:

Animals fasting			Animals fed		
	Oxygen used per gram per min.			Oxygen used per gram per min.	
Exp.	Portal organs	Liver	Exp.	Portal organs	Liver
1	·012 c.c.	—	8	·011 c.c.	·034 c.c.
2	·013 c.c.	·005 c.c.			·045 c.c.
3	·012 c.c.	·005 c.c.	9	·018 c.c.	·050 c.c.
4	·011 c.c.	·007 c.c.	10	·018 c.c.	·030 c.c.
5	·013 c.c.	·017 c.c.	11	·013 c.c.	·064 c.c.
6	·013 c.c.	·012 c.c.		·015 c.c.	·029 c.c.
7	·008 c.c.	·018 c.c.	12	·016 c.c.	·024 c.c.
Average	·012 c.c.	·011 c.c.		·015 c.c.	·035 c.c.

In our discussion of the call for oxygen we have reviewed the activity of many organs of the body, muscle, heart, kidney, secreting glands and absorbing epithelium; these organs are excited by the most diverse forms of stimulus, electrical stimuli, hormones, drugs, &c., and evince their activity by doing work of the most diverse kinds; in one respect only do they resemble one another, namely that in no organ excited by any form of stimulus can it be shown that positive work is done without the blood supply having to respond to a call for oxygen.

Page 105

Fig. 1.8 Page 105 copied from the book of Barcroft describing the increase in blood supply to an active tissue that consumes more oxygen (1914)

hemoglobin (HbO_2) in the small blood vessels, namely, in the microcirculation. The level of oxyhemoglobin in the microcirculation is affected by two factors, namely, oxygen consumption by the mitochondria and the microcirculatory blood flow. The demand for oxygen is affected by the specific activities taking place in each organ. The intracellular level of mitochondrial NADH (the reduced form) is a parameter related to oxygen balance.

1.3 The Discovery of Oxygen

In 2022, J.B. West reviewed the history of atmospheric oxygen as follows: *Many of us think a lot about oxygen. This includes how the normal body handles oxygen in health, but particularly how this is complicated by lung disease. Few of us are aware that as human inhabitants of the earth, we have a unique privilege. This is that as air breathers, we and most other animals on Earth, are the only living creatures in the known universe that have an unlimited supply of oxygen. This situation came about through one of the greatest miracles of nature, that is photosynthesis, the ability to release oxygen from water using the energy of sunlight. One consequence of this was that the first atmospheric oxygen came from the metabolism of microorganisms, the cyanobacteria, that used photosynthesis, but for which oxygen was an unwanted by-product. In fact, the oxygen had to be discarded for the organisms to thrive. When a major increase of oxygen concentration in the atmosphere occurred some 2 billion years ago, and the partial pressure of oxygen in the air rose to perhaps 200 mmHg, this Great Oxidation Event as it was called, was a death sentence for the large population of anaerobic animals for whom oxygen was toxic. Today much of the oxygen in the atmosphere comes from photosynthesis in microorganisms, including the cyanobacteria, and the recently discovered Prochlorococcus, that discard this unwanted by-product. The result is that the* PO_2 *in our atmosphere at sea level remains nearly constant at about 150 mmHg, although the factors responsible for this are not understood.*

The discovery of oxygen is a process that started more than 750 years ago (about 1250) and finished around 500 years later in 1780. There are at least nine names that are mentioned as related to this discovery and are presented in Figs. 1.9 and 1.10. The first five names contribute an indirect and minor part to the discovery process, while the last four scientists added the main information to the process of oxygen discovery within a 10-year period.

1.3.1 Ibn al-Nafis

The first one is Ibn al-Nafis, an Arab physician who was active in the thirteenth century (1213–1288). He was the first scientist who described the pulmonary circulation (West 2008; Haddad and Khairallah 1936; Loukas et al. 2008). West (2008) summarized the discovery as follows: *Ibn al-Nafis (1213–1288) was an Arab physician who made several important contributions to the early knowledge of the pulmonary circulation. He was the first person to challenge the long-held contention of the Galen School that blood could pass through the cardiac interventricular septum, and in keeping with this he believed that all the blood that reached the left ventricle passed through the lung. He also stated that there must be small communications or pores (manafidh in Arabic) between the pulmonary artery and vein, a prediction that preceded by 400 years the discovery of the*

Fig. 1.9 The pictures of the nine scientists who were involved and contributed to the discovery of oxygen (Severinghaus 2016a). [Modified]

pulmonary capillaries by Marcello Malpighi. Ibn al-Nafis and another eminent physiologist of the period, Avicenna (ca. 980-1037), belong to the long period between the enormously influential school of Galen in the 2nd Century, and the European Scientific Renaissance in the 16th Century. This is an epoch often given little attention by physiologists but is known to some historians as the Islamic Golden Age. Its importance is briefly discussed here.

1.3.2 Leonardo da Vinci

The second scientist who was cited as a discoverer of oxygen in 1515 is Leonardo da Vinci (Reti 1952; West 2017; Mitzner and Wagner 1992). As mentioned by West in (2017) regarding Leonardo, *Leonardo da Vinci (1452-1519) enjoys a reputation as one of the most talented people of all time in the history of science and the arts.*

Main Contributors to Oxygen Discovery	Discovery year
1. Ibn al Nafis (1213-1288).	1250
2. Leonardo da Vinci (1452-1519).	1515
3. Michael Servetus (1511-1553).	1552
4. Michael Sendivogius (1566- 1636).	1603
5. John Mayow (1641-1679).	1668
6. Carl Wilhelm Scheele (1742-1786).	1771
7. Joseph Priestley (1733-1804).	1774
8. Antoine Laurent Lavoisier (1743-1794).	1775
9. Henry Cavendish (1731-1810).	1783

Fig. 1.10 The list of 9 scientists who were involved in the discovery of oxygen according to Severinghaus (2016a)

However, little attention has been given to his contributions to physiology. One of his main interests was engineering, and he was fascinated by structural problems and the flow patterns of liquids. He also produced a large number of ingenious designs for warfare and a variety of highly original flying machines. But of particular interest to us are his contributions to bioengineering and how he used his knowledge of basic physical principles to throw light on physiological function. For example, he produced new insights into the mechanics of breathing including the action of the ribs and diaphragm. He was the first person to understand the different roles of the internal and external intercostal muscles. He had novel ideas about the airways including the mode of airflow in them. He also worked on the cardiovascular system and had a special interest in the pulmonary circulation. But, interestingly, he was not able to completely divorce his views from those of Galen, in that although he could not see pores in the interventricular septum of the heart, one of his drawings included them. Leonardo was a talented anatomist who made many striking drawings of the human body. Finally, his reputation for many people is based on his paintings including the Mona Lisa that apparently attracts more viewers than any other painting in the world.

1.3.3 Michael Servetus

Michael Servetus was the third person involved in the discovery of oxygen via description of the pulmonary circulation in 1552. Stefanadis et al. (2009) summarized his discovery as follows: "Michael Servetus was the first doctor ever to challenge and scientifically argue against the theories of Galen, which predominated for 14 centuries in medical schools worldwide. Even though he was relatively correct in scientific terms, Servetus was punished because of his boldness in challenging Galen's theories and was condemned to death by the Holy Inquisition. Yet, by publicly challenging Galen's and Hippocrates' predominant and unquestionable lessons on medicine for the first time, Servetus opened the door for other doctors to challenge and correct those theories and subsequently to bring about a new view of human anatomy and physiology. This article underlines the contribution of Servetus to the description of the pulmonary circulation."

1.3.4 Michael Sendivogius (1566–1636)

The fourth sage was a famous sixteenth-century Polish noble physician and alchemist, Michael Sendivogius. In 1604, he published that air contains what he named the "secret food of life." He had realized that this "food" was the same gas emitted by saltpetre (potassium nitrate) when heated. He called it part of the "saltnitre of the earth." Sendivogius's writings were frequently copied and read for over a century. However, no one grasped the chemical importance of his discovery, not even physicists Boyle and Newton at Oxford, who read his work 60 years later. Sendivogius' "food of life" story was only recently rediscovered. He discovered that air is not a single substance and contains a life-giving substance—later called oxygen—170 years before Scheele's discovery of the element.

1.3.5 John Mayow (1641–1679)

John Mayow was the fifth scientist who was involved in the discovery of oxygen. He was a chemist, physician, and physiologist who is remembered today for conducting early research into respiration and the nature of air. Mayow worked in a field that is sometimes called pneumatic chemistry. As a chemist and physiologist, he discovered, about a 100 years before Joseph Priestley and Antoine-Laurent Lavoisier, and identified spiritus nitroaereus (oxygen) as a distinct atmospheric entity. He also found out that animals take up the same gas in their blood during respiration and that the air exhaled by animals had a less amount of spiritus nitro-aereus gas than the one present in the open air.

Mayow's first experiment (as seen in Fig. 1.11) involved placing a candle in an inverted jar and placing the jar in a tub of water. The purpose of the water was to seal the vessel. To equalize the initial air pressure on both sides of the jar, Mayow used a straw, but quickly withdrew the straw once the pressures were equalized. As the candle burned, the water in the jar was drawn slightly upward until the candle was extinguished (because it had consumed the oxygen, not because the water had extinguished it). Mayow correctly explained that the air inside the jar had lost some of its "elasticity" and was no longer able to supply a force to the water equal to the atmospheric pressure. He suggested that this occurred because the air had been "deprived of nitro aerial and

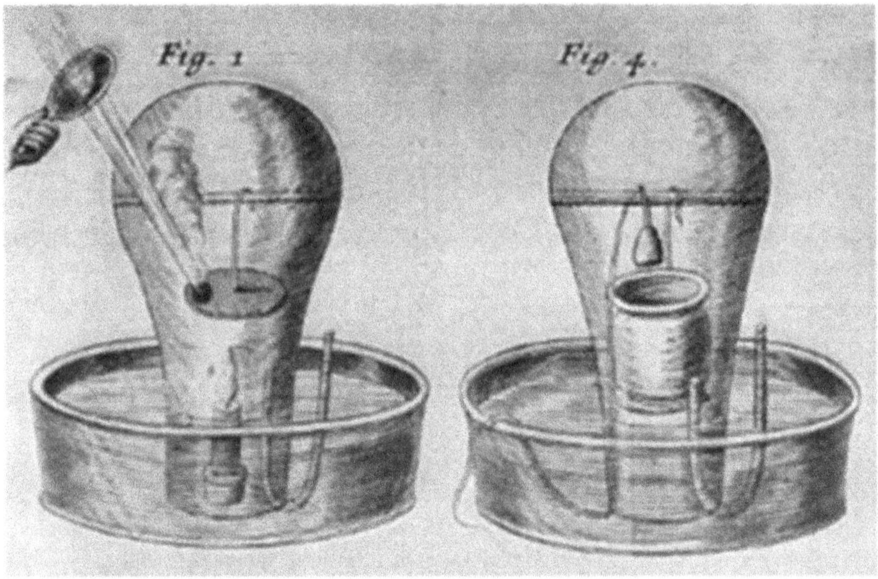

Fig. 1.11 From John Mayow experiments related to the discovery of oxygen (By John Mayow, 1674. Public domain…www.researchgate.net)

elastic particles" by the burning of the flame "so that the air there is not able as before to resist the pressure of the atmosphere."

Despite the critical need of oxygen and its significant use in body energy metabolism, the final steps of oxygen discovery in the atmosphere happened about 250 years ago (1771–1775). Until about 250 years ago, the story of oxygen discovery was described and discussed in terms of oxygen as a chemical element and the most important discovery of chemistry and science (Cassebaum and Schufle 1975; Smeaton 1992; Anon 1970; West 2014a; Marshall and Marshall 2005; Palmer 2006).

1.3.6 Carl Wilhelm Scheele (1742–1786)

The sixth discoverer of oxygen, Carl Wilhelm Scheele, was born in German-speaking Pomerania. Trained in Sweden, he became a very clever apothecary and chemist. In 1768, he published new studies of metal chemistry. In 1770, he became director of Locke Pharmacy in Uppsala. He worked at the university there with the famous chemist Torbern Bergman, who helped him publish his work in Latin in the journal Nova Acta. In 1774, Lavoisier sent his new chemistry textbook to Bergman, including a copy for Scheele, whose published work he much admired. Scheele promptly wrote Lavoisier to thank him for the book. He explained how, in 1771, he had generated a strange new gas by heating certain metallic earths. He had named it fire air because it greatly brightened a candle flame and supported life in mice. Because neither he nor Bergman could relate it to the phlogiston theory, Scheele had delayed publishing. He asked Lavoisier to repeat the experiment and then help him explain it.

The activity of Scheele was summarized by West (2014a) as follows: *Carl Wilhelm Scheele (1742–1786) has an important place in the history of the discovery of respiratory gases because he was undoubtedly the first person to prepare oxygen and describe some of its properties. Despite this, his contributions have often been overshadowed by those of Joseph Priestley and Antoine Lavoisier, who also played critical roles in preparing the gas and understanding its nature. Sadly, Scheele was slow to publish his discovery and therefore Priestley is rightly recognized as the first person to report the preparation of oxygen. This being said, the thinking of both Scheele and Priestley was dominated by the phlogiston theory, and it was left to Lavoisier to elucidate the true nature of oxygen. In addition to his work on oxygen, Scheele was enormously productive in other areas of chemistry. Arguably he discovered seven new elements and many other compounds. However, he kept a low profile during his life as a pharmacist, and he did not have strong links with contemporary prestigious institutions such as the Royal Society in England or the French Académie des Sciences. He was elected to the Royal Swedish Academy of Science but only attended one meeting. Partly as a result, he remains a somewhat nebulous figure despite the critical contribution he made to the history of respiratory gases and his extensive researches in other areas of chemistry. His death at the age of 43 may have been hastened by his habit of tasting the chemicals that he worked on.*

1.3.7 Joseph Priestley (1733–1804)

Oxygen's seventh discoverer was Unitarian minister Joseph Priestley. Two churches fired him as too radical, but he soon became famous while teaching grammar and science at Warrington Academy near Manchester, helping to make it a leading school for dissenters—in fact, the "cradle of Unitarianism," as one scholar called it. On August 1, 1774, Priestley, by heating red mercury calc, made a gas that caused a glowing splinter to burst into flame and supported life in a mouse in a bottle. He named it dephlogisticated air). In October 1774, in Paris, he described his method of making it to Lavoisier, who never acknowledged his help.

West in (2014b) summarized his activities as follows: *Joseph Priestley (1733–1804) was the first person to report the discovery of oxygen and describe some of its extraordinary properties. As such he merits a special place in the history of respiratory physiology. In addition, his descriptions in elegant 18th-century English were particularly arresting, and rereading them never fails to give a special pleasure. The gas was actually first prepared by Scheele (1742–1786) but his report was delayed. Lavoisier (1743–1794) repeated Priestley's initial experiment and went on to describe the true nature of oxygen that had eluded Priestley, who never abandoned the erroneous phlogiston theory. In addition to oxygen, Priestley isolated and characterized seven other gases. However, most of his writings were in theology because he was a conscientious clergyman all his life. Priestley was a product of the Enlightenment and argued that all beliefs should be able to stand the scientific scrutiny of experimental investigations. As a result, his extreme liberal views were severely criticized by the established Church of England. In addition, he was a supporter of both the French and American Revolutions. Ultimately his political and religious attitudes provoked a riot during which his home and his scientific equipment were destroyed. He therefore emigrated to America in 1794 where his friends included Thomas Jefferson and Benjamin Franklin. He settled in Northumberland, Pennsylvania although his scientific work never recovered from his forced departure. But the descriptions of his experiments with oxygen will always remain a high point in the history of respiratory physiology.*

1.3.8 Antoine Laurent Lavoisier (1743–1794)

Lavoisier had become the world's most brilliant chemist by the age of 30 years. In the spring of 1775, he began his studies of Priestley's newly published gas. He named it principe oxigene but continued to refer to it as "vital air." However, after 8 years of extensively examining its chemistry, he was still unable to prove whether vital air was a new element or a compound. And he doubted, but still couldn't disprove the phlogiston theory. By 1783, he was stalled by these dilemmas. In his Elements of Chemistry, Lavoisier falsely wrote that "This species of air was discovered almost at the same time by Mr. Priestley, Mr. Scheele, and myself," both a bold lie and proof that he had read Scheele's 1774 letter (Fig. 1.12b), using his method of making fire air while failing to acknowledge Scheele.

West in (2013) discussed and summarized the collaboration between Antoine and his wife Marie-Anne Lavoisier in the discovery of oxygen as well as the first measurements of human oxygen consumption as follows: *ANTOINE-LAURENT LAVOISIER (1743–1794) enjoys a reputation as one of the most eminent scientists of the late 18th Century* (Fig. 1.12a). *His wife, Marie-Anne Pierrette Paulze Lavoisier (1758–1836), collaborated in many of his studies, but her contributions have received relatively little attention. Antoine Lavoisier's interests ranged over a wide area of chemistry. One of his major advances was to clarify the concept of an element as a substance that could not be further broken down by chemical analysis. His extensive classification of known elements was described in his book Traité élémentaire de chimie (Elementary Treatise on Chemistry) published in 1789. In the life sciences Lavoisier was the first person to recognize the true nature of oxygen. This had been isolated by Joseph Priestley (1733–1804) and Carl Wilhelm Scheele (1742–1786), but its role in respiration was not understood because of confusion caused by the erroneous phlogiston theory. Antoine with the help of his wife demolished this. Lavoisier also clarified the similarities between respiration and combustion, and he was responsible for the first studies of human oxygen consumption under various conditions. Lavoisier was the first person to clearly state the role of oxygen, carbon dioxide, and nitrogen in respiration. Here he built on the previous work of other investigators particularly*

Marie Ann Lavoisier's nécessaires hid Scheele's letter for 219 years

Fig. 1.12 Antoine Laurent Lavoisier (1747–1794) with his wife Marie-Anne (1759–1836). On the far left is the drawing board used by Marie-Anne. On the far right is a glass jar containing water and resting on a porcelain dish. This jar was presumably used for collecting gas. Next to it is a narrow tube partly filled with mercury. This is an eudiometer for measuring the oxygen concentration of gases. To the left of that is a mercury receiver for collecting gas. On the floor, there is a large glass vessel that was used for experiments to make water from hydrogen and oxygen. Next to it is a hydrometer for measuring the specific gravity of fluids. This magnificent portrait by Jacques Louis David (1788) is in the Metropolitan Museum of Art, New York. Reproduced by permission. The Metropolitan Museum of Art, Purchase, Mr. and Mrs. Charles Wrightsman Gift, in honor of Everett Fahy, 1977 (1977.10). Image© The Metropolitan Museum of Art. [Modified]

Priestley. This English Nonconformist minister carried out an experiment in August 1774 when, on heating some red mercuric oxide, he found that a remarkable gas was produced. He stated "but what surprised me more than I can yet well express, was that a candle burned in this air with a remarkably vigorous flame… and a piece of red-hot wood sparkled in it" (Priestley, 1775). *Furthermore, he showed that a mouse was able to survive longer in this gas than in ordinary air, and he famously surmised that it might be useful for people with disease. Thus there was no doubt that Priestley had produced oxygen. However, unfortunately he did not understand its nature. Priestley was a follower of the phlogiston theory that had been promoted by Georg Ernst Stahl (1659–1734) and that stated that all combustible materials are composed of ash (calx) and phlogiston (Greek for inflammable), and that during burning, phlogiston escaped, leaving the dephlogisticated ash behind. This is in fact the reverse of what happens during combustion when oxygen combines with a combustible material, but the theory was enormously influential in the mid-18th Century. Priestley is often credited with "discovering" oxygen just as Columbus "discovered" America but neither Priestley nor Columbus correctly identified what they found. Carl Wilhelm Scheele (1742–1786) in Sweden had independently produced oxygen even before Priestley although the publication of his findings was delayed* (Scheele, 1777). *Scheele called the gas "fire-air" but again he was influenced by the phlogiston theory and did not recognize its true nature. Priestley visited Paris in October 1774 with his patron Lord Shelburne and they had dinner with Lavoisier and some other chemists. At that time Priestley described the experiments that he had carried out with mercuric oxide; naturally, these evoked great interest. As a result, Lavoisier repeated the experiments and was eventually able to understand the chemical processes. For example, in 1775 in what became known as the Easter Memoir, Lavoisier stated that "the substance which combines with metals during calcination, thereby increasing their weight, is nothing else than the pure portion of the air which surrounds us and which we breathe." This was the coup de grâce to the phlogiston theory. As we shall see later, Marie-Anne Lavoisier played an important part in this advance because it was she who translated the English texts of Priestley and another proponent of phlogiston, Richard Kirwan (1733–1812), so that Antoine could read them.*

1.3.9 Henry Cavendish (1731–1810)

In 1766, Cavendish discovered and published how to make an inflammable gas (H_2) by putting iron filings in strong acid. In the late 1770s, he and Priestley noted that burning inflammable air caused dew to form on glass walls. In 2014, J.B. West published a paper summarizing the activities of Gavendish as follows: *Henry Cavendish (1731–1810) was an outstanding chemist and physicist. Although he was not a major figure in the history of respiratory physiology he made important discoveries concerning hydrogen, carbon dioxide, atmospheric air, and water. Hydrogen had been prepared earlier by Boyle but its properties had not been recognized; Cavendish described these in detail, including the density of the gas. Carbon dioxide had also previously been studied by Black, but Cavendish clarified its properties and measured its density. He was the first person to accurately analyze atmospheric air and reported an oxygen concentration very close to the currently accepted value. When he removed all the oxygen and nitrogen from an air sample, he found that there was a residual portion of about 0.8% that he could not characterize. Later this was shown to be argon. He produced large amounts of water by burning hydrogen in oxygen and recognized that these were its only constituents. Cavendish also worked on electricity and heat. However, his main contribution outside chemistry was an audacious experiment to measure the density of the earth, which he referred to as "weighing the world." This involved determining the gravitational attraction between lead spheres in a specially constructed building. Although this was a simple experiment in principle, there were numerous complexities that he overcame with meticulous attention to experimental details. His result was very close to the modern accepted value. The Cavendish Experiment, as it is called, assures his place in the history of science.*

In many published papers and internet sites, the credit of oxygen discovery was given either to the American Joseph Priestley (Partington 1962; Miller 1987; Williams 2003) or to the German-Swedish Carl Wilhelm Scheele (Smeaton 1963; Partington 1962; Taylor 2019). The available information indicates that Scheele was the first one to discover oxygen in 1771–1772 but he published it officially a few years later in 1777. He claimed that in 1774, he sent a letter to the famous French chemist Antoine de Lavoisier, regarding his discovery, but Lavoisier claimed that he never received the letter. Priestley discovered oxygen and in 1774 he notified his findings to Antoine de Lavoisier who studied oxygen chemistry for a number of years. Priestley actually published his discovery in 1775. Wilkinson published a paper in (2004) describing the significant contribution of Lavoisier, Scheele and Priestley to the understanding of respiratory physiology in the eighteenth century. Here are his conclusions: "There appear to be a great desire in history to define who was the first to do something. In reality it makes no great difference to today's work who was first to define oxygen in the eighteenth century. What can be seen from the above is the ability of these three great men to set new paradigms and work to prove their theories. Some were more accurate than others but it is the ability to take that lateral step that sets these characters apart from their peers. It could be considered that if Scheele first isolated oxygen then Priestley defined its properties and Lavoisier showed it was an element but in fact they all added a significant set of pieces to the physiological jigsaw puzzle of which oxygen was just a part."

1.3.10 John Severinghaus (1922–2021)

In 1996, John Severinghaus (Fig. 1.13) became interested in the story of the discovery of oxygen and published his first paper on this subject (Severinghaus 2002). In the introduction to the cited paper, Severinghaus wrote the next paragraph: "AMERICANS are taught that the English Unitarian minister Joseph Priestley discovered oxygen in 1774. Scandinavians are taught that the Swedish apothecary Carl Wilhelm Scheele generated oxygen in Uppsala in 1771–1772, several years before Priestley. Scheele claimed that he wrote Lavoisier, describing the experiments in September 1774. However, Lavoisier denied seeing or receiving his letter. Among Nordic historians, Lavoisier has never been forgiven for this rebuff. Some historians have doubted whether such a letter was ever sent or received, because Scheele first published in 1777. Priestley brought the news of his discovery to Lavoisier in October and in publications overturning the conventional Phlogiston Theory, causing a revolution in chemistry."

One year later, Severinghaus published an extended paper on this topic (Severinghaus 2003) and came to the conclusion that Mary Ann Lavoisier received the original letter of Scheele and hid it until 219 years later the letter was found (Fig. 1.12b). Later, Severinghaus combined the discovery of Oxygen as an element in the eighteenth century to the description of the pulmonary circulation in the thirteenth century (Severinghaus, 2016a, 2016b). This subject was discussed in other papers describing the Islamic Golden age in the thirteenth century (West 2008; Haddad and Khairallah 1936).

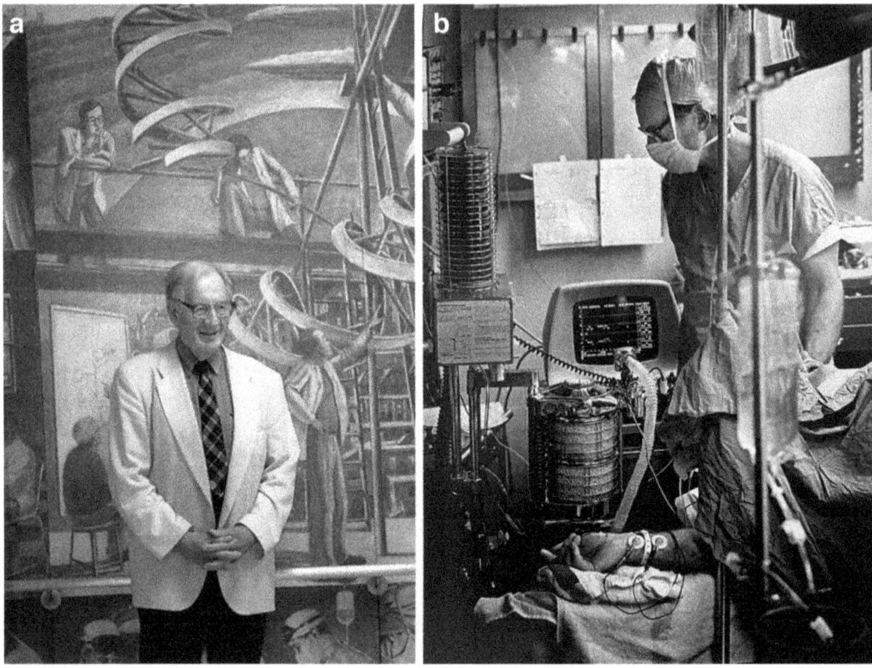

Fig. 1.13 (**a**) The "Master Tinker" and the blood–Gas analyzer. (**b**) John Severinghaus near a patient that was monitored in the hospital of the University of California, San Francisco, California, USA (From: John W. Severinghaus, M.D., 1922 to 2021, Anesthesiology. 2021; 135(4): 555-557.) [Modified]

1.4 The Discovery of the Mitochondrion

The understanding of mitochondrial function has been a challenge for various investigators, since its discovery more than 150 years ago. In 1981, Ernster and Schatz reviewed the history of mitochondrial structure and function studies (Scheele 1777). There is no real single answer regarding who discovered mitochondria. The process of discovery and identification was a gradual one, and four scientists were involved in the discovery of the mitochondria are presented in the next few figures.

The Swiss anatomist and histologist Albert von Kölliker (1856) shown in Fig. 1.14 described in 1856 what he called "granules" in the sarcoplasm of striated muscle (Fig. 1.14). According to A. L. Lehninger, "Kölliker was among the first to notice the arrangement of granules in the *sarcoplasm* of striated muscle over a period of years beginning around 1850. These granules were later called sarcosomes by *Retzius* (Fig. 1.15) in 1890. These sarcosomes have come to be known as the *mitochondria*—the power houses of the cell." In the words of Lehninger, "Kölliker should also be credited with the first separation of mitochondria from cell structure. In 1888, he teased these granules from insect muscle, in which they are very profuse, found them to swell in water, and showed them to possess a *membrane*." As mentioned,

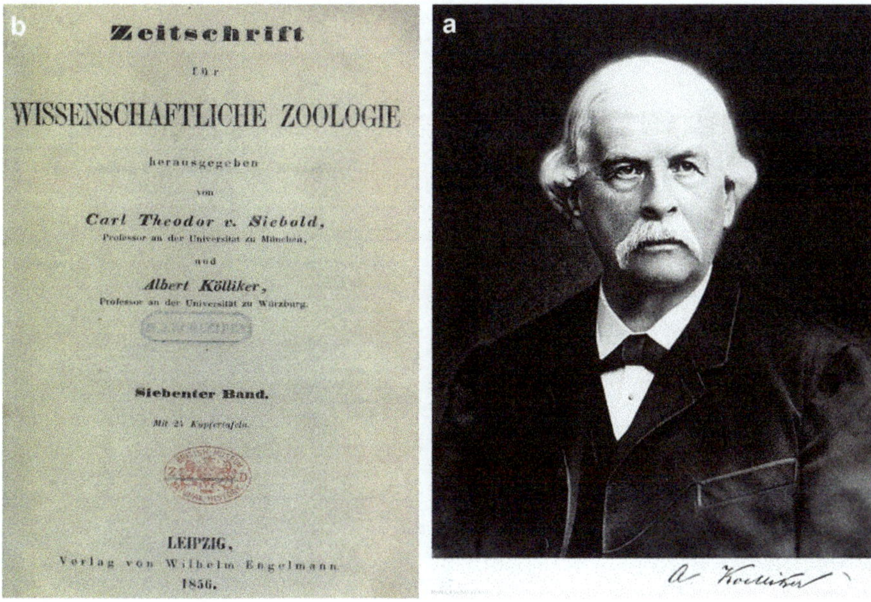

Fig. 1.14 Albert von Kölliker (**a**) who contributed initially to the discovery of the mitochondrion (**b**). [Modified]

Retzius in 1890 called the granules, Sarcosomes, which later on were identified as the mitochondria (Retzius 1890). Retzius published more than 300 scientific works in anatomy, embryology (Fig. 1.15), *eugenics*, *craniometry*, zoology and botany. He gave his name to the 60 μm-diameter Retzius cells in the central nervous system of the leech (*Hirudo medicinalis*). During his time at the Karolinska Institute, he made important contributions to anatomical descriptions of the muscles of the eardrum, the bones of the middle ear, and the Eustachian tube.

The discovery of mitochondria in general came in 1890 when Richard Altmann (Fig. 1.16a), a cytologist, identified the organelles and dubbed them "bioblasts" as presented in his publication seen in Fig. 1.16b (Altmann 1890). The fourth scientist involved in the discovery of the mitochondria was Carl Benda (Fig. 1.17), that in 1898, coined the term *mitochondria* from Greek thread, mitos, and granule, chondros (Benda 1898).

1.5 Mitochondrial Function and Its Monitoring Systems

In 2006, I (Avraham Mayevsky) asked Prof Britton Chance to write a short statement regarding the history of monitoring of NADH redox state. Here is Prof. Chance's response:

Fig. 1.15 Gustav Retzius and his publication that called the granules “sarcosomes.” [Modified]

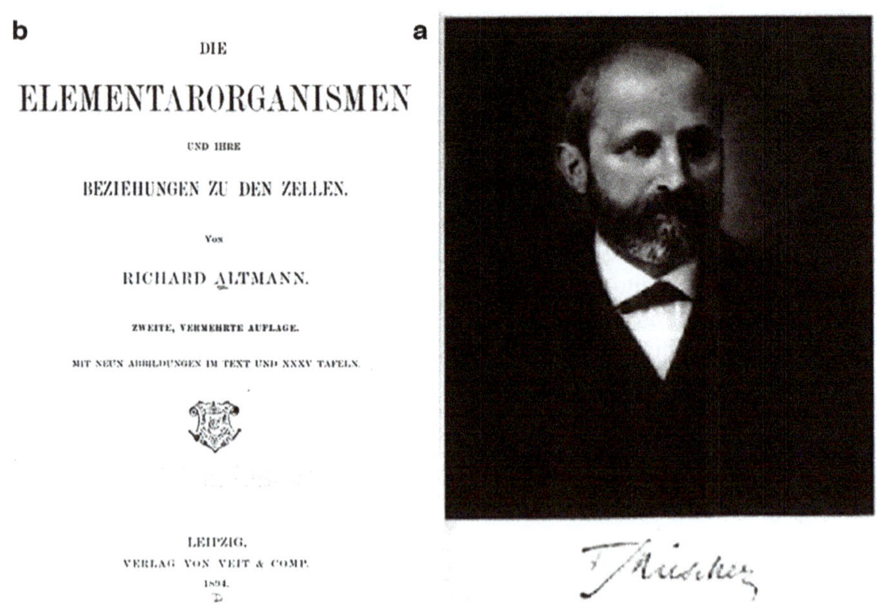

Fig. 1.16 (**a**) Richard Altman who gave the name “bioblasts” that later on were identified as the mitochondria. (**b**) The book was published by Altmann (1890). [Modified]

b

Carl Benda (30 December 1857 Berlin – 24 May 1932 Turin) was one of the first microbiologists to use a microscope in studying the internal structure of cells. In an 1898 experiment using crystal violet as a specific stain, Benda first became aware of the existence of hundreds of these tiny bodies in the cytoplasm of eukaryotic cells and assumed that they reinforced the cell structure. Because of their tendency to form long chains, he coined the name mitochondria ("thread granules"). These bodies had first been noted in 1857 by the physiologist and pioneer of the light microscope, Albert von Kölliker and were later termed "bioblasts" by Richard Altmann in 1886.

a

Fig. 1.17 Prof. Carl Benda (**a**) who coined the term mitochondria in 1898 (**b**). [Modified]

We will start by bringing a citation from Chance et al. in (1973) *"The accumulation of evidence since the pioneer work of Otto Warburg an "Atmungsferment"* (Warburg 1949*) and David Keilin on cytochromes (*Keilin, 1925, 1966*) as the keystones of cellular oxygen utilization led us to the study of the redox states of electron carriers in isolated mitochondria as a function of oxygen concentration and to develop techniques to measure the states of anoxia and normoxia in living tissues.*

*Although Warburg isolated the reduced di-phosphopyridine nucleotide (DPNH) and the reduced tri-phosphopyridine nucleotide (TPNH), he was apparently reluctant to imagine that those molecules were part of the respiratory chain. He also studied their optical properties and identified the 340 nm absorption band as being characteristic of the reduced form. The idea that these might also belong to Keilin's respiratory chain (*Keilin, 1925*) was not considered and in fact, was to some extent abhorrent to Warburg to imagine that he should support Keilin's views.*

Thus, the observation made spectroscopically in Chance's difference spectrum of the respiratory chain showing prominent peaks of what was soon to be called NADH and a trough due to flavoproteins and finding that they were respectively over 40 and over 10 times greater concentration than Keilin's cytochrome c was in a sense an overwhelming blow to Keilin. In my (Britton Chance) close contact with him, he never wished to discuss these two components as new and overwhelmingly present members of the "respiratory chain" (taking note that the Keilin and Hartree preparation (1945)*, was carried out in such a way that no trace of NADH survived the rigorous sand grinding and acid extraction). In fact, personal contact with Keilin subsequent to him being awarded the Copley Medal, evoked no response whatsoever of these novel components, which certainly must have been prominent in his mind. Nor in any conversation with him did he discuss Otto Warburg's discovery of these components.*

The blue autofluorescence of cells and tissues was recognized from the first studies of its fluorescence microscopy and was termed "autofluorescence." There was no attempt to characterize it or to do what was obvious to later researchers, which was to see whether oxygen or nitrogen affect the fluorescence, a study which Keilin relied upon to identify his cytochromes (the transition from trivalent to divalent caused a significant change in the electron configuration of the cytochromes). Since the oxidized form-NAD^+ is not absorbing light in this range, it was possible to evaluate the redox state of the mitochondrion by monitoring the UV absorbance or blue fluorescence of NADH.

Lubbers (1995) concluded that "the most important intrinsic luminescence indicator is NADH, an enzyme of which the reaction is connected with tissue respiration and energy metabolism."

Figure 1.18 presents the pictures of the five cited scientists that affected significantly the development of the theoretical and experimental technology for the monitoring of mitochondrial NADH function in vitro and in vivo.

The discovery of the optical properties of reduced Nicotine Amide Adenine Dinucleotide—NADH (earlier names: DPNH—diphosphopyridine nucleotide, or PN—pyridine nucleotide), has led to a very intensive research since the early 1950s. The reduced form of this molecule, NADH, shown in Fig. 1.19a, b (Kohen et al. 1969), absorbs light at 320–380 nm (Fig. 1.19c) and emits fluorescent light at 420–480 nm range (Fig. 1.19d) (Mayevsky and Barbiro-Michaely 2013a).

As the oxidized form NAD^+ does not absorb light in this range, it was possible to evaluate the redox state of the mitochondria by monitoring the UV absorbance or Blue fluorescence of NADH.

The foundations for the use of NADH fluorescence as a marker of mitochondrial activity were presented in detail by Chance and Williams (1955a, 1955b) for isolated mitochondria in vitro (see Fig. 1.20). The "resting" state of the mitochondria *in vitro* was defined as "State 4," where NADH was 99% in the reduced form, and ADP was the rate-limiting substance. If ADP is added to a suspension of mitochondria, ATP synthesis will be stimulated, oxygen consumption will increase and the rate-limiting factor will be the activity of the respiratory chain. During this State 3, or the "active state," the NADH redox state will decrease or become more oxidized (about 50%). When the "resting" mitochondria are deprived of oxygen, the activity of the mitochondria will stop and NADH will reach its maximum redox state (State 5).

The first detailed study using fluorescence spectrophotometry of NADH in intact Baker's yeast cells and Algae cells, was published in 1957 by Duysens and Amesz (1957) as seen in Fig. 1.21.

The use of NADH fluorometry for in vivo studies started at the end of the 1950s and has been presented in various publications. In a summative study, Chance compared spectra of NADH measured from rat kidney cortex to those of rat brain cortex (Fig. 1.22a, b) and found that the main effect of anoxic transition, both in the kidney and the brain, was a large increase in the fluorescence intensity with no detectable shift in the spectra (Chance 1967). Similar results were obtained by Chance (1966) when intact sartorius toad muscle was stimulated (Fig. 1.23a). Stimulation of the

Fig. 1.18 The five scientists who affected our understanding of mitochondrial function and NADH monitoring in vitro and in vivo situations. (**a**) Otto Heinrich Warburg (**b**) David Keilin (**c**) Britton Chance (**d**) Dietrich W. Lubbers (**e**) Frans F. Jobsis. [Modified]

muscle (State 4 to State 3 transitions) led only to a decrease in the intensity of the fluorescence emission spectrum.

Sundt and Anderson (1975a, 1975b) applied in vivo fluorometry to study the brain of the squirrel monkey (Fig. 1.23b), and recorded a greatly increased intensity in the dead brain as opposed to normal brain, with intermediate values in the ischemic brain. At the same time, Harbig et al. showed a clear increase in the NADH spectrum between normoxic and anoxic conditions in cat brain seen in Fig. 1.24a (Harbig et al. 1976).

Corderio et al. compared the excitation spectra in porcine Myo cutaneous flap with those of NADH in solution shown in Fig. 1.24b (Cordeiro et al. 1995). Exposing the flap to 6 h ischemia dramatically increased the intensity of 450 nm light. Furthermore, the present authors compared two sources of excitation light: the usual mercury arc and laser excitation at 324 nm (Mayevsky 1984). The main problem in applying such spectra, under in vivo conditions, is the effects of hemodynamic changes under anoxia. To overcome this problem, we induced repetitive cycles of anoxia that suppress the autoregulatory compensation mechanisms (Mayevsky 1984).

Figure 1.24c shows six scans obtained from the normoxic (A1, A2) and the anoxic brain (B1, B2) using an Hg arc, and comparative scans using Hg (C1) and the laser source (C2) in the dead brain. The results indicate that the location of the emission spectrum peak is identical in normoxia and anoxia, and is close when the laser is used. We also used the laser source in several in vivo studies and found that the light

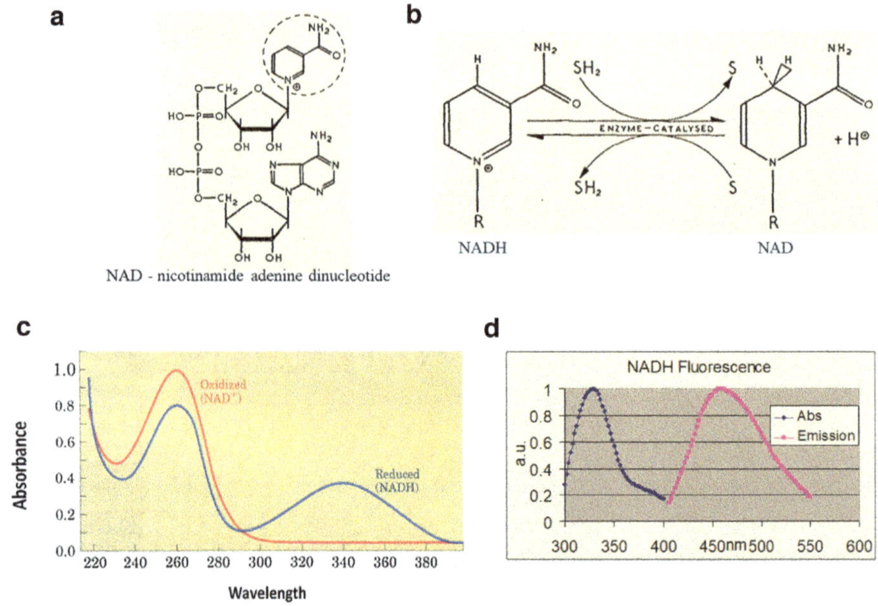

Fig. 1.19 (**a**) The structures of NAD^{+}. The nicotinamide group (broken ring) is the "functional" part of both molecules, i.e., the portion of the molecules where oxidation and reduction take place and (**b**) The transition between oxidized and reduced NADH. © Springer Science, reprinted with permission (Kohen et al. 1969). (**c**) Absorption Spectra of NAD^{+} and NADH. (**d**) Excitation and emission spectra of NADH. © Reprinted with kind permission of Springer Science + Business Media (Mayevsky and Barbiro-Michaely, 2013a, 2013b)

TABLE 2 · Metabolic States of Mitochondria and Associated Oxidation-Reduction Levels of Respiratory Enzymes

Characteristics						Steady state percentage reduction of components				
State	$[O_2]$	ADP level	Substrate level	Respiration rate	Rate-limiting substance	a	c	b	Flavo-protein[a]	$NADH_1$
1	>0	Low	Low	Slow	ADP	0	7	17	21	~90
2	>0	High	~0	Slow	Substrate	0	0	0	0	0
3	>0	High	High	Fast	Respiratory chain	4	6	16	20	53
4	>0	Low	High	Slow	ADP	0	14	35	40	99
5	0	High	High	0	Oxygen	~100	~100	~100	~100	~100

[a] These values are based on the amount of flavoprotein of the respiratory chain. Other fractions, reduced by anoxia, are present but disregarded here.
Source: After Chance and Williams [13].

Fig. 1.20 The definition of mitochondrial metabolic state, in vitro, in 1955, by Chance and Williams, opened up a new era in spectroscopic measurements of respiratory chain enzymes' redox state in vitro as well as in vivo (Chance and Williams 1956). [Modified]

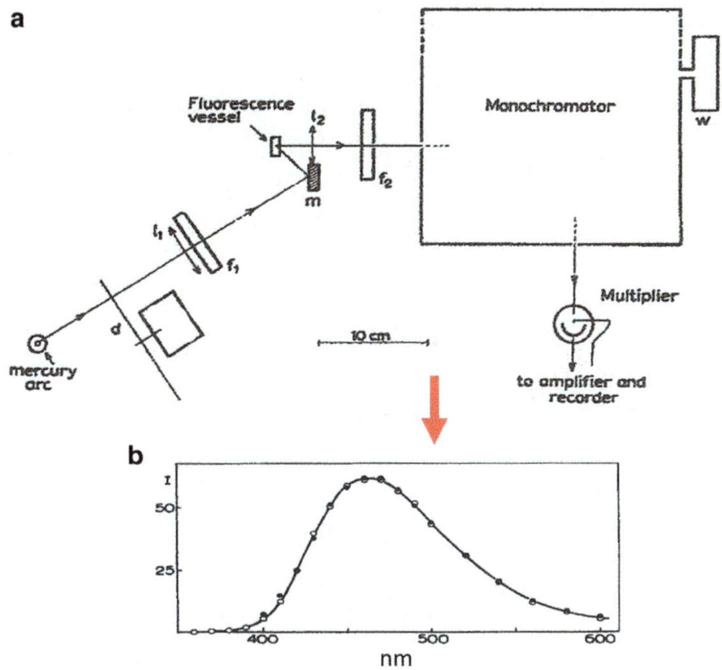

Fig. 1.21 (**a**) Apparatus for measurement of "fluorescence spectra." By means of a lens l_1 and a mirror in the mercury arc was imaged in a vessel containing the fluorescing solution or cell suspension. The fluorescing image was focused by lens l_2 on the entrance slit of the monochromator. Filter set l_1 isolated the spectral region of the exciting light, filter set l_2 absorbed the exciting radiation and transmitted light of longer wavelengths. The wavelength drum was driven by a synchronous motor. A sectioned disc *d*, mounted on another synchronous motor, modulated the incident light. The a.c. component of the current, caused by the light falling on the iP 21 multiplier, was amplified, rectified and fed into the recorder. (**b**) Fluorescence spectra of DPNH. The open circles are for excitation by wavelength 313 mμ, the black ones for excitation by 366 mμ. The spectrum for 313 mμ has been multiplied by a certain factor to make its maximum of equal height as the maximum of the spectrum excited by 366 mμ. The spectra appear to be identical; the maximum is at about 462 mμ (Duysens and Amesz 1957). [Modified]

intensity did not harm the brain during several hours of measurement. Under anoxia, a clear increase in NADH fluorescence spectra was recorded.

The first time that NADH was monitored in vivo in more than one site simultaneously was reported by Chance et al. (1962). They used two units of the same type of fluorometer placed on two organs in the same rat. Two organs were exposed in the same rat, namely the brain and the kidney. In this study they exposed the animal to various perturbations including anoxia and adrenaline injection. Figure 1.25a shows the probes located above the brain (right side) and the kidney (left side). Typical response of the kidney and the brain to anoxia is presented in Fig. 1.25b. The results were presented on one recorder therefore the quantitative comparison between the responses was possible.

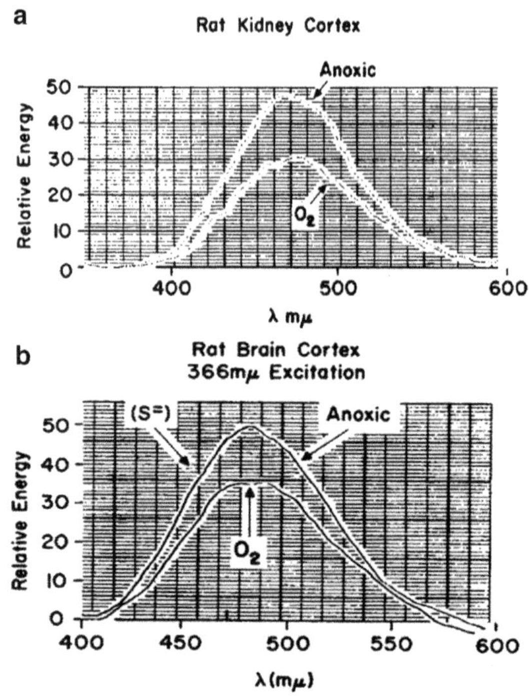

Fig. 1.22 (**a** and **b**) The emission spectra of rat kidney (**a**) and rat brain cortex (**b**), under aerobic conditions (lower trace) and under anoxic conditions (upper trace) (Chance 1967)

The responses of the brain to hypoxia is presented in Fig. 1.26b. According to the definition of Chance and Williams (1956, 1955a), a shift toward State 5 involves an increase in NADH proportional to a decrease in O_2 supply. In those days, the fluorescence was measured and displayed without correction for hemodynamic artifacts which was developed later in time. A clear increase in NADH fluorescence was recorded under the deprivation of oxygen.

In the early 1960s, Chance et al. (1963) tested the effect of irreversible ischemia on brain NADH using the decapitation model as seen in Fig. 1.26a. Here is the citation from the paper by Chance.

"One would expect that breathing nitrogen would lead to a very rapid transition from aerobiosis to anaerobiosis since it has already been pointed out in this symposium that breathing nitrogen would serve to deplete the blood of oxygen. As shown in Fig. 1.26a, about 8 s are required between the start of DPN reduction and the attainment of half-maximal reduction. In an attempt to observe the time for DPN reduction in ischemia, we have employed a decapitation technique with the mouse; the optical

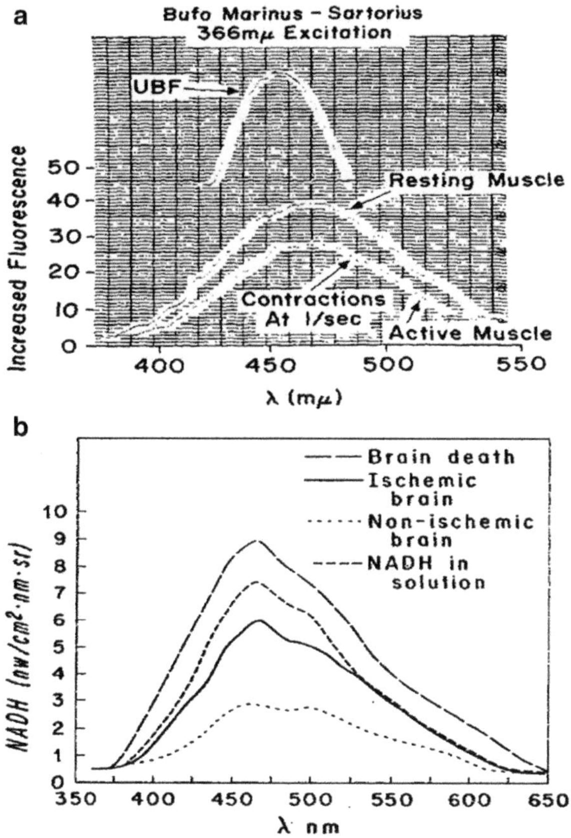

Fig. 1.23 (**a**) Emission spectra for the resting and stimulated sartorius muscle of the toad. Note that the spectrofluorometer is calibrated in a separate experiment by the emission of 4-methylumbelliferone shown at the top of the chart (UBF) (Chance, 1966). (**b**) Emission spectral fluorescence curves for NADH in solution, in normal brain, in ischemic cortex, and in cortex 5 min after death (Sundt and Anderson 1975a). [Modified]

system is arranged so that the slight mechanical artifact occurring on decapitation would not disturb the fluorescence excitation."

In order to enable the monitoring of NADH fluorescence in unanesthetized animals or other in vivo preparations, a flexible means was needed to connect the fluorometer with the tested organ, for example, the brain. This was achieved in 1972, when UV-transmitting quartz fibers became available (Schott Jena Glass, Germany). We have used the light-guide-based fluorometer for in vivo monitoring of the brain (Mayevsky and Chance 1973; Chance et al. 1973) subjected to anoxia or cortical spreading depression. The historical development of light-guide-based

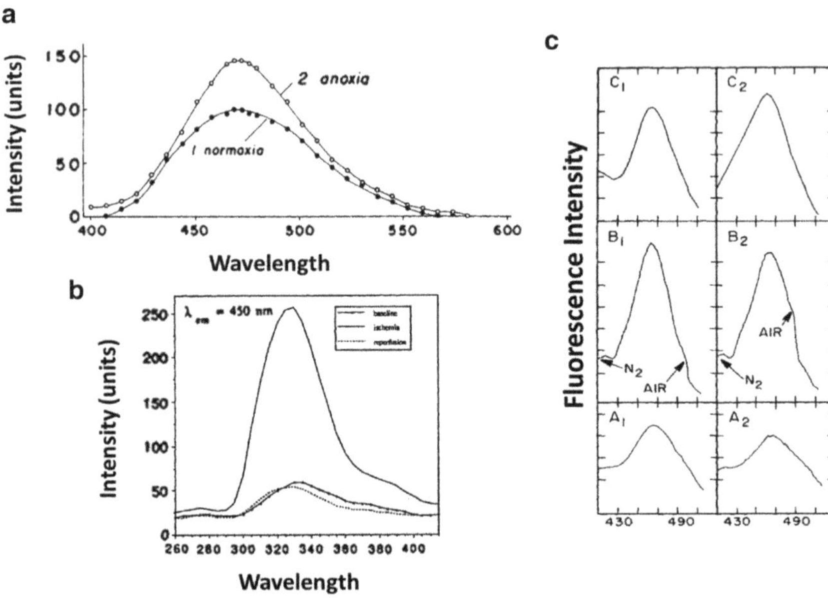

Fig. 1.24 (**a**) Pyridine nucleotide fluorescence emission spectra from cat brain cortex in vivo in (1) normoxia and (2) nitrogen anoxia. Both spectra peak at 469–473 nm. The spectra are not corrected for photocell sensitivity (Harbig et al. 1976). (**b**) Representative excitation spectra from an experimental myocutaneous flap at baseline, after 6 h of ischemia, and after 4 h of reperfusion. Fluorescence is recorded at a wavelength of 450 nm, and the excitation energy ranges from 260 to 430 nm. Excitation maximum of all spectra is 330–340 nm (Cordeiro et al. 1995). (**c**) Emission spectra of the brain under excitation of 366 nm light (A1, B2, B1, B2, and C1) or laser 324 nm light (C2). C1 and C2 were measured from a dead brain (Mayevsky 1984). [Modified]

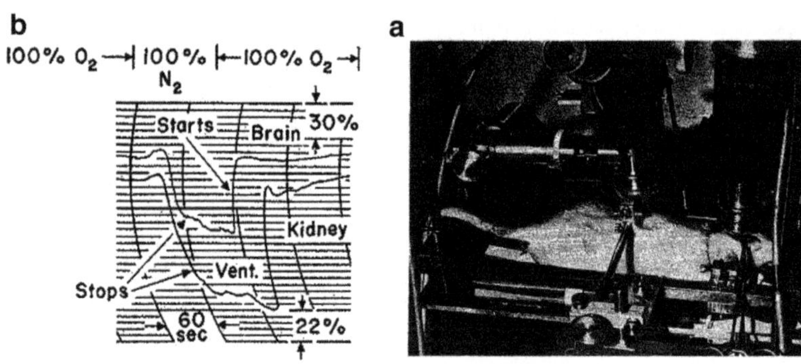

Fig. 1.25 (**a**) Experimental arrangement for simultaneous microfluorometry of brain and kidney cortex in the rat. Two microfluorometers are focused on the exposed surfaces of the brain and kidney. By means of a tracheal cannula the oxidation-reduction level of the intracellular pyridine nucleotide can be altered, and the corresponding fluorescence changes can be recorded by the two microfluorometers. For the kidney fluorometer, the water-cooled lamp housing attached to the Leitz "Ultrapak" illumination system is shown. Light choppers and photomultipliers are not shown (Chance et al. 1962). (**b**) Simultaneous recordings of fluorescence changes in rat brain and rat kidney in a cycle of anoxia. Fluorescence increases are indicated in a downward direction (Chance et al. 1963). [Modified]

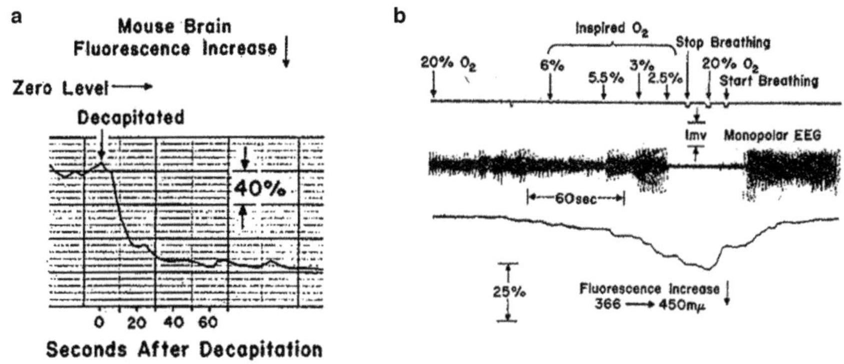

Fig. 1.26 (**a**) Record of the kinetics of fluorescence changes of the mouse brain cortex in ischemia caused by decapitation of a mouse (Chance et al. 1963). [Modified] (**b**) Correlation between EEG and pyridine nucleotide reduction in the slow transition from aerobiosis to anoxia. The top trace includes the measurements of inspired oxygen concentration according to the Pauling oximeter (Chance and Schoener 1962)

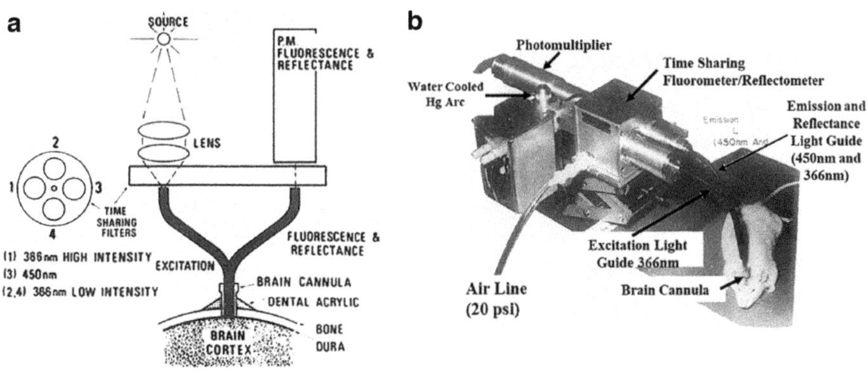

Fig. 1.27 (**a**) Time sharing fluorometer/reflectometer connected to a rat brain cannula by a flexible y-shape light guide (Mayevsky 1984). (**b**) Picture of the time-sharing fluorometer reflectometer used in the 1970th. The rat brain is connected to the device by a flexible fiber optic probe (Mayevsky 1984)

fluorometery-reflectometry is shown in Figs. 1.27 and 1.28. The original device functioned on the time-sharing principle (Fig. 1.27a), where four filters were placed in front of a two arms light guide. Filters one and three enabled the measurement of NADH fluorescence, while filters two and four were used to measure tissue reflectance at the excitation wavelength. The reflectance trace was used to correct the NADH signal for hemodynamic artifacts and to indicate changes in the blood volume of the sampled tissue.

In this original system, only one photomultiplier tube was used for the detection of the two signals. Figure 1.27b presents one of the first in vivo brain monitoring

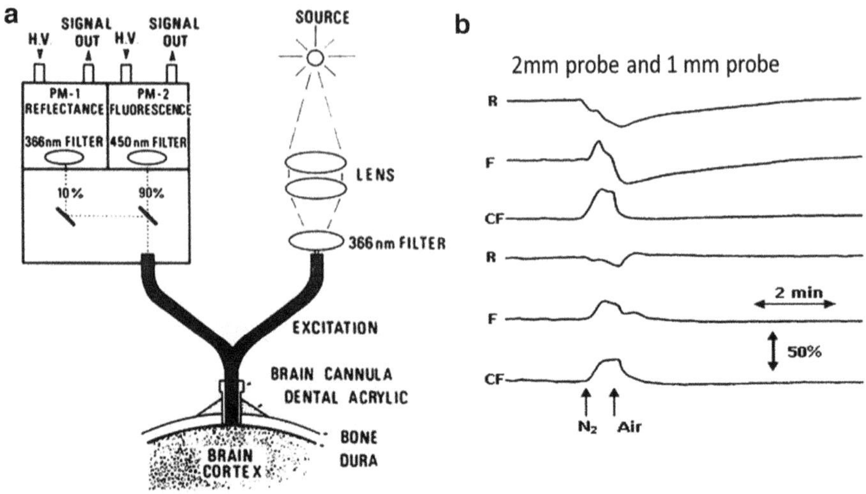

Fig. 1.28 (**a**) The structure of the standard DC fluorometer/ connected to the brain of a small animal (Mayevsky 1984). Excitation (right), Emission (left)-optical fibers for the monitoring of NADH redox state. HV high voltage, PM photomultiplier. (**b**) The effect of the monitored tissue volume (the diameter of the fiber optic probe) was tested under anoxia as shown in B, upper three traces (2 mm diameter) and lower three traces (1 mm diameter) (Mayevsky 1984)

time sharing setups, the time sharing approach (AC mode) was replaced by splitting the light emitted from the tissue into two unequal fractions for the measurement of fluorescence and reflectance signals. This model, named the DC type fluorometer, had originally a three-way light guide which was later replaced by a two arms light guide probe (Fig. 1.28a). In all three configurations, the reflectance signal was used for the correction of the fluorescence signal.

Our group developed and used the model shown in Fig. 1.28a in the late 1970s. This model is still being used in our laboratory to monitor the brain, heart, liver, and kidney, as also in multi-site or multi-organ monitoring. The responses to anoxia using different diameters of the fiber optic bundle are shown in Fig. 1.28b.

References

Altmann, R. 1890. *Die elementarorganismen und ihre beziehungen zu den zellen*. Leipzig: Veit & comp.

Anon. 1970. Carl Wilhelm Scheele (1742–1786) Swedish apothecary. *The Journal of the American Medical Association* 212: 2258–2259.

Barcroft, J. 1914. *The respiratory function of blood*. Cambridge: Cambridge University Press.

Benda, C. 1898. Ueber die spermatogenese der vertebraten und höherer evertebraten, II. Theil: die histiogenese der spermien. *Archiv für Anatomie und Physiologie* 73: 393–398.

Branson, R. D., and B. R. Robinson. 2011. Oxygen: When is more the enemy of good? *Intensive Care Medicine* 37: 1–3.

Cassebaum, H., and J. A. Schufle. 1975. Scheele's priority for the discovery of oxygen. *Journal of Chemical Education* 52: 442.

Chance, B., and B. Schoener. 1962. Correlation of oxidation-reduction changes of intracellular reduced pyridine nucleotide and changes in electro-encephalogram of the rat in anoxia. *Nature* 195: 956–958.

Chance, B., and G. R. Williams. 1955a. A method for the localization of sites for oxidative phosphorylation. *Nature* 176: 250–254.

Chance, B., and G. R. Williams. 1955b. Respiratory enzymes in oxidative phosphorylation (III-The steady state). *The Journal of Biological Chemistry* 217: 409–427.

Chance, B., and G. R. Williams. 1956. The respiratory chain and oxidative phosphorylation. In *Advances in enzymology*. XVII edition. Ed. Nord F. F. New York: Interscience Publisher, Inc.

Chance, B. 1966. The identification and control of metabolic states. In *Genootschap ter bevordering van natuur-, genees-, en heelkunde te Amsterdam,* 5–37.

Chance, B. 1967. *Biochemical studies of transitions from rest to activity*. Baltimore: The Williams & Wilkins Company.

Chance, B., P. Cohen, F. Jobsis, and B. Schoener. 1962. Intracellular oxidation-reduction states *in vivo*. *Science* 137: 499–508.

Chance, B., N. Oshino, T. Sugano, and A. Mayevsky. 1973. Basic principles of tissue oxygen determination from mitochondrial signals. *Advances in Experimental Medicine and Biology* 37A: 277–292.

Chance, B., B. Schoener, and F. Schindler, 1963. The intracellular oxidation-reduction state. In *Proceedings of a Symposium held under the joint auspices of the International Union of Biochemistry and the International Union of Physiological Sciences*, edited by Dickens F., and Neil E., 367–392. London: Pergamon Press.

Cordeiro, P. G., R. E. Kirschner, Q.-Y. Hu, J. J. C. Chiao, H. Savage, R. R. Alfano, L. A. Hoffman, and D. A. Hidalgo. 1995. Ultraviolet excitation fluorescence spectroscopy: A noninvasive method for the measurement of redox changes in ischemic myocutaneous flaps. *Plastic and Reconstructive Surgery* 96: 673–680.

Duysens, L. N. M., and J. Amesz. 1957. Fluorescence spectrophotometry of reduced phosphopyridine nucleotide in intact cells in the near-ultraviolet and visible region. *Biochimica et Biophysica Acta* 24: 19–26.

Haddad, S. I., and A. A. Khairallah. 1936. A forgotten chapter in the history of the circulation of the blood. *Annals of Surgery* 104: 1–8.

Harbig, K., B. Chance, A. G. B. Kovach, and M. Reivich. 1976. In vivo measurement of pyridine nucleotide fluorescence from cat brain cortex. *Journal of Applied Physiology: Respiratory, Environmental and Exercise Physiology* 41: 480–488.

Keilin, D., and E. F. Hartree. 1945. Purification and properties of cytochrome C. *The Biochemical Journal* 39: 289.

Keilin, D. 1925. On cytochrome, a respiratory pigment, common to animals, yeast, and higher plants. *Proceedings of the Royal Society of London. Series B, Containing Papers of a Biological Character* 98: 312–339.

Keilin, D. 1966. *The history of cell respiration and cytochrome*. Cambridge, UK: Cambridge University Press.

Kohen, E., C. Kohen, and B. Thorell. 1969. Use of microfluorimetry to study the metabolism of intact cells. *Biomedical Engineering* 554–565.

Kolliker, A. 1856. Einige Bemerkungen uber die endigungen der Hautnerven un deu bau der Muskelzellen. *Zeitschrift fur wissenschartliche Zoologie* 8: 311–325.

Longo, L. D. 2016. Sir Joseph Barcroft: One victorian physiologist's contributions to a half century of discovery. *The Journal of Physiology* 594: 1113–1125.

Loukas, M., R. Lam, R. S. Tubbs, M. M. Shoja, and N. Apaydin. 2008. Ibn al-Nafis (1210-1288): The first description of the pulmonary circulation. *The American Surgeon* 74: 440–442.

Lubbers, D. W. 1995. Optical sensors for clinical monitoring. *Acta Anaesthesiologica Scandinavica Supplementum* 39: 37–54.

Marshall, J. L., and V. R. Marshall. 2005. Rediscovery of the Elements: Carl Wilhelm Scheele. University of North Texas Libraries, UNT Digital Library. Spring 2005; Indianapolis, Indiana. Crediting UNT College of Arts and Sciences.

Martin, D. S., and M. P. Grocott. 2013. III. Oxygen therapy in anaesthesia: The yin and yang of O_2. *British Journal of Anaesthesia* 111: 867–871.

Mayevsky, A., and E. Barbiro-Michaely. 2013a. Shedding light on mitochondrial function by real time monitoring of NADH fluorescence: I. Basic methodology and animal studies. *Journal of Clinical Monitoring and Computing* 27: 1–34.

Mayevsky, A., and E. Barbiro-Michaely. 2013b. Shedding light on mitochondrial function by real time monitoring of NADH fluorescence: II: Human studies. *Journal of Clinical Monitoring and Computing* 27: 125–145.

Mayevsky, A., and B. Chance. 1973. A new long-term method for the measurement of NADH fluorescence in intact rat brain with implanted cannula. *Advances in Experimental Medicine and Biology* 37A: 239–244.

Mayevsky, A. 1984. Brain NADH redox state monitored *in vivo* by fiber optic surface fluorometry. *Brain Research Reviews* 7: 49–68.

Mayevsky, A. 2013. Shedding light on life: Optical assessment of mitochondrial function and tissue vitality in biology and medicine. *Journal of Clinical Monitoring and Computing* 27: 1–34.

Mayevsky, A. 2015. *Mitochondrial function in vivo evaluated by NADH fluorescence*. Switzerland: Springer International.

Miller, F. A. 1987. Joseph Priestley, preeminent amateur chemist. *Journal of Chemical Education*, 64:745.

Mitzner, W., and E. Wagner. 1992. On the purported discovery of the bronchial circulation by Leonardo da Vinci. *Journal of Applied Physiology: Respiratory, Environmental and Exercise Physiology (1985)* 73: 1196–1201.

Palmer, W. 2006. Carl Wilhelm Scheele (1742–1786): The life of a great chemist. *IPS-USA-2006 internet conference on advances in the internet, processing, systems and interdisciplinary research*. Palo Alto, California, USA.

Partington, J. R. 1962. The discovery of oxygen. *Journal of Chemical Education* 39: 123–125.

Priestley, J. 1775. *Experiments and observations on different kinds of air*. London: Johnson.

Reti, I. 1952. Leonardo da Vinci's experiments on combustion. *Journal of Chemical Education* 29: 590–596.

Retzius, G. 1890. Muskelfibrflle und Sarcoplasman. *Biologische Untersuchungen Stockholm, N.F.* 1: 51–88.

Roy, C. S., and C. S. Sherrington. 1890. On the regulation of the blood-supply of the brain. *The Journal of Physiology* 11: 85–158 17.

Scheele, C. W. 1777. *Chemische abhandlung von der luft und dem feuer (chemical studies on air and fire)*. Uppsala, Sweden and Leipzig, Germany: Verlegt von Magn. Swederus Buchhandler, zu finden bey.

Severinghaus, J. W. 2002. Priestley, the furious free thinker of the enlightenment, and Scheele, the taciturn apothecary of Uppsala. *Acta Anaesthesiologica Scandinavica* 46: 2–9.

Severinghaus, J. W. 2003. Fire-air and dephlogistication. Revisionisms of oxygen's discovery. *Advances in Experimental Medicine and Biology* 543: 7–19.

Severinghaus, J. W. 2016a. Eight sages over five centuries share oxygen's discovery. *Advances in Physiology Education* 40: 370–376.

Severinghaus, J. W. 2016b. The most important discovery of science. In *Oxygen transport to tissue XXXVII. Advances in experimental medicine and biology*, eds. C. E. Elwell, T. S. Leung, and D. K. Harrison.

Smeaton, W. A. 1963. New light on Lavoisier: The research of the last ten years. *History of Science* 2: 51–69.

Smeaton, W. A. 1992. Carl Wilhelm Scheele (1742–1786): Provincial Swedish pharmacist and world-famous chemist. *Endeavour* 16: 128–131.

Stefanadis, C., M. Karamanou, and G. Androutsos. 2009. Michael Servetus (1511–1553) and the discovery of pulmonary circulation. *Hellenic Journal of Cardiology: HJC = Hellenike kardiologike epitheorese* 50: 373–378.

Sundt, T. M., and R. E. Anderson. 1975a. Reduced nicotinamide adenine dinucleotide fluorescence and cortical blood flow in ischemic and nonischemic squirrel monkey cortex. I. Animal preparation, instrumentation, and validity of model. *Stroke* 6: 270–278.

Sundt Jr., T. M., and R. E. Anderson. 1975b. Reduced nicotinamide adenine dinucleotide fluorescence and cortical blood flow in ischemic and nonischemic squirrel monkey cortex. 2. Effects of alterations in arterial carbon dioxide tension, blood pressure, and blood volume. *Stroke* 6: 279–283.

Taylor, G. 2019. The chemical works of Carl Wilhelm Scheele. *Ambix* 66: 90–92.

Warburg, O. 1949. Heavy metal prosthetic groups and enzyme action. In ed. T. A Lawson. Oxford: Clarendon Press.

West, J. B. 2008. Ibn al-Nafis, the pulmonary circulation, and the Islamic Golden Age. *Journal of Applied Physiology: Respiratory, Environmental and Exercise Physiology* 105: 1877–1880.

West, J. B. 2013. The collaboration of Antoine and Marie-Anne Lavoisier and the first measurements of human oxygen consumption. *American Journal of Physiology. Lung Cellular and Molecular Physiology* 305: L775–L785.

West, J. B. 2014a. Carl Wilhelm Scheele, the discoverer of oxygen, and a very productive chemist. *American Journal of Physiology. Lung Cellular and Molecular Physiology* 307: L811–L816.

West, J. B. 2014b. Joseph Priestley, oxygen, and the enlightenment. *American Journal of Physiology. Lung Cellular and Molecular Physiology* 306: L111–L119.

West, J. B. 2017. Leonardo da Vinci: Engineer, bioengineer, anatomist, and artist. *American Journal of Physiology. Lung Cellular and Molecular Physiology* 312: L392–L397.

Wilkinson, D. J. 2004. Resuscitation great. The contributions of Lavoisier, Scheele and Priestley to the early understanding of respiratory physiology in the eighteenth century. *Resuscitation* 61: 249–255.

Williams, K. R. 2003. The discovery of oxygen and other Priestley matters. *Journal of Chemical Education* 80: 1129–1131.

Chapter 2
Brain Biochemical and Physiological Activities—In Vivo Markers of Brain Pathology

Abstract Normal mitochondrial function in the process of metabolic energy production is a key factor in maintaining cellular activities. Many pathological conditions in animals, as well as in patients, are directly or indirectly related to dysfunction of the mitochondria. Monitoring the mitochondrial activity by measuring the autofluorescence of NADH has been the most practical approach since the 1950s. This review presents the principles and technological aspects, as well as typical results, accumulated in our laboratory since the early 1970s. We were able to apply the fiber-optic-based NADH fluorometry to many organs monitored in vivo under various pathophysiological conditions in animals. These studies were the basis for the development of clinical monitoring devices as presented in another article. The encouraging experimental results in animals stimulated us to apply the same technology in patients after technological adaptations as described in the following article (Mayevsky and Barbiro-Michaely 2013b).

Keywords Cellular energy metabolism · In vivo monitoring · Intracellular NADH autofluorescence · Multiorgan monitoring · Tissue oxygen balance · Brain metabolism

2.1 Introduction

The aim of the present study is to summarize the principles used in order to determine the vitality and functional state of various tissues in vivo by measuring the tissue energy balance. Later on, the brain will be used as a representative organ of the body and more detailed review will demonstrate the various in vivo real-time techniques available for the brain in clinical as well as experimental animal studies. Basically, most of those techniques could be applied to almost any other organ or tissue in the

A. Mayevsky et al., *Physiological Mapping of the Cerebral Cortex in Neuropathology*,
https://doi.org/10.1007/978-3-032-13019-8_2

body. The same technology was also applied to monitor various organs simultaneously (Manor et al. 1995; Kraut et al. 1998; Mayevsky and Chance 1982).

2.1.1 Tissue Energy Balance: Tissue Vitality

Tissue vitality, which will be defined rigorously below, is related to the tissue's ability to perform its work. While metabolism refers to the chemical changes taking place within an organism, by far the greater proportion of metabolism arises, directly or indirectly, from the organism's need for the energy necessary to do work. As such, tissue vitality provides information that is central to a tissue's metabolic state. A brief description of how cells are able to execute "work" is presented in Fig. 2.1. Healthy tissue cells perform various types of work, such as contraction of muscle, absorption of nutrients, production of secretions, activities of the nervous system, and the synthesis of cell constituents. The energy needed to perform these types of work is derived from the degradation of foodstuffs, primarily through biological oxidation reactions. These oxidation reactions, however, are not a direct reaction of molecular oxygen with the substrate, but, rather, a transfer of electrons, mediated by a respiratory chain of several complex enzyme systems, in which oxygen is the ultimate electron acceptor. The electron transfers (oxidations) down the respiratory chain result in the production of energy-rich, pyrophosphate bonds of ATP. Concomitantly with the electron transfers, the respiratory chain components switch between reduced and

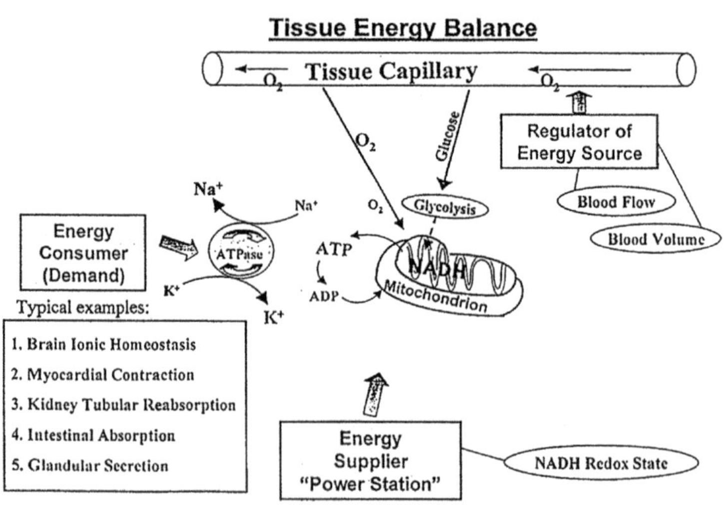

Fig. 2.1 Schematic presentation of the main activities related to tissue energy balance (see text for details). The demanding processes are different in various tissues, while the energy supply mechanisms are very similar. Each tissue produces most of its energy by producing ATP mainly in the mitochondria, depending on a continuous supply of O_2 via the blood stream (Mayevsky et al. 2000b)

oxidized states, each of which has different spectroscopic properties. Hydrolysis of the pyrophosphate bonds provides the energy necessary for the cell's work. The production of the pyrophosphate bonds, and thus the ability of the cell to do work, depends on the sufficient supply of sugar and oxygen to the tissue by the bloodstream. Without a sufficient supply, cells cannot function properly or, ultimately, survive. Since most of the energy consumed by tissues is dependent on the availability of oxygen, energy and oxygen will be used synonymously. In normal tissue, the ratio or balance between oxygen supply and oxygen demand is positive. That is, the supply mechanism (oxygenated blood circulation) is able to provide the amount of O_2 needed for the demand mechanisms (ATP production for various types of work). This balance is expressed in the following equation:

$$\text{Oxygen balance a } \frac{\text{Oxygen supply}}{\text{Oxygen demand}}$$

As can be seen, the O_2 balance reflects the cell's ability to do work. When supply exceeds demand, the vitality is greater than when the supply is insufficient to meet the demand. In this way, O_2 balance reflects the metabolic state of the tissue. Under normal conditions, a balance between O_2 demand and supply is kept by various mechanisms such as autoregulation. When an imbalance between O_2 demand and supply develops, due to insufficient supply, some organs will decrease O_2 demand by reducing activity. It is possible to monitor and provide parameters related to oxygen (energy) supply and oxygen (energy) demand. From the point of view of supply, oxygenated blood availability must be measured. These measurements can be made, since static and moving blood cells each have specific optical properties that are measurable. On the demand side, let us consider the respiratory chain located inside the mitochondria. ATP production is a demand-driven process carried out by oxidative phosphorylation that is, in turn, dependent on the supply of energy-rich substrate (metabolites of glucose) and molecular oxygen. The components of the respiratory chain (various enzymes) undergo continuous oxidation–reduction reactions, starting with NADH and continuing through the flavoproteins and finally reaching the various cytochromes. The terminal electron acceptor at the end of the respiratory chain is molecular O_2. Since the spectroscopic properties of the respiratory chain components are unique to their redox states, they can be used as internal markers of the state of oxidative phosphorylation and, hence, inform us regarding the state of oxygen demand. The intracellular concentration of NADH (reduced form) is a parameter related to a tissue's state of energy metabolism. Energy metabolism is dependent on pO_2. Information on pO_2 in tissue, therefore, is helpful for the evaluation of tissue metabolic activity. The need for an intracellular pO_2 indicator, as a parameter in the physiology and biochemistry of living tissue, emerged more than 50 years ago. Mitochondria are the intracellular organelles that use most of the consumed oxygen. As a result, the redox state of electron carriers in isolated mitochondria as a function of oxygen concentration was studied extensively. The components of the respiratory chain, roughly in order of increasing mid potential starting with NADH, at approximately −300 mV, at one end and cytochrome c, at +220 mV at the other end. Chance et al. (1973) concluded that, "For a system at

Metabolic States of Mitochondria and Associated Oxidation-Reduction Levels of Respiratory Enzymes

	Characteristics					Steady state percentage reduction of components				
State	$[O_2]$	ADP level	Substrate level	Respiration rate	Rate-limiting substance	a	c	b	Flavo-protein[a]	$NADH_1$
1	>0	Low	Low	Slow	ADP	0	7	17	21	~90
2	>0	High	~0	Slow	Substrate	0	0	0	0	0
3	>0	High	High	Fast	Respiratory chain	4	6	16	20	53
4	>0	Low	High	Slow	ADP	0	14	35	40	99
5	0	High	High	0	Oxygen	~100	~100	~100	~100	~100

[a] These values are based on the amount of flavoprotein of the respiratory chain. Other fractions, reduced by anoxia, are present but disregarded here.

Fig. 2.2 The classical table of Chance and Williams (1956) describing the various metabolic states of the isolated mitochondria are shown. The changes in the redox state of the various respiratory chain enzymes appear on the right side

equilibrium, NADH is at the extreme low potential end of the chain, and this may be the oxygen indicator of choice in mitochondria and tissue as well." In order to elucidate these concepts, Fig. 2.1 demonstrates an example of tissue vitality processes in a typical, normally functioning cellular utility, i.e., the Na^+/K^+-ATPase "pump," a membrane-bound enzyme that pumps the two ions against their concentration gradients. In excitable tissues, such as brain or muscle, as well as in other cells, the activity of the Na^+/K^+-ATPase is very sensitive to changes in ionic homeostasis. An increase in extracellular potassium ion concentration, $[K^+]$, will stimulate pumping activity in order to bring the extracellular $[K^+]$ back to normal levels (i.e., 3 mM range). The activation of Na^+/K^+-ATPase will increase the hydrolysis of ATP and thus the mitochondria will phosphorylate the ADP molecules that are released by the consuming processes, namely the Na^+/K^+-ATPase (see Fig. 2.1). The accelerated activity of the mitochondria will be accompanied by a more oxidized state of the respiratory chain components, as well as an increase in oxygen consumption (Erecinska and Silver 1989). In order to deliver more oxygen to the cells, oxygenated blood flow and blood volume will be increased and compensate for the extra oxygen needed by the tissue. This coupling between energy consumption and energy production is maintained as long as the O_2 supply is regulated well and tissue vitality remains in the normal range. Under conditions where the oxygen supply or delivery is limited, such as after stroke or heart attack, the energy supplier, i.e., the mitochondria, will not be able to produce the amount of ATP needed. As a result, energy-demanding processes will be limited. The net effect of the imbalance between energy demand and supply will be detected as a decrease in tissue vitality. This can lead to the development of various pathological changes. The possibility of monitoring mitochondrial function in vivo, in experimental animals and patients is of great importance and could contribute significantly to the understanding of various pathological processes.

Most of the information on mitochondrial function has been accumulated from in vitro studies. A relatively small portion of published papers dealt with the monitoring of mitochondrial function in vivo and in real-time. During the past 50 years, we have published over 200 papers in this very significant area, including the largest number of studies using NADH redox state monitoring in patients. In the present communication I summarize my research using fiber optic based fluorometry/reflectometry (Chance et al. 1973; Mayevsky and Chance 1973), based on the concept developed by Chance and Williams (1955a, 1955b). They defined the mitochondrial metabolic state in vitro as shown in Fig. 2.2. Detailed explanation of the table is given later in Sect. 2.1.4. The monitoring principles for one of the electron carriers, namely, Nicotinamide Adenine Dinucleotide (NADH) is presented. Since 1973, we accumulated enough in vivo data that can be compared to the in vitro studies, as shown in Fig. 2.3. As can be seen, there are differences between the definition of metabolic state in vitro and in vivo. The main difference is the definition of the resting state of the mitochondria. As seen in the left side of Fig. 2.2 under the resting state (State 4) in vitro, the NADH is 99% in the reduced form. As evaluated by many experiments in vivo, the resting state of all tissues is between State 4 and State 3. More details could be found in our past review paper (Mayevsky and Rogatsky 2007). In order to demonstrate the monitoring of NADH in a broad perspective, a summary of tissue energy metabolism and its evaluation by various techniques is given here.

Mitochondria In-Vitro				
State #	ADP level	Respiration Rate	Limiting Substrate	NADH %
5	High	0	Oxygen	~100
4	Low	Slow	ADP	99
3	High	Fast	Resp. Chain	53
2	High	Slow	Substrate	~0

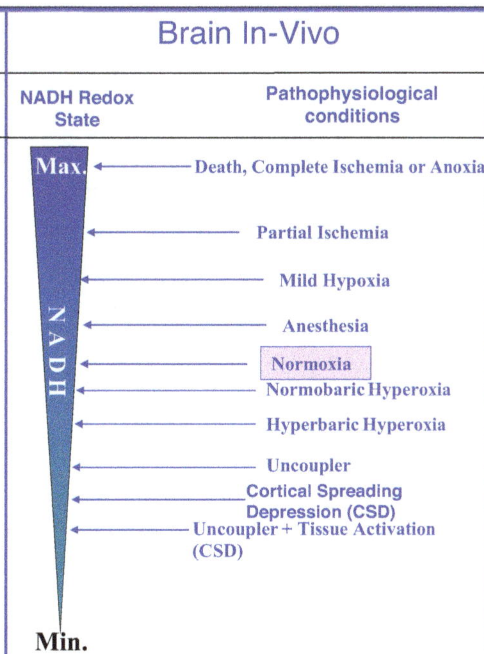

Fig. 2.3 This figure presents the comparison between mitochondrial metabolic states, defined by Chance and Williams (1956) and responses of the brain in vivo to changes in O_2 supply and metabolic activation (Mayevsky and Barbiro-Michaely 2013a)

2.1.2 Tissue Energy Metabolism

The functional capacity of any tissue is related to its ability to perform its work. It is possible to assess this ability through the knowledge of changes in the oxygen balance, i.e., the ratio of oxygen supply to oxygen demand in the tissue. Schematic presentation of the balance between oxygen supply and demand in a typical tissue is shown in Chap. 1, Figs. 1.3 and 1.4. The supply of oxygen is dependent upon microcirculatory blood flow and the level of oxygen bound to the hemoglobin in the small blood vessels. The saturation of hemoglobin is affected by the oxygen gradient between room air through the large arteries and small arterioles to the intracellular space and finally the mitochondrion. The demand for energy is affected by the specific activities taking place in each organ. The intracellular level of mitochondrial NADH (the reduced form) is a parameter related to the energy balance.

Figure 2.4 presents the involvement of the mitochondria in cellular energy metabolism. Substrates and O_2 are supplied and regulated by the blood in the microcirculation, namely from the very small arterioles and the capillary bed. The main function of the mitochondria is to convert the potential energy stored in various substrates (e.g. Glucose) into ATP. The inner membrane of the mitochondria contains five complexes of integral membrane proteins, including NADH dehydrogenase (Complex 1).

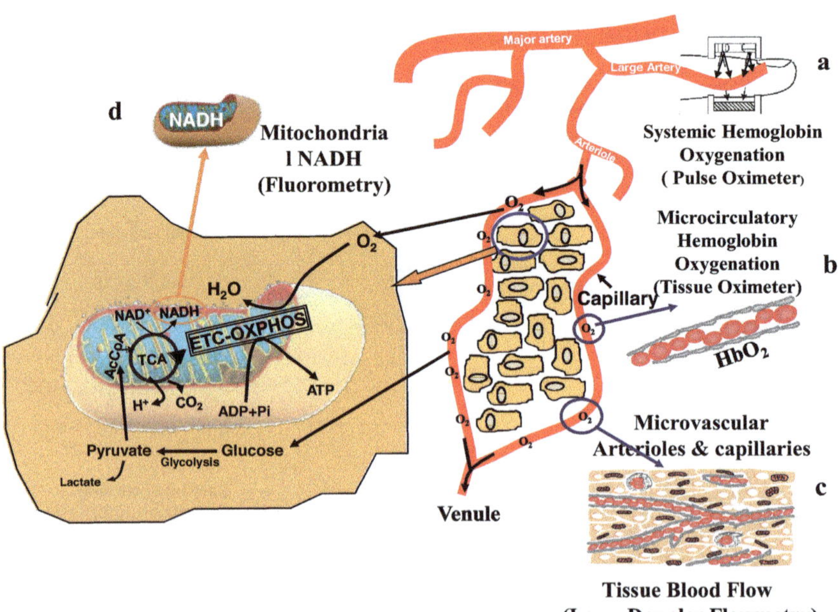

Fig. 2.4 The available techniques for real-time monitoring of energy metabolism at the tissue level. Part (**a**) shows the coupling between the macro-circulation monitored by the Pulse oximeter and the microcirculation. (**b–d**) Monitoring of vascular and intracellular compartments (see text for details) (Mayevsky and Barbiro-Michaely 2013a)

Three of those proteins are involved in the respiratory chain activity. The main function of the respiratory chain is to gradually transfer electrons from NADH and $FADH_2$ (originating from the Krebs cycle) to oxygen (O_2). With the addition of protons (H^+), water molecules (H_2O) are generated in Complex 4. NADH is a substrate or a coenzyme for the enzymatic activity of dehydrogenases that form part of the respiratory chain and reside in the inner membrane of the mitochondria. Further details on the biochemical properties of NADH can be found in publications by Chance and his collaborator (Chance and Thorell 1959a, 1959b). The formation of ATP depends on the sufficiency of substrate (i.e., glucose) and oxygen supply to the tissue by the blood flow in the microcirculation (Fig. 2.4). Without a sufficient supply, cells cannot function properly and can, ultimately, die. Since most of the energy consumed by tissues is dependent upon the availability of oxygen, the terms "energy" and "oxygen" will be used here synonymously.

The electron transfers (oxidations/reductions) down the respiratory chain result in the production of adenosine tri-phosphate (ATP). Concomitantly with the electron transport, the respiratory chain components switch between reduced and oxidized states, each of which has different spectroscopic properties. Hydrolysis of the pyrophosphate bonds provides the energy necessary for the cell's work. In order to assess the energy demand, it is necessary to measure different organ-specific parameters. For example, in the brain, the energy demand can be evaluated by measuring the extracellular levels of K^+ that reflect the activity of the major ATP consumer Na^+/K^+-ATPase (Mayevsky 1984a; Mayevsky and Chance 1982). In the heart, most of the energy is consumed by muscle contraction activity. On the other hand, the energy supply mechanism is the same for all tissues: oxygenated blood reaching the capillary bed releases O_2 that diffuses into the cells. Therefore, it is possible to evaluate tissue energy supply by monitoring the same four different parameters in all tissues. Energy exchange depends on pO_2 (partial oxygen pressure) in the tissue. Information regarding pO_2 in the tissue, therefore, is helpful for the evaluation of tissue metabolic activity. Oxygen levels can be measured by oxygen electrode however the sensitivity and accuracy of oxygen electrodes in the range of 1 mmHg (intracellular level) is not sufficient for the evaluation of mitochondrial function.

The need for an intracellular pO_2 indicator, as a physiological and biochemical parameter of living tissue, emerged more than 60 years ago. Mitochondria are the intracellular organelles that consume most of the oxygen. Therefore, the redox state of electron carriers in isolated mitochondria in vitro as well as in vivo as a function of oxygen concentration has been extensively studied. Chance et al. concluded that "For a system at equilibrium, NADH is at the extreme low potential end of the chain, and this may be the oxygen indicator of choice in mitochondria and tissue as well" (Chance et al. 1973). Lubbers (1995) concluded that "the most important intrinsic luminescence indicator is NADH, an enzyme of which the reaction is connected with tissue respiration and energy metabolism" (Lubbers 1995).

As of today, the ability to measure tissue energy metabolism at the microcirculation and cellular level is relatively limited.

2.1.3 *Real-Time Monitoring of Tissue Vitality*

Several real-time invasive and noninvasive techniques have been developed to determine tissue energy metabolism or tissue vitality in vivo. A short description of the available techniques for monitoring tissue energy balance described by various investigators is presented in Fig. 2.4. These techniques could be adapted and used in experimental animals as well as in patients. In order to present the linkage between the systemic and the tissue level oxygenation, the pulse oximeter technique is shown in Fig. 2.4a although the signal does not originate from the microcirculation.

2.1.3.1 Systemic Pulse Oximetry (A)

This standard of care device measures the oxygenation level of the hemoglobin dependent on the pulse passing in the large arteries. Therefore, this technique cannot provide information on hemoglobin saturation in the microcirculation.

2.1.3.2 Microcirculatory Hb Oxygenation (B)

Oxygenated hemoglobin can be monitored at the microcirculatory level using the absorption spectrum of hemoglobin, which is different in its oxygenated or deoxygenated state. One possibility is to illuminate the tissue with 585 nm light, which is an isosbestic point, and with 577 nm light, which is a non-isosbestic point, in which the oxy-hemoglobin absorb more light than the deoxy-hemoglobin form. By subtracting the 585 nm reflectance from the 577 nm reflectance, a parameter correlated to blood oxygenation is obtained. A detector collects the light reflected from the tissue and converts it into oxy-hemoglobin levels (Mayevsky et al. 1990; Rampil et al. 1992).

2.1.3.3 Microcirculatory Blood Flow (C)

Real-time monitoring of tissue blood flow (TBF) can be achieved using laser Doppler flowmetry (LDF) (Stern et al. 1977; Haberl et al. 1989; Dirnagl et al. 1989). The LDF measures relative changes, which correlate with the relative TBF alterations. The principle of LDF is to measure the Doppler shift, namely, the frequency change undergone by the light reflected from moving red blood cells. The results are presented as percentages of a full scale (0%–100%), providing relative perfusion values. In our laboratory, we used LDF to measure TBF (tissue blood flow) in the brain, kidney, and liver (Mayevsky et al. 1999b, 2001; Barbiro-Michaely et al. 1998). This method has been successfully employed in many animal models, including rodents, cats and baboons under stroke (Mayevsky and Rogatsky 2007).

2.1.3.4 Mitochondrial NADH (D)

NADH monitoring from the organ surface (brain, kidney, liver, testis, etc.) is performed by the fluorometric technique based on the original work by Chance and Williams (Chance and Williams 1955b; Mayevsky 1984a). The excitation light (366 nm) is passed from the fluorometer to the tissue through a bundle of quartz optical fibers. The emitted light (450 nm), together with the reflected light at the excitation wavelength, is transferred to the fluorometer through another bundle of fibers. The measured changes in fluorescence and reflectance signals are calculated as percent values relative to the calibrated signals under normoxic conditions. This type of calibration is not absolute, but it provides reliable and reproducible results for different animals and different laboratories (Mayevsky 1984a; Mayevsky and Chance 1982). The combination of NADH fluorescence and LDF techniques has been routinely utilized in our laboratory for the past 30 years (Mayevsky et al. 1999a, 2001, 2002; Sonn and Mayevsky 2000).

2.1.4 Monitoring of NADH: History, Principles, and Basic Technology

Since the discovery of pyridine nucleotides by Harden and Young (1906) more than 1000 papers have been published on the use of NADH as a marker for mitochondrial function. The present review will summarize almost 40 years of research using fiber optic-based fluorometry/reflectometry.

The discovery of the optical properties of reduced Nicotinamide Adenine Dinucleotide—NADH (earlier names: DPNH—diphosphopyridine nucleotide, or PN—pyridine nucleotide), has led to very intensive research since the early 1950s. The reduced form of this molecule, NADH, absorbs light at 320–380 nm (Fig. 1.19c, and d) and emits fluorescent light at 420–480 nm range (Fig. 1.19d). As the oxidized form NAD^+ does not absorb light in this range, it was possible to evaluate the redox state of the mitochondria by monitoring the UV absorbance or Blue fluorescence of NADH. The establishment of the method to monitor NADH in vivo was done between 1952 and 1962.

An intensive use of the in vivo NADH monitoring approach started in 1962. The "classical" paper on in vivo monitoring of NADH was published in 1962 by Chance et al. (1962a, 1962b). They were able to simultaneously monitor the brain and kidney of anesthetized rats using two microfluorometers. In 1962, Chance and collaborators elaborated on this kind of in vivo monitoring and used it in other rat organs (Chance et al. 1962c; Chance and Schoener 1962). The various instruments used in our laboratory to measure NADH were calibrated by measuring the fluorescence signal intensity as a function of NADH dissolved in solution. The typical calibration curve was presented in Fig. 4C in Mayevsky and Barbiro-Michaely (2013a), and the correlation between the concentration of NADH in solution and the fluorescence intensity was statistically highly significant. We also compared the presented calibration curve to a standard FDA-approved device and found no difference (Mayevsky et al. 2004).

The absorption and fluorescence spectra of NADH (the reduced form) have been well characterized at different levels of organization, i.e., in solution, mitochondria and cell suspensions, tissue slices, and organs in vitro and in vivo. There is a universal agreement that the intensity of the fluorescence band, independent of the organization level of the environment, is proportional to the concentration of mitochondrial NADH (the reduced form), particularly when measured in vivo from a tissue.

The biochemical and physiological significance of these spectral qualities is also universally accepted; that is, an increase in the fluorescence intensity indicates a more reduced state of NADH and of the rest of the mitochondrial electron transfer chain. Under various circumstances, changes in the redox state of the electron transport chain can be associated with various conditions. In Fig. 2.2, the metabolic states of the mitochondria in vitro (defined by Chance and Williams 1955b) are shown. The "resting" state of the mitochondria in vitro was defined as "State 4," where NADH was 99% in the reduced form, and ADP was the rate-limiting substance. If ADP is added to a suspension of mitochondria, ATP synthesis will be stimulated, oxygen consumption will increase, and the rate limit will be determined by the activity of the respiratory chain. During this State 3, or the "active state," the NADH redox state will decrease or become more oxidized (about 50%). When the "resting" mitochondria are deprived of oxygen, the activity of the mitochondria will stop, and NADH will reach its maximum redox state (State 5).

A definitive description of the mitochondrial metabolic state has never been given for in vivo conditions. Therefore, we described the in vivo mitochondria conditions as recorded by NADH fluorescence in a representative tissue or organ—e.g. the brain. While the range between minimal NADH (~0) and its maximal level was determined in vitro, it is almost impossible to determine in the intact brain or other organs in vivo. For example, State 2, with a substrate-free medium, could not be achieved in vivo since the tissue would die. On the other hand, the maximal level of NADH (State 5) could be monitored in vivo under complete deprivation of oxygen by anoxia or complete ischemia. We used the changes in NADH levels monitored in vivo to create a new scale ranging from a maximal definite point to the minimal level, as shown in Fig. 2.3. Details of this approach have been published (Mayevsky 1984a).

As seen in Fig. 2.3, the maximal NADH level is achieved under complete O_2 deprivation that can be induced both under in vitro and in vivo conditions. This signifies that this definitive point can be used to determine State 5 in vivo as well. The problem is to determine the metabolic state of a tissue in an in vivo situation. If we adopt the in vitro value of a resting state (State 4), this would signify that the increase in NADH during State 5, induced by anoxia (0% O_2), would be only 1%. According to all in vivo studies, this is not the case, and during anoxia the increase in NADH is larger than the decrease under State 4 to State 3 transitions. Figure 2.3 (right-hand panel) illustrates that the observed level of NADH increase is indeed larger than the decrease. Therefore, we concluded that, under in vivo conditions, the "resting" metabolic state of the brain is found between States 4 and 3 rather than in State 4 as defined in vitro (Mayevsky 1984a). In order to determine the maximal and minimal levels of NADH in vivo it is almost impossible to use cyanide or uncoupler (FCCP). Nevertheless, we were able to determine the maximal level by anoxia and the "minimal" level by a nonfluorescing uncoupler. We injected the uncoupler PCP (pentachlorophenol) into the ventricles of the rat's brain while monitoring the NADH responses to anoxia and spreading depression (Mayevsky 1976). If one wants to perform a reliable study with cyanide, the animal will have to die, and the results will not be helpful; therefore, we used the anoxia in order to measure the maximal level of NADH. Using fiber optic fluorometry, we were able to monitor the brain in anesthetized as well as awake rats (Mayevsky 1984a; Mayevsky and Rogatsky 2007).

In the following paragraph, the method and results of the specific organ monitored in the laboratory of Prof. Mayevsky are presented together. This will allow the potential user of the NADH monitoring device to understand the material in a better way. In our previous review (Mayevsky and Rogatsky 2007), we listed the papers published by various investigators in the field of NADH monitoring. It is important to note that this review summarizes almost 40 years of activity in our laboratory.

2.2 Methods

2.2.1 *Principles of NADH Monitoring Technology*

In order to monitor NADH fluorescence, it is possible to use one of the two principles available. At the early stage, it was necessary to measure and identify the fluorescence spectrum of NADH. Fluorescence spectra were compared in different in vitro and in vivo preparations. In parallel, the second approach was adopted, namely measuring the total fluorescence signal accumulated and integrated into a single intensity value using appropriate filters. This approach was necessary in order to measure NADH fluorescence continuously. The fluorescence spectra of NADH measured in various in vitro and in vivo models by different investigators were summarized in 2007 (Mayevsky and Rogatsky 2007). The origin of the Fluorescence signal was proven to come mainly from the mitochondria as well as the comparison between NADH fluorescence intensity and biochemical analysis of pyridine nucleotides were described in details recently (Mayevsky and Rogatsky 2007). Since the early 1950s, different groups have developed a number of fluorometers adapted to their specific experimental protocols. As of today, most of the groups using NADH fluorometers construct the devices in-house, since no suitable commercial products were available until recently.

The basic features of NADH fluorometers consist of the following:

1. A light source (including appropriate filters).
2. An optical path to the preparation and back to the detection unit.
3. Detection and signal processing units.
4. Signal recording and storage units.

In our earlier review published in 1984 and 2007, we extensively specified the light-guide-based fluorometry used in our studies (Mayevsky 1984a).

In order to enable the monitoring of NADH fluorescence in unanesthetized animals or other in vivo preparations, a flexible means was needed to connect the fluorometer with the tested organ, for example, the brain. This was achieved in 1972, when UV-transmitting quartz fibers became available (Schott Jena Glass, Germany). We have used the light-guide-based fluorometer for in vivo monitoring of the brain (Chance et al. 1973; Mayevsky and Chance 1973) subjected to anoxia, cortical spreading depression, and other pathophysiological conditions. Later on, the same technology was applied to monitor other organs as well.

2.2.1.1 Calibration and Correction of the Monitored Signals

In order to reduce the variation between different animals in a specific protocol, a standard procedure for signal calibration was used by various investigators. Since NADH could not be calibrated in absolute NADH concentrations, it was necessary to use a calibration procedure before each experiment. Different groups had developed standard procedures for calibration.

The Factors affecting NADH fluorescence and reflectance signals were described in detail in 2007 (Mayevsky and Rogatsky 2007).

The excitation and emission spectra of NADH are affected by the redox state of this fluorochrome and by other factors, leading to artifacts in the fluorescence measurements. In this section, various NADH-unrelated factors affecting the measured signal, will be discussed. Since most fluorometers involve the measurement of the total backscattered light at the excitation wavelength (i.e., 366 nm), the discussion will concern changes in NADH fluorescence as well as in tissue reflectance.

The following factors may affect the two **measured** signals, 366 nm reflectance (R) and 450 nm fluorescence (F):

1. Tissue movement due to mechanical or intracranial pressure changes.
2. Extracellular space events, such as volume changes or ion shifts between intra- and extracellular spaces.
3. Vascular and intravascular events, for example, oxy-deoxy Hb changes, and blood volume changes due to autoregulatory vasoconstriction under pathological conditions.
4. Intracellular space factors, such as O_2 level, ATP turnover rate, substrate availability, and mitochondrial redox state. This subject will be discussed later on.

In order to compensate for NADH-unrelated factors, various approaches have been developed. In the paper published by Ince et al. (1992) the various correction techniques are listed and discussed in detail. It appears that the correction technique is also dependent on the instrument configuration (Harbig et al. 1976; Balaban and Mandel 1988). Most of the published materials are based on the 1:1 ratio, when subtracting the 366 nm reflectance from the fluorescence signal. As of today, a new approach is still lacking to compensate for non-NADH factors affecting the NADH fluorescence signal. We have found that subtracting the reflectance from the fluorescence or dividing the two parameters provides similar net NADH changes. Very recently, Bradley and Thorniley (2005) published a review article dealing with the various correction techniques for tissue fluorescence. They summarized their review by the following conclusion: "even though research has been conducted into correction techniques for over thirty years, the development of a successful and practical correction technique remains a considerable challenge."

2.2.2 *Monitoring the Brain*

A special table and probe holding device was constructed in order to perform brain as well as other organs preparation for the monitoring period. The device and the tools for the operation procedure are shown in Fig. 2.5a, b. The brain is operated while the head is connected to a special head holder for the period of the operation (20–30 min) and then can be released, for the monitoring period, as shown in Fig. 2.6c. The other monitored organs, i.e, muscle, kidney, or liver have to be held by a micromanipulator during the monitoring period.

As mentioned before, the brain was the main organ monitored by other investigators as well as in our group. In order to be acquainted with the NADH monitoring device, it is recommended to start with brain monitoring and later on to move and use the selected organ. The reason for it is that the connection between the fiber optic probe

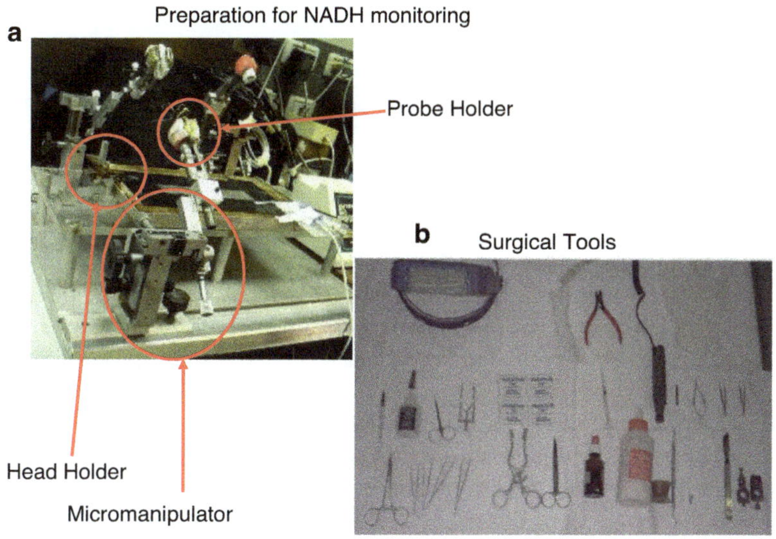

Fig. 2.5 The system used to prepare and monitor up to four organs simultaneously. (**a**) The same system enables to perform a craniotomy while the animal is connected to a special head holder. On the right side (**b**), the recommended surgical tools are shown (Mayevsky and Barbiro-Michaely 2013a)

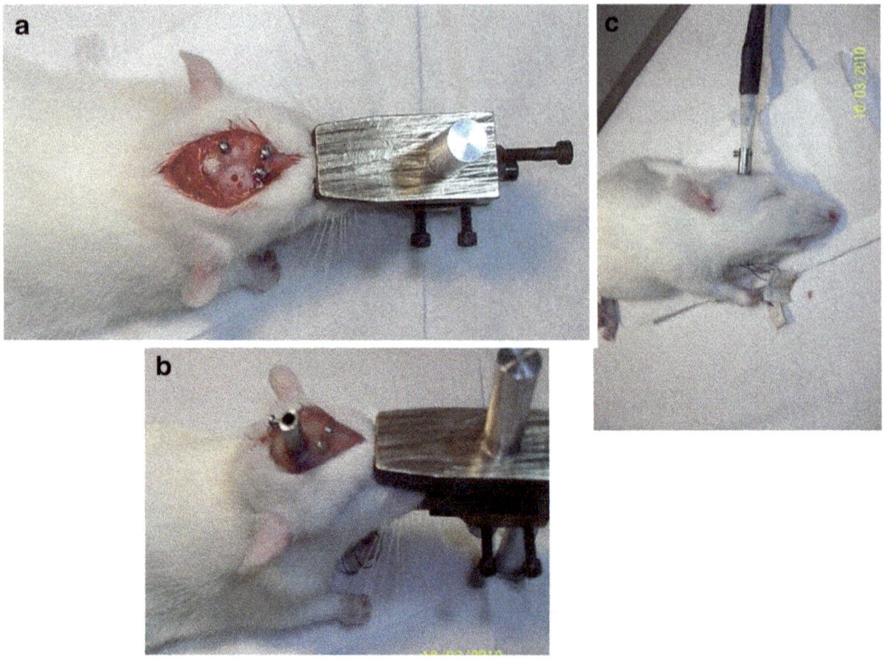

Fig. 2.6 Stages in the preparation of the rat brain for NADH monitoring. (**a**) Location of screws enabling the fixation of the light guide holder to the skull by dental acrylic cement. (**b**) The view of the skull after the end of the operation. (**c**) After the insertion of the fiber optic probe into its holder, the animal is ready for the monitoring (Mayevsky and Barbiro-Michaely 2013a)

and the monitored tissue must be constant during the monitoring session and in the brain it is easy to be achieved by connecting a probe holder to the skull using acrylic cement as shown in Fig. 2.6.

A midline incision is made in the skin, exposing the skull. Three holes were drilled in the skull and appropriate small screws were inserted into the skull (less than 1 mm in depth) as shown in Fig. 2.6a. An appropriate hole (3–5 mm in diameter) was drilled in the right or left parietal bone for the fixation of a light guide holder in which the monitoring probe was inserted later on. The light guide holder and the three screws were then fixated to the skull using dental acrylic cement (Fig. 2.6b). Ten minutes later the head of the animal was released from the head holder, and the probe was inserted to a predetermined depth and fixed by a set screw (Fig. 2.6c).

In Fig. 2.7, the structure of the standard DC fluorometer/reflectometer connected to the brain of a small animal is shown (Mayevsky and Rogatsky 2007). In Fig. 2.8, various types of light guide holders are presented (Mayevsky and Chance 1975). The thread outside the bottom of this cannula enables screwing it into the skull and also gives a better connection between the cannula and the cement. Type A was used in experiments in which animals were exposed to nitrogen/oxygen breathing cycles or to hyperbaric pressure of oxygen. The EEG was measured between the two electrodes, 1–2, cannula type B was used in all experiments in which spreading depression was

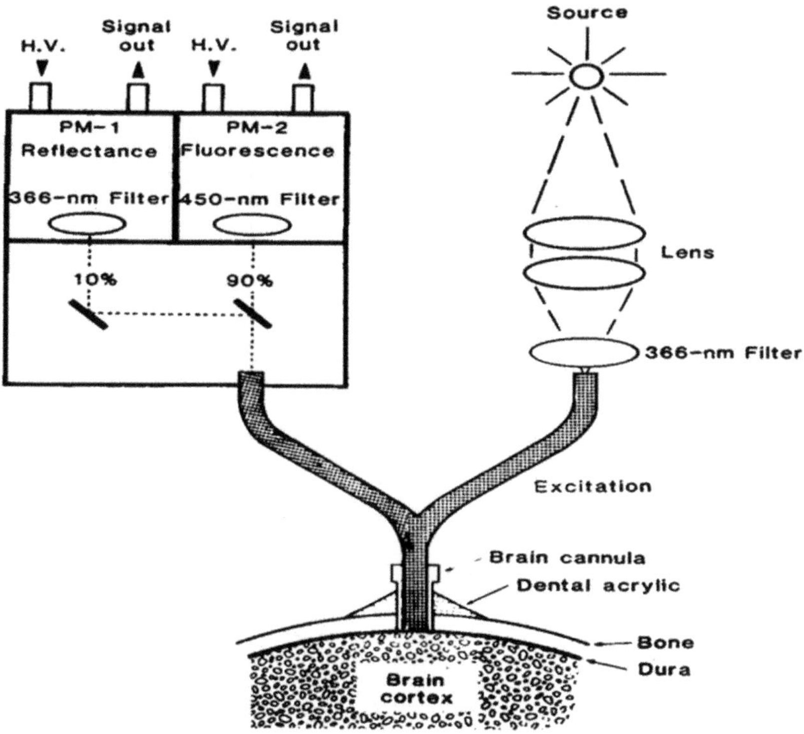

Fig. 2.7 The structure of the standard DC fluorometer/reflectometer connected to the brain of a small animal (Mayevsky and Barbiro-Michaely 2013a)

elicited by application of KCl solution above the Dura. The KCl electrodes 1, 2–3, 4. The third type, C, has two compartments in the bottom, the small one for KCl application 3–4 and the large compartment, 1–2, where chemicals such as Metrazol were applied and affected a larger area. A fifth electrode was located 180° to electrodes 3–4, so the EEG was measured at the same time.

2.2.3 Monitoring NADH Together with Other Parameters

The need for multiparametric monitoring of other parameters, in addition to NADH, results from the basic understanding that NADH is affected by two major factors. The redox state of NADH reflects not only the availability of oxygen inside the mitochondria but also the turnover rate of the ATP-ADP cycling activity (State 4–State 3 transition). The interaction between these two factors affects the nature of NADH response to various conditions. For example, an increase in brain energy consumption (e.g., cortical spreading depression) under oxygen restriction will be manifested as an increase in NADH rather than a decrease (oxidation) measured in

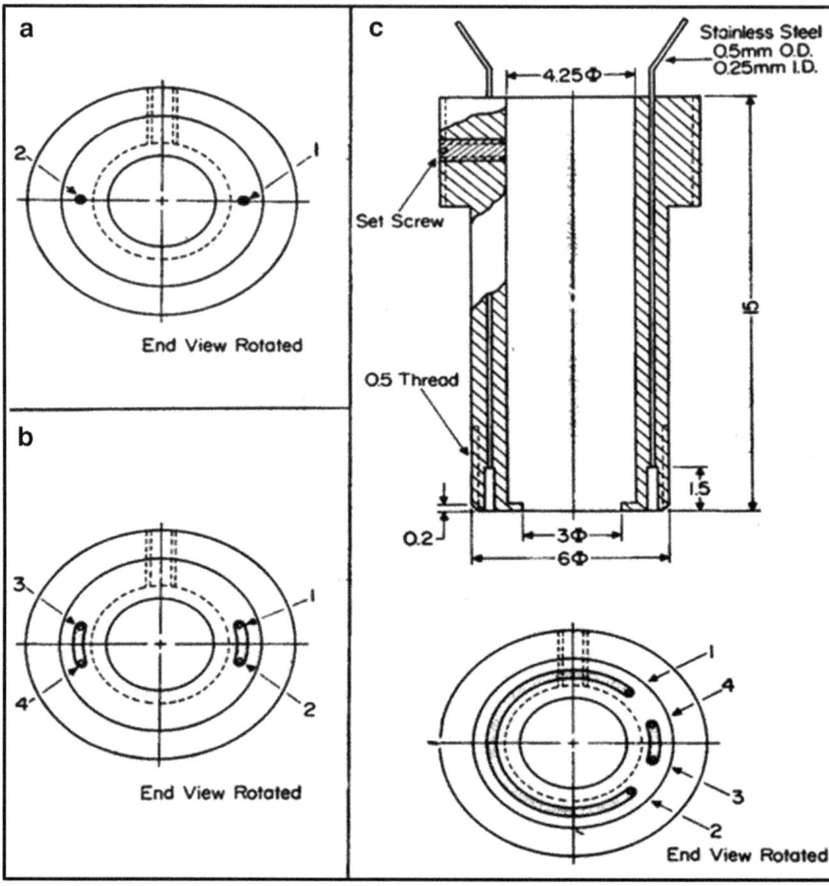

Fig. 2.8 The construction of the Light guide holders (cannulas) for measurement of NADH fluorescence from the surface of the brain of an awake rat in experiments in which topical application of drugs is needed (Mayevsky and Chance 1975). In this figure, various types of light guide holders are presented. The thread outside the bottom of this cannula enables screwing it into the skull and also provides a better connection between the cannula and the cement. Type a was used in experiments in which animals were exposed to nitrogen/oxygen breathing cycles or to hyperbaric pressure of oxygen. The EEG was measured between the two electrodes, 1–2, cannula type b was used in all experiments in which spreading depression was elicited by application of KCl solution above the Dura. The KCl electrodes 1, 2–3, 4. The third type, c, has two compartments in the bottom, the small one for KCl application 3–4 and the large compartment, 1–2, where chemicals such as Metrazol were applied, affecting a larger area. A fifth electrode was located 180° to electrodes 3–4, so the EEG was measured at the same time

normal, well-oxygenated brain tissue (Mayevsky 1978a). According to Chance and Williams, an increase in ATP production is always recorded as a decrease in NADH (Chance and Williams 1955b, 1956). Therefore, the "reduction cycle" measured by the NADH signal in response to CSD can be interpreted as an artifact of some kind. This phenomenon and the fact that the mitochondrial NADH signal cannot yet be calibrated in absolute values prompted us to develop a multi-parametric monitoring approach and a multiparametric probe that could be used in various tissues exposed to different pathophysiological conditions. By this approach, two major advantages were gained. First, it provided the possibility of a better interpretation of the recorded results. And second, non- physiological responses could also be more easily detected.

In all these studies, NADH was and is the most important parameter, while the inclusion of other parameters improves our ability to explain pathophysiological processes. The availability of fiber optic probe in 1972 opened up the possibility to attach various probes to NADH probe. The shift from single parameter monitoring to the highly developed assembly of up to 8–9 parameters was a gradual process. The milestones of the process of this development appear in our publication (Mayevsky and Barbiro-Michaely 2013a). In the early days, Chance et al. (1962b, 1964) and Jobsis et al. (Jobsis et al. 1971; Rosenthal and Jobsis 1971) incorporated the ECoG electrodes in the monitored brain. In 1973, we combined the ECoG electrodes with the fiber optic probe. Later on, electrodes for measurement of extracellular potassium ions and DC potential using micro (Mayevsky et al. 1974b) or surface electrode (Crowe et al. 1976, 1977) were added to our monitoring system. The combination of NADH probe and pO_2 probe was developed in the late 1970s (Mayevsky et al. 1980). In the 1980s, a few types of multiprobe systems were applied as seen in our paper (Mayevsky and Barbiro-Michaely 2013a).

The significant step was the embedding of a laser Doppler flowmeter sensor into the multiprobe assembly to monitor continuously the microcirculatory blood flow, in addition to NADH, and reflectance and various ions. The most developed multiprobe assembly was finalized in the 1990s. The effort to apply our developed multiparametric monitoring system to clinical situation led to the discovery of the cortical spreading phenomenon in the human brain (Mayevsky et al. 1996). The in vivo multiparametric monitoring approach has also applied to other organs in the body. Kedem's group supplemented NADH measurements of the dog heart, with the monitoring of tissue blood flow by heat clearance and of isometric contractile tension (Acad et al. 1987). Our group used the multiparametric monitoring system to study the function of the rat kidney in vivo (Amran-Cohen et al. 2003). The influence of ischemia on NADH and microcirculatory blood flow was studied in rat liver in vivo (Barbiro-Michaely et al. 1998). The various probes used in our multiparametric Assemblies (MPA) are presented in Fig. 2.9. One of the classical MPA used in our studies is shown in Fig. 2.10.

2.3 Results

2.3.1 Responses of the Brain to Anoxia, Hypoxia, and Ischemia

In this book, we define anoxia as a complete deprivation of O_2 caused by breathing 100% N_2. Hypoxia implies that the deprivation of O_2 from the breathing mixture is partial and ranges between 21% (normal air) and 0% (anoxia). Ischemia is defined as a decrease in O_2 supply due to a decrease in blood flow to the brain. The degree

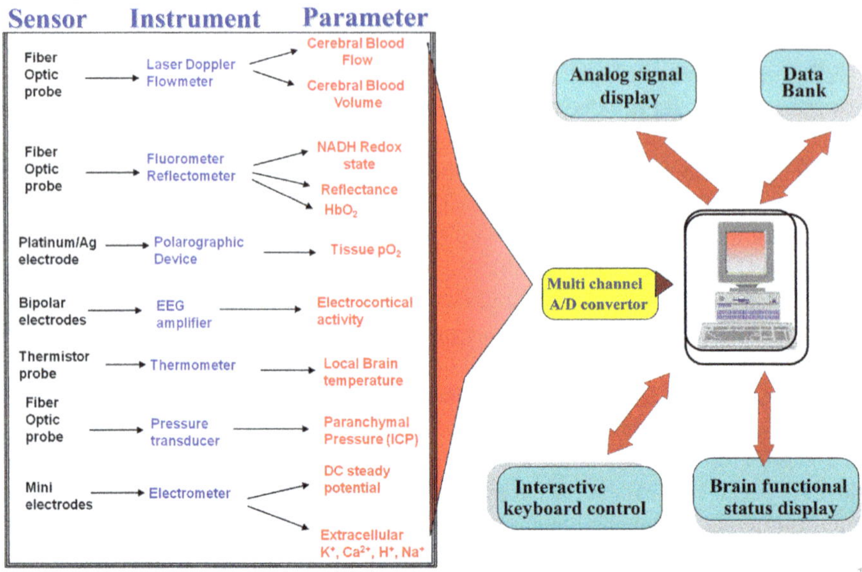

Fig. 2.9 The details of the various instruments and sensors used in the multiparametric monitoring system applied in our studies

of ischemia can vary from a complete absence of flow (complete ischemia) to various levels of blood flow (partial ischemia). Although oxygen deficiency is the main event in each of the three experimental conditions (anoxia, hypoxia, and ischemia), other physiological factors may differ. For example, microcirculatory blood flow is decreased under ischemia, but increases under brain hypoxia. Thus, changes in the tissue due to other blood flow-related factors are not identical.

The responses to hypoxia and anoxia are very similar; therefore, they will be discussed together. According to the definition of Chance and Williams (1955a, 1955b), a shift toward State 5 involves an increase in NADH proportional to a decrease in O_2 supply.

It is assumed that the response of NADH fluorescence to hypoxia or anoxia, induced in vivo, should be very similar to the response of isolated mitochondria. As shown in Fig. 2.11, left side, when the brain was exposed to 100% N_2, the fluorescence (F) showed a clear increase-decrease cycle depending on the availability of O_2. In an auto-regulated blood-perfused brain, it is expected that the lack of O_2 will trigger compensation mechanisms that may lead to an increase in the blood flow and volume, or a decrease in the R signal. Indeed, R exhibited a large decrease due to the increase in blood volume (vasodilatation of brain vessels).

Unilateral ischemia in this specific gerbil (Fig. 2.11, right side) led to an increase in NADH in the ipsilateral hemisphere very similar to the change under anoxia. The effect of the blood volume changes, measured by the reflectance trace, was very small in the gerbil, and as a result, the fluorescence Flu signal and the corrected fluorescence

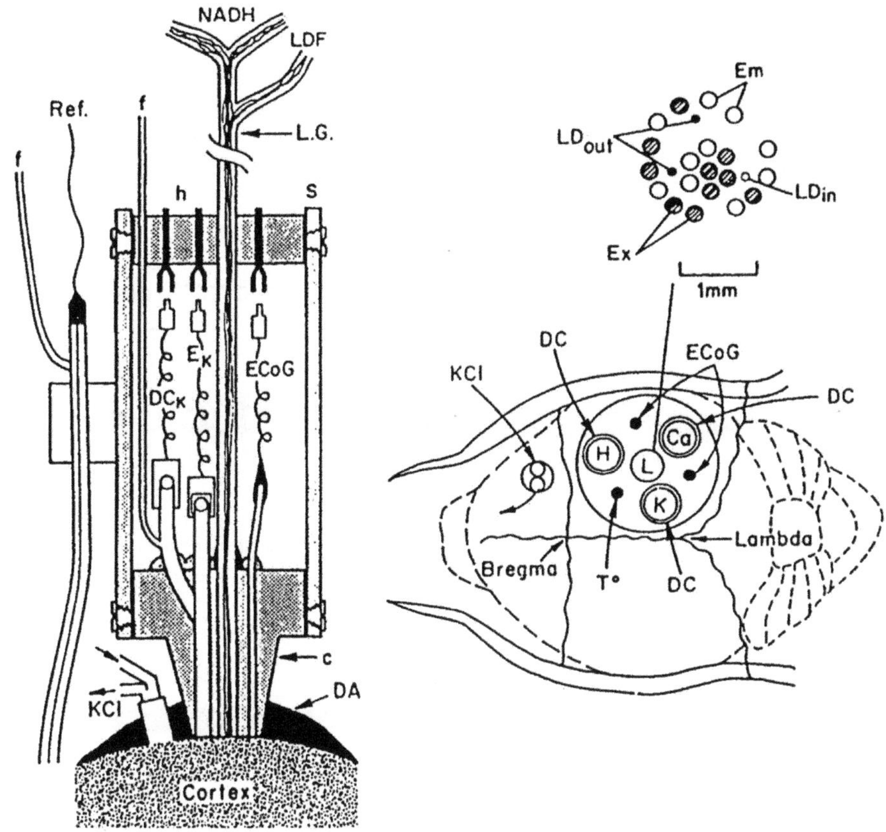

Fig. 2.10 Multiprobe assembly used to monitor hemodynamic, metabolic, ionic, and electrical activities from the brain. The left side shows a longitudinal section. In the right part, the location of the MPA and the skull is shown. C Plexiglas probe holder, DA dental acrylic cement, KCI push pull cannula for KCI, ECoG electrocortical electrodes, K, Ca, H ion specific electrodes, DC area of DC steady potential monitoring, NADH two arms of NADH monitoring light guide, h connectors holder, s aluminum sleeve, Ref reference electrode, f reeling tube of the reference or the other DC electrodes, T_0 thermistor for local temperature measurement, Ex, Em excitation and emission fibers for NADH monitoring, LD_{in}, LD_{out} optical fibers for monitoring blood flow and volume, LDF laser Doppler flowmeter fibers, L.G combined light guide (Mayevsky et al. 2000b)

signal (NADH) were very similar. In this specific gerbil, the connection between the two anterior cerebral arteries was minimal and, therefore, no blood compensation between the two hemispheres was noted (Mayevsky et al. 1983).

The responses to hypoxia presented in Fig. 2.12 are very similar to those found under anoxia. In the left part, brain response to 10% and 5% oxygen are presented in comparison to the effect of pure nitrogen. In this record, the brain oxygen level in the tissue, measured by an oxygen electrode, is presented too. In Fig. 2.13, the animal was exposed to 100% oxygen and then continuous decrease of oxygen down to 0% was induced. The correlation between the level of oxygen supply and the NADH levels in the brain is obvious.

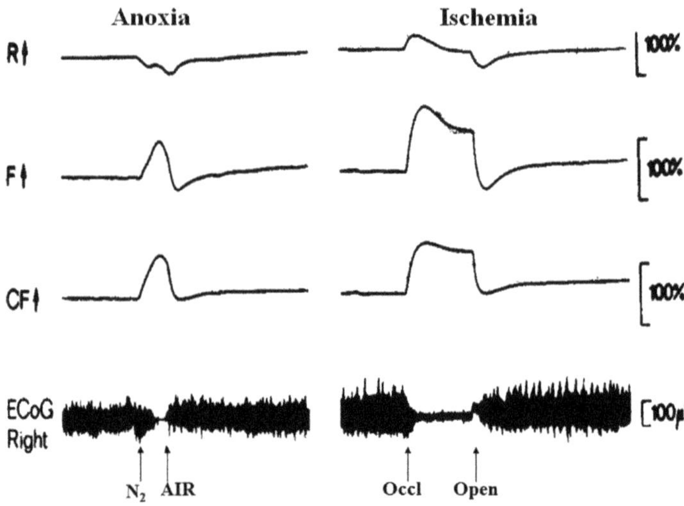

Fig. 2.11 Responses of a gerbil's brain to anoxia and ischemia induced by occlusion of the right carotid artery. Ref reflectance, Flu fluorescence, ECoG electrocorticography (Mayevsky and Barbiro-Michaely 2013a)

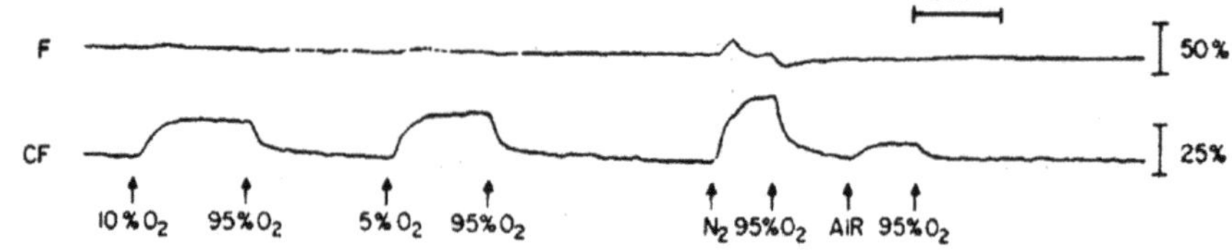

Fig. 2.12 Effects of hypoxia and anoxia on brain oxygen levels, NADH, and electrical activity

The effects of complete ischemia, induced by decapitation, are shown in Fig. 2.14b. Using this type of animal model enables testing the effects of various drugs, such as anesthetics or other experimental conditions, on mitochondrial function in vivo. As seen in Fig. 2.14b, the NADH response started 2–3 s after the decrease in blood supply and reached the maximal level within 10–15 s. In this animal, the two hemispheres were monitored simultaneously and the responses are very similar. The analysis of the

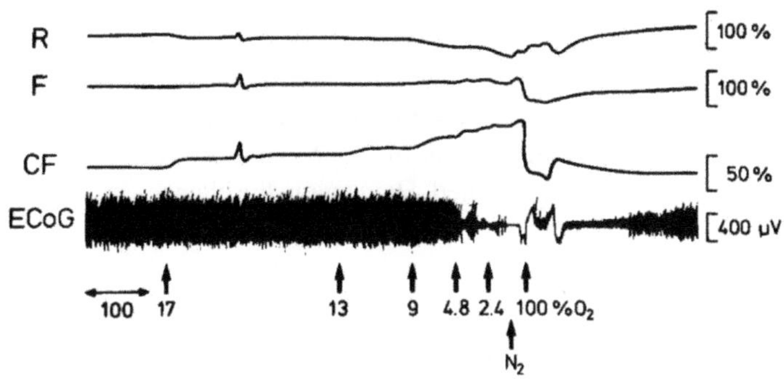

Fig. 2.13 Significant hypoxia was tested up to complete anoxia (Mayevsky and Barbiro-Michaely 2013a)

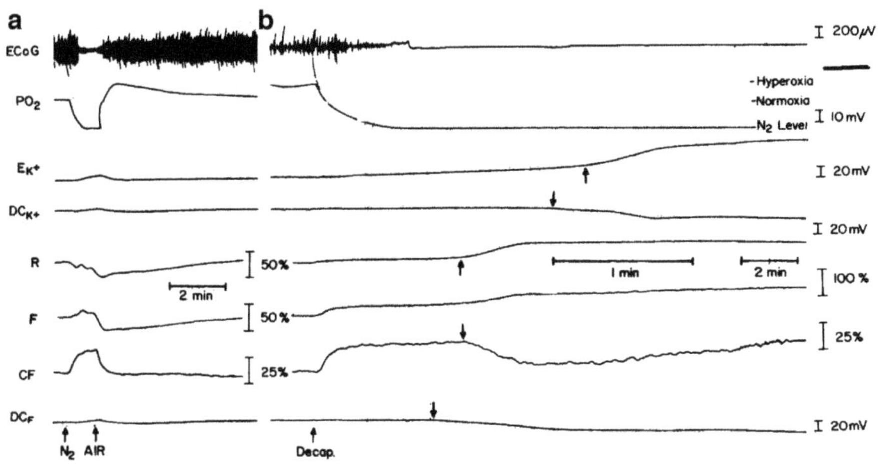

Fig. 2.14 Metabolic, ionic, reflected-light, and electrical responses to anoxia (**a**) and decapitation (**b**) in the rat brain are shown (Mayevsky and Sclarsky 1983)

data could be done by definition of few parameters representing the amplitude and the timing of various events after the induction of ischemia (Zarchin and Mayevsky 1981).

The reflectance trace (R) in Fig. 2.11 shows a typical decrease due to the increase in blood volume in the monitored site. The NADH (CF) showed a significant increase, which recovered to the baseline after resuming air breathing. Unilateral ischemia in this specific gerbil led to an increase in NADH in the ipsilateral hemisphere very similar to the change under anoxia. The effect of the blood volume changes, measured by the R trace, was very small in the gerbil, and as a result, the fluorescence F signal and the corrected fluorescence signal (CF) were very similar. In this specific gerbil, the connection between the two anterior cerebral arteries was minimal and, therefore, no blood compensation between the two hemispheres was noted (Mayevsky et al. 1983).

In Fig. 2.12, the correlation between NADH and tissue oxygen level is shown. The responses of NADH to anoxia are the same as in Fig. 2.11. The NADH was elevated in parallel with the decrease in oxygen. In Fig. 2.14b, the effects of decapitation on NADH and tissue oxygen are presented. Here again, the changes in NADH and tissue oxygen were very well correlated.

2.3.2 Responses to Cortical Spreading Depression and Epileptic Activity

As shown by Chance and Williams (1955b, 1956), the activation of the mitochondria by increased ADP is coupled with oxidation of NADH (decreased NADH levels), and is known as the State 4–State 3 transition in isolated mitochondria. We will describe the effects of tissue activation on NADH fluorescence under normoxic conditions as well as during limitation of O_2 supply in the tissue (hypoxia, ischemia).

Since the discovery of brain Cortical Spreading Depression (CSD) by Leao in 1944, it has become clear that the wave of depolarization passing through the tissue increases energy consumption (Leão 1944a). The effects of CSD on brain mitochondrial NADH were first described in 1973 and a clear oxidation wave was recorded (Rosenthal and Somjen 1973; Mayevsky and Chance 1973). A year later, we demonstrated the coupling between the elevated extracellular K^+ induced by CSD and the greater ATP requirement (NADH oxidation) needed to recover the normal ionic homeostasis (Somjen et al. 1976). This coupling suggests that upon stimulation of Na^+/K^+-ATPase, the released ADP will be recovered by the mitochondria to ATP. This stimulation of the respiratory chain will increase oxygen consumption and, consequently, blood flow will be increased to compensate for the extra oxygen needed (Mayevsky and Weiss 1991). Extra oxygen supply or compensation can be achieved by either an increase of CBF, enhanced O_2 extraction, or both. In a well-controlled study, we showed that concomitantly to NADH oxidation and extracellular K^+ changes, CBF doubled while oxygen extraction ratio remained the same (Mayevsky and Weiss 1991).

One of the leading groups, headed by Frans Jobsis, published a number of studies showing the effects of CSD on brain energy metabolism (LaManna and Rosenthal 1975; LaManna et al. 1981; Lothman et al. 1975; Rosenthal and Somjen 1973; Somjen et al. 1976; Sylvia and Rosenthal 1979). Several works have been published by Chance and collaborators (Kovach et al. 1984; Haselgrove et al. 1981), Dora's group published two papers in this field (Dora et al. 1984; Kovach et al. 1984).

Our group has published over 30 papers on the relationship between mitochondrial NADH redox state and CSD. Also, we had reported, for the first time, the development of CSD in the brain of a comatose patient (Mayevsky et al. 1996). In order to induce CSD, we used a specially designed light guide holder (Fig. 2.8b) enabling topical application of various chemicals or drugs. Typical responses to CSD monitored in

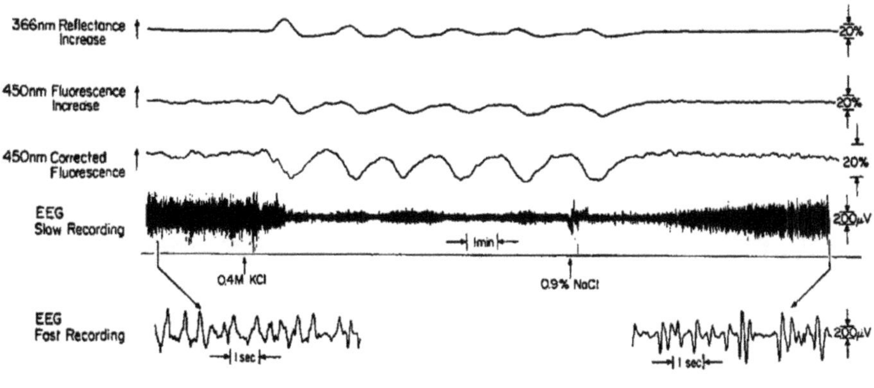

Fig. 2.15 Repetitive responses of the brain to cortical spreading depression evoked by application of 0.4 M KCI epidurally. The third trace is on an expanded amplitude scale. The arrow direction shows an increase in the optical signals (Mayevsky and Chance 1974)

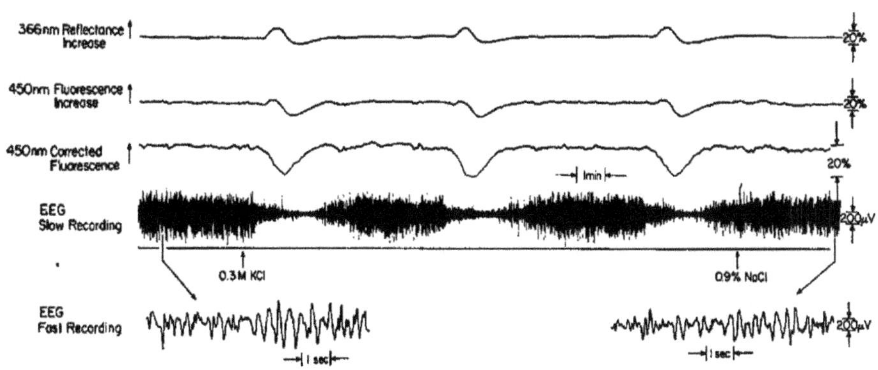

Fig. 2.16 Using a lower concentration of KCl (0.3 M) led to a lower frequency of SD responses (Mayevsky and Chance 1974)

the brain of an awake rat are shown in Figs. 2.15 and 2.16. In Fig. 2.16, it is possible to see the correlation between the depression of electrical activity (ECoG) and the oxidation (decrease) of NADH, in three repetitive cycles. The relationship between changes in various parameters developed during CSD is presented in Fig. 2.15. The initial events are changes in electrical (ECoG and DC potential) and ionic homeostasis (elevated extracellular potassium). The response to those changes is the activation of ion pumps that increase energy consumption. The NADH is more oxidized and blood flow to the brain is elevated.

In 1975, we published our first paper on mitochondrial responses to epileptic activity measured in non-anesthetized rats (Mayevsky and Chance 1975). A clear oxidation of NADH was found, even when Metrazol was applied epidurally, as shown in Fig. 2.18.

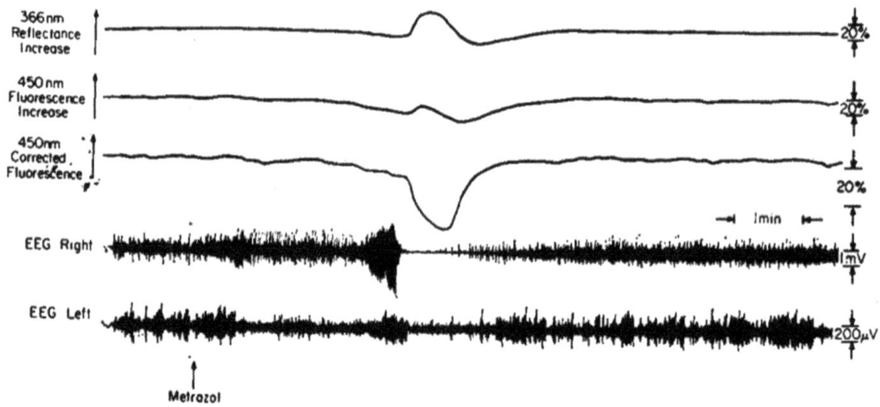

Fig. 2.17 The effects of Metrazol on brain NADH fluorescence, 366 nm reflectance, and the EEG are shown. Note that the amplitude of the EEG calibration in the right hemisphere is 1 mV

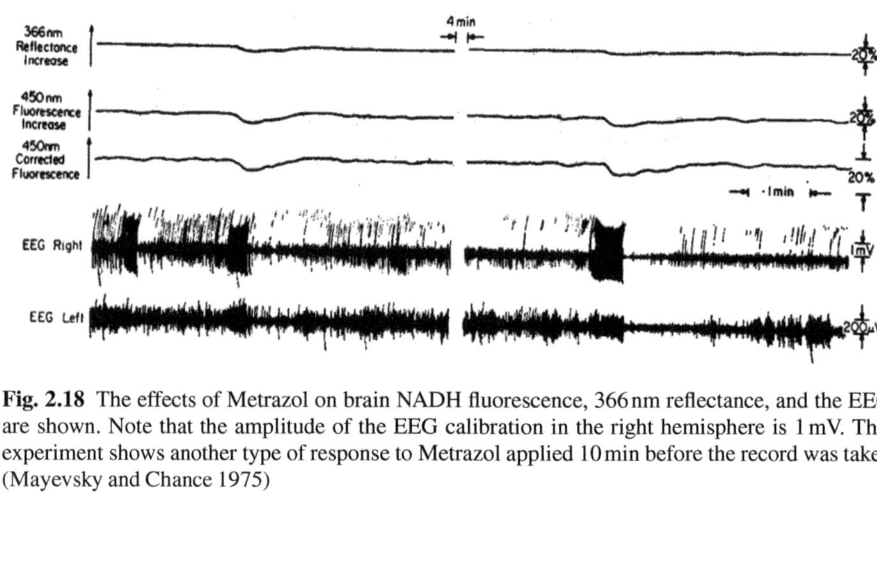

Fig. 2.18 The effects of Metrazol on brain NADH fluorescence, 366 nm reflectance, and the EEG are shown. Note that the amplitude of the EEG calibration in the right hemisphere is 1 mV. This experiment shows another type of response to Metrazol applied 10 min before the record was taken (Mayevsky and Chance 1975)

In most of our studies, we found that the response to epileptic activity was followed by a wave of cortical spreading depression, as can be seen in Fig. 2.17 (Mayevsky 1978b, 1983b, 1984a). In this case, we monitored a gerbil that developed epileptic activity after exposure to noise (Mayevsky et al. 1983). We found a similar coupling of NADH oxidation with CSD development, when epileptic activity was developed due to oxygen toxicity induced by exposing the rat to hyperbaric oxygenation (Mayevsky and Shaya 1980; Mayevsky 1983b, 1984b; Mayevsky et al. 1974a). In a small part of the animals tested, the response to the epileptiform drug was not followed by a CSD wave, as seen in Fig. 2.18.

2.3.3 *Experimental Multiparametric Results*

In order to demonstrate the monitoring capacity and potential of MPA, a typical long-term continuous monitoring of the normal brain is shown in Fig. 2.19. The segments A–E were taken at various intervals of time after 2 ml saline IV injection to the awake rat. The figure presents the stability of the monitoring system during more than 15 h. Most of the parameters monitored were very stable while the rat was located in a restraining cage. This kind of model could be used for new compound testing after injection into the body via IP or IV routine. According to the changes recorded in the various parameters, it is possible to hypothesize regarding the mechanism of action of a tested drug or treatment given. For example, if the CBF was the sole parameter to be affected by the compound while the NADH redox state and all other parameters were stable, one could conclude that the compound has a unique effect on the CBF. The fact that NADH was stable indicated that mitochondrial function was adequate. Another possibility is that a compound may specifically inhibit the Na^+/K^+-ATPase regardless of the energy state of the brain. Under such conditions, it is possible that the extracellular K^+ level will be elevated while all other parameters will be stable. In case the tested compound is a potential epileptic activity inducer, all the parameters will respond to the injection of the compound. It is important to note that due to the changes measured by the various sensors along the time axis, it is necessary always to use a sham control group of animals injected by saline or monitored without the treatment. Another option is to use the same type of experimental setup as presented in Fig. 2.10 and to use a known perturbation in order to stimulate the brain while monitoring all the parameters. One of the possible perturbations to be used is the cortical spreading depression (SD), a well-reversible known pathophysiological stimulus described over 50 years ago (Leão and Morison 1945; Leão 1944a, 1944b).

During the passage of an SD wave, a massive depolarization propagates from the stimulation site to the rest of the hemisphere (Bures et al. 1974). Due to the disturbance

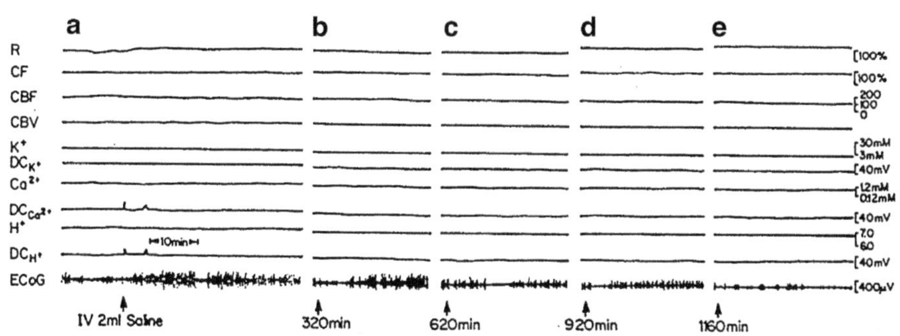

Fig. 2.19 Multiparametric recording of various brain physiological and biochemical activities measured from unanesthetized rat injected with 2 ml saline. Parts (**a–e**) represent vraious time points in the recording session. R, CF:reflected light (366 nm) and corrected NADH fluorescence (CF), CBF, CBV cerebral blood flow and volume, K^+, Ca^{2+}, pH extracellular levels of potassium, calcium, and H^+, DC_{K^+}, $DC_{Ca^{2+}}$, DC_{pH} DC steady potential around the ion-specific electrodes (Mayevsky et al. 2000b)

in ionic homeostasis (Mayevsky and Chance 1974, 1975, 1982), the energy metabolism is stimulated as indicated also by an increase in O_2 consumption and CBF (Mayevsky and Weiss 1991). Figure 2.20 shows a typical response of the brain to SD after IV infusion of alcohol to the awake rat. As can be seen in A, the control SD, a large increase in extracellular K^+ was correlated to the activated energy metabolism as seen by the oxidation of NADH (CF) and the increase in cerebral blood flow (RCBF). In this rat, the typical "oxidation cycle" was recorded 240 min after the end of alcohol infusion (Lothman et al. 1975). It appears that the autoregulation of CBF was affected. This may also be seen in the RCBF trace. There was a correlation between the responses of the CBF and NADH (CF) during the various stages, suggesting that alcohol induced abnormal microcirculation and energy metabolism. The situation appears quite complicated since the absolute CBF level at stage C is higher than at stages A + B, yet the mitochondria did not exhibit a typical oxidation cycle. It is possible to use the MPA in a gerbil brain ischemic model for compound testing. In order to test the potential protection activity of various drugs, the gerbil brain was exposed to bilateral or unilateral carotid artery occlusion (Mayevsky and Breuer 1992; Mayevsky 1990). Figure 2.21 presents, in a schematic way, the responses of the gerbil brain to unilateral carotid artery occlusion, which led to partial ischemia. In this case, the CBF decreased in two stages due to the partial ischemia that developed. The decrease in CBF was concomitant and correlated to the increase in the mitochondrial NADH and a slow increase in the extracellular level of K^+. At a certain point of the partial ischemic event, ischemic depolarization (ID) was developed. As seen, the CBF decreased to its minimal level

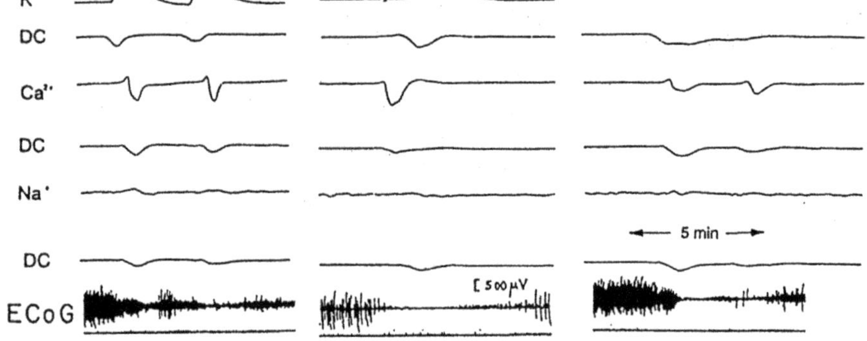

Fig. 2.20 Three responses to spreading depression recorded during control (**a**), immediately (**b**), as well as 240 min after the end of intravenous infusion of 3 ml 30% alcohol (**c**). In this figure, we are presenting the three DC traces recorded around the K, Ca^{2+}, and Na^+ electrodes exhibiting similar responses. RCBF relative CBF; other abbreviations are as in Fig. 2.15 (Mayevsky et al. 2000b)

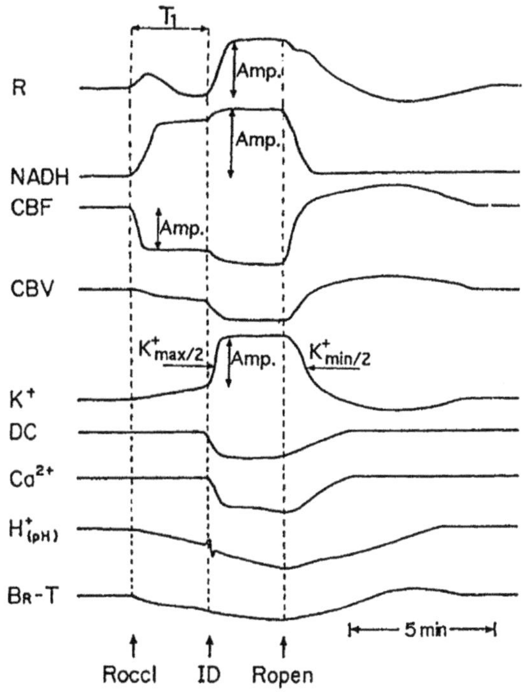

Fig. 2.21 Schematic representation of the typical responses to ischemia in the Mongolian gerbil. R reflectance, NAOH 450 nm corrected NAOH, CBF, CBV cerebral blood flow and volume, K^+, Ca^{2+}, H^+ extracellular levels of ions, DC steady potential, BR-T brain temperature, R_{occl}, R_{open} occlusion and opening of the right carotid artery, ID ischemic depolarization (Mayevsky et al. 2000b)

together with the increase in NADH. The ionic homeostasis was disturbed completely as seen by the steep increase in the K^+ and a rapid decrease in extracellular Ca^{2+}. Also, a large increase in the R trace was recorded due to a massive constriction of blood vessels in the microcirculation. After the reopening of the carotid artery all parameters returned to the preischemic levels within 5–10 min. In order to evaluate the protective effects of a tested compound, various parameters could be calculated regarding timing or amplitude values as seen in the figure. To demonstrate the potential use of the gerbil model for compound evaluation, two possible responses are shown in Figs. 2.22 and 2.23. In each gerbil, two ischemic episodes were induced, one before (A) and one after (B) the IP injection of the tested compound. All the responses shown in Fig. 2.21 could be seen in Fig. 2.22a. Namely, that due to the occlusion of the right carotid artery ischemic depolarization was developed. In Fig. 22b, most of the responses to the same length of ischemia were blocked by the protective drug. The decrease in CBF as well as the increase in NADH (CF) are the same as in Fig. 22a. The extracellular K^+ was slightly elevated but the second fast increase due to the lack of energy was prevented. Also, all other parameters were unchanged, including the ECoG. Figure 2.23 shows

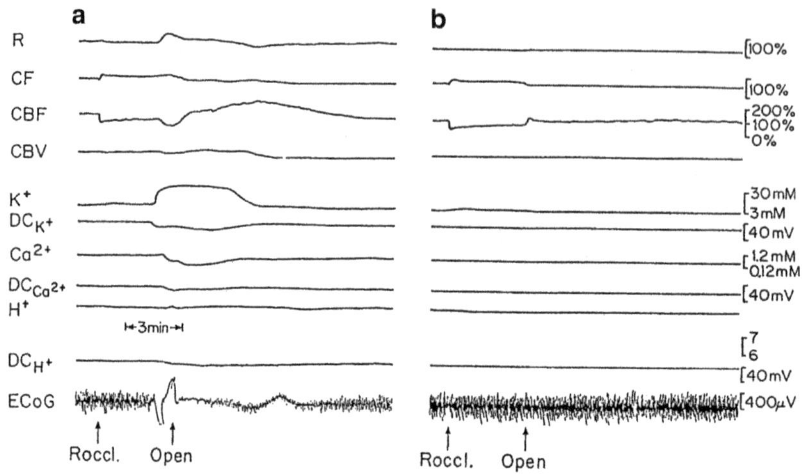

Fig. 2.22 Effects of unilateral carotid artery occlusion (R_{occl}) on various gerbil brain activities monitored before (**a**) and after (**b**) injection of anti-ischemic protection drug. All abbreviations are as in Fig. 2.19 (Mayevsky et al. 2000b)

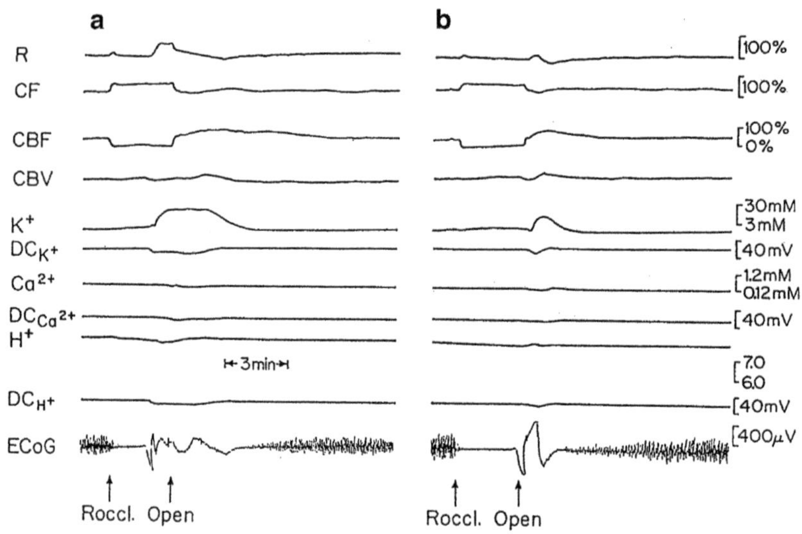

Fig. 2.23 Effects of unilateral carotid artery occlusion (R_{occl}) on various gerbil brain activities monitored before (**a**) and after (**b**) injection of anti-ischemic protection drug. All abbreviations are as in Fig. 2.15 (Mayevsky et al. 2000b)

the effects of the same tested compound in a second gerbil. Here again, in (A) all the expected responses to ischemia were recorded in all parameters. In this gerbil, the compound injected led to very clear effects on the responses (B) to ischemia, although the ID was prevented, as in Fig. 2.22b. The time to record the ID was prolonged and also the amplitude of changes in the R trace or the K^+ was smaller.

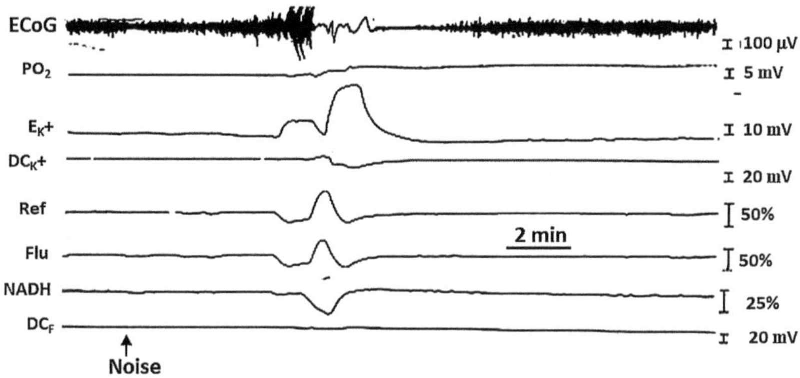

Fig. 2.24 The responses of the brain to an audio stimulation (noise) evoking epileptic seizures (Mayevsky and Rogatsky 2007)

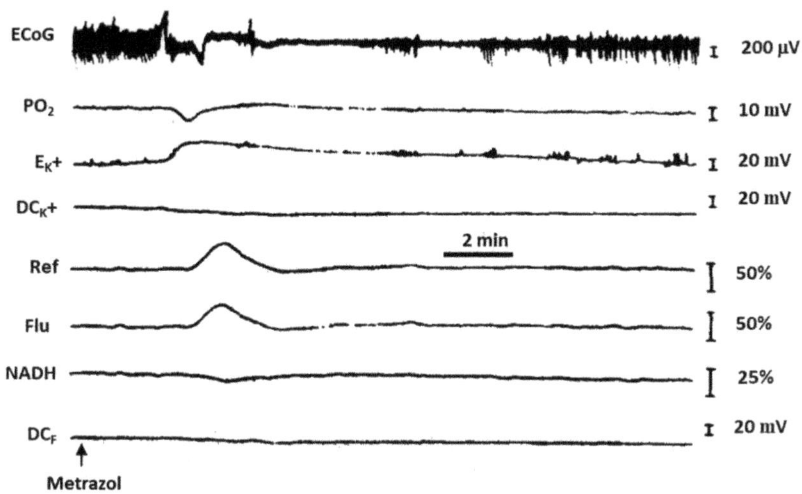

Fig. 2.25 Multiparametric responses of the brain to injection of Metrazol evoking epileptic seizures (Mayevsky 1983a)

The effects of epileptic activity while using the MPA system are presented in Figs. 2.24 and 2.25. Figure 2.24 presents the progression of seizure activity in a special strain of gerbils, developing epileptic effects as a result of monotonic noise or other factors (Mayevsky et al. 1986; Mayevsky and Rogatsky 2007). As seen, the exposure of the awake gerbil to noise resulted in the development of epileptic activity followed by a CSD wave. The coupling between the two pathological events is clear, and manifested by the electrical activity (ECoG) as well as extracellular K^+ levels and NADH redox state. During the epileptic stage, extracellular potassium increased and started to recover to the baseline, but then a larger elevation was recorded. There is a clear correlation between the various parameters during epilepsy and CSD. The decrease in NADH was smaller during the first stage, followed by an oxidation cycle typical for CSD.

The effects of partial brain ischemia on the responses to epileptic activity are shown in Fig. 2.25. In this rat, the two carotid arteries were ligated 20 h prior to the experiment. During the discharge activity, K^+ increased in the extracellular space and then spreading depression developed. The main effects of the ischemia are on the recovery processes due to the slow oxygen delivery to the cells.

The PO_2 showed a decrease in the cycle, indicating that the compensatory mechanism of increased blood flow is impossible as measured also by the increased reflectance. The short oxidation cycle of NADH found in the normoxic brain (Figs. 6.26 and 6.27 in paper by AM; Mayevsky 1983a) is very unclear in the ischemic brain. Due to the slow supply of energy, the recovery of K^+ into the cells was much slower as compared with normoxic brain. The electrical activity of the brain was also affected, as could be seen in the slow recovery of the ECoG and DC traces after the SD episode (Figs. 2.26 and 2.27).

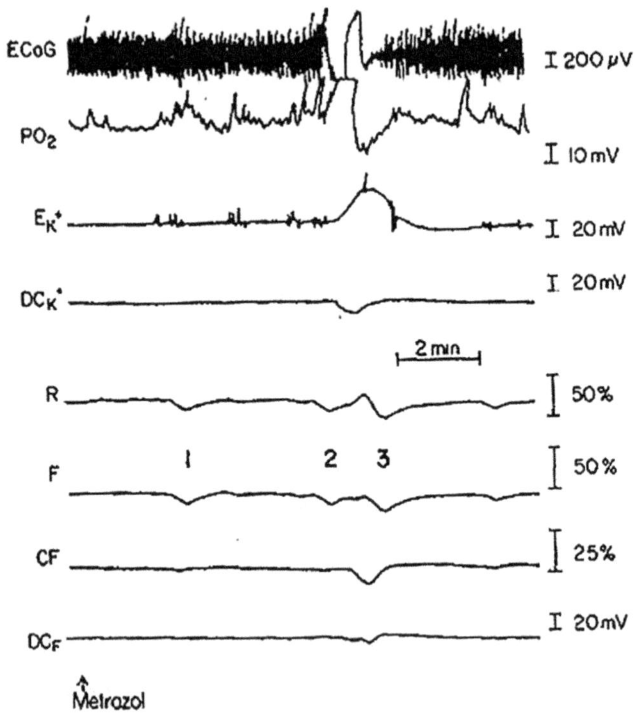

Fig. 2.26 The effects of Metrazol injection (intraperitoneal) on the development of seizure activity and the metabolic, ionic, and electrical responses of the awake rat

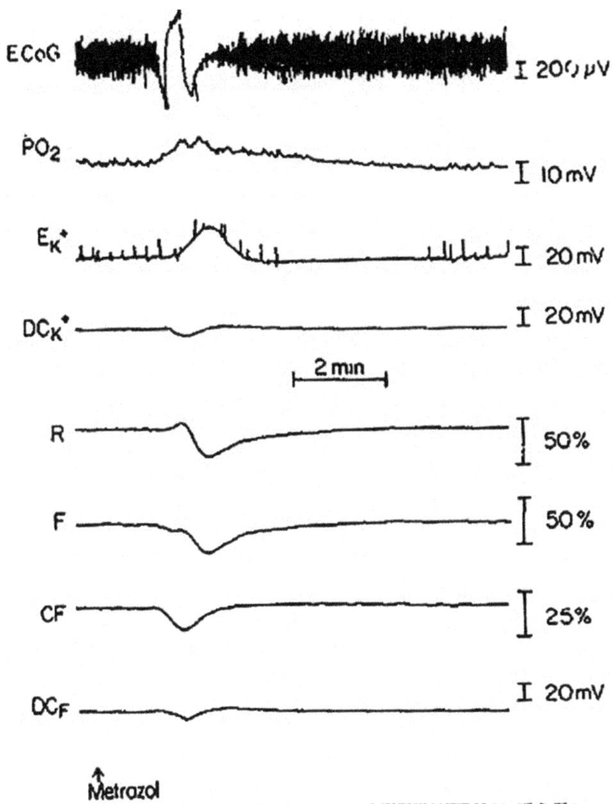

Fig. 2.27 A spreading depression phenomenon was recorded after the injection of Metrazol. The preceding epileptic activity is not clear in the record

2.4 Discussion and Conclusions

Most of the techniques developed for real-time evaluation of tissue and organ function are based on a single parameter monitoring. The MPA in the present study opened up new possibilities that enabled us to analyze and diagnose the physiological and biochemical state of various organs in real time. We applied the same approach to study various organs in the body exposed to pathophysiological situations as well as to various drugs. The brain was the main target organ tested, as shown, but other organs, such as the heart, liver, or kidney, were tested as well. Using the canine heart in vivo we have shown that the same MPA developed could be sutured to the epicaridium of the beating heart (Osbakken and Mayevsky 1996). The same basic approach was applied to the rat liver exposed to reversible ischemia (Barbiro-Michaely et al. 1998). Recently, we used the MPA approach to monitor the kidney of rats as well as during human transplantation procedure (Mayevsky et al. 2000a). The introduction of multiparametric monitoring devices is a natural trend after long years of brain monitoring (in head-injured patients) with single or dual parameter instruments (Sioutos et al. 1996; Miller et al. 1996). Indeed, few attempts have been made to increase the number of monitored parameters. Kirkpatrick et al. (1996) have shown that the multimodality monitoring system used provided a useful tool for assessing the brain of comatose patients. In their approach, only hemodynamic parameters from the brain tissue were recorded (ICP, CBF). In our approach, we combined a set of parameters representing hemodynamic, metabolic, ionic, and electrical activities measured simultaneously in severely head-injured patients (Mayevsky and Brabiro-Michaeli 2013b). The development of multiparametric monitoring systems for the diagnosis of compound effects will increase the efficiency of the evaluation of the mechanism of action of the specific compound tested. Not only the changes in the individual parameters but also the interrelation between them will serve as an indicator for the responses monitored under compound influence. The changes (increase or decrease) in CBF may identify the occurrence of a vascular event effecting perfusion, regardless of the efficiency of utilization of this blood by the tissue. At that point, if the coupling between the flow and O_2 consumption is intact and maximum O_2 extraction has been attained a decrease in CBF will be accompanied by an increase in the NADH redox state, indicating metabolic stress. On the other hand, when uncoupling occurs between CBF and O_2 consumption, no complementary change in the NADH redox state will be seen. When SD is induced under relatively normal conditions, complete coupling occurs between the increase in O_2 consumption and the large increase in CBF. On the other hand, in under-limited O_2 supply (ischemia, hypoxia), the responses of the brain to SD are altered, as shown also in Fig. 2.20 when alcohol was infused to the rat. The combination of the NADH probe with the CBF probe used in the present study indicated that the reversed NADH response (an increase instead of a decrease) is due to the limited CBF compensation, as indicated by the initial decreased CBF response (Fig. 2.20). Determination of critical values of flow decrease is not simple, and therefore the combination with the NADH redox state data and K^+ levels helps the identification of critical transition points in the development of a pathological state under compound effect. The extracellular K^+ level under normal conditions is in the range of 3–4 mM and any malfunction of the energy supply systems will be reflected by an increase in K^+ due to an inhibition of the Na^+/K^+-ATPase activity. The degree of K^+ increase during pathological states will enable the determination of the degree of brain damage as well as the recovery from such damage (decrease in K^+ levels). Animal experiments have shown that it is necessary to monitor as many parameters as possible simultaneously and in real time. The combination of monitoring ICP and ECoG, as well as the MPA, enables better diagnosis of the functional state of the brain. It is also well accepted that brain temperature is correlated with the metabolic usage of energy resources. Therefore, monitoring temperature has significant value in directing the treatment to avoid high brain temperature and brain damage. We want to emphasize that this MPA system was tested in many hundreds of animal experiments, including control, untreated animals monitored continuously for up to 24 h. It was found that all signals remained within the normal range during the entire monitoring period. None of the sensors penetrates the brain itself. Therefore, damage to the tissue is minimal and complete recovery is expected after the monitoring period and removal of the MPA and the holder.

References

Acad, B., J. Sonn, E. Furman, M. Scheinowitz, and J. Kedem. 1987. Specific effects of nitroprusside on myocardial O_2 balance following coronary ligation in the dog heart. *Journal of Cardiovascular Pharmacology* 9:79–86.

Amran-Cohen, D., J. Sonn, M. Luger-Hamer, and A. Mayevsky. 2003. The effect of ischemia and hypoxia on renal blood flow, energy metabolism and function *in vivo*. *Advances in Experimental Medicine and Biology* 540:93–101.

Balaban, R. S., and L. J. Mandel. 1988. Metabolic substrate utilization by rabbit proximal tubule. An NADH fluorescence study. *The American Journal of Physiology* 254:F407–F416.

Barbiro-Michaely, E., Y. Zurovsky, and A. Mayevsky. 1998. Real time monitoring of rat liver energy state during ischemia. *Microvascular Research* 56:253–260.

Bradley, R. S., and M. S. Thorniley. 2005. A review of attenuation correction techniques for tissue fluorescence. *Journal of the Royal Society Interface* 3:1–13.

Bures, J., O. Buresova, and J. Krivanek. 1974. *The mechanism and applications of Leao's spreading depression of electroencephalographic activity*. New York: Academic.

Chance, B., and B. Schoener. 1962. Correlation of oxidation-reduction changes of intracellular reduced pyridine nucleotide and changes in electro-encephalogram of the rat in anoxia. *Nature* 195: 956–958.

Chance, B., and B. Thorell. 1959a. Fluorescence measurements of mitochondrial pyridine nucleotide in aerobiosis and anaerobiosis. *Nature* 184:931–934.

Chance, B., and B. Thorell. 1959b. Localization and kinetics of reduced pyridine nucleotide in living cells by microfluorometry. *The Journal of Biological Chemistry* 234:3044–3050.

Chance, B., and G. R. Williams. 1955a. Respiratory enzymes in oxidative phosphorylation (I-Kinetics of oxygen utilization). *The Journal of Biological Chemistry* 217:383–393.

Chance, B., and G. R. Williams. 1955b. Respiratory enzymes in oxidative phosphorylation (III-The steady state). *The Journal of Biological Chemistry* 217:409–427.

Chance, B., and G. R. Williams. 1956. The respiratory chain and oxidative phosphorylation. In *Advances in enzymology*. XVII edition. ed. F. F. Nord. New York: Interscience Publisher, Inc.

Chance, B., P. Cohen, F. Jobsis, and B. Schoener. 1962a. Intracellular oxidation-reduction states *in vivo*. *Science* 137:499–508.

Chance, B., P. Cohen, F. Jobsis, and B. Schoener. 1962b. Localized fluorometry of oxidation-reduction states of intracellular pyridine nucleotide in brain and kidney cortex of the anesthetized rat. *Science* 136:325.

Chance, B., V. Legallias, and B. Schoener. 1962c. Metabolically linked changes in fluorescence emission spectra of cortex of rat brain, kidney and adrenal gland. *Nature* 195:1073–1075.

Chance, B., N. Oshino, T. Sugano, and A. Mayevsky. 1973. Basic principles of tissue oxygen determination from mitochondrial signals. *Advances in Experimental Medicine and Biology* 37A: 277–292.

Chance, B., B. Schoener, and S. Elsaesser. 1964. Control of the waveform of oscillations of the reduced pyridine uncelotide level in a cell-free extract. *Proceedings of the National Academy of Sciences United States of America* 52:337–341.

Crowe, W., A. Mayevsky, and L. Mela. 1977. Application of a solid membrane ion selective electrode to *in vivo* measurements. *The American Journal of Physiology* 233:C56–C60.

Crowe, W., A. Mayevsky, L. Mela, and I. Silver. 1976. Measurements of extracellular potassium, D.C. potential and ECoG in the cortex of the conscious rat during cortical spreading depression. In *Ion and enzyme electrodes in biology and medicine*, eds. M. Kessler, L. C. Clark, D. W. Lubbers, I. A. Silver, and W. Simon. Munich: Urban and Schwarzenberg.

Dirnagl, U., B. Kaplan, M. Jacewicz, and W. Pulsinelli. 1989. Continuous measurement of cerebral cortical blood flow by laser-Doppler flowmetry in a rat stroke model. *Journal of Cerebral Blood Flow & Metabolism* 9:589–596.

Dora, E., L. Gyulai, and A. G. B. Kovach. 1984. Determinants of brain activation-induced cortical NAD/NADH responses *in vivo*. *Brain Research* 299:61–72.

Erecinska, M., and I. A. Silver. 1989. ATP and brain function. *Journal of Cerebral Blood Flow and Metabolism* 9:2–19.

Haberl, R. L., M. L. Heizer, A. Marmarou, and E. F. Ellis. 1989. Laser-Doppler assessment of brain microcirculation: Effect of systemic alterations. *American Journal of Physiology. Heart and Circulatory Physiology* 256:H1247–H1254.

Harbig, K., B. Chance, A. G. B. Kovach, and M. Reivich. 1976. *In vivo* measurement of pyridine nucleotide fluorescence from cat brain cortex. *Journal of Applied Physiology: Respiratory, Environmental and Exercise Physiology* 41:480–488.

Harden, A., and W. Young. 1906. Alcoholic ferment of yeast-juice Part II Co-ferment of yeast-juice. *Proceedings of the Royal Society* B78:369–375.

Haselgrove, J., C. Barlow, E. Eleff, B. Chance, and S. Lebordais 1981. Correlation of electrical signals and mitochondrial redox state during spreading depression. In Advances in Physiological Science, 25 edition. eds. A. G. B. Kovach, E. Dora, M. Kessler, and I. A. Silver. Budapest: Pergamon Press.

Ince, C., J. M. C. C. Coremans, and H. A. Bruining.1992. *In vivo* NADH fluorescence. In Adv. Exp. Med.: Oxygen Transport to Tissue XIV. eds. W. Erdmann and D. F. Bruley. New York: Plenum Press.

Jobsis, F. F., M. O'Connor, A. Vitale, and H. Vreman. 1971. Intracellular redox changes in functioning cerebral cortex. I. Metabolic effects of epileptiform activity. *Journal of Neurophysiology* 3465: 735–749.

Kirkpatrick, P. J., P. Smielewski, S. Piechnik, J. D. Pickard, and M. Czosnyka. 1996. Early effects of mannitol in patients with head injuries assessed using bedside multimodality monitoring. *Neurosurgery* 39:714–720. Discussion 720–721.

Kovach, A. G. B., E. Dora, and L. Gyulai. 1984. Relationship between steady redox state and brain activation- induced NAD/NADH redox responses. *Advances in Experimental Medicine and Biology* 169:81–100.

Kraut, A., Y. Zurovsky, and A. Mayevsky. 1998. Simultaneous hemodynamic and metabolic responses to oxygen deficiency in the brain and other organs of the rat. *Neuroscience Letters. Supplement* 51:S24.

Lamanna, J. C., and M. Rosenthal. 1975. Effect of ouabain and phenobarbital on oxidative metabolic activity associated with spreading cortical depression in cats. *Brain Research* 88:145–149.

Lamanna, J. C., S. J. Peretsman, A. I. Light, and M. Rosenthal. 1981. GR; NADH. Oxygen sufficiency in the "working" brain. In *Advances in physiological sciences. Oxygen transport to tissue*, eds. A. G. B. Kovach, E. Dora, and I. A. Silver, 95–96. Budapest am: Akademiai Kiado.

Leão, A. A. P., and R. Morison. 1945. Propagation of spreading cortical depression. *Journal of Neurophysiology* 8:33–45.

Leão, A. A. P. 1944a. Pial circulation and spreading depression of activity in the cerebral cortex. *Journal of Neurophysiology* 7:391–396.

Leão, A. A. P. 1944b. Spreading depression of activity in cerebral cortex. *Journal of Neurophysiology* 7: 359–390.

Lothman, E., J. Lamanna, G. Cordingley, M. Rosenthal, and G. Somjen. 1975. Responses of electrical potential potassium levels, and oxidative metabolic activity of the cerebral neocortex of cats. *Brain Research* 88:15–36.

Lubbers, D. W.. 1995. Optical sensors for clinical monitoring. *Acta Anaesthesiologica Scandinavica Supplementum* 39:37–54.

Manor, T., S. Meilin, and A. Mayevsky. 1995. A new "multiprobe-multisite" system for *in vivo* monitoring of traumatic brain injury in the rat. In *4th Annual meeting of Israel Society for Neuroscience*, Eilat, Israel.

Mayevsky, A., and E. Barbiro-Michaely. 2013a. Shedding light on mitochondrial function by real time monitoring of NADH fluorescence: I. Basic methodology and animal studies. *Journal of Clinical Monitoring and Computing* 27:1–34.

Mayevsky, A., and E. Barbiro-Michaely. 2013b. Shedding light on mitochondrial function by real time monitoring of NADH fluorescence: II: human studies. *Journal of Clinical Monitoring and Computing* 27:125–145.

Mayevsky, A., and Z. Breuer. 1992. Brain vasculature and mitochondrial responses to ischemia in gerbils. I. Basic anatomical patterns and biochemical correlates. *Brain Research* 58:242–250.

Mayevsky, A., and B. Chance. 1973. A new long-term method for the measurement of NADH fluorescence in intact rat brain with implanted cannula. *Advances in Experimental Medicine and Biology* 37A:239–244.

Mayevsky, A., and B. Chance. 1974. Repetitive patterns of metabolic changes during cortical spreading depression of the awake rat. *Brain Research* 65:529–533.

Mayevsky, A., and B. Chance. 1975. Metabolic responses of the awake cerebral cortex to anoxia hypoxia spreading depression and epileptiform activity. *Brain Research* 98:149–165.

Mayevsky, A., and B. Chance. 1982. Intracellular oxidation-reduction state measured *in situ* by a multicannel fiber-optic surface fluorometer. *Science* 217:537–540.

Mayevsky, A., and G. Rogatsky. 2007. Mitochondrial function *in vivo* evaluated by NADH fluorescence: From animal models to human studies. *American Journal of Physiology. Cell Physiology* 292:C615–C640.

Mayevsky, A., and D. S. Sclarsky. 1983. Correlation of brain NADH redox state, K^+, PO_2 and electrical activity during hypoxia, ischemia and spreading depression. *Advances in Experimental Medicine and Biology* 159:129–141.

Mayevsky, A., and B. Shaya. 1980. Factors affecting the development of hyperbaric oxygen toxicity in the awake rat brain. *Journal of Applied Physiology: Respiratory, Environmental and Exercise Physiology* 49:700–707.

Mayevsky, A., and H. R. Weiss. 1991. Cerebral blood flow and oxygen consumption in cortical spreading depression. *Journal of Cerebral Blood Flow & Metabolism* 11:829–836.

Mayevsky, A. 1976. Brain energy metabolism of the conscious rat exposed to various physiological and pathological situations. *Brain Research* 113:327–338.

Mayevsky, A. 1978a. Ischemia in the brain: the effects of carotid artery ligation and decapitation on the energy state of the awake and anesthetized rat. *Brain Research* 140:217–230.

Mayevsky, A. 1978b. The responses of an awake brain to HPO under increased CO_2 concentration. *Advances in Experimental Medicine and Biology* 92:735–740.

Mayevsky, A. 1983a. Metabolic, ionic and electrical responses to experimental epilepsy in the awake rat. In *Proc. first internat. cong. cerebral blood flow, metabolism & epilepsy*, eds. M. Baldy, D. H. Moulinier, D. H. Ingvar, and B. S. Meldrum. England: John Libbey.

Mayevsky, A. 1983b. Multiparameter monitoring of the awake brain under hyperbaric oxygenation. *Journal of Applied Physiology: Respiratory, Environmental and Exercise Physiology* 54:740–748.

Mayevsky, A. 1984a. Brain NADH redox state monitored *in vivo* by fiber optic surface fluorometry. *Brain Research Reviews* 7:49–68.

Mayevsky, A. 1990. Level of ischemia and brain functions in the Mongolian gerbil *in vivo*. *Brain Research* 524:1–9.

Mayevsky, A. 1984b. Brain oxygen toxicity. Invited review. In *8th Symp. under-water physiol*, eds. A. J. Bachrach and M. M. Matzen, 69–89. Bethesda, Maryland: Undersea Medical Society Inc. 6097.

Mayevsky, A., W. Crowe, and L. Mela. 1980. The interrelation between brain oxidative metabolism and extracellular potassium in the unanesthetized gerbil. *Neurological Research* 1:213–226.

Mayevsky, A., A. Doron, T. Manor, S. Meilin, N. Zarchin, and G. E. Ouaknine. 1996. Cortical spreading depression recorded from the human brain using a multiparametric monitoring system. *Brain Research* 740:268–274.

Mayevsky, A., A. Doron, S. Meilin, T. Manor, E. Ornstein, and G. E. Ouaknine. 1999a. Brain viability and function analyzer: Multiparametric real-time monitoring in neurosurgical patients. *Acta Neurochirurgica. Supplementum* 75:63–66.

Mayevsky, A., K. H. Frank, S. Nioka, M. Kessler, and B. Chance, B. 1990. Oxygen supply and brain function *in vivo*: a multiparametric monitoring approach in the Mongolian gerbil. In *Oxygen transport to tissue XII*, eds. J. Piper, T. K. Goldstick, and M. Meyer. New York: Plenum Pub.

Mayevsky, A., D. Jamieson, and B. Chance. 1974a. Oxygen poisoning in the unanesthetized brain: Correlation of the oxidation-reduction state of pyridine nucleotide with electrical activity. *Brain Research* 76:481–491.

Mayevsky, A., H. Kaplan, J. Haveri, J. Haselgrove, and B. Chance. 1986. Three-dimensional metabolic mapping of the freeze-trapped brain: Effects of ischemia on the Mongolian gerbil. *Brain Research* 367:63–72.

Mayevsky, A., T. Manor, E. Pevzner, A. Deutsch, R. Etziony, N. Dekel, and A. Jaronkin. 2004. Tissue spectroscope: A novel *in vivo* approach to real time monitoring of tissue vitality. *Journal of Biomedical Optics* 9:1028–1045.

Mayevsky, A., S. Meilin, T. Manor, N. Zarchin, and J. Sonn. 1999b. Optical monitoring of NADH redox state and blood flow as indicators of brain energy balance. *Advances in Experimental Medicine and Biology* 471:133–140.

Mayevsky, A., R. Nakache, M. Luger-Hamer, D. Amran, and J. Sonn. 2001. Assessment of transplanted kidney vitality by a multiparametric monitoring system. *Transplantation Proceedings* 33: 2933–2934.

Mayevsky, A., R. Nakache, H. Merhav, M. Luger-Hamer, and J. Sonn. 2000a. Real time monitoring of intraoperative allograft vitality. *Transplantation Proceedings* 32:684–685.

Mayevsky, A., E. Ornstein, S. Meilin, N. Razon, and G. E. Ouaknine. 2002. The evaluation of brain CBF and mitochondrial function by a fiber optic tissue spectroscope in neurosurgical patients. *Acta Neurochirurgica. Supplement* 81:367–371.

Mayevsky, A., G. G. Rogatsky, and J. Sonn. 2000b. New multiparametric monitoring approach for real-time evaluation of drug tissue interaction *in vivo*. *Drug Development Research* 50:457–470.

Mayevsky, A., N. Zarchin, H. Kaplan, J. Haveri, J. Haselgrove, and B. Chance. 1983. Brain metabolic responses to ischemia in the Mongolian gerbil: *In vivo* and freeze trapped redox state scanning. *Brain Research* 276:95–107.

Mayevsky, A., T. Zeuthen, and B. Chance. 1974b. Measurements of extracellular potassium, ECoG and pyridine nucleotide levels during cortical spreading depression in rats. *Brain Research* 76:347–349.

Miller, J. D., I. R. Piper, and P. F. Statham. 1996. ICP monitoring indications and techniques. In *Neurotrauma*, eds. R. K. Narayan, J. E. Wilberger, and J. T. Povlishock. New York: McGraw-Hill.

Osbakken, M., and A. Mayevsky. 1996. Multiparameter monitoring and analysis of *in vivo* ischemic and hypoxic heart. *Journal of Basic and Clinical Physiology and Pharmacology* 7:97–113.

Rampil, I. J., L. Litt, and A. Mayevsky. 1992. Correlated, simultaneous, multiple-wavelength optical monitoring *in vivo* of localized cerebrocortical NADH and brain microvessel hemoglobin oxygen saturation. *Journal of Clinical Monitoring* 8:216–225.

Rosenthal, M., and F. F. Jobsis. 1971. Intracellular redox changes in functioning cerebral cortex. II. Effects of direct cortical stimulation. *Journal of Neurophysiology* 34:750–762.

Rosenthal, M., and G. Somjen. 1973. Spreading depression, sustained potential shifts, and metabolic activity of cerebral cortex of cats. *Journal of Neurophysiology* 36:739–749.

Sioutos, P. J., J. A. Orozco, and L.P. Carter. 1996. Regional cerebral blood flow techniques. In *Neurotrauma*, eds. R. K. Narayan, J. E. Wilberger, and J. T. Povlishock. New York: McGraw-Hill.

Somjen, G. G., M. Rosenthal, G. Cordingley, J. Lamanna, and E. Lothman. 1976. Potassium, neuroglia, and oxidative metabolism in central gray matter. *Federation Proceedings* 35:1266–1271.

Sonn, J., and A. Mayevsky. 2000. Effects of brain oxygenation on metabolic, hemodynamic, ionic and electrical responses to spreading depression in the rat. *Brain Research* 882:212–216.

Stern, M. D., D. L. Lappe, P. D. Bowen, J. E. Chimosky, G. A. Holloway Jr., H. R. Keiser, and R. L. Bowman. 1977. Continuous measurement of tissue blood flow by laser-Doppler spectroscopy. *The American Journal of Physiology* 232:H441–H448.

Sylvia, A. L., and M. Rosenthal. 1979. Effects of age on brain oxidative metabolism *in vivo*. *Brain Research* 165:235–248.

Zarchin, N., and A. Mayevsky. 1981. The effects of age on the metabolic and electrical responses to decapitation in the awake and anesthetized rat brain. *Mechanisms of Ageing and Development* 16: 285–294.

Chapter 3
Real-Time Brain Responses to Various Models of Ischemia

Abstract The metabolic state of the awake rat's cerebral cortex was evaluated by monitoring the intramitochondrial NADH redox state using the fiber optic-based surface fluorometry/reflectometry in real time. The blood supply to the brain was altered by occlusion of one or more blood vessels (arteries), for various intervals of time. Also, the evaluation of the energy state of the brain was tested by exposing the cerebral cortex to Cortical Spreading Depression (CSD) in normoxia and ischemia. The results show that in the rat even bilateral carotid arteries occlusion induced only partial ischemia due to the existence of the circle of Willis. In order to increase the level of ischemia in the rat, it was necessary to occlude three to four blood vessels providing blood to the brain. The metabolic response to CSD showed an oxidation cycle before occlusion and a few hours after the occlusion a reduction cycle of NADH was recorded. The effect of occlusion was reversible, and 7–8 days after the occlusion, the response to CSD was as before the occlusion. The effects of two types of anesthetics as well as partial ischemia on the response to complete ischemia (decapitation) were tested. The rate of NADH increase after decapitation was also significantly faster in the awake brain in comparison to the anesthetized one.

Keywords NADH redox state · Cortical spreading depression · Fiber optic fluorometry · Partial ischemia · Complete ischemia · Brain metabolism · In vivo monitoring

3.1 Introduction

Various animal models of brain ischemia were developed and used in order to study the responses of brain pathophysiology recorded in real time. Chance et al. (1963) tested the effect of decapitation on brain NADH redox state measured in the mouse.

A. Mayevsky et al., *Physiological Mapping of the Cerebral Cortex in Neuropathology*,
https://doi.org/10.1007/978-3-032-13019-8_3

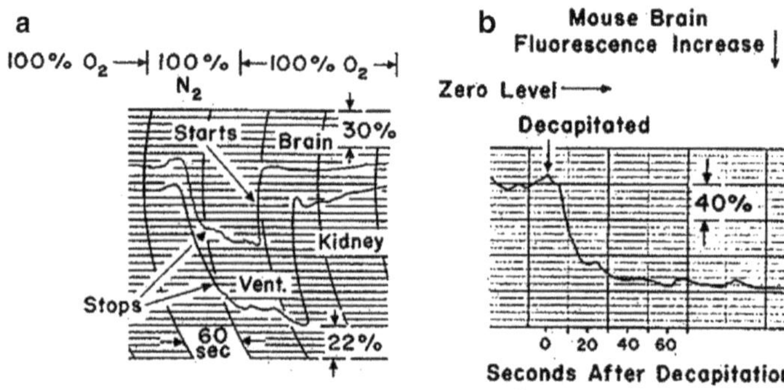

Fig. 3.1 (**a**) Simultaneous recordings of fluorescence changes in rat brain and rat kidney in a cycle of anoxia. Fluorescence increases are indicated in a downward direction (Chance et al. 1963). (**b**) Record of the kinetics of fluorescence changes of the mouse brain cortex in ischemia caused by decapitation of a mouse (Chance et al. 1963)

They compared the responses to ischemia induced by decapitation, seen in Fig. 3.1b (complete ischemia) to those measured in the rat's brain under anoxia (100% nitrogen) seen in Fig. 3.1a. Chance et al. (1963) discussed the results obtained in Fig. 3.1 as seen in the next paragraph:

> ***"Comparison with ischemia by decapitation.*** One would expect that breathing nitrogen would lead to a very rapid transition from aerobiosis to anaerobiosis since it has already been pointed out in this symposium that breathing nitrogen would serve to deplete the blood oxygen (Chance and Schoener 1963). As shown in Fig. 3.1a, about 8 s are required between the start of DPN reduction and the attainment of half maximal reduction. In an attempt to observe the time for DPN reduction in ischemia, we have employed a decapitation technique with the mouse; the optical system is arranged so that the slight mechanical artifact occurring on decapitation would not disturb the fluorescence excitation. A typical experimental result is shown in Fig. 3.1b. In this case, the time between the start of DPN reduction and reaching the half-maximal value is about 5 s. In other words, there is almost as rapid a transition from aerobiosis to anoxia in the nitrogen-breathing phenomena as in a decapitation procedure—with, of course, the advantage of reversibility in the case of anoxia."

In order to monitor the effects of brain ischemia it is important to use an animal model that will be easy and simple to prepare as well as will create a wide range of ischemia after occlusion of the related blood vessels. Figure 3.2 presents the arterial blood supply to the brain of a rat (a) and a gerbil (b). The "circle of Willis" in the rat is complete while in the gerbil it is variable between different animals. The meaning of this difference in the architecture of blood supply to the brain, between the rat and gerbil, is that the gerbil was adopted as a model in ischemia studies since the middle of the 1960s. Recently, I published a book on the Mongolian Gerbil brain (Mayevsky 2024) in which the various aspects of its unique blood circulation to the brain were described in detail. In the current chapter, the rat's brain as a model to ischemia is presented.

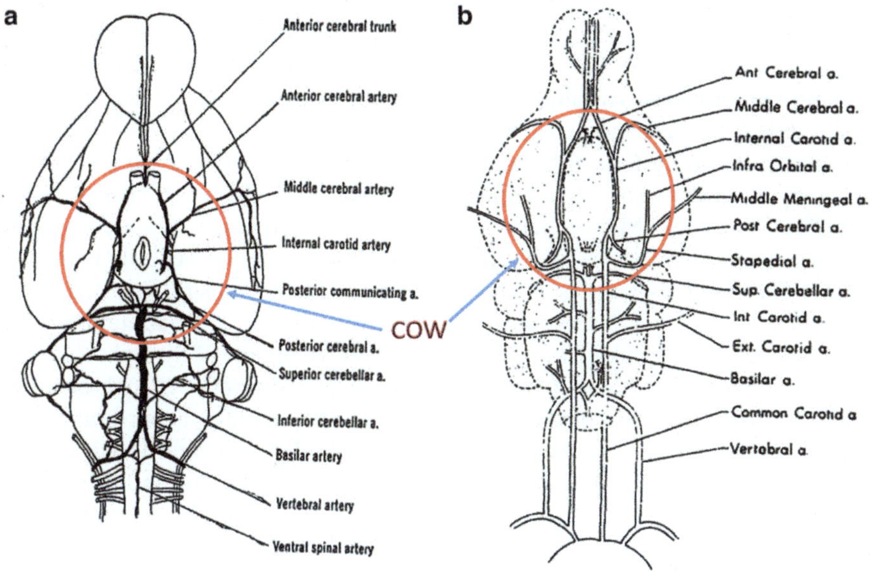

Fig. 3.2 (**a**) Ventral view of the rat's brain and its arterial system including the Circle of Willis (Zeman and Innes 1963). (**b**) General schematic presentation of the blood vessels originating from the aorta and supplying the gerbil head and brain (Berry et al. 1975). (Reprinted with permission from Elsevier)

3.2 Preliminary Study 1

In 1976, we developed a new rat model exposed to decapitation in the unanesthetized state (Mayevsky and Chance 1976).

3.2.1 Introduction and Methods

The effects of complete ischemia on brain energy metabolism have been studied in various animal preparation (Lowry et al. 1964; Folbergrová et al. 1970; Swaab and Boer 1972; Gercken and Preuss 1969). One of the approaches on studying those effects is to convert the brain into a closed system, by decapitation, and after fixation of the tissue to measure changes in various metabolites such as glucose, lactate, ATP, and Creatine phosphate. The freezing procedure avoids most of the postmortem changes of brain metabolites levels, but the events of the first few seconds after decapitation are obscure. The development of the surface fluorometry technique to measure oxidation-reduction state of NADH in the anesthetized and the awake rat (Mayevsky and Chance 1973; Chance et al. 1962, 1973) make the study of the effect of decapitation (a closed system) possible on NADH levels in the brain. In the present communication, we describe the effects of decapitation on brain NADH fluorescence and electrical activity of the awake rat. The modified time-sharing fluorometer-reflectometer was used in all experiments. This fluorometer enables the measurement of NADH fluorescence (450 nm) and changes in intensity of reflected light at the excitation wavelength (366 nm) from the surface of brain cortex. The coupling between the fluorometer and the brain was done by a fused light guide (2 mm active diameter) made of quartz fibers (Schott Comp., Mainz, Germany). Figure 3.3 shows the experimental setup. Twelve male Wistar rats were used (180–220 g) and were anesthetized with Equithesin (Jensen-Salsbery Laboratories, USA) 0.5–0.6 ml injected intraperitoneally. The animal was mounted on a head holder for the operation time. The parietal bone was exposed and a hole (6 mm diameter) was drilled in the middle of the right side. Two stainless steel screws were implanted above each hemisphere for ECoG recording. A Plexiglas cannula (light pipe holder) was screwed epidurally (about 1 mm deep). The cannula and the screws were cemented to the skull using dental acrylic cement. Figure 3.4a shows the location of the cannula and the screws. The animal was connected to the fluorometer (30 min after operation) by inserting the common part of the light pipe into the cannula. This procedure avoids any type of movement artifact during the decapitation. Three to four hours after operation, when the animal was fully awake, decapitation was done using a guillotine (Stoelting Co.) as seen in Fig. 3.4b.

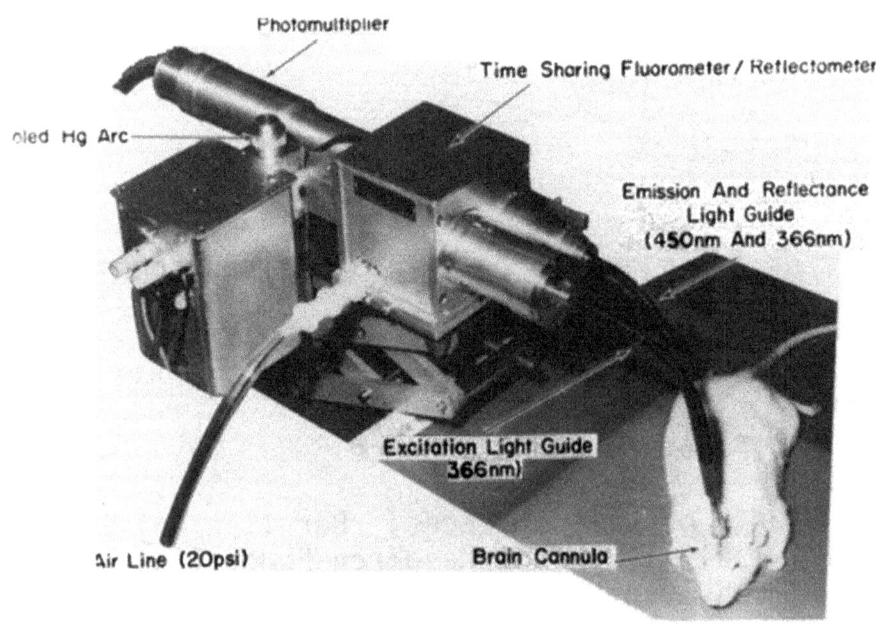

Fig. 3.3 An operated rat implanted with cannula connected to the time sharing fluorometer/reflectometer (Mayevsky and Chance 1976)

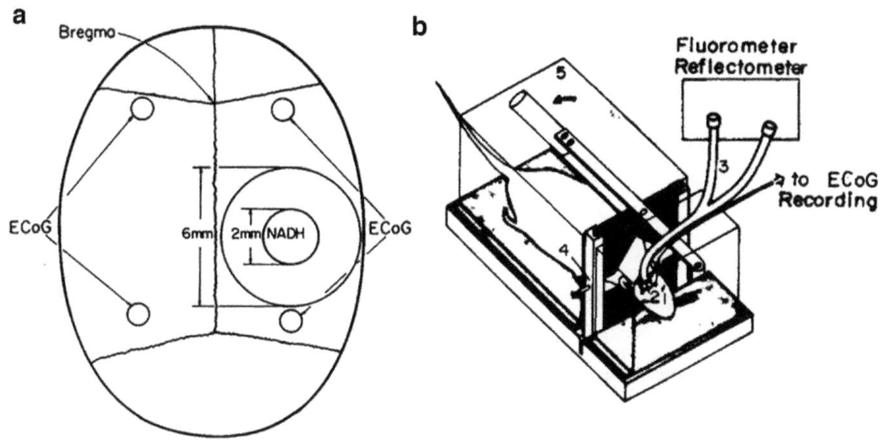

Fig. 3.4 (**a**) Schematic presentation of the location of the cannula and the ECoG electrodes implanted above the brain. (**b**) A special cage used for decapitation of a rat while continuous monitoring of brain NADH and electrical activity—ECoG

Measurements of ECoG, NADH fluorescence and reflectance were taken during, as well as after, decapitation (10–15 min).

3.2.2 *Results and Discussion*

The increase in NADH levels during N_2–O_2 breathing cycle was used as a test for the energetic state of the brain (Mayevsky and Chance 1975). All animals were exposed to N_2 for 1 min and the increase in NADH fluorescence under N_2 was 40%–50% in comparison to the normoxic level, which has the same range as reported previously (Mayevsky and Chance 1973,1975).

Figure 3.5 shows a typical response of the awake brain to decapitation. The upper trace shows that the reflectance of 366 nm light increases immediately after decapitation by 20% and reached a plateau within 10 s. In most animals, this increase was smaller than 20% and in few even undetectable. Two minutes later, a secondary increase in reflectance was recorded. This secondary increase in reflectance was recorded in ten out of 12 animals. The second trace from the top shows the response of the 450 nm fluorescence. In this animal, the maximum level of fluorescence was 130% above the normoxic level. The third trace shows the corrected fluorescence (the difference between the first and the second traces) response to decapitation. This correction technique was described earlier by others (Jobsis et al. 1971; Harbig and Rievich 1973) and us (Mayevsky and Chance 1973; Chance et al. 1973) and the correction ratio in all experiments was 1:1. At the same time that the reflectance trace showed the secondary increase, the corrected fluorescence trace showed a transient

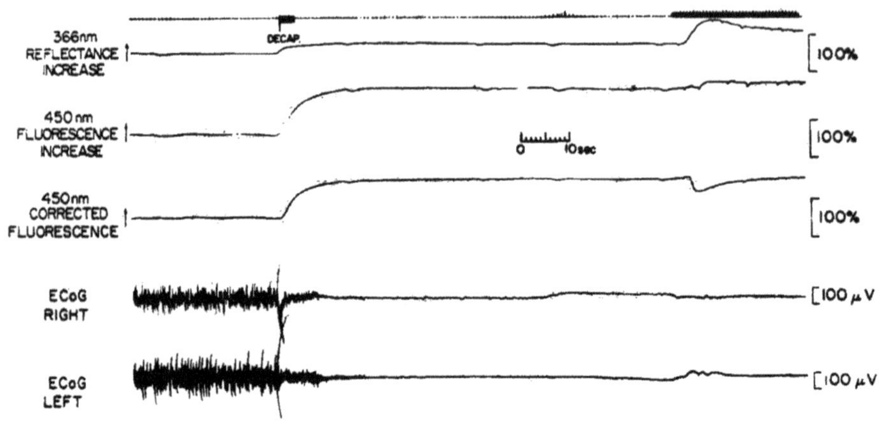

Fig. 3.5 The effects of decapitation on the oxidation–reduction state of NADH and ECoG recorded from the cortex of the awake rat (Mayevsky and Chance 1976)

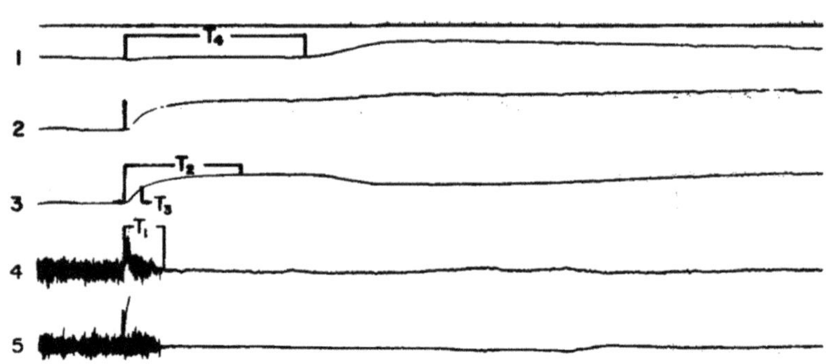

Fig. 3.6 Another example to show the effects of decapitation as in Fig. 3.5. The traces numbered 1–5 are the same and the parameters T_1–T_4 were calculated from the traces (Mayevsky and Chance 1976)

decrease which returned to its maximum level within few minutes. The lower two traces show the ECoG response to decaptation. The right hemisphere was the same side from which the NADH fluorescence was measured. In most animals, the ECoG traces showed a mechanical artifact when the decapitation was performed. In this animal the ECoG became flat within 10–15 s and two hemispheres showed an identical response. Figure 3.6 shows the response of another animal to decapitation which is similar to the one shown in Fig. 3.5.

In order to analyze the results obtained from all animals, we defined a few parameters calculated from data obtained from experiments (Fig. 3.5).

In order to analyze the results obtained from all animals, we defined a few parameters calculated from data obtained from experiments (Fig. 3.6).

T_1—Time when the ECoG amplitude was indistinguisable from the baseline level.
T_2—The time when corrected fluorescence reached its maximum level.
T_3—The time when corrected fluorescence reached 50% of the maximum level.
T_4—The time when the reflectance trace showed the secondary increase.

The calculation was done on data obtained from 12 rats. The means and their standard error were:

$$T_1 = 12.5 \pm 0.9 \text{ s} \qquad T_3 - 6.04 \pm 0.55 \text{ s}$$
$$T_2 = 32.8 \pm 3.2 \text{ s} \qquad T_4 = 66.3 \pm 9.5 \text{ s}$$

Changes in fluorescence intensity were calculated also. The maximum increase in corrected fluorescence (max ΔCF) at time T_2 was 81.6 ± 6% above the normoxic level measured before decaptitaion. The ΔCF at time T_1 was 65 ± 6.3% and by dividing the two values.

$$\frac{\Delta CFT_1}{\Delta CF\ \max} 78.4 \pm 2.8$$

which means that ECoG disappeared when NADH level reached about 80% of its maximum reduction. We also measured the time after decapitation when the corrected fluorescence started to increase, T_5, and the calculated value was 1.5 ± 0.14 s. This value shows that the oxygen stores in the tissue are very small and as soon as the oxygen supply was interrupted the NADH level increased.

The results presented in this preliminary presentation show that when the brain is transferred to a closed system by decapitation one can study the correlation between electrical activity and the metabolic state of the brain. This approach was described by Swaab and Baer (1972) and earlier by King et al. (1967). They described the effects of decapitation on the ECoG activity and found that at 10 s the activity disappeared from cortical area and, in our findings, it is in the same range of time (12.5 s). The main advantage of our approach is that we can measure the NADH oxidation-reduction state in the decapitated brain continuously and correlate it to the electrical activity. The ECoG disappeared when the NADH was reduced to 80% from its maximum level, and the same result was obtained when the animal was exposed to N_2 environment (Mayevsky and Chance 1973, 1975). We can assume that the rate of increase in NADH levels in the decapitated brain is dependent on the oxygen store in the tissue as well as the rate of consumption. This model will enable us to study the effects of various factors, such as anesthetics or hypoxia, on the rate of oxygen consumption in the decapitated brain.

3.3 Preliminary Study 2

In this second preliminary study, the preparation of the rat was the same as described in the first preliminary study.

In order to study the effects on the metabolic response of the brain resulting from the limitation of its blood supply, the carotid artery ligation technique was used (Eklof and Siesjo 1972). The common carotid arteries were dissected in the neck, and loops of ligature were placed around the vessels. The ends of the ligatures were taken out through the wound so that it was later possible to close the arteries while the rat was awake. Measurements of NADH oxidation–reduction state from the cortex were taken by implantation of a light guide holder above the right hemisphere of the brain (Mayevsky and Chance 1975). Initiation of spreading depression (SD) cycles was achieved by flushing KCl solution through a small push-pull cannula implanted above the right hemisphere so that NADH fluorescence measurements were taken at the same time. The distance between the application site of KCI and the light guide location was only 4–5 mm so that 0.5 M KCl solution flushing epidurally gave repeated cycles of SD. Before ligation of the arteries, the animal was exposed to 1–2 N_2-air breathing cycles and to S.D. cycles using 0.3–0.5 M KCl applied topically to the cortex (Mayevsky and Chance 1974, 1975). NADH fluorescence and ECoG measurements were taken, and these, representing the response of the normal brain, served as a control. Three hours after the operation, when the animals were fully awake, the carotid arteries were ligated. In some animals, both left and right arteries were ligated at the same time. In another group of animals only one artery was ligated, and in the third group the two arteries were ligated, but not at the same time. Figure 3.7 shows the response of NADH fluorescence and ECoG to carotid artery ligation. After ligation of the right artery, an increase of 25 % in NADH fluorescence was recorded (second and third traces from the top). The effect on 366 nm reflectance was very small. The ECoG of the right hemisphere showed a short transient depression, which recovered very quickly to normal (30 s). The ligation of the left artery (while the right artery remained closed) resulted in an increase in NADH fluorescence which remained higher compared to that in the pre-ligation period. A gradual increase in NADH fluorescence (about 10%) was recorded during the time interval of ligation of the right and left arteries. In all

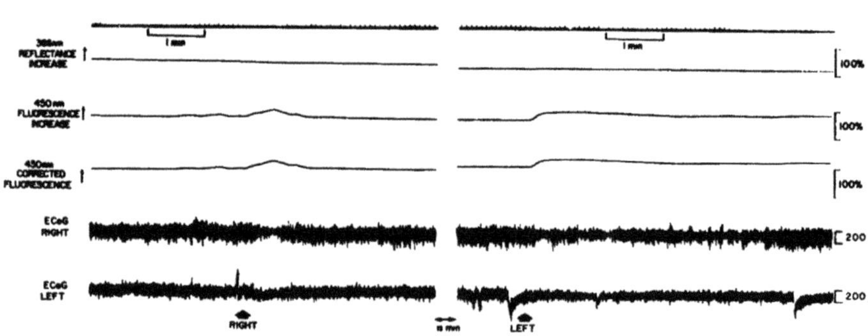

Fig. 3.7 Effects of unilateral and bilateral carotid artery ligation on NADH levels and ECoG. The arrows (right and left) at the bottom show the time of ligation (Mayevsky 1976b)

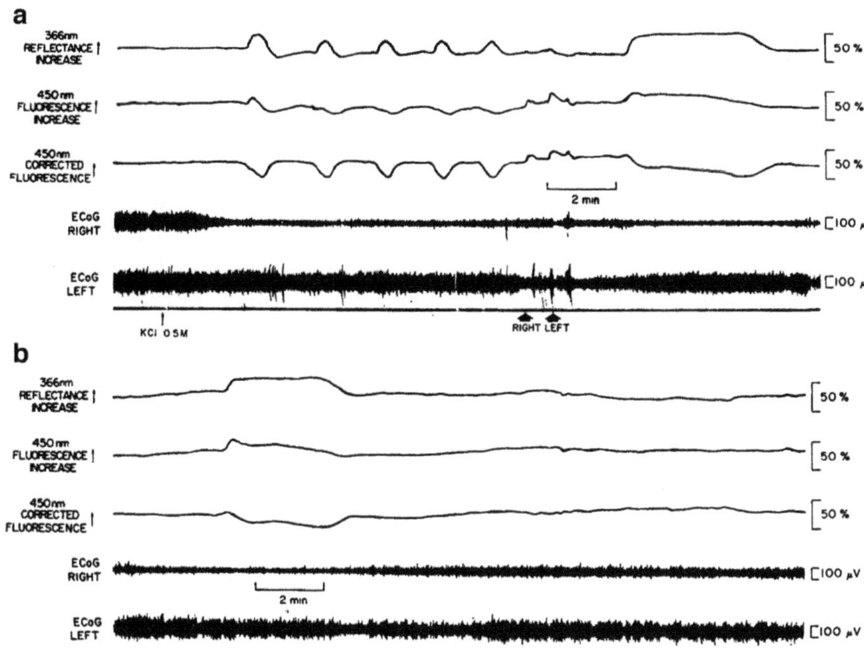

Fig. 3.8 The effect of carotid arteries ligation on the response of NADH levels and ECoG to cortical spreading depression. The arrows (right and left) show the time of ligation. Part (**b**) is a direct continuation of (**a**). (Mayevsky, 1976b)

animals, the metabolic state of the brain was tested by applying KCl solution to the cortex at various time intervals after the ligation. Immediately after the bilateral ligation a normal response of the brain to SD, namely the oxidation-reduction cycle of NADH, was observed. Within 2–3 h after ligation, as a result of exposure of the cortex to KCl, the ECoG became depressed, while the metabolic response of the NADH disappeared in most animals. Ligation of one carotid artery (left or right) did not affect significantly the response of the brain to SD. Figure 3.8 shows the response of an awake cortex to a bilateral carotid artery ligation. Before ligation, 0.5 M KCI was applied to the cortex and the normal metabolic response to spreading depression was recorded (Mayevsky and Chance 1974). The bilateral ligation of the carotid arteries caused an increase in NADH, and since the KCl remained on the cortex during this period (about 3 min), the response to SD was immediately apparent. In this particular animal, the duration of the oxidation cycle became very long after the ligation, as a response to spreading depression. A similar long oxidation cycle was recorded again later (lower part), but within 2 h the metabolic response to SD disappeared (not shown in Fig. 3.8).

3.4 Effects of Partial and Complete Ischemia on the Brain of Unanesthetized and Anesthetized Rats

3.4.1 Introduction

The effects of ischemia were studied in various animal models and were analyze by various physiological and biochemical techniques (Mayevsky and Chance 1976; Ginsberg et al. 1976; Rosenthal et al. 1976; Sundt and Anderson 1975; Eklof and Siesjo 1972; Molinari and Laurent 1976; Levy and Duffy 1975; Lowry et al. 1964; Harbig and Reivach 1973). The main disadvantage of all the techniques is in using anesthetized or paralyzed and artificially ventilated animals. One of the techniques to study ischemia is the artery occlusion approach (Molinari and Laurent 1976; Eklof and Siesjo 1972; Rosenthal et al. 1976; Ginsberg et al. 1976). Eklof and Siesjo (1972) described the effects of bilateral carotid artery occlusion on CBF and energy state of the rat brain. Decapitation is another technique widely used in biological experiments. Two types of experiments were done in brain research using this technique. In the first approach, by decapitation, brain is transferred into a closed system so that energy consumption can be measured under various experimental situations (Cohen 1973; Mayevsky 1976a; Maker and Lehrer 1971). Lowry et al. (1964) made a significant contribution to this question. The aim of the second attitude is to investigate the changes which take place in the brain after decapitation so that a better understanding of euthanasia is achieved (Peng et al. 1977; Swaab and Boer 1972). The availability of the new fluorometric technique to study NADH oxidation reduction state (Chance and Baltscheffsky 1958; Chance et al. 1962; Mayevsky 1976b; Mayevsky and Chance 1973, 1975) enables us to investigate the kinetic effects of carotid arteries ligation in the awake rat brain. In other experiments, we measured in a continuous way the changes in the awake and anesthetized brain after decapitation. We correlate the energy metabolism with the electrical activity of the cortex. Preliminary results were described previously (Hernández et al. 1978).

3.4.2 Methods

Time-sharing fluorometer/reflectometer was used in all experiments and was described previously in detail (Chance et al. 1975; Mayevsky and Chance 1975). Male Wistar rats 180–220 g were used. The operation of the rat included the implantation of a plexiglass cannula epidurally (above the right parietal cortex), two pairs of stainless steel screws and a push-pull cannula for application of KCI solution (Mayevsky and Chance 1975; Mayevsky 1976b). The cannulas and the electrodes were cemented to the skull by dental acrylic cement. Three types of experiments were performed: (a) acute ligation of the carotid arteries in which the two arteries were isolated in the neck and loops of ligature were placed around the vessels. The ligation was done while measuring NADH and ECoG; (b) chronic ligation of the two carotids was done by exposing the arteries and ligation at various intervals of time before NADH was measured. The measurement of NADH was started 30 min after the implantation of the cannula by inserting the common part of the light guide and fixing it epidurally. At this stage, the rat was anesthetized, but within 2–3 h the experiment started when the rat became fully awake. The rat was located in a small plexiglass cage for the duration of the experiments (up to 24 h); (c) in the decapitation experiments, the rat was located in a special cage combined with a guillotine. This procedure enables the performance of the decapitation of the awake rat with minimum disturbances; this was done 3–4 h after the operation. In the anesthetized group of rats, the procedure was the same as described and the injection was done 3–4 h after the operation. Thus, the effects of the operation and residual effect of the first injection of the anesthetics were minimal. The details of the calibration technique of NADH measuring system were identical to those described by us previously (Mayevsky and Chance 1975).

3.4.3 *Results*

3.4.3.1 The Effect of Acute Carotid Arteries Ligation

In this set of experiments, we ligate the carotid arteries in two ways: (1) Ligation of one carotid artery (left or right) and measuring the response of the brain to unilateral ligation for a few hours; (2) Simultaneous ligation of the two carotids and measuring NADH and ECoG up to 24 h postligation. The protocol of the experiment in this part was the following. The rat was anesthetized and operated on as described in Methods, and 30 min after the operation NADH, as well as ECoG measurements started while the rat was anesthetized. During the control period of the experiment, the brain was exposed to N_2 for 1 min in order to check the metabolic state of the brain. The response of the brain to spreading depression was tested by washing the surface of the dura with KCl solution (0.4–0.6 M) through the push-pull cannula. After control values were obtained and when the rat was awake the carotid arteries were ligated (one or two) while measuring the NADH and ECoG. At various intervals of time after ligation the brain was exposed to spreading depression and compared to the control values.

(1) ***The effect of unilateral ligation***. Preliminary results of this type of experiment were described by us previously (Mayevsky 1976b). In the present communication, we can summarize results obtained from eight rats in which unilateral ligation was performed. The main results are summarized in Table 3.1. The increase in NADH under N_2 exposure was almost 60 as compared to the normoxic steady state level. The effect of unilateral occlusion on the response to SD is in the amplitude of the oxidation cycle. This effect was significant ($p < 0.05$) but, since the cycles after occlusion are longer, we integrate the area below the oxidation cycle and there is no significant difference before and after unilateral occlusion. The steady level of NADH showed a transient response to the ligation as we described earlier (Mayevsky 1976b). The reason for this is the compensation

Table 3.1 The effects of unilateral carotid artery ligation on metabolic response of the brain

Animal no.	% Reduction under N_2	% Oxidation in SD	% Oxidation in SD after unilateral ligation
50	55	28	8
54	52	14	10
57	48	36	12
80	56	22	22
81	60	14	9
83	56	11	9
89	70	10	8
90	60	20	18
Mean ± SD	57 + 6.5	19 + 9.0	12 ± 5.2

in blood supply from the other hemisphere, so that the increase in NADH after occlusion recovers to normal a few minutes later.

(2)***The effects of acute bilateral ligation***. In this set of experiments, ten rats were involved. Figure 3.9 shows a typical response of the brain to acute bilateral carotid arteries ligation. Part A shows the response of the NADH to spreading depression and the effect of the ligation on the steady state level of NADH. Parts B, C, and D show the response of NADH to spreading depression elicited 15 min, 90 min, and 18 h, respectively, after the ligation. In this particular animal, the changes of the reflectance (upper trace in each part) were minimal during the spreading depression cycles in Part A. In Part D, one can see very clearly the very long biphasic cycle of the reflectance. The effect of the ligation on the oxidation–reduction state of NADH was transient, namely, that a small increase was recorded during the ligation, but within two minutes it recovered to the baseline. The main effect of the ligation is on the response of the brain to spreading depression, which increases oxygen utilization (Mayevsky and Chance 1974, 1975; Mayevsky et al. 1974; Rosenthal and Somjen 1973). During the first 60 min after ligation the oxidation cycle can be detected as a response to SD, but the amplitude of the oxidation (decrease) was diminished and the duration of the cycle became much longer (almost 4 min in B and only about 2 min in A). The qualitative change in the response of the brain to SD started several hours after the ligation. In Part C, one can see the response to SD 90 min after the ligation. The uncorrected trace, as well as the corrected one, shows a biphasic response of NADH to SD. At the beginning of the cycle an increase in NADH was recorded, while afterwards the NADH decreased below the baseline level. In Part D (18 h after ligation), the response of the brain was completely opposite to what was measured before the ligation (Part A). It seems that instead of an “oxidation cycle” as a response to SD, we found a “reduction cycle.” In this animal, the correction of the NADH for hemodynamic changes is not necessary, and the uncorrected trace also shows the same results. The effects of unilateral and bilateral carotid artery occlusion on the ECoG were recorded in all the experiments, but it was difficult to see any change without more sophisticated techniques of analysis.

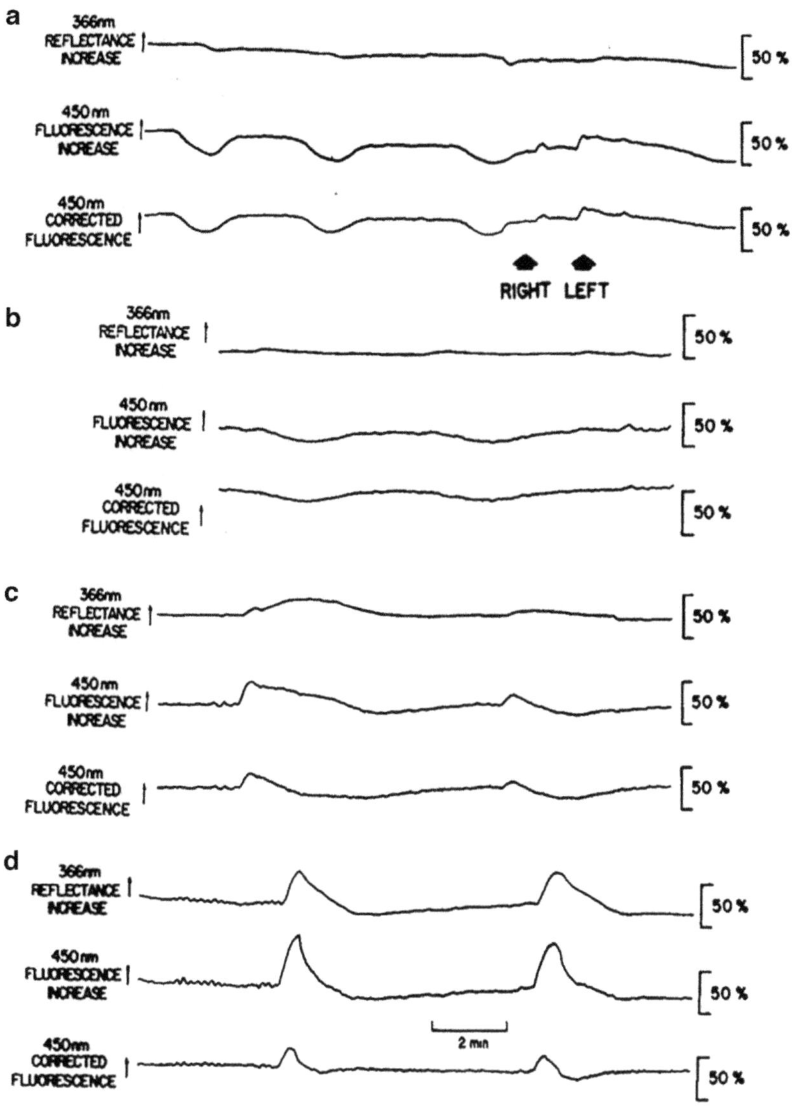

Fig. 3.9 The effect of acute bilateral carotids occlusion on the metabolic response to cortical spreading depression (SD). Part (**a**) was recorded before the carotid occlusion. Parts (**b**), (**c**), and (**d**) were measured 15 min, 90 min, and 18 h after occlusion, respectively (Mayevsky 1978a)

3.4.3.2 The Effects of Chronic Bilateral Carotid Occlusion

In this set of experiments, 22 rats were involved. The occlusion of the two carotids was done at different time intervals before the NADH and ECoG were measured. The various animals ranged between 15 h and 14 days of carotid occlusion. After operation

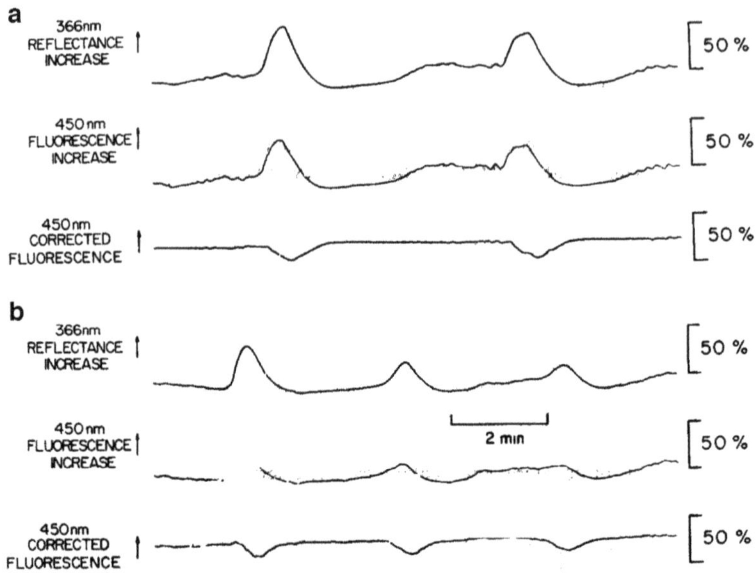

Fig. 3.10 The effects of 1O-day bilateral carotids occlusion on the metabolic responses to SD (Mayevsky 1978a). Part (**a**) was recorded 3 hours after the preparation od the animal to the brain monitoring and Part (**b**) was recorded 24 hours after the the operation of the brain to the monitoring

and implantation of the light guide holder and ECoG electrodes, the rat was connected to the fluorometer. In order to test the effects of the chronic occlusion, the brain was exposed to anoxia (N_2) and to SD elicited by washing KCl solution epidurally. According to the results we divided the 22 rats into three groups: (I) occlusion up to 3 days; (II) occlusion of 4–7 days; (III) occlusion of 8–14 days. The effect of the ligation on the response of the brain to SD was found to be a reversible effect. In rats of the first group (up to 3 days), the NADH fluorescence showed a "reduction cycle" after application of KCl epidurally as was measured after acute bilateral ligation (Fig. 3.9d). In the third group (8–14 days), the response to SD was as in the normal brain. In the second group (4–7 days), the results were not clear-cut and were seen as dependent upon the physiological and metabolic state of the animal and the brain. Figure 3.10 shows the results obtained from the brain after 10 days of bilateral occlusion. Part (a) was measured a few hours after the operation was done and Part (b) is from the same rat 24 h postoperation. The response of the brain to SD is the same in Parts (a) and (b), which shows that the preparation is stable up to 24 h after the operation.

3.4.3.3 The Effects of Decapitation in the Awake and Anesthetized Rat

A total of 37 rats were used in our study in the various treatments. The group of the unanesthetized (awake) rats served as a control group for comparison with other treatments (Table 3.2). The response of the awake rat brain to decapitation was described

Table 3.2 The effects of decapitation on ECoG, NADH fluorescence, and reflected light measured from the rat brain

Group	Treatment (number of rats)	T_0-time to ECoG = 0 (s)	T_1-time to CF increase (s)	T_2-time to CF_{max} (s)	CF_{max} % increase after decapitation	CF_0	CF_0/CF_{max}	T_2-time to CF (max/2) (s)	T_2-time to secondary reflectance increase	CF N_2 % increase after anoxia
A	Unanesthetized (7)	12.7 ± 1.4	0.93 ± 0.18	19.1 ± 3.4	74 ± 4.9	64 ± 5.0	88 ± 5.1	3.1 ± 0.39	45 ± 3.9	44 ± 4.4
N	Nembutal (7)	16.5 ± 0.95*	2.29 ± 0.18**	19.7 ± 2.5	81 ± 3.1	77 ± 2.4*	95 ± 1.7	4.7 ± 0.4**	76 ± 6.8**	56 ± 6.2
U	Urethane (8)	16.0 ± 0.62*	1.54 ± 0.33	19.9 ± 2.8	64 ± 5.8	61 ± 5.7	96 ± 1.3	4.0 ± 0.4	90 ± 10.3**	46 ± 4.4
C	Carotides occluded (9)	11.2 ± 0.55	1.56 ± 0.18*	21.6 ± 3.0	63 ± 5.0	55 ± 4.1	88 ± 2.2	4.4 ± 0.3*	64 ± 9.6	49 ± 5.2
CU	Carotides occluded (6)	15.2 ± 1.05	2.75 ± 0.34*	28.0 ± 5.4	43 ± 4.9**	37 ± 4.8**	85 ± 3.7	6.2 ± 0.92**	148 ± 10.1***	40 ± 3.8

ECoG electocortigram, *CF* corrected fluorescence, CF_0, % increase corrected fluorescence when ECoG = 0 ± SEM, standard error of the mean, *C* carotides occluded unanesthetized, *CU* carotids occluded urethane. * $p < 0.05$, ** $p < 0.01$, *** $p < 0.001$

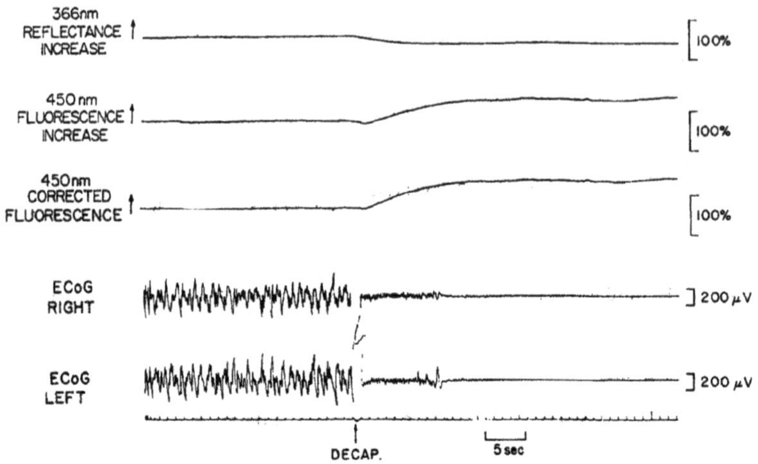

Fig. 3.11 The response of the brain to decapitation under urethane anesthesia (Mayevsky 1978a)

in our previous report. The anesthetics used were Nembutal (40 mg/kg i.p.) and urethane (1 g/kg i.p.). Figure 3.11 shows a typical response of the anesthetized brain with urethane to decapitation. The reflectance changes were very small as was found in most animals. The ECoG response was unique in comparison to the other treatments and appeared in all the eight rats of this group. Immediately after the decapitation a sharp decrease in the amplitude was recorded and this low amplitude (high frequency) remains until the complete disappearance. The response of a partial ischemic awake brain to decapitation was tested in nine rats. In these animals, both carotid arteries were ligated 40 h before the decapitation. According to previous studies, we knew that, at this time, the brain is not well perfused with blood. The NADH fluorescence and the ECoG responses are very similar to those in a normal awake rat and only accurate measurements and calculation can show the effect of the partial ischemia on the response of the brain to decapitation. In 36 out of 37 rats used in our studies, we found a large increase in reflectance a long time after the NADH reached its maximum value and the ECoG became flat. We called this response the "secondary reflectance increase" (SRI) in contrast to the initial or primary response, which appeared immediately after the decapitation. In order to analyze, the effect of decapitation in an accurate way, the following parameters were measured or calculated:

$T_0 =$ time (s) when electrical activity was very low and close to 0;
$T_1 =$ interval between the decapitation and the point that corrected fluorescence increased above baseline;
$T_2 =$ time to reach the maximum level of NADH after decapitation;
$T_3 =$ time to reach a level of NADH which is half of its maximum increase;
$T_4 =$ time when a large increase of the reflectance was measured;
$CF_{max} =$ maximum % increase of NADH above baseline after decapitation;
$CF_0 =$ % increase of NADH above baseline when ECoG = 0;

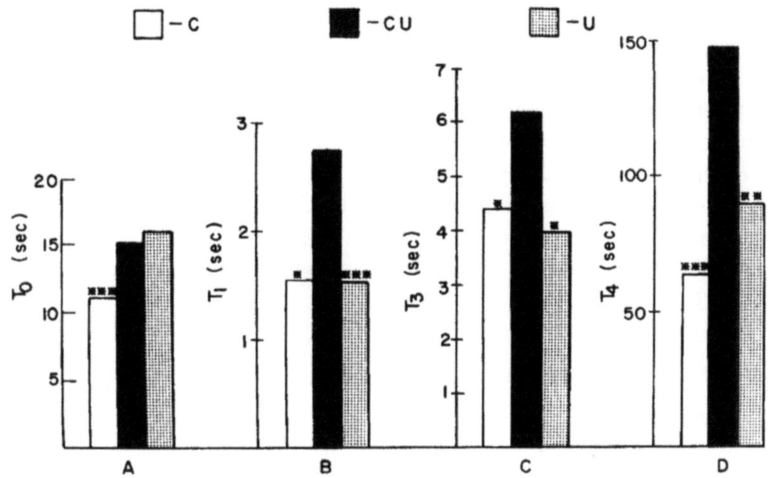

Fig. 3.12 The effects of urethane on various parameters measured from the brain of carotids occluded rats. $*p < 0.05$; $**p < 0.01$ (Mayevsky 1978a)

CF_0/CF_{max} = this value represents the percent change of NADH increase when ECoG = 0, in proportion to the maximum increase measured in the same rat;

CF_{N2} = % increase of NADH above baseline under nitrogen atmosphere (anoxia).

All parameters mentioned were calculated for all animals of each group and the mean as well as the standard error of the mean were calculated. The results are shown in Table 3.2. The statistical test for significance of the differences between the means was the Student *t*-test. The level of significance, which is shown in Table 3.2, was calculated and compared to group A, namely, the unanesthetized group. In order to test the effect of the metabolic state (partial ischemia) of the brain prior to the decapitation on its response, two groups of rats were prepared by bilateral carotid artery occlusion 20–50 h before decapitation (groups C and CU in Table 3.2). The calculation of *t*-test was also done between groups U, C, and CU, and the results are presented in Fig. 3.12. There are a few parameters which do not depend upon the treatment given to the rat before the decapitation. The T_2 value was in the same range and no significant differences were found between the five groups. The measurements showed that the maximum increase of NADH after decapitation was also the same in all groups except the CU group in which the CF_{max} was significantly lower. We found that the CF_0 cannot be used, as it is due to variability between the rats; therefore, we used the correction (CF_0/CF_{max}). There is no effect of the treatments on the relation between the ECoG and the NADH increase level. The effect of anoxia was the same in all groups of rats, and this shows that after the operation, while the rat was still under the Equithesin effect, the metabolic response was the same. The results show that the time until ECoG = 0 was shorter in the awake rats (A and C) in comparison to the anesthetized groups (N,

U, and CU). The occlusion of the carotid arteries did not affect this parameter, but the anesthetics have a significant effect. The value of T_1, which is the time until a change in CF was recorded, seems to be shorter in the unanesthetized (A) group. In a few of the rats of this group, the change started immediately after the decapitation, without any lag period. It seems that both carotid occlusion and anesthetics are affecting this parameter (Table 3.2). Another very sensitive parameter to decapitation is the slope of the increase of NADH as was estimated from the CF_{max}. Here, again, the lowest value was measured in the awake group. Another parameter which showed very interesting differences is the time taken to get a large increase of the 366 nm reflectance signal, as shown in Table 3.2.

3.4.4 Discussion

The aim of our present study was to investigate the kinetic effects of unilateral and bilateral carotid artery ligation on the metabolic state of the awake brain and the response of the awake and anesthetized brain to decapitation. The response of the brain to cortical spreading depression (SD) was used as an indicator to its metabolic state. It was shown previously by others and by us that SD may lead to an increased extracellular potassium level, and by an active process the K^+ ions will move back into the cells. This active pumping of K^+ consumes energy and can be detected as an oxidation of NADH (Rosenthal and Somjen 1973; Mayevsky et al. 1974). We have shown previously that a short exposure of the awake rat to hypoxia (Mayevsky and Chance 1975) may lead to a different response to SD, namely that the oxidation cycle became longer in time and the amplitude of the oxidation became smaller. By integrating the area below the oxidation cycle, it was found that the hypoxia alters it very little, which means that under reduced oxygen supply the system will recover to normal, but more slowly. In our present study, we extended the question to a state of reduced and limited blood flow to the brain, and it seems that the response depended on the time after the ligation was done. The supply of blood to the rat brain is done by two main pairs of arteries, the internal carotid and the vertebral, which forms an anastomotic ring on the base of the brain, the circle of Willis (Zeman and Innes 1963). By occluding one or two carotid arteries the cortical area may get a supply of blood via alternative vessels and only partial or incomplete ischemia is achieved. Our results demonstrate this very clearly, namely that, when unilateral occlusion was done, the response to SD was quite normal (Table 3.1), and the oxidation cycle was only longer in duration and smaller in amplitude, but the area below the cycle was not affected, as was described previously when hypoxia was induced (Mayevsky and Chance 1975). Salford et al. (1973) did not get an effect of unilateral occlusion on the energy state of the brain either, unless hypoxemia was combined with the occlusion. When bilateral occlusion was done the response to SD varied with time after the occlusion. During the first hour or so the response was similar to the one measured after unilateral occlusion (Fig. 3.9a, b), namely, that due to various factors, such as changes in blood pressure or increased circulation through the circle of Willis, the oxygen supply was sufficient to show an oxidation cycle. In their study, Eklof and Siesjo (1972) found that even a 50% reduction in CBF induced by bilateral carotid occlusion did not change the energy state of the brain. Only when hypotension was induced was the lower energy state found. They did not measure what happened after a few hours of occlusion in the normotensive rat. From our results it is clear that 2–3 h after bilateral ligation the oxygen supply to the cortex is very limited, and when an increase in O_2 consumption was induced by SD, the steady state of NADH became reduced (Fig. 3.9c, d). We can explain this in the following way. When SD was induced in a normal brain O_2 consumption and as a result blood flow is higher to supply more oxygen, so that NADH became more oxidized. However, when the blood flow is limited, the increase of O_2 consumptions due to SD will result in increased NADH or, as we called it, the "reduction cycle." The conditions are similar to hypoxia in which NADH will increase due to lack of oxygen. The reduction cycle as a response to SD was recorded in the acute experiment and in the chronic rats up to 7 days of occlusion. The oxidation cycle appeared again in the third group (8–14 days), which means that the effect of the bilateral occlusion is reversible. We can assume that new collateral circulation was developed within the first week of occlusion. This "correction" in the circulation seems to be quite slow, but since the only way to get new vessels is from the circle of Willis, it may take such a long time. In our previous work (Mayevsky 1976a), we showed that, in the brain, after 2–3 days of ligation, the reflectance response to anoxia was very small as compared to a normal brain. The reflectance change is in correlation to the blood volume changes in the tissue (Harbig and Reivach 1973; Rosenthal and Somjen 1973; Jobsis et al. 1971; Ginsberg et al. 1976). Therefore, we can conclude that in the first few days after occlusion the cortex has a very small capacity to increase its blood volume and a "reduction cycle" will be the response to SD. The effect on the response of the brain to SD during the first hour after ligation is quite surprising. One would expect that the compensation mechanism needs some time to be developed, but, in our case, within 1 h the brain was able to compensate blood flow to the cortex when SD was induced and an "oxidation cycle" was recorded. Most of the data accumulated on NADH fluorescence and ischemia deals with short-time effect and also not in the awake brain model. Sundt and Anderson (1975) found that, in the squirrel monkey, under ischemia NADH level went up under middle cerebral artery occlusion. Rosenthal et al. (1976) and Ginsberg et al. (1976) measured NADH response to ischemia from the brains of cats. They concluded that collateral circulation may be established after the occlusion. In our preparation, the effects of the occlusion are more pronounced, since both carotids were occluded and establishing collateral circulation will take more time (more than 1 week). In general, we feel that the rat brain can serve as a good model for partial ischemia, but not for a complete ischemia, unless one can occlude the vertebral artery in the awake rat as well. Decapitation, which is in use in order to study brain ischemia, converts the brain into a closed biological system (Cohen 1973; Folbergrová et al. 1970; King et al. 1967; Lowry et al. 1964). The main technique for studying brain energetics under such conditions is the freezing of the brain after various intervals of time and analysis of various metabolites. One of the main problems in this approach is the freezing rate of the tissue, as was discussed by Maker and Leher (1971). The availability of the fluorometric technique for the measurement of NADH oxidation-reduction state enables us to study brain energetics after decapitation of an awake animal, as well as to investigate the effects of anesthesia or partial ischemia on the response to decapitation. The analysis of the changes in NADH fluorescence after decapitation gives us the possibility to determine the time at which brain oxygen started to be limited (T_1, Table 3.2) as well as the rate by which oxygen is utilized by the brain cortex (T_3, Table 3.2). The measurement of ECoG together with NADH enables the study of the interrelation between energy state of the brain and its general electrical activity as well as the effects of various factors on this interrelation. The question arose as to whether, under our conditions, cytosolic NADH reduction might influence the results. It seems that, even if cytosolic NADH increases dramatically, it will occur very late after decapitation. Lowry et al. (1964) found a four- to sevenfold increase in cerebral glycolysis 10–20 rain after decapitation of 10-day-old mice and, even if one compares it to an adult mouse, the time of glycolysis increase is very long as compared to our time scale of NADH increase. In our studies we never saw any secondary increase of NADH in the range of 3–10 min postdecapitation, and the longest time it took to get maximum fluorescence increase of NADH was less than one minute. According to Chance et al. (1962), approximately 60% of NADH may be intramitochondrial and, as was calculated and suggested by others (Chance et al. 1973; Jobsis and Duffield 1967; Jobsis et al. 1971), the contribution of the cytosolic NADH to the measurement by surface fluorescence is minimal; it is possible to conclude that our kinetic changes of NADH are purely intramitochondrial.

The measurements of the ECoG under decapitation show that, in the awake rat, the disappearance time is shorter than in the anesthetized one. The same is true in the carotids occluded groups, namely that urethane delayed the disappearance of the ECoG under partial ischemic conditions (T_0 in Table 3.1 and Fig. 3.12). Swaab and Boer (1972) found that after 10 s the electrical activity was undetectable. Mikeska and Klemm (1975) recorded EEG activity up to 27 s after decapitation. This long-lasting EEG can be due to the use of a paralyzed and artificially respired rat model. King et al. (1967) found in their studies that the ECoG of a lightly anesthetized mouse disappeared 10–20 s after decapitation was done. It seems that our data is in the same range as was described by others. The correlation between the disappearance of the ECoG and the energy state CF_0 of the brain was very poor, as was shown in Table 3.2. The value of CF_0/CF_{max} was the same in all groups, namely that the ECoG disappeared when the NADH reached 85%–95% of its maximum value, and the anesthetics have a small effect on this value. These results confirm the data obtained when NADH and ECoG were measured under anoxic conditions. Chance et al. (1962, 1963) found that the electrical activity disappeared when NADH reached 80%–90% of its maximum value. In our previous studies (Mayevsky and Chance 1973, 1975; Mayevsky 1976b), we also showed the same relationship between the ECoG disappearance and the oxidation-reduction state of NADH.

The effect of decapitation on the metabolic activity is evaluated in our study by two parameters as shown in Table 3.2, namely, the T_1 and T_3 values. It seems that these two parameters are very sensitive to the state of the brain before decapitation was done. In the awake brain, the time before seeing any change in NADH is shorter, and the slope of the increase line of NADH is also larger in comparison to that in the anesthetized state. This is true in the normal as well as in the carotids occluded

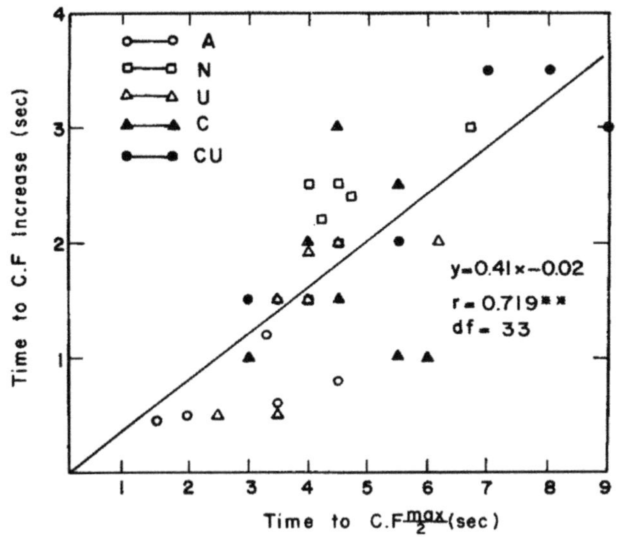

Fig. 3.13 The correlation between the time at which NADH increases after decapitation and the slope of the increase curve (Mayevsky 1978a)

groups. The correlation between these two parameters is statistically significant, as shown in Fig. 3.13. It seems that in the anesthetized brain the rate of change in energy consumption is slower than in the awake state. Lowry et al. (1964) also reported that in 10-day-old mice anesthesia slowed down the rate of decrease in high-energy phosphate. The secondary reflectance increase which was recorded in almost all rats can be explained by a spasm of the vessels occurred postdecapitation. This response will squeeze out the blood from the vessels and as a result the absorption of light will decrease, and increase of reflectance was recorded. It seems that anesthetics slowed this response of the reflectance.

3.5 Effects of Age and Anesthesia on the Responses to Complete Ischemia in Rats

3.5.1 Introduction

The effects of age on the metabolic activity of the brain have been tested on various animal models. Most of the studies described the differences either between the newborn and the adult (Gregoire et al. 1978; Hernández et al. 1978; Thurston and McDougal Thurston and McDougal, 1969) or between the adult and the old animal (Patel 1977; Peng et al. 1977; Pettegrew and Ewing 1971). In our present study, we compared rats of six age groups from 14 days to 2.5 years. Comparison of the tolerance to hypoxia between young and adult animals has been described by various investigators (Berlet 1976; Duffy et al. 1975; Mares et al. 1976; Stafford and Weatherall 1960). It was found that the young are more tolerant to hypoxia. This resistance cannot be due to the fact that glycolysis is the main source of energy in the young brain (Samson and Dahl 1975), nor to more energy reserves such as ATP and phosphocreatine (Berlet 1976; Thurston and McDougal Thurston and McDougal, 1969). It was suggested that the resistance was due to low oxygen consumption (Patel 1977; Mayevsky 1976b).

In the brains of old animals, some events relevant to energy metabolism have been described. A decrease in blood flow was reported (Fazekas et al. 1955; Meier-Ruge 1975), as well as a decrease in oxygen consumption (Patel 1977; Peng et al. 1977; Pettegrew and Ewing 1971). It is not clear whether those changes are primary or secondary factors in the aging process (Ferrendelli et al. 1971; Shenkin et al. 1953; Simard et al. 1971).

Studies of the effects of age on the brain energy metabolism have been made using various techniques. Lowry and his collaborators (King et al. 1967) described the effects of decapitation, which transferred the brain into a closed system, on the energy metabolism by measuring the levels of substrates and metabolites. The same approach was described by other investigators (King et al. 1967; Maker and Lehrer 1971; Swaab and Bear 1972).

In 1976, Mayevsky and Chance (1976) described a new approach by which the metabolic and electrical responses were correlated in the decapitated awake brain. Later, they showed the effects of anesthesia on the response to decapitation (Mayevsky 1978a). In the present communication, we will describe the effects of age (2 weeks to 2.5 years) on the metabolic, hemodynamic and electrical responses of the awake and anesthetized rat brain to decapitation. The aims of the present study were to correlate in detail the metabolic (NADH fluorescence) and electrical (electrocorticogram) responses to decapitation with the age of the animal tested. A further point was to study the effect of anesthetics on the mentioned correlation. By using the same technique for all ages, we assumed that variability between various ages due to technical reasons would be diminished. Preliminary results of this study were previously described (Zarchin and Mayevsky 1979).

3.5.2 Methods

Male Wistar rats were used in all experiments. The rats were divided into six age groups and each group contained awake and anesthetized animals (for detailed numbers, see Table 3.3). The young rats were kept with their mother until the time of the operation. Two types of DC fluorometer/reflectometers were used: one for unilateral measurement, and the other for bilateral records. The details of the instruments were recently described (Mayevsky 1978b). The basic approach used to study the changes in metabolic activity was to monitor continuously the oxidation-reduction state of pyridine-nucleotide (NADH) from the cortical surface. By illuminating the brain with 366 nm light, the intramitochondrial NADH is excited and fluoresces at 450 nm which

Table 3.3 The effects of age on the hemodynamic, metabolic and electrical responses[a] to decapitation of the awake and anesthetized rat brain

Body weight	Age (weeks)	Treatment[b] and no. of rats	T_0(s)	T_1 (s)	T_2 (s)	CF_{max} (%)	CF_0/CF_{max}	T_3 (s)	T_4 (s)	CF_{N2} (%)
20	2	Aw (10)	21 ± 1	1.9 ± 0.2	38 ± 3	90 ± 3	88 ± 1	7.0 ± 0.3	110 ± 15	68.3
		An (9)	63 ± 5***	4.3 ± 0.3***	68 ± 7***	95 ± 5	95 ± 2**	15 ± 1.7***	514 ± 66***	66 ± 4
60	4	Aw (9)	15 ± 1	1.1 ± 0.1	29 ± 3	99 ± 8	90 ± 1	4.6 ± 0.4	74 ± 10	82 ± 7
		An (7)	36 ± 5***	3.2 ± 0.4***	43 ± 3***	112 ± 5	94 ± 3	11.2 ± 0.9***	250 ± 26***	77 ± 6
100	6	Aw (7)	16	0.8 ± 0.1	34 ± 2	73 ± 6	91 ± 2	4.3 ± 0.4	66 ± 6	69 ± 4
		An (7)	29 ± 3***	1.4 ± 0.2**	39 ± 5	100 ± 5***	96 ± 1*	5.4 ± 0.6	122 ± 9***	71 ± 6
200	10	Aw (10)	16 ± 1	1.1 ± 0.2	18 ± 2	70 ± 4	96 ± 1	3.3 ± 0.2	82 ± 9	52 ± 2
		An (5)	38 ± 2***	1.3 ± 0.1	45 ± 4***	93 ± 5***	96 ± 1	5.5 ± 0.5***	1444 ± 12***	56 ± 6
400	48	Aw (5)	21 ± 2	0.5 ± 0.1	48 ± 5	68 ± 7	82 ± 4	6.1 ± 0.6	73 ± 7	38 ± 4
		An (4)	37 ± 4***	1.5 ± 0.4*	48 ± 4	68 ± 8	92 ± 4	7.1 ± 0.4	135 ± 36	50 ± 3*
300	120	Aw (4)	34 ± 2	1.1 ± 0.3	38 ± 6	65 ± 5	99 ± 1	5.1 ± 0.7	96 ± 29	48 ± 6
		An (4)	60 ± 9*	2.6 ± 0.4**	66 ± 5**	61 ± 9	95 ± 2	13.5 ± 1.5***	202 ± 26*	43 ± 6

Values are expressed as mean ± SEM [a]See text for explanation of parameters and abbreviations [b]Aw = awake; An = anesthetized $^*p < 0.05$, $^{**}p < 0.01$, $^{***}p < 0.001$; differences between awake and anesthetized were evaluated by *t*-test can be measured. Another parameter measured is the intensity of the reflected light at the excitation wavelength (366 nm), which is believed to be correlated to blood volume in the tissue monitored (for references, see Mayevsky and Chance 1975; Mayevsky 1976b, 1978a). The reflected light was used as a means of correcting the NADH fluorescence measurements.

Animals were anesthetized with Equithesin (ET—a mixture of chloral hydrate, pentobarbital, magnesium sulfate and propylene glycol) by intraperitoneal injection (0.3 ml per 100 g body weight). This dose was used for all age groups except the 2-week old group, for which diluted ET was used and the dose calculated accordingly. The injection kept the animal anesthetized for the duration of the operation and its connection to the measuring system. The skull was exposed by a midline incision of the skin, and one or two 4 mm diameter holes were drilled in the parietal bone for observation of metabolic changes. Two stainless steel screws were used as electrocorticogram (ECoG) electrodes on each hemisphere. A cannula for application of KCl solution was used in most experiments in order to test the response of the brain to spreading depression (Mayevsky 1976b). The cannulae and the ECoG electrodes were cemented to the skull using dental acrylic cement. Figure 3.14a shows schematically the localization of the various cannulae and ECoG electrodes on the skull.

The animal was connected to the fluorometer/reflectometer and placed in the specially designed plexiglass guillotine-cage for decapitation (Fig. 3.14b). After connecting the animal to the fluorometer, two different procedures were used to test the metabolic state of the cortical tissue monitored. The brain was exposed to anoxia by flushing nitrogen through a nasal mask. Only animals that showed a "normal" increase of NADH during the anoxia were used. If the response was much smaller

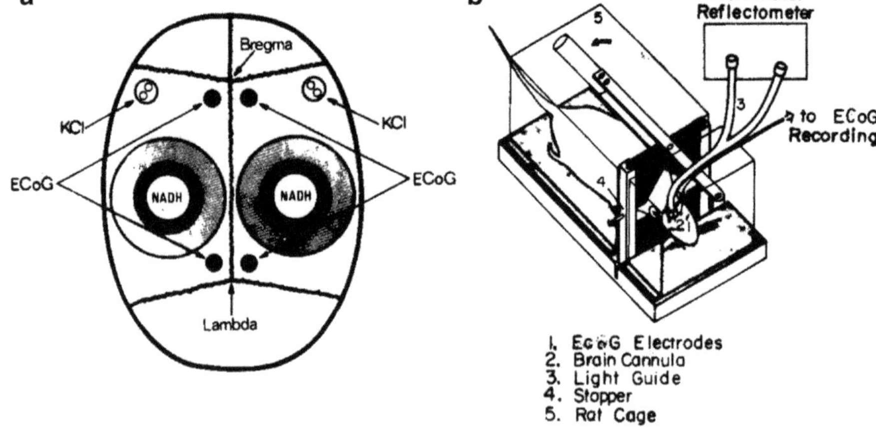

Fig. 3.14 Schematic representations of the location of various cannulae and electrodes on the skull of a rat (**a**) and the special cage used for decapitation (**b**). NADH was measured from one or two sites using a DC fluorometer/reflectometer. KCI (0.5 M) was applied topically in order to initiate spreading depression. The electrocorticogram (ECoG) was recorded using a bipolar technique (Zarchin and Mayevsky 1981)

(below the average of the group), the animal was discarded. The same test was done by exposing the brain to cortical spreading depression which should show a mitochondrial state 4→3 transition (oxidation or decrease in NADH levels). Decapitation was done in the fully awake rat, as indicated by the ECoG and motor activity. In the anesthetized groups, the rats were allowed to awaken and were anesthetized again by intraperitoneal injection of ET.

3.5.3 Results

A typical metabolic, hemodynamic and electrical response of the brain to decapitation is shown in Fig. 3.15. For each age group bilateral measurements were performed in some rats to test the variability between the brain's two hemispheres. After analyzing the results, we decided to use only unilateral measurements due to the very small variability between the two hemispheres. The four upper traces in Fig. 3.15 were measured from the right hemisphere and the lower four from the left hemisphere. It can be seen that the differences between the two hemispheres are insignificant in

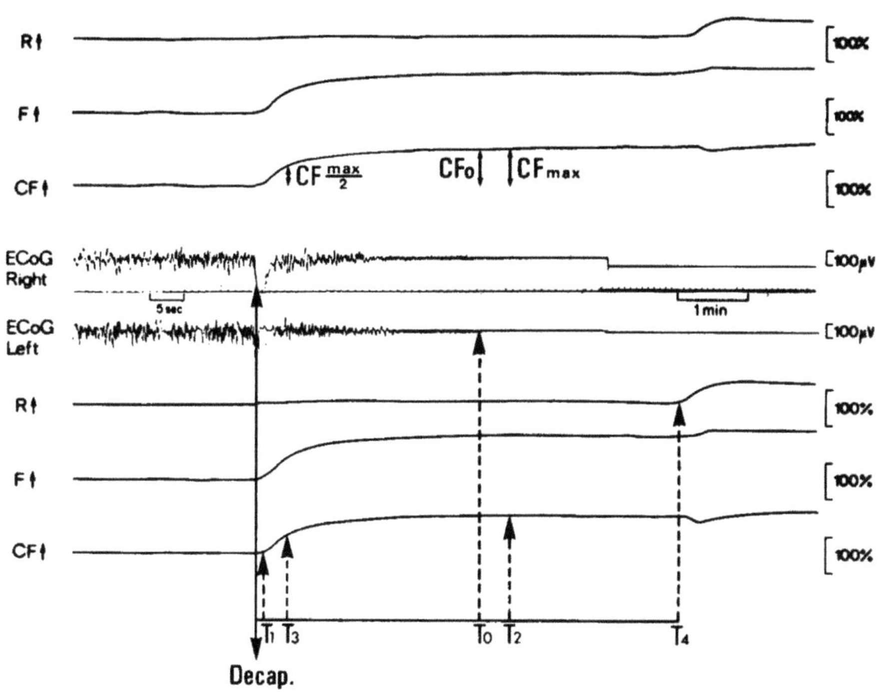

Fig. 3.15 Metabolic, reflected-light and electrical responses to decapitation (measured bilaterally). The upper four traces were measured from the right hemisphere and the lower four from the left hemisphere. R reflectance, F fluorescence, CF corrected fluorescence (F-R), ECoG electrocorticogram. See text for definitions of the measured parameters (Zarchin and Mayevsky 1981)

most cases. The 366 nm reflected light (R) shows a very small initial response to the decapitation. However, a very large secondary reflectance increase was recorded 1–2 min after ECoG = 0 when NADH reached its maximal level. The uncorrected (F) and corrected (CF) 450 nm fluorescence signals are similar to those described previously (Mayevsky and Chance 1976; Mayevsky 1978a). In order to analyze the effects of age on the responses to decapitation, various parameters were measured and calculated from the analog signals, as shown in Fig. 3.15. The definitions for the various parameters are as follows:

T_0 = time (s) when electrical activity was very low and close to 0.

T_1 = interval between decapitation and the point when corrected fluorescence started increasing.

T_2 = time when the maximum level of NADH was reached after decapitation.

T_3 = time when NADH reached a level which is half of its maximum increase ($CF_{max/2}$).

T_4 = time when a large increase of the reflectance was measured (SRI = secondary reflectance increase).

CF_{max} = maximum percentage increase of NADH above baseline after decapitation.

CF_0 = percentage increase of NADH above baseline when ECoG = 0.

$\frac{CF_0}{CF_{max}}$ = percentage of NADH increase when ECoG = 0 in proportion to the maximum increase measured in the same rat.

CF_{N2} = percentage increase of NADH above baseline in nitrogen environment (anoxia).

All correlation coefficients between T_0, T_1, T_2, T_3, and T_4 were significant ($p < 0.01$).

Table 3.3 summarizes the results obtained from 81 awake and anesthetized rats of various ages. The differences between the awake and anesthetized rats of each group were analyzed by Student's *t*-test and are shown in the table. For most of the parameters representing time, the anesthetic had a significant effect. The responses to anoxia and spreading depression (KCL 0.5 M) were measured in each animal after its connection to the measuring system as a standard test of the state of the brain. The response to anoxia (last column in Table 3.3) was not different between the rats of each group, with one exception (the 48-week-old group). The response to spreading depression recorded on the NADH trace showed a typical oxidation-reduction cycle as described previously (Mayevsky 1976b).

Figure 3.16 presents graphically the differences in four temporal parameters between awake and anesthetized rats at various ages. In all four parameters presented in Fig. 3.16 the effect of age on the response to decapitation is pronounced, and a "bell" shape is seen in the anesthetized groups. This shape indicates that in the young and old rats the response times are longer than in the intermediate ages. The same phenomenon was seen, though not as clearly in the awake animals. Table 3.4 shows the results obtained from a two-way analysis of variance for parameter T_3. Both age and anesthetic effects were significant ($p < 0.01$). Analysis of variance was also used in testing the other parameters, namely, T_0, T_1, and T_4. The interaction between the age of the animal and the treatment was found to be significant for the three parameters ($p < 0.01$).

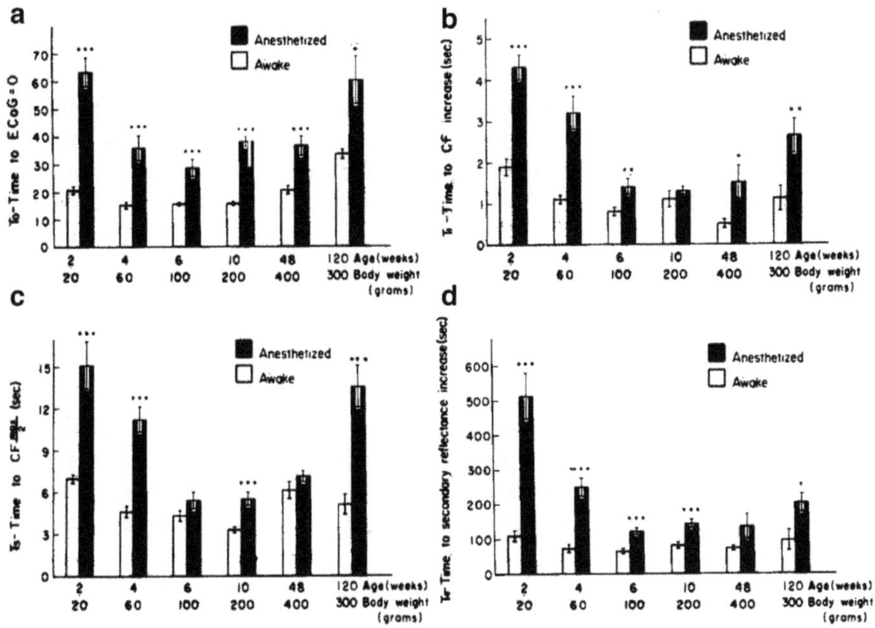

Fig. 3.16 The effects of age on the various responses of the awake and anesthetized rats to decapitation. Significance levels between awake and anesthetized rats: $*p < 0.05$; $**p < 0.01$; $***p < 0.001$ (Zarchin and Mayevsky 1981)

Table 3.4 Analysis of vasriance for the differences in parameter T_3 obtained from awake and anesthetized animals of various ages

Source of variation	SS	df	MS	*F*	*p*
Total	1499.1	80	18.7		
Age	397.9	5	79.6	16.5	<0.01
Anesthesia	378.2	1	378.2	78.5	<0.01
Age—anesthesia interaction	184.2	5	36.8	7.6	<0.01
Error	332.2	69	4.8		

3.5.4 Discussion

The advantages of our fluorometric technique for studying brain energy metabolism by monitoring the NADH oxidation-reduction state have been described previously (Mayevsky 1976b, 1977, 1978a, 1978b; Mayevsky and Chance 1975, 1976). In the present study we tested the effects of age (2 weeks to 2.5 years) on the awake and anesthetized rat brain during decapitation (complete ischemia).

The metabolic (NADH fluorescence), hemodynamic (reflectance) and electrical (ECoG) responses were obtained. Two groups of parameters on variables were calculated from the analog signals. The first group are temporal variables such as T_0–T_4 and the other one described the magnitude of redox changes during the decapitation. Longer times to obtain isoelectric recording (T_0) indicate that the energy needed to maintain electrical activity was available due to the slow consumption or utilization (production stopped due to the lack of O_2). The longer times regarding NADH changes also indicate a slower O_2 utilization and expressed resistance to the ischemia induced by the decapitation. This can be demonstrated in parameter T_1; namely, that when oxygen utilization is slow, the CF started to increase later suggesting a lower metabolic rate as well as resistance of the brain. Two of the parameters calculated from the analog signal measured after decapitation did not show any significant relation to age or treatme of the animal. CF_{max} was not affected by the anesthetic, except in two age groups (6 and 10 weeks) as shown in Table 3.3. This finding can be explained by the fact that the resting oxygen supply to the brain is not affected dramatically by the anesthetic, so that the oxidation-reduction state of NADH remained the same. The parameter CF_0/CF_{max} had a very high value (90–100) in all animals, and was also not dependent on age or anesthesia.

The electrical activity disappeared at a certain level of oxygen availability as was found previously. From the present results and together with what was published in the past (Mayevsky and Chance 1975). We can conclude that in order to maintain electrical activity the brain has to have certain amounts of available energy not depending upon the age or the treatment given to the animal. Three main parameters measured were affected by the state of the animal. Two are metabolic—T_1 and T_3—and the third represents electrical activity—T_0. All three parameters were related to the age of the animal and to the existence of anesthetics. Differences between the various ages were more pronounced in the anesthetized state, as can be seen from Fig. 3.16. In all parameters described in Fig. 3.16, the difference between the awake and anesthetized state was very noticeable in the young and old animals, so that a "bell" shape was formed. The effects of anesthesia were previously described by others who showed that the rate of oxygen consumption was lower in the anesthetized brain (Fink and Haschke 1973; Lowry et al. 1964), and that cerebral blood flow was also lower under anesthetics (Goldman and Sapirstein 1973).

The direct interrelation between energy availability and electrical activity indicates that the cortical tissue needs a constant supply of oxygen for normal electrical activity (Swaab and Bear 1972; Chance and Schoener 1963). The late disappearance of the ECoG after decapitation in the young (2 weeks) and the old (120 weeks) animals corresponds to the slower rate of utilization of high-energy phosphate compounds (Simard et al. 1971). Under anesthesia the time to disappearance of the ECoG was longer, due to the slower rate of energy utilization (Fink and Haschke 1973).

The results presented in this study are very similar to those described previously on the effect of anesthetics on the hemodynamic, metabolic and electrical responses to decapitation (Mayevsky 1978a). In two parameters measured (T_0 and T_4), the effects of anesthetics were smaller in our previous study (Mayevsky 1978a). Two reasons may explain the differences; (1) in the present study Equithesin was used as an anesthetic in contrast to Pentobarbital or Urethane used previously; (2) the difference in time between the two studies (2 years) may account for small variations in the strain of the rats. For this last reason, we always compared awake to anesthetized animals in the same set of experiments.

Another parameter which was sensitive to the age and treatment of the animal was the secondary increase in reflectance (SRI). This SRI was found in animals tested and described a unique response of light absorption change after the death of the brain. This phenomenon was found by us previously (Mayevsky 1978a; Mayevsky and Chance 1976), and we assume that it originates from the large decrease in blood content of the vessels, or a change in the size of the extracellular space. As was shown previously by various investigators (Hansen 1978; Van Harreveld 1972; Dora and Zeuthen 1976), complete depolarization induced a shift of water and electrolytes between the intra- and extracellular spaces. If such a change occurs after decapitation, then an increase in reflectance is possible.

Very recently, we measured the changes in NADH redox state, reflected-light, extracellular K^+, as well as the DC potential from the cortex of an awake animal (Mayevsky and Friedli, unpublished observation). It was found that the large increase in reflectance (SRI) appeared simulatanously with a large negative shift in the DC potential and a large increase in extracellular K^+. Therefore, we can interpret the SRI event as a response to the complete depolarization occurring after complete ischemia (decapitation).

3.6 Physiological Mapping of the Cerebral Cortex Exposed to Three and Four Blood Vessel Occlusion

3.6.1 Introduction

The "four-vessel occlusion" model, described by Pulsinelli and Brierley (1979) for producing brain ischemia is a useful model for investigators. The effects of the ischemia produced in this model have been described (Blomqvist et al. 1984; Capdeville et al. 1986; Dienel and Pulsinelli 1986; Onodera et al. 1986; Volpe et al. 1984). Harrison et al. (1985) described the mitochondrial redox responses in this model while optically monitoring one hemisphere of the ischemic brain. Due to the large variability in the ischemic level produced by this technique (Blomqvist et al. 1984), it is desirable to monitor online in real-time the level of ischemia produced in each hemisphere of the brain. The aim of the present study was to compare the mitochondrial NADH redox responses monitored simultaneously from the two cortical hemispheres after unilateral and bilateral carotid artery occlusions following electrocauterization of the vertebral arteries. The NADH redox signals were correlated with the hemodynamic responses measured by the 366 nm light reflectance trace (Chance et al. 1973; Mayevsky 1984), as well as the electrical activity (ECoG). To estimate the completeness of the ischemia, 100% N_2 was applied to the rat and the maximum level of NADH was monitored and used as a reference point.

The same animal model was used for studies where the metabolic responses to the four-vessel occlusion were compared with the extracellular levels of K^+ and Ca^{2+} using the multiporbe assembly (MPA) approach (Friedli et al. 1982; Mayevsky 1983; Mayevsky et al. 1985). The development of this model enabled us to compare the effects of ischemia withthose of cortical spreading depression (Leão 1944; Bures et al. 1974; Mayevsky et al. 1974) in the same animal, which was suggested as a good animal model for migraine (Lauritzen 1987a, 1987b).

3.6.2 Methods

3.6.2.1 NADH Redox State Fluorometry

Excitation light (366 nm) was passed to the brain via a bundle of quartz optical fibers. The light emitted at 450 nm, together with a light reflected at the excitation point, was transferred to the fluorometer via a second fiber bundle. The changes in the reflected light were correlated with changes in tissue blood volume and served to correct hemodynamic artefacts appearing in the NADH measurement. Details of this technique have been published (Mayevsky and Chance 1982, 1983; Mayevsky 1984). In the present work, one- or two-channel fluorometer reflectometry was used. The changes in fluorescence and reflectance signals were calculated relative to the calibrated signals under normoxic conditions. This type of calibration is not absolute, but provides reliable and reproducible results from different animals and between different laboratories. More details of this technique have already been published by our group (Mayevsky and Chance 1982, 1983; Mayevsky 1984).

3.6.2.2 Multiprobe Assembly

The MPA used in the present study has been described in detail in our previous publications (Friedli et al. 1982; Mayevsky 1983; Mayevsky et al. 1985). Thus, only major points will be described. The MPA has the following main features: (i) all components have the same type of noninvasive surface contact with the cortex, (ii) it can very easily be removed from the skull at the end of the experiment without any damage, so that repetitive experiments can be performed in a short period of time; (iii) all signals monitored have a very low sensitivity to movements of the animal, even while exposed to convulsions. In the present study, the pH electrode was substituted by a Ca^{2+} electrode (Mayevsky et al. 1989).

3.6.2.3 Animal Preparation

Male Wistar rats (250–280 g) were anesthetized by i.p. injection (0.3 ml 100 g–[1]) of Equithesin (each ml contains 9.72 mg pentobarbital sodium, 42.51 mg chloral hydrate, 21.25 mg magnesium sulphate, 44.34% wt/vol propylene glycol, 11.5% alcohol and water). An additional supplement of Equithesin was added (0.05–0.1 ml 100 g–[1]) every 30 min to maintain the anesthesia. After a midline scalp incision, the skull was exposed and a 5 mm hole was drilled in the two parietal bones for light-guide holders located epidurally. Four holes (1.5 mm hole diameter) were drilled in various points of the skull and stainless steel screws were used as ECoG electrodes and anchors for dental acrylic. The two common carotid arteries were isolated for later occlusion and the rat was connected to the measuring system by inserting the light guide into the implanted holder. Each rat was exposed to 100% N_2 (30–60 s) and the response was recorded from the two cerebral hemispheres. This test aimed to check the postoperative basal metabolic state of the brain.

While monitoring the NADH from the brain, the two vertebral arteries were cauterized using the regular procedure (Pulsinelli and Brierley 1979). Ten minutes later the effect of bilateral carotid occlusion was tested and electrocoagulation was repeated until severe ischemia was recorded (increased level of NADH compared with anoxia). This approach provided us with complete or near-complete ischemia in all animals as evaluated by the level of NADH. One can calculate the exact level of cortical ischemia by comparing the percentage change of NADH increase during the ischemia with the maximal level recorded under anoxia (0% oxygen). This normalization technique helped us to compare two hemispheres in various animals, although the absolute percentage changes were different.

In experiments where the MPA was used, the animal preparation was slightly modified. The skull was exposed through a midline incision and a hole of 6.0 mm diameter was drilled in the right parietal bone. Four holes (1.5 mm diameter) were drilled in various points of the skull, and stainless steel screws were used to hold the cement to the skull. The dura mater was removed carefully, and the MPA was held in a micromanipulator and located over the cortex without creating any extra pressure on the brain. To elicit spreading depression a push-pull surface cannula (2 mm diameter) was placed epidurally 2 mm from the MPA.

3.6.2.4 Results

In the present study, we used ten rats to verify the technique and the approach. A second group of seven rats was used in studies where the MPA was used to monitor the responses to four-vessel occlusion and spreading depression.

Figure 3.17 shows a typical record obtained after occlusion of the right carotid artery (R_{OCCL}) followed by the left carotid occlusion (L_{OCCL}). As can be seen, the right hemisphere became more reduced (increased CF_R) and partially re-oxidized later via the anterior circulation and the left carotid artery (still open). The signals of the left hemisphere remained unchanged under right carotid occlusion. Bilateral

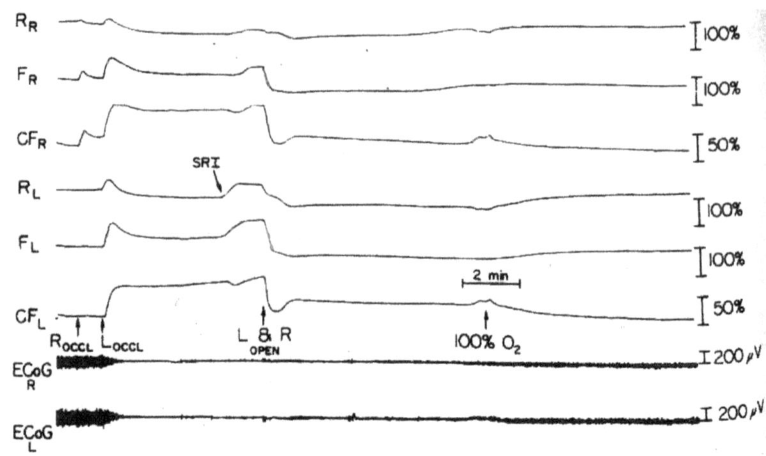

Fig. 3.17 Effects of ischemia on the metabolic, hemodynamic and electrical responses. R_R, F_R, and CF_R reflectance, fluorescence and corrected fluorescence measured from the right hemisphere. R_L, F_L, and CF_L the same three signals measured from the left hemisphere. $ECoG_R$, $ECoG_L$ electrocorticogram monitored from the right and left hemisphere. SRI secondary reflectance increase (Mayevsky et al. 1990)

occlusion induced a large increase in the CF signals of the two hemispheres, which recovered after reopening of the blood vessels (L + R open). The electrical activity (ECoG) disappeared 30 s after occlusion of the left carotid artery. The reflectance traces (R_R and R_L) show an initial small transient increase followed by a decrease due to hemodynamic changes. Four minutes after the occlusion, large increase in the reflectance was recorded (SRI) and has been described by us as a secondary reflectance increase (Friedli et al. 1982). This change in reflectance is not always corrected by the subtraction technique and a transient apparent oxidation of NADH can be seen in the corrected fluorescence trace (CF). During the reperfusion stage a large decrease in the reflectance was recorded (increased blood volume), which gradually returned to the pre-ischemic baseline. The large change in reflectance is due to massive vasoconstriction, which occurred under the ion shifts as recorded by the surface mini-electrodes. The SRI always appeared when a large increase in extracellular K was recorded together with a massive Ca^{2+} decrease. When 100% O_2 was supplied, oxidation of NADH was recorded. This indicates that the mitochondrial redox state had not completely recovered from ischemic episode. The same type of response was measured in a different rat (Fig. 3.18). The recovery from the 5 min of ischemia was faster as seen in the NADH traces (CF_R, CF_L), as well as in the ECoG traces. In this animal, the SRI appeared after 3.5 min in both hemispheres.

To determine the completeness of the ischemia induced by bilateral carotid occlusion, 100% N_2 was applied to the rat and a typical record is seen in Fig. 3.19. The responses to the bilateral occlusion were similar to those shown in Figs. 3.17 and 3.18. A pronounced decrease (recovery) in the CF trace during the occlusion suggested that

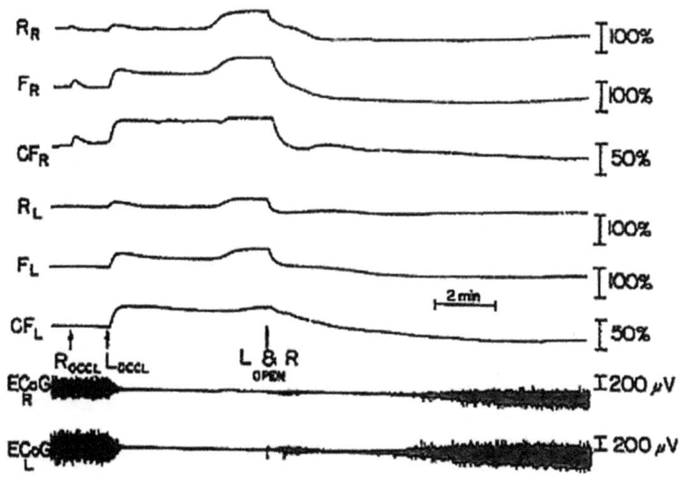

Fig. 3.18 Metabolic and electrical responses to severe ischemia induced by four-vessel occlusion in the rat. See Fig. 3.17 for abbreviations (Mayevsky et al. 1990)

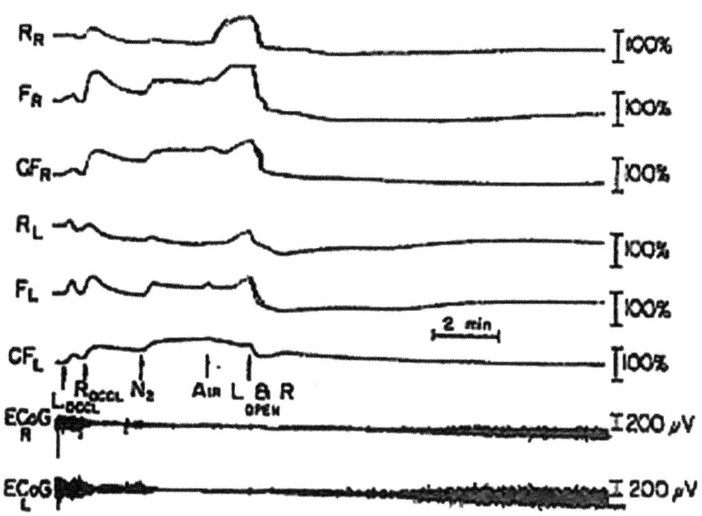

Fig. 3.19 Effects of ischemia and anoxia on the optical signals measured from the rat brain. NADH and reflected light were monitored from the two hemispheres. See Fig. 3.17 for abbreviations (Mayevsky et al. 1990)

the coagulation of the vertebral arteries was not 100% efficient. Application of N_2 led to an increase in NADH in both hemispheres, as expected under incomplete ischemia. The ECoG depression during the ischemia was not completed until N_2 was applied. In this animal also, the R trace showed a large increase (SRI) 4 min after the occlusion.

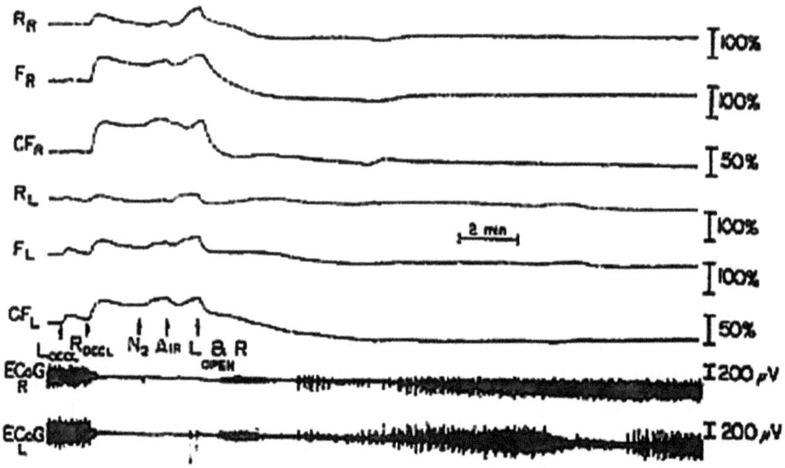

Fig. 3.20 Effects of incomplete ischemia induced by occlusion of the carotid arteries as measured by the mitochondrial redox state hemodynamic responses. See Fig. 3.17 for abbreviations (Mayevsky et al. 1990)

The same type of incomplete ischemia is shown in Fig. 3.20, as can be seen from the additional increase in CF signal under N_2 conditions. The ECoG was also further depressed under the anoxia. The partial recovery of the NADH under returning air to the four-vessel occluded rat is clear from Figs. 3.19 and 3.20, suggesting incompleteness of the ischemia.

In Fig. 3.21, an unsuccessful coagulation of the vertebral arteries is shown. The unilateral and bilateral occlusion led to a transient increase in NADH in both hemispheres. In this rat, a second cauterization was done which led to complete ischemia under bilateral carotid artery occlusion. Another example of incomplete coagulation of the vertebral arteries is shown in Fig. 3.22a in which right carotid occlusion was performed in the left carotid-artery-occluded rat. Here again, only a small transient increase in NADH was recorded. In this animal, terminal anoxia led to complete deprivation of O_2 (Fig. 3.22b) and a maximal NADH accumulation was noted. Also, the change in the reflected light was recorded (SRI).

The use of the four-vessel occlusion could replace the technique used to obtain complete ischemia by the decapitation procedure (Mayevsky 1978a; Mayevsky and Chance 1976). To test this idea, we performed the following experiment: one of the common carotid arteries was ligated permanently and occlusion of the second artery produced complete ischemia. To confirm this possibility, we compared the responses recorded after decapitation and a typical record is shown in Fig. 3.23. In Fig. 3.23a, the effects of left carotid occlusion are shown (the other three vessels were occluded). The same pattern of responses was recorded when decapitation was induced in the same rat (Fig. 3.23b). The ECoG disappeared 20–25 s after the induction of complete ischemia (left occlusion or decapitation). The NADH redox state increased shortly

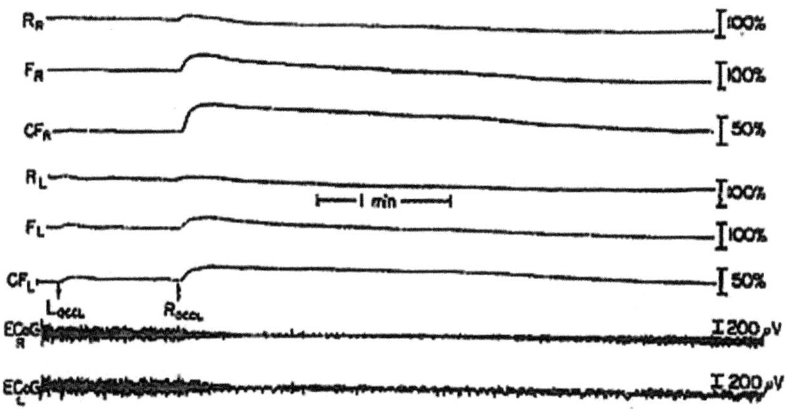

Fig. 3.21 Metabolic and electrical responses to carotid artery occlusion in the rat. The level of ischemia was very low. See Fig. 3.17 for abbreviations (Mayevsky et al. 1990)

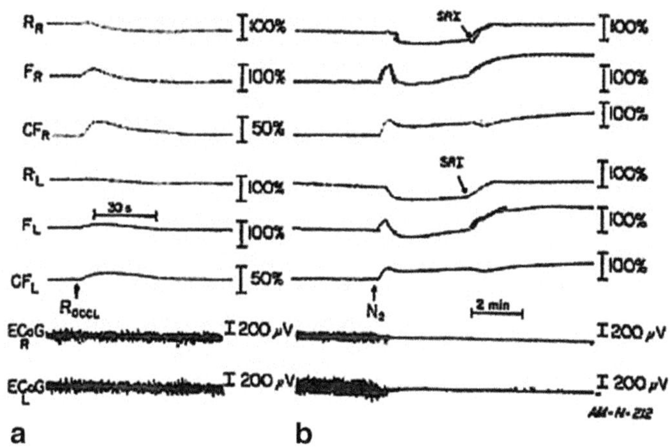

Fig. 3.22 Effects of right carotid occlusion in a three-vessel occlusion rat. (**a**) Due to the incompleteness of the vertebral cauterization only a small response was recorded. (**b**) The effects of anoxia on the metabolic and hemodynamic parameters. See Fig. 3.17 for abbreviations (Mayevsky et al. 1990)

after the induction of ischemia (Fig. 3.23a, b) and reached a maximal level within less than 1 min.

Although a large difference in NADH increase was noted between the two hemispheres, the normalizaition technique suggested has helped to solve the problem. All the changes are calculated in relation to the maximal levels under anoxia in the same monitoring site.

In experiments where the MPA was used, the responses of the brain to ischemia were compared with those recorded under spreading depression (SD). Both insults

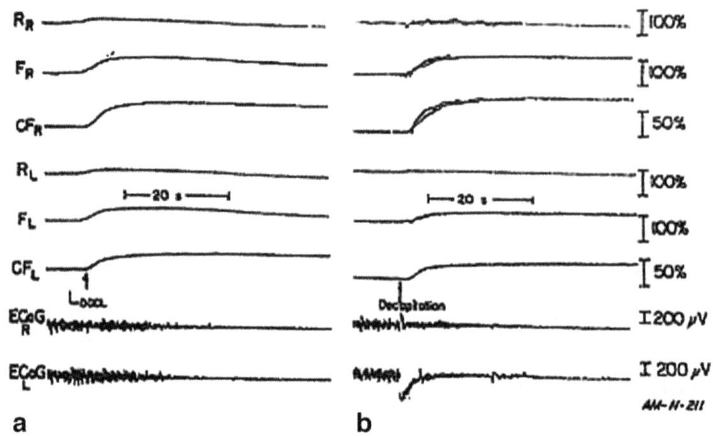

Fig. 3.23 Effects of complete ischemia induced by left carotid artery ligation (while the right one was occluded) in (**a**) and decapitation, (**b**) and the various parameters measured from the rat brain. See Fig. 3.17 for abbreviations (Mayevsky et al. 1990)

were induced in the same rat and were also repeated a few times alternately during each experiment. For the two examples of typical records (Figs. 3.24 and 3.25) only the qualitative responses will be discussed here. Figure 3.24 shows the responses of the brain to the two spreading depression waves passed below the MPA located 2–3 mm from the site of the SD initiation (KCl applied epidurally). The propagated wave of SD was detected initially by the Ca^{2+} electrode, followed by the NADH response and later on by the K^+ electrode. The ECoG response was recorded according to the spatial location of the electrodes. As expected, the wave of depolarization developed under SD led to an extracellular increase in K^+ and a decrease in Ca^{2+}, accompanied by an increase in energy utilization detected by the oxidation of NADH (decreased CF signal). The reflectance trace shows the typical biphasic response as described previously (Mayevsky and Chance 1974, 1976, 1982). The two cycles shown in Fig. 3.24 are very similar to each other in the changes detected by the various probes. The ECoG and the two DC potential traces show the typical depression and a negative shift, respectively, as expected.

Figure 3.25 shows the effects of bilateral carotid artery occlusion on the various activities of the brain after coagulation of the two vertebral arteries. The level of ischemia created by the four-vessel occlusion is evaluated by the NADH increase during the insult. As seen in the CF trace (NADH-corrected fluorescence), the level of ischemia was about 70%–80% of the maximum achieved by the addition of N_2 only to the animal. During the ischemic period, the ionic homeostasis of the brain was disturbed, as can be seen in the changes in extracellular levels of K^+ and Ca^{2+}. The leakage of K^+ started immediately after the occlusion of the two arteries. This initial slow efflux of K^+ was followed by a second fast leakage (Phase II) at the same time as a large influx of Ca^{2+} was recorded (decrease in extracellular measured level). The second

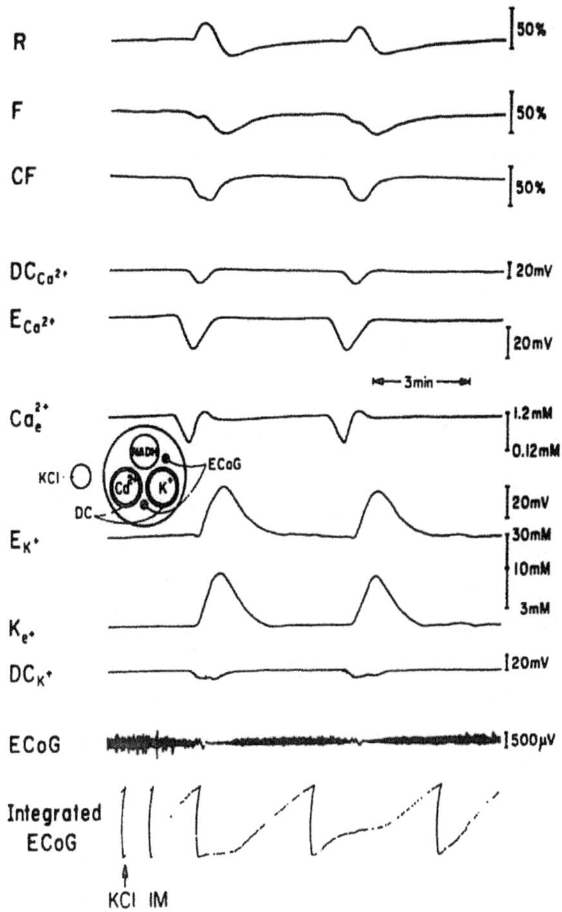

Fig. 3.24 Effects of cortical Spreading Depression on the various activities measured from the brain of a rat. R, F, and CF are reflectance, fluorescence and corrected fluorescence. ECa^{2+}, EK^{+} uncorrected calcium and potassium electrode potentials. Ca^{2+}_{e}, K^{+}_{e} corrected ion activities. $DCCa^{2+}$, DCK^{+} DC steady potential measured around the two electrodes. ECoG electrocorticogram. Spreading depression was initiated 2 mm from the monitoring site (Mayevsky et al. 1990)

phase of K^{+} leakage was accompanied by a negative shift in the DC steady potential (complete depolarization) and by a large increase in the reflectance trace—the SRI.

Most of the signals returned to the preischemic level within 5 min after reperfusion. The ECoG did recover, but after a longer period of time (not shown in Fig. 3.25).

3.6.2.5 Discussion

The four-vessel occlusion model developed by Pulsinelli and Brierly (1979) has been used in studies of complete ischemia. In this model, as well as in other ischemic

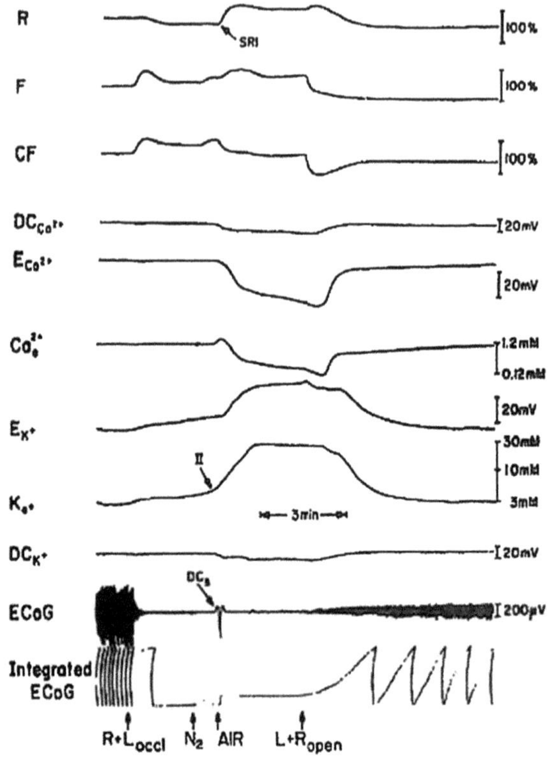

Fig. 3.25 Effects of four-vessel ischemia or the metabolic, ionic and electrical activities of the same rat as shown in Fig. 3.24. SRI secondary reflectance increase. II: Changes in rate of K^+ leakage. DC_R DC potential shift reorded on the ECoG trace. See Fig. 3.24 for abbreviations (Mayevsky et al. 1990)

models, the main problems are confirming that complete ischemia has occurred and verifying the level of ischemia in real-time. The best approach is to monitor the intracellular O_2 level. Harrison et al. (1985) used the NADH and cytochrome aa_3 redox state as a criterion of ischemia in the four-vessel occlusion model. They concluded that the surface fluorometry procedure is "indicative of intramitochondrial rather than cytoplasmic pyridene nucleotide." In previous studies (Mayevsky 1976b), we correlated brain responses to anoxia and spreading depression using a time-sharing fluorometry reflectometer for NADH and flavoproteins (FP) redox state and came to the same conclusion. In the present study, we were able to monitor the NADH redox state from the two cortical hemispheres, allowing the effect of unilateral carotid artery ligation to be studied.

The main advantage of the fibre-optic surface fluorometry technique is the potential use of an awake rat during the study, as well as behavioral assessment of the animal after the experiment. If needed, the light guide could be removed after the measurement and replaced by a plexiglass plug to protect the brain. As seen in the representative figures shown in the present study, the occlusion of one carotid artery (after cauterization of the vertebral arteries), led to a small ipsilateral increase in NADH level. The level of ischemia produced the unilateral or bilateral occlusion of the carotid arteries independently of the completeness of the vertebral arteries cauterization as well as the patency of the anterior cerebral arteries. By monitoring the NADH redox state bilaterally one can quantify the level of ischemia in each hemisphere, in a continuous manner and in real-time. This allows correlation of the insult level to other measured parameters such as extracellular K or Ca^{2+} and DC potential. The maximum NADH level, which is evaluated by applying 100% N_2 or decapitation of the animal at the end of the experiment, could serve as an internal standard. As shown in Fig. 3.22b, the addition of N_2 after bilateral carotid occlusion led to an additional increase in NADH, indicating that the vertebral cauterization was not 100% successful. The potential contribution of blood from other blood vessels such as the spinal artery is minimal in the rat.

The use of the MPA in our studies contributes significantly to the evaluation of the ischemic damage created by the four-vessel occlusion technique. Also, a comparison between ischemia and spreading depression was possible in the same animal.

Our studies on those two phenomena revealed that there are basic differences between ischemia and SD, although some changes in the ionic homeostasis state have some similarities (Hansen and Zeuthen 1981; Hansen, 1985). We monitored, for the first time, simultaneously and continuously the NADH redox state (by surface fluorometry), the extracellular ionic activities of K^+ and Ca^{2+} (by surface mini-electrodes) and the electrical activities (DC potential and electrocorticogram).

The only way one can correlate the ionic activities with the energy state is by monitoring in real-time most of the parameters related to it. Since it is preferable to measure NADH from the surface of the brain, we developed the surface mini-electrode to measure the extracellular ion activities. In our original description of the ion-selective K^+ mini-electrode, we showed that only small differences between microelectrodes and surface electrodes were recorded under spreading depression (Crowe et al. 1977). The advantage of the surface electrode technique is that the integrity of the tissue and microcirculaton remained intact compared with the "artificial" microenvironment created by the insertion of the microelectrode into the brain.

Under ischemia the primary change is in energy supply, while in SD it is the depolarization wave leading to changes in electrical activity and ionic homeostasis. This difference between ischemia and SD has a significant effect on the recovery processes of the two pathological situations. The difference is shown in Figs. 3.24 and 3.25 and only by simultaneous measurements of all the parameters involved in the various processes can one understand them. As we have shown, the NADH redox state, under these two pathological situations, is changing toward a highly reduced state (ischemia) or to a very oxidized one (SD). The net ourcome is a very fast recovery under SD while under ischemia it will be very slow and insult-duration dependent. During the ischemic episode, the Na^+-K^+-ATPase activity is inhibited while it is stimulated under spreading depression so that the recovery processes different kinetics.

References

Berlet, H. H. 1976. Hypoxic survival of normoglycaemic young adult and adult mice in relation to cerebral metabolic rates. *Journal of Neurochemistry* 26: 1267–1274.

Berry, K., H. M. Wisniewski, L. Svarzbein, and S. Baez. 1975. On the relationship of brain vasculature to production of neurological deficit and morphological changes following acute unilateral common carotid artery ligation in gerbils. *Journal of the Neurological Sciences* 25: 75–92.

Blomqvist, P., H. Mabe, M. Ingvar, and B. K. Siesjö. 1984. Models for studying long-term recovery following forebrain ischemia in the rat. 1. Circulatory and functional effects of 4-vessel occlusion. *Acta Neurologica Scandinavica* 69: 376–384.

Bures, J., O. Buresova, and J. Krivanek. 1974. *The mechanism and applications of Leao's spreading depression of electroencephalographic activity*. New York: Academic.

Capdeville, C., D. Pruneau, M. Allix, M. Plotkine, and R. G. Boulu. 1986. Model of global forebrain ischemia in the unanesthetized rat. *Journal de Pharmacologie* 17: 553–560.

Chance, B., and H. Baltscheffsky. 1958. Respiratory enzymes in oxidative phosphorylation (VII - binding of intramitochondrial reduced pyridine nucleotide). *The Journal of Biological Chemistry* 233: 736–739.

Chance, B., P. Cohen, F. Jobsis, and B. Schoener. 1962. Intracellular oxidation-reduction states *in vivo*. *Science* 137: 499–508.

Chance, B., V. Legallias, J. Sorge, and N. Graham. 1975. A versatile time-sharing multichannel spectrophotometer reflectometer and fluorometer. *Analytical Biochemistry* 66: 498–514.

Chance, B., N. Oshino, T. Sugano, and A. Mayevsky. 1973. Basic principles of tissue oxygen determination from mitochondrial signals. *Advances in Experimental Medicine and Biology* 37A: 277–292.

Chance, B., and B. Schoener. 1963. *Symp. on oxygen in animal organism, abstract*. 1972/02/01 ed.

Chance, B., B. Schoener, and F. Schindler. 1963. The intracellular oxidation-reduction state. In Oxygen in the Animal Organism Proceedings of a Symposium held under the joint auspices of the International Union of Biochemistry and the International Union of Physiological Sciences. eds. F. Dickens and E. Neil. London: Pergamon Press.

Cohen, M. M. 1973. Biochemistry of cerebral anoxia, hypoxia and ischemia. In *Monographs in neural sciences*, eds. Cohen, M. M. Basel: Karger.

Crowe, W., A. Mayevsky, and L. Mela. 1977. Application of a solid membrane ion selective electrode to *in vivo* measurements. *The American Journal of Physiology* 233: C56–C60.

Dienel, G. A., and W. A. Pulsinelli. 1986. Uptake of radiolabeled ions in normal and ischemia-damaged brain. *Annals of Neurology* 19: 465–472.

Dora, E., and T. Zeuthen. 1976. Brain metabolism and ion movements in the brain cortex of the rat during anoxia. In *Ion and enzyme electrodes in biology and medicine*, ed. M. Kessler. Baltimore: University.Park Press.

Duffy, T. E., S. J. Kohle, and R. C. Vannucci. 1975. Carbohydrate and energy metabolism in perinatal rat brain: Relation to survival in anoxia. *Journal of Neurochemistry* 24: 271–276.

Eklof, B., and B. K. Siesjo. 1972. The effect of bilateral carotid artery ligation upon the blood flow and the energy state of the rat brain. *Acta Physiologica Scandinavica* 86: 155–165.

Fazekas, J. F., J. Kleh, and F. A. Finnerty. 1955. Influence of age and vascular disease on cerebral hemodynamics and metabolism. *The American Journal of Medicine* 18: 477–485.

Ferrendelli J. A., W. G.,Sedgwick and V. Suntzeff. (1971) Regional energy metabolism and lipofuscin accumulation in mouse brain during aging. *Journal of Neuropathology and Experimental Neurology* 30:638–649

Fink, B. R., and R. H. Haschke. 1973. Anesthetic effects on cerebral metabolism. *Anesthesiology* 39: 199–215.

Folbergrová, J., O. H. Lowry, and J. V. Passonneau. 1970. Changes in metabolites of the energy reserves in individual layers of mouse cerebral cortex and subjacent white matter during ischaemia and anaesthesia. *Journal of Neurochemistry* 17: 1155–1162.

Friedli, C. M., D. S. Sclarsky, and A. Mayevsky. 1982. Multiprobe monitoring of ionic, metabolic, and electrical activities in the awake brain. *The American Journal of Physiology* 243: R462–R469.

Gercken, G., and H. Preuss. 1969. The effect of breathing oxygen on the metabolism of the rat brain under normal and ischaemic conditions. *Journal of Neurochemistry* 16: 761–767.

Ginsberg, M. D., M. Reivich, S. Frinak, and K. Harbig. 1976. Pyridine nucleotide redox state and blood flow of the cerebral cortex following middle cerebral artery occlusion in the cat. *Stroke* 7: 125–131.

Goldman, H., and L. A. Sapirstein. 1973. Brain blood flow in the conscious and anesthetized rat. *The American Journal of Physiology* 224: 122–126.

Gregoire, N. M., A. Gjedde, F. Plum, and T. E. Duffy. 1978. Cerebral blood flow and cerebral metabolic rates for oxygen, glucose, and ketone bodies in newborn dogs. *Journal of Neurochemistry* 30: 63–69.

Hansen, A. J. 1978. The extracellular potassium concentration in brain cortex following ischemia in hypo- and hyperglycemic rats. *Acta Physiologica Scandinavica* 102:324–329.

Hansen, A. J. 1985. Effect of anoxia on ion distribution in the brain. *Physiological Reviews* 65:101–148.

Hansen, A. J., and T. Zeuthen. 1981. Extracellular ion concentrations during spreading depression and ischemia in the rat brain cortex. *Acta Physiologica Scandinavica* 113: 437–445.

Harbig, K., and M. Reivach. 1973. The effect of ischemia and hypoxia on the pyridine nucleotide redox state of the cerebral cortex of cats. *Stroke* 4: 341.

Harbig, K., and M. Rievich. 1973. *Stroke* 4: 341. (Abstract VI-43).

Harrison, M., T. J. Sick, and M. Rosenthal. 1985. Mitochondrial redox responses to cerebral ischaemia produced by four-vessel occlusion in the rat. *Neurological Research* 7: 142–148.

Hernández, M. J., R. W. Brennan, R. C. Vannucci, and G. S. Bowman. 1978. Cerebral blood flow and oxygen consumption in the newborn dog. *The American Journal of Physiology* 234: R209–R215.

Jobsis, F. F., and J. C. Duffield. 1967. Oxidative and glycolytic recovery metabolism in muscle. Fluorometric observations on their relative contributions. *The Journal of General Physiology* 50: 1009–1047.

Jobsis, F. F., M. O'connor, A. Vitale, and H. Vreman. 1971. Intracellular redox changes in functioning cerebral cortex. I. Metabolic effects of epileptiform activity. *Journal of Neurophysiology* 3465: 735–749.

King, L. J., G. M. Schoepfle, O. H. Lowry, J. V. Passonneau, and S. Wilson. 1967. Effects of electrical stimulation on metabolites in brain of decapitated mice. *Journal of Neurochemistry* 14: 613–618.

Lauritzen, M. 1987a. Cerebral blood flow in migraine and cortical spreading depression. *Acta Neurologica Scandinavica. Supplementum* 113:1–40.

Lauritzen, M. 1987b. Cortical spreading depression as a putative migraine mechanism. *Trends in Neurosciences* 10:8–13.

Leão, A. A. P. 1944. Spreading depression of activity in cerebral cortex. *Journal of Neurophysiology* 7:359–390.

Levy, D. E., and T. E. Duffy. 1975. Effect of ischemia on energy metabolism in the gerbil cerebral cortex. *Journal of Neurochemistry* 24: 1287–1289.

Lowry, O. H., J. V. Passonneau, F. X. Hasselberger, and D. W. Schulz. 1964. Effect of ischemia on known substrates and cofactors of the glycolytic pathway in brain. *The Journal of Biological Chemistry* 239: 18–30.

Maker, H. S., and Lehrer, G. M. 1971) Effect of ischemia. In *Handbook of neurochemistry,* ed. A. Lajtha. New York: Plenum Press.

Mares, P., N. Kríz, G. Brozek, and J. Bures. 1976. Anoxic changes of extracellular potassium concentration in the cerebral cortex of young rats. *Experimental Neurology* 53: 12–20.

Mayevsky, A. 1976a. Brain energy metabolism of the conscious rat exposed to ischemic conditions. In *IXth World Conf. European Microcirculation Soc.*
Mayevsky, A. 1976b. Brain energy metabolism of the conscious rat exposed to various physiological and pathological situations. *Brain Research* 113:327–338.
Mayevsky, A. 1977. Brain energy metabolism of the conscious rat exposed to ischemic conditions. In Bibliotheca Anatomica, Vol 15: Recent Advances in Basic Microcirculatory Research, edn. 15, 361-364. ed. Lewis, D.H., Basel: Karger.
Mayevsky, A. 1978a. Ischemia in the brain: the effects of carotid artery ligation and decapitation on the energy state of the awake and anesthetized rat. *Brain Research* 140:217–230.
Mayevsky. 1978b. Shedding light on the awake brain. In *Frontiers in bioenergetics: From electrons to tissues*. II edition. eds. P. L. Dutton, J. Leigh, and A. Scarpa. New York: Academic Press.
Mayevsky, A. 1983. Multiparameter monitoring of the awake brain under hyperbaric oxygenation. *Journal of Applied Physiology* 54:740–748.
Mayevsky, A. 1984. Brain NADH redox state monitored *in vivo* by fiber optic surface fluorometry. *Brain Research Reviews* 7:49–68.
Mayevsky, A. 2024. *The Mongolian Gerbil brain: Mitochondrial function, vasculature, and pathophysiological states*. Switzerland: Springer International Publishing.
Mayevsky, A., and B. Chance. 1973. A new long-term method for the measurement of NADH fluorescence in intact rat brain with implanted cannula. *Advances in Experimental Medicine and Biology* 37A: 239–244.
Mayevsky, A., and B. Chance. 1974. Repetitive patterns of metabolic changes during cortical spreading depression of the awake rat. *Brain Research* 65: 529–533.
Mayevsky, A., and B. Chance. 1975. Metabolic responses of the awake cerebral cortex to anoxia hypoxia spreading depression and epileptiform activity. *Brain Research* 98: 149–165.
Mayevsky, A., and B. Chance. 1976. The effect of decapitation on the oxidation-reduction state of NADH and ECoG in the brain of the awake rat. *Advances in Experimental Medicine and Biology* 75: 307–312.
Mayevsky, A., and B. Chance. 1982. Intracellular oxidation-reduction state measured *in situ* by a multicannel fiber-optic surface fluorometer. *Science* 217: 537–540.
Mayevsky, A., and B. Chance. 1983. Multisite measurements of NADH redox state from cerebral cortex of the awake animal. *Advances in Experimental Medicine and Biology* 159: 143–155.
Mayevsky, A., R. B. Duckrow, E. Yoles, N. Zarchin, and D. Kanshansky. 1990. Brain mitochondrial redox state, tissue hemodynamic and extracellular ion responses to four-vessel occlusion and spreading depression in the rat. *Neurological Research* 12: 243–248.
Mayevsky, A., C. M. Friedli, and M. Reivich. 1985. Metabolic, ionic and electrical responses of the gerbil brain to ischemia. *The American Journal of Physiology* 248: R99–R107.
Mayevsky, A., E. Yoles, N. Zarchin, and D. Kaushansky. 1989. Brain vascular, ionic and metabolic responses to ischemia in the Mongolian gerbil. In *German-Israel physiol. meeting, Jerusalem.*
Mayevsky, A., T. Zeuthen, and B. Chance. 1974. Measurements of extracellular potassium, ECoG and pyridine nucleotide levels during cortical spreading depression in rats. *Brain Research* 76: 347–349.
Meier-Ruge, W. 1975. Experimental pathology and pharmacology in brain research and aging. *Life Sciences* 17: 1627–1635.
Mikeska, A., and W. R. Klemm. 1975. EEG evaluation of humaneness of asphyxia and decapitation euthanasia of the laboratory rat. *Laboratory Animal Science* 25: 175–179.
Molinari, G. F., and J. P. Laurent. 1976. A classification of experimental models of brain ischemia. *Stroke* 7: 14.
Onodera, H., K. Iijima, and K. Kogure. 1986. Mononucleotide metabolism in the rat brain after transient ischemia. *Journal of Neurochemistry* 46: 1704–1710.
Patel, M. S. 1977. Age-dependent changes in the oxidative metabolism in rat brain. *Journal of Gerontology* 32:643–646.

Peng, M. T., Y. I. Peng, and F. N. Chen. 1977. Age-dependent changes in the oxygen consumption of the cerebral cortex, hypothalamus, hippocampus, and amygdaloid in rats. *Journal of Gerontology* 32: 517–522.

Pettegrew, R. K., and K. L. Ewing. 1971. Life history study of oxygen utilization in the C57BL/6 mouse. *Journal of Gerontology* 26: 381–385.

Pulsinelli, W. A., and J. B. Brierley. 1979. A new model of bilateral hemispheric ischemia in the unanesthetized rat. *Stroke* 10: 267–272.

Rosenthal, M., D. Martel, J. C. Lamanna, and F. F. Jobsis. 1976. *In situ* studies of oxidative energy metabolism during transient cortical ischemia in cats. *Experimental Neurology* 50: 477–494.

Rosenthal, M., and G. Somjen. 1973. Spreading depression, sustained potential shifts, and metabolic activity of cerebral cortex of cats. *Journal of Neurophysiology* 36: 739–749.

Salford, L. G., F. Plum, and B. K. Siesjö. 1973. Graded hypoxia-oligemia in rat brain. I. Biochemical alterations and their implications. *Archives of Neurology* 29: 227–233.

Samson, F. E., and N. A. Dahl. 1975. Cerebral energy requirement of neonatal rats. *The American Journal of Physiology* 188: 277–280.

Shenkin, H. A., P. Novak, B. Goluboff, A. M. Soffe, and L. Bortin. 1953. The effects of aging, arteriosclerosis, and hypertension upon the cerebral circulation. *The Journal of Clinical Investigation* 32: 459–465.

Simard, D., J. Olesen, O. B. Paulson, N. A. Lassen, and E. Skinhoj. 1971. Regional cerebral blood flow and its regulation in dementia. *Brain* 94: 273–288.

Stafford, A., and J. A. Weatherall. 1960. The survival of young rats in nitrogen. *The Journal of Physiology* 153: 457–472.

Sundt, Jr., T. M., and R. E. Anderson. 1975. Reduced nicotinamide adenine dinucleotide fluorescence and cortical blood flow in ischemic and nonischemic squirrel monkey cortex. 2. Effects of alterations in arterial carbon dioxide tension, blood pressure, and blood volume. *Stroke* 6: 279–283.

Swaab, D. F., and K. Bear. 1972. The presence of biologically fabile compounds during ischemia and their relationship to EEG in rat cerebral cortex and hypothalmus. *Journal of Neurochemistry* 19: 2843–2853.

Swaab, D. F., and K. Boer. 1972. The presence of biologically labile compounds during ischemia and their relationship to the EEG in rat cerebral cortex and hypothalamus. *Journal of Neurochemistry* 19: 2843–2853.

Thurston, J. H., and D. B. McDougal Jr. 1969. Effect of ischemia on metabolism of the brain of the newborn mouse. *The American Journal of Physiology* 216: 348–352.

Van Harreveld, A. 1972. The extracellular space in the vertebrate central nervous system. In *The structure and function of nervous tissue*, ed. G. H. Bourne. New York: Academic Oress.

Volpe, B. T., W. A. Pulsinelli, J. Tribuna, and H. P. Davis. 1984. Behavioral performance of rats following transient forebrain ischemia. *Stroke* 15: 558–562.

Zarchin, N., and A. Mayevsky. 1979. The effects of age on cerebral metabolic and electrical responses to complete ischemia in the awake and anesthetized rat. *Israel Journal of Medical Sciences* 15: 17.

Zarchin, N., and A. Mayevsky. 1981. The effects of age on the metabolic and electrical responses to decapitation in the awake and anesthetized rat brain. *Mechanisms of Ageing and Development* 16: 285–294.

Zeman, W., and J. R. M. Innes. 1963. *Craigie's neuroanatomy of the rat.* New York: Academic Press.

Chapter 4
Oscillations of Brain Mitochondrial NADH and CBF Under Ischemia

Abstract A new in vivo model for studying brain metabolic and hemodynamic oscillatory phenomena during ischemia is described. In this model, acute or chronic occlusion of one or two carotid arteries in the rat is performed. Due to the partial ischemia developed, oscillations in the level of intramitochondrial pyridine nucleotides (NADH) as well as flavoproteins (Fp) were recorded from the brain by monitoring the fluorescence of these respiratory chain components. The two fluorescent signals (NADH and Fp) were measured by using the time sharing or DC fluorometer/reflectometer. The changes in the reflected light at the excitation wavelengths (366 and 450 nm) were recorded simultaneously. Bilateral carotid artery occlusion induced immediate oscillations (6–9 waves per min) in the mitochondrial redox state as well as in tissue blood volume in both hemispheres. To verify the accuracy of the NADH monitoring system, including the correction technique for hemodynamic and other artifacts, we used the intracarotid artery saline bolus injection approach. The results could be summarized as follows: (1) unilateral carotid artery occlusion resulted in delayed development of oscillations, particularly in the ipsilateral hemisphere; (2) the oscillation phenomenon was reversible if recirculation restarted within 5 min. Occlusion for more than 30 min resulted in irreversible oscillations; (3) the oscillation appearances and intensities were affected by various physiological conditions. Vasoconstriction, induced by hyperoxia, stimulated the oscillations while vasodilation, induced by hypercapnia, depressed them. Anoxia, hypoxia, and spreading depression (50) abolished the oscillations. Glucose injection was not effective. Anesthesia under certain conditions enhanced the oscillations; (4) a clear phase difference was recorded between blood volume (R signal) and NADH redox state (F signal) waves; (5) analysis by phase relationship showed that the oscillations in various locations on the same hemisphere or between hemispheres appeared at different timings, suggesting a local oscillator. Using laser Doppler flowmetry simultaneously with NADH redox state and ion selective extracellular K^+ and Ca^{2+} electrodes, it was found that the mitochondrial oscillations in the NADH redox state were inversely correlated to tissue blood flow and volume. No correlation was found between the K^+ and Ca^{2+} levels and the

A. Mayevsky et al., *Physiological Mapping of the Cerebral Cortex in Neuropathology*,
https://doi.org/10.1007/978-3-032-13019-8_4

metabolic and hemodynamic oscillations, excluding the possibility that membrane pump activity (Na^+/K^+-ATPase or Ca^{2+}-ATPase) are the oscillators. It was concluded that the metabolic oscillations recorded under ischemic conditions were dependent on local changes in blood supply as evaluated by blood flow and volume changes and pO_2 level measurements. These oscillations in O_2 supply led to the intramitochondrial oscillations of the NADH and Fp components of the chain.

Keywords Ischemic brain · Mitochondrial redox state · Laser Doppler flowmetry · Calcium antagonist · Metabolic oscillations · Extracellular K^+ and Ca^{2+}

4.1 Introduction

A few years after the initial studies in which NADH fluorescence was measured, Chance et al described oscillations in NADH fluorescence in various experimental models (Chance and Legallias 1963; Chance et al. 1964a, 1964b, 1965a, 1965b, 1973). A few more papers were published in this field (Richard et al. 1996; Ghosh et al. 1971; Bashford et al. 1980; Richter et al. 1975; Harrison 1970). Figure 4.1 shows the correlation between oscillations in NADH fluorescence and pO_2 in the perfused hearts in vitro. The oscillations started after initial depletion of endogenous substrate followed by the addition of glucose (Chance et al. 1965b).

In the next paragraph, the interpretation of the figures' results by Prof. B Chance is presented.

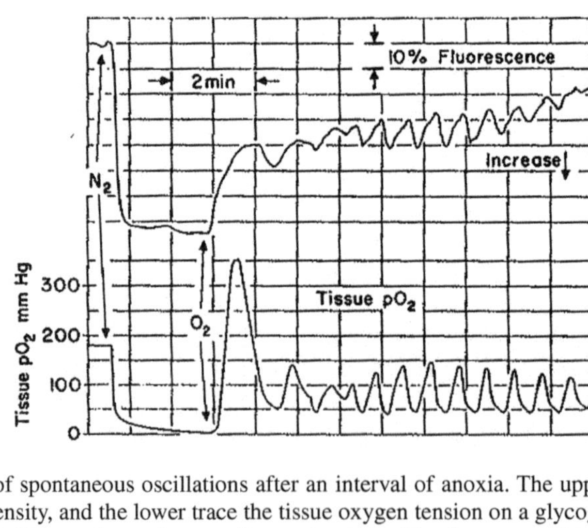

Fig. 4.1 Onset of spontaneous oscillations after an interval of anoxia. The upper trace shows the fluorescence intensity, and the lower trace the tissue oxygen tension on a glycogen-depleted heart. The perfusion fluid was changed to medium saturated with nitrogen, and back to oxygenated medium at the points shown in the figure (Chance et al. 1965b)

Spontaneous oscillations after anoxia. *A different type of instability of the pyridine nucleotide fluorescence has, on occasion, been observed in hearts first depleted of endogenous substrates by a combination of successive additions of epinephrine and periods of anoxia and later provided with glucose. Some preliminary results are presented here because of their intrinsic interest.* Figure 4.1 *shows an example of the onset of a series of oscillations of both the tissue fluorescence and oxygen tension following an interval of anoxia lasting 2.5 min. On perfusing the heart with oxygenated medium, the fluorescence decreases for about a minute to a value less than the fully aerobic level, and the oxygen concentration increases. Shortly thereafter, the oxygen concentration decreases and the heart was observed to beat more quickly and forcefully, although simultaneous measurements of the force of contraction were not made in this experiment. When the tissue oxygen tension falls to about 100 mmHg, the fluorescence starts to increase again, and a series of fluctuations in both the oxygen and fluorescence traces persists for at least 10 cycles. It is apparent from* Fig. 4.1 *that there is a slight phase shift between the two traces, fluorescence minimum slightly preceding oxygen concentration minimum. In each cycle, the onset of the fluorescence increase occurs when the* pO_2 *falls to between 70 and 100 mmHg, while the minimum* pO_2 *was 45 mmHg. It is difficult to know whether the fluctuations of oxygen are the casue of the pyridine nucleotide fluctuations or vice versa. Since decreased* pO_2 *has been shown to correspond either to an increased formation of DPNH in the cytoplasm (as with epinephrine) or in the mitochondria (as with Amytal or anoxia), it is not clear which DPN space is being affected in the above experiment. One possibility is that variations in the metabolic demand of the heart cause periodic hypoxia due to poor coronary perfusion with consequent mitochondrial DPN reduction. Another more interesting possibility is that we are observing a critical dependence of the mitochondrial respiration on the supply of exogenous glucose. In this case, the oscillations of fluorescence are more likely to be of glycolytic origin, possibly with fluctuating ADP or phosphate levels controlling the rate of the phosphofructokinase reaction.*

Oscillations in brain NADH fluorescence were described in 1976 by Dora et al. (1976) as seen in Figs. 4.2 and 4.3 where part of an oscillatory experiment is demonstrated. In Fig. 4.2, the beginning, and in Fig. 4.3, the end of oscillations are shown. Oscillation was elicited by decreasing the oxygen content of inspired gas mixture from 30% to 6% as is marked by the arrow (Fig. 4.2). As a consequence of hypoxia, arterial blood pressure dropped from the initial 120 to 25 mmHg. Corrected fluorescence increased by 30% and reflectance changed moderately (10% increase) and EEG showed a slight decrease as compared to the control activity. Following these abrupt changes blood pressure started to increase. The oxidation-reduction state of brain cortex shifted towards oxidation, and the minimal value of NADH fluorescence and the electrical activity were close to the control levels. Later on, in 1981, the same group published two more papers on this interesting subject (Dora and Kovach 1981a, 1981b).

In the next paragraph, the interpretation of the figure's results by Dora et al. is presented.

In Fig. 4.2, the beginning, and in Fig. 4.3, the end of oscillations are shown. Oscillation was elicited by decreasing the oxygen content of inspired gas mixture from 30% to 6% as is marked by the arrow. As a consequence of hypoxia, arterial

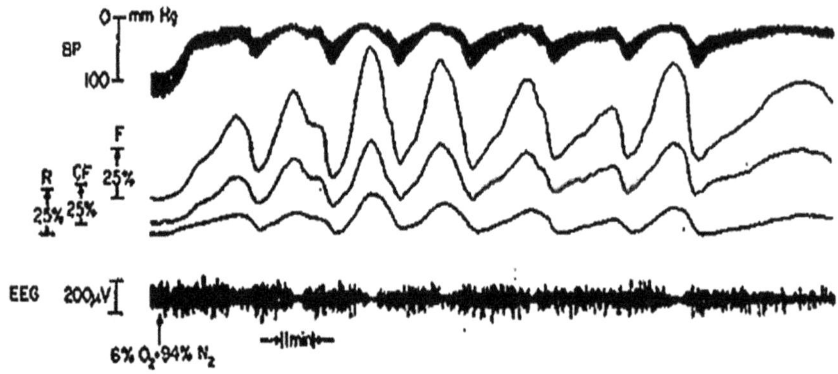

Fig. 4.2 In a typical experiment, the beginning of oscillations in arterial blood pressure (BP), EEG, and fluorescence (F), corrected fluorescence (CF), and reflectance (R) of the rat brain cortex are shown (Dora et al. 1976)

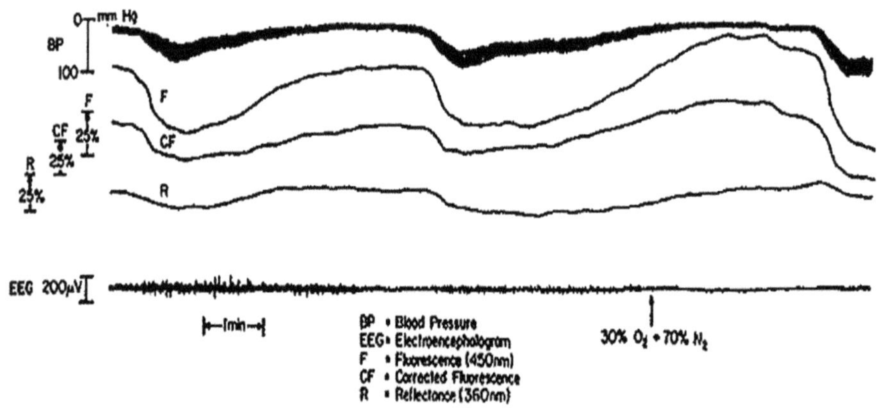

Fig. 4.3 In a typical experiment, the end of oscillations of arterial blood pressure (BP), EEG, and fluorescence (F), corrected fluorescence (CF), and reflectance (R) of the rat brain cortex are shown (Dora et al. 1976)

blood pressure dropped from the initial 120 to 25 mmHg. Corrected fluorescence increased by 30% and reflectance changed moderately (10% increase) and EEG showed a slight decrease as compared to the control activity. Following these abrupt changes blood pressure started to increase. Oxido-reduction state of brain cortex shifted towards oxidation and the minimal value of NADH fluorescence and the electrical activity were close to the control levels. It is interesting to note that the amplitudes of subsequent oscillations of reflectance and corrected fluorescence were not equal. First, they increased then decreased and again increased. In Fig. 4.3 the waves as compared to Fig. 4.2 have slowed down. At the beginning one wave lasted for 2 min, but before oscillations stopped, the waves had a periodicity of 5 min. The animal started to breath normal gas mixture at the time which is marked by the arrow, but the recovery of the rat from hypoxia was very slow. Arterial blood pressure and corrected fluorescence and reflectance recovered almost completely 30 min later but not the EEG. At this time activity had returned but it was considerably reduced as compared to the control. It is interesting to compare the anoxic reactions before and after oscillations.

Oscillations in cellular metabolism have been described in various in vitro and in vivo systems. Chance et al. (1964a) described damped oscillations in cytosolic pyridine nucleotides (NADH) in anaerobic yeast suspensions. These oscillations appeared after glucose addition and their wavelength varied from 0.6 to 0.25 min. The oscillations were ascribed to glycolytic activity, and it was suggested that the oscillations began when the mitochondrial control on the adenosine diphosphate (ADP) level decreased. It was also shown by Chance et al. (1964b) that a system that contains cytoplasmic components without respiratory activity demonstrates oscillations that vary from 0.2 min to 1.5 h and are inversely correlated with the temperature.

Some review articles have been written on oscillating enzyme reactions (Jobsis et al. 1971; Hess and Chance 1978) and on glycolytic systems by Harbig et al. (1976). Friesen and Block (1984) described a qualitative model to explain the mechanism of some oscillators, including the glycolytic oscillator. Although many publications have described oscillation systems in vitro, only a few published reports exist of in vivo conditions especially dealing with brain metabolic oscillations. Davies and Bronk (1957) demonstrated the oscillatory phenomenon in brain oxygen measured on the surface above the pial arteries and veins. The wavelength of the oscillations was approximately 12 s. Oscillations of the same wavelength were shown by Dora and Kovach (1981a). They measured NADH as well as blood volume from the brain cortex of anaesthetized, immobilized, and artificially ventilated cats. The oscillations described were sensitive to hypoxia, hypercapnia, and brain stimulation.

Recently, Vern et al. (1988) described low-frequency oscillations of cortical oxidative metabolism during waking and when asleep. A few more studies on the developed oscillations due to various factors were published (Doubovikov and Aksenov 2020; Shoemaker and Kamar 2025; Karkan et al. 1999; Klein et al. 2013; Hudetz et al. 1992).

Our work is aimed at investigating brain metabolic oscillation, which can be induced in a reproducible way in a small animal model. We describe oscillations in the intramitochondrial NADH level and in blood volume, as measured simultaneously from up to four sites in the rat's brain cortex, while exposed to partial ischemia. Also, we have correlated the NADH, blood flow, and volume changes to other local parameters such as tissue pO_2, extracellular K and Ca levels and electrical activities (DC potential and ECoG). It can be shown that the oscillations are vascular as well as metabolic, and that the relationship between the phases of the different parameters can be analyzed. Preliminary results have been published by Mayevsky et al. (1984). It is proposed that this ischemic brain, in which oscillations developed, may be used as an in vivo animal model for biological rhythm studies.

4.2 Methods

4.2.1 *Surface Fluorometry*

The intramitochondrial NADH redox state was monitored by a light guide DC fluorometer reflectometer having one, two, or four measurement sites on the brain (Mayevsky 1984; Mayevsky and Chance 1982). The source for the excitation light (366 nm) was a 100 W air-cooled mercury lamp. The emitted light from the brain was split in a 90:10 ratio for the measurement of the NADH fluorescence and 366 nm reflectance, respectively. Calibration and correction techniques have been described recently in detail by Mayevsky (1984). In experiments where NADH fluorescence was measured simultaneously with flavoprotein (Fp) fluorescence, the time-sharing fluorometer/reflectometer was used as described by Mayevsky (1976). We used "Y" shaped one-, two- and four-channel light guides made of quartz fibers having a 1 or 2 mm diameter at the common end of the one, two, and four channels, respectively.

4.2.2 *Multiparametric Monitoring System*

In experiments where NADH fluorescence was monitored together with tissue oxygen levels (pO_2), extracellular potassium (K^+), calcium (Ca^{2+}) levels, and DC steady potential, the multiprobe assembly (MPA) described in the references (Shapiro et al. 1982; Friedli et al. 1982) was used. In experiments where the microcirculation blood flow was evaluated, we used the laser Doppler flowmeter BPM 403a (TSI Inc., St Paul, MN, USA) described in detail (Haberl et al. 1989; Dirnagl et al. 1989) and shown to be correlated with H_2 clearance and autoradiography determination of cerebral blood flow (CBF). A 1.4 mm probe (model P-433 no. 6, TSI, Inc.) was introduced into the MPA and used simultaneously with the other parameters.

4.2.3 *Animal Preparation*

Male, Charles River rats were used in most experiments, except for a few studies done with the Wistar strain. The animal (120–220 g) was anaesthetized by an intraperitoneal (i.p.) injection of Equithesin (ET) (each ml contained 9.72 mg pentobarbital sodium, 42.51 mg chloral hydrate, 21.25 mg magnesium sulphate, 44.34% wt/vol propylene glycol, 11.5% alcohol and water) at a dose of 0.3 ml per 100 g body weight. After a midline incision of the skin, the skull was exposed and a hole (4 mm) or holes were drilled in one or two parietal bone areas. Four holes (1.5 mm diameter) were drilled for the stainless steel screws used as ECoG electrodes. The appropriate light guide holders were located epidurally and cemented to the skull using dental acrylic cement. In experiments where K_e^+, Ca^{2+}, and pO_2 were monitored, the dura mater was removed before the MPA was placed on the brain and cemented to the skull in the same way. To elicit spreading depression (SD) while monitoring NADH and flavoproteins (Fp), a push-pull cannula (for KCI application) was located near the light guide holder as described (Mayevsky and Chance 1975) previously. In experiments where a bolus of saline or glucose was injected, a catheter was introduced into one carotid artery towards the brain. Using our experimental approach, the entire area of the brain monitored was protected and the entire wound area of the skin (treated with a local anesthetic) was covered with dental cement. During most of the experiments, the animals were only slightly anaesthetized (0.1 ml ET per 100 g every 30 min) or were even awake and located in a small Plexiglas cage. Therefore, due to technical difficulties, the animal was not monitored for blood gases or blood pressure.

4.3 Results

Figure 4.4 demonstrates the classical results recorded in the brain of a rat 10 days after bilateral carotid artery occlusion. In all seven rats that were treated in the same way, oscillations were recorded. When only one carotid artery was occluded, oscillations were recorded in the two hemispheres. We have already shown in a preliminary study that unilateral or bilateral carotid artery ligation for 24–48 h led to an oscillatory phenomenon in brain blood volume (366 nm reflectance) and in the NADH uncorrected and corrected fluorescence (450 nm). As seen in Fig. 4.5, occlusion of the two carotid arteries for 24 h induced organized oscillations (waves

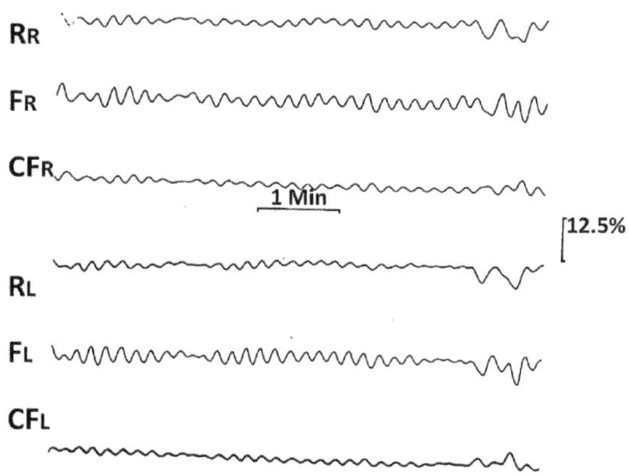

Fig. 4.4 Recording of brain oscillations 10 days after bilateral carotid artery occlusion in rats. RR, FR, and CFR are reflectance, fluorescence, and corrected fluorescence recorded in the right brain hemisphere. RL, FL, and CFL are the reflectance, fluorescence, and corrected fluorescence recorded from the left hemisphere

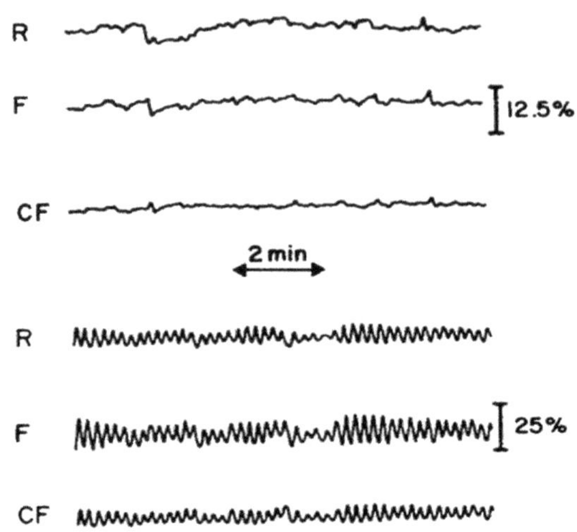

Fig. 4.5 Oscillations in the reflectance (R), fluorescence (F) and corrected NADH fluorescence (CF) recorded from a 24h bilateral carotid occluded rat (lower three traces) as compared with a control normal rat (upper three traces). The upper three traces were recorded from the unoccluded right hemisphere of the same rat (Mayevsky and Ziv 1991)

per minute) in NADH (F, CF) as well as in blood volume (R), which were not observed in a control animal (upper three traces) even when the scale sensitivity was increased in the control animal. To create a repeatable animal model for the various studies, we tested the appearance of the oscillations as a function of time after carotid artery occlusion, as well as the level of ischemia (unilateral or bilateral carotid occlusion).

Figure 4.6b1 shows the oscillations recorded from a 24 h left carotid artery ligated rat. As seen in part b1, the intensity of the oscillations in the ipsilateral occluded side (left) was much higher than in the unoccluded contralateral hemisphere (upper three traces). The record shown in Fig. 4.6b2 was taken from the same rat 2 h later when the animal became awake. As seen, the amplitude as well as the frequency of the oscillations increased. Also, some oscillations were recorded from the contralateral unoccluded side—the right side (upper three traces).

Table 4.1 summarizes the effects of ischemia duration as well as intensity on the development of the oscillations. Three examination times were used, namely, immediately after the occlusion, 24 h and 10 days later. For each examination time, we used different groups of rats. In each examination group, we have unilateral as well as bilateral carotid occluded subgroups. As seen, unilateral occlusion led to clear results either in the ipsilateral side within 24 h or in both sides after 10 days. This level of ischemia (unilateral occlusion) was not efficient when testing was done immediately after the occlusion. Bilateral occlusions have the best results after 24 h (15/15), and therefore in most of the other experiments, we used the bilateral occlusion for a 24 h period. As shown by Oldendorf (1989) bilateral carotid occlusion led to the development of

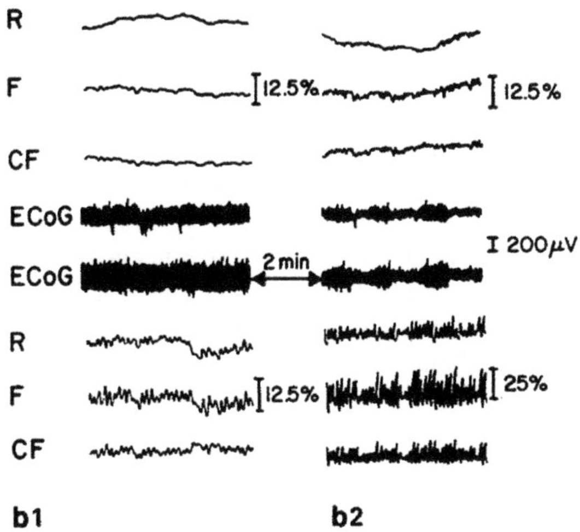

Fig. 4.6 (b1) The intensity of the oscillations measured bilaterally from a 24h left carotid artery occluded rat (lower three traces): (b2) this was measured 2h after Part b1 when the rat was fully awake. The upper three traces were recorded from the unoccluded right hemisphere of the same rat (Mayevsky and Ziv 1991)

Table 4.1 Development of metabolic oscillations under various occlusion and testing procedures (Mayevsky and Ziv 1991)

		Oscillations developed in		Reversibility (after reopening)	
Examination time	Occlusion side	Two hemispheres	One hemisphere	Immediate	After ½ h
Immediately	Unilateral	4/10		3/4	
	Bilateral	9/11		5/6	1/6
24 h	Unilateral	1/18	15/18		
	Bilateral	15/15	0/15		
10 days	Unilateral	5/5			
	Bilateral	7/7			

collaterals and increased perfusion through the enlarged basilar artery. The reversibility of the oscillation development was possible only if reopening of the carotid arteries was done within minutes after the beginning of the oscillations. After 30 min bilateral occlusion, reversibility was not possible.

After finding the best conditions for oscillation development, it was necessary to eliminate the possibility of correction artifacts as a source for the corrected fluorescence oscillation. In these experiments, we needed to determine whether or not the oscillations that appeared in the corrected fluorescence signal were due to a hemodynamic artifact or an inaccurate correction factor. Therefore, we used the approach that enabled us to test the correction factor of the fluorescence measurement in the same rat brain. The technique is based on the same method used originally by Jobsis et al. (1971) and Harbig et al. (1976).

In order to test the effect of various levels of ischemia on the development of the oscillation the following study was performed. In eight rats, the carotid arteries were isolated and monitoring of NADH fluorescent and reflectance were monitored in the two hemispheres. Figures 4.7–4.9 show typical results of this protocol. Figure 4.7a shows the bilateral typical responses to 100% nitrogen exposure. Figure. 4.7b shows the effects of unilateral and bilateral carotid occlusion. Oscillations in fluorescence

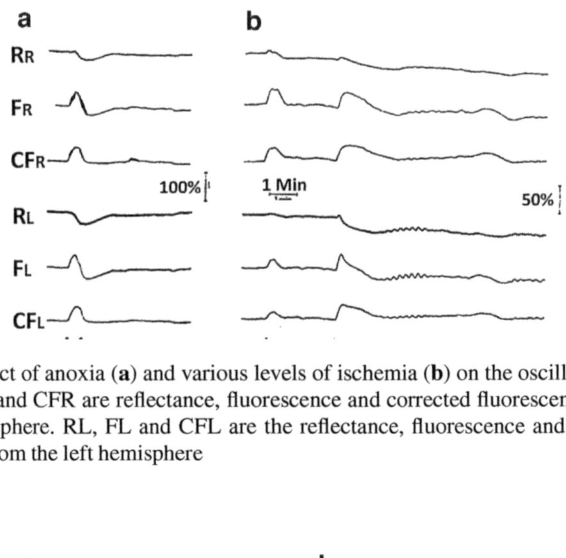

Fig. 4.7 The effect of anoxia (**a**) and various levels of ischemia (**b**) on the oscillation developed in the brain. R, FR and CFR are reflectance, fluorescence and corrected fluorescence recorded in the right brain hemisphere. RL, FL and CFL are the reflectance, fluorescence and corrected fluorescence recorded from the left hemisphere

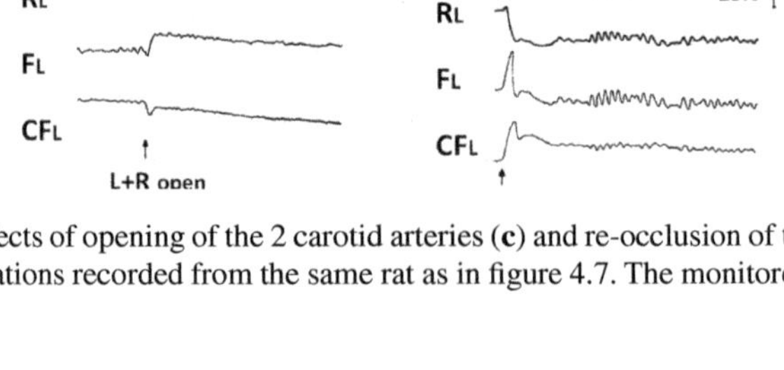

Fig. 4.8 The effects of opening of the 2 carotid arteries (**c**) and re-occlusion of the 2 carotid arteries (**d**) on the oscillations recorded from the same rat as in figure 4.7. The monitored parameters are as in figure 4.7

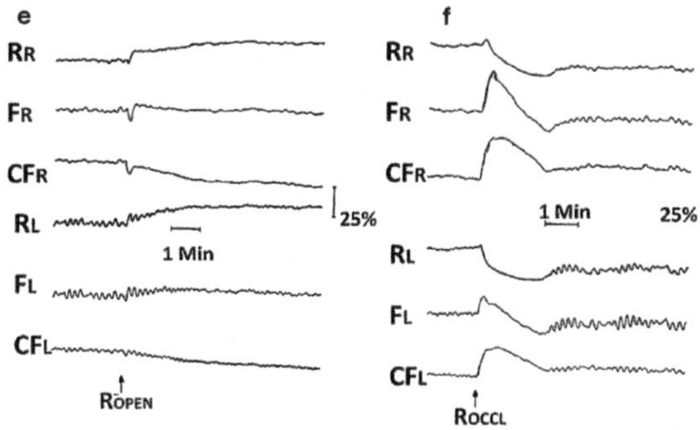

Fig. 4.9 The effects of opening of the Right carotid artery (**e**) and right carotid occlusion (**f**) on the oscillations recorded from the same rat as in figure 4.7. The monitored parameters are as in figure 4.7

and reflectance were recorded. After opening of the two carotid arteries (Fig. 4.8c), the oscillations disappeared, and appeared again after occlusion of the two carotid arteries (Fig. 4.8d). After opening of the right carotid artery (Fig. 4.9e), the intensity of the oscillations decreased in the two hemispheres and returned to the higher levels after occlusion of the right artery (Fig. 4.9f).

To differentiate between the "real" metabolic oscillations and the hemodynamic artifacts caused by inaccurate correction factors, we performed the following experiment: in five animals, a plastic cannula was inserted into one of the carotid arteries towards the brain and ischemia was induced by unilateral or bilateral carotid artery ligation. A series of 20 doses of 0.02–0.05 ml saline was injected into the carotid artery and an increase in the reflected light (R) as well as in the fluorescence light (F) was recorded (Fig. 4.10a). The increase in the reflectance light signal was due to the lower absorption properties of the tissue (less blood). Assuming that the small blood dilution did not cause tissue hypoxia, the increase in the fluorescence light was attributed solely to blood volume changes. Therefore, the F/R ratio represents the correction factor of the reflected light and that of the fluorescence light. The regression line was plotted and its slope represents the correction proportion for each animal tested (Fig. 4.10b).

The correction factor calculated by the saline flushing technique in four out of five animals tested was very close to 1 (0.82–1.1). Therefore, using such a factor in the fluorometer will provide very similar CF values to those occulated and measured during the experiments. Only in one animal (no. 2 in Table 4.1) did the value of the correction factor calculated by the saline flushing technique differ to a large extent. In this animal, we calculated the corrected fluorescence (CF) value by using the empirical correction factor (1.8) to obtain the net NADH CF values during the oscillation. Indeed, even using the 1.8 correction factor for calculation of the CF values in this animal showed a clear oscillation in NADH under these conditions, namely, that the oscillation was not generated by an artifact. The results presented in Table 4.2 show that only in one rat the CF calculated under saline flushing and oscillations was not

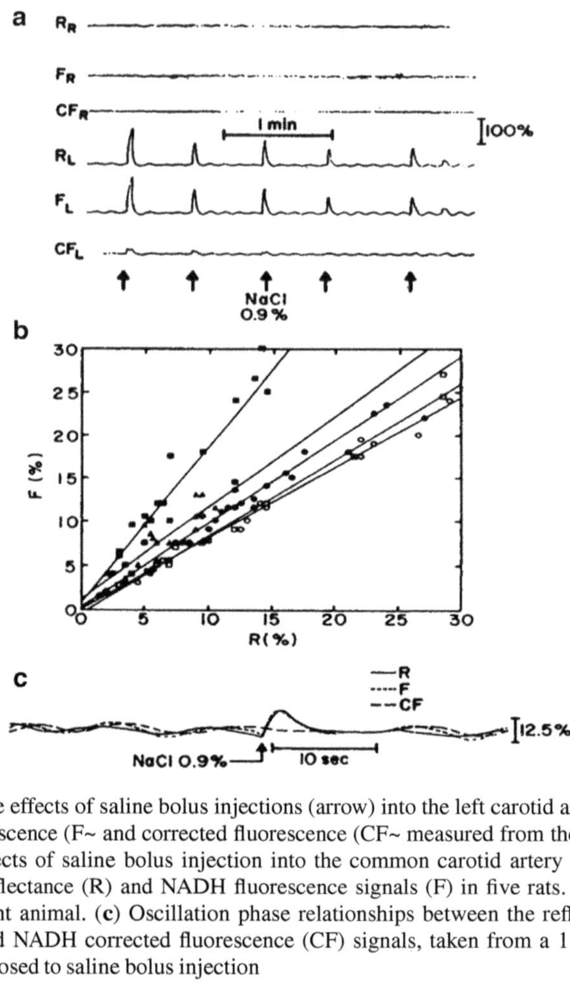

Fig. 4.10 (**a**) The effects of saline bolus injections (arrow) into the left carotid artery on the reflectance (R~, fluorescence (F~ and corrected fluorescence (CF~ measured from the two hemispheres of a rat. (**b**) Effects of saline bolus injection into the common carotid artery on the percentage change in the reflectance (R) and NADH fluorescence signals (F) in five rats. Each symbol represents a different animal. (**c**) Oscillation phase relationships between the reflectance (R), fluorescence (F), and NADH corrected fluorescence (CF) signals, taken from a 1 h bilateral carotid occluded rat, exposed to saline bolus injection

significantly different. These results indicate that the oscillations measured in the CF signal are real and were not created by the correction procedure.

Since the mitochondrial origin of the NADH signal is not accepted by all investigators, we verified that the oscillations in NADH signals are mitochondrial. To do so, we monitored simultaneously the fluorescence of NADH and Fp. The appearance of oscillations in the oxidized Fp fluorescence (only mitochondrial) can be seen in Fig. 4.11 (carotid arteries were occluded for 24 h). The effect of anoxia (Fig. 4.11) as well as of SD (Fig. 4.12) can be seen. NADH was monitored simultaneously with the Fp as well as two reflectance signals (366 and 460 nm) using a time sharing fluorometer (Mayevsky 1976). Due to the difference in the fluorescent properties, the NADH increased during anoxia while Fp decreased, but both changes have similar kinetics and the brain respiratory chain became more reduced. During SD both signals (NADH and

Table 4.2 Reflectance ® and fluorescence (F) relative changes recorded during saline injection into the carotid artery and during the oscillations (Mayevsky and Ziv 1991)

			F/R during		
Rat no.	Slope	Correlation coefficient	Oscillations	Saline injection	*P*
1	0.9	0.99	(21) 1.43 ± 0.11	(18) 0.89 ± 0.07	<0.001
2	1.8	0.94	(15) 1.69 ± 0.58	20 1.89 ± 0.07	<0.1
3	0.8	1.00	(18) 1.54 ± 0.51	(18) 0.83 ± 0.08	<0.001
4	1.0	0.98	(58) 1.20 ± 0.13	(18) 0.97 ± 0.06	<0.001
5	1.1	0.84	(36) 1.91 ± 0.48	(17) 1.25 ± 0.33	<0.05

Numbers in parentheses represent the number of flushes in the same animal. The numbers represent the mean ± SD

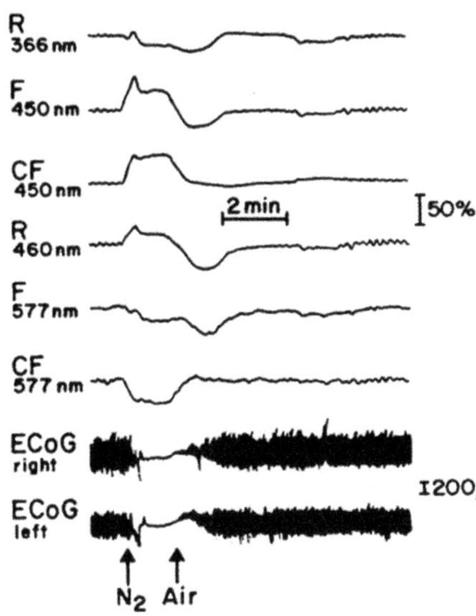

Fig. 4.11 Effects of anoxia on the oscillations in NADH and Fp fluorescence and reflectance. The records were obtained from a rat with bilateral carotid arteries occluded for 24h: channels one to three represent the NADH measurement while Fp fluorescence and reflectance are in channels four to six (Mayevsky and Ziv 1991)

Fp) showed an increase in redox state due to the limited amounts of oxygen available in the ischemic brain (Mayevsky 1978). Under both anoxia and SD the oscillations disappeared during the perturbation itself.

Another type of verification of the oscillation's nature came from studies where pO_2 was monitored simultaneously with NADH redox state as well as extracellular K (five different rats). Bilateral carotid arteries ligating for 24 h induced oscillations

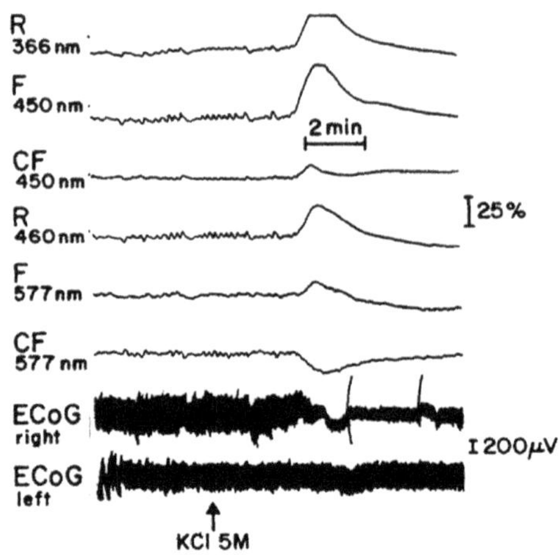

Fig. 4.12 Effects of spreading depression on the oscillations in NADH and Fp fluorescence and reflectance. The records were obtained from a rat with bilateral carotid arteries occluded for 24h: channels one to three represent the NADH measurement while Fp fluorescence and reflectance are in channels four to six (Mayevsky and Ziv 1991)

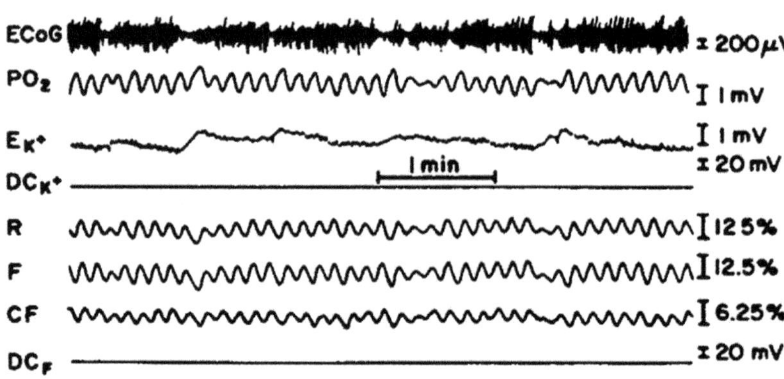

Fig. 4.13 Multiparametric recording from the brain of an awake rat having the two carotid arteries occluded for 45 h. ECoG electrocorticogram, pO_2 tissue oxygen pressure, E_{K^+} uncorrected potential measured from the valinomycin potassium electrode, DC_{K^+}, DC_F DC steady potential measured around the K^+ electrode and the light guide, R reflectance, F fluorescence, CF corrected fluorescence representing the net NADH signal (Mayevsky and Ziv 1991)

in the partial pressure of oxygen as measured with the oxygen electrode above the brain cortex (Fig. 4.13). The oscillations in the optical signals also appeared as changes in blood volume (R) as well as in the intramitochondrial NADH redox state (F, CF). The rat was anaesthetized for the operation period and then placed in a small Plexiglas cage connected to the monitoring system. The rat was awake during the measurements period which was 3–4 h after the operation and changes in the ECoG pattern were recorded simultaneously with the changes in extracellular K levels (K_e^+). These changes are in the range of ±1 mM. During the activation phase of the ECoG, the well-organized oscillations were somewhat distorted.

4.3.1 Phase Relationship Between Various Parameters or Different Brain Areas

In all experiments, a clear phase difference was found between the R, F, and CF signals. However, as saline was injected, the R and F signals showed a simultaneous increase due to the hemodilution effect appearing in both signals at the same time. A very interesting phase relationship was found between the $p0_2$ and the R signals, as shown in Fig. 4.14 tested in four different animals. It seems that whenever the R signals increased (blood volume decreased), the pO_2 decreased, showing that the changes in blood volume directly affected the oxygen availability (although the measurement sites of the various parameters were not identical).

To identify the type of oscillator that affects the continuation of the oscillations, we tested the effects of various metabolic perturbations as shown in Table 4.3. It was

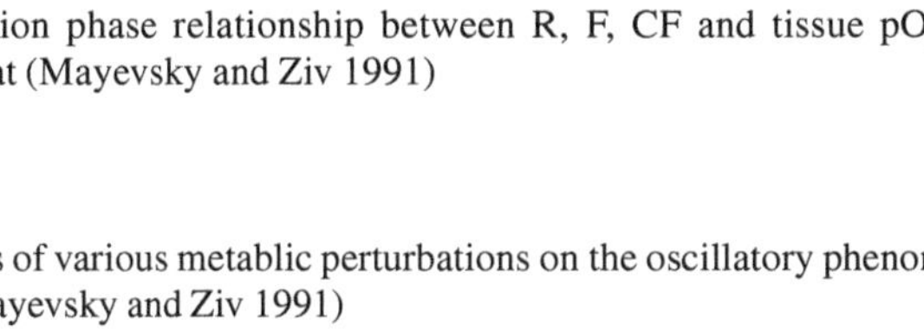

Fig. 4.14 Oscillation phase relationship between R, F, CF and tissue pO_2 in a 24h bilateral carotid occluded rat (Mayevsky and Ziv 1991)

Table 4.3 Effects of various metablic perturbations on the oscillatory phenomenon in the ischemic brain (Mayevsky and Ziv 1991)

Treatment	Responses developed in	Types of response
Hypercapnia (10% CO_2)	8/9	Abolished
Hyperoxia (100% O_2)	7/8	Initiated or stimulated
Hypoxia (10% O_2)	5/5	Abolished
Anoxia (0% O_2)	10/10	Abolished
Glucose (10%)	4/4	No effect
Anesthetics	5/5	Inhibited
Spreading depression	5/5	Abolished

found that changes in O2 supply to the brain affected the oscillations. As shown previously, the level of ischemia enhanced the development of the oscillations (unilateral versus bilateral carotid occlusion).

The effect of increased cerebral blood flow on the oscillations was tested by hypercapnia (10% CO_2 + 21% O_2 in N_2), as shown in Fig. 4.15a, increased blood flow, as indicated by the decrease in the R signal, abolished the oscillations until the rat breathed air again. The average NADH level remained approximately the same except for a transient short increase or decrease immediately after CO_2 application and removal.

The effects of changing the FiO_2 are shown in Fig. 4.16. This is a recording from a rat brain 24 h after bilateral carotid occlusion. Under 100% O_2 breathing, the well-defined oscillations in the pO_2, R, F, and CF signals were observed. However, when room air was given, the oscillations were abolished in all signals mentioned. Furthermore, it resulted in slight hypoxia as seen by the decrease in pO_2 and increase in CF signals. Evidently, the small decrease in energy supply (under air breathing) led to an increase in the extracellular K level, which returned to its original level as soon as the animal breathed 100% O_2 again.

Under hypoxic (10% O_2) or anoxic (0% O_2) conditions, the intensity of the oscillations diminished or even disappeared during the low O_2 supply period. In Fig. 4.11, we showed the anoxic effect tested in five rats. In a group of five rats, hypoxia led to a decrease in the intensity of the oscillations or abolished it during the insult period.

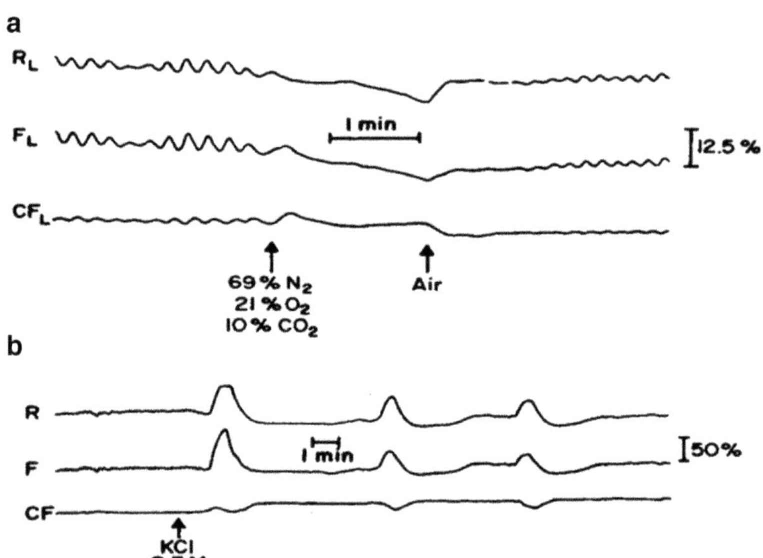

Fig. 4.15 (**a**) The response of the brain NAOH redox state to hypercapnia (10% CO^2). The two carotid arteries were occluded for 2 h and measurements were taken from the left hemisphere. (**b**) The effect of spreading cortical depression (SO) on the oscillations measured from the brain of a 7-day bilateral carotid occluded rat (Mayevsky and Ziv 1991)

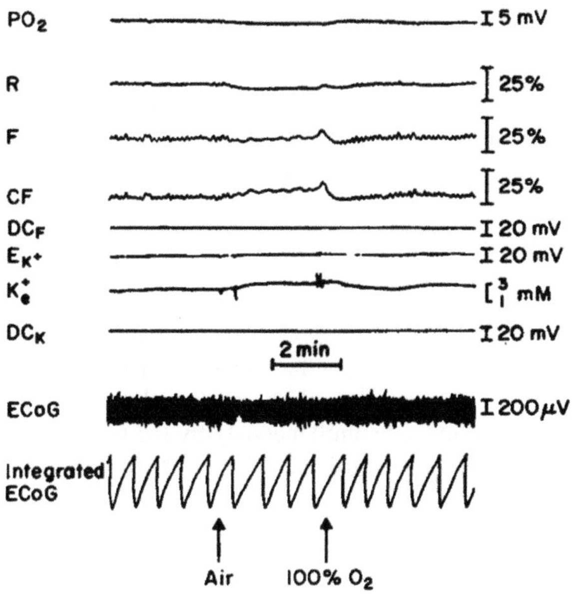

Fig. 4.16 Effects or hyperoxia 100% O_2 on the metabolic oscillations, ionic and electrical activities of the ischemic rat brain. The two carotid arteries were occluded for 24 h. The rat was breathing 100% O_2 before the exposure to air. ECoG electrocorticogram, pO_2 tissue oxygen pressure, E_{K^+} uncorrected potential measured from the valinomycin potassium electrode, DC_{K^+}, DC_F DC steady potential measured around the K^+ electrode and the light guide, R reflectance, F fluorescence, CF corrected fluorescence representing the net NADH signal (Mayevsky and Ziv 1991)

When glucose was injected into the carotid artery of four different animals, no significant change in the oscillation pattern was recorded (a total of 2 ml 10% glucose solution in 10–20 injections). NADH was monitored from the hemispheres and besides a transient change due to the hemodilution effect no significant changes were monitored (Table 4.3).

To change the energy demand by the brain, we tested the effects of spontaneous changes in activity (awake-sleep state), level of anesthesia, as well as induction of SD, which is known to change energy utilization (Mayevsky and Chance 1975; Mayevsky and Sclarsky 1983; Mayevsky 1978; Rosenthal and Somjen 1973; Siesjo 1978). As shown in Fig. 4.13, changes in the electrical activity (ECoG) representing waking responses of the animal resulted in a decrease in the intensity of the waves. In five other rats, we tested the effect of anesthetics on the oscillations and found that in the very deeply anaesthetized state (increased doses of ET by i.p. injection) the waves disappeared together with the depression or cessation of the electrical activity.

The effects of increased energy utilization were tested in rats in which SD was induced by KCI applied topically to the surface of the brain. As shown in Fig. 4.12, the SD wave led to the disappearance of the oscillations in NADH as well as Fp fluorescence. Due to the ischemic insult induced by the carotid occlusion, the NADH and Fp became reduced during the SD episode as was described previously (Rosenthal and Somjen 1973; Mayevsky 1978; Mayevsky and Sclarsky 1983). Under other conditions (10 days of bilateral carotid occlusion), the response to SD was an oxidation of NADH (Fig. 4.15b) and the oscillations also disappeared.

The phase relationships between different locations on the brain were analyzed in another group of five rats exposed to bilateral carotid ligation. As shown in Fig. 4.17, the phase of the waves appearing in the R, F, or CF signals varied between the right and left hemispheres. Furthermore, it can be seen that the phase relationships in the two hemispheres exhibited pattern changes in time, namely, that the oscillator is active, not in a fixed duration, at each location in the two hemispheres. The different wave pattern was studied also in the same hemisphere in which four measuring sites (1 mm diameter each) were monitored simultaneously by the four-channel fluorometer/reflectometer. As shown in Fig. 4.17b1, b2, which were recorded at 15 min intervals,

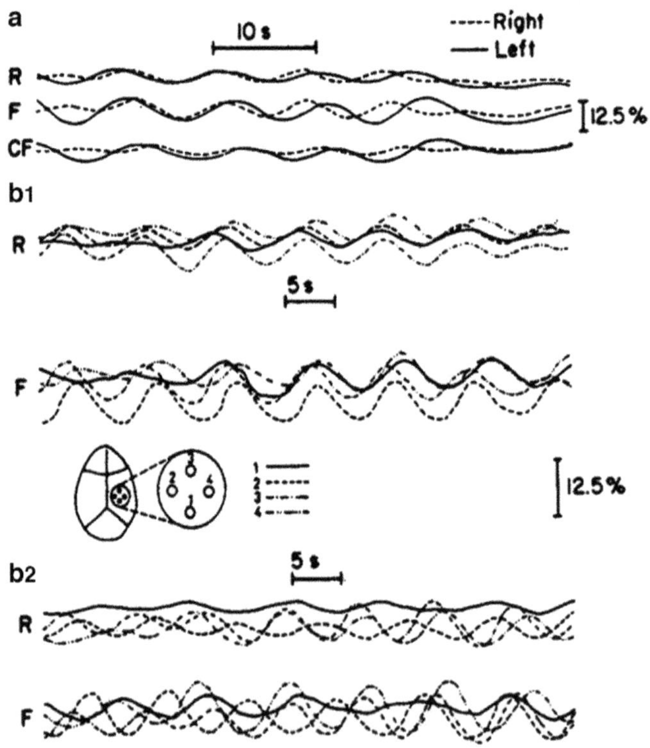

Fig. 4.17 Oscillation phase relationships between two cerebral hemispheres (**a**) or four different measuring sites in the same hemisphere in a bilateral carotid artery occluded rat (**b1** and **b2**) (Mayevsky and Ziv 1991)

each measuring site showed a different wave pattern at various times with regard to amplitude and phase relationships in the same brain hemisphere.

In a very recent set of experiments, we were able to incorporate into the MPA a second light guide connected to a laser Doppler flowmeter. Using this approach, we were able to correlate relative CBF (RCBF) to the NADH redox state as well as the other parameters monitored by the MPA during the oscillations. Figure 4.18 shows a record obtained from a rat in which the two carotid arteries were occluded for 20 h prior to the measurements. As expected, RCBF showed the same kind of oscillations as monitored by the NADH redox state. When the flow is increasing (LDF) then the NADH is more oxidized (CF) and vice versa. Also, the blood volume as measured by the laser Doppler (LDv) correlated very well with that evaluated by the 366 nm reflectance (R) trace changes, namely, an increase in blood volume corresponded to a decrease in the R trace. Oscillations were not recorded in the other parameters monitored, even when the sensitivity was increased five to tenfold (right side of Fig. 4.18). The extracellular level of Ca^{2+} was steady around 1.2 mM, and K^+ levels were even below 3 mM and steady. In this ischemic

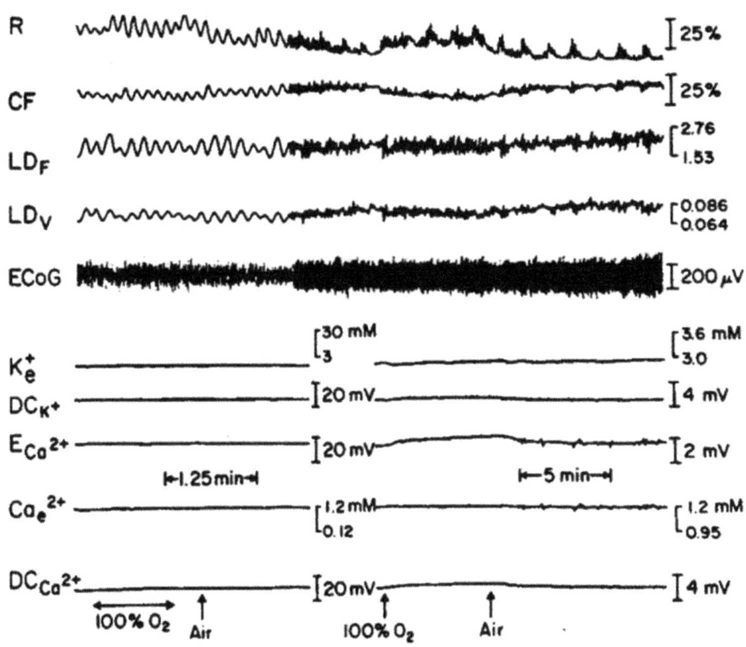

Fig. 4.18 Hemodynamic and metabolic oscillations measured from a rat brain in which the carotid arteries were occluded for 20 h. R reflectance, CF NADH corrected fluorescence, lD_F and lD_V laser Doppler flow and volume respectively, K_e^+, ECa^{2+}, Ca_i^{2+} corrected extracellular K^+, uncorrected and corrected extracellular calcium activities, respectively, DC_K^+, $DCca^{2+}$ DC steady potential measured concentric to the K and Ca electrodes, ECoG electrocorticogram (Mayevsky and Ziv 1991)

brain, changing the gas mixture from air to 100% O_2 led to a small oxidation of NADH as well as a decrease in blood volume. To test the hypothesis that Ca^{2+} was involved in the continuation of the oscillations (effect on smooth muscle), we exposed the oscillating brain to a Ca^{2+} channel blocker—nimodipine. Figure 4.19 shows the typical response to i.p. injection of nimodipine (0.1 ml containing 0.4 mg) from a rat with bilateral carotid occlusion for 20 h. The main effect of the drug was the decrease in RCBF and increase in NADH redox state as well as the blockage of the organized oscillations in the four traces (R, CF, LDF, and LDv)- The Extracellular K^+ and Ca^{2+} levels were not affected by the injection of nimodipine. The ECoG showed a mild depression probably due to a transient decrease in the available energy. This blockage of the oscillations by nimodipine was repeated similarly in four rats. When the same dose of nimopidine was injected into four normoxic control rats, no response was observed in all traces.

The effects of hypoxia (10% O_2) on the oscillations were tested in five rats 24 h after bilateral carotid artery occlusion. Figure 4.20 shows the clear oscillation developed in the ischemic brain. During the hypoxic event, the oscillations disappeared and returned after providing air to the animal. In another group of four rats, in which the two carotid arteries were occluded, injection of 2 ml 10% glucose solution into the left carotid artery was made. As seen in Fig. 4.21, the nature of the oscillations was not affected by the glucose even after another hour.

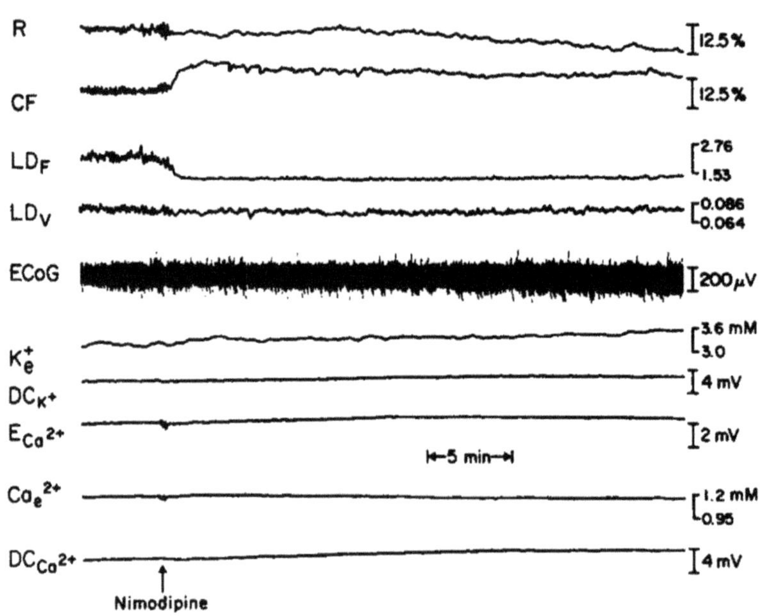

Fig. 4.19 The effects of Ca^{2+} channel blocker nimodipine on the various metabolic, hemodynamic, and ionic parameters monitored from a 20 h bilateral carotid occluded rat. All abbreviations are as in Fig. 4.18 (Mayevsky and Ziv 1991)

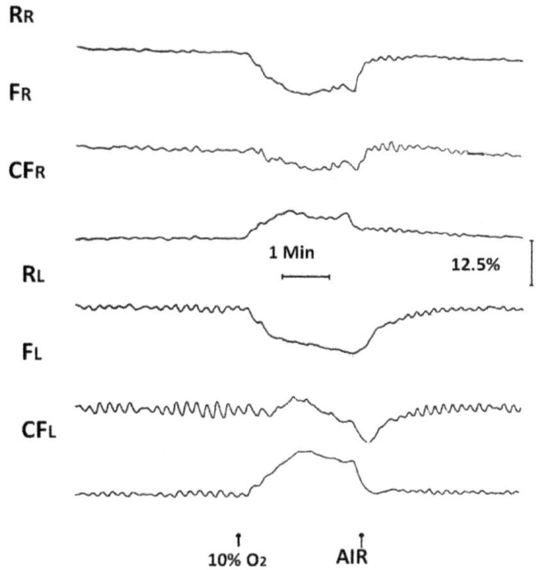

Fig. 4.20 Effects of hypoxia on the oscillations recorded after occlusion of the two carotid arteries for 1 h. RR, FR, and CFR are reflectance, fluorescence and corrected fluorescence recorded in the right brain hemisphere. RL, FL, and CFL are the reflectance, fluorescence, and corrected fluorescence recorded from the left hemisphere

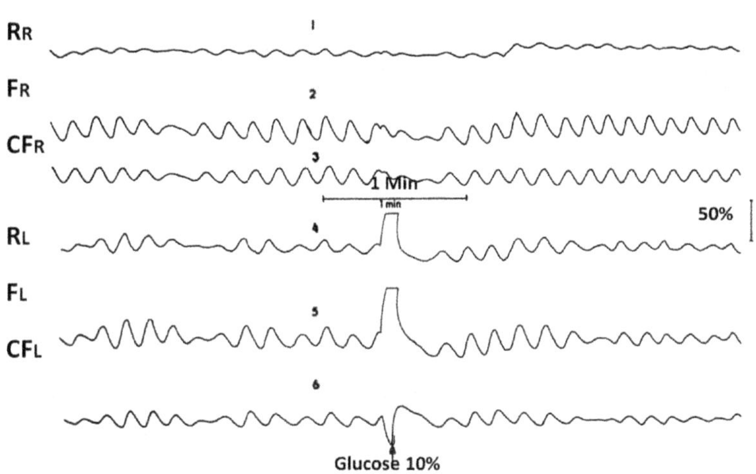

Fig. 4.21 Effects of glucose injection on the oscillations recorded after occlusion of the two carotid arteries for 1 h. RR, FR, and CFR are reflectance, fluorescence, and corrected fluorescence recorded in the right brain hemisphere. RL, FL, and CFL are the reflectance, fluorescence, and corrected fluorescence recorded from the left hemisphere

4.3.2 Metabolic and Hemodynamic Oscillation in the Non-Ischemic Brain

In a separate study, we investigated the appearance of oscillation in brain blood flow as well as mitochondrial NADH redox state (Meilin et al. 1999). The aim of this study was to characterize spontaneous oscillations of blood flow and mitochondrial NADH redox state in the cerebral cortex of rats, which underwent different treatments. Also, oscillation in CBF and NADH recorded from the human brain are presented.

At least 90 min after surgery, the animals underwent one of the various protocols:

Alcohol administration: The rats were injected IV with 30% alcohol, 3 ml 1200 g during 30 min.

CO exposure: The rats were placed in an airtight Plexiglas cage. The rats were exposed to 40 min of 1000 ppm CO followed by 20 min of 3000 ppm of CO (Mayevsky et al. 1995a).

N-Arg administration: The rats were injected with 50 mg kg of NO-nitro-L-arginine (N-Arg) **IP** (Meilin et al. 1995).

Ischemia: 24 h prior to brain surgery the common carotid arteries were isolated and occluded by 4-0 silk thread ligatures. 24 h later after, the brain surgery was completed and the MPA was placed. The animal was left to recover until fully awake.

Ischemia + Anesthesia: 24 h prior to brain surgery the common carotid arteries were isolated and occluded by 4-0 silk thread ligatures. During the monitoring period, the animal was anesthetized by additional **IP** injections of 0.3–0.5 ml equithesin every 20–30 min.

Trauma: A special modified MPA system was used for human brain monitoring in the Intensive-Care Unit as published previously (Mayevsky et al. 1996). The patient was hospitalized due to comatose state induced by severe head injury (GCS = 8). Preparation of the patient for brain monitoring was performed in the neurosurgical operating theater under general anesthesia. The patient was prepared for monitoring as follows:

1. A 15 mm diameter hole was carefully drilled in the frontal bone area selected according to CT scan. The hole was drilled in a "normal" looking section of the brain.
2. After cleaning the exposed area of bone residues, the dura mater was cut to enable direct contact between the MPA probes and the brain.
3. The MPA holder was screwed into the skull to a predetermined depth to avoid any undue pressure on the brain.
4. The operated area was closed by suturing the skin around the MPA holder (covered until the beginning of the monitoring period).
5. The patient was transferred to the intensive care unit and after stabilization and connection to the routine patient monitoring system the sterilized MPA was introduced into the holder (for contact with the brain) and fastened to it in order to ensure measurements for up to 72 h.

Continuous monitoring of the brain exposed to those pathological states reveals that spontaneous hemodynamic and metabolic oscillations were developed and recorded under the various treatments. The oscillations were recorded in the following parameters: CBF, reflected light (correlated to CBV), and mitochondrial NADH redox state. Figures 4.22–4.24 present records taken from one animal of each of the treated groups.

Figure 4.22a was recorded from an anesthetized rat monitored 24 h after two carotids occlusion.

Figure 4.22b was recorded from an awake rat monitored 24 h after two carotids occlusion.

Figure 4.23a was recorded from an animal that received an IV infusion of alcohol.

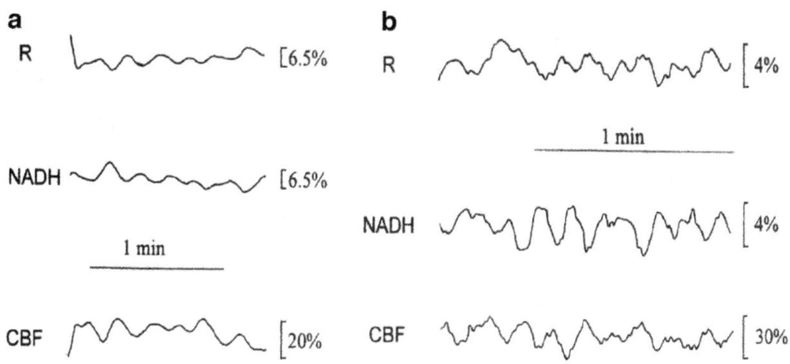

Fig. 4.22 Oscillations recorded from different animals under various treatments: (**a**) Anesthetized animal monitored 24 h after two carotids occlusion procedure. (**b**) Awake animal monitored 24 h after two carotids occlusion procedure. R reflectance, NADH mitochondrial NADH redox state, CBF cerebral blood flow, CBV cerebral blood volume (Meilin et al. 1999)

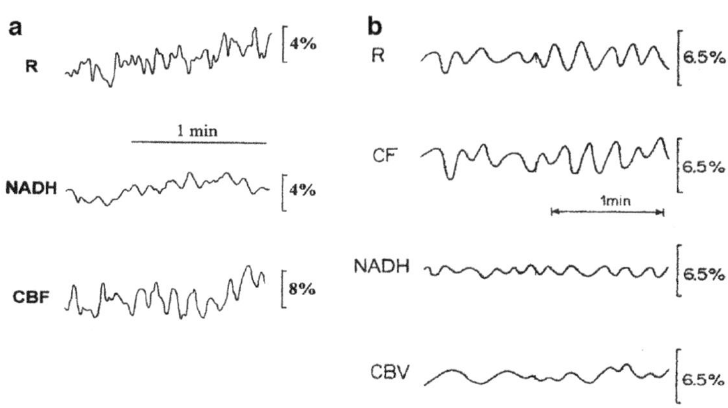

Fig. 4.23 Oscillations recorded from different animals under various treatments: (**a**) Animal receiving alcohol IV. (**b**) Oscillations monitored after N-Arg administration. R reflectance, NADH mitochondrial NADH redox state, CBF cerebral blood flow, CBV cerebral blood volume (Meilin et al. 1999)

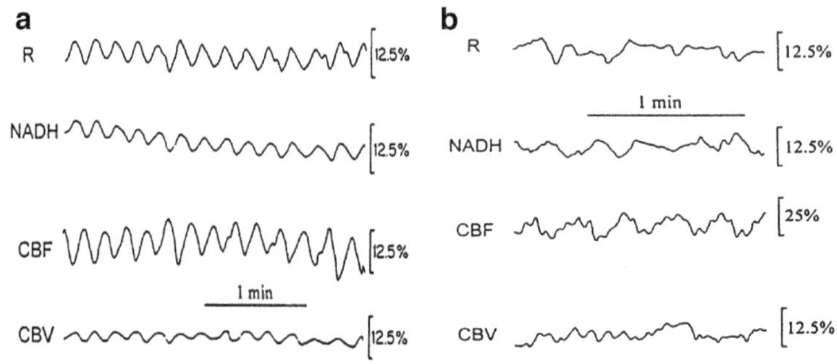

Fig. 4.24 Oscillations recorded from different animals under various treatments: (**a**) Animal exposed to 3000 ppm CO. (**b**) Oscillation recorded from the human cortex after trauma. R reflectance, NADH mitochondrial NADH redox state, CBF cerebral blood flow, CBV cerebral blood volume (Meilin et al. 1999)

Figure 4.23b was monitored after N-Arginine administration.

Figure 4.24a was recorded after exposing the rat to Carbon Monoxide.

Figure 4.24b was recorded from a patient after a head injury event.

It demonstrates typical changes in CBF reflected light and NADH during the oscillation period. The range of amplitude in CBF oscillation peaks is 6%–30% and NADH oscillations range is 5%–10%. The oscillation frequency was not significantly different among the various treatments and was found to be in the range of 3.6–5.5 waves/min (Table 4.4). Although NADH signal follows the CBF, a clear negative correlation was found between the two parameters in all treatments ($r = (-0.5)-(-0.8)$, $p<0.01$). Any increase or decrease in CBF during the oscillation wave was followed by an increase in the mitochondrial redox state oscillation wave. Although the frequency of the oscillation waves was found to be slower (2.5 waves/min) than those obtained in animals, the oscillations monitored from the injured patient were very similar to those recorded from the rat's cortex (Fig. 4.24b).

In this study, we observed similar oscillations in terms of frequency, correlation, and wave characterization under various treatments. The significant correlation between the hemodynamic and metabolic oscillations suggests that the trigger to this phenomenon could be the autoregulatory response to a minor lack of cellular metabolic energy. These findings are in agreement with a previous study carried out in our laboratory, which demonstrates Ischemia-related oscillations of CBV and NADH redox state (Mayevsky et al. 1984). We observed cortical hemodynamic and metabolic oscillations only in pathological states related to either a reduction in substrate (such as CBF reduction, oxygen depletion) or to a reduction in tissue O_2 extraction. In our previous study (Mayevsky and Ziv 1991), we showed that pO_2 oscillated at the same frequency as the NADH as compared to the CBF measured in the present study. In an original work of Dora and Kovach (1981a, 1981b) the oscillations were induced by either blood pressure fluctuations, bleeding or by deep anesthesia. In the present study, we were also able to detect more pathophysiological conditions that initiated the oscillations and also to correlate the NADH response to the microcirculatory oscillations (CBF).

Table 4.4 Frequency (wave/min) of the oscillations obtained at various treatments (Meilin et al. 1999)

Treatment	Ischemia	Anesthesia + ischemia	Alcohol	N-Arg	3000 ppm CO	Trauma (human)
Frequency (wave/min)	3.9 ± 0.4	4.1 ± 0.42	3.7 ± 0.8	3.7 ± 0.17	5.5 ± 0.62	2.5

In our previous studies, the initiation factor of the oscillations was partial ischemia, suggesting that a systemic factor was involved. According to the present study to a decrease in blood flow to the brain. The unique effect of carbon monoxide (CO) on the brain was an increase in CBF (Mayevsky et al. 1995b). The fact that oscillations were induced under CO suggests that the inducing factor is located along the line of microcirculation to the mitochondrial activity. Therefore, we conclude that the source of the oscillations is a change in brain regulation of microcirculation and the pathological state developed. It is also clear that the oscillator is not central but rather a local one affected by various perturbations of CO levels, depth of anesthesia as well as events such as Trauma.

4.4 Discussion

Rhythmic activity of the CNS has been investigated in various in vitro as well as in vivo; models (Lydic 1989). In most of the studies published, the naturally appearing rhythms were investigated. In the present study, we have developed an animal model in which brain microcirculatory and intracellular metabolic oscillations are induced in an in vivo situation. Such a model could be used as a tool for studying the triggering mechanism for the development of the oscillations as well as to identify the oscillator and its sensitivity to various factors, such as energy metabolism, ion homeostasis, or neuropharmacological treatment. The first question to be discussed is the mechanism of oscillation induction. In the work described by Dora and Kovach (Dora and Kovach 1981a), the oscillations were induced by either blood pressure fluctuations, bleeding, or by deep anesthesia, but the mechanism of initiation was not discussed. In our study, the initial factor in the development of the oscillations is the carotid artery occlusion (unilateral or bilateral) by itself and not only the resulting ischemia. We could postulate that since the occlusion occurred before the carotid body location, its involvement in the initiation process is an option. One possibility is that the carotid body is sending neural information (chemoreceptors) directly to the small arterioles as part of the autoregulatory and compensation mechanism to keep brain perfusion in the normal range. The fact that the phenomenon is irreversible after short periods of occlusion (Table 4.1), although blood flow and O_2 supply may return to normal, supports a neural triggering mechanism started by the carotid occlusion and not only the ischemic episode itself. Our way of initiation of the oscillation is a reproducible one and has not been described before by others. The second question is, are the recorded oscillations of a real metabolic nature, namely, are the changes in NADH redox state "real" and not created by correction artifacts? Using the saline bolus injection technique developed and adopted previously (Dora and Kovach 1981a; Harbig et al. 1976), it was possible to prove that the net NADH changes (oscillations) were metabolic and were not the result of a subtraction artifact. Using the calculated correction factor (by saline injection) during the oscillations did not change the appearance of the phenomenon in the calculated NADH trace. Furthermore, the phase difference between the R and F signals (Fig. 4.17a) supports the same conclusion. Also, the fact that Fp is oscillating at the same time supports the mitochondrial origin of the NADH oscillations.

The most critical and significant issue is the identification of the primary oscillator. To investigate it, we used the multiparametric monitoring approach, which provides information continuously and in real-time mode. We have proved that the oscillations were measured in the vascular tone (blood flow and volume), extracellular O_2 levels (tissue pO_2) as well as intracellular events (NADH and Fp fluorescence). It is the first time that so many physiological and biochemical parameters were monitored simultaneously in a real-time mode from the oscillating brain in vivo (Jobsis et al. 1977; Vern et al. 1988; Davies and Bronk 1957; Lydic 1989; Clark et al. 1958; Dora and Kovach 1981a; Dirnagl et al. 1989). The data presented proved that the oscillator is a local one rather than connected to a central mechanism controlling the oscillations in the entire brain. Using the four light guides above the same hemisphere, variations in the phase of the reflectance or NADH fluorescence oscillations were found. The size of the tissue volume monitored by the light guide was about 1 mm^3, so that assuming a distance between the light guides of 1 mm, we can conclude that the oscillating volume is not larger than the 1–2 mm^3 range. Two main oscillating elements are feasible: (1) smooth muscle cells in a vascular element wall; (2) neural or glial origin for the oscillations. The indirect evidence suggests that the oscillator is located in a vascular element rather than in neural or glial components. If the oscillator is located in neurons or glia, we may expect to see oscillations in the extracellular K^+ level due to changes in active pump activity, which consumes more than 50% of the energy produced. The results obtained by the MPA cannot support such an idea (Figs. 4.13, 4.16, 4.18, and 4.19). We did not record any fluctuations in K^+_e or Ca^{2+}_e during the metabolic oscillations induced by the carotid occlusion.

As shown in Fig. 4.13, changes in extracellular K^+ were correlated only with the variations in the ECoG activity in the unanaesthetized rat. This finding suggests that the changes in the electrical activity originating in the very complicated integrated processes between neuron induced variations in the Na^+/K^+-ATPase activity or in the rate of K^+ leakage or both (Korytova 1977; Melnikovova 1978). Indeed, under those conditions the well-organized oscillations in blood volume (R) and intramitochondrial NADH redox state (CF) were disturbed by the changes in alertness of the animal. We postulate then that the oscillations in the NADH redox state are due to changes in blood flow and volume, although the intracellular origin of an oscillator such as the interorganelle movement of Ca^{2+} cannot be excluded. Our measurement of extracellular Ca^{2+} does not support such an idea, but intracellular Ca^{2+} changes in the micromolar range (undetected by extracellular electrode) are feasible (Berridge and Galione 1988), and may affect the mitochondrial respiration. As shown by Uematsu et al. (1989), intracellular accumulation of Ca^{2+} under ischemia was blocked by pretreatment of the cat with nimodipine—a calcium channel blocker. They did not record any change in NADH redox state under the same treatment. The oscillations in our work were sensitive to changes in CBF; bilateral occlusion induced a higher intensity of the waves (compared with unilateral occlusion) due to the more severe limitation in CBF. However, when CBF was enhanced by increasing F; CO_2 or by low FiO_2 (hypoxia) the oscillations disappeared (see Fig. 4.15a). The addition of glucose did not change the pattern of the oscillation, suggesting that substrate for the mitochondria was not a limiting factor leading to the oscillations. Davies and Bronk (1957) also found that hypoxia abolished the oscillations they observed in the cat brain. The fact that hyperoxia activates the oscillations suggests that O_2 is not the direct and immediate oscillator driving a feedback loop to change the flow and mitochondrial respiration. Also, the localized nature of the oscillator in our studies (phase differences between various points in the same hemisphere or between two hemispheres) supports the hypothesis of a local feedback mechanism, linking cortical oxidative metabolism to vascular tone and changes in cerebral blood flow and volume (Jobsis et al. 1977; Melnikovova 1978; Moskalenko 1975; Severinghaus 1975; Clark et al. 1958).

A decrease in the activity level of the animal by injection of anesthetics resulted in the disappearance of the waves. When energy utilization was stimulated by SD, the oscillations disappeared regardless of the state of the brain ("oxidation cycle" or "reduction cycle"). Analysis of the phase difference between the various parameters measured suggests that the oscillations in R and CF are two interdependent but separate phenomena having phase differences. We have never observed oscillations in one of these parameters alone.

It seems that the oscillations described in the present study have a different origin from those of the glycolytic one described previously (Bioteaux and Hess 1974; Hess and Bioteux 1968; Hess et al. 1969; Hess and Chance 1978) or those described recently by Vern et al. (1988). There is a frequency similarity between our oscillations and those described by Davies and Bronk (1957) for variation in pO_2 and by Dora and Kovach (1981a) for NADH in cat brains.

4.5 Conclusions

It can be concluded that the change in CBF induced by ligation of one or two carotid arteries initiates oscillations in blood supply to the tissue, for example, by changing muscle tone in the blood vessel wall. Such a change was detected using the R signal and laser Doppler flowmeter as indicators. The oscillations in the tissue pO_2 may originate as a result of CBF fluctuations. Changes in intracellular O_2 supply to the mitochondria were also found to be oscillating as measured by the NADH and Fp signals.

References

Bashford, C. L., B. Chance, D. Lloyd, and R. K. Poole. 1980. Oscillations of redox states in synchronously dividing cultures of *Acanthamoeba castellanii* and *Schizosaccharomyces pombe. Biophysical Journal* 29:1–12.

Berridge, M. J., and A. Galione. 1988. Cytosolic calcium oscillators. *The FASEB Journal: Official Publication of the Federation of American Societies for Experimental Biology* 2:3074–3082.

Bioteaux, A., and B. Hess. 1974. Oscillations in glycolysis, cellular respiration and communication. *Faraday Symposia of the Chemical Society* 8:202–214.

Chance, B., R. W. Estabrook, and A. Ghosh. 1964a. Damped sinusoidal oscillations of cytoplasmic reduced pyridine nucleotinde in yeast cells. *Proceedings of the National Academy of Sciences* 51:1244–1251.

Chance, B., and V. Legallias. 1963. A spectrofluorometer for recording of intracellular oxidation-reduction states. *IEEE Transactions on Bio-medical Electronics* 10:40–47.

Chance, B., B. Schoener, and S. Elsaesser. 1965a. Metabolic control phenomena involved in damped sinusoidal oscillations of reduced diphosphopyridine nucleotide in a cell-free extract of *Saccharomyces carlsbergensis. The Journal of Biological Chemistry* 240:3170–3181.

Chance, B., B. Schoener, and S. Elsaesser. 1964b. Control of the waveform of oscillations of the reduced pyridine uncelotide level in a cell-free extract. *Proceedings of the National Academy of Sciences United States America* 52:337–341.

Chance, B., G. Williamson, I. Y. Lee, L. Mela, D. Devault, A. Ghosh, and E. K. Pye. 1973. Synchronization phenomena in oscillations of yeast cells and isolated mitochondria. *Biological and Biochemical Oscillators* 285–300.

Chance, B., J. R. Williamson, D. Jamieson, and B. Schoener. 1965b. Properties and kinetics of reduced pyridine nucleotide fluorescence of the isolated and *in vivo* rat heart. *Biochemische Zeitschrift* 341:357–377.

Clark Jr. L.C., G. Misrahy, and R. P. Fox. 1958. Chronically implanted polarographic electrodes. *Journal of Applied Physiology: Respiratory, Environmental and Exercise Physiology* 13:85–91.

Davies, P. W., and D. W. Bronk. 1957. Oxygen tension in mammalian brain. *Federation Proceedings* 16:689–692.

Dirnagl, U., B. Kaplan, M. Jacewicz, and W. Pulsinelli. 1989. Continuous measurement of cerebral cortical blood flow by laser-Doppler flowmetry in a rat stroke model. *Journal of Cerebral Blood Flow & Metabolism* 9:589–596.

Dora, E., and A. G. B. Kovach. 1981a. Metabolic and vascular volume oscillations in the cat brain cortex. *Acta Physiologica Academiae Scientiarum Hungaricae* 57:261–275.

Dora, E., and A. G. B. Kovach, 1981b. NAD-NADH and vascular volume oscillations in the cat brain cortex. In Advances in Physiology Science vol 25. Oxygen Transport to Tissue, eds. Kovach, A. G. B., Dora, E., Kessler, M., and Silver, I. A. Budapest: Pergamon Press.

Dora, E., K. Olaffson, B. Chance, and A. G. B. Kovach. 1976. Cortical NADH, pO_2, electrical activity and arterial blood pressure oscillations in hypoxaemia. *Advances in Experimental Medicine and Biology* 75:299–305.

Doubovikov, E. D., and D. P. Aksenov. 2020. Oscillations and concentration dynamics of brain tissue oxygen in neonates and adults. *Journal of Computational Neuroscience* 48:21–26.

Friedli, C. M., D. S. Sclarsky, and A. Mayevsky. 1982. Multiprobe monitoring of ionic, metabolic, and electrical activities in the awake brain. *The American Journal of Physiology* 243:R462–R469.

Friesen, W. O., and G. D. Block. 1984. What is a biological oscillator? *The American Journal of Physiology* 246:R847–R853.

Ghosh, A. K., B. Chance, and E. K. Pye. 1971. Metabolic coupling and synchronization of NADH oscillations in yeast cell populations. *Archives of Biochemistry and Biophysics* 145:319–331.

Haberl, R. L., M. L. Heizer, A. Marmarou, and E. F. Ellis. 1989. Laser-Doppler assessment of brain microcirculation: Effect of systemic alterations. *American Journal of Physiology. Heart and Circulatory Physiology* 256:H1247–H1254.

Harbig, K., B. Chance, A. G. B. Kovach, and M. Reivich. 1976. *In vivo* measurement of pyridine nucleotide fluorescence from cat brain cortex. *Journal of Applied Physiology: Respiratory, Environmental and Exercise Physiology* 41:480–488.

Harrison, D. E. F. 1970. UnDamped oscillations of pyridine nucleotide and oxygen tension in chemostat cultures of *Klebsiella aerogenes*. *The Journal of Cell Biology* 45:514–521.

Hess, B., and A. Bioteux. 1968. Control of glycolysis. In *Regulatory function of biological membranes*, ed. L. Jarnefelt. Amsterdam: Elsevier Publ, BBA Library.

Hess, B., A. Boiteux, and J. Kruger. 1969. Cooperation of glycolytic enzymes. *Advances in Enzyme Regulation* 7:149–167.

Hess, B., and B. Chance. 1978. In *Theoretical chemistry*, eds. H. Eyring and D. Hend. New York: Academic Press.

Hudetz, A. G., R. J. Roman, and D. R. Harder. 1992. Spontaneous flow oscillations in the cerebral cortex during acute changes in mean arterial pressure. *Journal of Cerebral Blood Flow and Metabolism* 12:491–499.

Jobsis, F. F., J. H. Keizer, J. C. Lamanna, and M. Rosenthal. 1977. Reflectance spectrophotometry of cytochrome aa3 *in vivo*. *Journal of Applied Physiology, Respiratory, Environmental and Exercise Physiology* 43:858–872.

Jobsis, F. F., M. O'Connor, A. Vitale, and H. Vreman. 1971. Intracellular redox changes in functioning cerebral cortex. I. Metabolic effects of epileptiform activity. *Journal of Neurophysiology* 3465: 735–749.

Karkan, D. M., C. Van Breemen, P. L. Skarsgard, G. J. Lagaud, and D. F. Hardwick. 1999. A Link between vasomotion and spontaneous oscillations of oxygen in rat brain. In *Oxygen transport to tissue XXI. Advances in experimental medicine and biology*, eds. Eke, A., and Delpy, D. T. Boston, MA: Springer.

Klein, K. U., S. Boehme, E. K. Hartmann, M. Szczyrba, L. Heylen, T. Liu, M. David, C. Werner, K. Markstaller, and K. Engelhard. 2013. Transmission of arterial oxygen partial pressure oscillations to the cerebral microcirculation in a porcine model of acute lung injury caused by cyclic recruitment and derecruitment. *British Journal of Anaesthesia* 110:266–273.

Korytova, H. 1977. Arousal-induced increase of cortical [K^+] in unrestrained rats. *Experientia* 33:242–244.

Lydic, R. 1989. Central pattern-generating neurons and the search for general principles. *The FASEB Journal: Official Publication of the Federation of American Societies for Experimental Biology* 3:2457–2468.

Mayevsky, A. 1976. Brain energy metabolism of the conscious rat exposed to various physiological and pathological situations. *Brain Research* 113:327–338.

Mayevsky, A. 1978. Ischemia in the brain: The effects of carotid artery ligation and decapitation on the energy state of the awake and anesthetized rat. *Brain Research* 140:217–230.

Mayevsky, A. 1984. Brain NADH redox state monitored *in vivo* by fiber optic surface fluorometry. *Brain Research Reviews* 7:49–68.

Mayevsky, A., and B. Chance. 1975. Metabolic responses of the awake cerebral cortex to anoxia hypoxia spreading depression and epileptiform activity. *Brain Research* 98:149–165.

Mayevsky, A., and B. Chance. 1982. Intracellular oxidation-reduction state measured *in situ* by a multicannel fiber-optic surface fluorometer. *Science* 217:537–540.

Mayevsky, A., A. Doron, T. Manor, S. Meilin, K. Salame, and G. E. Quaknine. 1995a. Repetitive cortical spreading depression cycle development in the human brain: A multiparametric monitoring approach. *Journal of Cerebral Blood Flow and Metabolism* 15:S34.

Mayevsky, A., A. Doron, T. Manor, S. Meilin, N. Zarchin, and G. E. Ouaknine. 1996. Cortical spreading depression recorded from the human brain using a multiparametric monitoring system. *Brain Research* 740:268–274.

Mayevsky, A., S. Meilin, G. G. Rogatsky, N. Zarchin, and J. Sonn. 1995b. Multiparametric monitoring of the awake brain exposed to carbon monoxide. *Journal of Applied Physiology: Respiratory, Environmental and Exercise Physiology* 78:1188–1196.

Mayevsky, A., and D. S. Sclarsky. 1983. Correlation of brain NADH redox state, K^+, PO_2 and electrical activity during hypoxia, ischemia and spreading depression. *Advances in Experimental Medicine and Biology* 159:129–141.

Mayevsky, A., and I. Ziv. 1991. Oscillations of cortical oxidative metabolism and microcirculation in the ischaemic brain. *Neurological Research* 13:39–47.

Mayevsky, A., I. Ziv, V., Ventura, and Y. Levy. 1984. Multiparametric oscillations in the rat brain in vivo. In: *Oscillations in physiological systems: dynamics and control*. London: The Institute of Measurement and Control.

Meilin, S., A. Mendelman, J. Sonn, T. Manor, N. Zarchin, and A. Mayevsky. 1999. Metabolic and hemodynamic oscillations monitored optically in the brain exposed to various pathological states. *Advances in Experimental Medicine and Biology* 471:141–146.

Meilin, S., N. Zarchin, A., Mayevsky, and S. Shapira. 1995. Multiparametric responses to cortical spreading depression under nitric oxide synthesis inhibition. In *Biochemical pharmacological and clinical aspects of nitric oxide*, eds. B. A. Weissman, N. Alon, and S. Shapira. New York: Plenum Press.

Melnikovova, H. 1978. Changes in the potassium ion concentration in the extracellular space of rat cerebral cortex during the arousal reaction. *Physiologia Bohemoslovenica* 27:131–138.

Moskalenko, Y. Y. 1975. Regional cerebral blood flow and its control at rest and during increased functional activity. In *Brain work*, eds. D. H. Lngvar and N. A. Lassen. CopenhagenMunksgaard.

Oldendorf, W. H. 1989. Trophic changes in the arteries at the base of the rat brain in response to bilateral common carotid ligation. *Journal of Neuropathology and Experimental Neurology* 48: 534–547.

Richard, P., B. M. Bakker, B. Teusink, K. Van Dam, and H. V. Westerhoff. 1996. Acetaldehyde mediates the synchronization of sustained glycolytic oscillations in populations of yeast cells. *European Journal of Biochemistry* 235:238–241.

Richter, O., A. Betz, and C. Giersch. 1975. The response of oscillating glycolysis to perturbations in the NADH/NAD system: a comparison between experiments and a computer model. *BioSystems* 7:137–146.

Rosenthal, M., and G. Somjen. 1973. Spreading depression, sustained potential shifts, and metabolic activity of cerebral cortex of cats. *Journal of Neurophysiology* 36:739–749.

Severinghaus, J. W. 1975. Coupling mechanisms flow to function and metabolism in brain, a general discussion. In *Brain work*, eds. Lngbar, D. H., and Lessen, N. A. Copenhagen: Munksgaard.

Shapiro, I. M., E. E. Golub, S. Kakuta, J. Hazelgrove, J. Havery, B. Chance, and P. Frasca. 1982. Initiation of endochondrial calcification is related to changes in the redox state of hypertrophic chondrocytes. *Science* 217:950–952.

Shoemaker, L. N., F. Kamar, D. Milej, L. Fitzgibbon-Collins, R. Eskandari, J. Bhangu, J. K. Shoemaker, and K. St. Lawrence. 2025. Two paths to one destination: Altered low-frequency oscillations of cerebral perfusion and oxygenation but not oxygen metabolism in mild cognitive impairment. *Journal of Applied Physiology (1985)* 138:1361–1371.

Siesjo, B. K. 1978. Anaesthesia, analgesia and sedation. In *Brain energy metabolism*, ed. B. K. Siesjo. Chichester, New York, Brisbane, Toronto: John Wiley & Sons.

Uematsu, D., J. H. Greenberg, W. F. Hickey, and M. Reivich. 1989. Nimodipine attenuates both increase in cytosolic free calcium and histologic damage following focal cerebral ischemia and reperfusion in cats. *Stroke* 20:1531–1537.

Vern, B. A., W. H. Schuette, B. Leheta, V. C. Juel, and M. Radulovacki. 1988. Low-frequency oscillations of cortical oxidative metabolism in waking and sleep. *Journal of Cerebral Blood Flow and Metabolism* 8:215–226.

Chapter 5
Mitochondrial Contribution to the Pathophysiology of Cerebral Ischemia

Abstract Cerebral ischemia is caused by damaged blood supply to the brain, leading to an interruption in the oxygen balance of supply vs. demand. As a result, the impaired tissue suffers from various metabolic alterations, which lead to the formation of a new microenvironment in the ischemic tissue. Lack of oxygen to the brain interrupts the mitochondrial electron transport chain activity, resulting in a decrease of ATP synthesis and elevation of mitochondrial NADH levels. Due to its sensitivity to cellular oxygen levels, mitochondrial NADH serves as an indication to various pathological conditions in the tissue. When focal cerebral ischemia occurs, only a limited area in the brain suffers from a decrease in blood supply. The ischemic tissue suffers from a gradual injury, which spreads from severe damage in the central core area and onto moderate damage in peripheral areas known as penumbra. While the core displays severe lack of blood supply, the penumbra holds partial supply of blood, due to the existence of collateral blood vessels. Correspondingly, oxygen supply and mitochondrial function are more severe in the core compared to the penumbra. Studies show that focal ischemic damage can be reduced by inducing another form of a more moderate ischemia as ischemic preconditioning or ischemic postconditioning treatments. Ischemic preconditioning is based on the assumption that mild degree of ischemia increases the tolerance of the tissue when facing additional ischemic insult. Ischemic postconditioning is induced during the primary ischemic duration or at reperfusion and aims to reduce reperfusion injury. The aims of the current study include the evaluation of mitochondrial function and blood flow before, during and following different states of cerebral ischemia, with emphasis on focal cerebral ischemia. The study investigated the effects of ischemic duration on cerebral blood flow (CBF) and mitochondrial NADH in different regions of the ischemic brain. A comparison between mitochondrial function and CBF during focal cerebral ischemia compared to global partial cerebral ischemia was held. The study also examined the variability in outcomes of mitochondrial function and CBF that were examined following focal cerebral ischemia. In addition, the ability of the brain to function under metabolic challenges, which include oxygen deprivation and

A. Mayevsky et al., *Physiological Mapping of the Cerebral Cortex in Neuropathology*,
https://doi.org/10.1007/978-3-032-13019-8_5

induced spreading depression (SD) waves, was examined. Finally, the study investigated the possibility of ischemia and reperfusion treatment using preconditioning and postconditioning ischemic procedures. Evaluation of mitochondrial function and the hemodynamic activity of the tissue was performed using a multisite multiparametric (MSMP) monitoring system, which combines NADH measurement using the fluorometric technique and CBF measurement using laser Doppler flowmetry, simultaneously and in real time in two regions of the brain. Focal ischemia was induced by middle cerebral artery occlusion (MCAO)—a common used mode of ischemic stroke. Global ischemia was induced by ligation of one or both carotid arteries. The results of the study introduce a continuous, multiparametric, in vivo and real-time monitoring of mitochondrial NADH and blood flow before, during and following cerebral ischemia. Results show that MCAO leads to reduction of tissue blood supply, accompanied by lack of oxygen and mitochondrial dysfunction. The hemodynamic and mitochondrial insult occurs in the hemisphere ipsilateral to the occluded artery. The ischemic lesion is gradual and spreads from severe damage in the core to moderate damage in the penumbra. Severity of mitochondrial and hemodynamic damage increases as ischemic duration increases, while the ability of the tissue to recuperate decreases. Study results demonstrate that after 24 h, transient focal ischemia leads to vast damage in the core alone, while permanent ischemia leads to equal damage to the core and the penumbra. The current study presents, for the first time, a comprehensive evaluation of different result profiles, which occur following a single model of focal ischemia. These profiles differ in CBF patterns, reperfusion reaction and survival of study animals. The wide variability in outcomes indicates the complexity of the MCAO model. Evaluation of this variability may assist other researchers using the same MCAO model as well as other models for focal ischemia. The current study presents a detailed comparison between the metabolic activity of the tissue during focal ischemia and global ischemia. Partial global ischemia, which occurs following unilateral carotid occlusion, has little immediate influence on the brain, resulting in differences between both hemispheres only 24 h following the occlusion. In contrast, partial global ischemia due to bilateral carotid occlusion leads to a more severe injury, which is demonstrated by the inability of the brain to cope with metabolic challenges immediately following the occlusion. Partial global ischemia, by unilateral or bilateral carotid occlusion, leads to a more moderate response compared to the core and even the penumbra of focal ischemia. Following the characterization of the metabolic activity in the ischemic tissue, the next step of the study included the evaluation of means to reduce the damage of focal ischemia. Ischemic preconditioning by unilateral and bilateral carotid occlusion did not meet the expectations of treatment to ischemic and reperfusion injury. The results rather show that unilateral or bilateral carotid occlusion prior to MCAO may serve as a model for emboli, in which damage in a main artery leads in a later stage to focal ischemia in a remote area. In contrast, unilateral or bilateral carotid occlusion which was used as ischemic postconditioning proved to be an effective intervention in order to reduce reperfusion injury in focal ischemia. Data obtained in this part of the study may be applicable in dealing with ischemic stroke at the clinic. The MSMP monitoring system, which was used in this study to monitor the function of the mitochondrial and blood supply in various ischemic conditions, has proved to be an effective means for evaluating the ischemic tissue and reveals a novel aspect of ischemic damage. The data collected in the current study may promote the current knowledge regarding the mitochondrial and hemodynamic contribution to the damage occurring due to cerebral ischemia.

Keywords NADH redox state · Cerebral blood flow · Mitochondrial function · Focal cerebral ischemia · Global cerebral ischemia · Rat's brain ischemia

5.1 Introduction

5.1.1 Brain Metabolism

5.1.1.1 Energy Production Processes in the Brain

The brain, the most vital organ in the body, has a high oxygen consumption. Oxygen and glucose are essential for the production of adenosine triphosphate (ATP), which serves as the energy reservoir of the tissue. All body tissues rely on this molecule as a critical energy source for various physiological functions. Glucose is the substrate from which most of the body's energy is derived. From one glucose molecule, 38 ATP molecules are produced. Two molecules are generated in the process of glycolysis, which takes place in the cytosol. Glycolysis is an anaerobic process in which glucose is broken down into two molecules of pyruvate. In the next stage, pyruvate molecules are converted into molecules of Acetyl-CoA, which serve as substrates for the citric acid cycle (Krebs cycle). The Krebs cycle occurs in the mitochondria and produces one molecule of GTP (later converted to ATP) and two molecules of CO_2. During this cycle, three molecules of Nicotinamide adenine dinucleotide (NAD^+) and one molecule of Flavin adenine dinucleotide (FAD^+) undergo reduction, becoming NADH and $FADH_2$, respectively. These reduced molecules donate their electrons (and protons) to the electron transport chain, where oxidative phosphorylation occurs. This process, which also takes place in the mitochondria, generates 32 ATP molecules. Electrons from NADH and $FADH^2$ are transferred through four protein complexes to a fifth complex, ATP synthase. This enzyme, embedded in the mitochondrial membrane, allows protons accumulated in the intermembrane space to flow through. The electrons pass through ATP synthase to an oxygen molecule, which, by accepting protons, becomes water (H_2O). The entire process produces energy, which the enzyme uses to attach a phosphate group to ADP, generating ATP (Berne and Levy 2000).

Under stress conditions, the body can synthesize ATP through two additional pathways. The enzyme creatine phosphokinase (PCr) transfers a phosphate group from phosphocreatine to ADP, producing one ATP molecule. The enzyme adenylate kinase combines two ADP molecules, yielding one ATP molecule and one AMP molecule. These alternative pathways provide an emergency energy reserve, generating a limited and low yield of ATP (Berne and Levy 2000).

Energy produced in brain cells is designated for two main purposes: intracellular functions such as neurotransmitter synthesis, and the activity of sodium-potassium pumps (Na^+/K^+-ATPase), which are essential for maintaining the neuronal resting potential and transmitting nerve signals (Mayevsky 1984).

Oxygen, the final electron acceptor in the chain, is indispensable for energy production. In its absence, ATP synthesis ceases. NADH, the first electron donor in the respiratory chain, is sensitive to tissue oxygen concentration. When oxygen is scarce, reduced NADH levels increase in accordance with tissue oxygen availability. Without oxygen, oxidative phosphorylation is impossible, and energy production halts after glycolysis. In this case, excess pyruvate produced in glycolysis is converted to lactate, which accumulates in the tissue, causing acidosis. Since the brain has no energy reserves, continuous blood flow is essential to provide glucose and oxygen to this tissue (Mayevsky and Rogatsky 2007).

5.1.1.2 Mitochondrial NADH

NAD^+, or its reduced form NADH, is the first electron donor in the mitochondrial electron transport chain. Under normal physiological conditions, reduced NADH molecules are generated in the mitochondria as part of glycolysis and the Krebs cycle. NADH is oxidized back to NAD^+ by donating a proton and an electron into the electron transport chain. Among the components of the respiratory chain, NADH is the most oxygen-sensitive, and therefore serves as an indicator of intracellular oxygen concentration and metabolic activity (Mayevsky 1984).

The ratio between reduced NADH and oxidized NAD^+ is determined by the mitochondrial metabolic state, inversely related to oxygen availability. Chance and colleagues, in a series of papers from 1955, described several mitochondrial metabolic states. In the resting state or deep anesthesia (State 4), 99% of NADH is in its reduced form. In the active state (State 3), reduced NADH levels decrease to only 53% (Chance et al. 1955). The normal physiological brain is typically in an intermediate state, between State 4 and State 3 (Mayevsky 1984). Under these conditions, with normal oxygen supply, the brain can meet increased ATP demand through autoregulatory mechanisms. In this state, the limiting factor for ATP production is the availability of substrate, namely the amount of ADP (Chance et al. 1955). When oxygen delivery to tissue is compromised, electron transfer in the respiratory chain is impaired. Consequently, ATP production decreases, and reduced NADH levels rise above their normal value (Mayevsky and Rogatsky 2007). Various studies have demonstrated correlations between NADH levels, blood flow, intra- and extracellular ion concentrations, and electrical activity in measured tissue (Mayevsky 1976, 1984; Mayevsky and Rogatsky, 2007). From this, it can be inferred that elevated levels of reduced mitochondrial NADH serve as an indicator of pathological conditions in tissue (Mayevsky 1984).

5.1.2 Cerebral Blood Flow

5.1.2.1 Anatomy of Cerebral Blood Vessels

The brain is supplied with blood through two common carotid arteries (CCA) and two vertebral arteries (VA). The vertebral arteries, located along the spinal column in the neck, merge to form the basilar artery (BA). The carotid arteries are located on either side of the trachea in the neck. Each carotid bifurcates into the internal carotid artery (ICA) and the external carotid artery (ECA). Unlike the ICA and VA, the ECA does not supply the brain itself but rather external regions: the tongue, thyroid gland, face, ears, pharynx, and salivary glands. Several arteries branch from the ECA, ultimately dividing into the superficial temporal artery (STA) and the maxillary artery (MA).

The cortex is supplied by three cerebral arteries. The posterior cerebral artery (PCA), branching from the basilar artery, supplies the occipital lobe and medial temporal lobe. The middle cerebral artery (MCA), the main branch of the ICA, supplies the lateral regions of the parietal, temporal, and frontal lobes. The anterior cerebral artery (ACA), another branch of the ICA, supplies the frontal lobe and medial parietal lobe.

The basilar artery, both ICAs, and the three cerebral arteries meet at the brainstem, forming a vascular ring called the Circle of Willis (Fig. 5.1). This structure allows for complete compensation in blood supply if one of the supplying vessels is impaired (Hartkamp et al. 1999; Lee 1995). The circle is completed by three communicating arteries: two posterior communicating arteries (PCoA) and one anterior communicating artery (ACoA).

5.1.2.2 Regulation of Cerebral Blood Flow—The Autoregulatory Mechanism

Continuous blood supply to the brain is essential for maintaining neuronal and glial metabolic activity. Since the brain lacks energy reserves, it requires a constant supply of oxygen for ATP synthesis (Mayevsky and Rogatsky 2007). The autoregulatory mechanism of the brain maintains stable blood supply across different mean arterial pressures (MAP) and cerebral perfusion pressures (CPP). This mechanism balances oxygen consumption with supply, so that during increased oxygen demand, blood flow to the tissue rises accordingly (Berne and Levy 2000).

Blood flow regulation in arterioles is achieved through several mechanisms: (a) Local regulation—determined by changes in pH, PO_2, PCO_2, and ion concentrations (K^+, Na^+, Ca^{2+}, Cl^-), depending on tissue metabolic status; (b) Neural regulation—affecting smooth muscle around arterioles, causing vasodilation or vasoconstriction depending on oxygen demand; (c) Vasoactive substances—molecules such as nitric oxide (NO) that influence vessel diameter. Through these mechanisms, vessels constrict or dilate in accordance with tissue metabolic and energetic state (Berne and Levy 2000).

The autoregulatory mechanism is effective within a mean arterial pressure range of 50–150 mmHg—the autoregulatory range. Outside this range, cerebral blood flow changes drastically, leading to altered CPP values and immediate biochemical and physiological damage to brain cells (Mayevsky 1984).

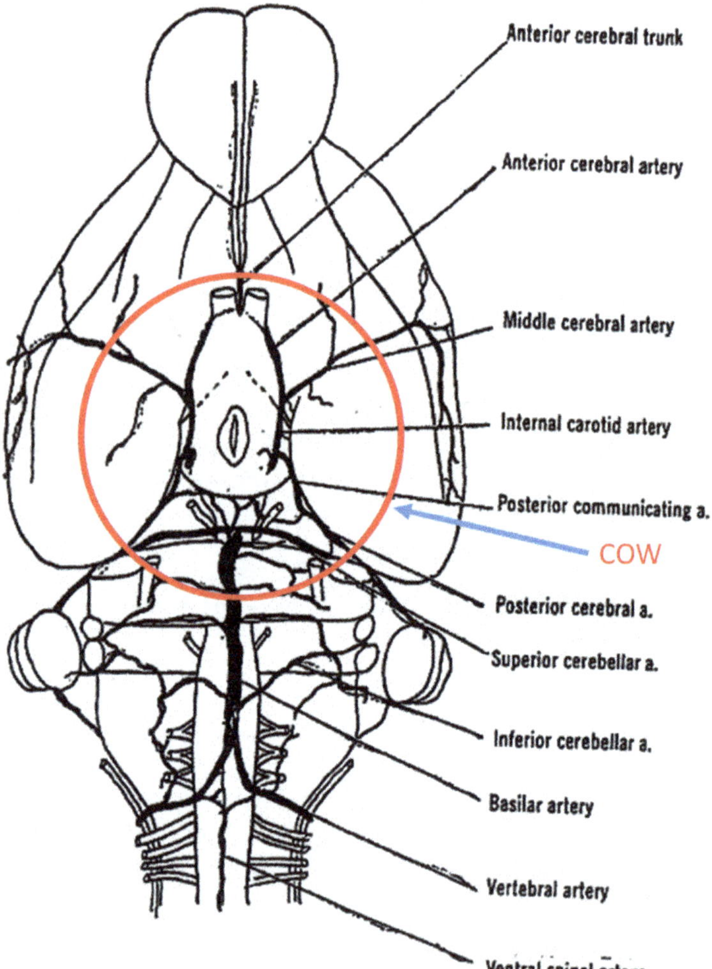

Fig. 5.1 Ventral view of the rat's brain and its arterial system including the Circle of Willis (Zeman and Innes 1963)

5.1.3 *Cerebral Ischemia*

Ischemia is defined as partial or complete cessation of blood flow to tissue, leading to oxygen deprivation. This results in reduced ATP levels in the tissue due to halted electron transport chain activity (Lipton 1999). Consequently, several changes occur, producing a new microenvironment in the ischemic tissue. These include suppression of electrical activity, ionic imbalance (both intra- and extracellular), decreased extracellular pH, and increased production of free radicals. Together, these factors halt protein synthesis in affected cells, functionally silencing the damaged tissue and potentially causing cell death in areas lacking blood supply (Lipton 1999).

Due to impaired respiration and ionic imbalance, the ischemic zone develops a microenvironment characterized by elevated extracellular potassium, excitatory amino acids such as glutamate, and acidic pH (Lipton 1999; Durukan and Tatlisumak 2007). The present work will focus on cerebral ischemia. Various types are classified based on location, extent, duration, and severity of brain damage.

5.1.3.1 Complete and Partial Ischemia

Complete ischemia refers to total cessation of blood flow to the affected area. Partial ischemia involves a sharp reduction in blood supply but not complete cessation, as collateral vessels compensate. Differences between complete and partial ischemia are reflected in extracellular pH: in partial ischemia, glucose is still available even though oxygen is limited. Respiration ceases, forcing neurons to rely on anaerobic glycolysis, producing lactate. The accumulation of lactic acid lowers pH (Lipton 1999). In both complete and partial ischemia, respiration is impaired, and reduced NADH levels rise (Mayevsky 1984).

5.1.3.2 Focal and Global Ischemia

Global ischemia occurs when the entire organ lacks proper blood supply, whereas focal ischemia affects a specific region. Global ischemia develops mainly in cardiac arrest or hemorrhagic stroke (Kuroiwa and Okeda 1994). Focal ischemia creates a gradient of damage severity, depending on collateral circulation. This produces a core region of maximal injury surrounded by penumbra areas with higher blood flow (Lipton 1999).

A condition of partial global ischemia may occur, for instance, following arrhythmias or significant blood pressure drops (Lipton 1999). Surgical occlusions of both carotid arteries (two-vessel occlusion, 2VO) induce partial global ischemia. In this case, minimal blood flow reaches the brain via the vertebral arteries, which primarily supply posterior regions. However, the Circle of Willis provides compensatory flow to other regions.

5.1.3.3 Acute and Chronic Ischemia

Another distinction is made between acute ischemia, where blood supply stops abruptly, and chronic ischemia, where blood supply deteriorates gradually, sometimes leading to complete ischemia. Acute ischemia may occur due to traumatic vascular injury or surgical procedures. Chronic ischemia is more common and often results from vascular calcification (Lipton 1999; Durukan and Tatlisumak 2007).

5.1.3.4 Reversible and Irreversible Ischemia

Laboratory models of ischemia are divided into reversible (transient ischemia), where blood flow is restored after occlusion removal, and irreversible (permanent ischemia), where occlusion persists without recovery (Lipton 1999).

5.1.3.5 Ischemia as Pathology—Stroke

Stroke is a condition in which the brain or a specific region suffers from impaired blood supply. This deprivation causes neuronal injury due to oxygen shortage, potentially leading to motor, sensory, or cognitive deficits depending on lesion location. Severe stroke effects may be permanent, as neuronal regeneration does not occur. Stroke is a medical emergency, potentially causing permanent neurological damage, complications, and death. It is the leading cause of adult disability in the Western world and the third leading cause of death worldwide (Cohen et al. 2011). Notably, 10% of reported strokes are due to MCA involvement (Mendel 2005).

Strokes are categorized as ischemic (80% of cases) or hemorrhagic. Ischemic stroke results from blocked blood supply, typically due to a thrombus or embolus (Kuroiwa and Okeda 1994). Thrombi commonly form in arteries damaged by atherosclerotic plaque. Other causes include hypoglycemia, tachycardia, hypertension, or tumors compressing blood vessels. Risk factors include high cholesterol, smoking, and advanced age. A milder form of ischemic stroke is the transient ischemic attack (TIA), in which vessel occlusion is temporary. Within 24 h of restored blood flow, no physiological or neurological damage remains. TIAs may precede ischemic strokes.

Hemorrhagic stroke occurs when vascular permeability increases, reducing blood supply to brain tissue due to bleeding. Hemorrhage may involve vessels or the ventricular system. Unlike ischemic stroke, which may be asymptomatic, hemorrhagic stroke is usually accompanied by headache (Cohen et al. 2011).

5.1.4 Focal Cerebral Ischemia

Focal cerebral ischemia is a condition in which there is a decrease in blood supply to a specific area of the brain. The extent of the damage depends on the extent of the blocked blood vessel and the duration of the blockage. Ischemic stroke is most often caused by a blood clot (thrombus) or embolus (Kuroiwa and Okeda 1994).

5.1.4.1 Spatial Damage in Focal Ischemia: Core and Penumbra

Focal ischemia results in a gradient of injury, ranging from the center of the injury to peripheral areas. The center of the injury is called the core, and the peripheral area is called the penumbra. The cells in the core area absorb the most severe injury, depending on the blood supply reaching the various areas. The blood supply to the core area is less than 20% of its initial level, and therefore a necrotic center is formed there. In contrast, the penumbra area has a partial blood supply, thanks to collateral blood vessels, and therefore the degree of damage in this area will be less than in the core area. Blood flow in the penumbra area is between 20% and 40% of baseline values (Lipton 1999). There are also studies that define a third area, the peri-infarct, which suffers from a decrease of up to 60% in blood flow following the creation of ischemia. The damage in this area is minimal (Lipton 1999; Ginsberg 2003; Hossmann 2006).

Another way to define core and penumbra regions is by electrical activity and metabolic activity. Electrical activity ceases in regions where blood flow is less than a threshold of 0.16–0.18 mL/g/min. When blood flow drops below a threshold of 0.1 mL/g/min, a massive increase in extracellular potassium occurs due to a decrease in tissue ATP levels, effectively indicating an impairment of the cell's energy metabolism. A penumbra region can be defined as the region between these two thresholds—regions where energy metabolism in the cell is maintained, but electrical activity is inhibited. Core regions are defined as regions where blood flow levels do not allow energy metabolism to occur (Lipton 1999). It should be noted that the threshold levels in question coincide with the blood flow percentages described earlier. Similar values have also been measured in various clinical studies in humans (Astrup et al. 1981).

The varying blood supply between the core and penumbra regions affects the energy balance essential for the tissue's ability to survive, and is the significant difference between the core and penumbra regions. The impaired energy balance has an impact on other differences between the regions, such as metabolism, ion balances, pH, and more. The degree of damage to the peripheral region depends on the total time of ischemia (Baron 2001). In addition, the core region undergoes further deterioration within 3 min of the onset of ischemia, as part of a phenomenon known as ischemic depolarization (ID), which will be discussed later.

Laboratory studies using the MCAO model to create focal ischemia show that the core region is located in the putamen and caudate nuclei (Astrup et al. 1981; Hossmann 2006; Pestalozza et al. 2002; Lipton 1999), and in cortical areas on the sides of the frontal and parietal lobes, including the somatosensory cortex. The penumbra regions surround the core region, and are located in the layer of cortex above it and in the posterior parts of the brain (Lipton 1999).

5.1.4.2 Duration of Ischemia

Increased duration of ischemia increases the severity of injury in all types of ischemia, both transient and permanent. During ischemia, energy supply to the tissue depends primarily on ATP generated from glycolysis. These metabolites provide energy for an additional 2 min from the onset of ischemia, after which ATP production ceases completely (Ginsberg 2003).

In cases of ischemia lasting a few minutes, there will be no long-term histological damage to the tissue. Initial histological damage can be observed as early as 5 min after the onset of ischemia (Lipton 1999). The main changes in ischemic tissue occur within the first hour. Brain cells undergo hypertrophy and a necrotic zone begins to form in the ischemic core. However, many cells in the ischemic area maintain a normal morphological shape during the first hours after the ischemic injury (Folbergrová et al. 1993).

The duration of ischemia has an impact on the brain's ability to recover. Studies show that the period of time in which ischemia damage can be repaired does not exceed 3 h (Bartus et al. 1995). The drug tPA (tissue plasminogen activator), which was developed in recent years, has led to an improvement in ischemic damage even after a period of 4.5 h from the moment of injury (Lipton 1999). The Mitochondria suffer from cumulative damage following local ischemia, which continues even after the blood flow is restored (Cohen et al. 2011).

5.1.4.3 Reperfusion Phase

Resumption of blood supply to the tissue after ischemia is called reperfusion (Duarte et al. 2003). This phase constitutes compensation for the damaged tissue, during which an increase in blood flow occurs to levels beyond the baseline levels before the ischemia was created. The duration of ischemia affects the degree of recovery of the damaged tissue—the longer the time the tissue did not receive a regular blood supply, the greater the degree of damage after the supply is restored, and the chance of recovery is reduced (Lipton 1999).

In focal ischemia, there is a variety of different responses during the reperfusion phase, including the absence of a hyperperfusion response (Ames et al. 1968). After reperfusion, a hyperemic response usually occurs in the ischemic area, followed by a return to normal physiological values in the core, while in the penumbra there is some decrease in blood flow (Lipton 1999). Immediately after reperfusion, extensive neuronal damage can still be observed, manifested by excess extracellular potassium, excess glutamate, and low pH (Tomlinson et al. 1993). Glucose levels can reach values of up to 50% of normal physiological values in the penumbra. The rate of ATP production in the penumbra does not return to its baseline level and remains low (Lipton 1999).

5.1.4.4 Reperfusion Injury

During the reperfusion phase, damage is usually added to the previous ischemic damage, a phenomenon known as reperfusion injury (Duarte et al. 2003). In this case, the resumption of blood flow leads to a sharp increase in the concentration of oxygen in the tissue, which is not fully utilized by the mitochondria, which leads to the formation of free radicals. The excess oxygen constitutes a substrate for oxidizing enzymes, which leads to the formation of free radicals at a higher level than they are formed under normal physiological conditions (Lipton 1999).

Many oxidation processes occur in the cell's mitochondria, and therefore, when the activity of the mitochondria is resumed, it constitutes a potential source of the formation of many free radicals, which damage the cell membranes, disrupt enzymatic activity and cause damage to DNA. An increase in the level of free radicals is one of the activation mechanisms of cell death, controlled or necrotic, in neurons (Lipton 1999). Therefore, the process of reperfusion injury can lead to the death of additional cells after the blockage is removed (Christophe and Nicolas 2006).

There are several antioxidant enzymes that neutralize free radicals, including superoxide dismutase (SOD), glutathione peroxidase (GSHPx), catalase, athione, ascorbic acid, and vitamin E. During the reperfusion phase, the ischemic tissue has not yet recovered to increase the production of these enzymes, a fact that, combined with the excess of free radicals, exacerbates tissue damage (Chan 1996). In focal ischemia, an increase in free radical levels is observed already during ischemia itself, starting from 30 min after the start of treatment and up to 12 h after it (Kinuta et al. 1989). The formation of free radicals mainly characterizes the penumbra region and not the core region, both during the ischemic phase and after the resumption of blood flow (Solenski et al. 1997).

5.1.4.5 Blood Flow Changes in Focal Ischemia

Ischemia is defined as a condition where the blood supply to tissue is impaired. In addition, ischemia affects the blood vessels in the affected tissue, particularly the capillary walls. During ischemia, the permeability of blood vessels increases, which can lead to edema and enlargement of the brain ventricles (Lipton 1999). In focal ischemia, blood flow in the core region drops to values less than 20% of the baseline level. Blood flow does not resume in the core during ischemia and may even deteriorate over time. Blood flow in the penumbra is between 20% and 40% of baseline values due to collateral supply (Lipton 1999; Durukan and Tatlisumak 2007; Ginsberg 2003).

The restoration of blood flow to the tissue is usually accompanied by a wave of hyperperfusion to compensate for the loss of flow during ischemia. Initially, higher-than-normal CBF values are observed, but within about 10 min, these values stabilize at 70% of the baseline value (Schmid-Elsaesser et al. 1998). Nevertheless, it has been found that in the first few minutes after blood flow is restored, no-flow areas may exist where flow does not resume. These areas are usually located at the ischemic lesion's epicenter, and recovery is relatively slow (Bengtsson and Siesjo 1990).

5.1.4.6 Mitochondrial Activity in the Ischemic Brain

The brain is entirely dependent on the blood supply of oxygen and **glucose**. Therefore, an impairment in blood supply causes, among other things, mitochondrial damage in neurons (Mayevsky 1984). The mitochondria are exposed to oxidation processes, a gradual increase in intracellular calcium levels, and a lack of essential enzymes (Tomlinson et al. 1993). These factors, combined with the lack of oxygen, impair mitochondrial function and disrupt the ATP synthesis process (Mayevsky 1984). Additionally, calcium entry into the mitochondria leads to the entry of water molecules, causing mitochondrial swelling (Halestrap 2006). Another injury that occurs during ischemia is an increase in membrane permeability due to the opening of ion channels, which damages the membrane potential (Sims and Muyderman 2010). Another indication of mitochondrial damage during ischemia is the release of cytochrome C in the damaged area, which indicates damage to the respiratory chain (Sims and Muyderman 2010). As a result of mitochondrial damage, the electron transport chain's activity ceases, and a decrease in tissue ATP levels can be observed.

The decrease in ATP levels is equivalent to the cell being depleted of its energy source. This decrease has numerous effects on processes that occur in the tissue under normal physiological conditions, such as the activity of free radical scavengers, protein phosphorylation, membrane transporter activity, and sodium-potassium pump activity (Lipton 1999). Nevertheless, the tissue's ATP level is not necessarily an indicator of the severity of the ischemic injury or the extent of the decrease in blood flow. Due to the decrease in blood flow and the functional damage to the electron transport chain, ATP production is significantly reduced but not completely stopped. The ischemic injury leads to the utilization of existing tissue glucose to create ATP through glycolysis. It also leads to the utilization of emergency energy stores, i.e., ATP creation using the enzymes creatine phosphokinase and adenylate kinase (Berne and Levy 2000). The severity of the injury, therefore, is a function of the availability of ATP in the tissue, combined with the activity of the electron transport chain (Lipton 1999). Mitochondrial NADH levels, which change according to the activity of the electron transport chain, are an accurate and effective indicator of the severity of the ischemic injury.

NADH reduction is characteristic of all types of cerebral ischemia and is observed in various animal models (Lipton 1999; Mayevsky et al. 1990; Mayevsky 1984). The increase in NADH levels occurs due to the decrease in oxygen concentration, as a result of the disruption in blood supply to the tissue. The reduction of mitochondrial NADH occurs in parallel with changes in ion concentrations and electrical activity during ischemia (Mayevsky 1984).

In focal ischemia, ATP levels in the penumbra decrease more moderately, to a level of 50%–70% compared to a normal physiological state, versus 18%–32% in the core (Sims and Muyderman 2010; Lipton 1999). A decrease was also observed in PCr levels, to values of 16%–28% in the core and 70% in the penumbra. In the core, these values were slightly higher than the decrease observed in full global ischemia (Sims and Muyderman 2010). Simultaneously, an immediate increase of up to 50% was observed in NADH levels in the ischemic tissue (Dora et al. 1986). The maximum percentage changes were observed after just 15 min and did not change significantly for another 2 h, but after 2 h of ischemia, a decrease in the total amount of adenosine metabolites, i.e., ATP + ADP + AMP, was observed, to a level of 40% in the core and 60% in the penumbra (Sims and Muyderman 2010).

When blood flow was restored after 2 h of MCA occlusion, ATP levels returned to their initial values in the penumbra but remained at 50%–70% in the core (Sims and Muyderman 2010; Hossmann 2006; Kristián and Siesjö,1998). Immediately after blood flow was restored, mitochondrial NADH was reduced to a level that varied according to the severity of the ischemic injury (Dora et al. 1986). After 4 h, a further decrease in ATP levels occurred. This finding, combined with lactate levels that remain high after reperfusion, indicates continuous mitochondrial damage in the tissue even after blood flow is restored (Hossmann 2006). After initial recovery, a further decrease in mitochondrial function occurs after 2–4 h of reperfusion (Canevari et al. 1997; Sims and Muyderman 2010).

Reperfusion injury, which occurs after blood flow is restored, has a destructive nature in cases of focal ischemia. 3–5 h of permanent ischemia cause mitochondrial swelling while maintaining the integrity of the mitochondrial membrane. After 24 h of ischemia, cells that have undergone necrosis can be observed, in which the mitochondrial membrane appears to be destroyed. In contrast, just 3 h of transient ischemia followed by 2 h of reperfusion are enough to demonstrate similar mitochondrial destruction to that which occurs after 24 h of permanent ischemia (Solenski et al. 2002).

5.1.4.7 Ion Concentrations in Brain Tissue During Focal Ischemia

Due to the disruption of the energy balance following the decrease in blood supply, ischemia causes changes in the balance of ions between the interstitial space and the cell cytoplasm.

- **Potassium:** The decrease in available ATP in the tissue leads to the cessation of sodium-potassium pump activity, which results in the accumulation of potassium in the interstitial space (Mayevsky et al. 1990). Potassium is the first ion to react to the change that occurs in the tissue, and its values change before changes are observed in the levels of other ions in the extracellular space. In focal ischemia, extracellular potassium levels rise immediately after ischemia is induced. In the core, a higher increase is observed compared to the penumbra, up to a level of 60 mM/L, and potassium values remain higher than normal during the occlusion. In the penumbra, decreases in potassium levels are observed for short periods, which can be attributed to the formation of depolarization waves. Following the restoration of blood flow after 2 h of ischemia, potassium values return to a normal level of 2.5 mM/L within about 5 min. After 6 h of reperfusion, a further increase in extracellular potassium levels is observed, to a value of 5 mM/L (Gidö et al. 1997).

Astr ocytes are responsible for regulating extracellular potassium levels by bringing potassium into the cell space in exchange for sodium and chlorine. During ischemia, the astrocytes cannot cope with the high potassium levels. Following the entry of many potassium ions, water molecules also enter the cytoplasm of the astrocytes, which leads to their swelling and eventually their cessation of activity (Lipton 1999).

- **Calcium:** In all cases of ischemia, an increase in intracellular calcium levels is observed at the expense of its extracellular level (Lipton 1999). The entry of calcium into the cells is accompanied by the entry of sodium ions and water molecules, as a result of which the cells swell. Additionally, calcium is released from the endoplasmic reticulum into the cytoplasm, which raises the intracellular calcium level even more (Kristián and Siesjö 1998).

The increase in calcium levels damages, among other things, the cell membrane, as phospholipase enzymes are activated and cause the breakdown of membrane phospholipids, resulting in an increase in free fatty acids in the cell (Lipton 1999). The destruction of the membrane causes additional calcium ions to enter the cell, and in fact, an intensifying process of membrane damage is created. Similar damage also occurs in the mitochondrial membrane (Sun and Gilboe 1994).

The increase in intracellular calcium levels leads to calcium entry into the mitochondria. As a result, the mitochondria swell and their function is impaired. Calcium causes damage to the mitochondrial membrane by activating channels of the type mitochondrial permeability transition pore (MPTP), thereby increasing the membrane's permeability to ions. The activation of MPTP channels damages the mitochondrial membrane potential, which is created under normal physiological conditions from the accumulation of H^+ ions in the intermembrane space. As long as these channels remain open, during ischemia and even during the reperfusion stage, the activity of the electron transport chain is impaired (Halestrap 2006).

Occlusion of the MCA for a short period (10 min) or longer (120 min) leads to a decrease in the extracellular calcium level in the core. In 2-h focal ischemia, a decrease in the extracellular calcium level was observed from 1.2 mM to values of 0.1 mM. The calcium level does not return to values greater than 50% of its initial level even after 6 h from the moment of arterial occlusion (Kristián et al. 1998). The decrease in the extracellular calcium level parallels the increase in the intracellular calcium level and is accompanied by the accumulation of calcium in the mitochondria (Kristián and Siesjö 1996). In the penumbra, potassium is observed leaving the cell and calcium entering it, in a recurring manner as part of the formation of depolarization waves. After blood flow is restored, the calcium level returns to its baseline values (Kristián and Siesjö 1996). The total calcium level in the tissue (the sum of the extracellular and intracellular levels) increases after 24 h following MCAO treatment. The increase was observed in both transient and permanent ischemia, but not in cases of global ischemia (Kristián et al. 1998).

- **Sodium:** From the moment ischemia is applied, the intracellular Na^+ level rises. One of the reasons for this is the cessation of sodium-potassium pump activity, which makes it impossible to remove excess sodium from the cell. Excess glutamate released in the ischemic tissue causes additional sodium ions to enter the cell, thereby contributing to the disruption of its balance (Lipton 1999). An increase in the concentration of free radicals in the tissue causes damage to the sodium channels, which as a result remain open for prolonged periods and further increase the sodium concentration inside the cell. The intracellular sodium level in the core is higher than its level in the penumbra, in accordance with the glutamate level and the decrease in energy metabolism that characterize these areas (Lipton 1999).

5.1.4.8 Glucose Metabolism and pH Level During Focal Ischemia

Glucose utilization is the main way to create ATP in the brain during ischemia. The defective oxygen supply and the cessation of the electron transport chain cause the utilization of glucose to create ATP through glycolysis. This process continues as long as glucose is present in the tissue. The main product of glycolysis is **lactate**, which accumulates in the damaged tissue during ischemia. Additionally, the cessation of respiratory chain activity leads to the accumulation of excess CO_2 in the tissue. The excess lactate and CO_2 lead to a decrease in the tissue pH level in all cases of ischemia. As a result, the severity of the ischemic injury worsens and the chances of recovery after blood flow is restored decrease (Wass and Lanier 1996). As the glucose level in the ischemic tissue rises, more ATP molecules are created, but the tissue becomes more acidic and is actually more damaged. A high glucose level has a destructive effect on mitochondrial activity during ischemia (Anderson et al. 1999).

In focal ischemia, the pH in the core drops to values of up to 6.1. In the penumbra, a pH gradient is created, from areas with a low pH to areas where a pH level of 7.0 is maintained (Lipton 1999). In the core, glucose utilization processes in the tissue are impaired, while in the penumbra, a normal physiological rate of glucose utilization is maintained throughout the period of ischemia (Ginsberg 2003). In contrast, about an hour after blood flow is restored following 2 h of ischemia, lower-than-normal glucose metabolism was observed in the penumbra. The initial glucose level in the tissue has an effect on the size of the damaged area in focal ischemia, as well as on mitochondrial activity. As the glucose level rises, the core area grows at the expense of the penumbra area (Anderson et al. 1999).

The disruption of the ion balance during ischemia leads to an excess release and prevention of reuptake of neurotransmitters, including dopamine, serotonin, GABA, and glutamate. High levels of **glutamate**, the most excitatory neurotransmitter in the nervous system, lead to cell destruction through apoptosis or necrosis. Therefore, during the ischemic injury, the excess glutamate worsens the damage caused by the ischemia (Paxinos and Watson 1997). It is known that astrocytes cease to function following ischemia. One of the functions of astrocytes is to regulate glutamate levels, which is prevented during an ischemic injury (Lipton 1999).

In focal ischemia, there is an increase in the level of extracellular glutamate within a few minutes from the start of ischemia, from values of 1–5 μM to a level of up to 35 μ M (Camacho and Massieu 2006). In the core, a further increase in glutamate occurs after about an hour from the moment of ischemia, to values of up to 82 μM. Glutamate levels in the core remain high, while in the penumbra there is a gradual recovery until they return to a baseline level (Camacho and Massieu 2006; Morimoto et al. 1996). After the occlusion is removed, glutamate values return to a baseline level within a few minutes, but after a few hours, a further increase in the extracellular glutamate level may occur (Nishizawa 2001).

5.1.4.9 Cell Death Following Focal Ischemia

Following ischemia, cell death can occur through necrosis or apoptosis. Apoptosis is programmed cell death that occurs in cases of tissue stress and originates from intracellular processes. In contrast, necrosis is cell death through cell destruction, which is caused by the influence of an external factor (Berne and Levy 2000). Following focal ischemia, neuron death via apoptosis occurs mainly in the core (Lipton 1999).

The leading causes of cell death following ischemia include morphological changes in the cell, damage to DNA, accumulation of glutamate, cessation of protein synthesis, or dysfunction of the cell's proteins (Lipton 1999). Another factor is mitochondrial damage, which is caused by damage to cytochrome C, following the accumulation of calcium in the mitochondria. This damage usually leads to apoptosis, but may also lead to necrosis, by destroying the cell membrane (Brustovetsky et al. 2002; Sims and Muyderman 2010).

During focal ischemia, the cells undergo edematous changes as a result of the cessation of sodium-potassium pumps. These changes are manifested by the swelling of the cells and mitochondria and are accompanied by slight membrane damage. Subsequently, the cells undergo a number of structural changes, such as the contraction of the nucleus, deformation of the cell and nuclear membranes, and the accumulation of vesicles in the cytoplasm. In the penumbra, cells can remain in this state for several days. This state is attributed mainly to cases of transient ischemia and is observed after the reperfusion stage, and is not observed in prolonged permanent ischemia (Lipton 1999). In the core, additional changes occur, including the destruction of the cell membrane, mitochondrial contraction, and the destruction of other organelles. This stage occurs after 6 h in the core of permanent focal ischemia, and after a few days in the penumbra as well (Lipton 1999).

Focal ischemia leads to the destruction of endothelial cells in the core, 30 min after the reperfusion stage in 1-h transient ischemia, or after 6 h of permanent ischemia (Lipton 1999). The destruction of endothelial cells leads to an increase in BBB (blood-brain barrier) permeability, a disruption in the ion balance in the tissue, and the formation of edema in the extracellular space. Studies prove that edema has an effect on neuron death in the core, depending on the duration of the ischemia (Albayrak et al. 1997). The destruction of endothelial cells is more prominent in permanent ischemia compared to transient ischemia, as expressed by the penetration of more leukocytes in this ischemia compared to transient ischemia (Lipton 1999).

5.1.4.10 The Involvement of Nitric Oxide (NO) in the Ischemic Brain

Nitric Oxide (NO) is a diffusive gas that takes part in regulating blood flow in the brain. NO is synthesized by the enzyme nitric oxide synthase (NOS). There are three main types of NOS enzymes, which originate from neurons (nNOS), epithelial cells (eNOS), and immune system cells (**iNOS**). The mechanism of action of NO includes raising the intracellular cGMP level and consequently enzymatic activity in the cell (Garcia and Stein 2006). An increase in NO concentration leads to vasodilation of blood vessels, among other things by preventing calcium from entering the smooth muscle cells in the blood vessel wall, which leads to their relaxation. Additionally, NO serves as a neurotransmitter in the CNS, is a regulating factor in the immune system, and affects the aggregation of blood platelets (Garcia and Stein 2006).

In focal ischemia, the NO concentration rises from 10-8M to 10-6M. A further increase occurs during the reperfusion stage (Malinski et al. 1993). During ischemia, there is an increase in the level and activity of all three types of enzymes that synthesize NO. The activity of NOS enzymes is higher in the core compared to the penumbra during ischemia and after reperfusion (Ashwal et al. 1998). An increase in intracellular calcium levels during ischemia leads to an increase in nNOS activity, mainly in the core. nNOS has a toxic effect on the tissue, since it increases the damage in the environment of the nerve cells by exposure to free radicals. The iNOS enzyme is present at a very low level in the normal brain. About 12 h after the ischemic injury, iNOS begins to increase, which reaches its peak after 48 h and causes additional destruction in the neurons. The enzyme returns to its initial level 7 days after ischemia. In contrast, eNOS has a protective effect, as it widens the blood vessels in the damaged tissue, thereby helping to maintain a minimal level of energy metabolism. Excess NO in the core contributes to the severity of the ischemic injury in this area. Disabling the activity of the nNOS enzyme in various ways has led to a decrease in the volume of the damaged area (Lipton 1999; Samdani et al. 1997; Garcia and Stein 2006; Astrup et al. 1981; Mayevsky 1976, 1984).

5.1.4.11 Electrical Activity in the Brain During Focal Ischemia

A decrease in tissue electrical activity occurs in all types of ischemia. Suppression of electrical activity occurs in the tissue before the disruption of the cell's ion balance or an increase in NADH levels, meaning that electrical activity is stopped before the energy metabolism is disrupted (Astrup et al. 1981).

In cases of focal ischemia, electrical activity ceases in areas where blood flow is below a threshold of 0.16–0.18 mL/g/min. This is characteristic of both the core and penumbra regions (Astrup et al. 1981). In the core, there is a complete electrical suppression. In the penumbra, depolarization waves are observed in the DC potential, similar to spreading depression (SD) waves in ischemic tissue. These waves demonstrate a temporary cessation of cell activity, which occurs due to the exit of potassium and the entry of sodium and calcium into the cells. At some point, the level of electrical activity in the penumbra also drops to a fixed threshold, and irreversible tissue damage occurs. Therefore, it can be said that based on electrical activity, the core region expands at the expense of the penumbra region (Nallet et al. 1999).

5.1.4.12 Ischemic Depolarization

Ischemic depolarization (ID), also called SD-like hypoxic depolarization (HSD), is a state of electrical suppression that occurs during a blood or oxygen supply irregularity to the brain, such as in cases of hypoxia, anoxia, or ischemia. Following the increased demand for ATP during ID, electrical suppression occurs in the tissue (expressed in EEG and DC potential), along with an increase in extracellular potassium levels, a decrease in extracellular sodium levels, swelling of the cells at the expense of the intercellular space volume, and neuron death (Li et al. 2000; Sukhotinsky et al. 2008).

ID is observed in various cases of complete ischemia, such as four-vessel occlusion or decapitation in rats, as well as in cases of bilateral carotid occlusion in gerbils (Mayevsky 1976; Mayevsky and Rogatsky 2007). In the core, about 3 min after the cessation of blood flow, a decrease in DC potential occurs, which is accompanied by an increase in extracellular potassium concentration to values of up to 70 mM. This phenomenon is not observed in the penumbra (Lipton 1999). During ID, there is also an increase in the tissue's reflection level. This increase occurs in parallel with the increase in the extracellular potassium level and the cessation of the electrical signal in the tissue (Mayevsky and Rogatsky 2007).

5.1.4.13 Models for Creating Focal Ischemia

The study of the effects of cerebral ischemia on the brain and its functions is conducted in a research context using various animal models. The most common way to create focal ischemia is by occluding the middle cerebral artery (MCA), known as middle cerebral artery occlusion (MCAO) (Lipton 1999; Hossmann 2006). This procedure creates complete focal ischemia in a large area of the cortex on the ipsilateral side of the occlusion (Prieto et al. 2005). There are several models for MCA occlusion, which are divided into two main categories: distal occlusion and proximal occlusion (Lipton 1999; Durukan and Tatlisumak 2007).

Distal occlusion affects the distant branches that split from the MCA. This occlusion can be achieved through a surgical procedure that is considered invasive to the brain (Buchan et al. 1992). Another method is through photothrombosis occlusion (Kalesnykas et al. 2008; Lee et al. 2007).

In contrast, in proximal occlusion, the MCA is occluded at its base, immediately after the point where it splits from the ICA. This method is more common, and the brain remains intact during the procedure (Durukan and Tatlisumak 2007; Lipton 1999). The common model for proximal ischemia is the MCA occlusion model using a nylon filament with a rounded tip. This model was developed by Tamura (1981) and has undergone several basic adaptations over the years by Koizumi et al. (1986) and Longa et al. (1989). Further updates to this model included coating the nylon filament with Poly-L-Lysine (Belayev et al. 1996; Bouley et al. 2007) or silicone (Schmid-Elsaesser et al. 1998). There is a high variability in the results of studies that used the filament model to create MCAO. This variability is expressed in CBF values, neurological test results, and the size of the damaged area (Belayev et al. 1996; Bouley et al. 2007; Gerriets et al. 2004; Laing et al. 1993; Mhairi Macrae 1992; Lipton 1999). Part of this variability is explained by differences in animal weight and the diameter of the occluding filament (Lipton 1999; Lourbopoulos et al. 2008; Belayev et al. 1996;

Bouley et al. 2007), as well as differences resulting from different animal strains (Dittmar et al. 2006; Prieto et al. 2005; Walberer et al. 2006).

Recently, an in vitro model for creating focal ischemia has been developed, based on growing cells on a medium with varying degrees of oxygen and glucose deprivation (oxygen-glucose-deprived medium, OGD) (Richard et al. 2010).

5.1.5 Partial Global Cerebral Ischemia

Partial global cerebral ischemia is a condition in which the blood supply to the entire brain decreases but is not completely cut off. This condition can occur as a result of an irregular heart rhythm or a significant drop in blood pressure (Tomlinson et al. 1993). An experimental model for partial global cerebral ischemia is the occlusion of a single carotid artery or both carotid arteries.

5.1.5.1 Bilateral Carotid Occlusion (BCO)

Following the occlusion of both carotid arteries, also known as bilateral carotid occlusion (BCO) or two-vessel occlusion, a minimal blood flow to the brain is maintained through the vertebral arteries. These arteries mainly supply the back of the brain, but blood flow is also compensated to other parts due to the Circle of Willis. It's important to note that BCO causes partial ischemia in mice and rat models, but not in gerbils, where this occlusion creates full ischemia due to the absence of compensation by the Circle of Willis (Lipton 1999).

During arterial occlusion, there is a temporary blood flow reduction of up to 80%, primarily in the front part of the brain (Tomlinson et al. 1993). Blood flow values return to their initial level within minutes, thanks to the blood supply from the vertebral arteries, which reach the entire brain due to the Circle of Willis (Mayevsky 1976). In cases of transient global ischemia, blood flow values return to their initial level within a few minutes after blood flow is restored (Mayevsky et al. 1990).

Similar to focal ischemia, cells swell in cases of global ischemia. There is a time window of 12 h to several days between the period of restricted blood supply to the tissue and cell death (Lipton 1999). Neuron death via apoptosis is less observed in global ischemia (Lipton 1999).

BCO disrupts the brain's energy balance, which is reflected in a decrease in ATP and GTP levels and an increase in AMP levels. Additionally, a decrease in the overall levels of adenosine nucleotide metabolites and PCr was observed, indicating damage to energy metabolism (Kozuka 1995; Vagnozzi et al. 1997). The levels of ATP and PCr continue to drop as the duration of ischemia lengthens (Vagnozzi et al. 1997). After blood flow is restored following 2 h of ischemia, these parameters return to their baseline values (Kozuka 1995). Mitochondrial NADH levels rise immediately after the arteries are occluded. The NADH returns to its initial level a few minutes later, due to the residual blood supply (Mayevsky 1976). Along with the rise in NADH, an increase in extracellular potassium levels is observed (Mayevsky et al. 1990). These phenomena are caused by the cessation of electron transport chain activity, which leads to a lack of ATP in the tissue (Mayevsky 1984).

During ischemia, electrical activity ceases for a few minutes, followed by electrical activity that is lower than normal. After blood flow is restored, electrical activity returns to its initial value within a few minutes (Mayevsky et al. 1990).

In the intercellular space, potassium levels rise, and calcium and sodium levels fall (Mayevsky et al. 1990; Kristián and Siesjö 1996). Simultaneously, the extracellular pH drops within a few seconds, from 7.0 to 6.4. This decrease is likely due to the accumulation of lactic acid in the tissue. Following this decrease, intracellular pH also drops within a few minutes from the start of ischemia (Lipton 1999; Tomlinson et al. 1993).

During ischemia, an increase in tissue lactate levels and a decrease in glucose levels were observed, indicating glycolysis activity. Two hours after blood flow was restored following 2 h of ischemia, a higher-than-normal level of glucose was observed due to the renewed blood supply. In contrast, lactate levels remained higher than normal, indicating difficulty in tissue recovery after ischemia (Kozuka 1995). Extracellular glutamate levels rise from 1–5 μM to up to 35 μM, similar to what happens during focal (Camacho and Massieu 2006). After the occlusion is removed, glutamate values return to their initial level within a few minutes (Nishizawa 2001).

5.1.5.2 Unilateral Carotid Occlusion (UCO)

Unilateral carotid occlusion (UCO) is used as a model for atherosclerosis, a common phenomenon that increases the risk of brain injuries, including stroke (Dieter and Laird 2005). Studies show that blocking a single carotid artery doesn't create a significant change in blood flow in the ipsilateral hemisphere compared to both the contralateral hemisphere and the blood flow before or after the occlusion (Bluestone et al. 2004; Mayevsky 1976). The brain maintains a normal blood supply thanks to the anatomical structure of the Circle of Willis, which allows blood to be supplied through the opposing carotid artery and the vertebral arteries. The electrical pattern, blood flow, and energetic activity of the tissue either don't change at all or undergo a minimal change and return to a normal level within a few seconds (Mayevsky 1976).

In contrast, some research models combine a single carotid occlusion with additional treatment, which causes damage in the ipsilateral hemisphere relative to the contralateral hemisphere. In the hypoxia-ischemia (HI) model, also called the Levine method (Levine 1960), one of the carotid arteries is occluded in addition to ventilating the animal with a gas mixture containing 8%–10% oxygen (Homola et al. 2006). In histological examinations, more severe necrotic damage was found in the ipsilateral hemisphere compared to the contralateral side (Park et al. 2006). Similar to ischemic tissue, HI recorded increases in glucose and lactic acid values and a decrease in the volume of the intercellular space. The changes were observed in the hemisphere supplied by the occluded carotid artery but not in the opposite hemisphere (Homola et al. 2006).

Other possible combinations are carotid occlusion and a drop in blood pressure (Bluestone et al. 2004), or a controlled increase in intracranial pressure (ICP) while maintaining a constant blood pressure (Busto and Ginsberg 1985). In all cases, the ipsilateral hemisphere showed a decrease in CBF (cerebral blood flow) values, an increase in lactate production, a decrease in glucose levels, and high levels of NADH, compared to the contralateral hemisphere. In the case of hypercapnia, a single carotid occlusion led to a decrease in CBF values in both hemispheres, but the decrease was sharper in the ipsilateral hemisphere (De Ley et al. 1985).

There are a number of phenomena that are characteristic of UCO, such as minimal damage to the blood supply to the eye cornea in rats (Slakter et al. 1984), or the appearance of metabolic oscillations in CBF and NADH recordings in the occluded side's hemisphere compared to the opposite hemisphere (Mayevsky and Ziv 1991). Oscillations are defined as regular fluctuations as a function of time, with a frequency and amplitude that change according to the different processes in which they are observed. A previous study by Mayevsky and colleagues showed that in the case of UCO, more oscillations were created in the ipsilateral hemisphere compared to healthy tissue, but fewer oscillations compared to partial global ischemia created by occluding both carotid arteries (Mayevsky and Ziv 1991).

5.1.6 Ischemic Preconditioning and Postconditioning

Studies show that damage from focal ischemia can be improved by applying an additional ischemic treatment either before or after the main ischemic event. Ischemic preconditioning is defined as an intervention that precedes ischemia and helps to lessen the severity of the ischemic injury. This treatment is based on the assumption that exposing the brain to relatively low-intensity ischemia will result in a less severe injury during a later ischemic event, compared to ischemia that was not preceded by a pretreatment. Following preconditioning, the brain develops a tolerance that is expressed in the next ischemic episode (Pignataro et al. 2009). In addition to using various types of ischemia, using hyperoxia as a pretreatment for ischemia also has a beneficial effect (Bigdeli 2011).

Ischemic postconditioning is defined as a mechanical interruption of the reperfusion stage by applying ischemia for an additional period during or after focal ischemia. The goal of postconditioning is to reduce reperfusion injury (Pignataro et al. 2009). The use of ischemic postconditioning as a possible treatment for cerebral ischemia began in recent years, following the success of the treatment in cases of heart muscle ischemia (Zhao 2009).

Many studies use a variety of methods for postconditioning: occlusion of the MCA, or occlusion of a single carotid artery or both carotid arteries after the reperfusion stage. In addition, there is variability regarding the duration of the posttreatment and the number of times this treatment is repeated. Studies show that compared to ischemia alone, the ischemic injury is less severe following the use of ischemic preconditioning, and even less severe following the use of ischemic postconditioning. The current study will use occlusion of one or two carotid arteries as a pre- or postischemia treatment.

5.1.7 Anoxia

Anoxia is a specific case of hypoxia where there is a complete absence of oxygen in the tissues (Berne and Levy 2000). This condition can be caused by a lack of air, suffocation, or drowning. During anoxia, the blood supply to the brain remains normal and even increases to meet the heightened demand for oxygen. The lack of oxygen leads to the cessation of the electron transport chain's activity in the mitochondria. As a result, the level of reduced NADH rises, and the ATP level decreases in all body tissues (Mayevsky 1984).

Brief exposure to nitrogen (N_2) is an experimental model for anoxia in laboratory animals. Studies show that brief exposure to nitrogen does not cause brain damage (Mayevsky 1984).

5.1.8 Spreading Depression (SD)

Spreading depression (SD), or cortical spreading depression (CSD), is a well-known phenomenon of a reversible response in the central nervous system to a local or systemic stimulus. It is an "all-or-nothing" phenomenon, in the form of a wave of regional electrical suppression in the gray matter of the cortex, which spreads from the wave's focus to its surroundings at a slow rate of 3–5 mm/min. The wave spreads in all directions from its point of origin within the hemisphere where it was created, without crossing to the contralateral hemisphere (Dreier 2011; Somjen 2001). The phenomenon was first observed in 1944 by Leão, who documented SD waves in a rabbit's cortex (Leão 1944). Over the years, the phenomenon was also studied in additional animal models (Somjen 2001). In 1995–1996, Mayevsky et al. first documented SD waves in humans (Mayevsky et al. 1995, 1996).

SD can occur spontaneously in various pathological conditions, such as ischemia, anoxia, migraines, epilepsy, and head injuries (Gorji 2001; Somjen 2001; Pietrobon 2005). In addition, SD waves can be induced by several external stimuli, such as rinsing with a KCl solution, electrical stimulation, needle puncture, thermal stimulation, or metabolic inhibitors (Dreier 2011).

There are several theories regarding the factors that influence the creation of SD waves in the brain. According to the potassium hypothesis, the electrical suppression that occurs during SD leads to an increase in the extracellular potassium level. During SD, the cells in the immediate vicinity of the passing wave are excited, causing them to release higher amounts of potassium into the intercellular space. In this way, the potassium level continues to rise as the wave progresses (Somjen 2001). Another hypothesis suggests that the waves are triggered by an increase in the level of the excitatory neurotransmitter glutamate. Studies show that excess glutamate can trigger SD waves. Additional studies prove that when SD waves are created by means other than glutamate, its level in the intercellular space increases (Somjen 2001; Smeaton 1963; Basarsky et al. 1999).

5.1.8.1 Tissue Changes During the Wave's Passage

During the wave's propagation, a decrease in electrical activity occurs for 20–30 s, and it returns to a baseline level after about 10 min (Somjen 2001). Similar to what happens during ischemia, in parallel with the decrease in electrical activity, there is an increase in potassium ions in the interstitial space. Simultaneously, a decrease is observed in sodium, calcium, and chlorine ions in the extracellular fluid. As a result, fluids enter the cells, causing them to swell and also leading to a decrease in the volume of the extracellular space (Somjen 2001). The accumulation of potassium in the extracellular space causes the release of glutamate (Mayevsky et al. 1998).

In the suppressed areas, there is a decrease in intracellular oxygen concentration (Marshall 1959), as well as an increase in oxygen consumption after recovering from the wave (Lukyanov and Bures 1967). In addition, during SD, an increase in glucose consumption and an increase in lactic acid production were measured. After about 10 min, these values returned to their initial level before the wave was created (Lauritzen et al. 1990). The increase in lactic acid, in addition to the accumulation of CO_2 in the tissue, leads to a decrease in pH, which returns to a normal level after the wave passes (Somjen 2001). These changes lead to a decrease in the tissue's ATP level, which creates an energy demand. In response to the high demand for ATP, the activity of the mitochondrial electron transport chain increases, and as a result, NADH is oxidized (Meilin et al. 1998; Mayevsky et al. 1990). The stage of the decrease in the NADH level can be preceded by an initial increase in its level (Hashimoto et al. 2000). The NADH oxidation waves occur in parallel with the electrical depolarization of the tissue (Sonn and Mayevsky 2000).

During an SD wave, there is a significant increase in tissue oxygen consumption, as a result of increased activity of sodium-potassium channels that require ATP (Sonn and Mayevsky 2000). The demand for oxygen is compensated by widening blood vessels and increasing blood flow to the depolarized area (Mayevsky and Weiss 1991). During an SD wave, NO is released, causing vasodilation in the blood vessels, which leads to an increase in CBF (Sonn and Mayevsky 2000). The injection of L-NAME after inducing SD waves prevented an increase in blood flow during the wave (Meilin et al. 1995).

Despite the variety of changes that SD waves create in the tissue, they do not create functional or histological damage in the brain. After the wave passes, all values return to their prewave state, so SD waves have no long-term effect on the brain (Sonn and Mayevsky 2000). A *summary* of these findings presents the SD phenomenon as a transient metabolic challenge, manifested by synchronized waves of hyperemia in CBF and oxidation in NADH, following the increased demand for oxygen.

In recent years, it has become accepted to include SD waves as part of the term spreading depolarization, which also includes phenomena of changes in the structure and function of neurons (Dreier 2011).

5.1.8.2 SD Waves in an Ischemic Brain

The brain's ability to compensate for itself by increasing blood supply during the propagation of SD waves is impaired in ischemic tissue. Accordingly, during an SD wave in ischemia, an initial decrease in CBF occurs, followed by a moderate increase. In conjunction with the decrease in blood flow, there is also an increase in the level of reduced NADH, in contrast to the decrease observed under normoxic conditions (Mayevsky et al. 1990). The wave duration in ischemia is significantly longer, due to a lack of oxygen, which causes slower recovery after the depolarization stage (Mayevsky 1976; Iijima et al. 1998). In contrast to SD waves in healthy tissue, SD waves during ischemia have a destructive effect on the tissue—each wave leads to further deterioration in the tissue, until a state of complete electrical suppression is reached (Nakamura et al. 2010).

In focal ischemia, the core region undergoes electrical suppression, so no spontaneous SD waves are observed there. These waves are observed only in the penumbra, where electrical activity is observed during the ischemia. Most of the waves have a nature and duration similar to induced ischemic SD waves, but some are longer and have the nature of electrical activity suppression. As the duration of the ischemia lengthens, the periods during which electrical activity in the tissue is suppressed also increase. From a certain point, electrical activity with a wave nature is no longer observed, and the penumbra enters a state of ID, i.e., complete electrical suppression (Nallet et al. 1999). The SD waves induced during ischemia have an effect on the size of the damaged area in focal ischemia, as demonstrated by the growth of the damaged area after each wave is created (Nakamura et al. 2010).

Ischemic depolarization (ID) is also called SD-like hypoxic depolarization (HSD), due to the similarity between the phenomena that occur in the brain during the two phenomena: electrical suppression, an increase in the extracellular potassium level, a decrease in the sodium level, swelling of the cells, and an increased demand for ATP (Somjen 2001). However, ID, unlike SD, is a state of complete electrical suppression that the entire ischemic tissue undergoes, without a wave pattern. The main difference between a state of SD and ID is the amount of oxygen available in the two cases. In an SD wave in healthy tissue, the amount of available oxygen is not limited. During the electrical suppression, there is an increase in blood flow, and with it an increase in the hemoglobin level, which raises the level of available oxygen. In contrast, in the case of ID, the amount of available oxygen decreases, either due to a lack of blood reaching the area or due to hypoxia or ischemia (Somjen 2001).

Following the occlusion of both carotid arteries, SD waves can be created spontaneously throughout the cortex (Mayevsky 1976). The spontaneous waves are similar in nature to induced SD waves during ischemia, i.e., they have a pattern of a decrease in CBF, an increase in the level of reduced NADH, and a long duration relative to the waves observed under normoxic conditions (Somjen 2001; Mayevsky et al. 1990). Studies show that single carotid occlusion has a minimal effect on SD waves compared to control animals (Mayevsky 1976).

5.1.8.3 SD Waves in Both Brain Hemispheres

The vast majority of studies show significantly that SD waves are a unilateral phenomenon, with no effect whatsoever on the opposite hemisphere, regardless of the method of wave induction or measurement (Somjen 2001; Mayevsky 1984; Leão 1944). For this reason, most studies document SD waves in only one hemisphere (unilateral spreading depression, USD), while protocols that include SD in both hemispheres (bilateral spreading depression, BSD) are rare.

Studies show that there is no difference in the tissue's response to USD waves versus BSD waves in rats. These studies include the response of rats to electrical shock on one side of their body (Best et al. 1975), the effect on appetite (Huston and Bures 1970), and responses related to the sense of taste (Buresová and Bures 1985). In several studies, the effect of BSD on memory and learning processes was examined. Inducing BSD in a rat led to the disruption of these processes, such as maze navigation (Brozek et al. 1974, Winn et al. 1975), learning to avoid electrical shock (Pianka 1974), motor performance, and depth perception (Islam and Buresová Islam and Buresová, 1975). The purpose of these studies was to create a stimulus using a double BSD in order to cause the suppression of parallel areas in both hemispheres. In these studies, USD caused minimal learning problems or had no effect at all.

A summary of these studies shows that during bilateral SD waves, there is no effect of waves in one hemisphere on waves in the opposite hemisphere. Similar to the current study, the aforementioned studies use BSD in order to create a metabolic challenge in both hemispheres of the brain, without an inter-hemispheric effect.

5.1.9 Research Objectives and Importance

5.1.9.1 Working Hypotheses

Mitochondrial damage during ischemia is a significant component of the pathophysiology of stroke and other brain injuries. Monitoring NADH in an in vivo model in real time serves as an indicator of the mitochondria's condition during an ischemic event. Therefore, ischemic tissue damage can be assessed through multiparametric monitoring of mitochondrial function and cerebral blood flow.

Previous studies show a fundamental difference between the core and penumbra regions that form during focal ischemia, based on the severity of the damage. The ischemic core region is more severely damaged than the penumbra region, which suggests that a greater increase in NADH levels and a decrease in blood flow are expected at the site of injury compared to the peripheral areas. Additionally, it's expected that the nature of the parameter changes will differ depending on the duration of the ischemia. It is estimated that as the duration of ischemia increases, the extent of brain injury will also increase, leading to a larger core region relative to the penumbra region, and a greater increase in mitochondrial NADH levels in these areas.

Following reperfusion, a decrease in NADH levels is expected, along with a restoration of blood flow to the tissue. The degree of recovery will depend on the intensity and duration of the ischemia. The extent of ischemia influences the recovery time, and it is estimated that partial global ischemia will cause less severe mitochondrial and hemodynamic damage compared to focal ischemia. Studies show that an additional ischemic event as a preconditioning treatment (ischemic preconditioning) or as a late treatment (ischemic postconditioning) for focal ischemia can alleviate ischemic damage. It can be hypothesized that global ischemia as a preconditioning treatment will lead to a decrease in the severity of both the ischemic and reperfusion injuries. Global ischemia as a late treatment would only lead to a decrease in reperfusion injury.

5.1.9.2 Research Objectives

The research objectives are to:

1. Examine the presence and nature of the relationship between different physiological parameters, as expressed in various cortical regions under different ischemic conditions.
2. Build a model to monitor mitochondrial NADH levels and blood flow in tissue during focal ischemia caused by MCA occlusion.
3. Identify the core and penumbra regions that form as a result of focal ischemia based on their response during monitoring, and examine the differences in parameter behavior between these regions.
4. Study the effect of ischemia duration on cerebral blood flow and mitochondrial NADH levels, both after MCA occlusion and during the reperfusion phase.
5. Compare mitochondrial function and blood flow in focal ischemia versus partial global ischemia, which is created by occluding one or both carotid arteries.
6. Examine mitochondrial function and cerebral blood flow in the ischemic brain following metabolic challenges, including oxygen deprivation and coping with induced spreading depression (SD) waves.
7. Develop methods for treating the damage from ischemic and reperfusion injury following focal ischemia.

5.1.9.3 Importance of the Research

Cerebral ischemia is one of the main causes of damage that occurs in various pathophysiological conditions, such as stroke, arteriosclerosis, and other brain injuries. The MCAO model is a well-established and common model for focal ischemia in humans.

Numerous studies on focal and global ischemia models have examined electrical activity, ion balance, cerebral blood flow, and other parameters.

The current study aims to examine the level of mitochondrial activity by directly measuring NADH levels and the relationship between this activity and the blood supply to the brain. The MSMP system, which shows changes in blood flow and NADH levels in ischemic tissue, provides real-time feedback about the changes occurring in the damaged tissue. To date, focal ischemia using the MCAO model has not been studied with multiparametric measurement. This type of measurement, with an emphasis on mitochondrial activity, could add to the knowledge about focal ischemia and other ischemic conditions. The MSMP system will also provide information regarding the connection between focal ischemia and other pathophysiological conditions.

Due to the widespread prevalence of stroke, finding ways to reduce damage during ischemia and following reperfusion is crucial. This research proposes a way to reduce the damage from focal ischemia by using global ischemia as a postischemia treatment (ischemic postconditioning). The results of this study may be important for adding to the knowledge of focal ischemia and its clinical applications.

5.1.9.4 Scientific Articles and Their Integration into the Work

This work integrates results from four articles published in scientific journals in the fields of neuroscience and optical instrumentation. The article titled "Real-Time Monitoring of Mitochondrial Function and Cerebral Blood Flow following Focal Ischemia in Rats" (Livnat et al. 2008) was the first to present the combination of the MCAO method with the MSMP monitoring system, to create a simultaneous, real-time monitoring of blood flow and mitochondrial function following focal ischemia.

The article titled "Real-Time Monitoring of Spatial and Temporal Metabolic Changes during Focal Cerebral Ischemia in Rats" (Livnat et al. 2011) deals with mitochondrial function and cerebral blood flow in response to different durations of focal ischemia as measured in different monitoring regions. The article titled "Mitochondrial Function and Cerebral Blood Flow Variable Responses to Middle Cerebral Artery Occlusion" (Livnat et al. 2010) discusses the high variability in the results obtained when performing the MCAO model.

The article titled "Mitochondrial Function and Cerebral Blood Flow Responses under Unilateral Carotid Occlusion in Rats" (Livnat et al. 2009) focuses on global ischemia: comparing unilateral carotid occlusion (UCO) to bilateral carotid occlusion (BCO) and their effects on mitochondrial function and cerebral blood flow in the brain.

The articles are integrated into the work at relevant points, with additional results and explanations regarding the discussed topics provided as needed. All the articles deal with different aspects of cerebral ischemia, aligning with the current research topic. Each article is innovative in its field and adds to the existing professional literature on cerebral ischemia.

5.2 Methods

5.2.1 Multisite Multiparametric System

Monitoring the different parameters in the animal's brain is done using the multisite multiparametric (MSMP) system, which was developed in Prof. Mayevsky's lab (Fig. 5.2). The system combines NADH measurement using the fluorometric method with the measurement of cerebral blood flow (CBF) using a laser Doppler (Mayevsky et al. 1998). The system includes a bundle of optical fibers, which are bundled together into a probe with a black elastic coating to prevent the penetration of external light. In this work, two probes were used, each with a diameter of 2 mm, which were placed in different locations on the surface of the cerebral cortex. The probes are placed inside special cannulas, with a diameter of 3 mm, which are fixed to the animal's head using dental cement and reinforcing screws. These cannulas prevent the movement of the fibers relative to the monitored tissue. The multiparametric measurement system allows for simultaneous hemodynamic and metabolic measurements from the same area in the cortex, continuously and in real time (Mayevsky et al. 1998).

5.2.1.1 Fluorometric Measurement of NADH

NADH levels were measured using the fluorometric method developed by Chance and his colleagues. This method is based on the differences between reduced NADH

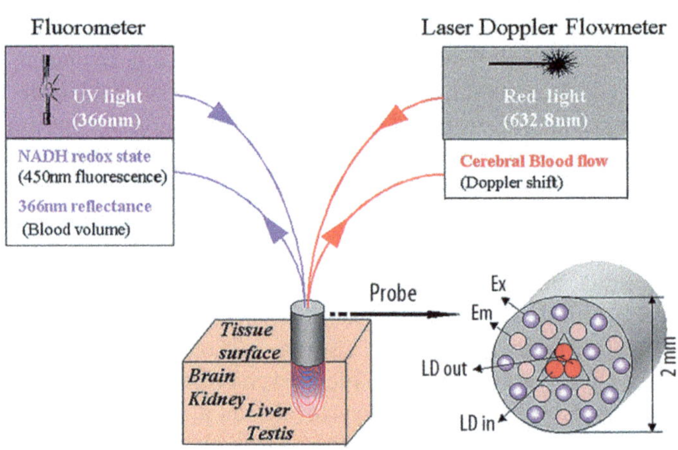

Fig. 5.2 Schematic illustration of the multiparametric measurement system. The measurement probe is connected to an LDF device for CBF monitoring, and to a photomultiplier for NADH monitoring. Em, Ex = fibers for excitation and emission of light, to and from the tissue, respectively, for mitochondrial NADH monitoring. LD in, LD out = fibers for LDF blood flow monitoring

and oxidized NAD^+. A reduced NADH molecule absorbs light in the UV spectrum (315–400 nm) and emits fluorescent light at wavelengths of 420–480 nm. In contrast, an oxidized NAD^+ molecule does not emit fluorescent light, which allows for the measurement of the ratio between these molecules using fluorescent methods.

In the fluorometric measurement method, the animal's brain is exposed by drilling a hole in the skull without removing the dura layer. The exposed brain area is illuminated with light at a wavelength of 366 nm. The light, from a mercury lamp, travels through an optical fiber system, lenses, and a filter. It is directed to the tissue using Y-shaped plastic fibers, where one arm transmits the light to the tissue and the other collects the light reflected from it. The light reflected from the tissue is amplified and converted into a digital signal. Some of the light is reflected from the tissue at a wavelength of 450 nm; this is fluorescence (Flu), which originates from light striking NADH molecules. The light that does not strike NADH molecules is reflected as reflectance (Ref) at the same wavelength as the excitation light (366 nm). The intensity of the reflectance light is inversely proportional to the tissue's absorption intensity and primarily reflects changes in blood volume.

Hemodynamic changes in the tissue cause changes in both the reflectance and fluorescence light absorption properties. To obtain a more accurate mitochondrial NADH measurement that is not affected by the tissue's absorption properties, the NADH measurement was corrected by subtracting the reflectance value from the fluorescence at a 1:1 ratio. This subtracted value (Flu-Ref) is called Corrected Fluorescence (CF). The CF value is a relative measure and a reliable indicator of the reduced NADH level in the monitored tissue at a given time. However, in experiments where a high reflectance level was recorded for an extended period, an artifact was noted in the CF measurement. In these cases, the CF value was not a reliable measure of mitochondrial NADH levels, and as a result, the uncorrected fluorescence (Flu) value served as a more reliable indicator.

5.2.1.2 CBF Measurement Using Laser Doppler

Laser Doppler Flowmetry (LDF) is a method for continuously measuring local blood flow in real time. The measurement is performed using a PeriFlux PF 2B LDF device from Perimed, Sweden. The device emits light from a helium-neon laser with an intensity of 2 mV and a wavelength of 632.8 nm. The light is projected onto the tissue via optical fibers. When the light strikes static tissue, it is reflected at the same wavelength. However, when the light hits red blood cells, its frequency changes, and it is reflected at a different wavelength according to the Doppler shift principle. The light returns from the tissue through additional optical fibers to a photodetector, which converts the photons into electrical signals and quantifies them. The speed of the red blood cells is expressed in the wavelengths of the reflected rays, while the number of shifted rays indicates the quantity of blood cells. The more blood cells there are in the tissue, the more rays hit them and are shifted as a result. The speed of the blood cells increases with blood flow, which in turn shortens the wavelengths of the reflected light.

From the average shift of the light rays, the tissue blood flow (TBF) can be calculated. The LDF method allows for TBF measurement in various tissues. In this study, cerebral blood flow (CBF) was measured.

5.2.2 *Surgical Procedures*

5.2.2.1 Preparation for Surgery

The experiments were conducted in accordance with the regulations of the Bar-Ilan University Animal Experimentation Ethics Committee and the Ministry of Health guidelines. Adult male Wistar rats, weighing 220–300 g, were anesthetized with an intraperitoneal (IP) injection of Equithesin (Eth). The anesthetic's composition included pentobarbital (9.72 mg), chloral hydrate (42.51 mg), magnesium sulfate (21.25 mg), propylene glycol (44.34%), and alcohol (11.5%) per mL of Eth, with the remainder being water. The animals were injected with Eth at a dose of 0.3 mL/100 g at the beginning of the experiment, with a supplement of 0.1 mL every hour until the end. Throughout the experiment, the animals were placed on a heating plate to maintain a constant body temperature of 37°C.

5.2.2.2 Induction of Focal Ischemia by MCAO

The induction of local ischemia by occluding the MCA was performed according to the protocol by Longa and colleagues (1989), with relevant adaptations (Fig. 5.3). The right CCA (common carotid artery) was exposed through a longitudinal incision along the neck and the overlying muscles (Digastric, Sternomastoid and Omohyoid Muscles) were retracted. The artery was carefully isolated, taking care to avoid damaging the adjacent vagus nerve. The carotid was then pulled with a silk thread to prevent blood flow through it during the surgery. Next, the ECA (external carotid artery) was exposed and isolated, with coagulation of its branching arteries up to its bifurcation into the STA (superficial temporal artery) and MA (maxillary artery). At this location, two silk threads were tied tightly, permanently stopping blood flow through these arteries. In the next step, the ICA (internal carotid artery) was isolated, with coagulation of its branches, up to the area where the pterygopalatine artery (PPA) branches off. The isolated ICA was then lifted with a silk thread to temporarily prevent blood flow through it.

After isolating all relevant blood vessels, another silk thread was loosely tied around the ECA, close to its bifurcation from the CCA. The isolated section of the ECA was then punctured with fine scissors. A preprepared nylon filament was inserted through the puncture. The filament was made from a 4/0 nylon suture, which was heated at its tip to create a rounded protrusion. The protruding tip of the thread was then coated with poly-L-Lysine, an adhesive material that has been shown to improve success rates in this model (Belayev et al. 1996; Bouley et al. 2007; Shimamura et al. 2006).

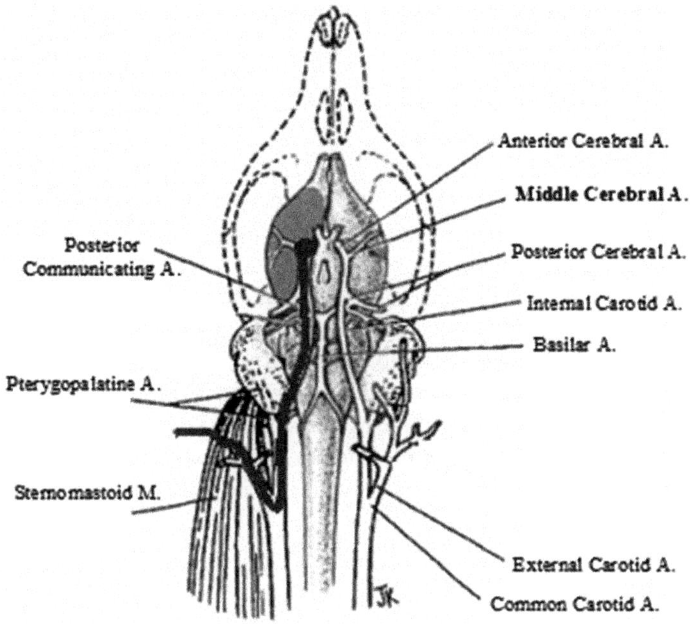

Fig. 5.3 Schematic diagram of the arteries supplying blood to the brain. During MCAO surgery, the nylon wire is inserted through the ECA into the ICA (Longa et al. 1989)

The filament was inserted from its protruding tip through the puncture in the ECA into the CCA. At this point, the loose thread prepared earlier was tied in a way that would prevent blood leakage through the puncture in the ECA but would also allow the nylon filament to move within the blood vessel. After the filament was inserted, blood flow was restored in the ICA and CCA by untying the temporary knots. The ECA was then detached from its location by cutting between the two silk threads tied near the PA and STA bifurcation, in a way that prevented bleeding.

The next step in the MCA occlusion was performed later, after the measurement system was connected to and monitoring the brain. At this stage, the nylon filament was gently withdrawn through the puncture, pulled out of the CCA, and left at the free end of the ECA. Simultaneously, the cut ECA was rotated to create a straight line between it and the ICA. While rotating, the nylon filament was pushed forward so its protruding tip entered the ICA. The thread was gently pushed along the ICA until after the PPA bifurcation. At this stage, it was verified that the filament continued its path through the ICA and did not enter the PPA. After this bifurcation, it is no longer possible to visually track the filament's progress. The filament was pushed further through the ICA until resistance was felt, from which it could be concluded that the filament was stopped at the beginning of the MCA, occluding it. Another indication of a successful occlusion is the length of the path the filament traveled inside the blood vessel, approximately 21 mm from its insertion point. At this stage, the nylon filament was left in place for a certain period according to each protocol, where it occluded the MCA. Later, the filament was withdrawn backward to restore blood flow through the MCA.

5.2.2.3 Carotid Artery Occlusion

Occlusion of one or both common carotid arteries (CCA) was included in different protocols as a standalone procedure or in combination with MCA occlusion. To expose the carotid arteries, an incision was made in the animal's neck. In experiments where global ischemia was induced for 24 h before measurement, one or two carotid arteries were tied with 4/0 silk thread. The neck area of these animals was then sutured, and they were left in a cage for a day with free access to food and water. In experiments where the arteries were occluded during monitoring, silk threads were looped around one or two arteries during the surgical stage before monitoring, but not tied. The threads were tied, and the arteries were occluded at a later stage, during the monitoring itself.

5.2.2.4 Head Surgery

The skull was exposed by making an incision in the skin and removing the connective tissue above it. Holes were drilled into the skull to anchor cannulas for the MSMP system's probe. The diameter of each cannula and hole was 2.8 mm. The holes were drilled in the parietal lobe in different locations according to the various experiment protocols (Fig. 5.4). In experiments where two measurement fibers were placed on two different hemispheres, one hole was drilled in each hemisphere, 1 mm posterior and 5 mm lateral to the bregma point. In experiments where two measurement fibers were placed on one hemisphere, two adjacent holes were drilled on the right hemisphere, 1 mm posterior and 1 and 5 mm lateral to the bregma point. The location of the cannulas was determined based on studies showing hemodynamic changes in the brain during focal ischemia (Schmid-Elsaesser et al. 1998) and with the help of a stereotactic atlas (Paxinos and Watson 1997). In experiments where SD waves were induced, two

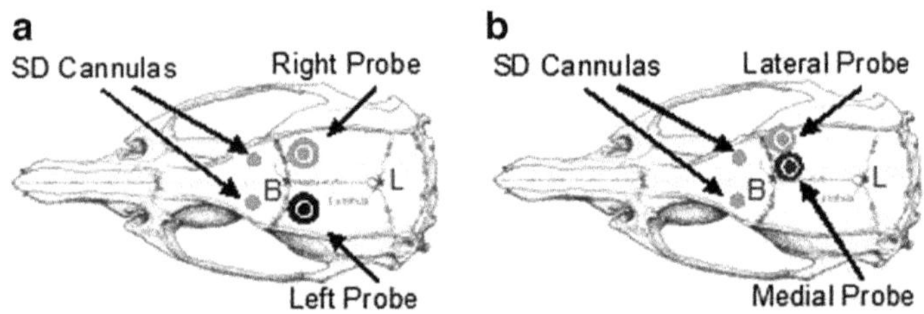

Fig. 5.4 Position of the measurement fibers (probes) and SD cannulas on the surface of the brain. (**a**) Placement of a measurement fiber over the entire hemisphere. (**b**) Placement of both measurement fibers over the right hemisphere

additional mm diameter holes were drilled in the frontal lobe for cannulas to inject KCl. The various cannulas were placed on the cortex without applying pressure on the brain tissue. To fix the cannulas to the animal's head, several screws were placed in the skull and dental cement was used.

5.2.2.5 Nitrogen Exposure

At the beginning of each experiment, after the MSMP system fibers were fixed to the brain, a brief exposure to nitrogen was performed for about 20–30 s, until respiration stopped. After a few seconds, the animal resumed breathing, either spontaneously or with ventilation assistance. Nitrogen exposure is used to check the integrity of the measurement system and the tissue after surgery. In this study, the initial nitrogen exposure also served as a control group for subsequent nitrogen exposures in the same animals at later stages of the experiment. Studies show that brief exposure to nitrogen does not cause brain damage (Mayevsky 1984). In each nitrogen exposure treatment, minimum or maximum reflectance values (Ref_peak) were sampled, depending on the observed response. The maximum NADH level (NADH_max), minimum blood flow (CBF_min), and maximum blood flow (CBF_max) were also sampled.

5.2.2.6 Inducing SD Waves

In various protocols, SD waves were induced by washing the tissue with a 1M concentration of KCl solution. To stop the waves, the tissue was washed with a 0.9% NaCl solution. The SD waves were induced in both hemispheres simultaneously, in each hemisphere separately, using two special cannulas for injecting KCl and NaCl. Each cannula was constructed from two PE-50 polyethylene tubes, one connected to a syringe and the other placed inside a solution bottle. The other ends of each tube (closer to the brain) were anchored in a shorter, thicker tube (PE-205), with a height difference of about 2 mm from each other and 2 mm from the end of the thicker tube. The thick tube was placed on the dura layer in a window drilled for the washing cannula. Using the syringe, fluids were drawn from the solution bottle through the tubes, creating localized circulation on the dura without fluid accumulation or pressure on the tissue. To stop the waves, the tissue was washed with a NaCl solution.

In each set of induced SD waves, the first wave was examined and statistically analyzed. In each wave, reflectance (Ref), NADH level, and tissue blood flow (CBF) were measured. In each wave, the minimum or maximum NADH values (NADH_peak) were sampled, depending on the nature of the wave. The minimum reflectance (Ref_min), maximum reflectance (Ref_max), and maximum CBF (CBF_max) values were also sampled.

5.2.2.7 Neurological Assessment

In animals where the effect of ischemia was examined after 24 h, a postural reflex test was performed to assess their neurological and motor function after ischemia (Bederson et al. 1986). This test included a forelimb extension check, during which the animals were held by their tail in the air. In this position, a healthy animal should extend its forelimbs straight toward the floor. In contrast, an injured animal should extend its forelimbs to the side contralateral to the injury. Resistance to light pressure was also examined: the animals were held by their tail with their feet touching a surface, and light pressure was applied to the right shoulder and then the left shoulder. Healthy animals should respond by resisting the light pressure and not moving from their position. Injured animals, however, lose the ability to resist pressure applied to the contralateral shoulder and consequently tilt to the side. In addition, the animals' ability to move was examined by placing them on a smooth surface. Healthy animals should move in a straight line, while injured animals are expected to move in a circular motion toward the affected side (Bederson et al. 1986).

5.2.3 Experimental Protocols

5.2.3.1 General Protocol—Beginning and End of Each Experiment

At the beginning of each experiment, after securing the MSMP system to the brain, a brief nitrogen exposure (anoxia) was performed as a control. Following this exposure, the animals were given 20–30 min to recover before starting the different treatments detailed for each protocol.

At the end of each experiment, the animals were euthanized by exposure to nitrogen for several minutes. Brain monitoring continued for several minutes after the anoxia to confirm the animal's death.

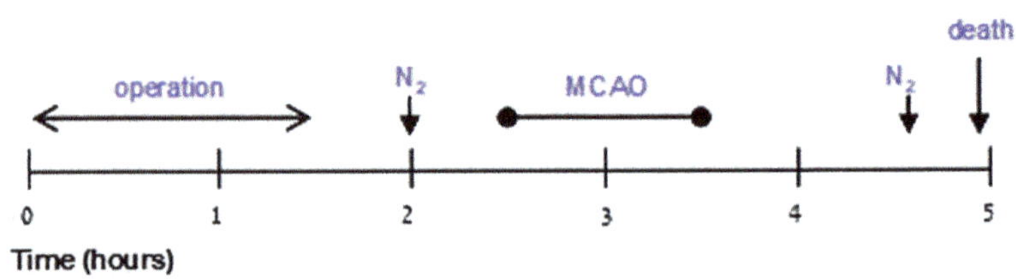

Fig. 5.5 Schematic presentation of the experimental protocol described in Sect. 5.3.3.2

5.2.3.2 Mitochondrial Function and Blood Flow Following MCAO—Monitoring in Both Hemispheres (Fig. 5.5)

A group of eight rats underwent MCA occlusion preparation and were then connected to the monitoring system, with each measurement fiber placed on a different cerebral hemisphere. After monitoring began and a brief nitrogen exposure was performed, the right MCA was occluded for 1 h, followed by the restoration of blood flow (reperfusion). One hour after reperfusion, a brief anoxia was performed, and the animal was then euthanized.

5.2.3.3 Effect of Ischemia Duration on Mitochondrial Function and Cerebral Blood Flow (Fig. 5.6)

Three groups of rats underwent MCA occlusion preparation and were then connected to the monitoring system, with both measurement fibers placed on the right hemisphere. After monitoring began and a brief nitrogen exposure was performed, the right MCA was occluded for 10 min ($n = 9$), 30 min ($n = 8$), or 60 min ($n = 8$), and then blood flow was restored. One hour after reperfusion, another nitrogen exposure was performed, and the animal was then euthanized.

5.2.3.4 Mitochondrial Function and Cerebral Blood Flow 24 h After Focal Ischemia (Figs. 5.7 and 5.8)

This protocol examined brain function 24 h after MCA occlusion. Some animals underwent transient occlusion (Fig. 5.7), in which the MCA was occluded for 60 min ($n = 8$). Another group of animals underwent permanent ischemia (Fig. 5.8), where the MCA was occluded for 24 h ($n = 8$). After the surgery, the rats were left to recover in a cage with free access to food and water. After 24 h, the animals' neurological and motor responses were tested to assess ischemic damage. Subsequently, the rats were anesthetized and connected to the monitoring system. The rats were exposed to brief anoxia, followed by induced SD waves.

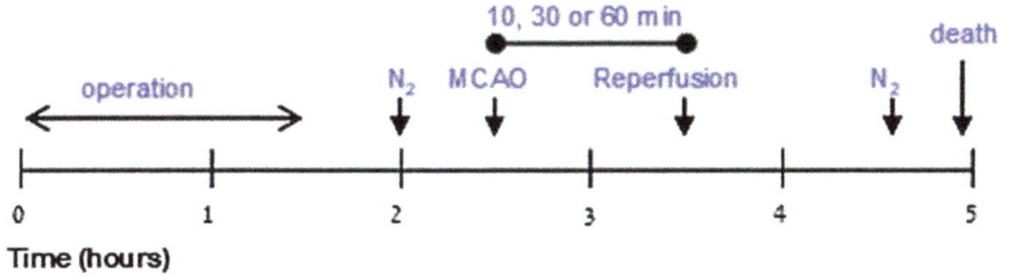

Fig. 5.6 Schematic presentation of the experimental protocol described in Sect. 5.3.3.3

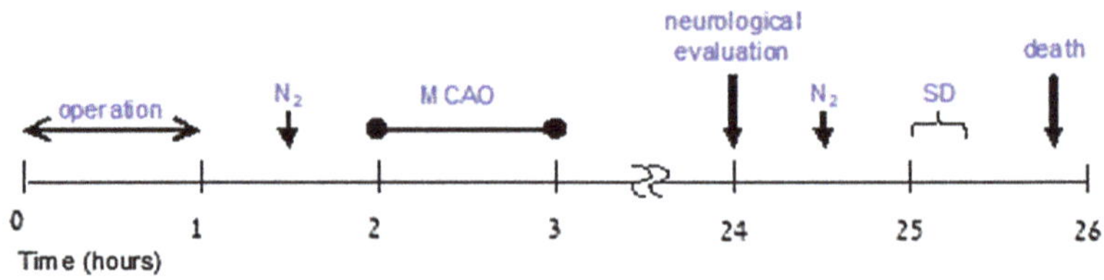

Fig. 5.7 Schematic presentation of the experimental protocol described in Sect. 5.3.3.4a

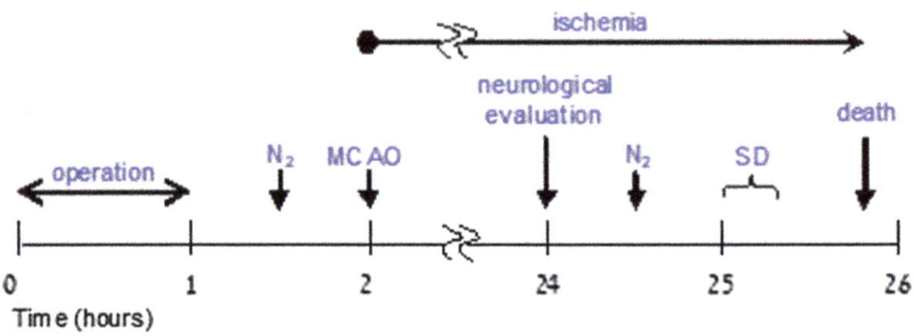

Fig. 5.8 Schematic presentation of the experimental protocol described in Sect. 5.3.3.4b

5.2.3.5 Examining the Degree of Variability in the MCAO Model Results

Animals that underwent MCA occlusion in various protocols throughout the work (a total of 96 rats) were divided into profile groups according to different CBF patterns observed in the experiments. Due to the different nature of the profile groups, only a qualitative comparison was generally made between the groups. A quantitative comparison was made between a group of animals in which an increase in blood flow was observed following MCA occlusion and a control group.

5.2.3.6 The Relationship Between Focal and Global Ischemia

This protocol is divided into four experimental groups: in each group, one or two carotid arteries were occluded, and monitoring was performed either near the time of occlusion or 24 h later. Tissue function was evaluated based on responses to nitrogen exposure and induced SD waves at different time points. In all groups, one measurement fiber was placed on each hemisphere.

Monitoring Near the Time of Carotid Artery Occlusion (Fig. 5.9)

In this group, three sets of **SD** waves (three waves per set) were induced in animals either near the time of a single carotid artery occlusion ($n = 10$) or near the time of a bilateral carotid artery occlusion ($n = 8$). After anesthesia and surgery, the rats underwent nitrogen exposure, followed by a first set of induced **SD** waves, which served as a control for the subsequent **SD** waves. One hour after the waves stopped, one or two carotid arteries were occluded. The recovery time from the artery occlusion (until the measured parameters stabilized) was 1 min for a single carotid artery occlusion and 10 min for a bilateral occlusion. After the parameters stabilized, a second set of **SD** waves was induced, and an hour later, a third set was induced. After the last **SD** waves stopped, the rats underwent a brief anoxia and were then euthanized.

Monitoring 24 h After Carotid Artery Occlusion (Fig. 5.10)

The rats underwent a single carotid artery occlusion ($n = 10$) or a bilateral carotid artery occlusion ($n = 8$), and were then left in a cage with free access to food and water. One day later, the animals underwent a brief anoxia, followed by a single set of three SD waves. The animal was then euthanized.

5.2.3.7 Effect of Ischemic Preconditioning on Ischemic and Reperfusion Damage (Fig. 5.11)

Two groups underwent occlusion of either the right carotid artery ($n = 8$) or both carotid arteries ($n = 8$) and were returned to a cage with free access to food and water.

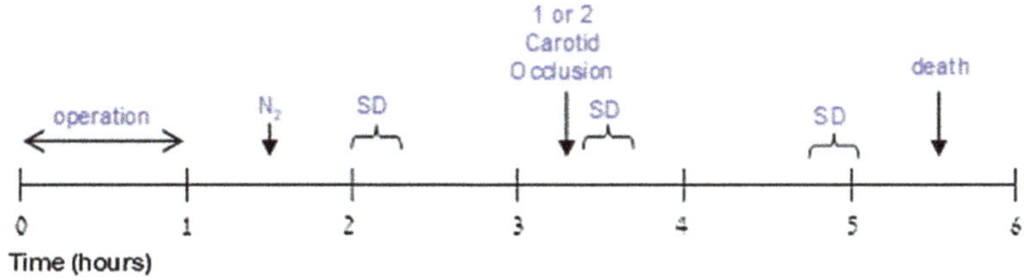

Fig. 5.9 Schematic presentation of the experimental protocol described in Sect. 5.3.3.6a

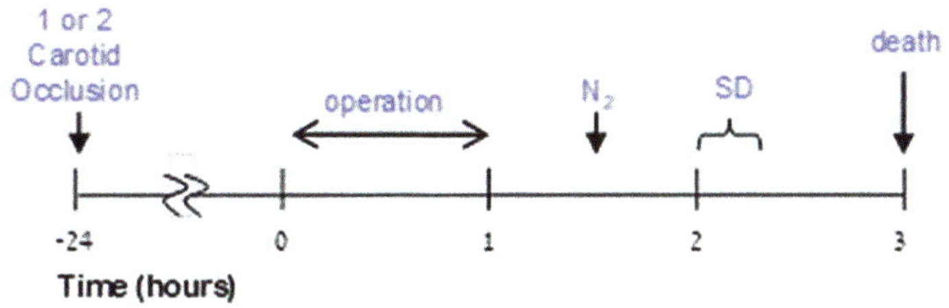

Fig. 5.10 Schematic presentation of the experimental protocol described in Sect. 5.3.3.6b

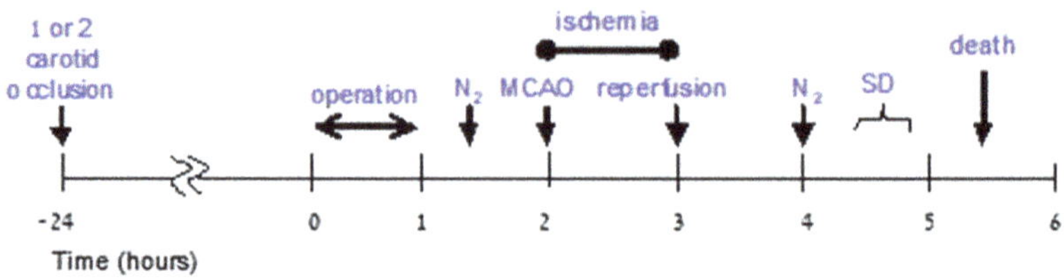

Fig. 5.11 Schematic presentation of the experimental protocol described in Sect. 5.3.3.7

After 24 h, the rats were anesthetized again and underwent the surgical preparation for MCAO. The animals were monitored for 1 h during which the MCA was occluded and for another hour after the occlusion was removed. Before the ischemia stage and after 1 h of reperfusion, the animals were exposed to nitrogen. After the second nitrogen exposure and before euthanizing the animals, SD waves were induced. These groups were compared to control groups that underwent either focal ischemia only or global ischemia only.

5.2.3.8 Effect of Ischemic Postconditioning on Ischemic and Reperfusion Damage

This section involved a total of four protocols, in which one or two carotid arteries were occluded for either 45 or 15 min.

Carotid Occlusion for 45 min (Fig. 5.12)

The animals underwent MCA occlusion. After 45 min, before removing the MCA occlusion, one ($n = 8$) or two ($n = 8$) carotid arteries were also occluded. After an additional 15 min, the MCA occlusion was removed (which had been occluded for a total of 1 h). Thirty minutes later, the carotid artery occlusion was removed (they had been occluded for a total of 45 min). The animals were then monitored for an additional hour. Before the ischemia stage and after 1 h of reperfusion, the rats were exposed to nitrogen, and then SD waves were induced.

Carotid Artery Occlusion for 15 min (Fig. 5.13)

The animals underwent MCA occlusion for 60 min. Simultaneously with the removal of the MCA occlusion, one ($n = 8$) or two ($n = 8$) carotid arteries were occluded. After an additional 15 min, the carotid artery occlusion was removed, and the animals were then monitored for another hour. Before the ischemia stage and after 1 h of reperfusion, the rats were exposed to nitrogen, and then SD waves were induced.

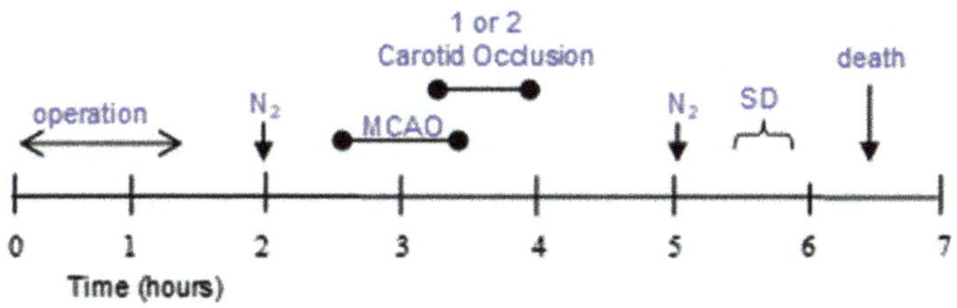

Fig. 5.12 Schematic presentation of the experimental protocol described in Sect. 5.3.3.8a

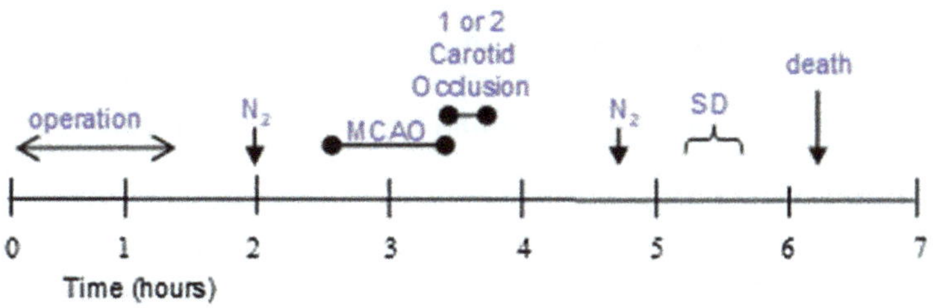

Fig. 5.13 Schematic presentation of the experimental protocol described in Sect. 5.3.3.8b

In both cases (carotid artery occlusion for 45 or 15 min), the groups were compared to control groups that underwent either focal ischemia only or global ischemia only (of a single carotid or both, depending on the protocol).

5.2.4 Data Collection and Processing

5.2.4.1 Data Collection

The data obtained during each experiment were saved on a computer using LabView software from National Instruments, USA. The data acquisition frequency was 100 Hz, from which the software sampled two data points per second. The software allows for simultaneous input from multiple channels and their conversion into digital signals, while creating an analog display during the experiment. In addition to the computerized data, information from each experiment was also recorded in an analog format on a Grass Instruments Model 7D polygraph.

5.2.4.2 Data Processing

The raw data from the computer were collected from all animals and sorted by the different parameters. Reflectance, fluorescence, and CF values were received in mV and converted to percentages, where the initial value was defined as 100% and a change of 500 mV was calculated as a 100% change relative to the initial level. CBF values were also received in mV and converted to percentages, based on a range between 0% at death and 100% at the beginning of the measurement. The data underwent statistical processing with the help of a certified statistician. For each data point, mean and standard error (SE) values were calculated.

The determination of differences between the baseline value of each parameter and the values later in the experiment was done using a paired t-test. A comparison between two measurement sites was done using an unpaired t-test. For comparing two groups or two measurement sites over a defined period (ischemia or reperfusion time), a repeated measures ANOVA with a Huynh-Feldt correction was used, with time serving as a covariant. If significant differences were found in the ANOVA test, an unpaired t-test was used between each time point in one group and the corresponding point in the other group to find the specific time periods in which the differences occurred.

The determination of differences between extreme values was done using paired t-test and unpaired t-test, depending on the data being compared. In addition, correlation coefficients between different parameters were calculated using the Pearson correlation coefficient. In all statistical comparisons, three significance levels were defined: $p < 0.05$ was marked with one asterisk or "x," $p < 0.01$ was marked with two asterisks or "xx," and $p < 0.001$ or less was marked with three asterisks or "xxx."

5.3 Results

5.3.1 Planning and Building a Multiparametric Monitoring Model in Combination with MCAO

5.3.1.1 Adapting the MCAO Model for Monitoring with the MSMP System

The current work developed a model that allows for continuous, real-time brain monitoring before, during, and after MCA occlusion. To achieve this, the MCAO model by Longa et al. (1989) was adapted for the MSMP monitoring system.

Naturally, MCA occlusion is performed on the ventral side of the animal, requiring the rat to be placed on its back, while brain surgery for monitoring requires the animal to be placed on its stomach. The development of the model's first stage required a method that would allow both actions to be performed simultaneously. The MCAO model was split into two stages. In the first stage, the animal was placed on its back for the initial MCA occlusion preparation, which included inserting the filament into the ECA and then into the right carotid artery. After this, the animal was placed on its stomach, the brain was exposed, the cannulas were secured to its head, the measurement fibers were connected, and the monitoring stage began. After a few minutes of monitoring, including a brief nitrogen exposure, the rat was again placed on its back for the second stage of the MCAO, in which the filament was moved into the ICA and then the MCA, all while monitoring the parameters.

To allow the animal to be turned onto its back for the final MCA occlusion stage while connected to the measurement system, a special surgical table was designed. This table included a groove through which the fibers could pass downwards. The table legs were adjustable to different heights, allowing for convenient surgery at a low height and then elevated monitoring to prevent excessive bending and damage to the optical fibers.

5.3.1.2 Selecting Measurement Regions

To find the ideal measurement sites for real-time brain monitoring during MCAO, several preliminary experiments were conducted where each measurement fiber was placed on parallel locations in both hemispheres. These experiments found that the ideal location to observe changes in both blood flow and NADH levels is 1 mm posterior to the bregma point. In the next stage, two measurement fibers were placed on the right hemisphere simultaneously, and another series of experiments was conducted to achieve stable, long-term monitoring at both sites (Fig. 5.14). Subsequently, several experiments were performed in which animals were monitored at both of these measurement areas under MCA occlusion. The results of these experiments showed a trend of differences in blood flow between the two measurement areas. Data collection for the current study began after these objectives were achieved.

5.3.2 Mitochondrial Function and Blood Flow in Different Brain Monitoring Regions

5.3.2.1 Introduction

This section is based on the article (Livnat et al. 2008) "**Real-Time Monitoring of Mitochondrial Function and Cerebral Blood Flow following Focal Ischemia in Rats**," published in the Journal of Innovative Optical Health Sciences (JIOHS) in 2008 (Livnat et al. 2008). The article presents reflectance, fluorescence, NADH, and CBF as measured before, during, and after 60 min of focal ischemia. The article shows results from monitoring both hemispheres and from monitoring two regions on a single hemisphere. The results presented in the article demonstrate that with MCA occlusion, CBF values decreased while NADH levels increased in the right hemisphere only (Figs. 5.15–5.17).

In contrast, no changes were recorded in the left hemisphere following the occlusion or after its removal. When monitoring two areas in the right hemisphere, lower CBF levels and higher NADH levels were observed in the medial measurement area (closer to the bregma point) compared to the lateral measurement area. After the occlusion was removed, the parameter values returned to their initial levels.

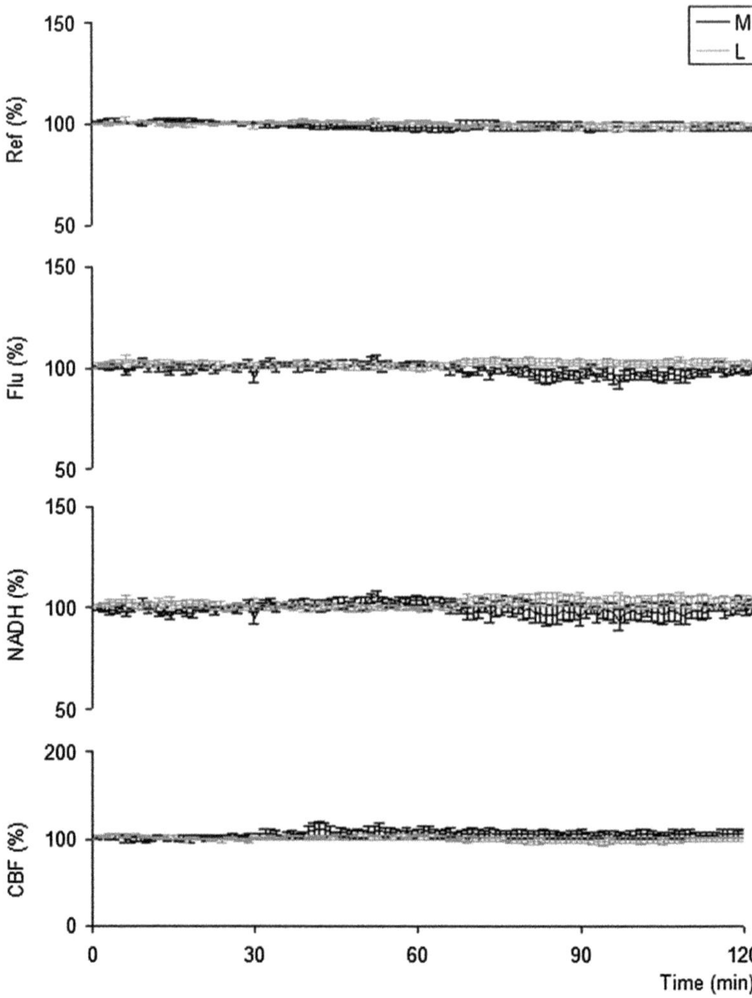

Fig. 5.14 Reflectance (Ref), fluorescence (Flu), NADH, and blood flow (CBF) levels during 2 h monitoring without treatment, as measured at two measurement sites in the right hemisphere - medial (M, black) and lateral (L, gray) relative to the bregma point. Results are presented as mean ± SE ($n = 5$).

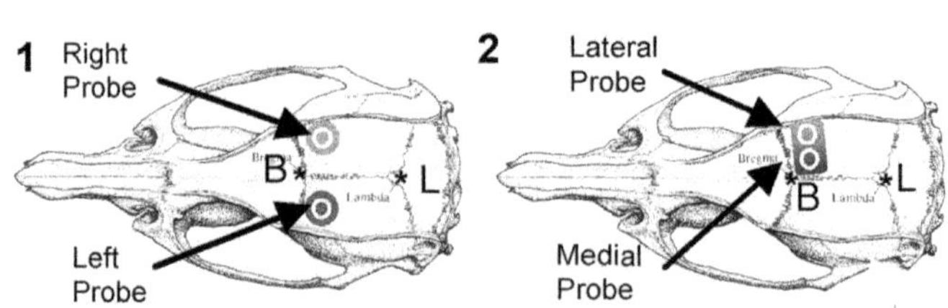

Fig. 5.15 Schematic representation of the location of the MSMP system fibers on the monitored sites in the bilateral model (1) and unilateral model (2). B Bregma, L Lambda

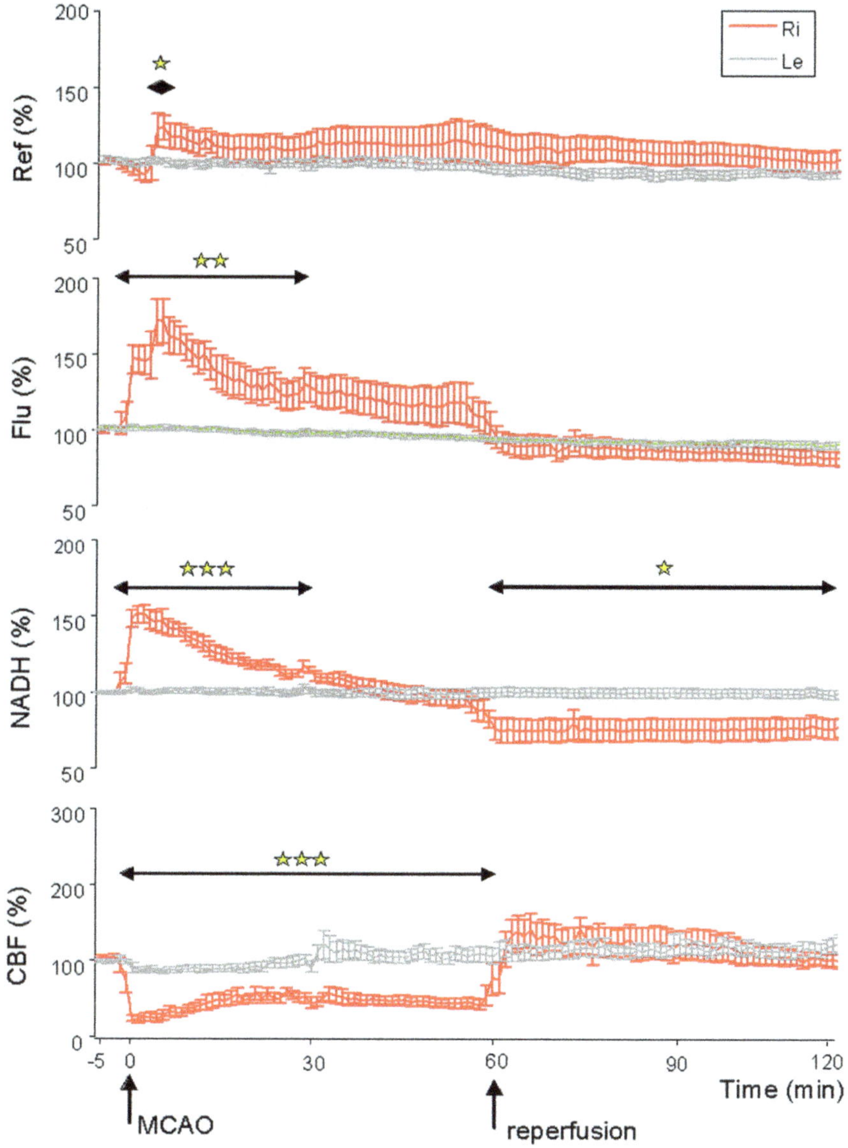

Fig. 5.16 Hemodynamic and metabolic changes in the bilateral monitoring model, following right MCAO for 60 min and reperfusion. Right hemisphere, Ri; Left hemisphere, Le. The results are presented as mean ± SE. ($n = 8$). Significant differences between the two hemispheres (in each minute) are marked by asterisks: $*p < 0.05$; $**p < 0.01$; $***p < 0.001$

Comparison of the data obtained with the results of other studies showed that the lateral region is defined as the core, the center of the ischemic injury, and the medial region is defined as the penumbra, the peripheral region that suffers from less severe injury. Consequently, in the following results and in subsequent articles, the terms core and penumbra will be used to characterize these regions.

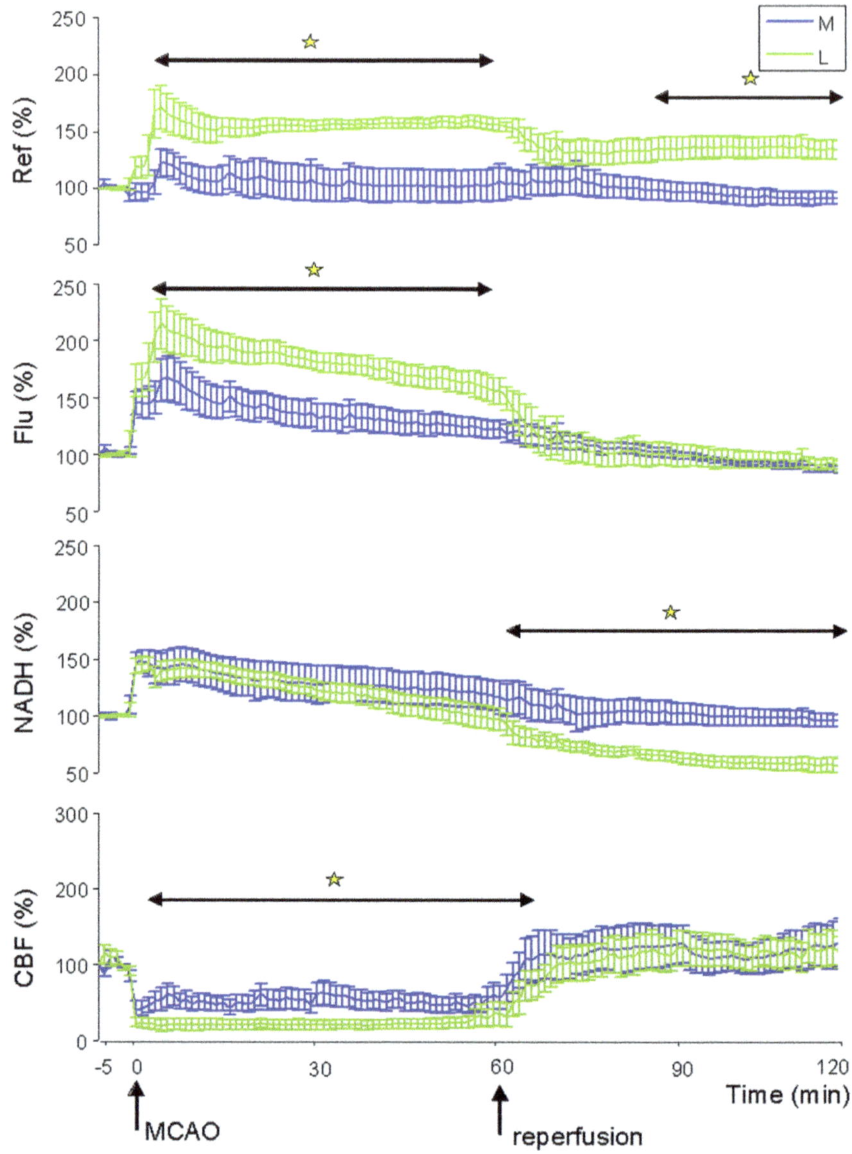

Fig. 5.17 Hemodynamic and metabolic changes in the unilateral monitoring model following right MCAO for 60 min and reperfusion. Lateral monitoring location, L; Medial monitoring location, M. The results are presented as mean ± SE. ($n = 9$). Significant differences between the two hemispheres are marked with asterisks, $*p < 0.05$

5.3.2.2 Nitrogen Exposure During Bilateral Monitoring

The following results supplement those published in the article mentioned previously. Figure 5.18 summarizes the group results for a brief exposure to nitrogen in normoxic

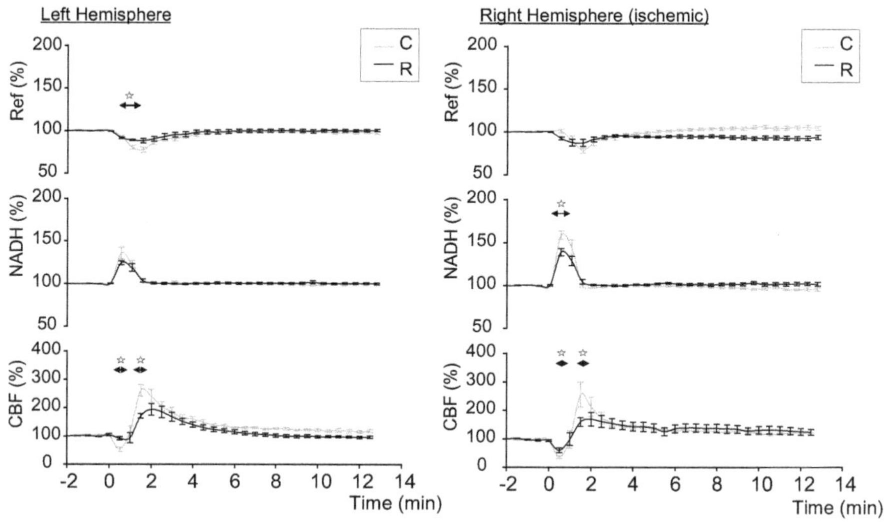

Fig. 5.18 Changes in parameters following brief exposure to nitrogen, as observed in the right and left hemispheres. Two treatments are shown: control treatment a few minutes before occlusion of the right MCA for 60 min (C), and 1 h after removal of the occlusion during the reperfusion phase (R). The time of onset of nitrogen exposure is marked as point 0 on the time axis. Results are shown as mean ± SE ($n = 8$). Significant differences between two nitrogen treatments are indicated by asterisks, $^{*}p < 0.05$

tissue (control nitrogen, C), compared to nitrogen exposure approximately 1 h after blood flow was restored following a 60-min right MCA occlusion (reperfusion nitrogen, R). In both treatments, one measurement fiber was placed in the parietal lobe of the right hemisphere, and another was placed in a corresponding area of the left hemisphere.

During nitrogen exposure, a significant decrease in reflectance levels was recorded relative to the baseline, in both the control nitrogen and the reperfusion nitrogen treatments. In the right hemisphere, the minimum level was 75% ± 3% during the control nitrogen exposure ($p < 0.001$), compared to 84% ± 4% during the reperfusion nitrogen exposure ($p < 0.05$), with no significant difference between the two treatments. In contrast, a significant difference was recorded in the left hemisphere between the control nitrogen and the reperfusion nitrogen ($p < 0.05$). The minimum reflectance value during control nitrogen exposure was 75% ± 3% ($p < 0.001$), versus only 87% ± 2% during reperfusion nitrogen exposure ($p < 0.001$) (Figs. 5.19–5.22).

NADH levels increased in the left hemisphere to a level of 134% ± 6% during the control nitrogen exposure ($p < 0.001$ relative to baseline), compared to only 125% ± 2% during the reperfusion nitrogen exposure ($p < 0.001$). In the right hemisphere, an increase to a value of 157% ± 7% was recorded during control nitrogen exposure ($p < 0.001$), compared to 139% ± 4% during reperfusion nitrogen exposure ($p < 0.001$). The increase in NADH levels was greater in the control treatment compared to the reperfusion stage, with the difference in the right hemisphere being significant

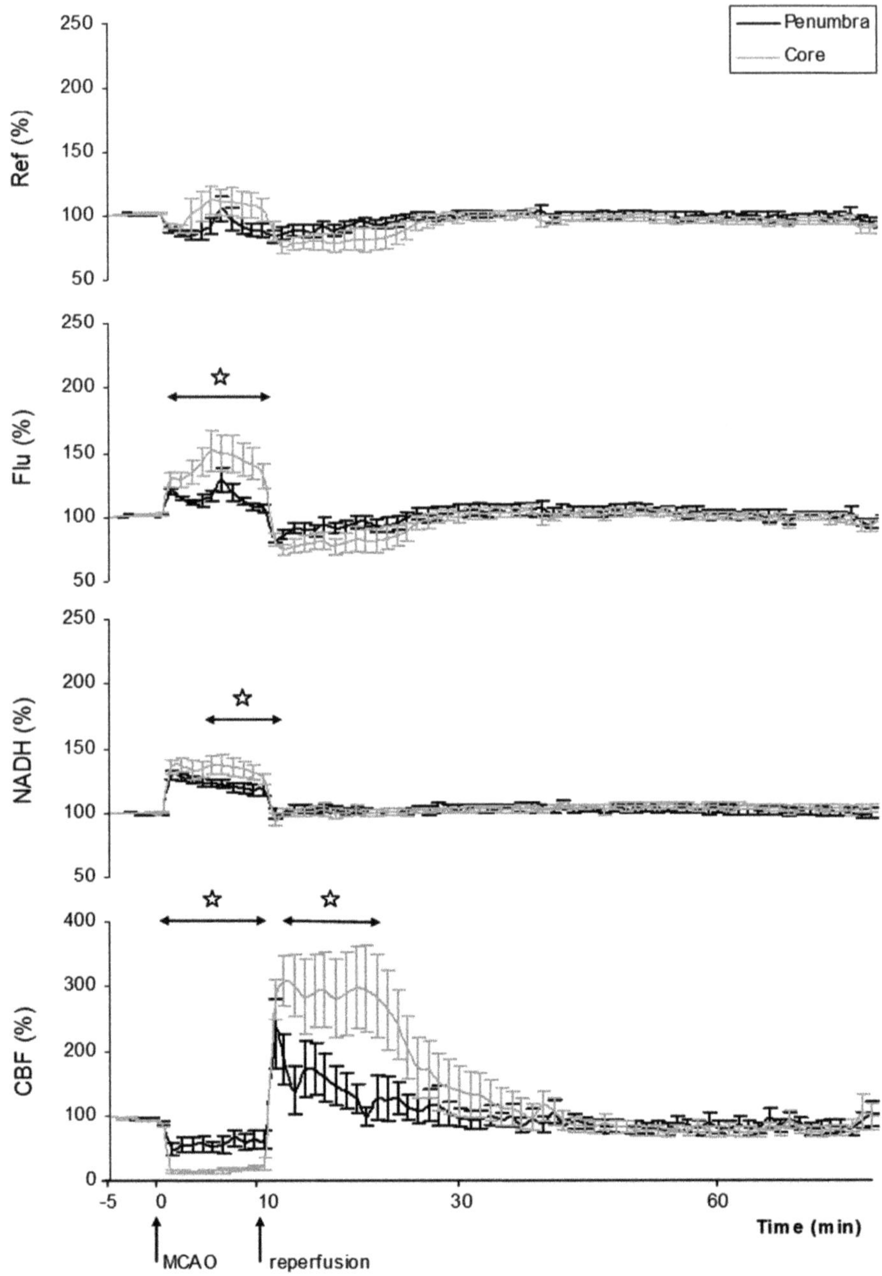

Fig. 5.19 Changes in reflectance (Ref), fluorescence (Flu), NADH and CBF following right MCAO for 10 min and reperfusion for 60 min. The results are represented as mean ± SE ($n = 9$). Significant differences between the two hemispheres are marked with asterisks, $*p < 0.05$

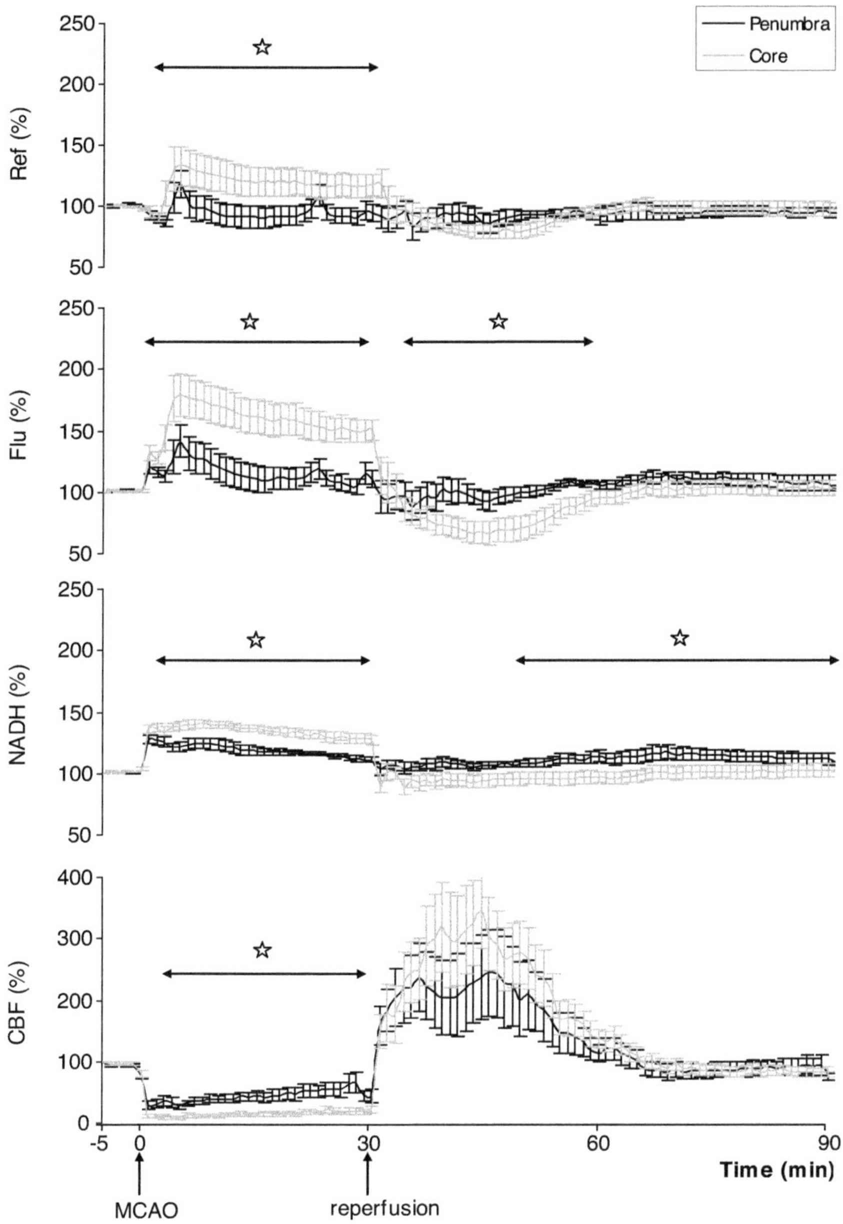

Fig. 5.20 Hemodynamic and metabolic changes following right MCAO for 30 min and reperfusion for 60 min. The results are represented as mean ± SE ($n = 8$). Significant differences between the two hemispheres are marked with asterisk, $^*p < 0.05$

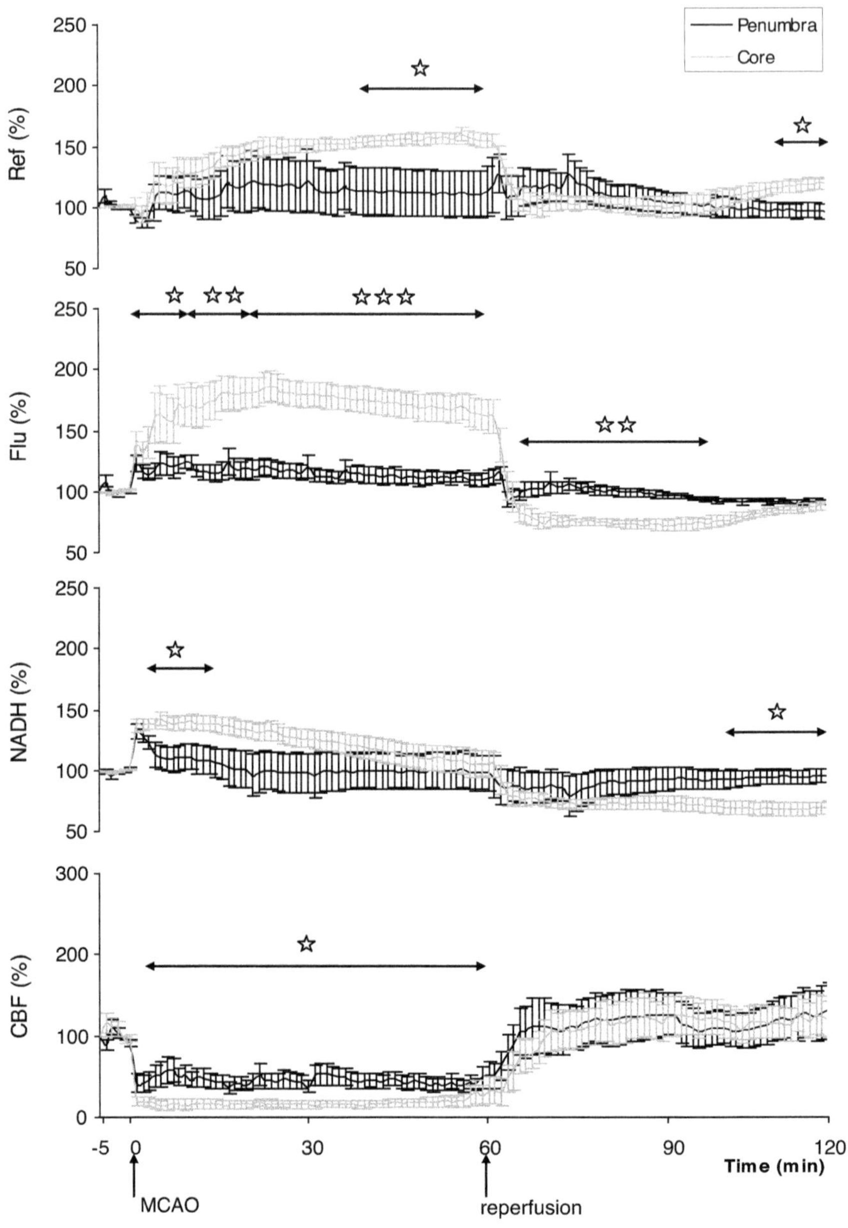

Fig. 5.21 Hemodynamic and metabolic changes following right MCAO for 60 min and reperfusion. The results are represented as mean ± SE ($n = 8$). Significant differences between the two hemispheres are marked with asterisks, $*p < 0.05$, $**p < 0.01$, $***p < 0.001$

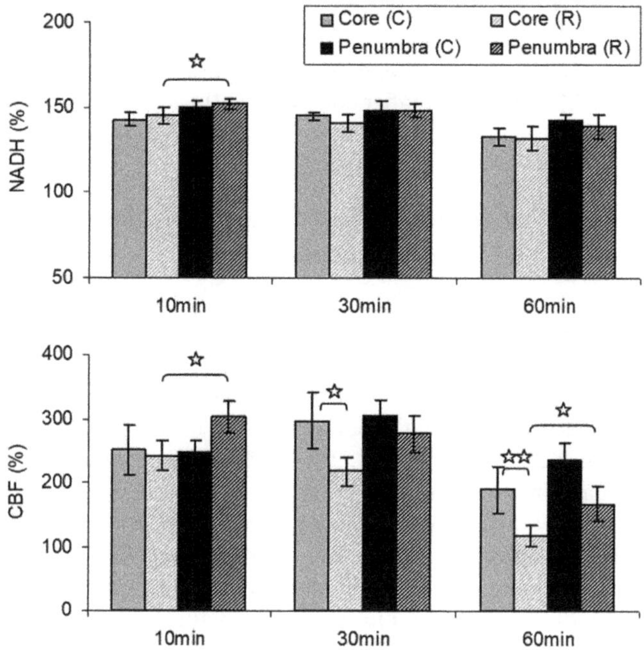

Fig. 5.22 Maximal CBF and NADH levels during short anoxia before the occlusion of the MCA (C, control) and 60 min after reperfusion (R) following three ischemic durations: 10, 30, and 60 min. The results are represented as mean ± SE ($n = 8$). Significant differences between the two hemispheres are marked with asterisks, $^{*}p < 0.05$

($p < 0.05$). Furthermore, the increase in NADH levels was greater in the right hemisphere relative to the left hemisphere, in both the control nitrogen ($p < 0.05$) and the reperfusion nitrogen ($p < 0.001$) treatments.

Blood flow showed a pattern of an initial decrease followed by an increase, both of which were significant relative to the starting point. In the right hemisphere, blood flow decreased to a value of 35% ± 5% in the control treatment ($p < 0.001$), compared to 60% ± 10% during reperfusion nitrogen exposure ($p < 0.001$). Blood flow then increased to a maximum value of 265% ± 53% in the control nitrogen treatment ($p < 0.001$) and 170% ± 24% in the reperfusion nitrogen treatment ($p < 0.01$). In the left hemisphere, blood flow decreased to a value of 55% ± 8% in the control treatment ($p < 0.001$), compared to only 92% ± 6% during reperfusion nitrogen exposure ($p < 0.05$). Blood flow then increased to a maximum value of 259% ± 25% in the control nitrogen treatment ($p < 0.001$) and 195% ± 20% in the reperfusion nitrogen treatment ($p < 0.01$).

The differences between the control nitrogen and reperfusion nitrogen in both hemispheres were significant, both during the blood flow decrease ($p < 0.05$) and at the maximum value ($p < 0.05$). It was also found that blood flow decreased to lower values in the right hemisphere compared to the left, in both the control nitrogen ($p < 0.001$) and the reperfusion nitrogen ($p < 0.05$) treatments.

Reflectance and NADH values returned to baseline within 2 min of the nitrogen treatment. Blood flow in the left hemisphere returned to baseline within about 6 min, whereas in the right hemisphere, blood flow remained at a value 30% higher than its initial value ($p < 0.01$) and also higher compared to the left hemisphere ($p < 0.01$).

5.3.3 *The Effect of Ischemia Duration on Mitochondrial Function and Cerebral Blood Flow*

5.3.3.1 Introduction

This section is based on the article "Real-Time Monitoring of Spatial and Temporal Metabolic Changes during Focal Cerebral Ischemia in Rats," which was published in Brain Research in 2011 (Livnat et al. 2011). The article examined blood flow and mitochondrial function in the core and penumbra as observed in focal ischemia for 10, 30, and 60 min. Following MCA occlusion, a decrease in CBF levels and an increase in NADH levels were observed in all groups, with a return to baseline after blood flow was restored. These changes were more intense in the core compared to the penumbra. The intensity of the ischemic injury increased with the duration of ischemia, as demonstrated by a progressively smaller hyperemia wave and reflectance levels that remained high as the ischemia duration increased. Mitochondrial damage is manifested by the absence of a compensatory response during reperfusion, as observed in the fluorescence levels of the different groups. The results from nitrogen exposure during the reperfusion stage indicate that the tissue's ability to cope with oxygen deprivation decreases depending on the duration of the ischemia and the location of the injury.

5.3.3.2 The Effect of Ischemia Duration on the Brain's Response to Brief Anoxia

The following results supplement the findings presented in the article above. Figure 5.23 shows a group summary of the brain's response to control nitrogen (C) compared to nitrogen exposure during the reperfusion phase (R) following a 10-min right MCA occlusion. In both measurement sites, a decrease in reflectance to 76% ± 3% ($p < 0.001$) was recorded during control nitrogen exposure and 69% ± 4% during reperfusion nitrogen exposure ($p < 0.001$). NADH levels increased in the penumbra to 150% ± 4% during control nitrogen exposure ($p < 0.001$) and to 152% ± 3% during reperfusion nitrogen exposure ($p < 0.001$). In the core, NADH increased to 143% ± 4% during control nitrogen exposure ($p < 0.001$) and to 145% ± 5% during reperfusion nitrogen exposure ($p < 0.001$). A significant difference ($p < 0.05$) was found in the maximum NADH values during reperfusion nitrogen treatment between the penumbra (152% ± 3%) and the core (145% ± 3%). Blood flow significantly decreased in the penumbra to 67% ± 6% during control nitrogen exposure ($p < 0.001$), while there was

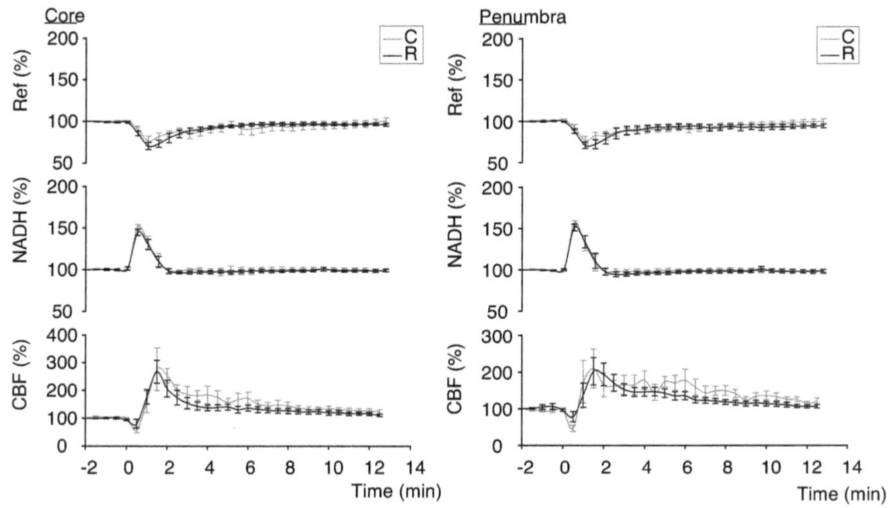

Fig. 5.23 Average pattern of response to nitrogen exposure, as observed in the right hemisphere at two measurement sites - core and penumbra. Two nitrogen treatments are shown: a few minutes before right MCA occlusion for 10 min (C), and 1 h after occlusion (R). Results are presented as mean ± SE ($n = 9$)

a nonsignificant decrease to 71% ± 15% during reperfusion nitrogen exposure. In the core, nonsignificant decreases in blood flow were recorded to a value of 84% ± 15% during control nitrogen exposure and 88% ± 15% during reperfusion nitrogen exposure. Following the decrease, an increase in blood flow was recorded, reaching 248% ± 19% during control nitrogen exposure ($p < 0.001$) and 303% ± 29% during reperfusion nitrogen exposure ($p < 0.001$) in the core, and 251% ± 49% during control nitrogen exposure ($p < 0.05$) and 243% ± 34% during reperfusion nitrogen exposure ($p < 0.05$) in the penumbra. A significant difference ($p < 0.05$) was found in the maximum blood flow values during reperfusion nitrogen treatment between the penumbra (303% ± 29%) and the core (243% ± 34%). No significant differences were found between the control nitrogen and reperfusion nitrogen for any of the parameters. Reflectance and NADH values returned to their baseline levels within about 2 min of the start of the treatment, while blood flow values returned to baseline after about 5 min.

Figure 5.24 shows a group summary of control nitrogen (C) compared to reperfusion nitrogen (R) following a 30-min right MCA occlusion. In the penumbra, reflectance decreased to a minimum of 71% ± 4% ($p < 0.001$) during control nitrogen exposure and 76% ± 2% during reperfusion nitrogen exposure ($p < 0.001$). In the core, reflectance decreased to a minimum of 72% ± 3% ($p < 0.001$) during control nitrogen exposure and 69% ± 3% during reperfusion nitrogen exposure ($p < 0.001$). NADH levels increased in the penumbra to 148% ± 6% during control nitrogen exposure ($p < 0.001$) and to 148% ± 4% during reperfusion nitrogen exposure ($p < 0.001$). In

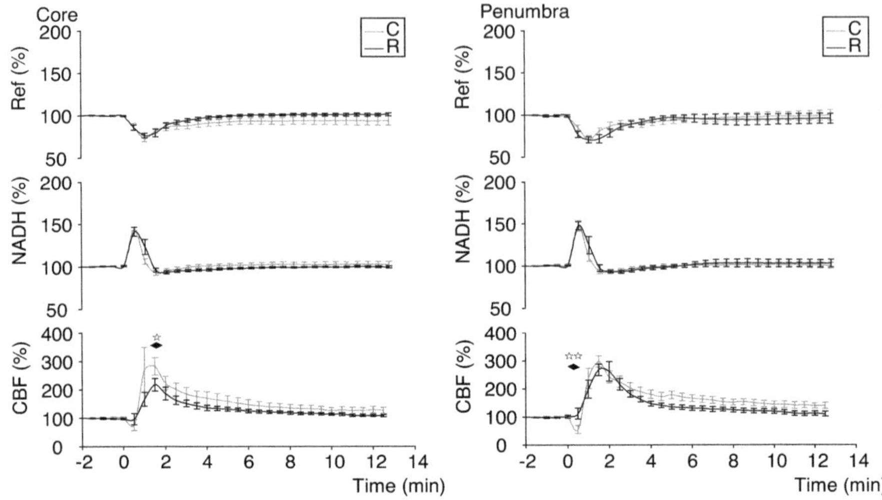

Fig. 5.24 Average pattern of response to nitrogen exposure, as observed in the right hemisphere at two measurement sites - core and penumbra. Two nitrogen treatments are shown: a few minutes before occlusion of the right MCA for 30 min (C), and 1 h after removal of the occlusion (R). Results are presented as mean ± SE ($n = 8$). Significant differences between two nitrogen treatments are indicated by asterisks, *$p < 0.05$; **$p < 0.01$

the core, a rise was recorded to 145% ± 2% during control nitrogen exposure ($p < 0.001$) and to 141% ± 5% during reperfusion nitrogen exposure ($p < 0.001$). Blood flow significantly decreased in the penumbra to a level of 40% ± 6% during control nitrogen exposure ($p < 0.001$), while there was a nonsignificant decrease to a level of only 93% ± 11% during reperfusion nitrogen exposure—a significant difference between the two nitrogen exposures ($p < 0.01$). In the core, significant decreases in blood flow were recorded to a value of 51% ± 9% during control nitrogen exposure ($p < 0.001$) and a nonsignificant decrease to a value of 78% ± 12% during reperfusion nitrogen exposure. Following the decrease, an increase in blood flow was recorded. In the penumbra, CBF values increased to 307% ± 22% during control nitrogen exposure ($p < 0.001$) and 277% ± 29% during reperfusion nitrogen exposure ($p < 0.001$). In the core, an increase to 297% ± 49% was recorded during control nitrogen exposure ($p < 0.001$) and 218% ± 23% during reperfusion nitrogen exposure—a significant difference between the two nitrogen exposures ($p < 0.05$). After several minutes, all parameters returned to baseline values.

Figure 5.25 shows a group summary of control nitrogen (C) compared to a brief nitrogen exposure approximately 1 h after blood flow was restored following a 60-min right MCA occlusion (R). In the penumbra, a decrease in reflectance to a minimum value of 81% ± 4% was recorded during control nitrogen exposure ($p < 0.001$) and 81% ± 5% during reperfusion nitrogen exposure ($p < 0.05$). In the core, a decrease to a minimum value of 79% ± 4% was recorded during control nitrogen exposure

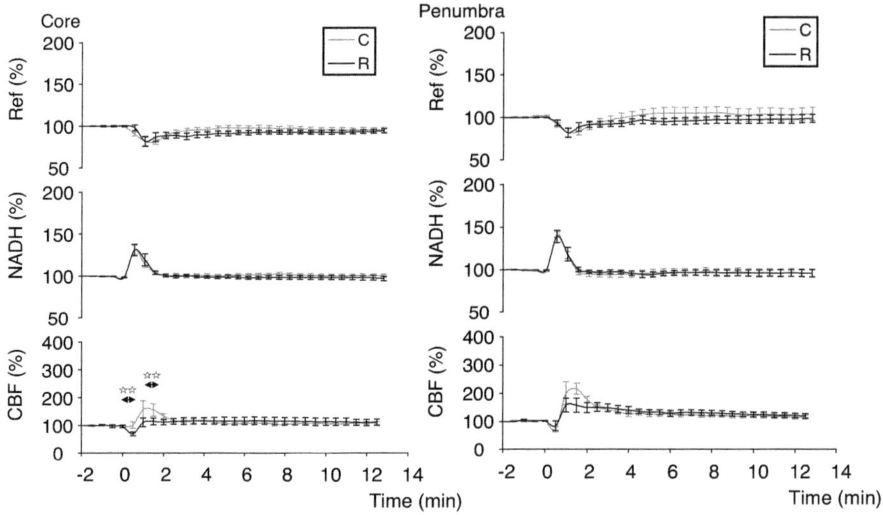

Fig. 5.25 Average pattern of response to nitrogen exposure, as observed in the right hemisphere at two measurement sites - core and penumbra. Two nitrogen treatments are shown: a few minutes before occlusion of the right MCA for 30 min (C), and 1 h after removal of the occlusion (R). Results are presented as mean ± SE ($n = 8$). Significant differences between two nitrogen treatments are indicated by asterisks, $^{*}p < 0.05$; $^{**}p < 0.01$.

($p < 0.001$) and 81% ± 5% during reperfusion nitrogen exposure ($p < 0.05$). NADH levels increased in the penumbra to 143% ± 3% during control nitrogen exposure ($p < 0.001$) and to 139% ± 7% during reperfusion nitrogen exposure ($p < 0.001$). In the core, an increase to 133% ± 5% was recorded during control nitrogen exposure ($p < 0.001$) and to 132% ± 7% during reperfusion nitrogen exposure ($p < 0.001$). Blood flow significantly decreased in the penumbra to 69% ± 7% during control nitrogen exposure ($p < 0.001$), while there was a nonsignificant decrease to 80% ± 13% during reperfusion nitrogen exposure. In the core, an inverse pattern was recorded: a nonsignificant decrease to 86% ± 12% during control nitrogen exposure, but a significant decrease to 66% ± 4% during reperfusion nitrogen exposure ($p < 0.001$). Following the decrease, an increase in blood flow was recorded. In the penumbra, CBF levels increased to 236% ± 27% during control nitrogen exposure ($p < 0.001$) and to 168% ± 27% during reperfusion nitrogen exposure ($p < 0.05$). In the core, CBF levels increased to 190% ± 56% during control nitrogen exposure ($p < 0.01$) and to only 117% ± 16% during reperfusion nitrogen exposure. The minimum blood flow in the core was significantly lower ($p < 0.01$) and the maximum blood flow was significantly higher ($p < 0.01$) during control nitrogen exposure relative to reperfusion nitrogen exposure. A similar trend was observed in the penumbra, but the differences were not significant. In addition, during reperfusion nitrogen exposure, a significant difference was observed between the two measurement sites in the maximum CBF level ($p < 0.05$). Blood flow returned to baseline values about 5 min after treatment in the core, but remained about 20% higher than its pretreatment value in the penumbra ($p < 0.05$). All parameters returned to their initial values a few minutes after the start of nitrogen exposure.

5.3.3.3 Comparison Between Different Monitoring Sites

Figure 5.26 compares two groups in which the MCA was occluded for 60 min. In one group, the measurement fibers were placed on the right and left hemispheres, and in the second group, two fibers were placed at two sites on the right hemisphere (core and penumbra). In both groups, the measurement fibers were placed 1 mm posterior to the bregma point, at different lateral positions. The location of the measurement fiber on the right hemisphere in the first group (referred to as "right") was between the two fibers in the second group. The results are presented as the mean only, without standard error, to facilitate the distinction between the different measurement areas and allow for a qualitative comparison. The CBF values measured in the "right" area decreased upon MCA occlusion to values similar to those observed in the core, but remained higher. After a few minutes, the CBF values in the "right" area slightly increased to a level close to the penumbra values. From that point until the ischemia was removed, a significant difference was observed between the "right" area and the core ($p < 0.01$). The reflectance and fluorescence of the "right" area were significantly different from those in the core, but not from the penumbra, starting about 5 min after the occlusion until its removal ($p < 0.01$). With the removal of the occlusion, the CBF values in the "right" area returned to their initial values, with no significant difference relative to the core or penumbra. No change was observed in the reflectance values in the "right" area at this stage, while the fluorescence values returned to baseline immediately and remained lower than the core and penumbra values. As a result, the NADH level in the "right" area decreased and remained low relative to the core and penumbra in the hour following ischemia. The differences in NADH levels between the "right" area and the penumbra were significant throughout that time ($p < 0.05$).

5.3.3.4 Spontaneous SD Waves During Focal Ischemia

Figure 5.27 shows the average pattern of spontaneous SD waves observed following MCA occlusion in the penumbra ($n = 8$), compared to control waves that were actively induced in a normoxic brain ($n = 8$). A rise in reflectance to a value of $160\% \pm 13\%$ was recorded. The increase in reflectance was significantly higher both relative to the initial level ($p < 0.001$) and relative to the increase observed in the control group ($118\% \pm 7\%, p < 0.05$). After the rise, a decrease to baseline was recorded. This pattern was significantly different ($p < 0.05$) from the control group, where a decrease to values lower than the initial level ($83\% \pm 1\%$) was recorded. In NADH, an increase to a value of $109\% \pm 2\%$ was recorded. This increase was not observed in the control group, where a decreasing trend in NADH was observed ($p < 0.01$). CBF levels were significantly lower relative to the control group ($p < 0.001$).

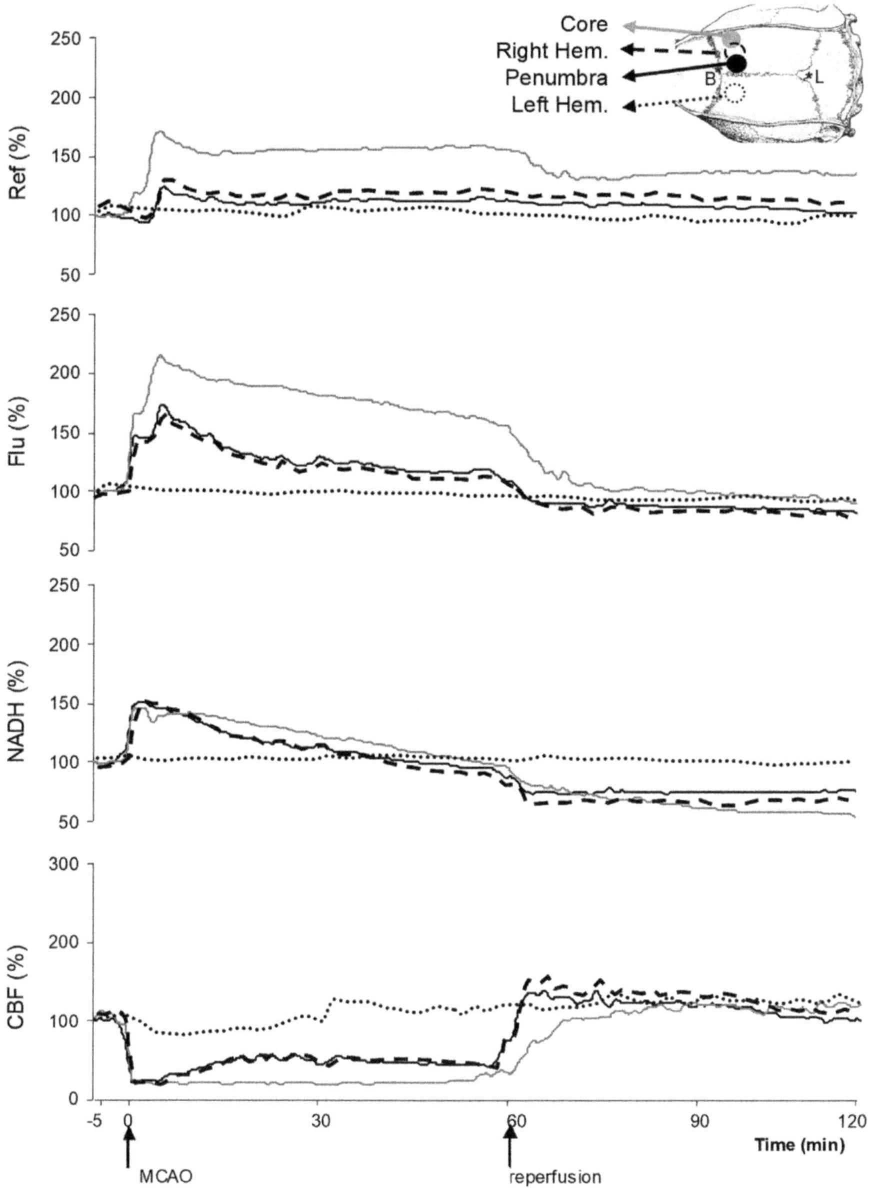

Fig. 5.26 Changes in parameters in response to right MCA occlusion for 1 h, followed by resumption of blood flow, as measured in two groups across four different points of the rat brain: in the left hemisphere (Left Hem., dotted line) and the right hemisphere (Right Hem., dashed line) in one group ($n = 8$), and at two measurement sites across the right hemisphere: core (solid gray line) and penumbra (solid black line) in the second group ($n = 8$). Results are presented as mean only, without standard error

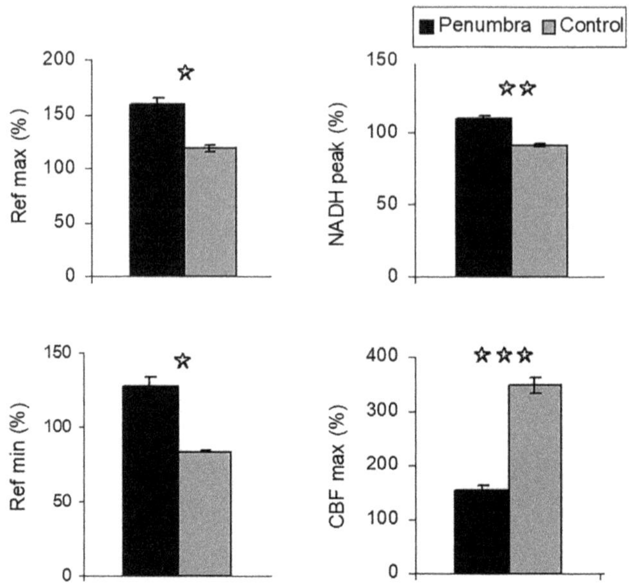

Fig. 5.27 Changes in parameters during spontaneously generated SD waves in the penumbra during MCAO (black), compared to a control group of SD waves induced in the normoxic brain (gray). Results are presented as mean ± SE ($n = 8$ in both groups). Significant differences between groups are indicated by asterisks, $^{*}p < 0.05$; $^{**}p < 0.01$; $^{***}p < 0.001$

5.3.3.5 Mitochondrial Function and Cerebral Blood Flow 24 h After Transient Focal Ischemia

In this part of the study, several tests were conducted to assess the long-term effect of focal ischemia. The animals in this part underwent MCA occlusion 24 h before monitoring its effects. In some animals, transient occlusion was performed, where the MCA was occluded for 60 min and then blood flow was restored. In other animals, permanent occlusion was performed, where the MCA was occluded without restoring blood flow.

Twenty-four hours after a 60-min transient MCA occlusion, the animals ($n = 8$) underwent a neurological evaluation to test their stability. When the ischemic animals were held by their tails, they extended their limbs with a tendency to the left side, which is contralateral to the injury. In a group of animals that did not undergo occlusion ($n = 8$), which served as a control for this part of the experiment, a response of forward limb extension was observed at this stage, without any leaning to a side. During the light pressure resistance test, which involved applying pressure to the right or left shoulder, the ischemic animals leaned to the left side, whereas the control animals showed stable posture when pressure was applied. When moving on a surface, the ischemic animals moved in a circle toward the affected side, while the control animals moved in a straight line.

Table 5.1 shows the minimum and maximum values for reflectance, NADH, and CBF during a brief nitrogen exposure performed 24 h after a 60-min MCA occlusion. The results are presented in comparison to nitrogen exposure in the same animals before MCA occlusion, which served as a control.

In the core area, no significant changes were recorded in the reflectance level, NADH, or maximum blood flow. As shown in Fig. 5.28, the minimum reflectance values in the nitrogen exposure performed 24 h after MCA occlusion were higher than the control values in both the core ($p < 0.001$) and the penumbra ($p < 0.05$). In addition,

Table 5.1 Nitrogen exposure before MCAO (for 60 min) and 24 h after deblocking

	Before MCAO		24 h after MCAO	
	Penumbra	Core	Penumbra	Core
Ref_peak	79±0.4	83±0.43	89±0.66	95±0.36
NADH_max	128±0.45	129±0.48	120±1.39	105±0.31
CBF_min	63±4.16	77±2.64	79±0.73	55±3.37
CBF_max	234±4.81	243±3.81	163±2.52	116±1.22

Minimum and maximum values of reflectance, NADH, and CBF during nitrogen exposure, as measured in the core and penumbra. Nitrogen exposure was performed before MCA occlusion and 24 h after reperfusion ($n = 8$).

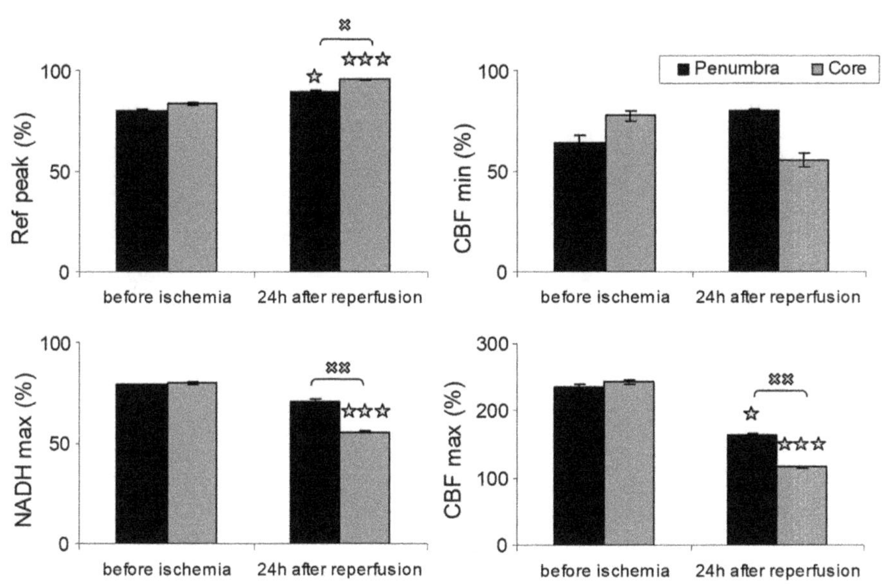

Fig. 5.28 Response to nitrogen exposure before MCA occlusion and 24h after MCA occlusion for 60 min, as measured in the core and penumbra. Results are presented as mean ± SE ($n = 8$). Significant differences between two nitrogen treatments are indicated by asterisks, $^{*}p < 0.05$, $^{***}p < 0.001$. Significant differences between the two measurement areas are indicated by x-marks, $^{x}p < 0.05$, $^{xx}p < 0.01$

the reflectance values in the core at this stage were higher relative to the penumbra ($p < 0.05$). The maximum NADH level in the core 24 h after occlusion was significantly lower both relative to the baseline group ($p < 0.001$) and relative to the penumbra at this stage ($p < 0.01$). The maximum blood flow was lower than the control values in both the core ($p < 0.001$) and the penumbra ($p < 0.05$). In addition, the blood flow values in the core at this stage were lower relative to the penumbra ($p < 0.01$).

Table 5.2 shows the maximum and minimum values for reflectance, NADH, and CBF as measured in SD waves that were induced 24 h after a 60 min MCA occlusion, in comparison to waves induced in normoxic (control) animals. In the core, no significant changes in the SD waves were recorded relative to the initial level. According to Fig. 5.29, it can

Table 5.2 SD waves 24 h after MCAO for 60 min compared to control group

	Control		24 h after MCAO	
	Penumbra	Core	Penumbra	Core
Ref_max	119 ± 2.37	108 ± 0.74	103 ± 1.48	101 ± 0.13
Ref_min	77 ± 0.48	85 ± 1.11	90 ± 0.39	97 ± 0.14
NADH_peak	87 ± 0.81	86±	108 ± 0.15	101 ± 0.11
CBF_max	252 ± 10.71	222 ± 11.87	161 ± 4.33	105 ± 0.59

Minimum and maximum values of reflectance, NADH, and CBF during induced SD waves, as measured in the core and penumbra. SD waves were induced 24 h after MCAO for 60 min ($n = 8$) and in animals that did not undergo MCAO (control, $n = 8$).

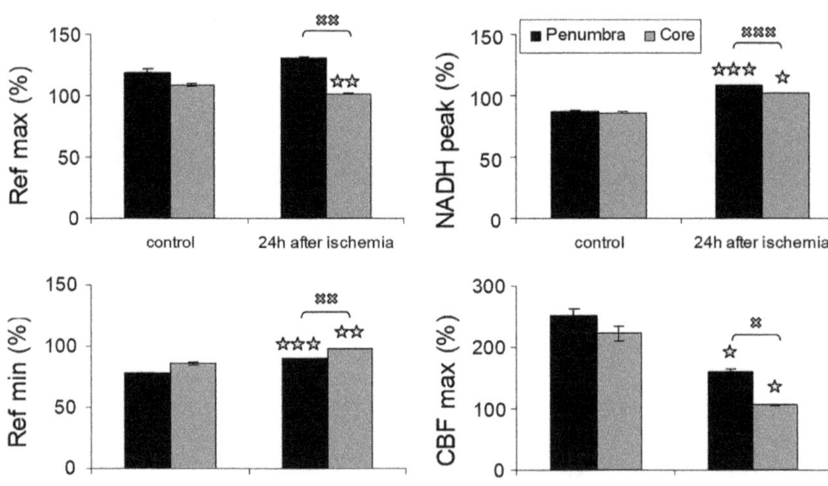

Fig. 5.29 Induced SD waves - measured in the core and penumbra 24 h after 60 min MCA occlusion ($n = 8$) compared to a control group ($n = 8$). Results are presented as mean ± SE. Significant differences between groups are indicated by asterisks, $*p < 0.05$, $**p < 0.01$, $***p < 0.001$. Significant differences between the two measurement areas are indicated by x-marks, $^{x}p < 0.05$, $^{xx}p < 0.01$, $^{xxx}p < 0.01$

be seen that in the waves induced 24 h after the occlusion was removed, lower maximum reflectance values were observed in the core both relative to the control group ($p < 0.01$) and relative to the penumbra ($p < 0.01$). The minimum reflectance level was higher in the core ($p < 0.01$) and the penumbra ($p < 0.001$) relative to the control group. Also, at this stage, the minimum reflectance level in the core was higher relative to the penumbra ($p < 0.01$). In the NADH level, higher maximum values were recorded relative to the control group in both the core ($p < 0.01$) and the penumbra ($p < 0.001$). Also, at this stage, lower NADH values were recorded in the core relative to the penumbra ($p < 0.001$). The maximum blood flow at both measurement sites was lower relative to the control group ($p < 0.05$), with significant differences between the sites ($p < 0.05$).

5.3.3.6 Mitochondrial Function and Cerebral Blood Flow 24 h After Permanent Focal Ischemia

In this section, animals ($n = 8$) were monitored 24 h after MCA occlusion without its removal, so the MCA remained occluded during the monitoring. A neurological evaluation of the animals' stability showed that there were differences between them and the control animals that did not undergo MCA occlusion ($n = 8$). These differences included leaning to the left when held by the tail or when light pressure was applied, and moving in a circle when placed on a surface, similar to the differences described following transient MCA occlusion.

Table 5.3 shows the minimum and maximum values for reflectance, NADH, and CBF during a brief nitrogen exposure performed 24 h after permanent MCA occlusion. The results are presented in comparison to nitrogen exposure in the same animals before MCA occlusion, which served as a control. With the exception of minimum CBF values, no changes were observed in the measured parameters relative to the control group, in both the core and the penumbra. As shown in Fig. 5.30, a decreasing trend in reflectance was observed before the occlusion, while 24 h after the occlusion was removed, the reflectance values remained almost unchanged, which led to differences between the nitrogen exposures in both the core ($p < 0.01$) and the penumbra ($p < 0.001$). The maximum NADH level after the occlusion was significantly lower

Table 5.3 Nitrogen exposure before MCAO and 24 h after permanent occlusion

	Before MCAO		24 h after MCAO	
	Penumbra	Core	Penumbra	Core
Ref_peak	79 ± 0.55	82 ± 0.89	101 ± 0.22	102 ± 0.75
NADH_max	129 ± 0.47	129 ± 0.96	104 ± 0.43	104 ± 0.36
CBF_min	61 ± 7.78	80 ± 5.31	59 ± 5.24	55 ± 6.75
CBF_max	237 ± 10.01	228 ± 6.66	109 ± 1.1	107 ± 0.43

Minimum and maximum values of reflectance, NADH, and CBF during nitrogen exposure, as measured in the core and penumbra. Nitrogen exposure was performed before MCA occlusion and 24 h after the onset of occlusion ($n = 8$).

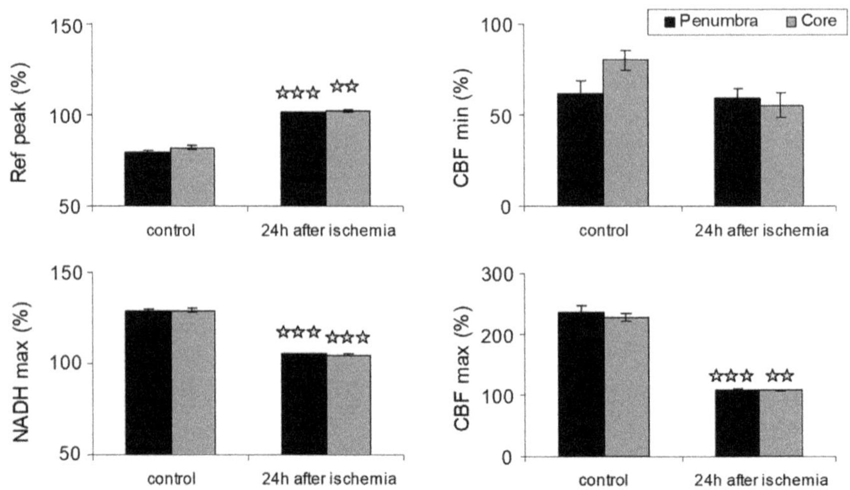

Fig. 5.30 Response to nitrogen exposure before MCA occlusion and 24 h after permanent occlusion, as measured in the core and penumbra. Results are presented as mean ± SE ($n = 8$). Significant differences between two nitrogen treatments are indicated by asterisks, $^{**}p < 0.01$, $^{***}p < 0.001$

Table 5.4 SD waves 24 h after permanent MCAO and in the control group

	Control		24 h after MCAO	
	Penumbra	Core	Penumbra	Core
Ref_max	119 ± 2.37	108 ± 0.74	101 ± 0.17	101 ± 0.26
Ref_min	77 ± 0.48	85 ± 1.11	98 ± 0.16	97 ± 0.26
NADH_peak	87 ± 0.81	86 ± 0.75	101 ± 0.1	102 ± 0.2
CBF_max	252 ± 10.7	222 ± 11.86	103 ± 0.31	104 ± 0.41

Minimum and maximum values of reflectance, NADH, and CBF during induced SD waves, as measured in the core and penumbra. SD waves were induced after MCA occlusion for 24 h ($n = 8$) and in animals that did not undergo MCA occlusion (control, $n = 8$).

relative to the control group in both measurement sites ($p < 0.001$). The maximum blood flow level after the occlusion was lower in both the core ($p < 0.01$) and the penumbra ($p < 0.001$).

Table 5.4 shows the maximum and minimum values for reflectance, NADH, and CBF as measured in SD waves induced 24 h after permanent MCA occlusion, in comparison to waves induced in normoxic (control) animals. In both measurement sites, no changes in the parameters were observed 24 h after occlusion relative to the baseline values measured before the wave induction (Fig. 5.31). Consequently, lower minimum reflectance values ($p < 0.01$), higher minimum NADH levels ($p < 0.05$),

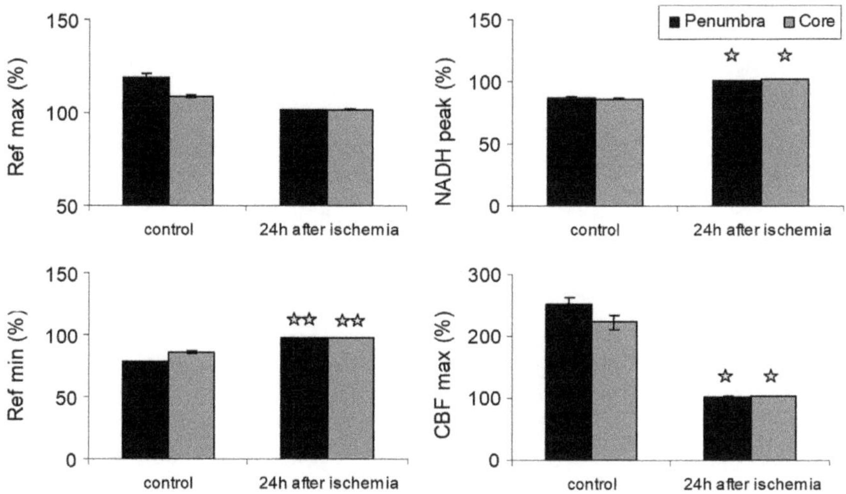

Fig. 5.31 Induced SD waves—measured in core and penumbra 24 h after permanent MCA occlusion ($n = 8$) versus control group ($n = 8$). Results are presented as mean ± SE. Significant differences between waves before and after occlusion are indicated by asterisks, $*p < 0.05$, $**p < 0.01$

and lower maximum NADH levels ($p < 0.05$) were recorded in the waves induced after the occlusion relative to the control waves.

5.3.4 Variability in the Brain's Response to MCAO

5.3.4.1 Introduction

This section is based on the article "Mitochondrial Function and Cerebral Blood Flow Variable Responses to Middle Cerebral Artery Occlusion," published in the *Journal of Neuroscience Methods* in 2010 (Livnat et al. 2010). The article discusses the variety of possible outcomes when performing the MCAO model. In the article, a uniform MCAO model was performed on 96 rats, in which reflectance, fluorescence, NADH, and CBF levels were monitored. The research results show that 58% of the animals exhibited the response pattern known in the literature and presented in this work: a decrease in blood flow and an increase in NADH levels at the beginning of the occlusion, and a return to baseline values after its removal. In contrast, in 42% of the experimental animals, six different response patterns were observed: (1) ischemia without reperfusion, (2) death following blood flow restoration, (3) low impact on the measured parameters following MCA occlusion, (4) increase in blood flow following MCA occlusion, (5) early spontaneous reperfusion, and (6) late reperfusion (Figs. 5.32–5.36).

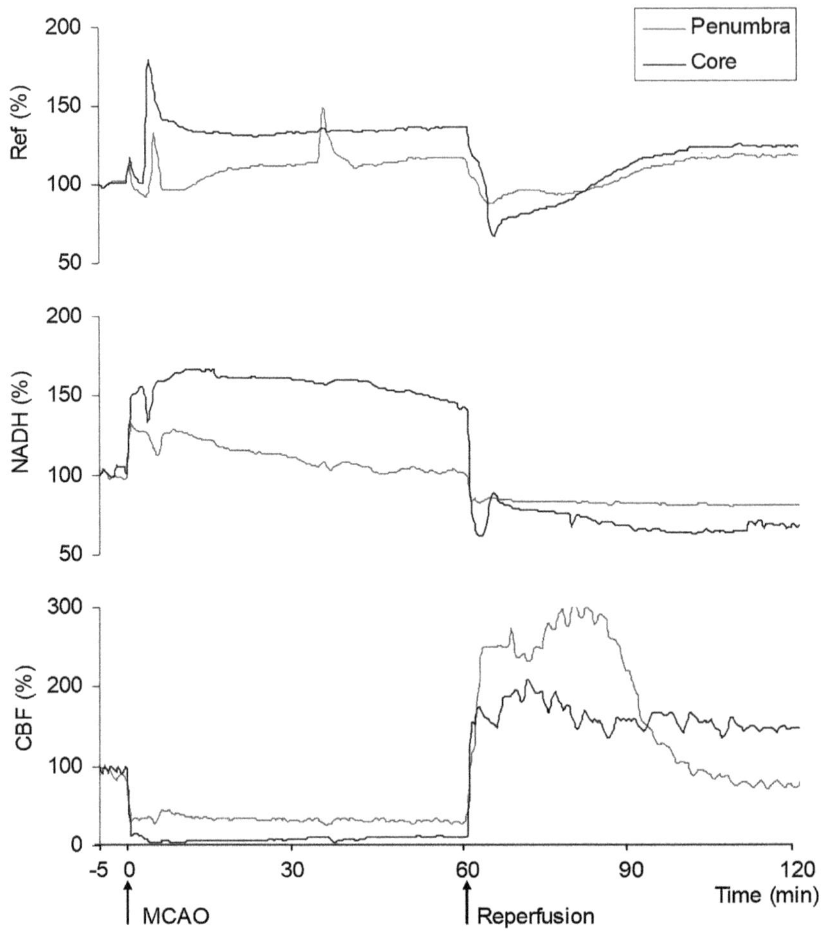

Fig. 5.32 Responses of reflectance (Ref), NADH (NADH) and CBF monitored in the core (black) and penumbra (gray) of the ischemic region, to 60 min of transient MCAO, followed by 60 min of reperfusion (P1)

5.3.4.2 Increase in Blood Flow in the Penumbra Following MCAO

Figure 5.37 shows a group average of one of the patterns discussed in the article cited above—animals in which blood flow in the penumbra increased immediately after MCA occlusion. In the core of this group, a significant decrease in CBF was recorded relative to the baseline level, to a value of 23 ± 3% ($p < 0.001$). Subsequently, blood flow stabilized at a value of around 35%. In contrast, in the penumbra, a significant increase in CBF was recorded immediately after occlusion ($p < 0.01$), to a value of 197 ± 25%. After about 8 min, a further increase was recorded, to a value of 220 ± 32%. The difference between the measurement areas was significant for the first 10 min from the moment of occlusion ($p < 0.001$). At the same time, there was a significant

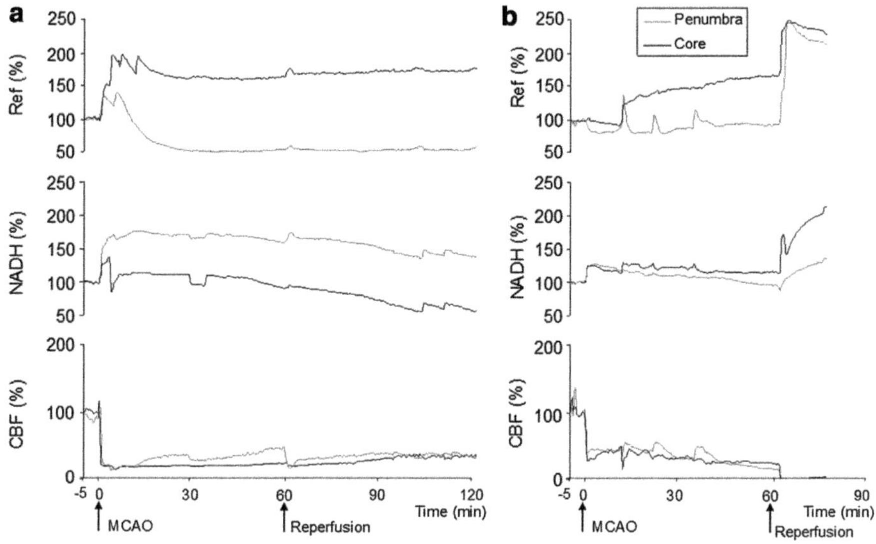

Fig. 5.33 Responses of reflectance (Ref), NADH (NADH) and CBF monitored in the core (black) and penumbra (gray) in P2 (**a**) and P3 (**b**)

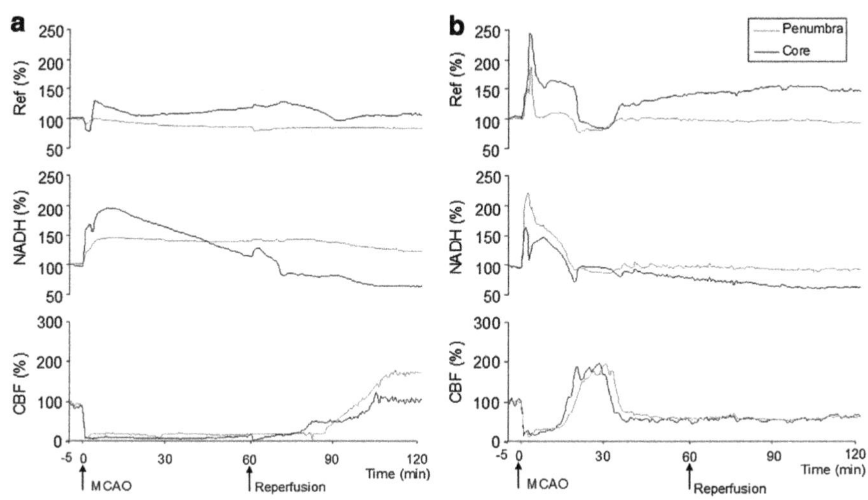

Fig. 5.34 Responses of reflectance (Ref), NADH (NADH) and CBF monitored in the core (black) and penumbra (gray) in P4 (**a**) and P5 (**b**)

decrease in reflectance relative to the baseline, to a value of $89 \pm 3\%$ in the penumbra ($p < 0.05$) and $78 \pm 1\%$ in the core ($p < 0.001$). The difference between the regions was significant in the first 5 min from the moment of occlusion ($p < 0.05$), after which a gradual recovery in the parameters began. Fluorescence showed a (nonsignificant)

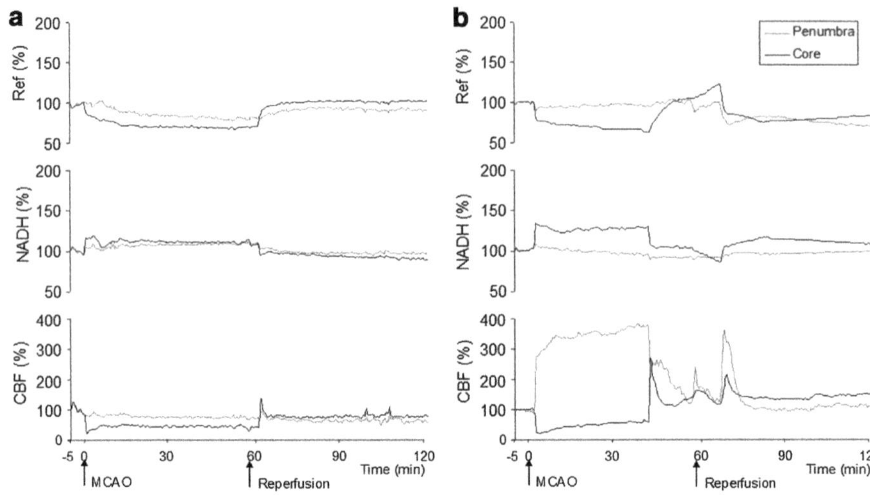

Fig. 5.35 Responses of reflectance (Ref), NADH (NADH) and CBF monitored in the core (black) and penumbra (gray) in P6 (**a**) and P7 (**b**)

upward trend to a value of 112 ± 6% in the penumbra, compared to a significant increase relative to the baseline to a value of 125 ± 5% in the core ($p < 0.05$). After a few minutes, a decrease was recorded in both measurement regions, with a return to baseline values observed in the penumbra and fluorescence remaining at a level of 112 ± 5% in the core ($p < 0.05$). The changes between the regions were found to be significant during the first 3 min of occlusion ($p < 0.05$). The NADH level increased significantly in both measurement regions compared to the baseline, reaching a level of 132 ± 3% in the core ($p < 0.001$) compared to 114 ± 4% in the penumbra ($p < 0.05$). These values were maintained for at least 10 min from the onset of ischemia. The differences between the measurement areas were significant for at least 10 min from the onset of ischemia ($p < 0.01$).

NADH values in the core were significantly lower than those observed in the core in cases where the pattern of results known from the literature was observed ($p < 0.05$). No significant differences were found between the penumbra in these animals and the core in the current group. The NADH level in the current group showed a trend of an intermediate pattern, being lower than that measured in the core and higher than that measured in the penumbra in the expected pattern of results, but without significant differences relative to these two areas (Table 5.5).

5.3.4.3 Variability in the Intensity of Mitochondrial Damage Following MCAO

In addition to the patterns reviewed in the article, differences were also observed in the intensity of change in fluorescence and NADH levels. Figure 5.38 presents a comparison between two identical protocols, which included a 60 min MCA occlusion, performed nearly 3 years apart—one in an early stage of the study (February–March 2007) and

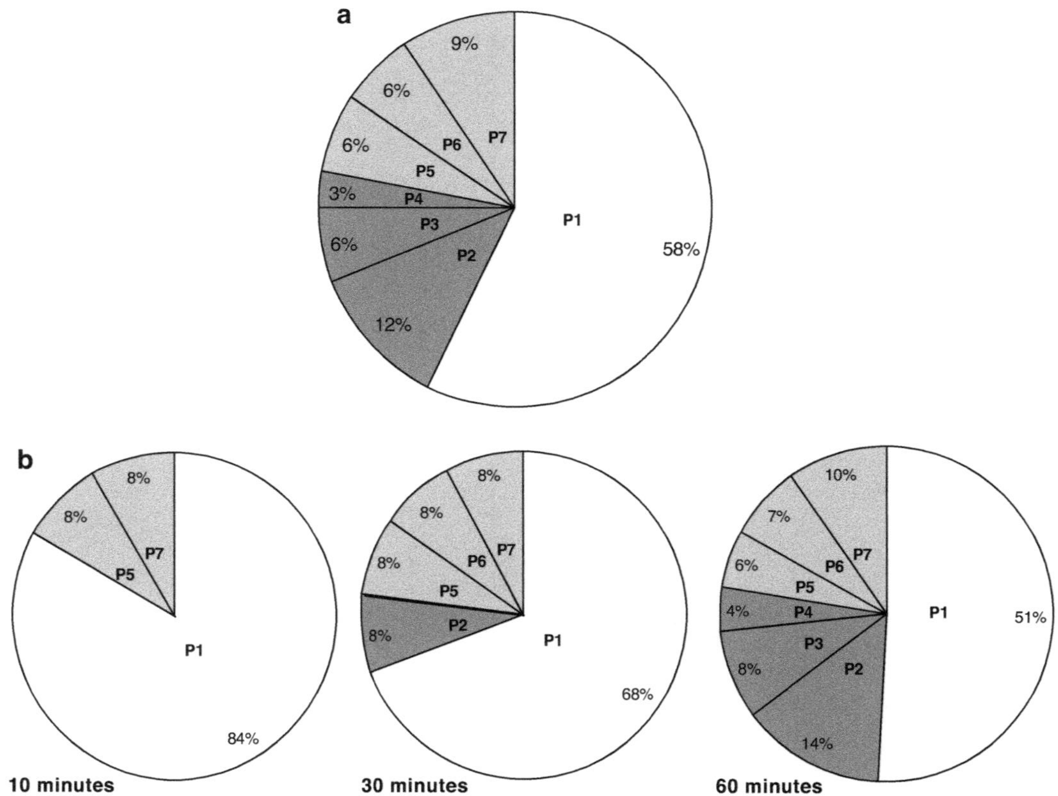

Fig. 5.36 **(a)** Distribution of all 96 experiments according to profiles P1–P7. **(b)** Distribution according to ischemic durations (10, 30, and 60 min). P2–P4 represent responses which occurred following the removal of the ligating filament (dark gray). P5–P7 represent responses which are related to the onset of the occlusion (bright gray)

one in a later stage (January 2010). This comparison shows significant differences between the groups in the fluorescence ($p < 0.01$) and NADH ($p < 0.05$) levels during ischemia in the core area. In the penumbra area, no significant changes were observed. Additionally, there were no significant differences between the groups during nitrogen

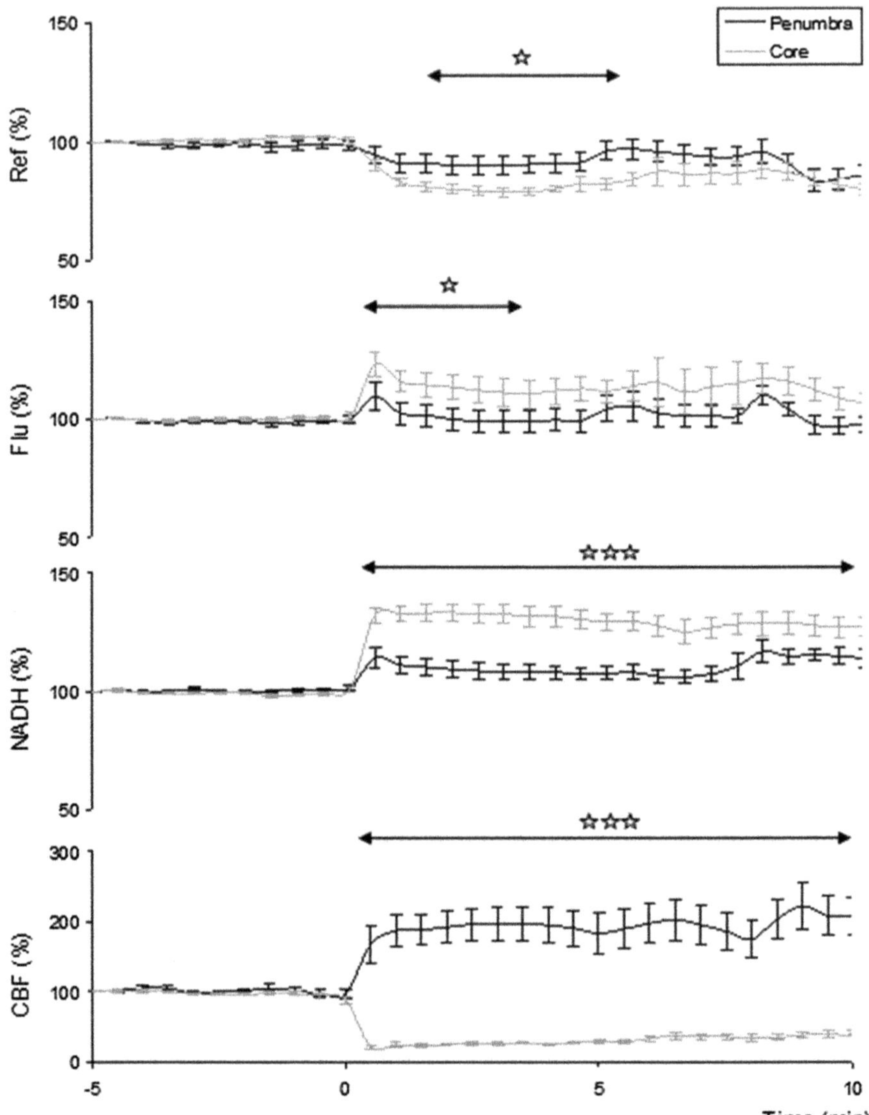

Fig. 5.37 Reflectance, fluorescence, NADH and CBF patterns, as measured in the core and penumbra in the first 10 min after MCAO. Results are presented as mean ± SE ($n = 6$). Significant changes at measurement sites are indicated by asterisks, *$p < 0.05$, ***$p < 0.001$

Table 5.5 Experimental profiles of responses following middle cerebral artery occlusion

Experimental profiles of responses following middle cerebral artery occlusion.

#	Profile	Ischemia		Reperfusion	
P1	Ischemia followed by reperfusion	↓↑			↑↓
P2	Ischemia without reperfusion	↓↑			==
P3	Death following reperfusion	↓↑			=↑
P4	Late reperfusion	↓↑			=↑=↓
P5	Spontaneous early reperfusion	↓↑↑↓			==
P6	Minor CBF responses in penumbra	Core↓↑	Penumbra==	Core↑↓	Penumbra==
P7	CBF elevation in the penumbra	Core↓↑	Penumbra↑=	Core↑↓	Penumbra↓=

Responses of CBF (black) and mitochondrial NADH (grey) to middle cerebral artery occlusion in seven experimental profiles. When needed, further discrimination between core and penumbra was added. ↑-increase; ↓-decrease; =-no change.

exposures—neither in the anoxia treatments performed before ischemia nor in the treatments performed 1 h after the reperfusion stage (this information is not presented).

This protocol of 60 min MCA occlusion was planned to serve as a control group for the subsequent models in the study. Following the differences in results between the two groups described above, it was decided that for the first part of the work, which has been reviewed so far, the first group would be used as a control group, and the updated group—which was worked on close to the time the protocols to be discussed from now on were performed—would serve as the control for the rest of the work (Figs. 5.39–5.42).

5.3.5 *The Relationship Between Focal and Global Ischemia*

5.3.5.1 Introduction

At the beginning of this section, the article "Mitochondrial Function and Cerebral Blood Flow Responses under Unilateral Carotid Occlusion in Rats" is presented, which was published as a proceedings article of SPIE (the International Society for Optics and Photonics) in (2009). This article compares the effects of a single carotid occlusion (UCO) to a bilateral carotid occlusion (BCO) on reflectance, fluorescence, NADH, and CBF levels, as observed during nitrogen exposure and induced SD waves up to 24 h after the occlusion. Both UCO and BCO led to a temporary decrease in CBF and a temporary increase in NADH. The hyperemic response in SD waves induced 24 h after a single carotid occlusion was lower than the control group in the ipsilateral hemisphere to the occlusion and higher than the control group in the contralateral hemisphere. The hyperemic response in SD waves induced 24 h after bilateral carotid occlusion was lower in both hemispheres relative to the control group, but was the same during nitrogen exposure.

5.3.5.2 Unilateral vs. Bilateral Carotid Artery Occlusion

Figure 5.43 shows a comparison between occlusion of one carotid artery versus two carotid arteries. The changes following unilateral occlusion of a carotid artery were significantly smaller compared to the changes in the same parameters following bilateral occlusion. The maximal level of fluorescence following single carotid occlusion was $108 \pm 3\%$, compared with up to $125 \pm 5\%$ in the group where both carotid arteries were occluded ($p < 0.001$). The maximal increase in NADH during unilateral

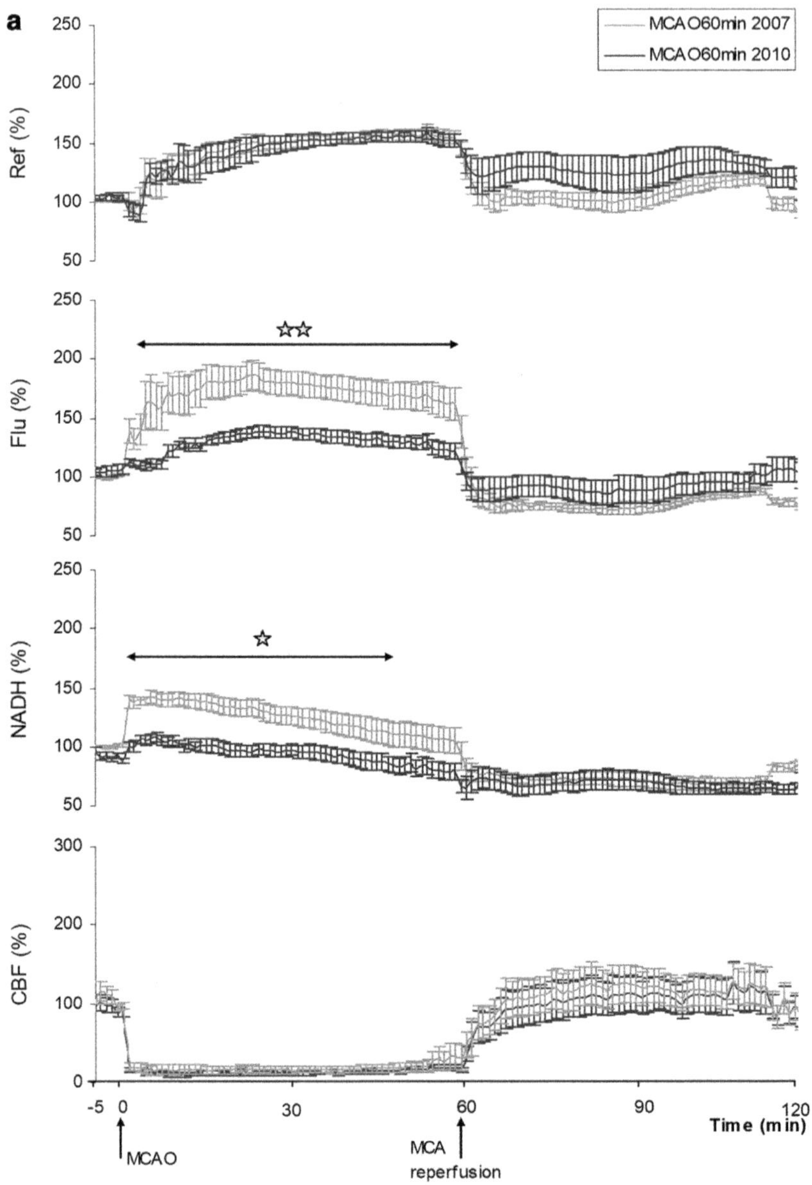

Fig. 5.38 Comparison of two identical protocols, in which the MCA was blocked for 60 min and then blood flow was restored. The protocols were performed 3 years apart (2007, $n = 8$ and 2010, $n = 8$). Measurements were made in two areas: the core (**a**) and the penumbra (**b**). Results are presented as mean ± SE. Significant differences between groups are indicated by asterisks, $*p < 0.05$, $**p < 0.01$

occlusion was 110 ± 2%, whereas during bilateral occlusion it reached 125 ± 4% ($p < 0.001$). A significant difference ($p < 0.01$) was also observed in the extent of blood flow reduction: 72 ± 8% in single carotid occlusion versus 24 ± 4% in bilateral carotid occlusion. In addition, the tissue recovery time following occlusion (UNI, black, $n = 10$) was less than 2 min after a single carotid occlusion, compared with 9 min after occlusion of both carotid arteries.

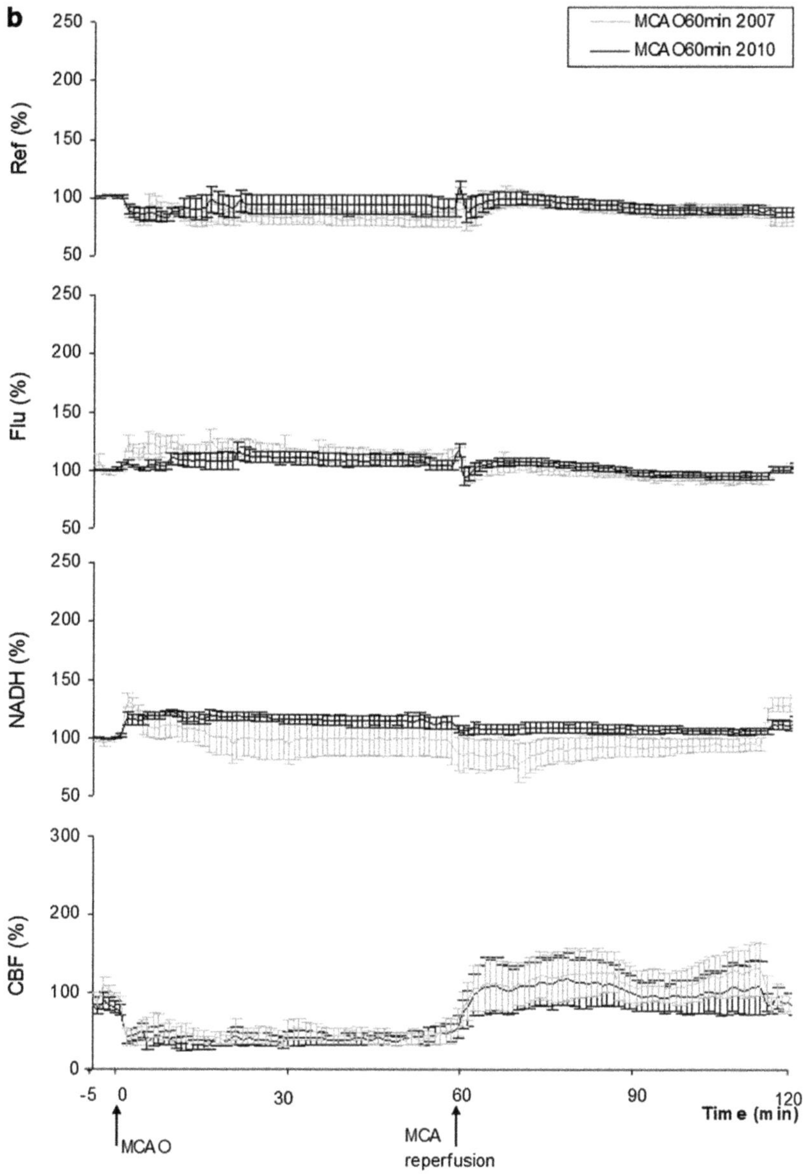

Fig. 5.38 (continued)

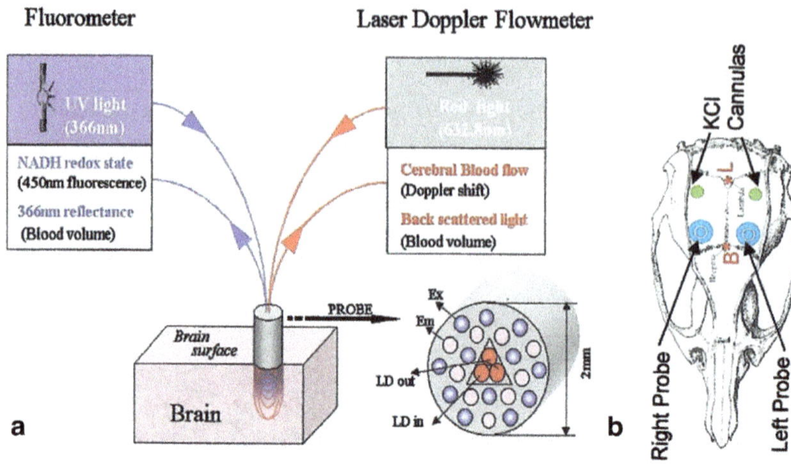

Fig. 5.39 Schematic representation of a single probe of the MSMP monitoring system (**a**), and the monitoring locations on the cerebral surface (**b**). The measuring probe is connected to an LDF device for CBF monitoring and to a photomultiplier for NADH measurement. Ex, Em, excitation and emission light for NADH fluorometry; LD_{in}, LD_{out}, fibers for LDF monitoring; B Bregma, L Lambda

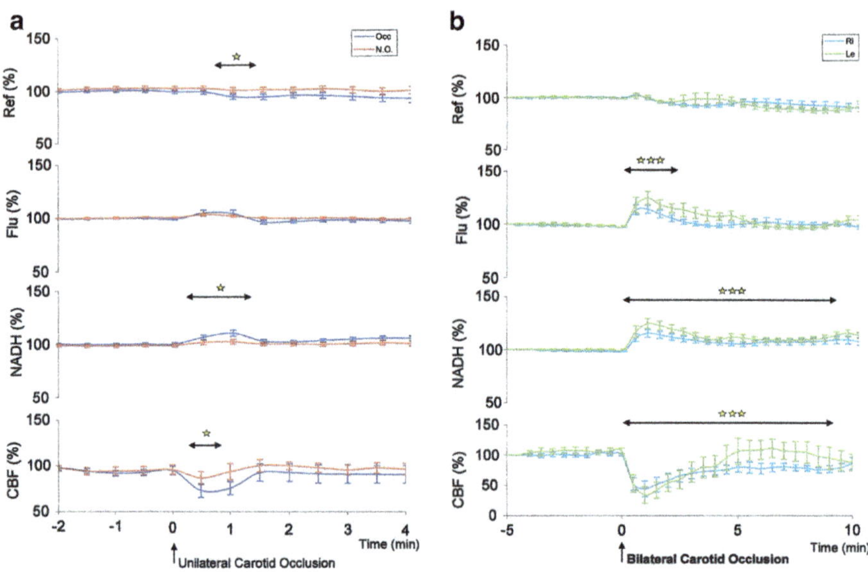

Fig. 5.40 The effects of UCO (**a**) and BCO (**b**) on the reflectance (Ref). NADH and CBF levels, as measured in both hemispheres the hemisphere ipsilateral to the occluded carotid (O_{cc}) and in the contralateral nonoccluded hemisphere (N.O.) in the UCO model, and in the left (L) and right (R) hemispheres in the BCO model. Significant differences between the two hemispheres are marked by asterisks, *$p < 0.05$

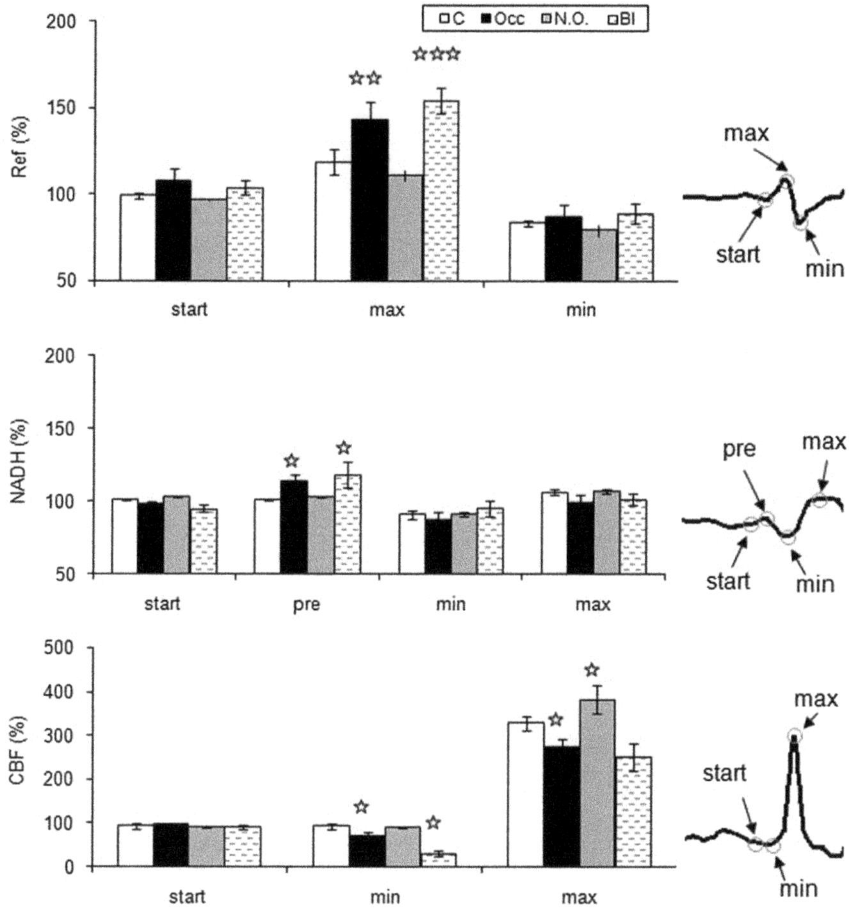

Fig. 5.41 The effects of an induced SD wave 24 h after different types of ischemia. The graphs represent the hemisphere on the side of the occluded carotid (O_{cc}) and the nonoccluded hemisphere (N.O.) in the UCO model, one hemisphere in the BCO model (BI) and a control wave (C). A sketch of each parameter change in a single SD wave is shown on the right. Significant differences in comparison to the control group in each phase are marked by asterisks, $*p < 0.05$, $**p < 0.01$, $***p < 0.001$

5.3.5.3 Induced Spreading Depression Waves Before and During Unilateral Carotid Artery Occlusion

Table 5.6 presents the minimum and maximum values for reflectance, NADH, and CBF, as observed in SD waves induced before, immediately after, and 1 h after unilateral carotid occlusion (UCO). The results presented represent the parameter values as measured in the first wave induced in each set of waves. Figure 5.44 presents results

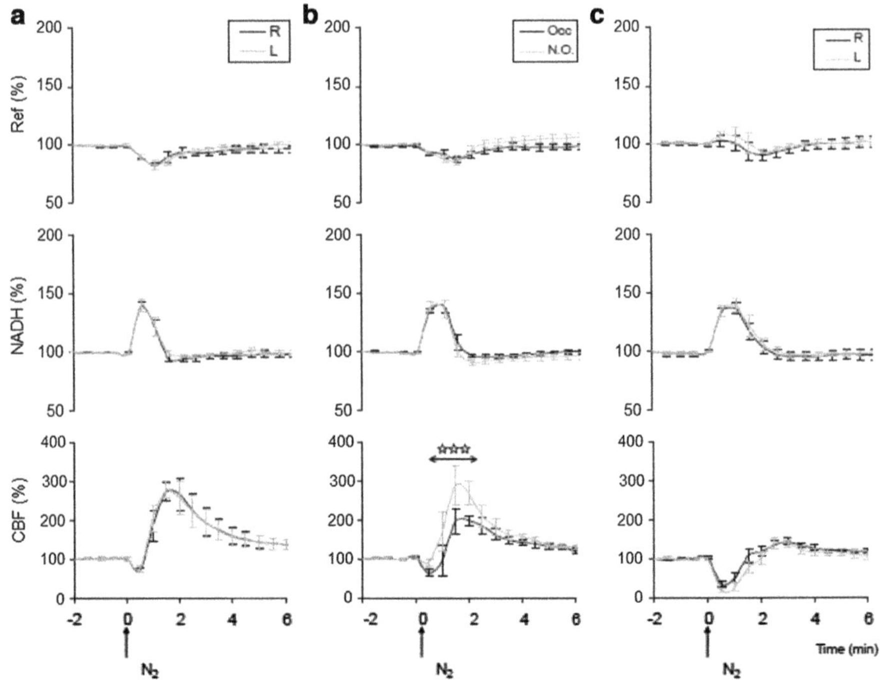

Fig. 5.42 The effects of Short Anoxia before (**a**) and 24 h after UCO (**b**) and BCO (**c**). R right hemisphere; L left hemisphere; O_{cc} the hemisphere on the side of the occluded carotid following UCO; N.O. the nonoccluded hemisphere. Significant differences between the two hemispheres are marked by asterisks, $^{***}p < 0.001$

from a single experiment in which SD waves were induced in a normoxic brain. During the passage of SD waves in normoxic tissue before carotid artery occlusion (control), a pattern of an increase followed by a decrease in reflectance was recorded ($p < 0.001$ relative to the start of measurement), a decrease in NADH levels ($p < 0.001$), and an increase in CBF levels ($p < 0.001$). No significant differences were found between the two hemispheres throughout the measurement. The waves induced 1 min and 1 h after the carotid artery occlusion were not significantly different from the control waves, and no differences between the hemispheres were recorded in them either.

Table 5.7 shows the minimum and maximum values for reflectance, NADH, and CBF as observed in SD waves induced 24 h after unilateral carotid occlusion. A pattern of an increase and then a decrease was recorded in reflectance, both significant relative to the starting point ($p < 0.001$). The maximum values of the reflectance wave were significantly higher in the hemisphere on the side of the occluded carotid compared to the opposite hemisphere ($p < 0.01$). In addition, the maximum reflectance values in the hemisphere ipsilateral to the occlusion were significantly higher relative to waves in a normoxic brain ($p < 0.05$), and also relative to the waves induced 1 min after the occlusion.

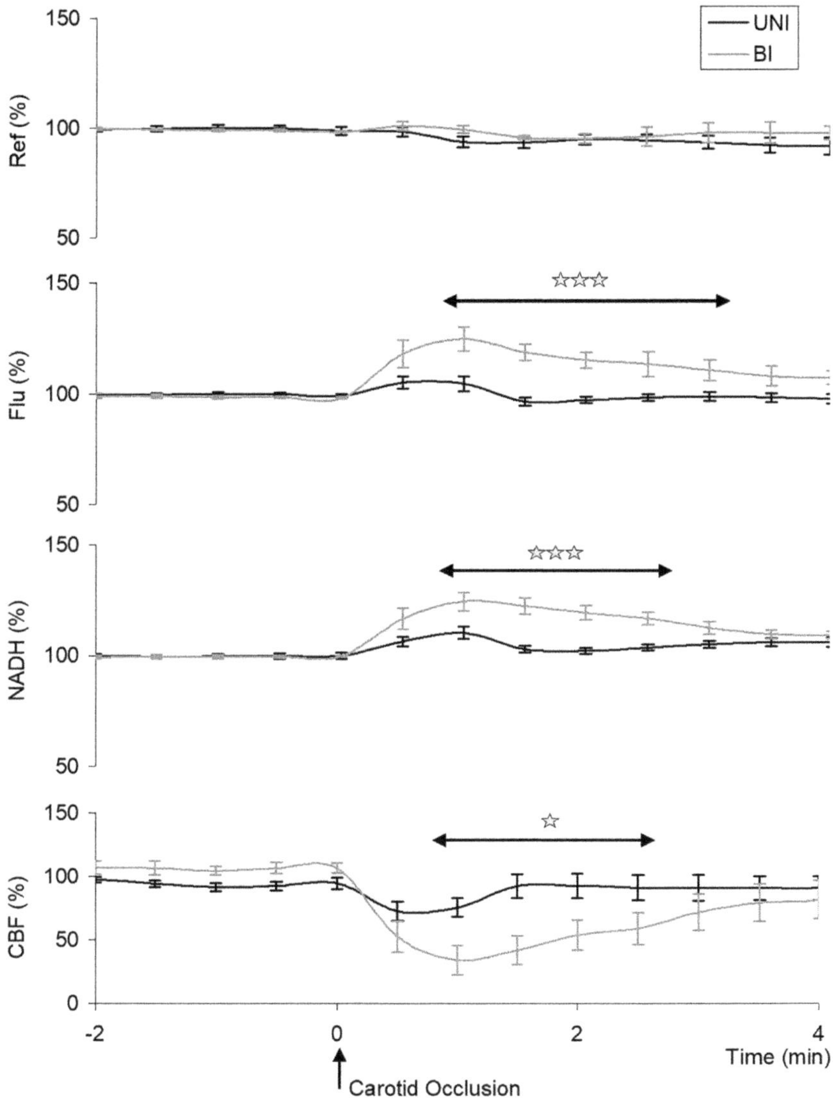

Fig. 5.43 Effect of unilateral occlusion of a single carotid artery (UNI, black, $n = 10$), compared with bilateral occlusion of both carotid arteries (BI, gray, $n = 8$) on hemodynamic and metabolic parameters in the rat brain. Significant differences between the two groups are indicated by asterisks, $^{*}p < 0.05$; $^{***}p < 0.001$

A pattern of oxidation followed by reduction was observed in NADH in both hemispheres. A premature reduction was observed in the hemisphere on the ipsilateral side of the occlusion only. In this hemisphere, an increase of 15% $\pm$4% was observed, which was significant both relative to the starting point ($p < 0.05$) and relative to the

Table 5.6 SD waves induced before, 1 min after, and 1 h after UCO

	Before UCO (control)		1 min after UCO		1 h after UCO	
	Occluded	Nonoccluded	Occluded	Nonoccluded	Occluded	Nonoccluded
Ref_max	119±3.5	114±3.9	113±5.9	113±5.0	120±3.6	114±4.0
Ref_min	83±2.5	83±3.1	75±3.1	81±2.0	84±2.0	84±3.0
NADH_ peak	88±1.5	89±1.7	94±4.3	89±2.9	89±1.5	89±1.8
CBF_max	328±39.1	299±33.3	328±58.4	314±28.6	328±39.2	291±33.4

Minimum and maximum values of reflectance, NADH, and CBF during an induced SD wave, as measured in the cortex of the hemisphere ipsilateral to the occlusion (occluded) and contralateral to the occlusion (nonoccluded). Waves were recorded before, immediately after, and 1 h after single carotid artery occlusion ($n = 10$).

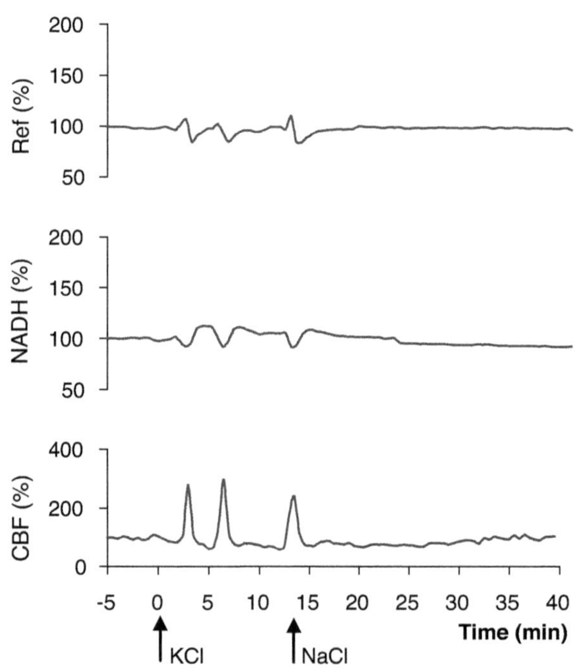

Fig. 5.44 Analog recording from a single experiment of SD waves in the normoxic brain. The waves were induced by rinsing the cortex with KCl solution and terminated by rinsing with NaCl

opposite hemisphere ($p < 0.05$). During the oxidation phase, a significant decrease relative to the baseline was observed in both hemispheres ($p < 0.01$). A decrease in blood flow was observed only in the hemisphere on the occluded side. The decrease was significant relative to the baseline level ($p < 0.05$) and relative to the opposite

Table 5.7 SD wave induced 24 h after UCO

	24 h after UCO	
	Occluded	Nonoccluded
Ref_max	143 ± 10.3	111 ± 3.0
Ref_min	87 ± 6.9	79 ± 3.1
NADH_peak	113 ± 4.4	103 ± 0.6
CBF_max	275 ± 17.6	383 ± 32.9

Minimum and maximum values of reflectance, NADH, and CBF during an induced SD wave, as measured in the cortex of the hemisphere ipsilateral to the occlusion (occluded) and contralateral to the occlusion (nonoccluded). Waves were induced 24 h after single carotid artery occlusion ($n = 10$).

hemisphere ($p < 0.05$). After the blood flow decrease, an increase was recorded, which was significant relative to the starting point in both hemispheres ($p < 0.001$).

The differences between the hemispheres at this stage were significant ($p < 0.05$). The maximum values in the hemisphere contralateral to the occlusion were significantly higher relative to the control group's values ($p < 0.05$), while the values in the ipsilateral hemisphere were significantly lower than this group's values ($p < 0.05$). After the waves stopped, all parameters returned to their initial values.

5.3.5.4 Induced Spreading Depression Waves Before and During Bilateral Carotid Occlusion

Table 5.8 summarizes the minimum and maximum values for reflectance, NADH, and CBF, as observed in SD waves induced before, immediately after, and 1 h after bilateral carotid occlusion (BCO). The control wave, induced before the artery occlusion, was not significantly different in any of the parameters compared to the control wave presented earlier in animals that underwent a single carotid artery occlusion. In the waves induced 10 min and 1 h after the occlusion, a different pattern was recorded relative to the control waves: an increase followed by a return to baseline was recorded in reflectance; an initial reduction followed by oxidation was recorded in NADH; and a decrease followed by an increase was observed in CBF.

In the wave induced after 10 min, reflectance levels were significantly higher relative to the baseline ($p < 0.001$), but with no significant difference relative to the control group. During the initial reduction, NADH levels increased significantly relative to the baseline ($p < 0.001$) and relative to the control group ($p < 0.05$). The decrease and increase in blood flow were significantly different both relative to the starting point ($p < 0.001$) and relative to the control group ($p < 0.01$). After the waves stopped, all parameters returned to their initial values. A similar pattern was also recorded in the wave induced 1 h after the carotid artery occlusion.

Table 5.9 shows the minimum and maximum values for reflectance, NADH, and CBF as observed in SD waves induced 24 h after bilateral carotid occlusion. The increase in

Table 5.8 Induced SD waves before, 1 min after, and 1 h after BCO

	Before BCO (control)		10 min after BCO		1 h after BCO	
	Right	Left	Right	Left	Right	Left
Ref_max	119 ± 7.5	120 ± 4.5	113 ± 8.5	117 ± 6.5	117 ± 4.5	117 ± 12.9
Ref_min	84 ± 1.5	82 ± 3.7	81 ± 7.5	85 ± 6.2	95 ± 2.5	93 ± 9.9
NADH_peak	91 ± 3.4	94 ± 1.2	112 ± 4.4	117 ± 2.6	105 ± 1.4	104 ± 3.9
CBF_max	348 ± 37.5	302 ± 39.0	181 ± 22.5	198 ± 33.8	176 ± 15.7	155 ± 8.7

Minimum and maximum values of reflectance, NADH, and CBF during an induced SD wave, as measured in the cortex of the right and left hemispheres. The waves were recorded before, 10 min after, and 1 h after occlusion of both carotid arteries ($n = 8$).

Table 5.9 SD wave induced 24 h after BCO

	24 h after BCO	
	Occluded	Nonoccluded
Ref_max	154 ± 7.5	165 ± 10.0
Ref_min	89 ± 5.6	118 ± 16.4
NADH_peak	119 ± 8.6	118 ± 2.8
CBF_max	252 ± 30.3	223 ± 33.6

Minimum and maximum values of reflectance, NADH, and CBF during an induced SD wave, as measured in the cortex of the right and left hemispheres. The waves were induced 24 h after occlusion of both carotid arteries ($n = 8$).

reflectance was significantly higher relative to the initial level ($p < 0.001$), relative to the increase in the normoxic state ($p < 0.001$), and also significantly different relative to the increase in reflectance observed 10 min ($p < 0.01$) and 1 h ($p < 0.01$) after the carotid occlusion. NADH values increased before a decrease was observed in them ($p < 0.001$). These maximum values were significantly higher than the NADH level of the control group at the same stage in both hemispheres ($p < 0.05$). In blood flow, a pattern of a decrease and then an increase was observed, both significant relative to the starting point ($p < 0.001$). The blood flow decrease was to significantly lower values relative to the CBF decrease observed in the control group ($p < 0.05$). Immediately after, a nonsignificant increasing trend was observed in blood flow, which was low relative to the control group and high relative to the values obtained in the waves induced 10 min and 1 h after the artery occlusion. In all parameters, no significant differences were recorded between the two hemispheres. After the waves stopped, all parameters returned to their initial values.

5.3.5.5 Tissue Function Following Unilateral vs. Bilateral Carotid Occlusion

Figure 5.45 shows the time differences between the first and second SD waves, as determined by measuring the peak times in CBF values. Figure 5.45a shows the time differences between waves induced before, 1 min after, 1 h after, and 24 h after a single carotid occlusion. In all the waves examined, no significant differences were observed between the two hemispheres in the time differences between the waves. The time difference between adjacent SD waves in the control measurement was about 5 min, while

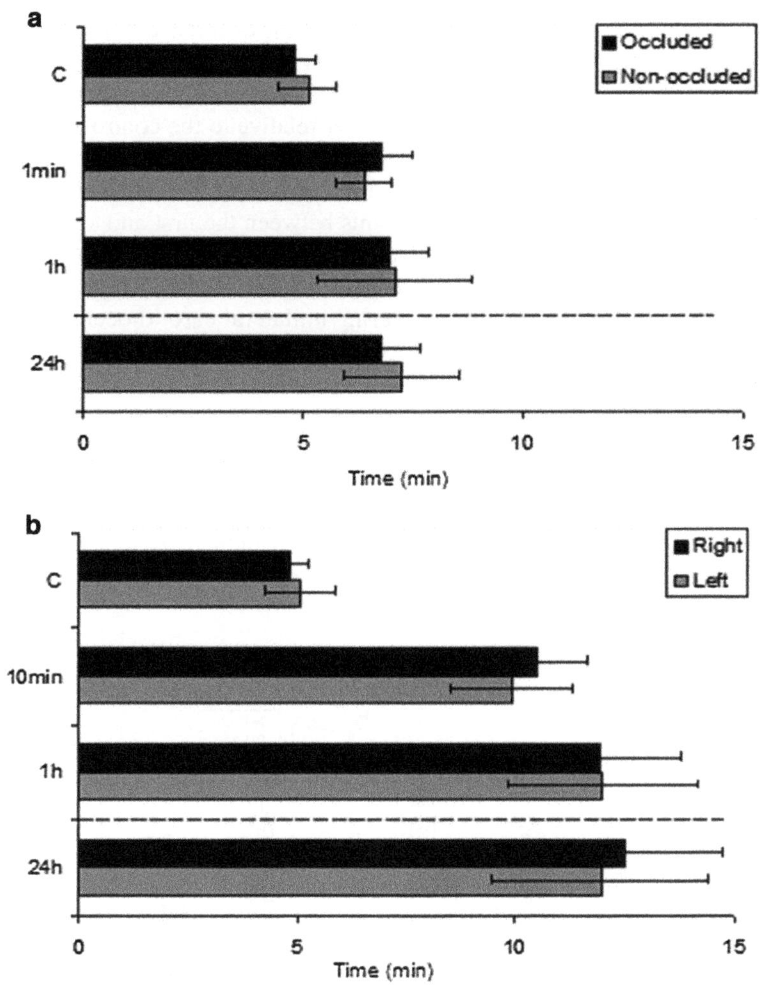

Fig. 5.45 The time differences between consecutive SD waves, determined by maximum CBF values. The results include four experimental groups: In the first group (**a**, top) three sets of waves were induced in both hemispheres before (control, C), 1 min after and 1 h after occlusion of a single carotid artery ($n = 10$). In the second group (**a**, bottom) waves were induced 24 h after occlusion of a single carotid artery ($n = 10$). In the third group (**b**, top) three sets of waves were induced in both hemispheres before (control, C), 10 min after and 1 h after occlusion of both carotid arteries ($n = 8$). In the fourth group (**b**, bottom) waves were induced 24 h after occlusion of the carotid arteries ($n = 8$)

the time differences observed 1 min, 1 h, and 24 h after carotid occlusion were between 6.5 and 7 min. No significant differences were found between the time differences of the waves at the three time points after carotid occlusion, nor relative to the control group.

Figure 5.45b shows four sets of waves: a control wave, and at three different time points (10 min, 1 h, and 24 h) after bilateral carotid occlusion. In all the waves examined, no significant differences were observed between the two hemispheres in the time differences between the waves. The time difference between adjacent SD waves in the control waves was about 5 min. The time difference 10 min after carotid occlusion was about 10 min. The time difference 1 and 24 h after artery occlusion was about 12 min. The time differences between the successive waves at the three time points after ischemia were significantly higher relative to the control waves before carotid occlusion ($p < 0.05$), and with no significant difference between the three groups of waves after artery occlusion. Similar results were obtained by comparing the time differences between common points between the first and second wave in other parameters (these data are not presented).

Figure 5.46 shows the hyperemia values recorded in the CBF level in the first SD wave of each series of waves induced during unilateral carotid occlusion (A) and bilateral carotid occlusion (b). Following unilateral ischemia, no significant changes were observed in the extent of hyperemia, even after 1 h. In contrast, with bilateral carotid occlusion, a significant decrease in hyperemia values can be seen starting from 10 min ($p < 0.01$) and 1 h after the occlusion ($p < 0.05$). Conversely, 24 h after bilateral carotid occlusion, a recovery trend can be seen, characterized by higher CBF values. In the case of unilateral carotid occlusion, however, a significant change relative to the baseline was observed at this stage, expressed as an increase in blood flow in the hemisphere contralateral to the occluded carotid ($p < 0.05$), accompanied by a decrease in blood flow in the ipsilateral hemisphere ($p < 0.05$).

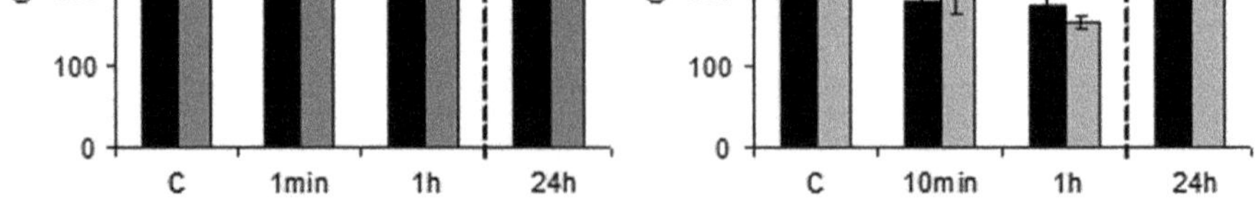

Fig. 5.46 Maximum CBF values in the SD wave in different global ischemia conditions: occlusion of a single carotid artery (**a**, $n = 10$) and occlusion of both carotid arteries (**b**, $n = 8$). C, control; occluded, hemisphere on the ipsilateral side to the occlusion; nonoccluded, hemisphere on the contralateral side to the occlusion; right, hemisphere; left, hemisphere. Results are presented as mean ± SE. Significant differences between groups relative to the control group are indicated by asterisks, $^*p < 0.05$; $^{**}p < 0.01$

5.3.5.6 Comparison Between Global and Focal Ischemia

Figure 5.47 shows the minimum CBF values measured following global ischemia in comparison to focal ischemia. Following focal ischemia, CBF values were significantly lower in the core relative to the penumbra at each of the ischemia durations—10 min ($p < 0.01$), 30 min ($p < 0.05$), and 60 min ($p < 0.05$). No significant differences were found among the three groups in minimum CBF values in the two measurement sites.

Following bilateral carotid occlusion, no differences were observed between the hemispheres. Blood flow during this occlusion decreased to values lower than the minimum level in the penumbra during MCAO, but higher than the blood flow recorded in the core. During single carotid occlusion, a significant decrease in CBF relative to the starting level ($p < 0.05$) was observed only in the hemisphere ipsilateral to the occluded artery. The minimum values during this occlusion were significantly higher relative to the minimum value observed following bilateral carotid occlusion ($p < 0.05$), and also relative to the CBF value in the penumbra ($p < 0.05$) and the core ($p < 0.001$) following MCA occlusion.

Figure 5.48 shows the maximum NADH values measured during ischemia in different models. Following 10-min or 30-min MCA occlusion, significant differences were observed between the core and the penumbra in the maximum NADH levels ($p < 0.05$).

These differences were not observed during a 1-h occlusion. Bilateral carotid occlusion led to NADH levels that were low relative to those observed in the core in all the groups where MCAO protocols were performed ($p < 0.05$). NADH levels in the penumbra showed a greater (but not significant) increasing trend relative to bilateral carotid occlusion. Single carotid occlusion resulted in maximum NADH levels that were significantly lower relative to the other types of ischemia ($p < 0.05$).

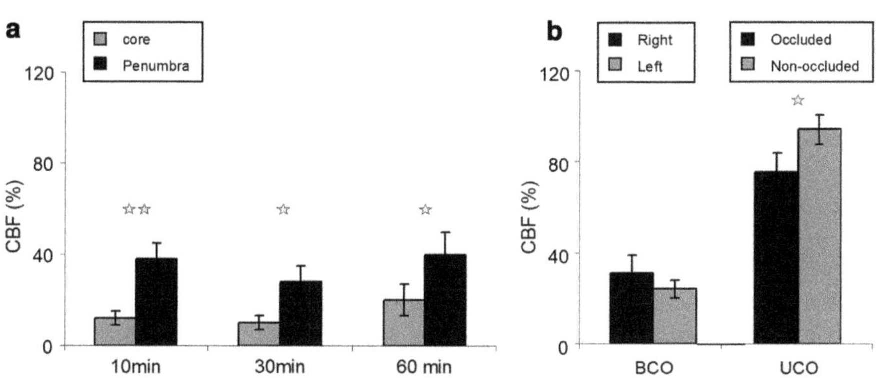

Fig. 5.47 Minimum CBF values in different ischemia models. (**a**) Focal ischemia (core and penumbra) for 10 min ($n = 9$), 30 min ($n = 8$), or 60 min ($n = 8$). (**b**) Occlusion of one (UCO, $n = 10$) or two (BCO, $n = 8$) carotid arteries. Results are presented as mean ± SE. Significant differences between two measurement sites are indicated by asterisks, $*p < 0.05$, $**p < 0.01$

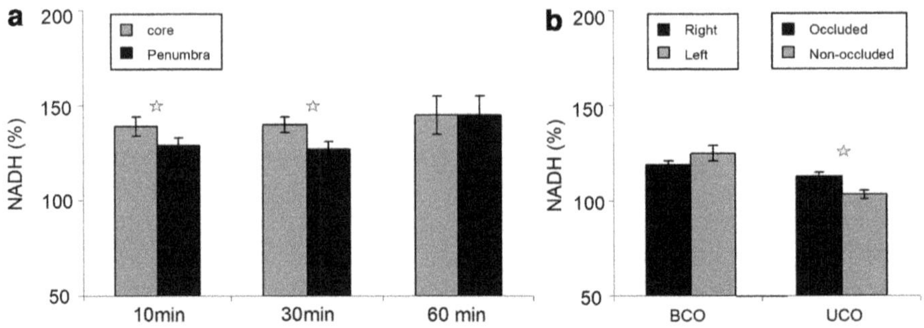

Fig. 5.48 Maximum NADH values in different ischemia models. (**a**) Focal ischemia (core and penumbra) for 10 min ($n = 9$), 30 min ($n = 8$) or 60 min ($n = 8$). (**b**) Occlusion of one (UCO, $n = 10$) or two (BCO, $n = 8$) carotid arteries. Results are presented as mean ± SE. Significant differences between two measurement sites are indicated by asterisks, $*p < 0.05$, $**p < 0.01$

5.3.6 *Global Ischemia as a Preconditioning Treatment for Focal Ischemia*

5.3.6.1 Unilateral Carotid Artery Occlusion

Figure 5.49 shows the parameter values in response to a 60-min MCA occlusion, which was preceded by a right carotid occlusion 24 h earlier. Upon MCA occlusion, blood flow decreased in both measurement areas, to 10% in the core area ($p < 0.001$ relative to baseline) and 32% in the penumbra area ($p < 0.001$). The CBF difference between the two areas was significant for the first 20 min of ischemia ($p < 0.05$). After a few minutes, a partial spontaneous recovery began, expressed as a moderate increase in blood flow in both areas, up to values of 45% in the core ($p < 0.001$) and 65% in the penumbra ($p < 0.05$).

Following MCA occlusion, a 7% decrease in reflectance values was recorded at both measurement sites ($p < 0.01$). Three minutes later, reflectance increased to values of 155% in the core and 105% in the penumbra. Consequently, significant differences in the reflectance level between the two areas were recorded throughout the occlusion period ($p < 0.01$). The penumbra's reflectance returned to baseline and remained at a similar level throughout the occlusion. Upon artery occlusion, an increase in fluorescence level was observed in the core, to a value of 132% ($p < 0.05$ relative to baseline), followed by a gradual decrease to a value of 126% at the end of the ischemia ($p < 0.05$). In the penumbra, a temporary increase to a value of 113% was recorded ($p < 0.05$), followed by a decrease to baseline after a few minutes. Fluorescence in the core was significantly higher relative to the penumbra for the first 20 min of ischemia ($p < 0.05$). The NADH level increased in both measurement areas, to a value of 111% in the core ($p < 0.001$ relative to baseline) and 114% in the penumbra ($p < 0.001$). A moderate and gradual decrease in the core was then recorded, down to a level of 85%

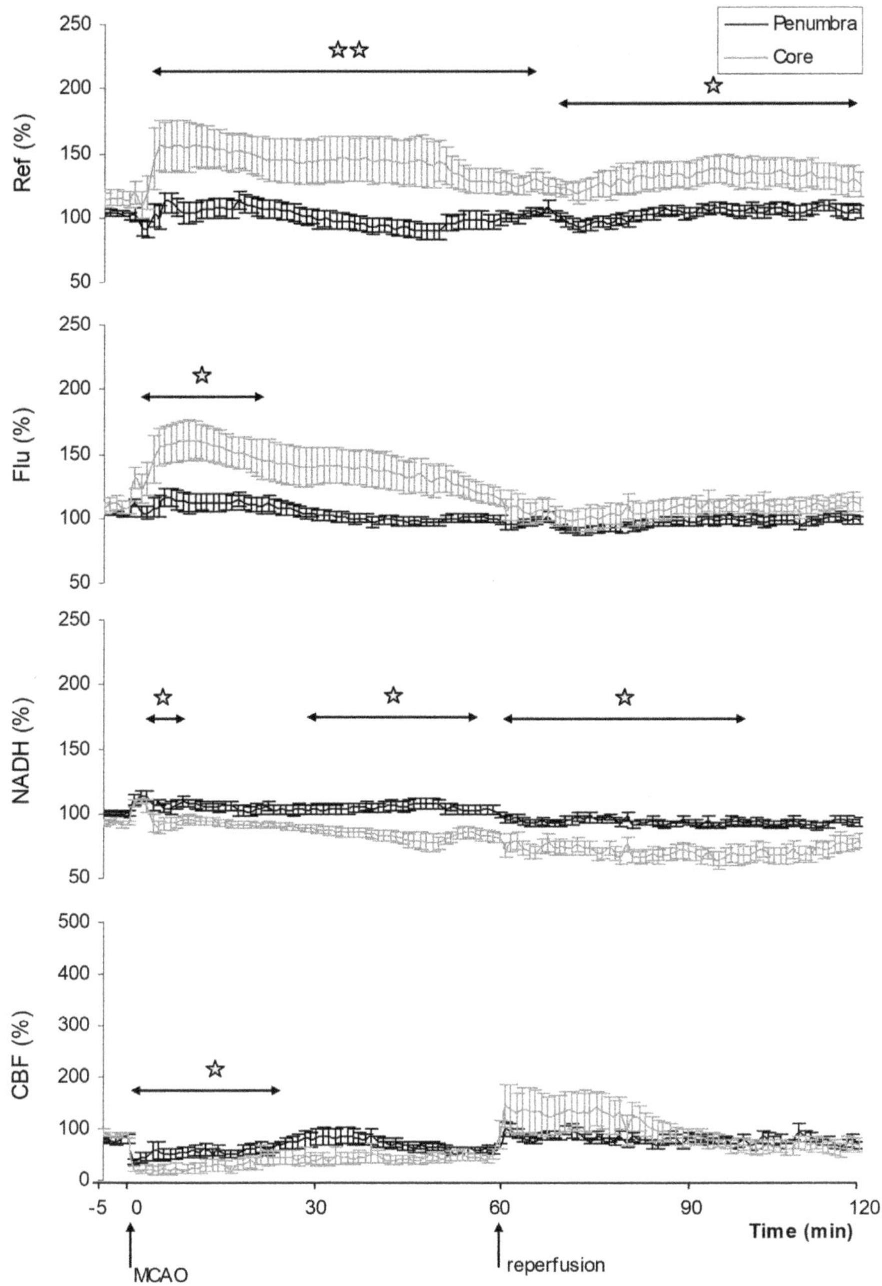

Fig. 5.49 Hemodynamic and metabolic changes following right MCA occlusion for 1, 24 h after right carotid occlusion, as measured in the core and penumbra. Results are presented as mean ± SE ($n = 8$). Significant differences between the two measurement sites are indicated by asterisks, $*p < 0.05$, $**p < 0.01$, $***p < 0.001$

at the end of the hour of ischemia, while in the penumbra, the NADH level remained at values higher than 105% throughout the entire hour. Consequently, significant differences between the two areas were recorded in the half-hour before the occlusion was removed ($p < 0.05$).

Following the restoration of blood flow, an increase in blood flow was recorded to a value of 140% in the core and 102% in the penumbra. About half an hour after the restoration of blood flow, CBF values in both measurement areas stabilized at 75%. Reflectance in the core decreased to a value of 124%, and after about half an hour, it increased again to a value of 135%. Throughout this time, reflectance values remained significantly higher than the initial reflectance values ($p < 0.05$). No change in reflectance was observed in the penumbra during this time. In fluorescence, a decrease to baseline was recorded in the core, and after about half an hour, an additional increase to a value of 110% was recorded, while no change was recorded in the penumbra relative to the baseline. The NADH level decreased in both measurement areas, to 65% in the core ($p < 0.05$) and 95% in the penumbra, which created a significant difference between the two areas for 54 min after the occlusion was removed ($p < 0.05$).

Figure 5.50 shows the changes in the different parameters following MCA occlusion with and without a right carotid occlusion 24 h earlier. In the group where a single carotid was occluded 24 h before MCAO, higher blood flow values were observed relative to the group with MCA occlusion only: for 20 min after the occlusion until its removal in the core area ($p < 0.05$), and for about 20 min after the occlusion in the penumbra area ($p < 0.05$). In this group, significantly higher reflectance and fluorescence values were observed in the core at the beginning of the occlusion ($p < 0.05$) relative to the second group. After a few minutes of occlusion, no more significant changes were observed between the groups in the reflectance, fluorescence, and NADH levels. The reflectance level in the core in the carotid artery occlusion group showed a decreasing trend, compared to an increasing trend in the second group, which led to significant differences between the groups about 5 min before the occlusion was removed ($p < 0.05$). In the first 15 min after the occlusion was removed, differences in the fluorescence level between the groups were observed in the penumbra ($p < 0.05$). Other than that, no further differences were recorded between the groups at this stage.

Table 5.10 and Fig. 5.51 show the responses to nitrogen exposure before and after a 24 h right carotid occlusion followed by a 60 min MCAO. The minimum reflectance level in the core was significantly higher both relative to the reflectance level in the penumbra ($p < 0.01$) and relative to the reflectance level in the previous treatment ($p < 0.01$). The maximum NADH level in the penumbra was significantly lower relative to the nitrogen treatment before artery occlusion ($p < 0.05$). The maximum blood flow values in the core area were significantly lower both relative to the penumbra ($p < 0.01$) and relative to the corresponding value in the core recorded during the nitrogen exposure before ischemia ($p < 0.05$).

Table 5.11 and Fig. 5.52 present induced SD waves after a 24 h right carotid artery occlusion followed by a 60 min MCAO, compared to a control group of animals

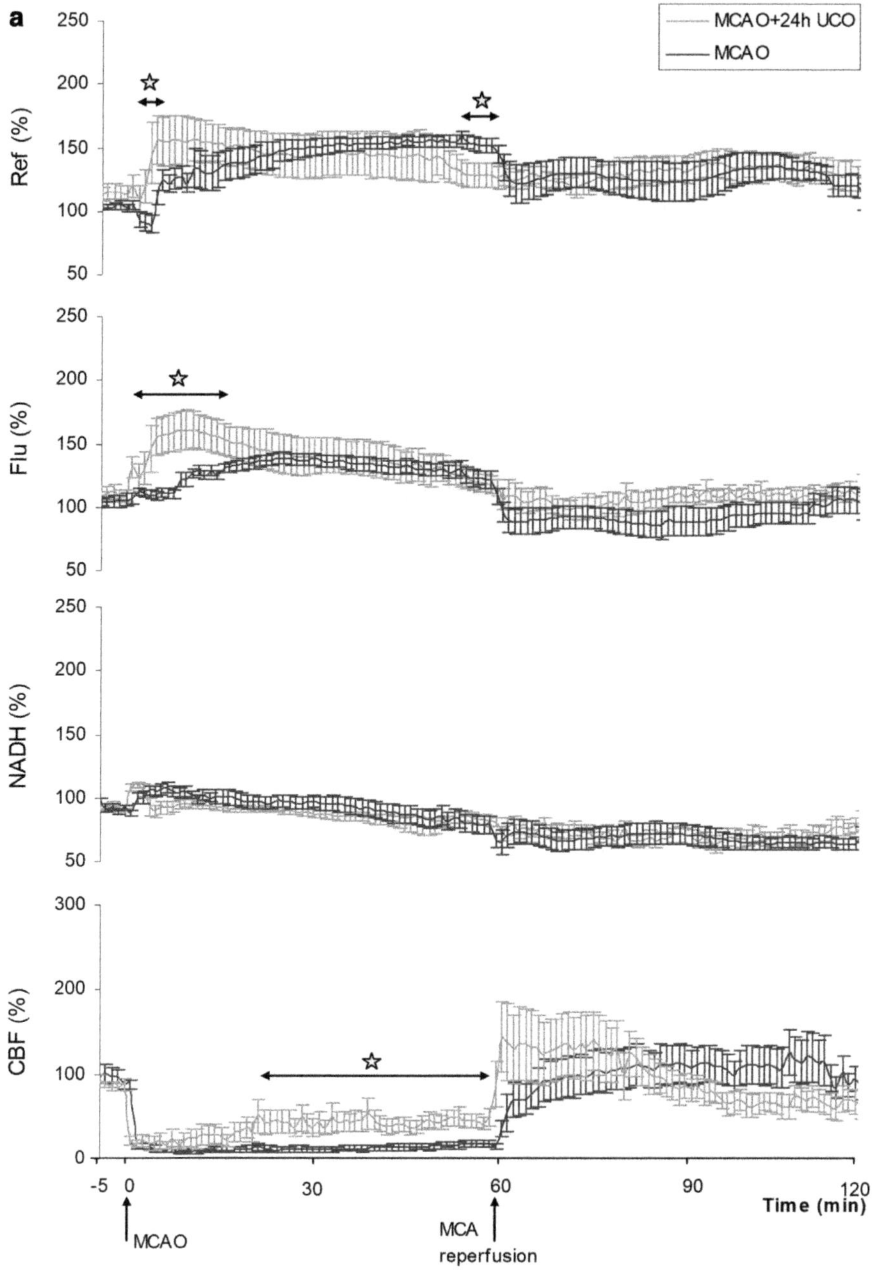

Fig. 5.50 Changes in various parameters following 1 h MCAO with ($n = 8$) or without ($n = 8$) 24 h prior carotid artery occlusion (UCO), as measured in the core (**a**) and penumbra (**b**). Results are presented as mean ± SE. Significant differences between the two measurement sites are indicated by asterisks, *$p < 0.05$, **$p < 0.01$, ***$p < 0.001$

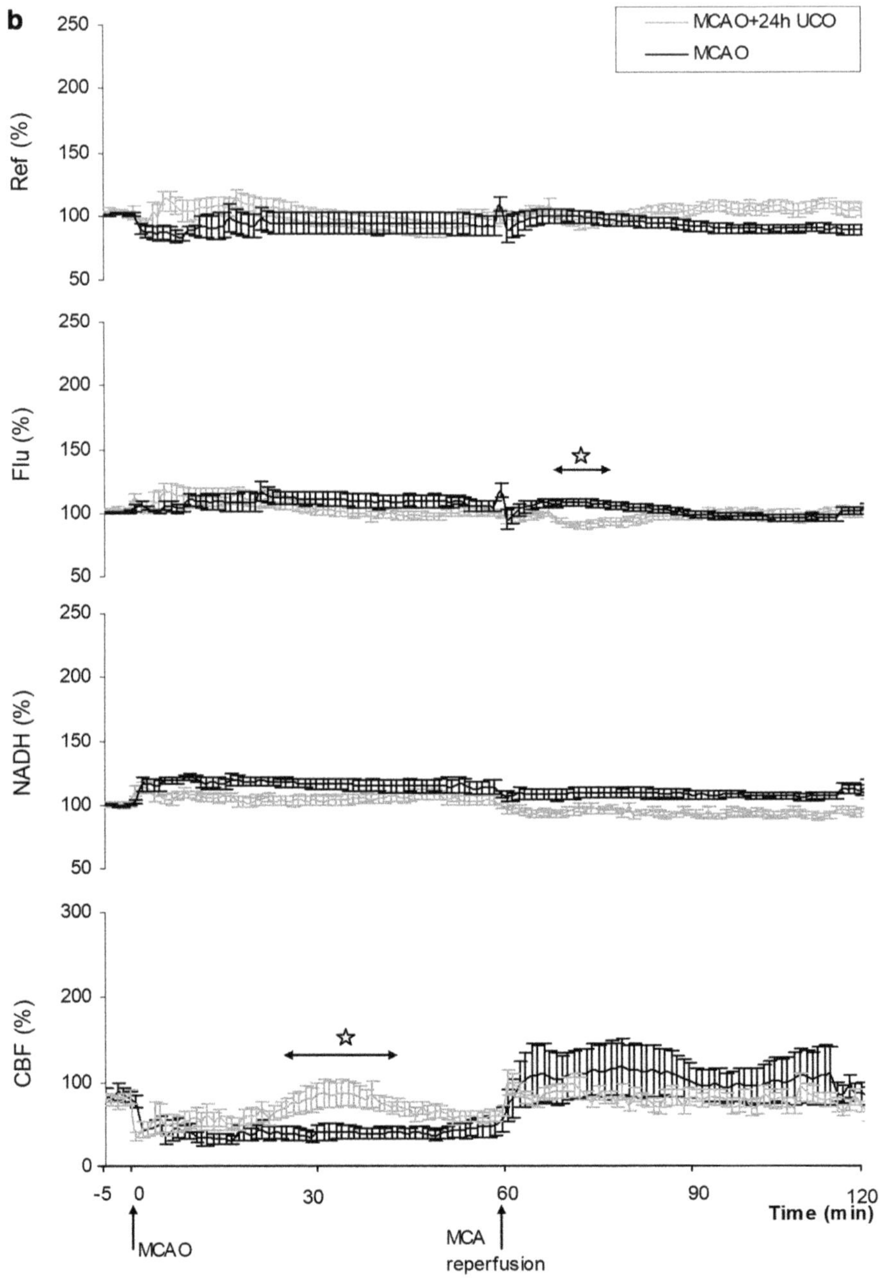

Fig. 5.50 (continued)

Table 5.10 Nitrogen exposure before and after UCO for 24 h along with MCAO

	Before UCO (control)		After 24 h UCO + 60 min MCAO	
	Penumbra	Core	Penumbra	Core
Ref_peak	65 ± 1.74	63 ± 1.6	75 ± 0.6	83 ± 0.6
NADH_max	147 ± 2.78	141 ± 1.8	126 ± 2.2	119 ± 3.9
CBF_min	48 ± 2.6	35 ± 3.5	64 ± 4.9	49 ± 4.7
CBF_max	300 ± 25.5	245 ± 12.37	218 ± 8.81	179 ± 11.3

Minimum and maximum values of reflectance, NADH, and CBF during nitrogen exposure, as measured in the core and penumbra. Nitrogen exposure was performed in the same animals before and after UCO occlusion for 24 h accompanied by MCA occlusion for 60 min ($n = 8$).

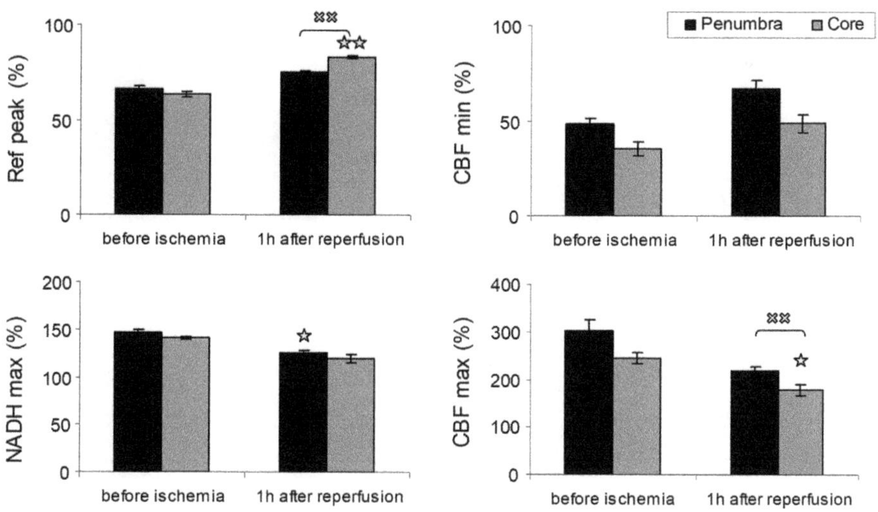

Fig. 5.51 Response to nitrogen exposure 24 h after unilateral carotid occlusion and before MCA occlusion, and 1 h after reperfusion, as measured in the core and penumbra. Results are presented as mean ± SE ($n = 8$). Significant differences between two nitrogen treatments are indicated by asterisks, *$p < 0.05$, **$p < 0.01$. Significant differences between the two measurement areas are indicated by x-marks, $^{xx}p < 0.01$

that underwent carotid artery occlusion only. The increase in reflectance level in the penumbra was significantly higher relative to the corresponding increase recorded in the SD wave before artery occlusion ($p < 0.001$). In the NADH level, a trend of increase was recorded after MCAO compared to a trend of decrease after UCO only, which created significant differences between the groups in the core ($p < 0.001$) and the penumbra ($p < 0.001$).

Table 5.11 SD waves after UCO for 24 h and MCAO for 60 min compared to control group

	Control		24 h after MCAO	
	Penumbra	Core	Penumbra	Core
Ref_max	111 ± 1.11	116 ± 1.06	112 ± 2.84	123 ± 5.23
Ref_min	61 ± 1.07	72 ± 1.4	73 ± 1.85	84 ± 2.5
NADH_peak	86 ± 0.66	85 ± 0.4	110 ± 1.29	107 ± 0.31
CBF_max	309 ± 8.86	328 ± 6.58	320 ± 23.1	411 ± 38.7

Minimum and maximum values of reflectance, NADH, and CBF during induced SD waves, as measured in the core and penumbra. SD waves were induced after single carotid occlusion for 24 h followed by MCA occlusion for 60 min ($n = 8$) and in control animals that underwent carotid artery occlusion alone, without MCAO (control, $n = 8$).

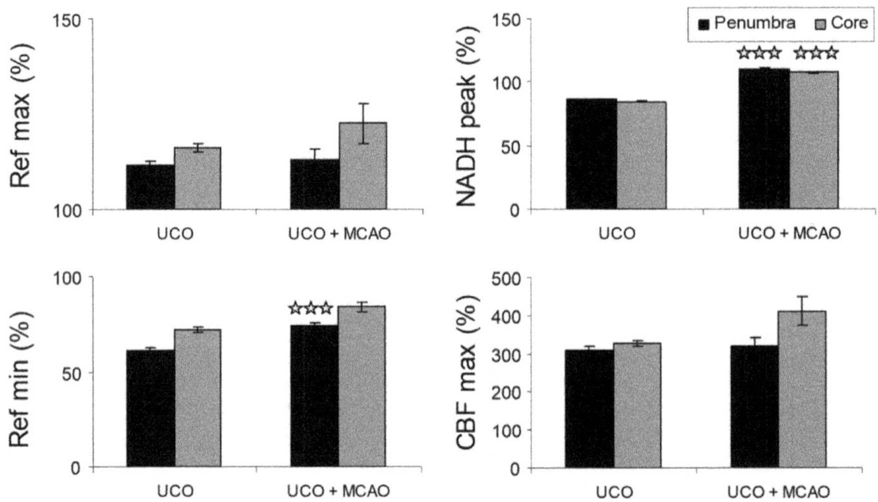

Fig. 5.52 Changes in various parameters as observed in the core and penumbra in SD waves that were induced after 24 h of UCO followed by 60 min of MCAO ($n = 8$), compared to waves that were induced after 24 hs of UCO without MCAO ($n = 8$). Results are presented as mean ± SE. Significant differences between groups are indicated by asterisks, $^{***}p < 0.001$

5.3.6.2 Bilateral Carotid Artery Occlusion

Figure 5.53 shows the reflectance, fluorescence, NADH, and CBF responses to a 60 min MCA occlusion. The MCA occlusion was performed 24 hafter a permanent occlusion of both carotid arteries. With MCA occlusion, blood flow decreased in both measurement areas, to 28% in the core area ($p < 0.001$ relative to baseline) and 44% in the penumbra area ($p < 0.01$) and remained at similar levels for the next hour. Following MCA occlusion, a 6% decrease in reflectance values was recorded at both

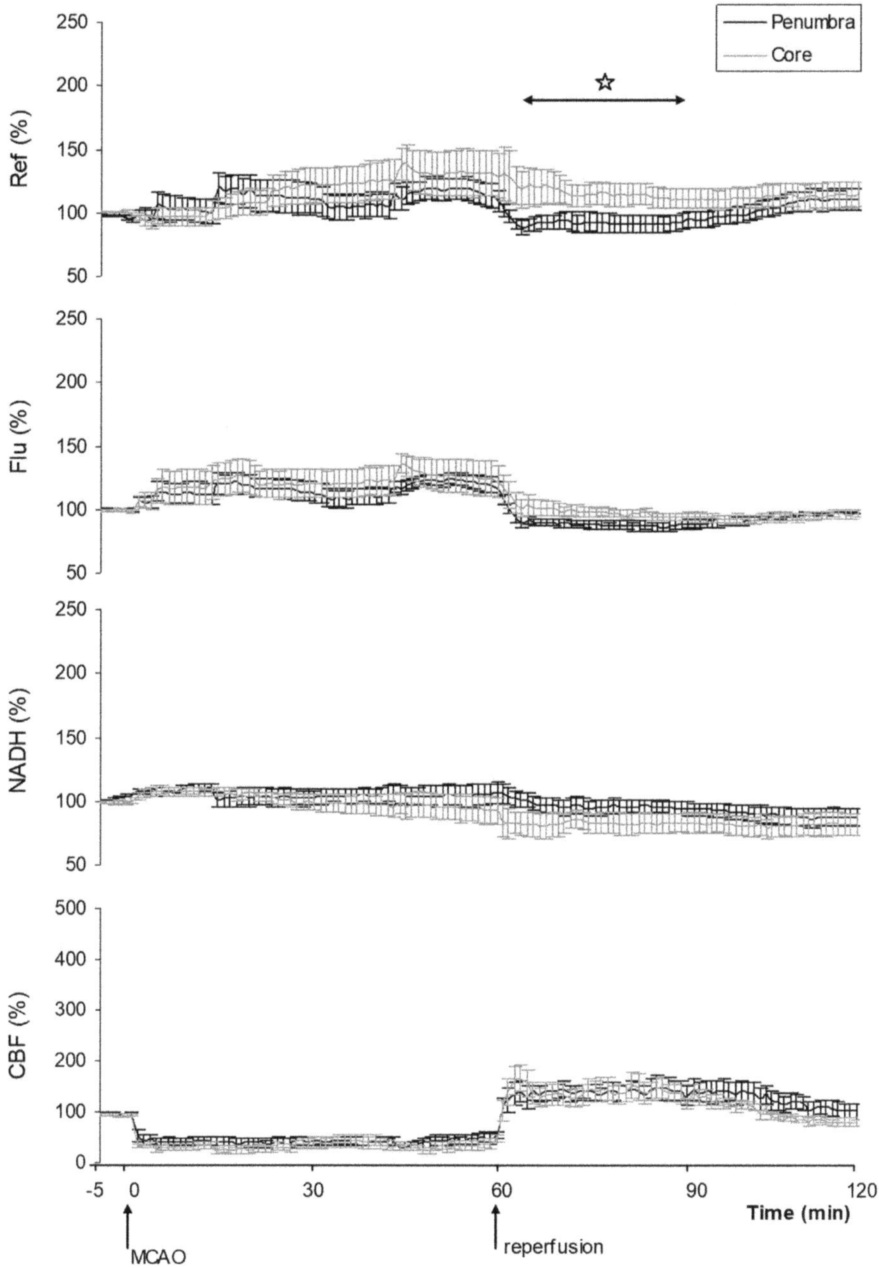

Fig. 5.53 Hemodynamic and metabolic changes in the core and penumbra following 1 h occlusion of the right MCA in rats that had undergone bilateral carotid occlusion 24 h earlier. Results are presented as mean ± SE ($n = 8$). Significant differences between the two measurement sites are indicated by asterisks, $^{*}p < 0.05$

measurement sites. After 3 min, an increasing trend in reflectance was recorded, up to maximum values of 135% in the core and 117% in the penumbra. In the fluorescence values, an increasing trend was recorded up to maximum values of 130% in the core and 122% in the penumbra (in both areas $p < 0.05$ relative to the baseline level in the last 15 min of ischemia). In the NADH level, an increase of 9% in the core and 6% in the penumbra was recorded (in both areas $p < 0.05$ relative to the baseline level in the first 20 min of the occlusion).

Following the restoration of blood flow, an increase in blood flow was recorded to a value of 166% in the core and 140% in the penumbra. About 1 h after the restoration of blood flow, CBF in the penumbra returned to its initial value, while in the core, CBF values stabilized at 80%. Reflectance decreased in the core to a value of 121% and to 90% in the penumbra, which created a significant difference between the two areas for 30 min after the occlusion was removed ($p < 0.05$). After about 1 h, no significant differences were observed in the reflectance level at both sites relative to the initial level.

Fluorescence in the core returned to baseline, while in the penumbra, a decrease to a value of 90% was recorded ($p < 0.05$ relative to baseline). After about 1 h, the reflectance values at both sites returned to their initial levels. In the NADH level, a decreasing trend was recorded, returning to baseline values after about 1 h.

Figure 5.54 shows the changes in the different parameters following MCA occlusion with and without bilateral carotid occlusion 24 h earlier. In the group where both carotid arteries were occluded, significantly lower fluorescence levels ($p < 0.05$) were observed in the core relative to the group with MCA occlusion only, for about 30 min after the occlusion was removed. In the other parameters, no significant differences were observed between the groups after the occlusion was removed.

Table 5.12 shows the minimum and maximum values of reflectance, NADH, and CBF during a brief nitrogen exposure performed 24 h after carotid artery occlusion before a 60 min MCA occlusion and 1 h after reperfusion. As shown in Fig. 5.55, the increase in reflectance values in the nitrogen exposure after MCAO was significantly higher in both areas relative to the increase in reflectance in the nitrogen treatment before MCAO ($p < 0.05$). The maximum increase in NADH level was lower in the nitrogen exposure after ischemia in the core ($p < 0.05$ relative to the previous nitrogen exposure). The maximum blood flow was lower in the nitrogen exposure after ischemia ($p < 0.05$).

Table 5.13 shows the maximum and minimum values for reflectance, NADH, and CBF as measured in SD waves in two groups. The maximum and minimum reflectance values were significantly higher ($p < 0.01$) at both measurement sites following MCAO combined with BCO compared to BCO alone (Fig. 5.56). In the NADH level, an increase was recorded in both treatments, which is characteristic of ischemic conditions. In addition, the increase in CBF was significantly lower following MCAO at both measurement sites ($p < 0.05$).

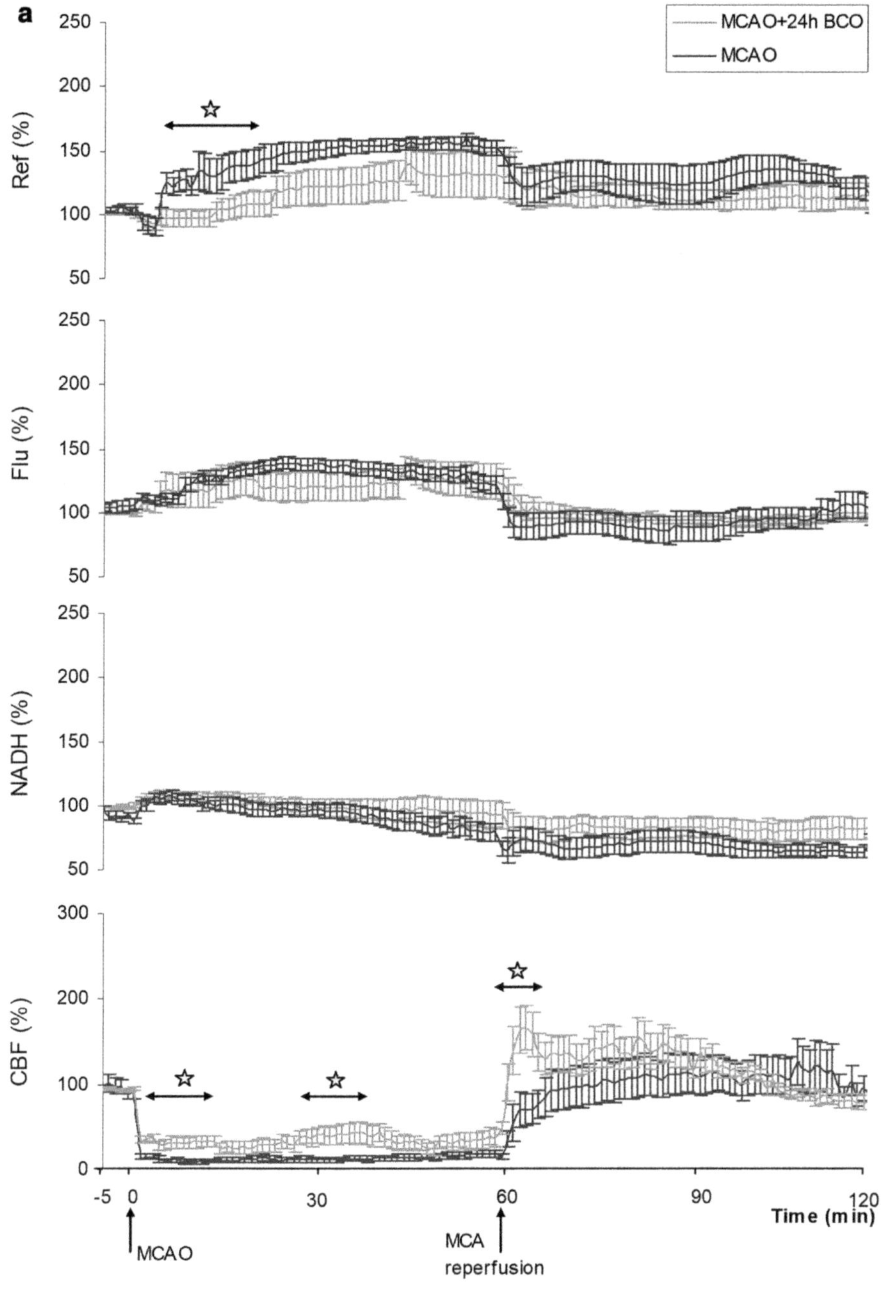

Fig. 5.54 Changes in various parameters in the core (**a**) and penumbra (**b**) following 1 h MCAO with (n = 8) or without (n = 8) bilateral carotid occlusion 24 h earlier (BCO). Results are presented as mean ± SE. Significant differences between groups are indicated by asterisks, $*p < 0.05$, $**p < 0.01$

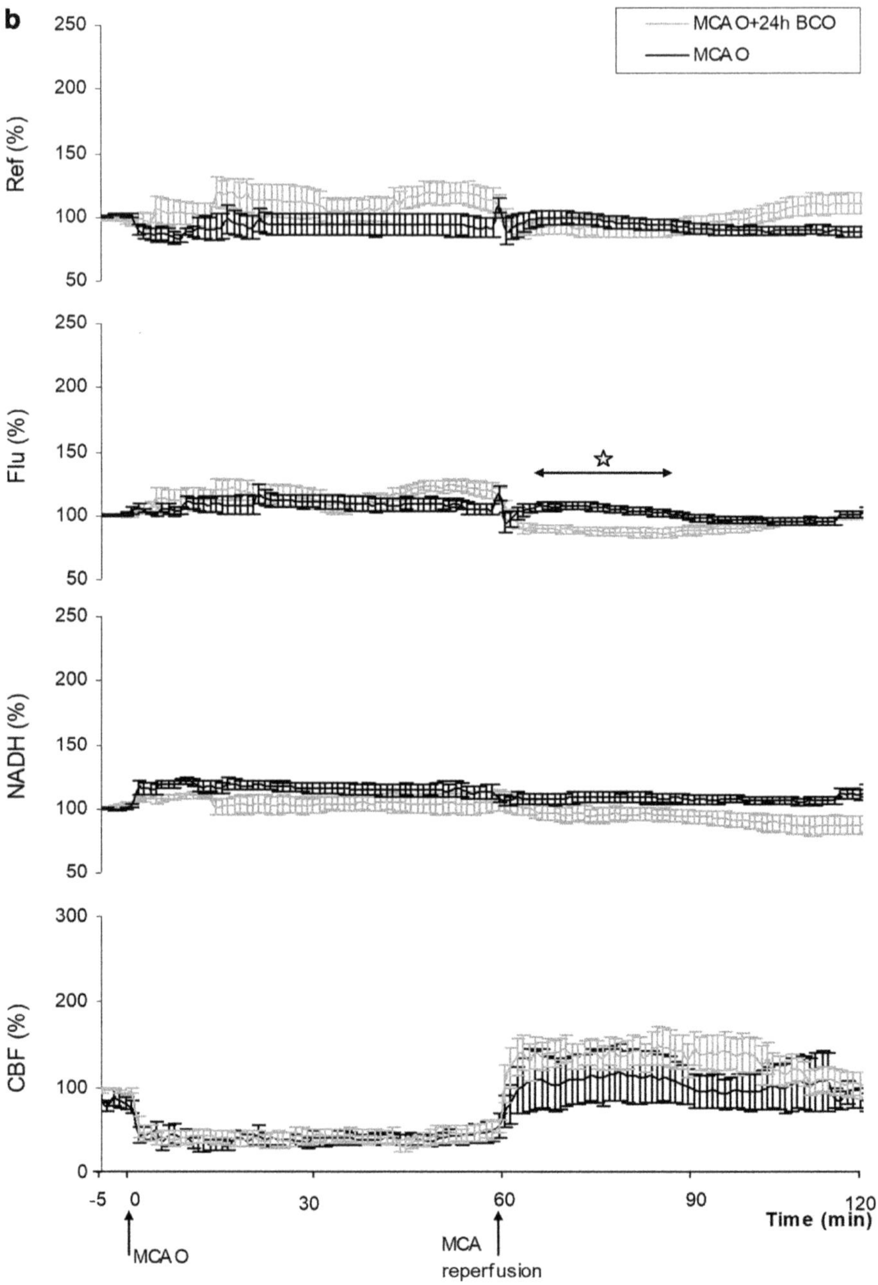

Fig. 5.54 (continued)

Table 5.12 Nitrogen exposure before MCAO and 1 h after MCAO in rats undergoing BCO

	BCO 24 h		BCO 24 h + MCAO 60 min	
	Penumbra	Core	Penumbra	Core
Ref_peak	86 ± 0.9	85 ± 0.6	87 ± 0.8	90 ± 0.5
NADH_max	141 ± 4.5	130 ± 0.3	119 ± 2.6	122 ± 1.7
CBF_min	60 ± 4.1	73 ± 3.1	79 ± 0.6	84 ± 0.6
CBF_max	233 ± 5.2	240 ± 4.1	221 ± 3.8	225 ± 3.1

Minimum and maximum values of reflectance, NADH and CBF during nitrogen exposure, as measured in the core and penumbra. Nitrogen exposure was performed before MCA occlusion and 1 h after reperfusion ($n = 8$) in rats that had undergone BCO the previous day.

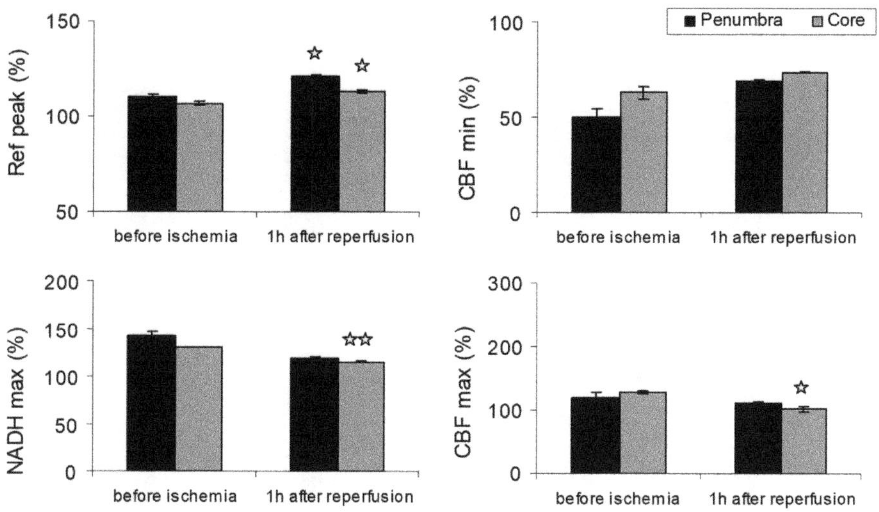

Fig. 5.55 Response to nitrogen exposure before MCA occlusion (left) and 1 h after reperfusion (right), as measured in the core and penumbra 24 h after occlusion of both carotid arteries. Results are presented as mean ± SE ($n = 8$). Significant differences between the two nitrogen treatments are indicated by asterisks, *$p < 0.05$, **$p < 0.01$

5.3.7 *Global Ischemia as a Postconditioning Treatment for Focal Ischemia*

5.3.7.1 Unilateral Carotid Occlusion for 45 min

Figure 5.57 shows the values of reflectance, fluorescence, NADH, and CBF in a 60-min MCA occlusion combined with a 45-min unilateral carotid occlusion that was initiated 15 min before the MCA occlusion was removed. With MCA occlusion,

Table 5.13 SD waves in animals that underwent BCO without MCAO and 1 h after MCAO

	BCO 24 h		BCO 24 h + MCAO 60 min	
	Penumbra	Core	Penumbra	Core
Ref_max	139 ± 2.4	134 ± 0.7	164 ± 2.6	170 ± 3.5
Ref_min	87 ± 0.5	85 ± 1.1	104 ± 1.3	99 ± 0.5
NADH_peak	110 ± 0.8	108 ± 0.7	112 ± 0.7	112 ± 0.6
CBF_max	234 ± 10.7	223 ± 11.9	139 ± 5.8	136 ± 4.1

Minimum and maximum values of reflectance, NADH, and CBF during induced SD waves, as measured in the core and penumbra. SD waves were induced 1 h after reperfusion following MCAO for 1 h, combined with permanent occlusion of the carotid arteries 24 h earlier ($n = 8$). In another group, SD waves were induced after occlusion of both carotid arteries only ($n = 8$).

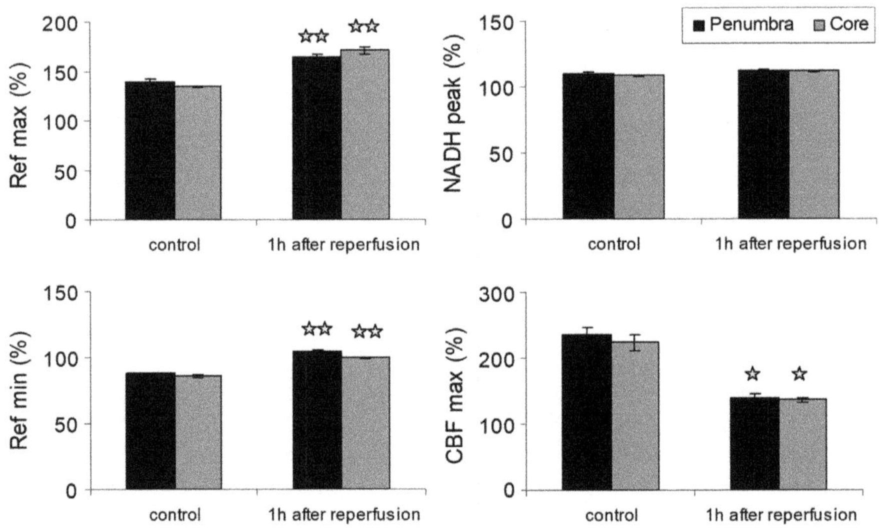

Fig. 5.56 Changes in the various parameters of SD waves recorded 24 h after occlusion of both carotid arteries (control, $n = 8$), and 1 h after reperfusion following MCAO in addition to carotid occlusion for 24 h (1 h after reperfusion, $n = 8$), as observed in the core and penumbra. Results are presented as mean ± SE. Significant differences between the two treatments are Hemodynamic and metabolic changes in the core and penumbra following 1 h right MCA occlusion combined with 45 min single carotid occlusion performed 15 min before MCA deblockage. Results are presented as mean ± SE ($n = 8$). Significant differences between the two measurement sites are indicated by asterisks, $*p < 0.05$. indicated by asterisks, $*p < 0.05$, $**p < 0.01$

blood flow decreased in both measurement areas, to 10% in the core area ($p < 0.001$ relative to baseline) and 40% in the penumbra area ($p < 0.001$). The differences in CBF between the core and penumbra were significant for the time the MCA remained occluded ($p < 0.05$). Following MCA occlusion, a 6% decrease in reflectance values

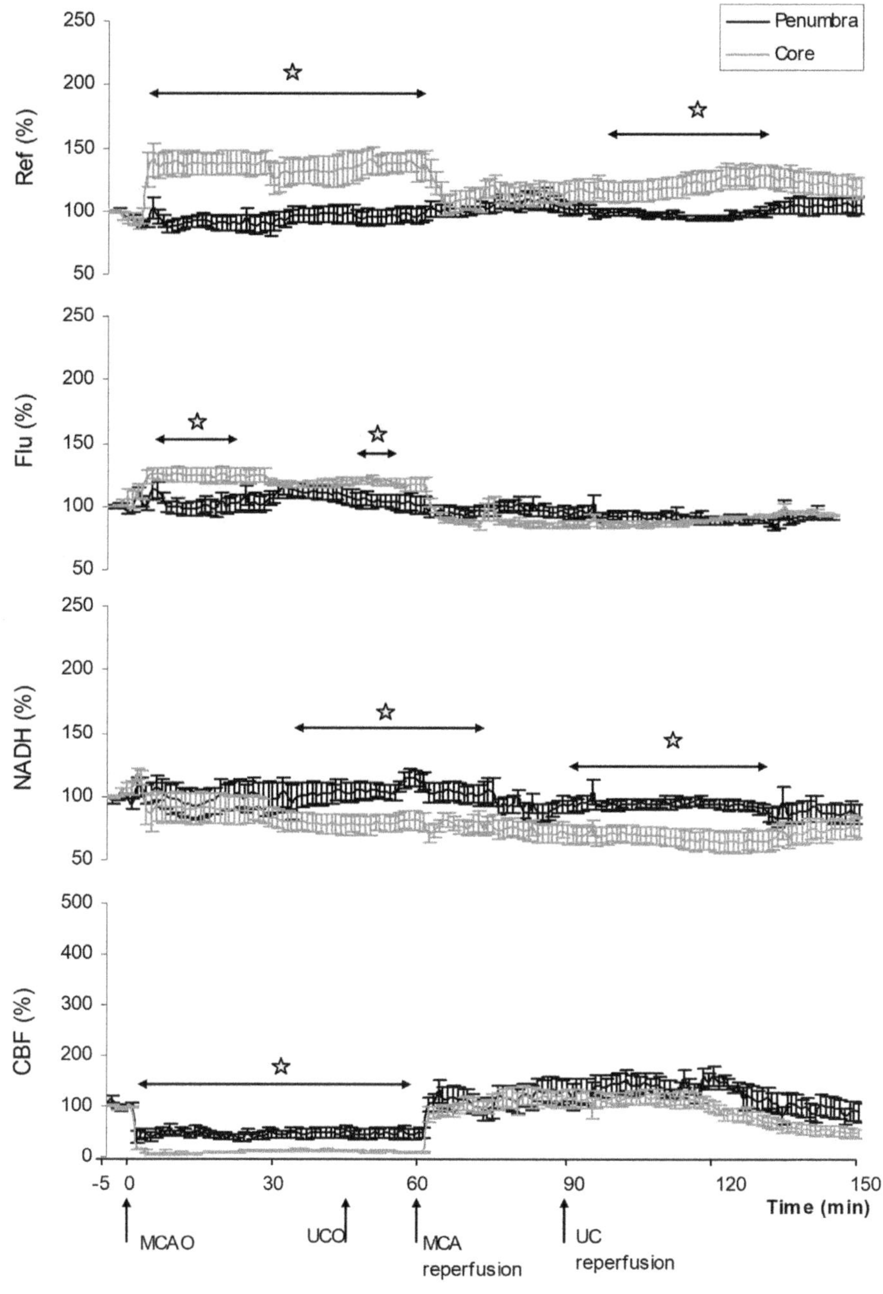

Fig. 5.57 Hemodynamic and metabolic changes in the core and penumbra following 1 h right MCA occlusion combined with 45 min single carotid occlusion performed 15 min before MCA deblockage. Results are presented as mean ± SE ($n = 8$). Significant differences between the two measurement sites are indicated by asterisks, $^*p < 0.05$

was recorded at both measurement sites. After 3 min, an increase was recorded up to maximum values of 135% in the core ($p < 0.05$) and 103% in the penumbra ($p < 0.05$). Reflectance values remained high in the core area until the local occlusion was removed, and returned to baseline values in the penumbra. As a result, significant differences between the two measurement areas were recorded until the MCA occlusion was removed ($p < 0.05$).

In fluorescence values, an increase was recorded up to maximum values of 126% in the core and 110% in the penumbra (in both areas $p < 0.05$ relative to baseline). Fluorescence remained high in the core, and returned to baseline values in the penumbra. A significant difference between the two areas was observed in the first 20 min of the local occlusion. In the NADH level, a moderate decrease was recorded in the core, which created significant differences between it and the penumbra about half an hour after MCA occlusion ($p < 0.05$). The right carotid artery occlusion 45 min after MCA occlusion did not affect CBF or reflectance values. In contrast, in fluorescence, significant differences between the core and penumbra were recorded from the moment of occlusion ($p < 0.05$).

Following the restoration of blood flow in the MCA, CBF and reflectance values in both measurement areas returned to their initial levels. Fluorescence and NADH in the core decreased to levels of 88% and 75%, respectively ($p < 0.05$). This response indicates a return of blood supply to a normal level, and consequently, excess mitochondrial activity to compensate for the energy deficit caused by the ischemia. The removal of the carotid artery occlusion led to a moderate decrease in CBF in the core, down to a level of 60% ($p < 0.01$). An increase in reflectance was recorded in the core, up to a value of 126% ($p < 0.05$). The NADH level in the core at this stage decreased to a value of 65% ($p < 0.01$ relative to baseline), which created a significant difference between the two measurement areas ($p < 0.01$).

Figure 5.58 shows the changes in the different parameters following MCA occlusion with and without ipsilateral carotid artery occlusion 15 min before the reperfusion stage. In the group where the carotid artery was occluded before the MCA occlusion was removed, a low fluorescence level was recorded in the core relative to the control group about 30 min after MCA occlusion for about 10 min ($p < 0.05$). Other than that, no significant differences were recorded between the groups in both the core and the penumbra.

In a control group ($n = 5$) in which the carotid artery was occluded for 45 min, no long-term changes (over 30 s) were observed following the occlusion in any of the measured parameters relative to baseline. Consequently, when comparing the combination of MCAO and UCO to this group, similar results were obtained as compared to the comparison of MCAO and UCO relative to baseline. For this reason, these data are not presented.

Table 5.14 and Fig. 5.59 show the minimum and maximum values for reflectance, NADH, and CBF during a brief nitrogen exposure performed before a 60 min MCA occlusion combined with a 45-min unilateral carotid occlusion, and 1 h after the MCAO was opened. Figure 5.59 demonstrates that after reperfusion, the CBF decrease in the core area was lower relative to the nitrogen exposure before the local ischemia ($p < 0.05$).

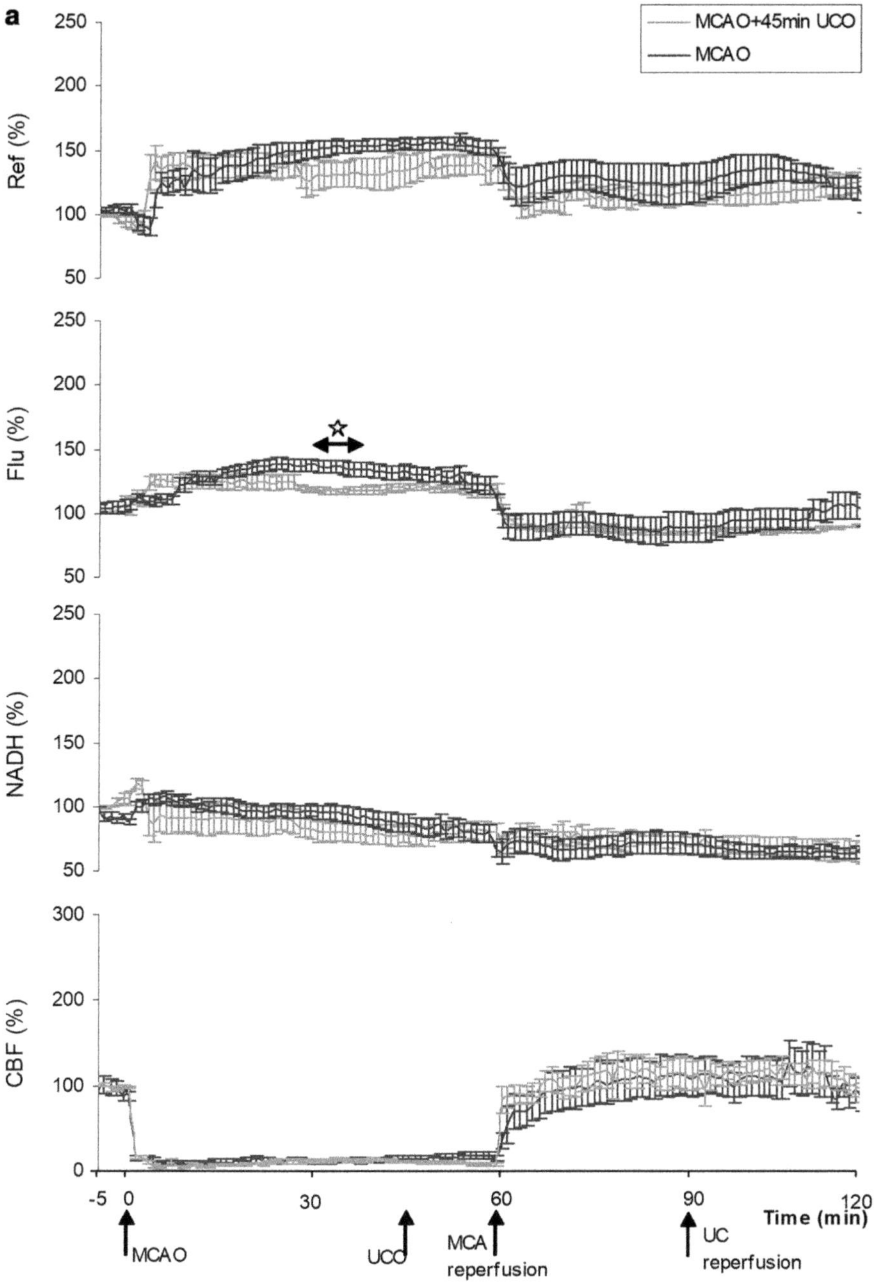

Fig. 5.58 Changes in various parameters in the core (**a**) and penumbra (**b**) following 1 h MCAO with ($n = 8$) or without ($n = 8$) ipsilateral carotid artery occlusion 15 min before the reperfusion phase (UCO). Results are presented as mean ± SE. Significant differences between groups are indicated by asterisks, *$p < 0.05$

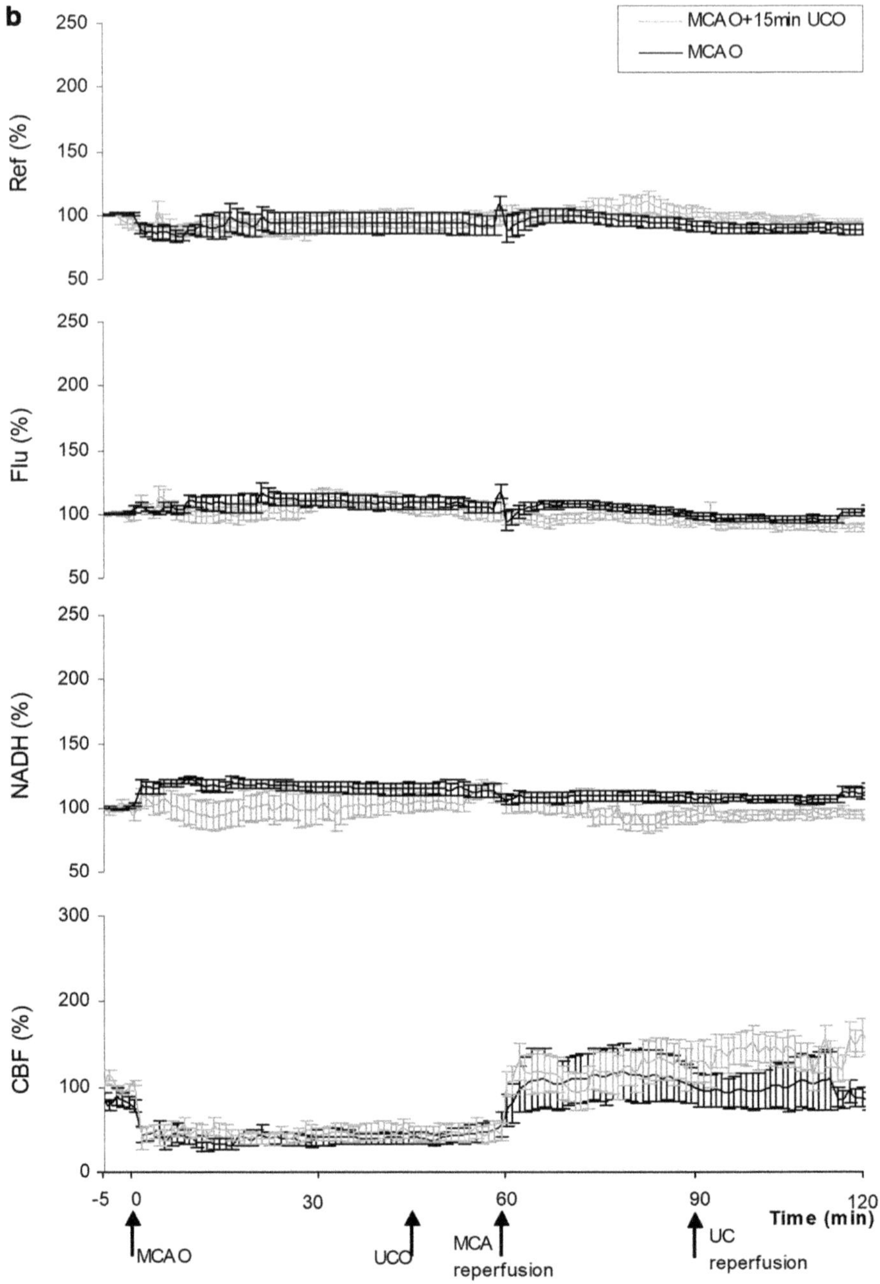

Fig. 5.58 (continued)

Table 5.14 Nitrogen exposure 1 h after MCAO combined with UCO versus control

	Control		MCAO 60 min + UCO 45 min	
	Penumbra	Core	Penumbra	Core
Ref_peak	86 ± 0.9	87 ± 0.7	87 ± 0.8	90 ± 0.5
NADH_max	133 ± 0.9	131 ± 0.3	128 ± 1.1	123 ± 1.7
CBF_min	74 ± 2.1	72 ± 2.1	79 ± 0.6	85 ± 0.7
CBF_max	270 ± 5.6	258 ± 4.6	260 ± 6.4	225 ± 3.1

Minimum and maximum values of reflectance, NADH, and CBF during nitrogen exposure, as measured in the core and penumbra. Nitrogen exposure was performed before MCA occlusion for 1 h in combination with UCO for 45 min (control), and 1 h after MCA opening ($n = 8$).

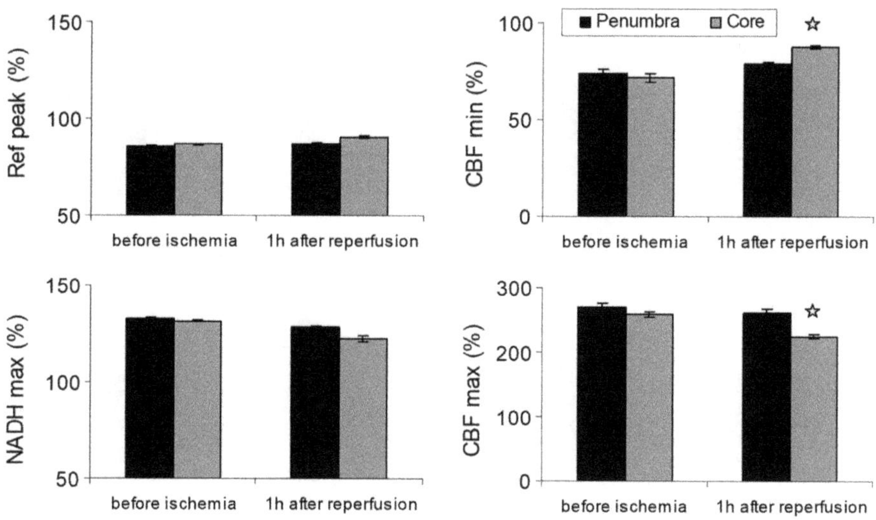

Fig. 5.59 Response to nitrogen exposure in the core and penumbra before MCA and UCO occlusion (left) and 1 h after reperfusion (right). Results are shown as mean ± SE ($n = 8$). Significant differences between two nitrogen treatments are indicated by asterisks, *$p < 0.05$

Table 5.15 and Fig. 5.60 show the maximum and minimum values for reflectance, NADH, and CBF as measured in SD waves induced 1 h after blood flow was restored following a 60-min MCA occlusion combined with a 45-min unilateral carotid occlusion, compared to SD waves in a normoxic state (Control). As can be seen in Fig. 5.60, no significant differences were observed between the waves induced before and after ischemia.

Table 5.15 SD waves 1 h after MCAO combined with UCO compared to control waves

	Control		MCAO 60 min + UCO 45 min	
	Penumbra	Core	Penumbra	Core
Ref_max	119 ± 2.4	108 ± 0.7	119 ±2.7	124 ± 3.5
Ref_min	78 ± 0.5	86 ± 1.1	84 ± 1.4	91 ± 0.8
NADH_peak	87 ± 0.8	86 ± 0.7	88 ± 1.9	91 ± 0.9
CBF_max	250 ± 10.7	221 ± 11.9	165 ± 5.8	162±

Minimum and maximum values of reflectance, NADH, and CBF during induced SD waves, as measured in the core and penumbra. SD waves were induced 1 h after reperfusion following MCAO for 60 min and single carotid occlusion for 45 min (n = 8). In another group, SD waves were induced under normoxic conditions (n = 8).

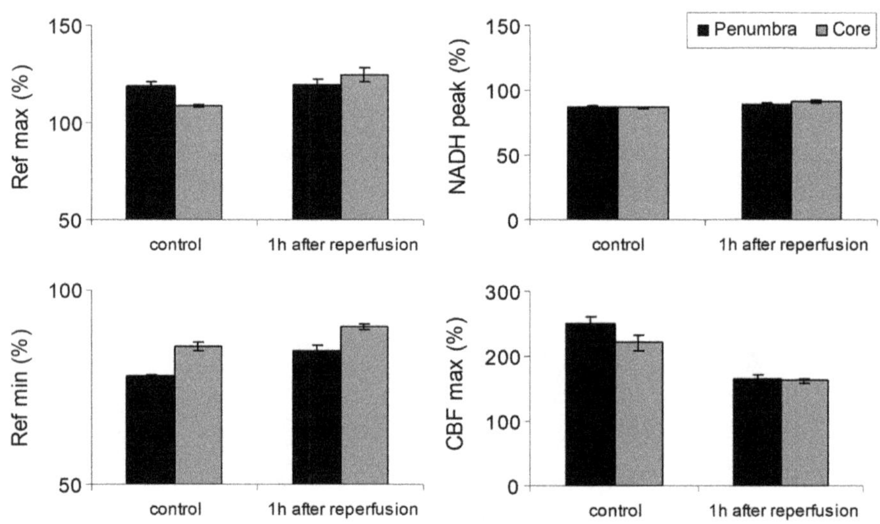

Fig. 5.60 Changes in various parameters in SD waves, evoked 1 h after reperfusion following MCAO for 60 min and UCO for 45 min (1 h after reperfusion, n = 8), compared to SD waves evoked in normoxic conditions (control, n = 8), as observed in the core and penumbra. Results are presented as mean ± SE

5.3.7.2 Bilateral Carotid Occlusion for 45 min

Figure 5.61 shows the responses of reflectance, fluorescence, NADH, and CBF to a 60-min MCA occlusion combined with a 45-min bilateral carotid occlusion that was initiated 15 min before the MCA occlusion was removed. Upon MCA occlusion, blood flow decreased in both measurement areas, to 15% in the core area ($p < 0.001$ relative to baseline) and 40% in the penumbra area ($p < 0.001$). The differences in CBF between the core and penumbra were significant for most of the time the MCA remained occluded ($p < 0.05$).

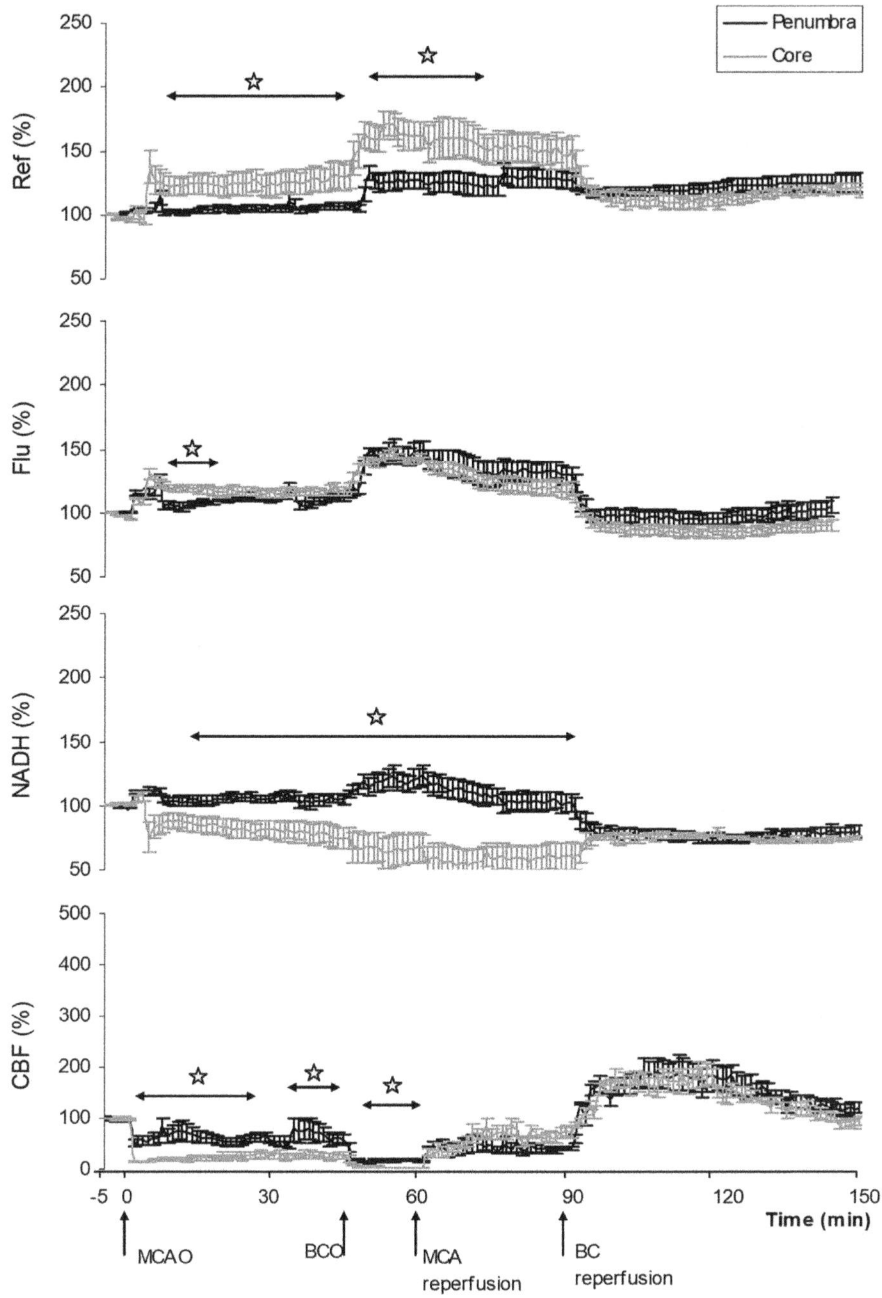

Fig. 5.61 Hemodynamic and metabolic changes in the core and penumbra following 1 h right MCA occlusion combined with 45 minbilateral carotid occlusion, performed 15 min before MCA deblockage. Results are presented as mean ± SE ($n = 8$). Significant differences between the two measurement sites are indicated by asterisks, $*p < 0.05$

Following MCA occlusion, a 6% decrease in reflectance values was recorded in the core area. After 3 min, an increase was recorded up to maximum values of 137% in the core ($p < 0.05$) and 113% in the penumbra ($p < 0.05$). Reflectance values remained high in the core area during the MCA occlusion time, and returned to baseline values in the penumbra. As a result, significant differences between the two measurement areas were recorded at this stage ($p < 0.05$). In fluorescence values, an increase was recorded up to maximum values of 129% in the core and 121% in the penumbra (in both areas $p < 0.05$ relative to baseline).

Fluorescence remained high in the core. In the penumbra, a decrease to baseline was recorded, but then an increase and stabilization at a level of 110%. A significant difference between the two areas was observed from the 5th minute of occlusion until the 20th minute. In the NADH level, a decrease of up to 30% was recorded in the core ($p < 0.01$ relative to baseline), which created significant differences between it and the penumbra 15 min after MCA occlusion ($p < 0.05$).

The carotid artery occlusion 45 min after MCA occlusion led to a decrease in blood flow to values of 5% in the core ($p < 0.001$ relative to baseline and $p < 0.05$ relative to blood flow before carotid occlusion) and 15% in the penumbra ($p < 0.001$ relative to baseline, $p < 0.05$ relative to blood flow before carotid occlusion). A significant difference ($p < 0.05$) was maintained between the two measurement areas. In the reflectance level, an additional increase was recorded, to levels of 170% in the core ($p < 0.001$ relative to baseline and before carotid occlusion) and 133% in the penumbra ($p < 0.01$ relative to baseline and before carotid occlusion), a significant difference between the two areas ($p < 0.05$). In the reflectance level, an additional increase was recorded, to values of 146% in both areas ($p < 0.001$ relative to baseline and before carotid occlusion). In the NADH level, an increase was recorded in the penumbra to a level of 115% ($p < 0.01$ relative to baseline and before carotid occlusion), while in the core, NADH levels remained low due to the high reflectance level, which led to a significant difference between the two areas ($p < 0.05$).

The removal of MCA occlusion led to an increase in blood flow to values of 50% in both measurement areas ($p < 0.01$ relative to baseline and relative to the previous stage). The other parameters were not affected by the occlusion removal at this stage. The restoration of blood flow in the carotid arteries 30 min later led to an increase in blood flow up to values of 180% in both areas ($p < 0.01$ relative to baseline). After about 1 h, blood flow values returned to baseline in both areas. In the reflectance level, a decrease to values of 115% was recorded in both measurement areas ($p < 0.05$ relative to baseline). The reflectance level remained higher than its baseline value for the next hour. Fluorescence values decreased to values of up to 85% in the core ($p < 0.05$). The fluorescence level in the penumbra returned to baseline values. The NADH level decreased to a level of up to 78% in both areas.

Figure 5.62 shows the changes in the different parameters following a 60 min MCA occlusion with and without bilateral carotid occlusion 15 min before the reperfusion stage. In the group where the carotid arteries were occluded, significantly lower fluorescence levels ($p < 0.05$) were observed in the core relative to the group with MCA occlusion only, 15 min after occlusion.

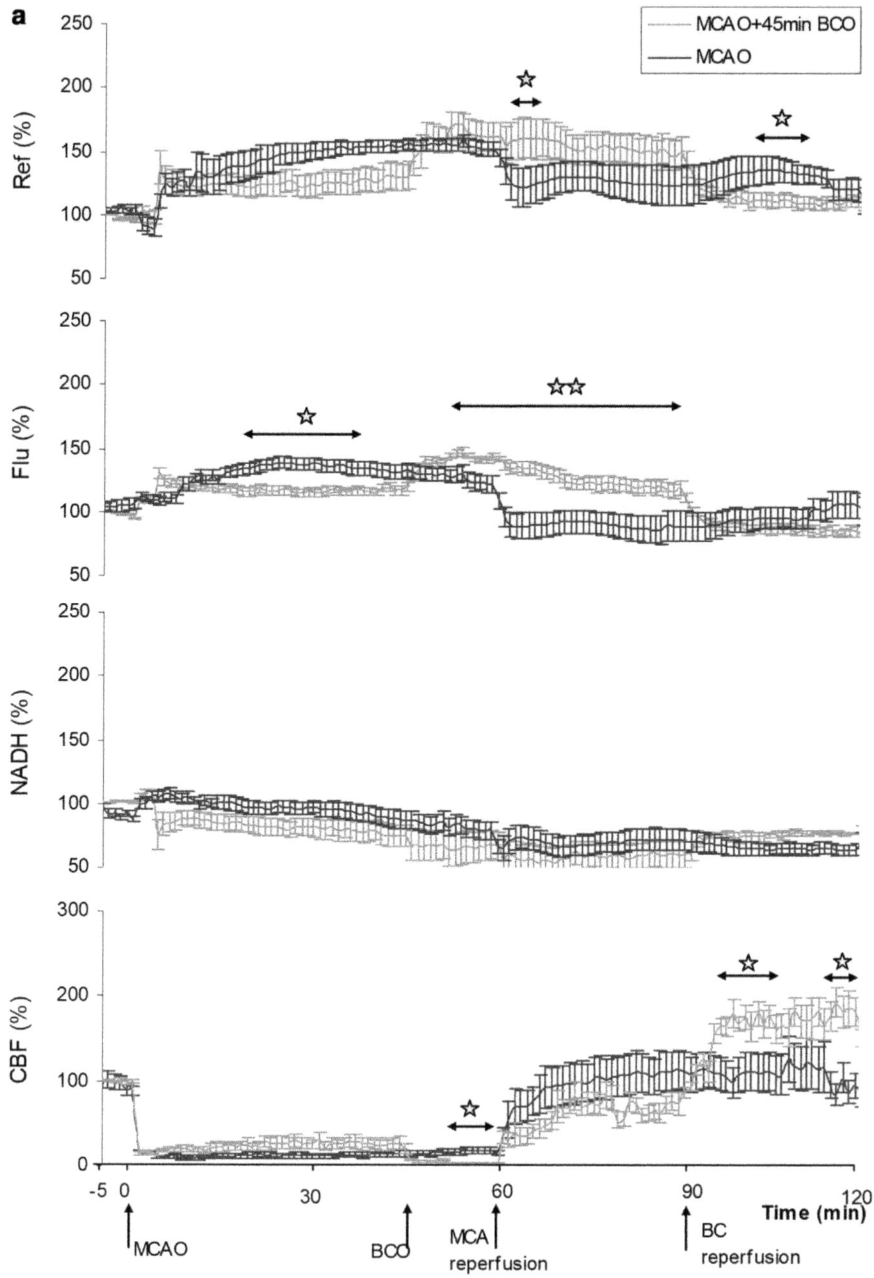

Fig. 5.62 Changes in various parameters following MCAO for 1 h with ($n = 8$) or without ($n = 9$) occlusion of both carotid arteries 15 min before the reperfusion phase (BCO), as measured in the core (**a**) and penumbra (**b**). Results are presented as mean ± SE. Significant differences between groups are indicated by asterisks, *$p < 0.05$, **$p < 0.01$

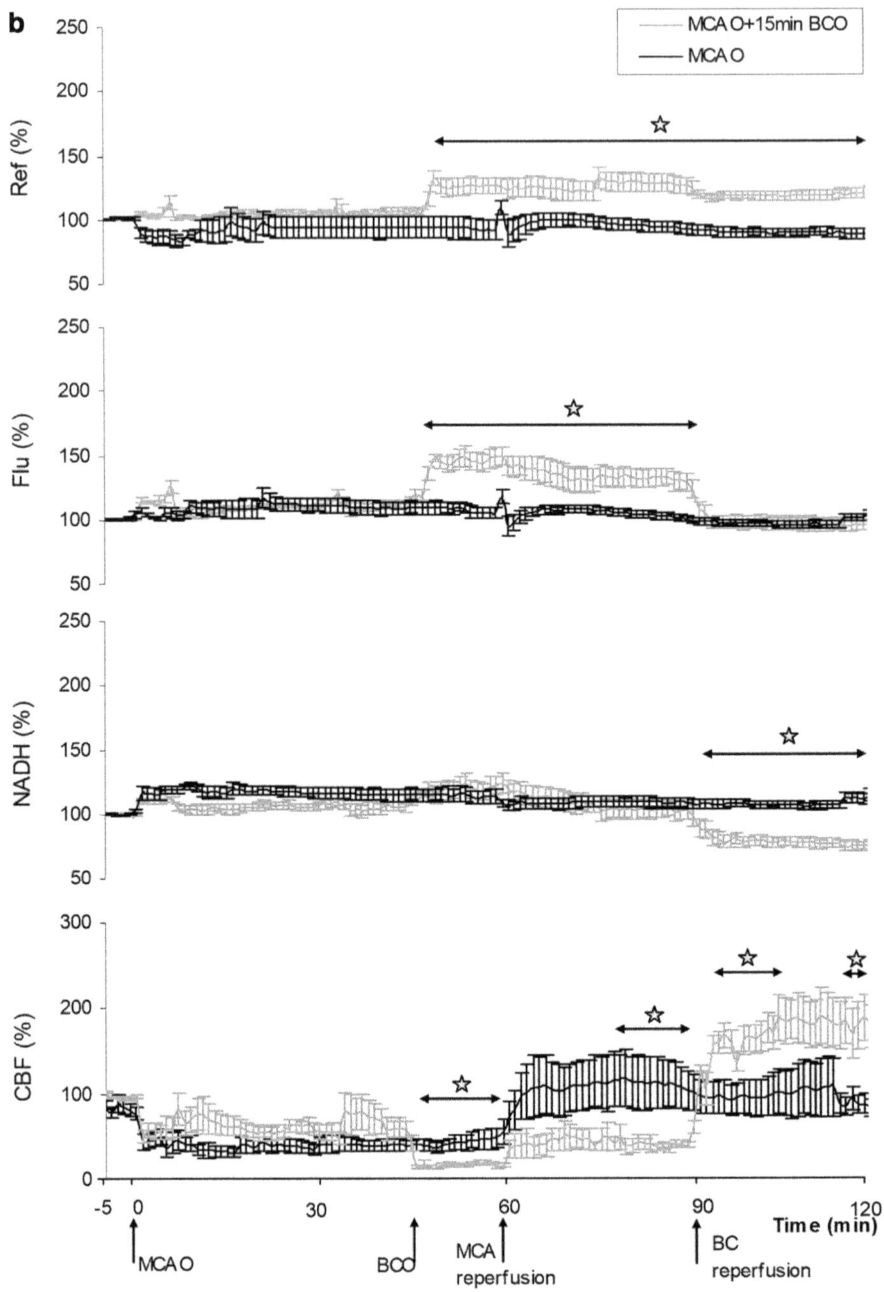

Fig. 5.62 (continued)

The bilateral carotid occlusion led to significantly lower blood flow ($p < 0.05$) and significantly higher fluorescence values ($p < 0.05$) at both measurement sites relative to the group where the carotid arteries were not occluded. In addition, in the penumbra, high reflectance levels were observed following carotid occlusion ($p < 0.05$). Following the removal of MCA occlusion while the carotid arteries remained occluded, fluorescence values remained higher than usual relative to the second group, and high reflectance values were also observed in the core ($p < 0.05$). Blood flow in the penumbra in the group where the carotids were occluded was lower at this stage relative to the group where the carotids were not occluded ($p < 0.05$). The removal of the carotid artery occlusion led to high CBF levels ($p < 0.05$) and high reflectance levels ($p < 0.05$) at both measurement sites relative to the other groups. In addition, significant differences between the groups were recorded in the NADH level in the penumbra at this stage ($p < 0.05$).

In a control group ($n = 5$) in which only the two carotid arteries were blocked for 45 min, an initial decrease of approximately 50% in blood flow was observed ($p < 0.01$) and after approximately 2 min stabilization at a value of 80% was observed until the blockage was removed. An increase of approximately 15% ($p < 0.05$) was observed in the fluorescence and reflectance levels and after approximately 2 min a return to baseline levels was observed. When comparing the combination of MCAO and BCO to this group, similar results were obtained as compared to the comparison between MCAO and BCO with respect to the initial measurement level. For this reason, the data are not shown.

Table 5.16 and Fig. 5.63 show the minimum and maximum values for reflectance, NADH, and CBF during a brief nitrogen exposure performed before a 60-min MCA occlusion combined with a 45-min bilateral carotid occlusion, and 1 h after blood flow was restored in both. No significant differences were observed between the two treatments.

Table 5.17 and Fig. 5.64 show the maximum and minimum values for reflectance, NADH, and CBF as measured in SD waves induced 1 h after blood flow was restored following a 60-min MCA occlusion combined with a 45-min bilateral carotid occlusion, compared to SD waves in a normoxic state. No significant differences were observed between the waves induced before and after ischemia.

5.3.7.3 Unilateral Carotid Occlusion for 15 min

Figure 5.65 shows the values for reflectance, fluorescence, NADH, and CBF for a 60 min MCA occlusion, combined with a 15-min unilateral carotid occlusion that was performed simultaneously with the removal of the MCA occlusion. Upon MCA occlusion, blood flow decreased in both measurement areas, to 14% in the core area ($p < 0.001$ relative to baseline) and 39% in the penumbra area ($p < 0.001$). The differences in CBF between the core and penumbra were significant for the time the MCA remained occluded ($p < 0.05$). Following MCA occlusion, a decrease in reflectance values was recorded—4% in the core and 8% in the penumbra. After 3

Table 5.16 Nitrogen exposure 1 h after MCAO combined with BCO versus control

	Control		MCAO 60 min + BCO 45 min	
	Penumbra	Core	Penumbra	Core
Ref_peak	86 ± 0.9	86 ± 0.7	87 ± 0.8	91 ± 0.5
NADH_max	141 ± 4.5	130 ± 0.3	119 ±2.6	122 ± 1.7
CBF_min	60 ± 4.1	73 ± 3.2	79 ± 0.6	84 ± 0.6
CBF_max	233 ± 5.2	240 ± 4.1	221 ± 3.8	225 ± 3.1

Minimum and maximum values of reflectance, NADH, and CBF during nitrogen exposure, as measured in the core and penumbra. Nitrogen exposure was performed before MCA occlusion for 1 h, in combination with BCO for 45 min, and 1 h after reperfusion ($n = 8$).

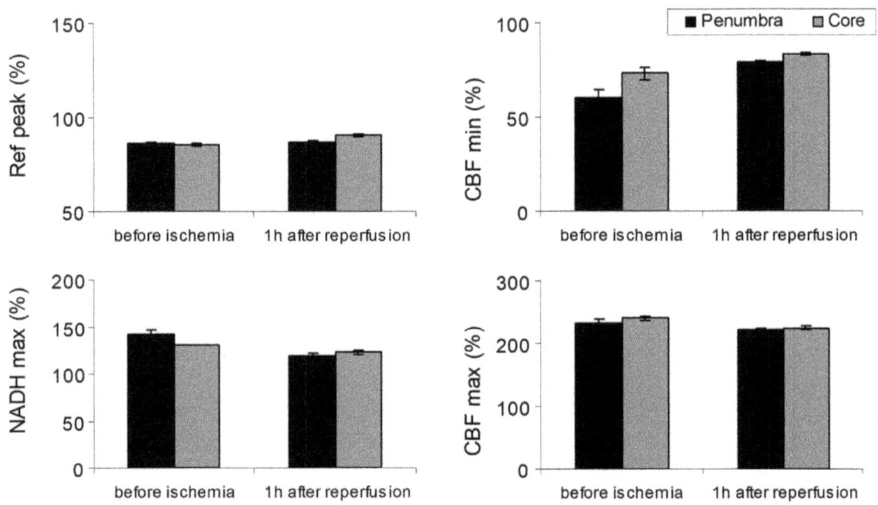

Fig. 5.63 Response to nitrogen exposure before MCA and BCO occlusion (left) and 1 h after reperfusion (right), as measured in the core and penumbra. Results are shown as mean ± SE ($n = 8$). Significant differences between two nitrogen treatments are indicated by asterisks, *$p < 0.05$

min, an increase was recorded in the core up to maximum values of 152% ($p < 0.001$ relative to baseline, $p < 0.01$ relative to penumbra). Reflectance values remained high in the core area for the rest of the experiment, and returned to baseline values in the penumbra. In fluorescence values, an increase was recorded up to maximum values of 134% in the core and 112% in the penumbra (in both areas $p < 0.05$ relative to baseline). A significant difference between the two areas was observed for up to 40 min from the moment of occlusion. The NADH level decreased moderately in the core and increased in the penumbra throughout the entire experiment.

The removal of MCA occlusion simultaneous with the unilateral carotid occlusion led to an increase in CBF, to maximum levels of 85% in the core and 106% in the

Table 5.17 SD waves 1 h after MCAO combined with BCO compared to control waves

	Control		MCAO 60 min + BCO 4 5 min	
	Penumbra	Core	Penumbra	Core
Ref_max	119 ± 2.4	108 ± 0.7	119 ± 2.7	124 ± 3.5
Ref_min	78 ± 0.5	86 ± 1.1	85 ± 1.4	79 ± 0.5
NADH_peak	87 ± 0.8	86 ± 0.7	8 9 ± 1.9	91 ± 0.9
CBF_max	250 ± 10.7	221 ± 11.9	165 ± 5.8	162 ± 4.1

Minimum and maximum values of reflectance, NADH, and CBF during induced SD waves, as measured in the core and penumbra. SD waves were induced 1 h after reperfusion following MCAO for 60 min and bilateral carotid occlusion for 45 min ($n = 8$). In another group, SD waves were induced under normoxic conditions ($n = 8$).

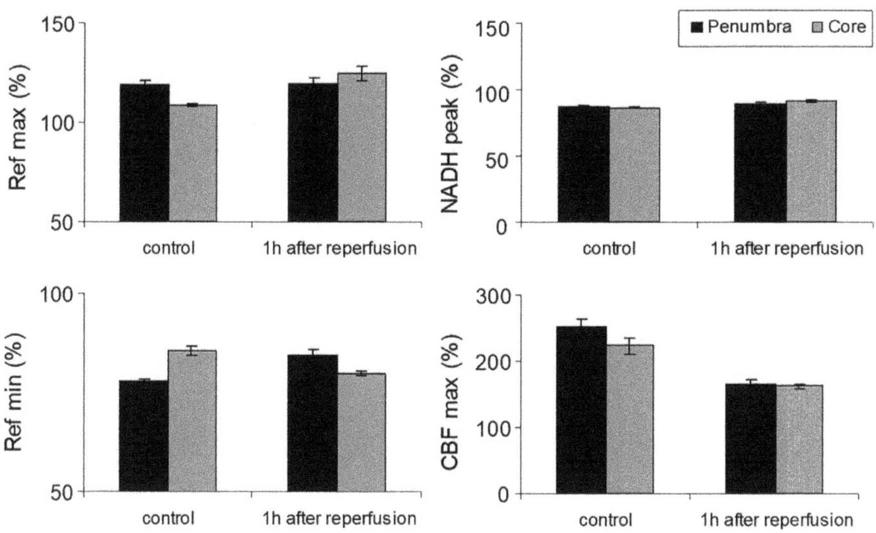

Fig. 5.64 Changes in various parameters in SD waves, evoked 1 h after reperfusion following MCAO for 60 min and UCO for 45 min (1 h after reperfusion, $n = 8$), compared to SD waves evoked in normoxic conditions (control, $n = 8$), as observed in the core and penumbra. Results are presented as mean ± SE

penumbra. Reflectance values in the core remained significantly higher relative to their baseline and relative to the penumbra ($p < 0.01$). The fluorescence level returned to baseline in both areas, and in NADH, values of about 72% were recorded in the core at this stage—values that were low both relative to the beginning of the experiment ($p < 0.001$) and relative to the penumbra at this stage ($p < 0.05$). The removal of the carotid artery occlusion 15 min after its occlusion did not lead to significant changes in the measured parameters.

Figure 5.66 shows the changes in the different parameters following a 60-min MCA occlusion with and without ipsilateral carotid artery occlusion for 15 min simultaneous

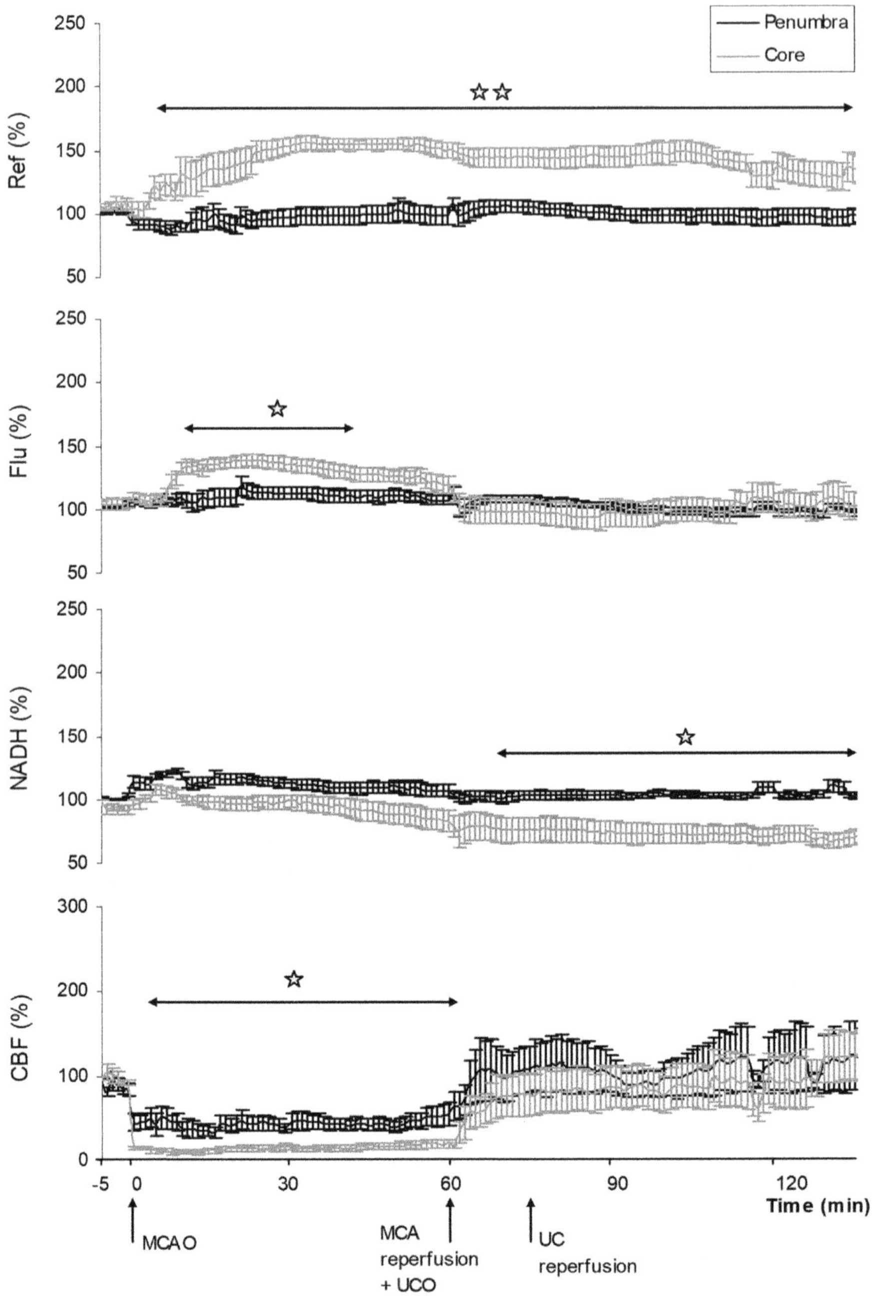

Fig. 5.65 Hemodynamic and metabolic changes in the core and penumbra following 1 h right MCA occlusion combined with 15-min single carotid artery occlusion performed concurrently with MCA decompression. Results are presented as mean ± SE ($n = 8$). Significant differences between the two measurement sites are indicated by asterisks, *$p < 0.05$

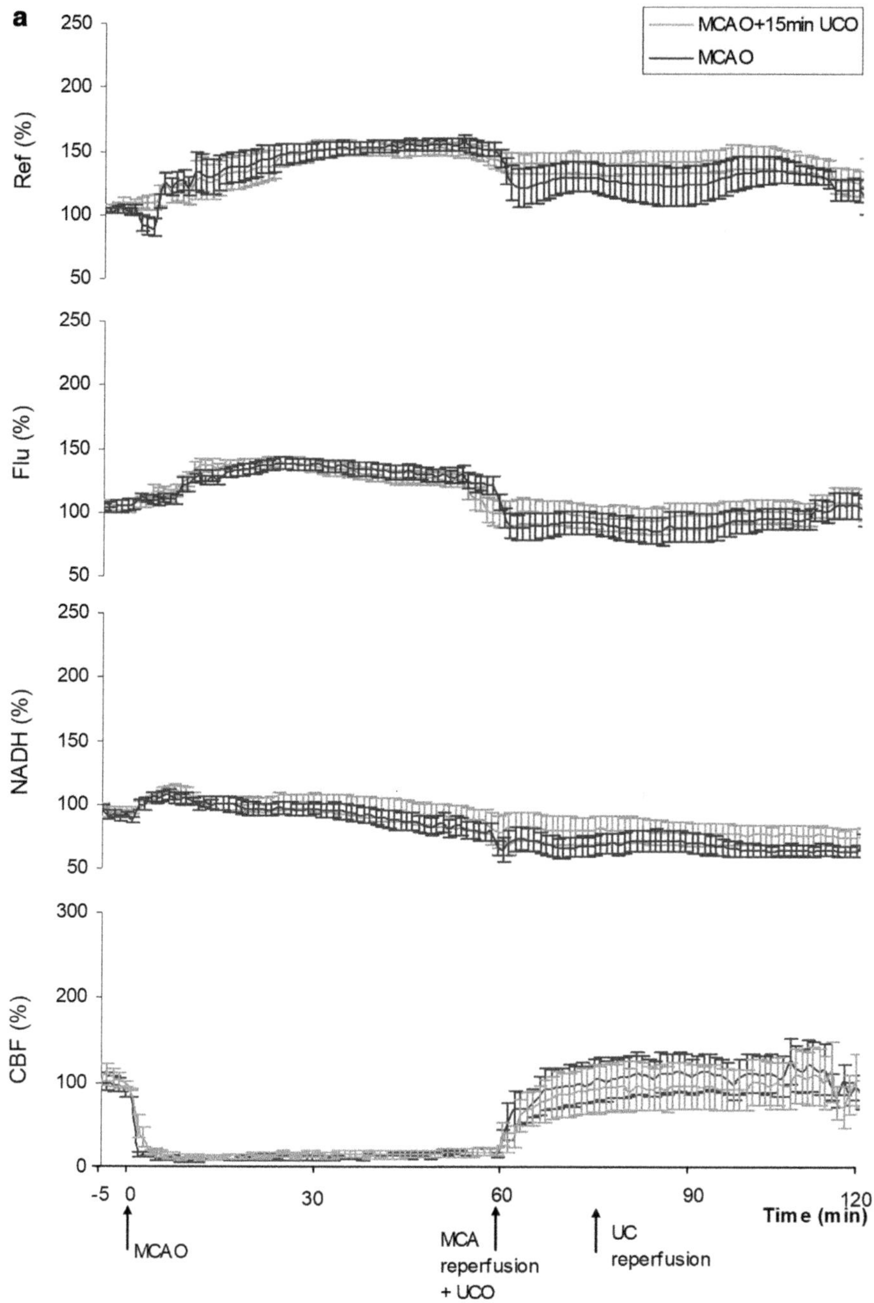

Fig. 5.66 Changes in various parameters in the core (**a**) and penumbra (**b**) following 1 h MCAO with ($n = 8$) or without ($n = 8$) 15-min ipsilateral carotid artery occlusion (UCO) concomitant with MCA deblockage. Results are presented as mean ± SE. Significant differences between groups are indicated by asterisks, $*p < 0.05$

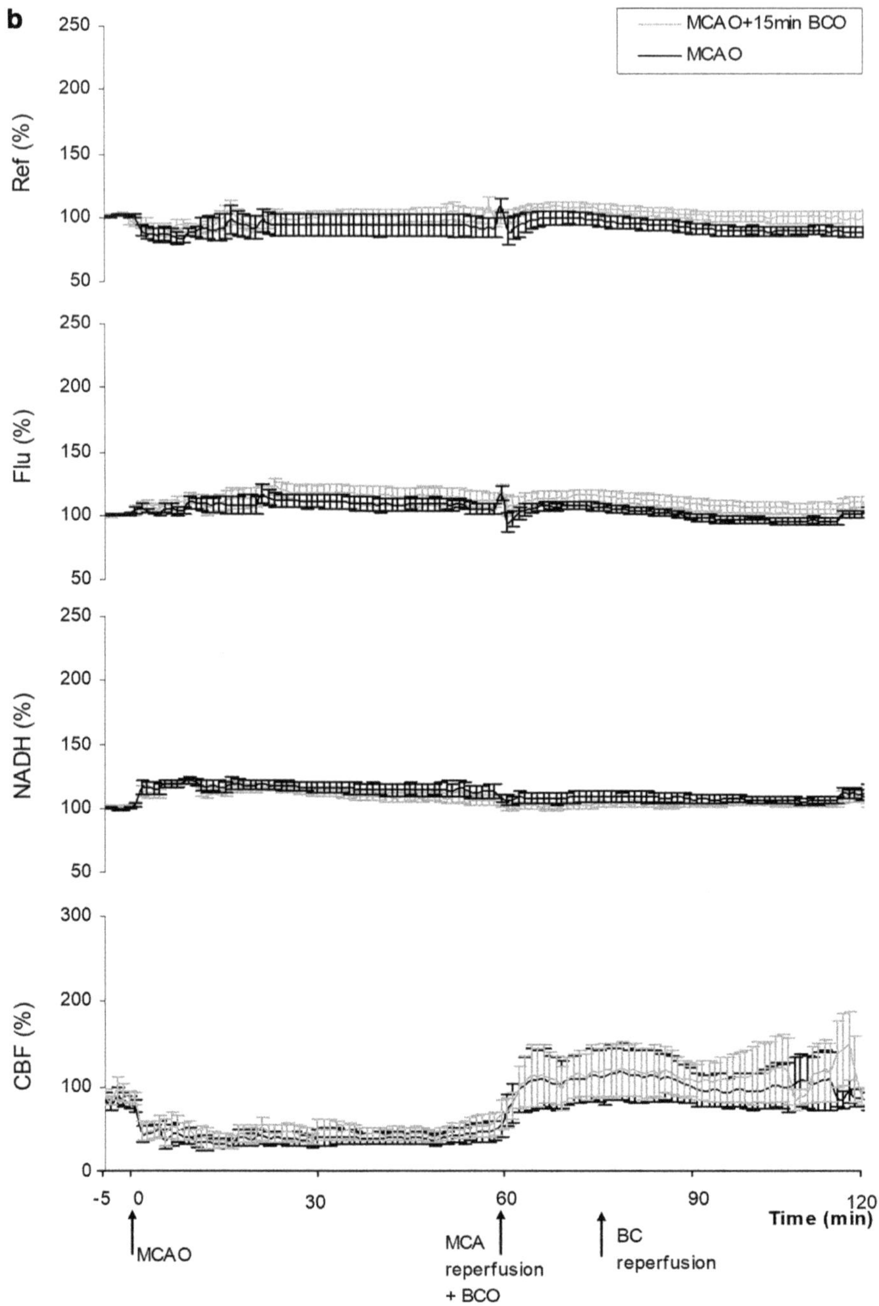

Fig. 5.66 (continued)

with the MCA reperfusion stage. No significant differences were recorded between the groups in both the core and the penumbra.

In a control group ($n = 5$) in which the carotid artery was occluded for 45 min, no long-term changes (over 30 s) were observed following occlusion in any of the measured parameters relative to baseline. Consequently, when comparing the combination of MCAO and UCO to this group, similar results were obtained as compared to the comparison of MCAO and UCO relative to baseline (data not shown).

Table 5.18 and Fig. 5.67 show the minimum and maximum values for reflectance, NADH, and CBF during a brief nitrogen exposure performed before a 60-min MCA occlusion combined with a 45-min unilateral carotid occlusion, and 1 h after the

Table 5.18 Nitrogen exposure 1 h after MCAO combined with UCO versus control

	Control		MCAO 60 min + UCO 45 min	
	Penumbra	Core	Penumbra	Core
Ref_peak	86 ± 0.49	82 ± 1.27	8 8 ± 0.78	90 ± 0.51
NADH_max	132 ± 0.53	131 ± 0.37	129 ± 1.06	121 ± 1.72
CBF_min	75 ± 1.79	75 ± 2.12	80 ± 0.57	83 ± 0.63
CBF_max	304 ± 10.05	267 ± 3.13	260 ± 6.39	240 ± 3.86

Minimum and maximum values of reflectance, NADH and CBF during nitrogen exposure, as measured in the core and penumbra. Nitrogen exposure was performed before ischemia (control) and after MCA occlusion for 1 h in combination with UCO for 15 min ($n = 8$).

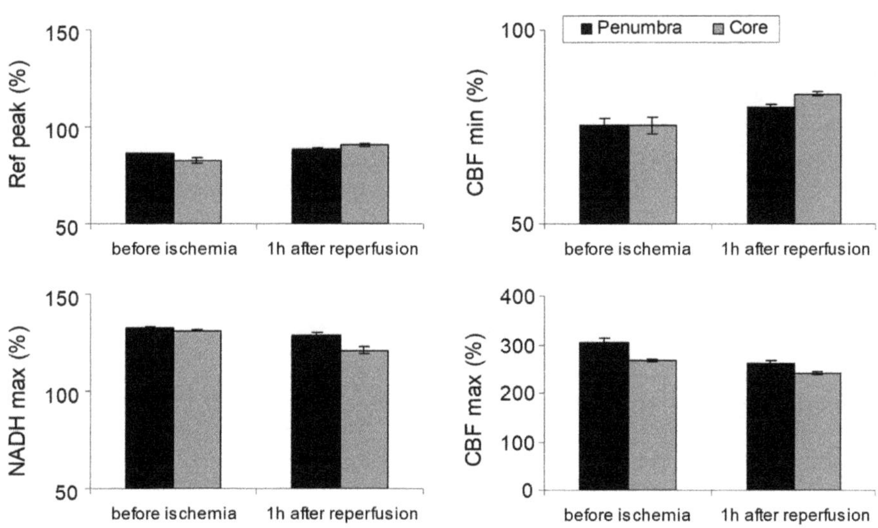

Fig. 5.67 Response to nitrogen exposure in the core and penumbra before MCA and UCO occlusion and 1 h after reperfusion. Results are presented as mean ± SE ($n = 8$). Significant differences between two nitrogen treatments are indicated by asterisks, $*p < 0.05$

MCAO was opened. No changes were observed between nitrogen exposure before MCA occlusion and UCO and nitrogen exposure after reperfusion.

Table 5.19 and Fig. 5.68 show the maximum and minimum values for reflectance, NADH, and CBF as measured in SD waves induced 1 h after blood flow was restored following a 60-min MCA occlusion combined with a 45-min unilateral carotid occlusion, compared to SD waves in a normoxic state (Control). No significant differences were observed between the waves induced before and after ischemia.

Table 5.19 SD waves 1 h after MCAO combined with UCO compared to control waves

	Control		MCAO 60 min + UCO 45 min	
	Penumbra	Core	Penumbra	Core
Ref_max	117 ± 2.3	105 ± 1.01	117 ± 2.69	122 ± 3.53
Ref_min	76 ± 0.53	84 ± 1.09	83 ± 1.29	89 ± 0.81
NADH_peak	85 ± 0.8	85 ± 0.64	88 ± 1.98	90 ± 1.02
CBF_max	242 ± 9.29	216 ± 11.65	162 ± 5.61	160 ± 4.1

Minimum and maximum values of reflectance, NADH, and CBF during induced SD waves, as measured in the core and penumbra. SD waves were induced 1 h after reperfusion following MCAO for 60 min and single carotid occlusion for 15 min ($n = 8$). In another group, SD waves were induced under normoxic conditions ($n = 8$).

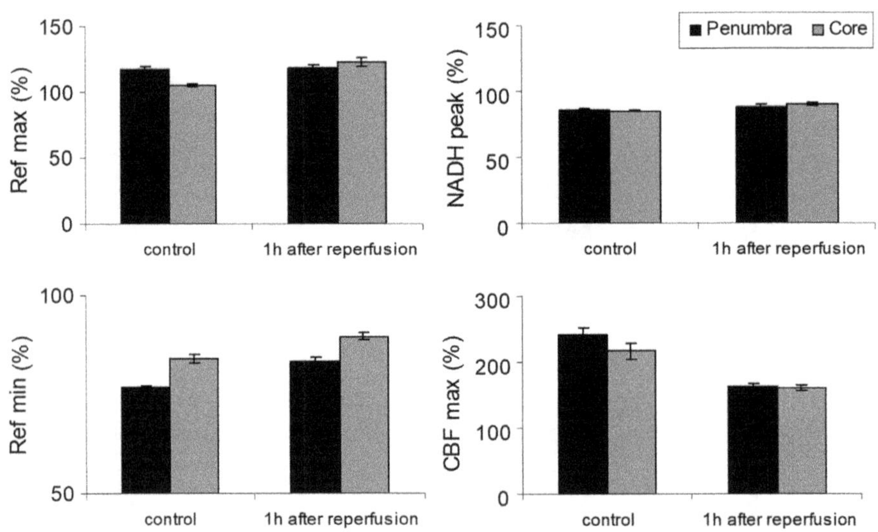

Fig. 5.68 Changes in various parameters in SD waves, evoked 1 h after reperfusion following MCAO for 60 min and UCO for 15 min (1 h after reperfusion, $n = 8$), compared to SD waves evoked in normoxic conditions (control, $n = 8$), as observed in the core and penumbra. Results are presented as mean ± SE

5.3.7.4 Bilateral Carotid Occlusion for 15 min

Figure 5.69 shows the responses of reflectance, fluorescence, NADH, and CBF to a 60 min MCA occlusion, combined with a 15-min bilateral carotid occlusion that was performed simultaneously with the removal of the MCA occlusion. During MCA occlusion, blood flow decreased in both measurement areas, to 10% in the core area ($p < 0.001$ relative to baseline) and 39% in the penumbra area ($p < 0.001$). The differences in CBF between the core and penumbra were significant for most of the time the MCA remained occluded ($p < 0.05$). In fluorescence values, an increase was recorded up to maximum values of 137% in the core ($p < 0.001$ relative to baseline) and 111% in the penumbra ($p < 0.05$). Fluorescence remained at a high level in the core until the reperfusion stage, while in the penumbra, a decrease to baseline was recorded. A significant difference between the two areas was observed until the MCA occlusion was removed.

In the NADH level, a decrease of up to 21% was recorded in the core ($p < 0.01$ relative to baseline), which led to significant differences between it and the penumbra in the first 30 min after MCA occlusion ($p < 0.05$). Three minutes after occlusion, an increase in reflectance was recorded, up to maximum values of 147% in the core ($p < 0.01$ relative to baseline and $p < 0.05$ relative to penumbra). Reflectance values remained high in the core area during the MCA occlusion time.

Following the restoration of blood flow in the MCA simultaneous with the bilateral carotid occlusion, significant differences between the measurement areas were recorded in all measured parameters ($p < 0.05$). With the removal of the carotid artery occlusion, CBF values increased to 220% in the core ($p < 0.01$ relative to baseline). CBF values in the penumbra returned to their initial levels, which caused significant differences between the areas for 15 min after the occlusion was removed ($p < 0.05$). In the reflectance level, a return to baseline values was recorded in both areas, but an additional increase in the core led to a significant difference between the areas ($p < 0.05$). Fluorescence values decreased in both measurement areas to their initial level. In the NADH level, a decrease of 30% was recorded in the core ($p < 0.01$ relative to baseline and $p < 0.05$ relative to penumbra).

Figure 5.70 shows the changes in the different parameters following a 60-min MCA occlusion with and without bilateral carotid occlusion simultaneous with the reperfusion stage. Upon restoration of blood flow in the MCA, significant differences were recorded in the core between the groups in CBF ($p < 0.05$), reflectance ($p < 0.05$), and fluorescence ($p < 0.01$). After the restoration of blood flow in the carotid arteries, a high level of CBF was recorded relative to the group where the carotids were not occluded for about 10 min ($p < 0.05$), while no significant differences were recorded in the other parameters at this stage. In the penumbra, no differences were recorded between the groups throughout the entire experiment.

In a control group ($n = 5$) in which only the two carotid arteries were blocked for 15 min, an initial decrease of approximately 50% in blood flow was observed ($p < 0.01$) and after approximately 2 min stabilization at a value of 80% was observed until the blockage was removed. An increase of approximately 15% ($p < 0.05$) was observed

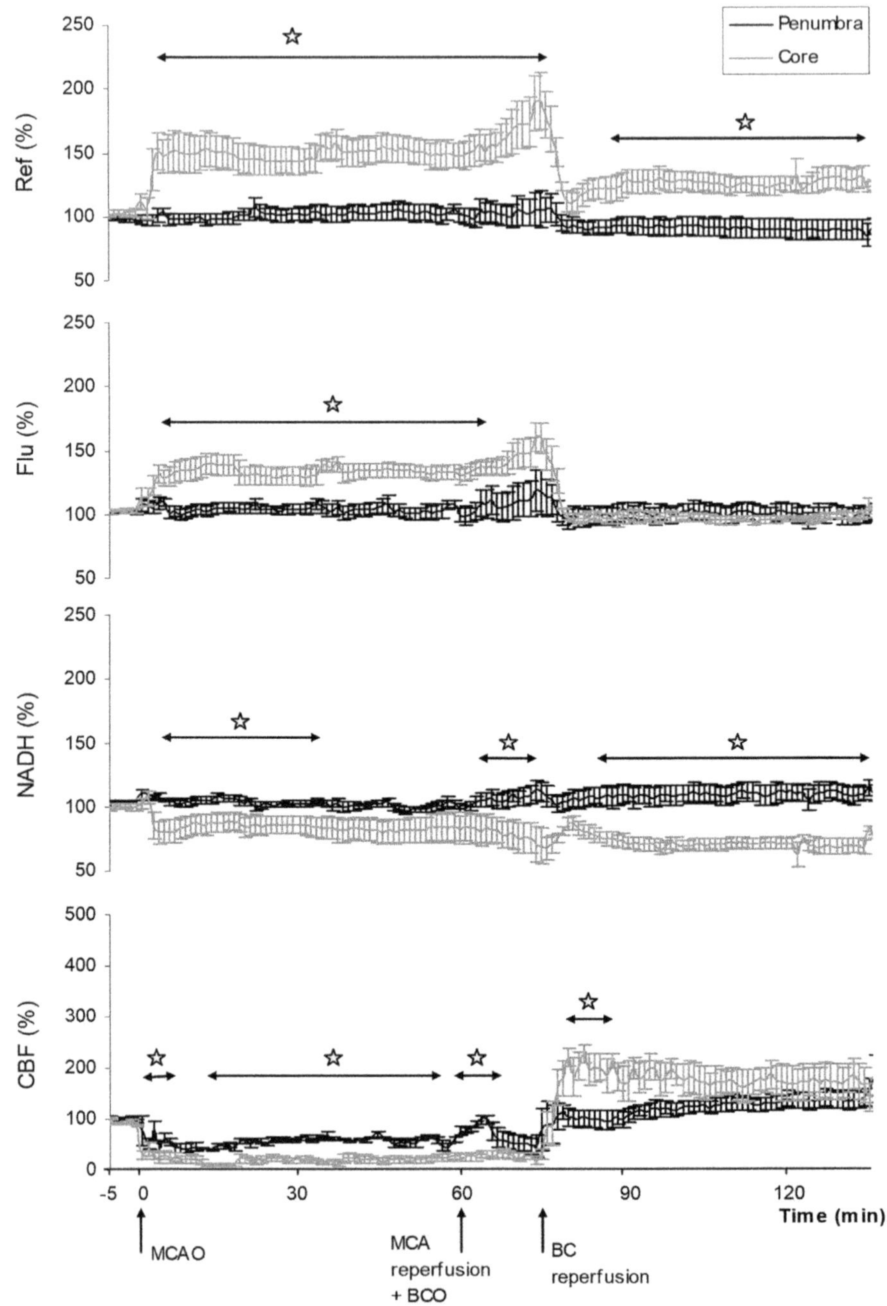

Fig. 5.69 Hemodynamic and metabolic changes in the core and penumbra following 1-h occlusion of the right MCA combined with 15-min occlusion of both carotid arteries, performed simultaneously with MCA decompression. Results are presented as mean ± SE ($n = 8$). Significant differences between the two measurement sites are indicated by asterisks, $*p < 0.05$

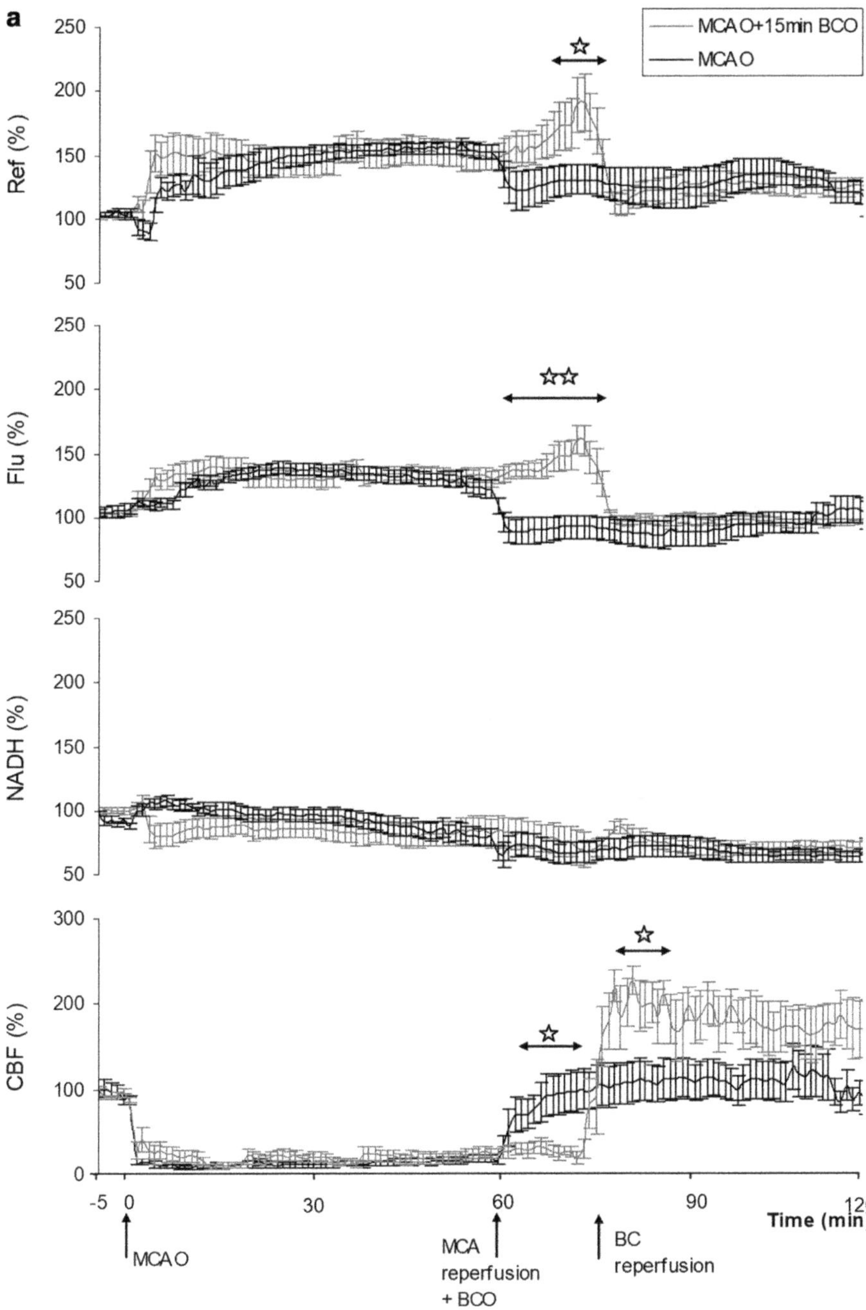

Fig. 5.70 Changes in various parameters following MCAO for 1 h with ($n = 8$) or without ($n = 8$) bilateral carotid occlusion during the reperfusion phase (BCO), as measured in the core (**a**) and penumbra (**b**). Results are presented as mean ± SE. Significant differences between groups are indicated by asterisks, $*p < 0.05$, $**p < 0.01$

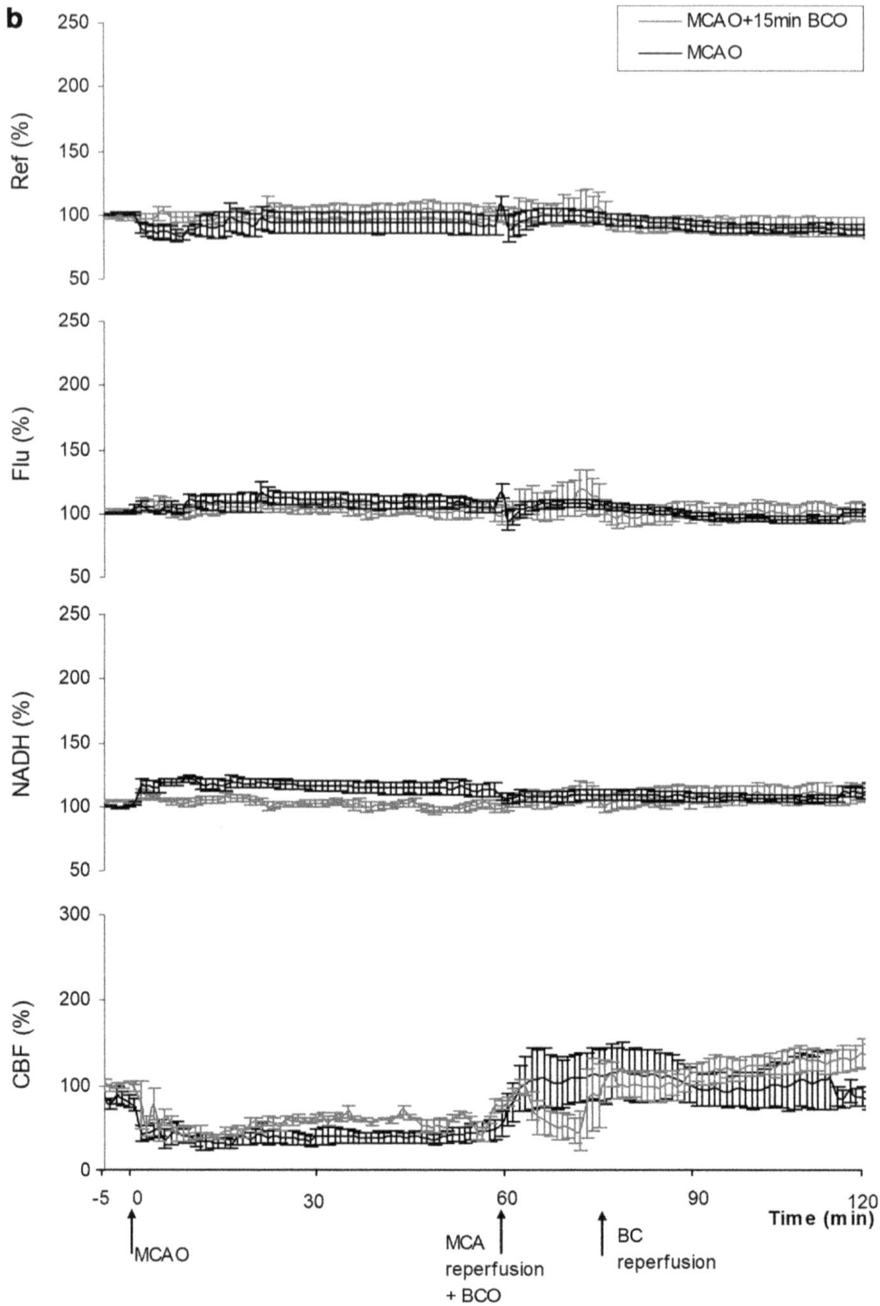

Fig. 5.70 (continued)

in the fluorescence and reflectance levels and after approximately 2 min a return to baseline levels was observed. When comparing the combination of MCAO and BCO to this group, similar results were obtained as compared to the comparison between MCAO and BCO with respect to the initial measurement level. For this reason, the data are not shown.

Table 5.20 and Fig. 5.71 show the minimum and maximum values of reflectance, NADH, and CBF during brief nitrogen exposures. The first exposure was performed before the ischemia stage, and the second exposure was performed after a 60-min MCA occlusion followed by a 15-min bilateral carotid occlusion. No significant differences were observed between the two treatments.

Table 5.21 and Fig. 5.72 show the maximum and minimum values for reflectance, NADH, and CBF as measured in SD waves induced 1 h after blood flow was restored following a 60-min MCA occlusion combined with a 15-min bilateral carotid occlusion, compared to SD waves in a normoxic state. No significant differences were observed between the waves induced before and after ischemia.

5.4 Discussion

5.4.1 Introduction

This study examined the effect of various states of cerebral ischemia on the brain's energy metabolism by continuously and in real time measuring mitochondrial function in combination with blood flow. Although much research has been conducted using CBF measurements during cerebral ischemia (Lipton 1999; Nagasawa and Kogure 1989; Prieto et al. 2005; Schmid-Elsaesser et al. 1998), there are not many studies that have measured NADH levels (Sims and Muyderman 2010) which indicate mitochondrial activity (Mayevsky 1984), in combination with CBF. The current study aims to address this point, with an emphasis on focal ischemia, using the middle cerebral artery occlusion (MCAO) model. This work used a unique measurement system that allows for the measurement of these parameters in two hemispheres or at two measurement points in the same hemisphere.

Table 5.20 Nitrogen exposure 1 h after MCAO and BCO for duration versus control

	Control		MCAO 60 min + BCO 45 min	
	Penumbra	Core	Penumbra	Core
Ref_peak	86 ± 0.9	85 ± 0.69	87 ± 0.78	90 ± 0.5
NADH_max	141 ± 4.5	130 ± 0.34	119 ± 2.56	122 ± 1.67
CBF_min	60 ± 4.11	72 ± 3.15	79 ± 0.57	83 ± 0.55
CBF_max	233 ± 5.25	240 ± 4.12	221 ± 3.82	224 ± 3.14

Minimum and maximum values of reflectance, NADH, and CBF during nitrogen exposure, as measured in the core and penumbra. Nitrogen exposure was performed before ischemia, and 1 h after reperfusion following 1 h of MCA occlusion combined with 15 min of BCO ($n = 8$).

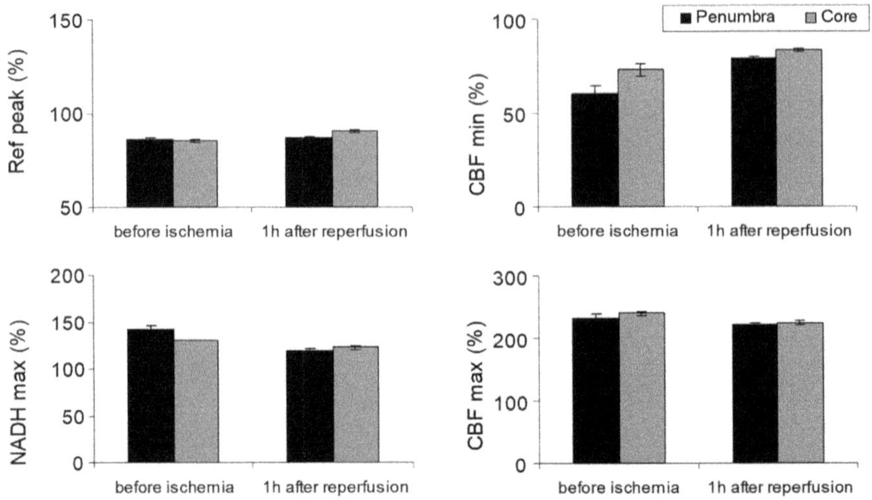

Fig. 5.71 Response to nitrogen exposure before MCA occlusion for 60 min and BCO for 15 min and 1 h after reperfusion, as measured in the core and penumbra. Results are presented as mean ± SE ($n = 8$).

Table 5.21 SD waves 1 h after MCAO and BCO for 15 min compared to control waves

	Control		MCAO 60 min + BCO 45 min	
	Penumbra	Core	Penumbra	Core
Ref_max	119 ± 2.37	108 ± 0.74	119 ± 2.26	124 ± 3.52
Ref_min	77 ± 0.48	85 ± 1.11	84 ± 1.35	79 ± 0.57
NADH_peak	87 ± 0.81	86 ± 0.74	88 ± 1.44	91 ± 1.04
CBF_max	252 ± 10.7	222 ± 11.87	165 ± 1.94	162 ± 4.05

Minimum and maximum values of reflectance, NADH, and CBF during induced SD waves, as measured in the core and penumbra. SD waves were induced 1 h after reperfusion following MCAO for 60 min and bilateral carotid occlusion for 15 min ($n = 8$). In another group, SD waves were induced under normoxic conditions ($n = 8$).

5.4.2 Spatial Damage During Focal Ischemia

5.4.2.1 Core vs. Penumbra

This study examined the effect of right MCA occlusion on both hemispheres. The results show that focal ischemia via MCAO causes brain damage in the ipsilateral hemisphere only, with no effect on the contralateral hemisphere, as described in other studies (Hossmann 2006; Lipton 1999; Sims and Muyderman 2010; Longa et al. 1989).

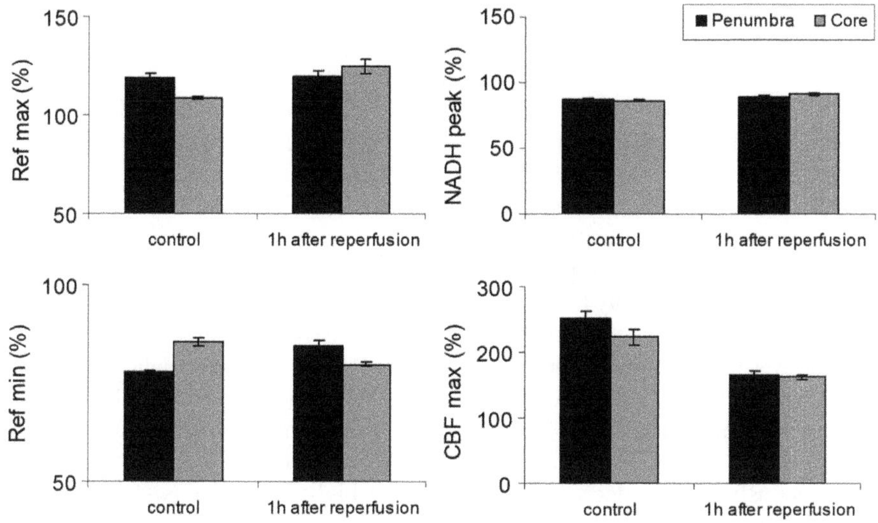

Fig. 5.72 Changes in various parameters in SD waves, evoked 1 h after reperfusion following MCAO for 60 min and BCO for 15 min ($n = 8$), compared to SD waves evoked in normoxic conditions (control, $n = 8$), as observed in the core and penumbra. Results are presented as mean ± SE

This pattern of damage indicates the successful execution of the MCAO model in this study, as well as the ability of the multisite multiparametric (MSMP) system, which was used in this work, to reliably monitor focal ischemia.

The study then focused on two measurement sites on the right hemisphere: a lateral and a medial area relative to the Bregma point. Blood flow and blood volume in the lateral area following MCA occlusion in the different models were low relative to the medial area. In both measurement areas, some blood flow recovery was observed after reaching the minimum level, which was expressed as an increase in CBF and a decrease in reflectance, likely due to an increase in blood volume in the tissue (Mayevsky 1976). The recovery in the lateral area, while the MCA was still occluded, was very low, in contrast to the higher degree of recovery in the medial area. This recovery increased the gap in blood supply between the areas.

The findings in the lateral area are characteristic of the ischemic core, which forms at the center of the ischemic injury. This area is characterized by a decrease in blood flow to a level of 20% of its initial value or less. Following the decrease in blood supply, a necrotic center forms, which is the focus of the ischemic injury (Nagasawa and Kogure 1989; Lipton 1999). The documented blood flow in the penumbra area ranges between 20% and 40% of baseline values (Lipton 1999; Nagasawa and Kogure 1989; Ginsberg 2003), which is consistent with the blood flow values we measured in the medial area. In this area, there is a partial blood supply due to collateral vessels, and therefore the extent of damage in this area is less than in the core area (Lipton 1999). Based on the blood flow patterns, the lateral monitoring area can be characterized as the core and the medial area as the penumbra. Existing anatomical knowledge supports this conclusion, as it associates the core with deeper and more lateral areas and the penumbra with the areas surrounding it (Lipton 1999).

Scientific literature shows that in studies using the MCAO model, the common duration for creating transient ischemia is up to 3 h, with most of the significant changes in the tissue occurring in the first hour after occlusion (Lipton 1999). The results observed in the different models in the current study indicate a decrease in blood flow following MCA occlusion. The blood supply was more severely affected in the core area relative to the penumbra, as observed in other studies (Lipton 1999). The damage to the blood supply to the ischemic tissue leads to an energy crisis, which is expressed as greater mitochondrial damage in the core area compared to the penumbra (Mayevsky 1984). Following the decrease in blood flow, an oxygen deficiency is created in the tissue, which disrupts the activity of the respiratory chain, and as a result, a reduction of mitochondrial NADH occurs in the ischemic tissue (Mayevsky 1984; Mayevsky et al. 1998). The reduction of NADH in the current study's groups occurs simultaneously with the decrease in blood flow. A higher degree of NADH reduction was observed in the core area relative to the penumbra. The difference between the NADH reduction patterns in the different areas corresponds to the different blood flow patterns between the two areas. However, in the penumbra, NADH oxidation was observed after its reduction, i.e., recovery, while in the core area, high levels of NADH remained throughout the entire occlusion time. The combination of blood flow and mitochondrial NADH levels, which is made possible by the MSMP system, presents a picture of greater mitochondrial damage in the core relative to the penumbra (Lipton 1999; Sims and Muyderman 2010).

During monitoring of both hemispheres, blood flow in the right hemisphere decreased to levels that were lower than those observed in the penumbra but higher than those observed in the core, as observed in the groups in which monitoring was performed at two sites on the right hemisphere. In addition, NADH reduction in this area was high relative to the reduction observed in the penumbra. From these results, it can be concluded that the monitoring area on the right hemisphere in the bilateral monitoring model constitutes an intermediate area, between the core and the penumbra. The blood flow in this area had characteristics similar to the penumbra, but the mitochondrial damage in this area was greater relative to the penumbra, as indicated by the NADH reduction level (Mayevsky 1984; Mayevsky et al. 1990). From a comparison between the three measurement areas, a picture of gradual brain damage following focal ischemia caused by MCA occlusion is obtained, as previously documented (Lipton 1999). The focus of the damage is in the core, which is located at a lateral point on the cortex. As one moves away from it toward the longitudinal fissure, a gradual decrease in the severity of the ischemic injury is observed, which is expressed in a greater supply of blood and mitochondrial NADH oxidation.

5.4.2.2 Ischemic Depolarization (ID)

Studies show that during focal ischemia, the core area undergoes a complete cessation of electrical activity (Nallet et al. 1999). This phenomenon is called ischemic depolarization (ID), or SD-like hypoxic depolarization (HSD) (Lipton 1999; Duarte et al. 2003; Somjen 2001). This phenomenon occurs as a result of an increase in extracellular potassium levels, which stems from an irregularity in blood or oxygen supply to the brain, as in cases of extreme ischemia (Mayevsky and Cohen 1990). Studies show that during ID, a rise in reflectance level (secondary reflectance increase, SRI) occurs within about 3 min, which happens simultaneously with a rise in fluorescence levels. These phenomena parallel the increase in extracellular potassium levels and the electrical suppression that occur in the ischemic tissue. The reflectance and fluorescence reach a maximum level, and then decrease to a level that remains higher than the baseline for the entire duration of the ischemia (Lipton 1999).

Similar phenomena were observed in the current study in the core of focal ischemia, indicating that a series of events characteristic of ID occurs in this area after 3 min. The phenomenon occurs in the core and is not observed in the penumbra, as documented in other studies (Mayevsky and Cohen 1990; Duarte et al. 2003). The ID phenomenon indicates more severe ischemic damage and tissue failure in the core area compared to the penumbra (Mayevsky and Rogatsky 2007; Mayevsky 1984). The mitochondrial NADH level in the current study was calculated as the subtracted value of reflectance from fluorescence. A rise to particularly high values in the reflectance level may cause an artifact in the NADH measurement, as observed in previous studies during ischemia and also at the time of death. Due to the calculation of NADH level as a subtracted value, a simultaneous change in reflectance and fluorescence levels leads to a situation where no change in NADH level is seemingly observed, a phenomenon that is characteristic of cases where SRI is documented (Nallet et al. 1999). The SRI phenomenon, which is an artifact in the NADH measurement using the subtraction method, may explain the gradual decrease in NADH level that occurs in ischemia for 60 min even before the occlusion is removed. In such situations, it is appropriate to examine the uncorrected fluorescence level, and indeed, it can be seen that following MCA occlusion, an increase in fluorescence values occurs to higher levels in the core compared to the penumbra for the entire duration of the ischemia.

5.4.2.3 Spontaneous SD Waves During Ischemia

Studies show that while the core area undergoes complete electrical shutdown in the form of ID, short periods of electrical shutdown are observed in the penumbra, with a pattern of spreading depression (SD) waves that are characteristic of ischemic tissue (Nallet et al. 1999). In the current study, at least one SD wave was observed in the penumbra area in 74% of the animals that underwent MCA occlusion for different durations. The first wave usually appeared after about 3–5 min, concurrent with the start of ID in the core area. The SD waves that were observed had an ischemic pattern, characterized by low hyperemia and NADH reduction, unlike waves in normoxic tissue (Kobayashi et al. 1995). SD waves in ischemic tissue may reduce long-term damage caused by the disruption in blood supply, as long as blood flow is restored after a short period (Nakamura et al. 2010; Takano et al. 1996). However, as the duration of ischemia lengthens and the frequency of the waves increases, the waves cause an exacerbation of the damage, which intensifies after each wave (Lipton 1999).

5.4.3 *Effect of Ischemia Duration on Mitochondrial Function and Cerebral Blood Flow*

5.4.3.1 Mitochondrial Function and Cerebral Blood Flow in the First Hour After Focal Ischemia

The results of the current study demonstrate that the duration of ischemia has an effect on the tissue's ability to recover from ischemic injury, in the stage where blood flow is restored (reperfusion). Following the restoration of blood flow, a hyperemic response was observed, the intensity of which is inversely proportional to the duration of ischemia. Hyperemia occurs following a 10 or 30 min occlusion, but not for 60 min, which indicates the tissue's ability to compensate after short periods of ischemia, and the absence of this ability in longer periods (Lipton 1999; Hossmann 2006). It is known that hyperemia following MCAO occurs in both the core and the penumbra and depends on the severity of the damage (Lipton 1999). In short ischemia durations, up to half an hour, higher hyperemia is observed in the core relative to the penumbra, indicating a greater need for blood supply to this area (Lipton 1999; Duarte et al. 2003). In contrast, after a 1-h ischemia is removed, the differences in the hemodynamic response between the areas become blurred. In this case, a prolonged lack of blood supply led to the tissue's inability to recover after flow was restored (Lipton 1999).

The decay of the hyperemia wave and the return of blood flow to a normal level also change depending on the severity of the damage. As the ischemia duration lengthens, the duration of blood supply recovery to the tissue also lengthens. In addition, a more prolonged hyperemic response is observed in the core area relative to the penumbra. The duration of the hemodynamic changes in the tissue indicates the severity of the ischemia, and the increased oxygen demand during the reperfusion stage (Lipton 1999; Berne and Levy 2000). The greater the damage, the longer the recovery attempt, as measured by the hyperemia wave in CBF, lasts. It is possible that a prolonged disruption in blood supply leads to damage to the activity of these autoregulatory mechanisms, and as a result, the tissue's ability to recover from ischemic damage is prevented (Lipton 1999).

In the current study, nitrogen exposure was used as a tool to check the tissue's condition in response to a temporary lack of oxygen. Normal tissue responds to nitrogen with an initial decrease in CBF, followed by a wave of hyperemia. In experiments where the MCAO model was performed, a trend of decrease in the intensity of hyperemia was found during nitrogen exposure 1 h after blood flow was restored. As the ischemia duration increases, a lower hyperemic response is observed during nitrogen exposure. After half an hour of ischemia, this phenomenon is observed only in the core area, but after 1 h of ischemia, a lower than normal hyperemia wave is also observed in the penumbra. The summary of these data indicates the spread of damage in the tissue in direct proportion to the ischemia time, as described in previous studies (Sims and Muyderman 2010).

Tissue recovery after MCAO is expressed by the return of the mitochondria to normal activity (Mayevsky 1984; Mayevsky et al. 1990). Following the restoration of blood flow, a recovery from the damage to mitochondrial function can be observed, as can be inferred from the oxidation of mitochondrial NADH (Mayevsky et al. 1998). As the occlusion time increases, a lower degree of NADH oxidation is recorded in the recovery stage, depending on the oxygen demand after blood supply is restored (Mayevsky et al. 1990). In addition, the extent of ischemic damage affects the rate of mitochondrial recovery, which is expressed as a more prolonged NADH recovery as the ischemia duration increases.

The more severe damage in the core area relative to the penumbra during ischemia leads to greater oxidation of NADH molecules in this area during the reperfusion stage. With the restoration of blood flow, mitochondrial activity returns to its initial level in the penumbra, but the increased oxygen demand in the core leads to excess oxidation of NADH (Sims and Muyderman 2010), which indicates increased mitochondrial activity in this area. That is, the greater the extent of tissue damage, the greater the mitochondrial activity during the reperfusion stage, as expressed by the oxidation of NADH. Despite the normal functioning of the mitochondria in the first hour after blood flow is restored, secondary damage to its function may occur in the following hours (Sims and Muyderman 2010), due to cumulative ischemic damage combined with reperfusion injury, which stems from an excess of free radicals in the tissue following a sharp increase in oxygen supply (Lipton 1999; Duarte et al. 2003).

The restoration of blood flow after a 1-h focal ischemia leads to a decrease in NADH to low levels in the core area. These changes are not accompanied by corresponding changes in CBF, i.e., an increase in blood flow; on the contrary, the core area is characterized by the absence of a hyperemic response, which indicates the tissue's inability to recover from the ischemia (Lipton 1999; Duarte et al. 2003). In addition, studies show that after blood flow is restored, low ATP levels remain in the core area, which indicates irreversible damage to mitochondrial function (Sun and Gilboe 1994; Sims and Muyderman 2010). It is likely that the high reflectance level observed in the reperfusion stage caused an artifact in the NADH measurement, which at this stage is not a measure of mitochondrial function. As mentioned, a more accurate evaluation of mitochondrial activity is done in these situations based on fluorescence values (Mayevsky and Rogatsky 2007), in which complete recovery is observed in the reperfusion stage after 1 h of ischemia. Another reason for the high reflectance level in the reperfusion stage may be the sinking of the brain, which occurs following a decrease in the volume of the interstitial space (Lipton 1999), and leads to the accumulation of fluids between the tissue and the measurement fiber, so that the light projected onto the tissue returns to the photodetector in a mirror effect (Mayevsky and Rogatsky 2007). On the other hand, an increase in reflectance may also result from the development of an edematous area as a result of ischemic damage

(Lipton 1999; Mayevsky and Rogatsky 2007). The increase in reflectance, combined with the fluorescence pattern, indicates an increase in mitochondrial damage during the reperfusion stage in the core area, depending on the duration of the ischemia (Mayevsky and Rogatsky 2007; Mayevsky 1984). Mitochondrial function can be damaged due to the accumulation of lactate, glutamate, and free radicals in the tissue (Lipton 1999; Kristián and Siesjö 1998), and may deteriorate even further due to secondary damage in the hours following the restoration of blood flow (Canevari et al. 1997; Sims and Muyderman 2010).

5.4.3.2 Mitochondrial Function and Cerebral Blood Flow 24 hAfter Focal Ischemia

In this part of the work, several tests were conducted to assess the long-term effect of focal ischemia—24 h after the occlusion was performed. In some of the animals, transient occlusion was performed, in which the MCA was occluded for 60 min and then blood flow was restored. In other animals, permanent occlusion was performed, in which the MCA was occluded without restoring blood flow. The results of the experiment show that 24 h after ischemia, whether transient or permanent, a different neurological behavior pattern was observed relative to the normoxic state. This behavior included forelimb flexion to the side contralateral to the injury when the rats were held by their tails, leaning to the side when pressure was applied to the forelimbs, and circular movement toward the ipsilateral side of the injury. These behavior patterns are characteristic of animals that have undergone brain damage (Gerlai et al. 2000; Zhang et al. 2000). In addition, permanent or transient occlusion led to a decrease in the animals' appetite and consequently to an average weight loss of about 15 g within 24 h, which was not observed in control animals that underwent a similar surgery without ischemia.

Mitochondrial function and blood supply to the tissue 24 h after MCA occlusion were evaluated according to the tissue's response to nitrogen exposure, in which the tissue's coping with a temporary lack of oxygen was examined, and during induced SD waves, in which the tissue's response to metabolic challenge was examined. Twenty-four hours after a 60 min MCA occlusion (transient occlusion), no response was observed in the core in any of the measured parameters, either during a brief nitrogen exposure or during the passage of induced SD waves. This indicates a very high level of damage in this area and is in fact evidence of the tissue's metabolic dysfunction. In the penumbra area, however, responses were observed during nitrogen exposure and SD waves, but these were of a different nature relative to the control treatments performed 24 h earlier in the same animals, as well as relative to similar treatments in normoxic animals. The high reflectance levels and low CBF levels indicate damage to the blood supply to the penumbra area 24 h after transient ischemia (Mayevsky and Rogatsky 2007; Chen et al. 2006). The maximum reduction of NADH during nitrogen exposure was low relative to the normoxic state, which indicates a higher than normal baseline NADH level, which is an indication of damage to mitochondrial function (Mayevsky 1984; Mayevsky and Chance 2007). In addition, the NADH reduction trend during induced SD waves, in contrast to the oxidation trend observed in normoxic tissue, is an indication of the tissue's inability to cope with metabolic challenges (Mayevsky et al. 1995). The differences between the monitoring areas in the responses to nitrogen and SD waves indicate more severe brain damage in the core relative to the penumbra, as also documented in other studies.

In contrast to the difference between the core and the penumbra observed following transient occlusion, no differences were observed between the monitoring areas 24 h after permanent MCA occlusion. In both areas, no changes were observed relative to the baseline in any of the measured parameters, either during a brief nitrogen exposure or during the passage of induced SD waves, which indicates the metabolic dysfunction of the tissue (Mayevsky et al. 1995). This pattern, which was observed in the core and penumbra following a 24 h permanent occlusion, is similar in its characteristics to the pattern observed only in the core 24 h after a 60 min transient occlusion.

From this, it can be concluded that 24 h after a transient occlusion, severe damage occurs in the core, to the point of mitochondrial and hemodynamic dysfunction, while in the penumbra, a more moderate ischemic damage occurs, which allows for blood supply and mitochondrial function. In contrast, 24 h after a permanent MCA occlusion, the mitochondrial and hemodynamic dysfunction covers the penumbra area in addition to the core. In other words, a deterioration occurs in the penumbra, and more parts of it join the core area.

5.4.4 Variability in the Brain's Responses Following MCAO

5.4.4.1 Variability in the Response Pattern

In scientific literature, the damage that occurs following MCA occlusion is characterized by an immediate decrease in blood flow in both the core and the penumbra (Hossmann 2006). These results, accompanied by mitochondrial damage, were also observed in the current study. However, within the study, six other response patterns were also observed that differed from the pattern described above. These patterns can be divided into three sub-categories: (1) ischemia without reperfusion, following a decrease in blood flow or death following the removal of the occlusion; (2) untimely reperfusion, which occurs before or sometime after the occlusion is removed; and (3) an unusual pattern in the penumbra, characterized by no change or an increase in blood flow.

According to the study's results, the probability of a different response pattern increases in direct proportion to the duration of ischemia. This fact may explain some of the profiles obtained, but here one must distinguish and examine at which stage of the ischemia the difference in results was observed. A hyperemic response before the occlusion is removed, nonsignificant changes in blood flow in the penumbra, or an increase in blood flow in the penumbra, which correspond to profiles P2-P4 in the article, are all phenomena that occur while the MCA is occluded. In contrast, the absence of a reperfusion response, death upon occlusion removal, or a delayed hyperemic response, which correspond to profiles P2-P4 in the article, all occur during the reperfusion stage.

The study's results demonstrate that responses related to reperfusion are more common in 60-min ischemia, and therefore are affected by the ischemia duration. The reason for this may lie in a decrease in the elasticity of the occluded artery, which may increase its vulnerability (Coulson et al. 2002). In addition, an increase in the ischemia duration may increase the chance of blood clots (thrombus) forming around the occluding filament, which may cause a difference in the intensity of the ischemia (Schmid-Elsaesser et al. 1998).

The absence of a reperfusion response was in all cases accompanied by mitochondrial damage, which was characterized by an increase in the NADH level (Mayevsky and Rogatsky 2007). This phenomenon can be attributed to mechanical damage to the occluded blood vessel, which occurs as a result of pulling the nylon filament backward (Coulson et al. 2002). This response can lead to subarachnoid hemorrhage (SAH), as demonstrated in other studies (Dittmar et al. 2006; Gerriets et al. 2004; Hungerhuber et al. 2006; Schmid-Elsaesser et al. 1998; Woitzik and Schilling 2002). In contrast to these studies, the current study demonstrates, in addition to the damage to blood supply, the accompanying mitochondrial damage, as expressed in the NADH levels, as a response to the decrease in oxygen supply to the tissue (Mayevsky and Rogatsky 2007). In addition, the study's results demonstrate the ability of the MSMP system to identify SAH through monitoring a single hemisphere. In some cases, the removal of the occlusion was accompanied by the immediate death of the animal. A possible explanation for this phenomenon may be damage to the function of the heart's myocardium layer, and consequently, damage to the autonomic nervous system (Cechetto et al. 1989).

A hyperemic response usually occurs in the ischemic area following the removal of the occlusion (Hossmann 2006). In the current study, in some of the animals, a hyperemic response was observed even before the occlusion was removed, accompanied by a recovery in mitochondrial function. The reason for this pattern may be an incomplete MCA occlusion due to a movement of the nylon filament (Gerriets et al. 2004; Zhao et al. 2008) or due to variability in the anatomical structure of the arteries supplying blood to the occluded area. According to the distribution of the results, the duration of ischemia has no effect on an early hyperemic response. In contrast, a delayed hyperemic response was observed in animals that underwent MCA occlusion for 60 min, but not in animals that underwent occlusion for shorter periods. This response may indicate mechanical damage to the MCA, as a result of which it is unable to immediately restore blood flow to the area that was previously occluded (Coulson et al. 2002).

In some cases, damage to the blood supply and mitochondrial function was observed in the core only, while in the penumbra, no changes were observed or an increase in blood flow was observed. In these cases, the damage observed in the core is lower relative to the levels of damage generally observed following MCAO, as observed in the current study, as well as known from the literature (Ginsberg 2003; Hossmann 2006; Lipton 1999). According to these results, it can be hypothesized that in these cases, a complete focal ischemia was not created following the insertion of the filament, or alternatively, that the blood supply to the tissue continued to exist due to the presence of collateral blood vessels (Zhao et al. 2008; Gerriets et al. 2004). This phenomenon can be explained as part of the "steal phenomenon," in which an additional decrease in blood supply occurs to damaged areas and blood is diverted to areas where perfusion is already maintained. This phenomenon is characteristic of the margins of the penumbra area in focal ischemia, and may increase the damage caused in the center of the ischemic tissue and improve the condition of the penumbra (Pinard et al. 2002). In addition, in these cases, a massive increase in reflectance was not observed, which was generally observed in the study in the core. The absence of this response may indicate a decrease in the intensity of the ischemia relative to the other groups (Mayevsky and Rogatsky 2007; Mayevsky and Cohen 1990). It is likely that in the current group, the MCA was not completely occluded. Blood flow through the artery did not completely stop as in the other groups, which created only partial ischemia in the core, and even increased blood flow in the penumbra.

Various studies indicate variability in the results following MCA occlusion, which is expressed in the size of the damaged area, results of neurological tests, or the survival of the experimental animals (Gerriets et al. 2004; Laing et al. 1993; Lourbopoulos et al. 2008; Schmid-Elsaesser et al. 1998; Zhao et al. 2008). The current study was the first to present comprehensive documentation of the variability in results obtained when performing a single model of focal ischemia using MCA occlusion. The MSMP system was found to be able to characterize different patterns of mitochondrial function and blood flow following the MCAO model already in the initial stages of ischemia.

5.4.4.2 Variability in Mitochondrial NADH Levels

In addition to the profiles described above, differences were also observed in the intensity of change in the fluorescence and NADH levels in the current study. An examination of the reasons for the differences revealed that they were due to differences between animals that were used in different periods during which the study was conducted. For a more precise examination of the differences, a comparison of all the measured parameters was performed between animals that were used nearly 3 years apart—in an early stage of the study (February–March 2007) and in a later stage (January 2010). This comparison found that there were indeed significant differences in the fluorescence and NADH levels between the two groups. Despite the differences, a similar pattern of increase in fluorescence and NADH levels was observed in both groups following MCA occlusion, to higher levels in the core compared to the penumbra. At the same time, no significant differences were found between the groups in the parameters indicating perfusion, namely CBF and reflectance. In addition, no differences were observed between the groups in all parameters during nitrogen exposure. It seems that the reason for these differences lies in differences between different breeding batches, which differ from each other in the nature of the mitochondrial damage following local ischemia, but not in the existence of such damage itself. The differences between the batches in their coping with ischemic conditions are expressed only in the mitochondrial response, and not in the perfusion pattern or the response to oxygen deprivation. Since the 60-min MCA occlusion was planned to serve as a control group for the subsequent models in the study, it was decided following the differences described above that in the first part of the work, one group of animals, close in time to the groups reviewed so far, would be used, and the group on which the work was done close to the time when the protocols to be discussed from now on were performed would serve as a control for the rest of the work.

5.4.5 The Relationship Between Focal and Global Ischemia

5.4.5.1 The Effect of Global Ischemia on Cerebral Blood Flow and Mitochondrial Function

The current study examined the effect of occluding one or both carotid arteries on mitochondrial function and cerebral blood flow. From the moment one or both carotid arteries are occluded, a decrease in blood flow accompanied by a reduction in mitochondrial NADH can be observed. The changes observed in the different parameters following the occlusion of both carotid arteries were significantly greater relative to the occlusion of a single artery. In bilateral occlusion, these changes were observed in both hemispheres, while in unilateral occlusion, changes were recorded only in the hemisphere on the side of the occluded carotid. In addition, the rat's brain showed a quicker recovery following a single carotid occlusion, in which the different parameters returned to baseline values after a few seconds, compared to the occlusion of both carotid arteries, in which recovery was recorded after about 10 min. As documented in other studies (Bluestone et al. 2004; Mayevsky 1976), a single carotid occlusion does not cause immediate brain damage. In this case, a normal blood supply to the brain is maintained, thanks to compensation from the other carotid artery and the vertebral arteries, which allow blood to be supplied to the entire brain through the Circle of Willis. The electrical pattern, blood flow, and energy activity of the tissue do not change at all, or undergo a minimal change and return to a normal level within a few seconds of the occlusion (Mayevsky 1976).

5.4.5.2 The Effect of Global Ischemia Duration on Brain Metabolism

The effect of carotid artery occlusion on brain tissue was examined, among other things, in a state of metabolic challenge through the induction of SD waves. In the first minutes after occlusion, differences can be observed in the SD waves induced in animals that underwent single carotid occlusion versus those that underwent bilateral occlusion. The SD waves induced in the brains of animals with a single carotid occlusion did not show significant changes relative to the control group's waves, which were induced in the same animals before the occlusion. In contrast, following bilateral occlusion, hemodynamic changes were observed during the passage of the wave, which were expressed as a decrease in blood flow that was not observed in the control group. The metabolic hyperemia, which characterizes SD waves in normoxia, was significantly lower following carotid occlusion, which indicates the inability of the ischemic tissue to meet the oxygen demand (Meilin et al. 1995; Sonn and Mayevsky 2000). Simultaneously, in reflectance, no decrease was observed after the increase phase, but a return to baseline levels. The low reflectance level in the control group is equivalent to an increase in blood flow and results from an increase in blood volume in the tissue. A combination of the absence of a decrease in reflectance and low blood flow levels is characteristic of ischemic tissue, in which a blood deficiency is felt (Mayevsky 1976).

Simultaneously with these phenomena, a pattern of mitochondrial NADH reduction was recorded in the waves induced in the ischemic tissue, compared to its oxidation in the control group. NADH oxidation results from an overactivity of the electron transport chain. This activity leads to a high oxygen demand, and as a result, an increase in blood flow to the brain (Mayevsky et al. 1998). In contrast, NADH reduction indicates ischemic damage and is characteristic of tissue suffering from an oxygen deficiency, due to a disruption in blood supply (Sonn and Mayevsky 2000). In addition, a slowing of the wave rate was recorded, as can be inferred from an increase in the time difference between two successive SD waves (Sonn and Mayevsky 2000). This wave pattern is characteristic of ischemic damage in the tissue, as documented in other studies (Iijima et al. 1998; Mayevsky 1976). Single carotid artery occlusion did not lead to changes in the immediate time frame after the occlusion. The SD waves after the occlusion were not significantly different from the control waves in any of the parameters, which indicates normal metabolism and a full compensatory ability of the brain (Mayevsky 1984; Sonn and Mayevsky 2000).

Sixty minutes after bilateral carotid artery occlusion, the brain tissue remains in a damaged state, as can be inferred from the pattern of the SD waves at this stage, especially in the blood flow changes. On the other hand, it seems that a single carotid occlusion for the same period causes only a small decrease in blood flow in the hemisphere on the ischemic side, which is expressed as a decrease in CBF in the SD waves (Sonn and Mayevsky 2000; Meilin et al. 1995).

Bilateral carotid occlusion for 24 h caused irreversible brain damage, as expressed in the response to nitrogen. During nitrogen exposure, the decrease in blood flow was greater than that observed in the control group, and was accompanied by a hyperemia wave, which was also significantly lower than is common in normal anoxia. In addition, the rate of NADH oxidation in the recovery phase from anoxia was slower in the ischemic group relative to the control group. These findings indicate metabolic damage in the tissue, which results from damage to the oxygen balance in the tissue following ischemia (Mayevsky 1984; Mayevsky et al. 1998). The tissue damage in this model is also expressed in the SD wave pattern, which is characterized by a reversal in NADH, a massive increase in reflectance, and a decrease in the extent of hyperemia relative to normoxic tissue. On the other hand, SD waves in this state showed a greater increase in CBF compared to SD waves in shorter ischemia periods, of 10 or 60 min. From this, it can be concluded that 24 h of ischemia is indeed longer, but it is not necessarily more severe than short ischemia, since there is enough time for the brain to stabilize and create a new hemodynamic state. On the other hand, short ischemia may appear more severe, due to the lack of time and ability of the tissue to stabilize (Lipton 1999; Mayevsky 1984).

5.4.5.3 Spatial Damage Following Global Ischemia

Unlike bilateral carotid artery occlusion, in which a decrease in the function of the entire brain was observed, a single carotid occlusion leads to a deterioration in the condition of the tissue only in the hemisphere on the side of the occluded carotid. The decrease in tissue function is expressed in the response to a brief nitrogen exposure, in which a decrease in CBF and then a smaller hyperemia was observed, as well as in SD waves, in which CBF decreased and a reversal in the NADH wave occurred. Simultaneously with the decrease in blood flow in the SD wave in the ischemic hemisphere, an increase in blood flow occurs in the hemisphere contralateral to the occluded artery. These changes indicate tissue damage in the hemisphere on the side of the occlusion, which leads to a decrease in mitochondrial function (Meilin et al. 1995; Sonn and Mayevsky 2000).

It seems that there is a gradualness in the extent of damage to brain tissue depending on the type of ischemia. In a state of bilateral carotid artery occlusion, the brain damage is more severe and is observed in earlier stages, compared to single carotid occlusion. The tissue undergoes metabolic and energy deterioration immediately after bilateral carotid artery occlusion, but not after single carotid artery occlusion, as can be inferred from the pattern of SD waves in the tissue (Sonn and Mayevsky 2000; Somjen 2001). The effect of a single carotid occlusion is expressed only after 24 h, when an increased blood supply to the hemisphere opposite the occluded carotid is observed, at the expense of the hemisphere on the side of the occlusion, which is expressed in both SD waves and a brief nitrogen exposure. These changes are also accompanied by mitochondrial damage in the same hemisphere, which is expressed as NADH reduction during an SD wave (Mayevsky 1984; Sonn and Mayevsky 2000).

Despite the decrease in blood supply, the entire brain manages to maintain mitochondrial and tissue function after a 24-h single carotid occlusion, as can be learned from the rate of SD wave propagation and the NADH reduction during nitrogen exposure, which are identical to those observed in normoxia. Another piece of evidence for the maintenance of tissue function is the behavior of the animals after 24 h. While in animals in which both carotid arteries were occluded, a clouding of the senses can be observed, accompanied by slow reactions and an average weight loss of about 15 g within 24 h, these phenomena did not occur in rats in which only one carotid was occluded for the same period. In summary, a single carotid occlusion can be characterized as an intermediate state, with characteristics of both healthy tissue and ischemic tissue.

5.4.5.4 Global vs. Focal Ischemia

Focal ischemia, which is observed following MCA occlusion, shows a limited area of damage, which is concentrated mainly in the parietal lobe of only one hemisphere. On the other hand, following bilateral carotid artery occlusion, global ischemia occurs, which encompasses most of the cortical areas. In contrast to these types of ischemia, during a single carotid occlusion, only partial damage to the mitochondria is observed, which is limited to the hemisphere on the side of the occluded artery only.

The blood supply through the MCA to its distribution area is exclusive, so that when it is occluded, complete ischemia is created in the core area (Lipton 1999), which leads to extensive mitochondrial damage that is also observed in the reperfusion phase. As one moves away from the core area toward the penumbra, the severity of the damage decreases, and only partial ischemia is observed, thanks to compensation through the opening of collateral blood vessels (Lipton 1999; Mayevsky 1984). A similar phenomenon occurs following global occlusion of one or both carotids, in which the mitochondrial damage is only partial, due to the compensation in blood supply through the Circle of Willis (Mayevsky 1976).

Partial ischemia allows for tissue recovery over time, even before the occlusion is removed, as was observed during global ischemia and in the penumbra area created following focal ischemia. In the case of bilateral global ischemia, the tissue reaches stabilization, which is expressed as a hyperemic response in SD waves that was higher after 24 h of ischemia relative to 1 h after the occlusion. In the penumbra area created following MCAO, the recovery trend is already observed in the first hour after the occlusion, as indicated by the gradual oxidation in NADH over the occlusion time. In contrast, in the core area, a lower degree of recovery is observed, which is characteristic of complete ischemia (Mayevsky et al. 1990). Recovery from complete ischemia is possible only after blood flow is restored, but the extent of improvement in mitochondrial function depends on the ischemia time (Mayevsky et al. 1990; Sims and Muyderman 2010).

It seems that the damage in the penumbra area of focal ischemia is greater than the extent of damage observed during partial global ischemia, as can be inferred from the hyperemic response in SD waves, which was very low in focal ischemia compared to global ischemia, and as shown by previous studies (Sims and Muyderman 2010; Cohen et al. 2011). In summary, it seems that partial global ischemia, caused by the occlusion of both carotid arteries, leads to more moderate mitochondrial damage relative to that observed in the core and penumbra of focal ischemia. In contrast, the lowest degree of damage is observed in partial global ischemia, which occurs following a single carotid artery occlusion.

5.4.6 Can Focal Ischemia Damage Be Reduced?

5.4.6.1 Introduction

Following the mapping of spatial damage and the examination of the effect of focal ischemia duration on mitochondrial function and cerebral blood flow, the current study attempted to reduce the intensity of ischemic damage. To achieve this, two approaches were taken that use partial global ischemia as a means to reduce the damage from focal ischemia. The first approach used global ischemia as a preconditioning treatment for focal ischemia. In this case, a single carotid artery or both carotid arteries were permanently occluded 24 h before MCA occlusion. The second approach used global ischemia as a postconditioning treatment for focal ischemia. In this case, a transient occlusion of one or two carotid arteries was performed after MCA occlusion and before or at the time of blood flow restoration in the MCA.

5.4.6.2 Global Ischemia as a Preconditioning Treatment

Studies show that preconditioning treatment for ischemia may alleviate the intensity of ischemic damage. Various studies use different types of preconditioning treatments (Pignataro et al. 2009). Consequently, the current study attempted to examine the effect of carotid occlusion as a preconditioning treatment for ischemia.

Unilateral Carotid Artery Occlusion

Following MCA occlusion 24 h after a single carotid artery occlusion, a more moderate decrease in blood supply was observed relative to MCA occlusion alone, as expressed by high CBF levels in both the core and the penumbra. These values are high relative to what is known in the literature about these areas (Lipton 1999). During ischemia, a partial spontaneous recovery occurred in blood flow, results that are consistent with similar studies describing blood flow in the core and penumbra following MCAO (Lipton 1999; Nagasawa and Kogure 1989; Duarte et al. 2003). In contrast, the high fluorescence level during ischemia indicates greater mitochondrial damage in the core area following the use of carotid occlusion as a preconditioning treatment for focal ischemia. It can be assumed that the low NADH values in the core do not result from increased activity of the electron transport chain, but are an artifact caused by high levels of reflectance (Mayevsky and Rogatsky 2007). These results seemingly present a contradictory situation, where despite the improvement in blood supply, damage to mitochondrial function occurs. In addition, during the stage of blood flow restoration in the MCA, a supposed recovery occurs in blood supply and mitochondrial function, as indicated by CBF and fluorescence values, respectively. Consequently, it is possible that the carotid artery occlusion led to a state of tissue resistance to local ischemia 24 h later (Kitagawa et al. 2005), as was also observed in other cases of ischemia (Kirino 2002; Zhang et al. 2003).

Nevertheless, it seems that 1 h after blood flow is restored in the MCA, the tissue loses its ability to cope with metabolic challenges such as a brief exposure to nitrogen and SD waves. The parameter pattern during nitrogen exposure that occurs about 1 h after reperfusion is significantly different from nitrogen exposure in normoxic tissue. The minimum reflectance level observed in the core was high both relative to the reflectance level in the penumbra and relative to the reflectance level during nitrogen exposure in normoxic tissue, likely as a result of the inability of the blood vessels in the core to constrict and dilate as needed (Mayevsky and Rogatsky 2007). The change in the reflectance pattern is consistent with the low levels of change observed in CBF, which indicate a deficiency in blood supply to the tissue according to its needs (Mayevsky and Chance 2007). In addition, a smaller than usual increase in NADH level was observed during nitrogen exposure. From this, it follows that the baseline NADH levels were higher than usual, which indicates impaired mitochondrial function about 1 h after ischemia (Mayevsky and Rogatsky 2007). These results are consistent with the NADH reduction pattern observed during the passage of SD waves, unlike the oxidation pattern observed in normoxic tissue (Sonn and Mayevsky 2000). From the summary of the results, it appears that a permanent occlusion of the carotid artery 24 h before MCA occlusion leads to an improvement in blood supply during focal ischemia, mainly in the core area. However, the tissue is unable to cope properly with metabolic challenges such as oxygen deprivation and the activation of cortical tissue. Therefore, a single carotid occlusion 24 h before MCA occlusion does not lead to an improvement in the tissue's metabolic function, but causes damage at least similar to focal ischemia without preconditioning treatment. The mitochondrial damage observed during ischemia and during exposure to nitrogen and SD waves in the reperfusion stage indicates that carotid occlusion as an ischemic preconditioning treatment preceding MCA occlusion is not effective in treating the damage from focal ischemia.

Bilateral Carotid Artery Occlusion

Following a permanent occlusion of both carotid arteries 24 h before MCA occlusion, there were almost no differences between the core and penumbra after MCA occlusion. This is in contrast to previous results obtained in the current research, in which more severe damage was caused in the core area relative to the penumbra, expressed as higher levels of reflectance and fluorescence and lower levels of CBF in the core relative to the penumbra. The absence of these differences indicates a similar degree of damage in both measurement areas, which is caused by the occlusion of the carotid arteries 24 h before the MCAO was performed.

Similar to single carotid occlusion, bilateral carotid occlusion as a preconditioning treatment for MCA led to an improvement in blood supply to the tissue during local ischemia that occurs after 24 h. The CBF values in the core after MCA occlusion, which are higher than what is known in the literature at this stage (Lipton 1999), as well as the lower reflectance level, indicate an improvement in the tissue's ability to supply blood. Furthermore, a higher than usual hyperemic response is observed in the core in the reperfusion stage, and mitochondrial function returns to a normal level in the penumbra following the use of bilateral carotid occlusion as a preconditioning treatment for focal ischemia. In light of these results, it is possible that the carotid artery occlusion led to tissue resistance to local ischemia 24 h later, as was also observed in other cases of ischemia (Kirino 2002; Kitagawa et al. 2005; Zhang et al. 2003).

Nevertheless, and similar to a single carotid artery occlusion, the tissue has difficulty coping with metabolic challenge in the form of SD waves and with a temporary lack of oxygen following nitrogen exposure. Also, following a bilateral carotid occlusion combined with MCA 24 h later, high reflectance levels, low maximum NADH, and a low hyperemic wave are observed during nitrogen respiration and during the passage of SD waves relative to normoxic tissue. Consequently, it appears that 1 h after blood flow is restored in the MCA, the tissue is unable to cope with metabolic challenges such as oxygen deprivation or SD waves (Sonn and Mayevsky 2000).

Summary

Due to the tissue's inability to withstand metabolic challenges of oxygen deprivation and SD waves, it appears that a permanent occlusion of one or two carotid arteries 24 h before MCAO does not lead to an improvement in the tissue's mitochondrial and hemodynamic function, but rather increases the existing damage.

Following these findings, it seems that the combination of a permanent occlusion of one or two carotid arteries with a later MCA occlusion can be considered a model for creating an embolus, in which the calcification of a large blood vessel leads to additional ischemia in an area distant from the MCA occlusion (Koizumi et al. 1986). Studies demonstrate that 20%–30% of stroke cases are caused by arteriosclerosis (Dieter and Laird 2005). An additional research model that links global ischemia with focal ischemia may contribute greatly to the study of the relationship between arteriosclerosis and stroke.

5.4.6.3 Global Ischemia as a Postconditioning Treatment

In recent years, studies have been conducted that show that delayed treatment for ischemia (ischemic postconditioning) has an alleviating effect on the degree of ischemic damage (Pignataro et al. 2009; Zhang et al. 2003). According to this approach, the current study examined the effect of a single carotid occlusion or bilateral carotid occlusion as a delayed treatment for ischemia. The carotid arteries were occluded before or simultaneously with the removal of MCA occlusion.

Unilateral Carotid Artery Occlusion

The occlusion of a single carotid artery, as well as the later removal of this occlusion, had no visible effect on the mitochondrial function and blood supply to the tissue, whether it was performed for 45 min starting before the removal of MCA occlusion or for 15 min simultaneously with the restoration of blood flow in the MCA. In real-time measurement, no differences were observed in the measured parameters with and without a single carotid occlusion as a delayed treatment for ischemia in both the core and the penumbra. Although this comparison cannot conclude about the beneficial effects of carotid occlusion as a delayed treatment, it does show that this treatment does not have a detrimental effect on tissue function. In contrast, a single carotid occlusion led to an improvement in the tissue's ability to cope with oxygen deprivation and metabolic excitation. Following a single carotid occlusion for 15 min immediately after blood flow was restored in the MCA, no differences were found in any of the measured parameters between nitrogen exposure that occurred in a normoxic state and another exposure 1 h after blood flow was restored. In addition, no differences were found in the comparison between SD waves that were induced in normoxic tissue and waves that were induced 1 h after blood flow was restored following a combination of MCAO and a delayed UCO treatment for 15 or 45 min. As demonstrated in the previous chapters, differences were observed during these metabolic challenges before ischemia and after reperfusion. From these results, it appears that following a single carotid occlusion as a delayed treatment for ischemia, the brain is able to cope with metabolic challenges such as oxygen deprivation and excitation using SD waves, similar to a normoxic brain. This is in contrast to what was observed after focal ischemia alone without the delayed treatment, as well as following an attempt at preconditioning treatment. Therefore, a single carotid occlusion as a delayed treatment for ischemia has a beneficial effect on mitochondrial function and the ability of blood to be supplied to the postischemic tissue.

Bilateral Carotid Artery Occlusion

Similar to a single carotid occlusion, the effect of a bilateral carotid occlusion 15 min before the removal of MCA occlusion for 45 min, or together with the removal of MCA occlusion for 15 min, was examined. In contrast to a single carotid occlusion, following a bilateral carotid occlusion, immediate changes were observed in mitochondrial function and blood flow. Bilateral carotid artery occlusion 15 min before the removal of MCA occlusion led to an additional decrease in blood supply to the brain, which was accompanied by an increase in reflectance and fluorescence levels, in both the core and the penumbra. Compared to MCA occlusion without carotid occlusion, blood flow was lower and the reflectance level was higher in these two areas as long as the carotid arteries remained occluded (for 45 min). After the removal of the carotid occlusions, a hyperemic response was recorded in the core and penumbra. A similar response was not observed following MCA occlusion alone. This indicates that the gradual manner in which blood flow was restored to the tissue led to an improvement in its ability to supply blood in the reperfusion stage. In contrast, the reflectance level in the penumbra continued to be higher than normal after the removal of the carotid occlusions, which led to an artifact in the measurement of the NADH level, as was also observed in other cases (Xing et al. 2008; Zhao 2007, 2009; Bigdeli 2011).

Following a 15-min bilateral carotid artery occlusion simultaneous with the removal of MCA occlusion, blood flow values remained low relative to the state where the MCA occlusion was removed without carotid occlusion. At the same time, an increase in reflectance and fluorescence levels was observed in the core. After blood flow was restored in the carotid arteries 15 min later, a hyperemic wave was observed in the core, which was not observed following MCAO alone. This was accompanied by a decrease in fluorescence and reflectance levels. The hyperemic response indicates the tissue's ability to compensate for the lack of blood supply following ischemia, which is made possible by carotid occlusion as a delayed treatment for focal ischemia. In the penumbra area, no different response was observed following carotid artery occlusion after MCA compared to MCA occlusion alone.

When coping with metabolic challenges—oxygen deprivation and SD waves—no differences were observed in mitochondrial activity and blood flow about 1 h after blood flow was restored in the MCA and the carotid arteries compared to normoxic tissue. Therefore, similar to what happens after a single carotid occlusion, bilateral carotid occlusion as a delayed treatment for focal ischemia led to a better coping with metabolic challenges such as nitrogen exposure and induced SD waves.

Carotid Artery Occlusion as a Treatment for Ischemic Damage

In summary, it appears that the occlusion of one or both carotid arteries that occurs close to the restoration of blood flow after focal ischemia leads to an improvement in mitochondrial function and blood supply to the tissue, which is expressed in the tissue's ability to cope with metabolic challenges such as SD waves and oxygen deprivation. The study's findings demonstrate that the addition of partial global ischemia, which serves as a postconditioning treatment, is a means of improving the metabolic function of ischemic brain tissue and may be an effective treatment for improving the damage from focal ischemia.

The occlusion of one or two arteries, for a period of 15 or 45 min, led to similar results in the brain's response to metabolic challenges such as excitation following SD waves or oxygen deprivation following nitrogen exposure. During bilateral carotid artery occlusion, temporary damage to blood supply and mitochondrial function was observed, a phenomenon that was not observed during a single carotid occlusion. Therefore, a single carotid occlusion is a more effective treatment compared to a bilateral carotid occlusion even at the moment the occlusion is removed. In addition, a single carotid occlusion for 45 min led to minimal damage to blood supply during nitrogen respiration, which is not observed during a 15-min occlusion. Therefore, among the procedures reviewed, a single carotid occlusion for 15 min has a beneficial effect on the damage from focal ischemia combined with the lowest temporary damage.

The results of the current study join a series of previous studies that indicate an improvement in reperfusion damage through delayed ischemic treatment. These studies demonstrate that delayed treatments against ischemia have a beneficial effect in cases of ischemia in the brain and heart, as well as against conditions of hyperperfusion syndrome and inflammation (Pignataro et al. 2009). Seemingly, ischemic postconditioning is a temporary addition of damage that is added to the damage already caused by ischemia, but in fact, the intervention during ischemia leads to a gradual return of blood to the ischemic tissue, and thereby leads to a decrease in reperfusion injury.

The study's results, which demonstrate the effectiveness of carotid occlusion as a delayed treatment for ischemic and reperfusion damage, may indicate a possible direction for treating stroke damage in the clinic. Additional research in this field, with or without the use of pharmacological means for treating stroke (such as pTA), is needed to determine the effectiveness of such treatments.

This study investigated mitochondrial function and blood flow before, during, and after focal ischemia, in conjunction with other pathologies. It provides extensive profiles of various outcomes that can result from a single model of focal ischemia. These results highlight the complexity of the MCAO model and the high variability in the nature of focal ischemia it produces.

5.4.7 *Summary and Conclusions*

The current study examined mitochondrial function and blood flow before, during, and after focal ischemia, in combination with other pathologies (Fig. 5.73). In the first stage, the changes occurring in mitochondria and blood supply to the tissue during ischemia have varying degrees of severity. The results of the study indicate that MCA occlusion leads to a decrease in blood supply to the tissue, which leads to oxygen deficiency and, as a result, to impaired mitochondrial function. The hemodynamic and mitochondrial damage occur in the hemisphere ipsilateral to the occlusion. The ischemic damage is gradual in nature and is more severe in the core region, which is the center of the damage, relative to the penumbra regions located at its edges. As the duration of ischemia increases, more severe ischemic damage is observed, and accordingly the tissue's ability to recover decreases. Accordingly, it was demonstrated in this study that after 24 h, transient focal ischemia leads to extensive damage in the core only, while permanent focal ischemia damages both the core and the penumbra.

This work comprehensively presents different profiles of outcomes that are all caused by a single model of focal ischemia. These results are evidence of the great complexity of the MCAO model, which results in a high variability in the nature of the focal ischemia that results from it.

The current study presents a comparison of the metabolic function of the tissue during focal ischemia compared to global ischemia. Partial global ischemia resulting from a single carotid occlusion has only a minor effect on the brain, manifested in differences between the hemispheres 24 h after the occlusion. In contrast, partial global

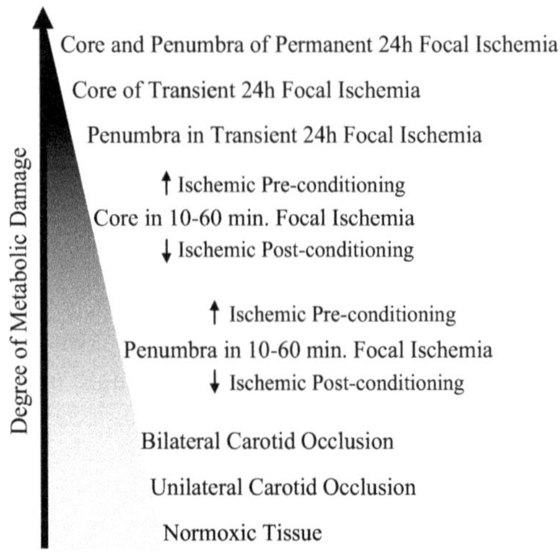

Fig. 5.73 Ranking the various models presented in the study according to the intensity of metabolic damage to the tissue

ischemia resulting from the occlusion of both carotid arteries results in more severe damage, manifested in the inability of the brain to cope with metabolic challenges from the moment of injury. Partial global ischemia, whether due to occlusion of one or both carotid arteries, results in mild damage relative to that observed in the core and even penumbra of focal ischemia.

After characterizing mitochondrial and hemodynamic function following focal ischemia and comparing them to global ischemia, several ways to reduce the damage of focal ischemia were examined: through ischemic preconditioning and ischemic postconditioning. Occlusion of one or both carotids as a pretreatment to reduce ischemic and reperfusion damage was found to be ineffective.

The results of the present study indicate that occlusion of one or two carotid arteries may serve as a model for embolus, in which damage to a major artery later leads to ischemic damage in a distant area. In contrast, occlusion of one or two carotid arteries as a late treatment has been shown to be an effective way to reduce the damage of reperfusion injury following focal ischemia. The information obtained in this part of the study may be useful for implementing the method of late treatment of ischemic damage as a method of treating stroke in patients.

The results of the present study were all obtained using the MSMP system, which allows in vivo monitoring of several parameters at multiple sites simultaneously, continuously and in real time. Using the MSMP system to measure mitochondrial function and blood flow during ischemia reveal another aspect of the tissue damage that occurs following various ischemic conditions. This information, combined with additional characteristics of mitochondrial activity, such as gene expression, membrane activity, and histological findings, may advance what is known about the mitochondrial contribution to tissue damage resulting from cerebral ischemia.

References

Albayrak, S., Q. Zhao, B. K. Siesjö, and M. L. Smith. 1997. Effect of transient focal ischemia on blood-brain barrier permeability in the rat: correlation to cell injury. *Acta Neuropathologica* 94:158–163.

Ames, A. I., R. I. Wright, M. Kowada, M. Thurston, and G. Majno. 1968. Cerebral ischemia. II. The noreflow phenomenon. *The American Journal of Pathology* 52:437–453.

Anderson, R. E., W. K. Tan, H. S. Martin, and F. B. Meyer. 1999. Effects of glucose and PaO_2 modulation on cortical intracellular acidosis, NADH redox state, and infarction in the ischemic penumbra. *Stroke* 30:160–170.

Ashwal, S., B. Tone, H. R. Tian, D. J. Cole, and W. J. Pearce. 1998. Core and penumbral nitric oxide synthase activity during cerebral ischemia and reperfusion. *Stroke* 29:1037–1046. discussion 1047.

Astrup, J., B. K. Siesjo, and L. Symon. 1981. Thresholds in cerebral ischemia—the ischemic penumbra. *Stroke* 12:723–725.

Baron, J. C. 2001. Perfusion thresholds in human cerebral ischemia: historical perspective and therapeutic implications. *Cerebrovascular Diseases* 11(Suppl 1):2–8.

Bartus, R. T., R. L. Dean, K. Cavanaugh, D. Eveleth, D. L. Carriero, and G. Lynch. 1995. Time-related neuronal changes following middle cerebral artery occlusion: implications for therapeutic intervention and the role of calpain. *Journal of Cerebral Blood Flow and Metabolism* 15:969–979.

Basarsky, T. A., D. Feighan, and B. A. Macvicar. 1999. Glutamate release through volume-activated channels during spreading depression. *The Journal of Neuroscience* 19:6439–6445.

Bederson, J. B., L. H. Pitts, M. Tsuji, M. C. Nishimura, R. L. Davis, and H. Bartkowski. 1986. Rat middle cerebral artery occlusion: evaluation of the model and development of a neurologic examination. *Stroke* 17:472–476.

Belayev, L., O. F. Alonso, R. Busto, W. Zhao, and M. D. Ginsberg. 1996. Middle cerebral artery occlusion in the rat by intraluminal suture. Neurological and pathological evaluation of an improved model. *Stroke* 27:1616–1622. discussion 1623.

Bengtsson, F., and B. K. Siesjo. 1990. Cell damage in cerebral ischemia: physiological, biochemical and structural aspects. In *Cerebral ischemia and resuscitation*, eds. A. Schurr and B. M. Rigor. Boca Raton: CRC Press.

Berne, R. M. and M. N. Levy. 2000. *Principles of physiology*. Philadelphia, PA: Elsevier, Mosby.

Best, P. J., J. Orr, and J. E. Pointer. 1975. Differential effects of cortical ablation and spreading depression on sensitivity to footshock: implications for the role of the cortex in learning. *Physiology & Behavior* 14:801–807.

Bigdeli, M. R. 2011. Neuroprotection caused by hyperoxia preconditioning in animal stroke models. *The Scientific World Journal* 11:403–421.

Bluestone, A. Y., M. Stewart, B. Lei, I. S. Kass, J. Lasker, G. S. Abdoulaev, and A. H. Hielscher. 2004. Three-dimensional optical tomographic brain imaging in small animals, part 2: unilateral carotid occlusion. *Journal of Biomedical Optics* 9: 1063–1073.

Bouley, J., M. Fisher, and N. Henninger. 2007. Comparison between coated vs. uncoated suture middle cerebral artery occlusion in the rat as assessed by perfusion/diffusion weighted imaging. *Neuroscience Letters* 412:185–190.

Brozek, G., O. Buresová, and J. Bures. 1974. Effect of bilateral cortical spreading depression on the hippocampal theta activity induced by oral infusion of aversive gustatory stimulus. *Experimental Neurology* 42:661–668.

Brustovetsky, N., T. Brustovetsky, R. Jemmerson, and J. M. Dubinsky. 2002. Calcium-induced cytochrome c release from CNS mitochondria is associated with the permeability transition and rupture of the outer membrane. *Journal of Neurochemistry* 80:207–218.

Buchan, A. M., D. Xue, and A. Slivka. 1992. A new model of temporary focal neocortical ischemia in the rat. *Stroke* 23: 273–279.

Buresová, O., and J. Bures. 1985. Unilateral and bilateral cortical spreading depression interferes with radial maze performance in rats. *Brain Research* 344:365–368.

Busto, R., and M. D. Ginsberg. 1985. Graded focal cerebral ischemia in the rat by unilateral carotid artery occlusion and elevated intracranial pressure: hemodynamic and biochemical characterization. *Stroke* 16:466–476.

Camacho, A., and L. Massieu. 2006. Role of glutamate transporters in the clearance and release of glutamate during ischemia and its relation to neuronal death. *Archives of Medical Research* 37:11–18.

Canevari, L., S. Kuroda, T. E. Bates, J. B. Clark, and B. K. Siesjö. 1997. Activity of mitochondrial respiratory chain enzymes after transient focal ischemia in the rat. *Journal of Cerebral Blood Flow and Metabolism* 17:1166–1169.

Cechetto, D. F., J. X. Wilson, K. E. Smith, D. Wolski, M. D. Silver, and V. C. Hachinski. 1989. Autonomic and myocardial changes in middle cerebral artery occlusion: stroke models in the rat. *Brain Research* 502:296–305.

Chan, P. H. 1996. Role of oxidants in ischemic brain damage. *Stroke* 27:1124–1129.

Chance, B., G. R. Williams, W. F. Holmes, and J. Higgins. 1955. Respiratory enzymes in oxidative phosphorylation (V- A mechanism for oxidative phosphorylation). *The Journal of Biological Chemistry* 217:439–451.

Chen, S., Z. Feng, P. Li, S. L. Jacques, S. Zeng, and Q. Luo. 2006. *In vivo* optical reflectance imaging of spreading depression waves in rat brain with and without focal cerebral ischemia. *Journal of Biomedical Optics* 11:34002.

Christophe, M., and S. Nicolas. 2006. Mitochondria: a target for neuroprotective interventions in cerebral ischemia-reperfusion. *Current Pharmaceutical Design* 12:739–757.

Cohen, J. E., E. Itshayek, S. Moskovici, J. M. Gomori, S. Fraifeld, R. Eichel, and R. R. Leker. 2011. State-of-the-art reperfusion strategies for acute ischemic stroke. *Journal of Clinical Neuroscience* 18:319–323.

Coulson, R. J., N. C. Chesler, L. Vitullo, and M. J. Cipolla. 2002. Effects of ischemia and myogenic activity on active and passive mechanical properties of rat cerebral arteries. *American Journal of Physiology. Heart and Circulatory Physiology* 283:H2268–H2275.

De Ley, G., J. B. Nshimyumuremyi, and I. Leusen. 1985. Hemispheric blood flow in the rat after unilateral common carotid occlusion: evolution with time. *Stroke* 16:69–73.

Dieter, R. S., and J. R. Laird. 2005. Carotid artery stenting: update. *International Journal of Cardiovascular Interventions* 7: 126–133.

Dittmar, M. S., B. Vatankhah, N. P. Fehm, G. Schuierer, U. Bogdahn, M. Horn, and F. Schlachetzki. 2006. Fischer-344 rats are unsuitable for the MCAO filament model due to their cerebrovascular anatomy. *Journal of Neuroscience Methods* 156:50–54.

Dora, E., K. Tanaka, J. H. Greenberg, N. H. Gonatas, and M. Reivich. 1986. Kinetics of microcirculatory, NAD/NADH, and electrocorticographic changes in cat brain cortex during ischemia and recirculation. *Annals of Neurology* 19: 536–544.

Dreier, J. P. 2011. The role of spreading depression, spreading depolarization and spreading ischemia in neurological disease. *Nature Medicine* 17:439–447.

Duarte, S. G., A. D. Campos, and B. O. Colli. 2003. Functional evaluation of temporary focal cerebral ischemia: experimental model. *Arquivos de Neuro-psiquiatria* 61:751–756.

Durukan, A., and T. Tatlisumak. 2007. Acute ischemic stroke: overview of major experimental rodent models, pathophysiology, and therapy of focal cerebral ischemia. *Pharmacology, Biochemistry, and Behavior* 87:179–197.

Folbergrová, J., Y. Kiyota, K. Pahlmark, H. Memezawa, M. L. Smith, and B. K. Siesjö. 1993. Does ischemia with reperfusion lead to oxidative damage to proteins in the brain? *Journal of Cerebral Blood Flow and Metabolism* 13:145–152.

Garcia, X., and F. Stein. 2006. Nitric oxide. *Seminars in Pediatric Infectious Diseases* 17:55–57.

Gerlai, R., H. Thibodeaux, J. T. Palmer, M. Van Lookeren Campagne, and N. Van Bruggen. 2000. Transient focal cerebral ischemia induces sensorimotor deficits in mice. *Behavioural Brain Research* 108:63–71.

Gerriets, T., E. Stolz, M. Walberer, C. Müller, C. Rottger, A. Kluge, M. Kaps, M. Fisher, and G. Bachmann. 2004. Complications and pitfalls in rat stroke models for middle cerebral artery occlusion: a comparison between the suture and the macrosphere model using magnetic resonance angiography. *Stroke* 35:2372–2377.

Gidö, G., T. Kristián, and B. K. Siesjö. 1997. Extracellular potassium in a neocortical core area after transient focal ischemia. *Stroke* 28:206–210.

Ginsberg, M. D. 2003. Adventures in the pathophysiology of brain ischemia: penumbra, gene expression, neuroprotection: The 2002 Thomas Willis lecture. *Stroke* 34:214–223.

Gorji, A. 2001. Spreading depression: a review of the clinical relevance. *Brain Research. Brain Research Reviews* 38: 33–60.

Halestrap, A. P. 2006. Calcium, mitochondria and reperfusion injury: a pore way to die. *Biochemical Society Transactions* 34:232–237.

Hartkamp, M. J., J. Van Der Grond, K. J. Van Everdingen, B. Hillen, and W. P. Mali. 1999. Circle of Willis collateral flow investigated by magnetic resonance angiography. *Stroke* 30: 2671–2678.

Hashimoto, M., Y. Takeda, T. Sato, H. Kawahara, O. Nagano, and M. Hirakawa. 2000. Dynamic changes of NADH fluorescence images and NADH content during spreading depression in the cerebral cortex of gerbils. *Brain Research* 872:294–300.

Homola, A., N. Zoremba, K. Slais, R. Kuhlen, and E. Syková. 2006. Changes in diffusion parameters, energy-related metabolites and glutamate in the rat cortex after transient hypoxia/ischemia. *Neuroscience Letters* 404:137–142.

Hossmann, K. A. 2006. Pathophysiology and therapy of experimental stroke. *Cellular and Molecular Neurobiology* 26: 1055–1081.

Hungerhuber, E., S. Zausinger, T. Westermaier, N. Plesnila, and R. Schmid-Elsaesser. 2006. Simultaneous bilateral laser Doppler fluxmetry and electrophysiological recording during middle cerebral artery occlusion in rats. *Journal of Neuroscience Methods* 154:109–115.

Huston, J. P., and J. Bures. 1970. Drinking and eating elicited by cortical spreading depression. *Science* 169:702–704.

Iijima, T., C. Shimase, Y. Iwao, and H. Sankawa. 1998. Relationships between glutamate release, blood flow and spreading depression: real-time monitoring using an electroenzymatic dialysis electrode. *Neuroscience Research* 32: 201–207.

Islam, S., and O. Buresová. 1975. The effect of cortical spreading depression on motor performance and depth avoidance in rats. *Activitas Nervosa Superior* 17:8–14.

Kalesnykas, G., T. Tuulos, H. Uusitalo, and J. Jolkkonen. 2008. Neurodegeneration and cellular stress in the retina and optic nerve in rat cerebral ischemia and hypoperfusion models. *Neuroscience* 155:937–947.

Kinuta, Y., H. Kikuchi, M. Ishikawa, M. Kimura, and Y. Itokawa. 1989. Lipid peroxidation in focal cerebral ischemia. *Journal of Neurosurgery* 71:421–429.

Kirino, T. 2002. Ischemic tolerance. *Journal of Cerebral Blood Flow and Metabolism* 22:1283–1296.

Kitagawa, K., Y. Yagita, T. Sasaki, S. Sugiura, E. Omura-Matsuoka, T. Mabuchi, K. Matsushita, and M. Hori. 2005. Chronic mild reduction of cerebral perfusion pressure induces ischemic tolerance in focal cerebral ischemia. *Stroke* 36:2270–2274.

Kobayashi, S., V. A. Harris, and F. A. Welsh. 1995. Spreading depression induces tolerance of cortical neurons to ischemia in rat brain. *Journal of Cerebral Blood Flow and Metabolism* 15: 721–727.

Koizumi, J., Y. Yoshida, T. Nakazawa, and G. Ooneda. 1986. Experimental studies of ischemic brain edema, I: a new experimental model of cerebral embolism in rats in which recirculation can be introduced in the ishemic area. *Japanese Journal of Stroke* 8:1–8.

Kozuka, M. 1995. Changes in brain energy metabolism, neurotransmitters, and choline during and after incomplete cerebral ischemia in spontaneously hypertensive rats. *Neurochemical Research* 20:23–30.

Kristián, T., G. Gidö, S. Kuroda, A. Schütz, and B. K. Siesjö. 1998. Calcium metabolism of focal and penumbral tissues in rats subjected to transient middle cerebral artery occlusion. *Experimental Brain Research* 120:503–509.

Kristián, T., and B. K. Siesjö. 1996. Calcium-related damage in ischemia. *Life Sciences* 59:357–367.

Kristián, T., and B. K. Siesjö. 1998. Calcium in ischemic cell death. *Stroke* 29:705–718.

Kuroiwa, T., and R. Okeda. 1994. Neuropathology of cerebral ischemia and hypoxia: recent advances in experimental studies on its pathogenesis. *Pathology International* 44:171–181.

Laing, R. J., J. Jakubowski, and R. W. Laing. 1993. Middle cerebral artery occlusion without craniectomy in rats. Which method works best?*Stroke* 24:294–297. discussion 297-8.

Lauritzen, M., A. J. Hansen, D. Kronborg, and T. Wieloch. 1990. Cortical spreading depression is associated with arachidonic acid accumulation and preservation of energy charge. *Journal of Cerebral Blood Flow and Metabolism* 10: 115–122.

Leão, A. A. P. 1944. Spreading depression of activity in cerebral cortex. *Journal of Neurophysiology* 7:359–390.

Lee, J. K., M. S. Park, Y. S. Kim, K. S. Moon, S. P. Joo, T. S. Kim, J. H. Kim, and S. H. Kim. 2007. Photochemically induced cerebral ischemia in a mouse model. *Surgical Neurology* 67: 620–625. discussion 625.

Lee, R. M. 1995. Morphology of cerebral arteries. *Pharmacology & Therapeutics* 66:149–173.

Levine, S. 1960. Anoxic-ischemic encephalopathy in rats. *The American Journal of Pathology* 36:1–17.

Li, J., Y. Takeda, and M. Hirakawa. 2000. Threshold of ischemic depolarization for neuronal injury following four-vessel occlusion in the rat cortex. *Journal of Neurosurgical Anesthesiology* 12:247–254.

Lipton, P. 1999. Ischemic cell death in brain neurons. *Physiological Reviews* 79: 1431–1568.

Livnat, A., E. Barbiro-Michaely, and A. Mayevsaky. 2009. Mitochondrial function and cerebral blood flow responses under unilateral carotid occlusion in rats. In *SPIE BiOS: biomedical optics*, eds. A. Mahadevan-Jansen and E. D. Jansen. Bellingham, WA: International Society for Optics and Photonics.

Livnat, A., E. Barbiro-Michaely, and A. Mayevsky. 2010. Mitochondrial function and cerebral blood flow variable responses to middle cerebral artery occlusion. *Journal of Neuroscience Methods* 188:76–82.

Livnat, A., E. Barbiro-Michaely, and A. Mayevsky. 2011. Real-time monitoring of spatial and temporal metabolic changes during focal cerebral ischemia in rats. *Brain Research* 1389:125–132.

Livnat, A., M. Tolmasov, E. Barbiro-Michaely, and A. Mayevsky. 2008. Real-time monitoring of mitochondrial function and cerebral blood flow follwing focal ischemia in rats. *Journal of Innovative Optical Health Sciences* 1:63–69.

Longa, E. Z., P. R. Weinstein, S. Carlson, and R. Cummins. 1989. Reversible middle cerebral artery occlusion without craniectomy in rats. *Stroke* 20:84–91.

Lourbopoulos, A., D. Karacostas, N. Artemis, I. Milonas, and N. Grigoriadis. 2008. Effectiveness of a new modified intraluminal suture for temporary middle cerebral artery occlusion in rats of various weight. *Journal of Neuroscience Methods* 173:225–234.

Lukyanov, L. D., and J. Bures. 1967. Changes in pO_2 due to spreading depression in cortex and nucleus caudatus of rat. *Physiologia Bohemoslovenica* 16:449–455.

Malinski, T., F. Bailey, Z. G. Zhang, and M. Chopp. 1993. Nitric oxide measured by a porphyrinic microsensor in rat brain after transient middle cerebral artery occlusion. *Journal of Cerebral Blood Flow and Metabolism* 13:355–358.

Marshall, W. H. 1959. Spreading cortical depression of Leão. *Physiological Reviews* 39:239–279.

Mayevsky, A. 1976. Brain energy metabolism of the conscious rat exposed to various physiological and pathological situations. *Brain Research* 113:327–338.

Mayevsky, A. 1984. Brain NADH redox state monitored *in vivo* by fiber optic surface fluorometry. *Brain Research Reviews* 7:49–68.

Mayevsky, A., and B. Chance. 2007. Oxidation-reduction states of NADH *in vivo*: from animals to clinical use. *Mitochondrion* 7:330–339.

Mayevsky, A., and S. Cohen. 1990. Brain ischemic depolarization and vasospasm in the Mongolian gerbil: The dependence on energy depletion levels. *Advances in Experimental Medicine and Biology* 277:315–322.

Mayevsky, A., A. Doron, T. Manor, S. Meilin, K. Salame, and G. E. Quaknine. 1995. Repetitive cortical spreading depression cycle development in the human brain: A multiparametric monitoring approach. *Journal of Cerebral Blood Flow and Metabolism* 15:S34.

Mayevsky, A., A. Doron, T. Manor, S. Meilin, N. Zarchin, and G. E. Ouaknine. 1996. Cortical spreading depression recorded from the human brain using a multiparametric monitoring system. *Brain Research* 740:268–274.

Mayevsky, A., R. B. Duckrow, E. Yoles, N. Zarchin, and D. Kanshansky. 1990. Brain mitochondrial redox state, tissue hemodynamic and extracellular ion responses to four-vessel occlusion and spreading depression in the rat. *Neurological Research* 12:243–248.

Mayevsky, A., S. Meilin, T. Manor, E. Ornstein, N. Zarchin, and J. Sonn. 1998. Multiparametric monitoring of brain oxygen balance under experimental and clinical conditions. *Neurological Research* 20:S76–S80.

Mayevsky, A., and G. Rogatsky. 2007. Mitochondrial function *in vivo* evaluated by NADH fluorescence: from animal models to human studies. *American Journal of Physiology. Cell Physiology* 292:C615–C640.

Mayevsky, A., and H. R. Weiss. 1991. Cerebral blood flow and oxygen consumption in cortical spreading depression. *Journal of Cerebral Blood Flow & Metabolism* 11:829–836.

Mayevsky, A., and I. Ziv. 1991. Oscillations of cortical oxidative metabolism and microcirculation in the ischaemic brain. *Neurological Research* 13:39–47.

Meilin, S., J. Sonn, N. Zarchin, G. Rogatsky, E. Guggenheimer-Furman, and A. Mayevsky. 1998. Responses of rat brain to induced spreading depression following exposure to carbon monoxide. *Brain Research* 780:323–328.

Meilin, S., N. Zarchin, A. Mayevsky, and S. Shapira. 1995. Multiparametric responses to cortical spreading depression under nitric oxide synthesis inhibition. In Biochemical pharmacological and clinical aspects of nitric oxide, eds. B. A. Weissman, N. Alon, and S. Shapira. New York: Plenum Press.

Mendel, T. 2005. ["Malignant" middle cerebral artery territory infarction]. *Neurologia i Neurochirurgia Polska* 39: 366–371.

Mhairi Macrae, I. 1992. New models of focal cerebral ischaemia. *British Journal of Clinical Pharmacology* 34:302–308.

Morimoto, T., M. Y. Globus, R. Busto, E. Martinez, and M. D. Ginsberg. 1996. Simultaneous measurement of salicylate hydroxylation and glutamate release in the penumbral cortex following transient middle cerebral artery occlusion in rats. *Journal of Cerebral Blood Flow and Metabolism* 16:92–99.

Nagasawa, H., and K. Kogure. 1989. Correlation between cerebral blood flow and histologic changes in a new rat model of middle cerebral artery occlusion. *Stroke* 20:1037–1043.

Nakamura, H., A. J. Strong, C. Dohmen, O. W. Sakowitz, S. Vollmar, M. Sué, L. Kracht, P. Hashemi, R. Bhatia, T. Yoshimine, J. P. Dreier, A. K. Dunn, and R. Graf. 2010. Spreading depolarizations cycle around and enlarge focal ischaemic brain lesions. *Brain* 133:1994–2006.

Nallet, H., E. T. Mackenzie, and S. Roussel. 1999. The nature of penumbral depolarizations following focal cerebral ischemia in the rat. *Brain Research* 842:148–158.

Nishizawa, Y. 2001. Glutamate release and neuronal damage in ischemia. *Life Sciences* 69:369–381.

Park, W. S., D. K. Sung, S. Kang, S. H. Koo, Y. J. Kim, J. H. Lee, Y. S. Chang, and M. Lee. 2006. Therapeutic window for cycloheximide treatment after hypoxic-ischemic brain injury in neonatal rats. *Journal of Korean Medical Science* 21:490–494.

Paxinos, G. and C. Watson. 1997. *The rat brain in stereotaxic coordinates*. Australia, Centercourt: Academic Press.

Pestalozza, I. F., S. Di Legge, M. Calabresi, and G. L. Lenzi. 2002. Ischaemic penumbra: highlights. *Clinical and Experimental Hypertension* 24:517–529.

Pianka, M. J. 1974. Active avoidance learning during bilateral spreading depression. *Physiology & Behavior* 12:725–732.

Pietrobon, D. 2005. Migraine: new molecular mechanisms. *The Neuroscientist* 11:373–386.
Pignataro, G., A. Scorziello, G. Di Renzo, and L. Annunziato. 2009. Post-ischemic brain damage: effect of ischemic preconditioning and postconditioning and identification of potential candidates for stroke therapy. *The FEBS Journal* 276: 46–57.
Pinard, E., H. Nallet, E. T. Mackenzie, J. Seylaz, and S. Roussel. 2002. Penumbral microcirculatory changes associated with peri-infarct depolarizations in the rat. *Stroke* 33:606–612.
Prieto, R., F. Carceller, J. M. Roda, and C. Avendaño. 2005. The intraluminal thread model revisited: rat strain differences in local cerebral blood flow. *Neurological Research* 27:47–52.
Richard, M. J., T. M. Saleh, B. El Bahh, and J. A. Zidichouski. 2010. A novel method for inducing focal ischemia in vitro. *Journal of Neuroscience Methods* 190:20–27.
Samdani, A. F., T. M. Dawson, and V. L. Dawson. 1997. Nitric oxide synthase in models of focal ischemia. *Stroke* 28: 1283–1288.
Schmid-Elsaesser, R., S. Zausinger, E. Hungerhuber, A. Baethmann, and H. J. Reulen. 1998. A critical reevaluation of the intraluminal thread model of focal cerebral ischemia: evidence of inadvertent premature reperfusion and subarachnoid hemorrhage in rats by laser-Doppler flowmetry. *Stroke* 29:2162–2170.
Shimamura, N., G. Matchett, T. Tsubokawa, H. Ohkuma, and J. Zhang. 2006. Comparison of silicon-coated nylon suture to plain nylon suture in the rat middle cerebral artery occlusion model. *Journal of Neuroscience Methods* 156:161–165.
Sims, N. R., and H. Muyderman. 2010. Mitochondria, oxidative metabolism and cell death in stroke. *Biochimica et Biophysica Acta - Molecular Basis of Disease* 1802:80–91.
Slakter, J. S., A. D. Spertus, S. S. Weissman, and P. Henkind. 1984. An experimental model of carotid artery occlusive disease. *American Journal of Ophthalmology* 97:168–172.
Smeaton, W. A. 1963. New light on lavoisier: the research of the last ten years. *History of Science* 2:51–69.
Solenski, N. J., C. G. Dipierro, P. A. Trimmer, A. L. Kwan, and G. A. Helm. 2002. Ultrastructural changes of neuronal mitochondria after transient and permanent cerebral ischemia. *Stroke* 33:816–824.
Solenski, N. J., A. L. Kwan, H. Yanamoto, J. P. Bennett, N. F. Kassell, and K. S. Lee. 1997. Differential hydroxylation of salicylate in core and penumbra regions during focal reversible cerebral ischemia. *Stroke* 28:2545–2551. discussion 2551-2.
Somjen, G. G. 2001. Mechanisms of spreading depression and hypoxic spreading depression-like depolarization. *Physiological Reviews* 81:1065–1096.
Sonn, J., and A. Mayevsky. 2000. Effects of brain oxygenation on metabolic, hemodynamic, ionic and electrical responses to spreading depression in the rat. *Brain Research* 882:212–216.
Sukhotinsky, I., E. Dilekoz, M. A. Moskowitz, and C. Ayata. 2008. Hypoxia and hypotension transform the blood flow response to cortical spreading depression from hyperemia into hypoperfusion in the rat. *Journal of Cerebral Blood Flow and Metabolism* 28:1369–1376.
Sun, D., and D. D. Gilboe. 1994. Ischemia-induced changes in cerebral mitochondrial free fatty acids, phospholipids, and respiration in the rat. *Journal of Neurochemistry* 62:1921–1928.
Takano, K., L. L. Latour, J. E. Formato, R. A. Carano, K. G. Helmer, Y. Hasegawa, C. H. Sotak, and M. Fisher. 1996. The role of spreading depression in focal ischemia evaluated by diffusion mapping. *Annals of Neurology* 39:308–318.
Tamura, A., D. I. Graham, J. Mcculloch, and G. M. Teasdale. 1981. Focal cerebral ischaemia in the rat: 1. Description of technique and early neuropathological consequences following middle cerebral artery occlusion. *Journal of Cerebral Blood Flow and Metabolism* 1:53–60.
Tomlinson, F. H., R. E. Anderson, and F. B. Meyer. 1993. Brain pH_i, cerebral blood flow, and NADH fluorescence during severe incomplete global ischemia in rabbits. *Stroke* 24:435–443.
Vagnozzi, R., B. Tavazzi, D. Di Pierro, B. Giardina, B. Fraioli, S. Signoretti, S. Distefano, M. Galvano, and G. Lazzarino. 1997. Effects of increasing times of incomplete cerebral ischemia upon the energy state and lipid peroxidation in the rat. *Experimental Brain Research* 117:411–418.
Walberer, M., E. Stolz, C. Müller, C. Friedrich, C. Rottger, F. Blaes, M. Kaps, M. Fisher, G. Bachmann, and T. Gerriets. 2006. Experimental stroke: ischaemic lesion volume and oedema

formation differ among rat strains (a comparison between Wistar and Sprague-Dawley rats using MRI). *Laboratory Animals* 40:1–8.
Wass, C. T., and W. L. Lanier. 1996. Glucose modulation of ischemic brain injury: review and clinical recommendations. *Mayo Clinic Proceedings* 71:801–812.
Winn, Jr, F. J., M. A. Kent, and T. M. Libkuman. 1975. Learned taste aversion induced by cortical spreading depression. *Physiology & Behavior* 15:21–24.
Woitzik, J., and L. Schilling. 2002. Control of completeness and immediate detection of bleeding by a single laser-Doppler flow probe during intravascular middle cerebral artery occlusion in rats. *Journal of Neuroscience Methods* 122:75–78.
Xing, B., H. Chen, M. Zhang, D. Zhao, R. Jiang, X. Liu, and S. Zhang. 2008. Ischemic post-conditioning protects brain and reduces inflammation in a rat model of focal cerebral ischemia/reperfusion. *Journal of Neurochemistry* 105:1737–1745.
Zeman, W. and J. R. M. Innes. 1963. *Craigie's neuroanatomy of the rat*. New York: Academic Press.
Zhang, H. X., G. H. Du, and J. T. Zhang. 2003. Ischemic pre-conditioning preserves brain mitochondrial functions during the middle cerebral artery occlusion in rat. *Neurological Research* 25:471–476.
Zhang, L., J. Chen, Y. Li, Z. G. Zhang, and M. Chopp. 2000. Quantitative measurement of motor and somatosensory impairments after mild (30 min) and severe (2 h) transient middle cerebral artery occlusion in rats. *Journal of the Neurological Sciences* 174:141–146.
Zhao, H. 2007. The protective effect of ischemic postconditioning against ischemic injury: from the heart to the brain. *Journal of Neuroimmune Pharmacology* 2:313–318.
Zhao, H. 2009. Ischemic postconditioning as a novel avenue to protect against brain injury after stroke. *Journal of Cerebral Blood Flow and Metabolism* 29:873–885.
Zhao, H., W. G. Mayhan, and H. Sun. 2008. A modified suture technique produces consistent cerebral infarction in rats. *Brain Research* 1246:158–166.

Chapter 6
The Involvement of Nitric Oxide (NO) in the Regulation of Brain Physiological Activities

Abstract This work examined the activity of the brain in rats exposed to hypercapnia (increase in $PaCO_2$) and hypoxia (decrease in PaO_2). In addition, the involvement of the NO molecule in hypercapnia and hypoxia was examined by administering an inhibitor of its synthesis (L-NAME) when exposing the animals to the above conditions. In this work, the brain response to hypercapnia and hypoxia was examined in animals injected with L-NAME, IP at a concentration of 50 mg/kg. To induce hypercapnia and hypoxia, the animals were ventilated with the following mixtures: 69% N_2 + 21% O_2 + 10% CO_2 and 1% CO_2 + 87% N_2 + 12% O_2, respectively. The results of this work showed that inhibition of NO synthesis by L-NAME caused an increase in arterial blood pressure and a decrease in cerebral blood flow. Therefore, it seems that NO has an important role in maintaining constant blood pressure and normal blood flow in the brain, similar to what has been reported in many other works. Hypercapnia induction in rats caused an increase in cerebral blood flow and mitochondrial NADH oxidation, while the extracellular potassium level did not change. These results suggest an improvement in oxygen balance due to hypercapnia. Exposing the rats to hypercapnia during NOS induction (by administering L-NAME) caused similar results (increased cerebral blood flow and NADH oxidation), except for the increased extracellular potassium level. Therefore, it is possible that during hypercapnia exposure, the NO molecule plays some role in improving the energy balance in the cell, which could contribute to the efficiency of the Na^+/K^+-ATPase pumps. Induction of hypoxia in rats caused an increase in cerebral blood flow and an increase in mitochondrial NADH level. The increase in cerebral blood flow probably failed to compensate for the oxygen deficiency due to hypoxia, and therefore the NADH level increased. Although The increase in NADH level, the extracellular potassium level did not change, which suggests that there was no disruption in the activity of Na^+/K^+-ATPase pumps and probably no ATP deficiency was created. Exposure of the rats to hypoxia during NOS induction caused similar results (an increase in cerebral blood flow and NADH level, without a change in extracellular potassium level). From the results presented in this work, it appears that the NO molecule does not participate in the vasodilation mechanism

A. Mayevsky et al., *Physiological Mapping of the Cerebral Cortex in Neuropathology*,
https://doi.org/10.1007/978-3-032-13019-8_6

during hypercapnia and hypoxia, and there are probably other regulatory pathways of cerebral blood flow that participate in the above reactions unrelated to the NO pathway.

Keywords Hypoxia · Hypercapnia · Cerebral blood flow · NADH redox state · Mitochondrial function · Extracellular K^+

6.1 Introduction

6.1.1 Brain Energy Metabolism

Brain tissue consumes a lot of energy, even though no mechanical work or external secretory activity occurs in it. This fact makes brain cells very sensitive to disturbances in the supply of energy sources (Siesjo 1978d). Since the energy reserves in the brain are small (Erecinska and Silver 1989), its normal activity is completely dependent on the supply of oxygen and glucose by the bloodstream. Moreover, since nerve cells have little capacity for regeneration, any decrease in the energy supply to the brain may have long-term consequences for the function and integrity of the organism. The energy required for the work of brain cells is obtained by oxidative catabolism of glucose (Siesjo 1978d). A small portion of ATP is produced in the cytosol by glycolysis (2 moles of ATP per mole of glucose), while most of the ATP is produced in the mitochondria by oxidative phosphorylation (36 moles of ATP per mole of glucose), with oxygen serving as the final electron acceptor in the respiratory chain (Go 1991). The energy is used to sustain two major activities of the brain:

1. Maintaining the resting potential of nerve cells, which allows for the transmission of nerve impulses. Resting potential is maintained by sodium-potassium pumps, which consume over 50% of the total energy consumed in neurons (Erecinska and Silver 1989).
2. Biosynthesis of neurotransmitters and proteins (Lehninger 1982). There is a correspondence between ATP utilization (by Na^+/K^+-ATPase pumps) and mitochondrial activity (see Fig. 6.1). Any change in oxygen supply to the brain will disrupt the balance between energy supply and demand and can result in the development of a pathological condition (Mayevsky 1984). Any decrease in oxygen supply to the brain will cause a slowdown in mitochondrial activity (Mayevsky 1976) and the reduction of respiratory chain components, therefore the NADH level will increase. Na^+/K^+-ATPase pumps will be inhibited due to the decrease in ATP production (Mayevsky et al. 1984) and as a result potassium will accumulate in the extracellular space. Any increase in energy demand will increase the flow of electrons through the respiratory chain and the oxidation of the various carriers (Mayevsky 1984). Figure 6.1 schematically demonstrates the dependence between blood flow, oxygen supply, pump activity and ionic changes occurring in the brain (Figs. 6.2 and 6.3).

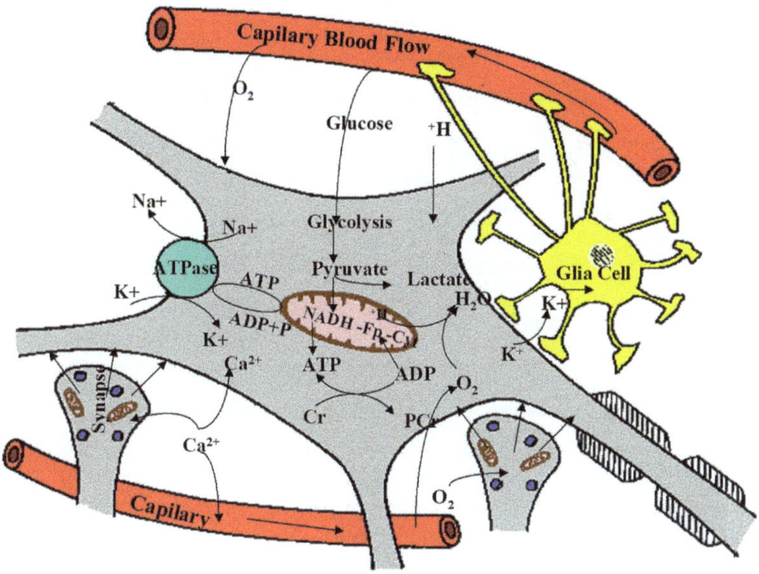

Fig. 6.1 Schematic presentation of the interconnection between energy metabolism and ionic homeostasis in the CNS

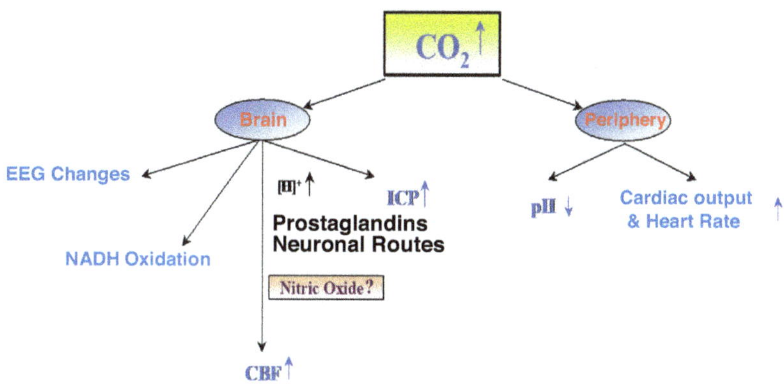

Fig. 6.2 The effect of CO_2 on the various parameters in brain and periphery

6.1.1.1 Autoregulation of Blood Flow in the Brain

One of the mechanisms by which the proper balance between oxygen consumption and supply is maintained is the autoregulatory mechanism of blood flow. When tissue function increases, the demand for oxygen and substrate will increase and blood flow will increase accordingly. Arterial blood flow is regulated by two mechanisms:

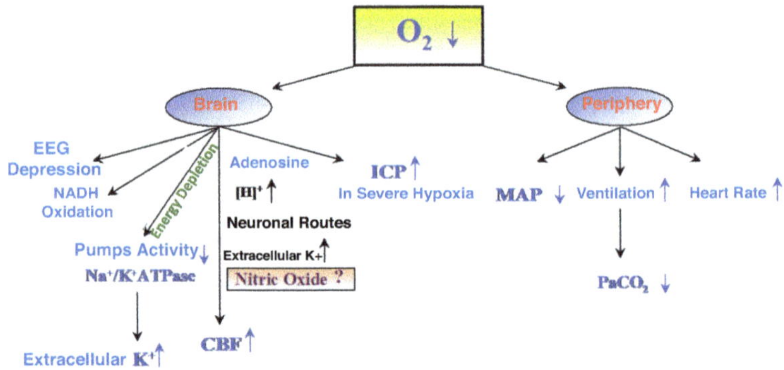

Fig. 6.3 The effect of hypoxia on the various parameters in brain and periphery

1. Local regulation through local changes in pH, pCO_2, pO_2, extracellular ion concentrations, adenosine, and lactate, which vary as a function of the metabolic state of the tissue (Betz 1972).
2. Neural regulation of arteriolar smooth muscle, which affects the diameter of the blood vessels (Harper et al. 1972).

In addition, the term autoregulation includes the fact that blood flow remains constant while perfusion pressure varies, within certain limits. The autoregulation mechanism dilates blood vessels when cerebral perfusion pressure (CPP) decreases, and constricts blood vessels when CPP increases (Johnson 1986). Cerebral blood flow (CBF) in humans is constant within a CPP range of 60–70 mmHg to 150–160 mmHg. Under conditions of hypercapnia, hypoxia, trauma, and local ischemia, autoregulation is reduced or absent (Siesjo 1978b). It is now known that the molecule Nitric Oxide (NO) has a specific role in maintaining normal cerebral blood flow and regulating systemic blood pressure. Inhibition of NOS (the enzyme that produces NO) causes a 20%–50% decrease in cerebral blood flow and an increase in systemic blood pressure (Backman et al. 1991).

6.1.1.2 Nitric Oxide

It is now known that this molecule participates in various processes in the brain, blood vessels, immune system, liver, pancreas, peripheral nervous system and lungs (Culotta and Koshland 1992; Iadecola and Zhang 1994). NO has been found to inhibit platelet aggregation and adhesion to the endothelium (Ignarro 1991; Moncada et al. 1988). In the brain, NO has been found to take an active part in the transmission of information between neurons, between neurons and glial cells, and is responsible for motor learning in the cerebellum, memory, behavior, and maintaining normal blood flow by regulating vascular tone, among other things (Palmer et al. 1988b; Stamler et al. 1992; Szabo 1996; Culotta and Koshland 1992). In the clinical field, it has been found that breathing newborns with pulmonary hypertension with NO gas reduces pulmonary vascular resistance and increases systemic oxygenation (Ehrenkranz and Neonatal Inhaled Nitric Oxide Study 1997; Roberts et al. 1997). The NO molecule has three forms: cation (NO^+), anion (NO^-), and radical (NO*). Each of the forms reacts differently with oxygen, metals, and other NO molecules (Stamler et al. 1992). The molecule of biological interest is the radical form NO*. NO synthesis from L-Arginine in mammals is carried out by the enzyme NO Synthase (NOS), with NADPH and oxygen serving as cofactors in the process. The byproduct of this reaction is the amino acid L-Citrulline (Palmer et al. 1988a; Marletta 1993).

There are three isoenzymes of NOS:

1. (eNOS) ENDOTHELIAL CONSTITUTIVE NOS—found in vascular endothelial cells and stimulated by acetylcholine.
2. (nNOS) NEURONAL CONSTITUTIVE NOS—synthesizes NO in neurons.
3. (iNOS) INDUCIBLE NOS—found in activated macrophages (Knowles and Moncada 1994).

eNOS is activated by Ca^{+2} and calmodulin, while iNOS is induced by endotoxins and cytokines, in a process that requires DNA transcription and protein synthesis. NO produced by iNOS is considered a vasodilator observed in the inflammatory process, and it also has antimicrobial activity and is cytotoxic to a large number of intracellular pathogens. The main cytotoxic effect is due to the final product of NO and superoxide, the radical peroxynitrite (ONOO–) (Young 1997; Beckman et al. 1990). NO is a molecule that diffuses easily through cell membranes and has a short half-life, so it is difficult to measure in vivo. Therefore, to study the role of NO, substances that inhibit the activity of NOS are used. When different inhibitors have been used, the degree of inhibition seems to differ from enzyme to enzyme and depends on:

1. The type of inhibitor.
2. The dose and route of administration of the inhibitor.
3. The time between administration of the inhibitor and the test of the response (Iadecola et al. 1994).

It was found that (NIO) *N*-iminoethyl-L-ornithine is the most effective inhibitor of endothelial NOS, while *N*ω-Arginine-Methyl-Ester (L-NAME) has been found to be the most effective for brain NOS (although it also has a strong inhibitory effect on the endothelial enzyme) (Rees et al. 1990). Other substances that are used as effective inhibitors are: Nitro-L-Arginine (L-NA); *N*-monomethyl L-Arginine (L-NMMA). They prevent the conversion of L-Arginine to L-Citrulline, because they are analogs of L-Arginine and the NOS enzyme binds to them instead of L-Arginine (Iadecola et al. 1994). Recently, another inhibitor, 7-nitroindazole (7-NI), has been found, which is specific for the nNOS enzyme (Wang et al. 1995).

6.1.1.3 Role of NO in the Brain

NO actively participates in the process NO is involved in the transmission of information between brain cells and has been shown to play an important role in the vasodilatory responses of cerebral blood vessels to a wide range of stimuli, including muscarinic antagonists, neuronal activation, hypoxia, and hypercapnia (Iadecola et al. 1994). NO as a neurotransmitter transmits information from nonadrenergic noncholinergic (NANC) neurons to smooth muscle in cerebral blood vessels (Toda and Okamura 1992). It is not stored and readily diffuses across membranes into cells (Culotta and Koshland 1992). NO has a specific role in maintaining normal cerebral blood flow. NOS inhibitors have been shown to reduce resting cerebral blood flow but do not affect glucose utilization or oxygen consumption in the brain, so their effect on resting blood flow cannot contribute to the suppression of cerebral energy metabolism (Iadecola et al. 1994). In another work, the hypothesis was raised that the NO molecule is responsible for the regulation of mitochondrial respiration by inhibiting cytochrome oxidase. According to this hypothesis, inhibition of NO production can cause an increase in oxygen consumption. Indeed, Brown presents works that have shown an increase in oxygen consumption after inhibition of NO production (Brown 1995). Regulation of brain cell activity and communication between them by NO is carried out by the release of neurotransmitters such as glutamate, acetylcholine and dopamine (Guevara-Guzman et al. 1994; Zhu and Luo 1992). The action of endothelial NOS is associated with acetylcholine, and therefore inhibition of the enzyme activity caused an inhibition of the relaxation induced by Ach (Moncada et al. 1991). Other substances that act by inducing NO release from the vascular endothelium in the brain include: ATP, ADP, bradykinin, thrombin, serotonin, histamine, substance p and norepinephrine (Szabo 1996). In addition to maintaining normal cerebral blood flow, NO has been found to play a role in regulating systemic blood pressure. Following administration of an inhibitor of the NOS enzyme, blood pressure and vascular resistance increased in rats, rabbits, dogs and monkeys. When L-Arginine was injected, blood pressure returned to normal levels (Rees et al. 1989; Toda and Okamura 1992). In a study in halothane-anesthetized rats, administration of L-NAME (40 mg/kg) reduced resting CBF by 42% and blood pressure increased. Blood pressure reached 70% of maximum within 10 min (28% increase) and rose further after 60 min (40% increase). CBF decreased to 70% of maximum within 10 min (37% decrease), and continued to decrease after 45 min (53% decrease) (Iadecola and Zhang 1994). In another study in awake rats, L-NMMA administered IV at doses ranging from 10 to 100 mg/kg produced dose-dependent increases in blood pressure and vascular resistance and decreases in CBF. Administration of L-NAME (30 mg/kg) produced similar effects (Sokoloff et al. 1992). In awake sheep, administration of L-NA (100 μg/kg) increased mean arterial pressure from 82 mmHg to 88 mmHg and CPP from 72 mmHg to 79 mmHg, while CBF was unchanged and cerebral lactate was increased (Iwamoto et al. 1992). In a study in isofluranesed monkeys, administration of L-NAME (60 mg/kg) reduced CBF and increased cerebral vascular resistance (McPherson et al. 1995). Administration of L-NAME (30 mg/kg, IV) to anesthetized cats caused significant vasoconstriction in arteries larger than 100 μm in diameter, but not in smaller arteries, while PaO_2 was not significantly changed (Ishimura et al. 1996). In addition, pulsatile blood flow has been shown to stimulate endothelial NO production (Szabo 1996). In a study in rats ventilated and anesthetized with sodium pentobarbital, 30 min after IV infusion of L-NAME (0 mg/kg^2) an increase in MAP (Mean Arterial Pressure) from 105 mmHg to 132 mmHg occurred while blood flow in the parietal cortex decreased and oscillations appeared. Administration of L-Arg stopped the oscillations at a constant MAP. Therefore, it is assumed that NOS activity inhibits the oscillations of blood flow in the cortex (Hudetz et al. 1995).

In addition, NO has been implicated in mediating tissue injury during cerebral ischemia. Release of glutamate from the presynaptic terminal causes a postsynaptic influx of calcium ions into the cell via the NMDA (*N*-Methyl-D-Aspartate) receptor, which triggers the initiation of neurochemical responses involving NOS activation (Garthwaite 1991). On the other hand, it has been reported that NO can reduce tissue damage during ischemia, especially in the early stages, by increasing cerebral blood flow to the ischemic area and inhibiting platelet aggregation (Zhang et al. 1994). The effect of NO on blood vessels is probably mediated by cGMP. NO activates guanylate cyclase in vascular smooth muscle, and increased activity of this enzyme leads to an increase in cGMP. cGMP acts as a messenger and causes various reactions, including the inhibition of smooth muscle contractions by lowering intracellular calcium concentrations (Ignarro 1991; Brown 1995). Recent studies indicate several mechanisms responsible for the NO and cGMP system: (A) The brain NOS enzyme is controlled by various substances, such as NADPH and O_2, which control the rate of its activity, and thus the rate of NO production. (B) NO production is regulated by cGMP, which depends on intracellular Ca^{2+} and calmodulin levels. (C) The activity of cGMP as a mediator depends on the expression and activity of guanylate cyclase, while the activity of this enzyme is controlled by intracellular calcium (Palmer et al. 1988b; Mayer et al. 1992).

Nitric oxide (NO) plays essential roles in the modulation of vascular tone, neurotransmission and the immune system (Rifkind et al. 2007). It is synthesized in all brain cells (Ignarro et al. 1999) from L-arginine by the enzyme nitric oxide synthase (NOS) (Nagababu et al. 2003). There are at least three sources of NO in the CNS: Endothelial isoform (eNOS), neuronal (nNOS), and inducible isoform (iNOS), which is generated by glial cells (Ignarro et al. 1999). Endothelial NO is synthesized in peripheral blood vessels and controls the extent of blood flow through these blood vessels by its relaxant effect on vascular smooth muscles (Cosby et al. 2003).

Nitric Oxide can also affect mitochondrial function mainly by inhibition of cytochrome C oxidase activity as was presented in the heart, brain, synaptosomes and intact astrocytes (Nohl et al. 2000; Millar et al. 1998; Li et al. 2001). NO binds in a competitive manner with O_2, and reversibly inhibits mitochondrial O_2, consumption. Under low O_2, concentrations (below 30 μM), cytochrome C oxidase is more sensitive to NO, suggesting that neuronal energy metabolism could actually be compromised under conditions associated with reduced blood supply to the brain (Ignarro et al. 1999).

Under normal physiological circumstances, there is a microvascular adaptation to local requirements (Boughton-Smith et al. 1993). However, excessive production of NO leads to a loss of this gentle control, resulting in generalized vasodilated state, accompanied by hypotension that is often refractory to fluid and vasopressors (Boughton-Smith et al. 1993). These circulatory abnormalities are accompanied by blood flow redistribution and are considered to be the putative cause of altered oxygen metabolism during sepsis (Crouser et al. 2002; Gutierrez 1993; Maginniss et al. 1994; Nelson et al. 1987). The decreased blood flow to the intestine, for example

(Spain et al. 1994; Weiss and Cohen 1978), has been consistently documented in experimental septic shock, whereas the skeletal muscle showed normal or even increased flows (Hershey and Bond 1993; Lubbe et al. 1992). In addition to the inter-organ redistribution of the blood flow, intra-organ redistribution also occurs, as demonstrated in the intestine (Revelly et al. 1996; Vallet et al. 1994). These alterations may favor the development of localized zones of ischemia, resulting in an imbalance between regional O_2 demand and delivery. The purpose of the present study (Tolmasov et al. 2009), was to evaluate the involvement of NO in the brain—a vital organ versus the small intestine—a less vital organ using L-NAME and nitrite.

Nitric Oxide has a short half-life and needs to be produced/released in close proximity to the site where it reacts. Recent data reports that the reduction of nitrite to NO may play an important role in supplying NO for the regulation of blood flow (Kader et al. 1993; Macrae et al. 1993).

Although mammals do not possess enzymes specifically designed for nitrite reduction, it has been shown that nitrite is reduced to NO in mammals. Nitrite can be reduced by mitochondria (Brown 2003) as well as by xanthine oxidase (Bolanos and Almeida 2007; Buerk 2007). However, most of the recent interest in nitrite reduction has emphasized the reduction by deoxygenated hemoglobin (Kader et al. 1993; Macrae et al. 1993; Gursoy-Ozdemir et al. 2000; Rao et al. 1998; Bolanos and Almeida 1999; Faraci and Heistad 1998; Zhang et al. 1995; Danielisova et al. 2004) that is present in red blood cells (RBCs) as oxygen is being delivered to the tissues. A physiological role for RBC nitrite reduction in regulating blood flow is supported by the observation that nitrite infusion results in an increase in RBC NO in conjunction with improved forearm blood flow, inhibition of hypoxic pulmonary vasoconstriction and in the prevention of delayed cerebral vasospasm (Macrae et al. 1993; Jung et al. 2006; Dezfulian et al. 2007; Friedli et al. 1982). Nitric Oxide plays an important role in the autoregulation of CBF and mitochondrial function in the partial ischemic brain (Eibeshitz et al. 2008).

Earlier studies on improved blood flow associated with nitrite infusion have been limited to the peripheral circulation. The cerebral blood flow (CBF) is, however, highly regulated under conditions of changing blood pressure, changes in carbon dioxide concentration and varying neuronal activity (Rampil et al. 1992) to maintain an adequate supply of glucose and oxygen for brain function. NO has also been shown to regulate CBF (Tonnesen et al. 2005; Mayevsky et al. 2002). To determine whether nitrite infusion and the resultant production of NO by nitrite reduction can influence tightly regulated CBF, we used a rat model.

Cortical spreading depression (SD), described initially by Leao (1944b), is a multifactorial event affecting the electrical, ionic, metabolic and hemodynamic activities in the brain (Vyskocil et al. 1972; Bures et al. 1974; Mayevsky et al. 1974; Mayevsky and Weiss 1991). Due to disturbances in the ion homeostasis, the Na^+/K^+-ATPase activity and energy metabolism are stimulated in order to restore the normal extracellular ion levels (Mayevsky and Weiss 1991; Mayevsky et al. 1974; Hansen 1985). The hemodynamic response to SD was a challenge to many investigators since the initial observation of dilation of pial vessels (Leão 1944a). He concluded that the vascular responses are secondary to the local changes in the activity of neural elements. The changes in cerebral blood flow (CBF) just before, during and after the depolarization wave of SD were described by various investigators (Hansen et al. 1980; Mies and Paschen 1984; Buresova 1957; Van Harreveld and Ochs 1957; Van Harreveld and Stamm 1952; Lauritzen 1984; Lauritzen et al. 1982), and have been reviewed by Lauritzen (1987a, 1987b). In all studies, a large increase in cerebral blood flow was recorded during the wave (Meilin et al. 1995). Lauritzen and collaborators descried a post-spreading depression wave hypoperfusion, while a preceding vasoconstriction (immediately before the wave) was not established or proved. The mechanism behind the changes in CBF due to the SD wave is not clear although recently nitric oxide NO was proposed to be involved (Duckrow 1993; Goadsby et al. 1992). NO was suggested as an important factor in CBF regulation (Iadecola et al. 1994; Irikura et al. 1994) as well as having direct effects on neuronal (Culotta and Koshland 1992). In order to clarify the role of NO in this multifactorial event (SD) we adopted the multiparametric monitoring system (MPA) by which the hemodynamic, metabolic, ionic and electrical activities were measured simultaneously from the surface of the brain (Mayevsky et al. 1992). Energy state and metabolism were evaluated by monitoring the cerebral blood flow and volume (Laser Doppler flowmetry) as well as intramitochondrial redox state (surface NADH fluorometry reflectometry). Ion homeostasis was evaluated by the extracellular K^+, Ca^{2+} and H^+ levels (surface mini-electrodes). Electrical activities were assessed by the spontaneous electrocortical activity—ECoG as well as by DC steady potential.

In recent years, it was indicated that changes in CBF play a major role in the pathogenesis of epilepsy. Since NO is one of the major regulators of CBF under various situations, its role in the case of epilepsy was vastly investigated. Except for its smooth muscle relaxant effect, NO has been implicated in the neuronal hyperexcitability which characterizes the pathophysiology of epilepsy (Mayevsky et al. 1992; De Sarro et al. 1991; Fedele and Raiteri 1999; Snyder 1992). Several studies have suggested that NO is an anticonvulsant agent, due to its antagonist NMDA receptor effect (Hosseinzadeh et al. 2005; Lancelot et al. 1998; Lei et al. 1992; Manzoni et al. 1992) making its role in epileptogenesis as either an anticonvulsive or a proconvulsive agent unclear (Rifkind et al. 2006). In one of the presented studies, we monitored the hemodynamic, metabolic and electrical changes in the cortex of Metrazol induced Status Epilepticus—SE rats, with and without the inhibition of NO synthase (Barbiro-Michaely et al. 2011).

6.1.2 *Effects of Hypercapnia*

CO_2 is considered one of the main factors that induce vasodilation of blood vessels in the brain, therefore cerebral blood flow is very sensitive to changes in the partial pressure of CO_2 in the arteries. This sensitivity is an expression of the autoregulation mechanism. Since blood flow should match the metabolic demands of the tissue, it is natural that CO_2, which is the main product of oxidative metabolism, will serve as a vasodilator (Siesjo 1978a). Hypercapnia occurs when the PCO_2 in arterial blood is above normal concentration. An increase in PCO_2 causes a decrease in pH and a rightward shift in the oxygen dissociation curve (Feihl and Perret 1994). Hypercapnia can be produced by hypoventilation or by breathing a gas mixture containing high concentrations of CO_2. In humans, ventilation with 5% CO_2 increased CBF by about 50% and ventilation with 27% CO_2 increased CBF by 100% (Kety and Schmidt 1948). When $PaCO_2$ is changed from −4 to over 400 mmHg, a maximal increase in CBF of about 150 mmHg occurs (Reivich 1964). These changes in CBF during hypercapnia are reversible. Anesthetized animals exposed to various concentrations of CO_2 were affected to a lesser extent than awake animals, probably due to the inhibitory effect of anesthesia on cerebral metabolism and CO_2 production (Traystman 1997). Hypercapnia causes the following changes: decrease in cellular and intercellular pH, increase in blood flow and blood volume in the brain, increase in ICP, change in EEG activity, changes in neuronal excitability, oxygen consumption was usually unchanged in most studies, while CO_2 at high concentrations can cause anesthesia (Siesjo 1978a; Kogure et al. 1975; Duelli and Kuschinsky 1993). The hemodynamic responses to exposure to hypercapnia (by CO_2 ventilation or hypoventilation) are characterized by an increase in heart rate, an increase in cardiac output, an increase in stroke volume and a decrease in vascular resistance (Feihl and Perret 1994). In an article summarizing the measurement of NADH by the fluorometric method in the brain, it was found that an increase in PCO_2 in arterial blood in hyperbaria caused NADH oxidation and a decrease in reflected light (Mayevsky 1984). These facts suggest an increase in cerebral blood flow. Further work in chloralose-anesthetized cats showed that hypercapnia caused NADH oxidation during 10%, 15%, and 30% CO_2 ventilation. The decrease in corrected fluorescence (CF) induced by 30% CO_2 ventilation was greater than the decrease induced by 10% and 15% CO_2 ventilation. The electrical activity in the cortex was reversibly decreased during 30% CO_2 ventilation. Local blood volume was increased during ventilation with the different mixtures and the reflectance was decreased (Gyulai et al. 1982). In work in halothane-anesthetized dogs, increasing $PaCO_2$ did not result in significant changes in PaO_2, brain temperature, or MAP while EEG amplitude and frequency were decreased (Artru and Michenfelder 1980). In other studies, hypoventilation caused an increase in capillary diameter and hyperventilation caused a decrease in capillary diameter in the brain (Duelli and Kuschinsky 1993, Muizelaar et al. 1988). In the study of Sato et al. (1992) ventilation with 5% CO_2 and 0% CO_2 caused an increase in CBF and systemic blood pressure according to the level of hypercapnia, in rats anesthetized with Urethane. In another study in rats anesthetized with Chloralose, ventilation with 6% CO_2 for 3 min caused an increase in cerebral blood flow of about 8.7%. After about 20 min, CBF and $PaCO_2$ returned to their baseline levels (Schmitz et al. 1996). Feihl and Perret (1994) concluded in their article that hypercapnia at a level of $PaCO_2 \leq 80$ mmHg when pH Arterial pH <7.15 is reversible and does not pose a risk. Under these conditions, oxygenation is maintained and increased susceptibility to respiratory acidosis is prevented. Several mechanisms have been proposed for the effects of CO_2 on cerebral vasodilation, including excess [H^+] ions in the extracellular fluid of the brain, prostaglandins, neuronal pathways, and Nitric Oxide:

(1) Excess [H^+] ions in the extracellular fluid of the brain—Most studies show an increase in [H^+] ions in the cerebrospinal fluid during hypercapnia. It is hypothesized that cerebral vasodilation is mediated primarily by changes in acidity around and/or within the smooth muscle cells of the cerebral vessels during hypercapnia (Wang et al. 1992). Large changes in $PaCO_2$ and bicarbonate ion

concentrations in the cerebrospinal fluid did not affect the diameter of arterioles in the pia unless a change in pH occurred. Therefore, it is concluded that the CO_2 molecule and bicarbonate ion are not vasoactive (Traystman 1997).

(2) Prostaglandins—Prostaglandins can mediate the response of CBF to CO_2. They have been shown to be important in vasodilation during hypercapnia in some species (gerbils, rats, and monkeys), but not in others (rabbits and cats). There may be an interaction between the prostaglandin system and nitric oxide production, such that prostaglandins accelerate and facilitate the release of nitric oxide (Traystman 1997).

(3) Neuronal pathways—years ago it was suggested that hypercapnia causes the excitation of cholinergic reflex pathways that are involved in vasodilation (James et al. 1969). The above mechanism is still controversial and there is no real evidence for neuronal involvement in vasodilation during hypercapnia.

(4) Involvement of nitric oxide—the importance of NO in the mechanism of cerebral vasodilation during hypercapnia is still unclear. Many studies have found that NOS inhibitors blocked the increase in CBF during hypercapnia, on the other hand other studies have not shown changes in cerebral vasodilation after administration of a NOS inhibitor in hypercapnia (see Chap. 6.1.3.). In conclusion, it is possible that an increase in hydrogen ions in the fluid, the extracellular fluid in the brain, increases NOS activity or prostaglandin production and release. There are probably many mechanisms involved in the effect of CO_2 on *vasodilation*.

6.1.3 Role of NO in Hypercapnia

In order to understand the role of NO in the vasodilation response of blood vessels in the brain to hypercapnia, few studies used NOS inhibitors. In most studies, it was found that NOS blockage decreased the elevation of CBF that was found under hypercapnia. In the control groups of those studies, using rats, cats, dogs and rabbits. This effect was reversible by injection of L-Arginine. The decreased in blood flow was in the range of 36%–94% depending on the dosage and site of injection (Iadecola et al. 1994; Wang et al. 1992).

Study performed in anesthetized rats, local Injection of NOS inhibitors L-NA or L-NMA (N^w-Methyl-L-arginine) decreased the response of CBF under hypercapnia (Iadecola 1992). Injection of NOLAG (N^g-Nitro-L-arginine) into the carotid artery led to the same results (Wang et al. 1992). Treatment with another inhibitor-7-NI which is specific to nNOS, decreased significantly (60%) the CBF under hypercapnia (70–80 mmHg).

The inhibition of the response to hypercapnia is similar to the results obtained in other studies. Therefore, it was assumed that most of the NO involved in the response to hypercapnia originates from nNOS and not from eNOS (Wang et al. 1995). Injecting L-NAME (IV-20 mg/kg) to anesthetized rats decreased the response of the CBF by 47% when the animal breath 7.5% CO_2. The effect was reversible in part after injecting L-arginine (Estevez and Phillis 1997). Providing L-NAME (30 mg/kg IV) to anesthetized rats blocked completely the increase in CBF under moderate hypercapnia ($PaCO_2$~60 mmHg)

(Horvath et al. 1994). In another study in anesthetized rats, injecting of similar dose of L-NAME decreased significantly (by 42%) the increase in CBF when breathing the rat by 10% CO_2 and also decreased the time for recovery by 89% (Buchanan and Phillis 1993). Injecting L-NAME directly to the cortex of anesthetized rats stopped the vasodilation of brain vasculature by 60%–70% during hypercapnia ($PaCO_2 = 65$ mmHg). Providing L-arginine IV blocked partially those effects (Reid et al. 1995).

In another study, in anesthetized rats, L-NNA (N^G-nitro L-arginine) was given locally as well as IV. Under those two situations, the increased in CBF was inhibited under breathing 5% CO_2, but given locally L_NNA was less effective in the inhibition (Fabricius and Lauritzen 1994). A study performed in anesthetized cats giving L-NAME (30 mg/kg IV) led to a significant decrease in CBF under breathing 5% CO_2, while a lower dosage of L-NAME (3 mg/kg IV) led to smaller changes (Sandor et al. 1994). On the other hand, there are studies that found NOS inhibitors did not decrease the vasodilation of brain blood vessels during hypercapnia. Adachi et al. (1992) found in anesthetized rats that giving L-NOArg (L-N^G-nitroarginine) did not lower the increase in CBF, under respiration with 10% CO_2. In another study performed in awake rats, giving L-NMMA (10–100 mg/kg IV) did not change the increase in CBF obtained under breathing 10% CO_2. In another study performed in awake rats, giving L-NMMA 10–100 mg/kg IV did not alter the increase in CBF under breathing 5% CO_2. Giving L-NAME (30 mg/kg IV) decreased the elevation in CBF under hypercapnia but the effect was not statistically significant (Sokoloff et al. 1992). Study performed in anesthetized Monkeys given L-NAME (60 mg/kg IV) did not change CBF in various areas in the brain (besides the cortex) under moderate hypercapnia (PCO_2~50 mmHg) or severe (PCO_2~70 mmHg) (McPherson et al. 1994). Giving L-NMMA to newborn pigs did not change the increase of arteriolar diameter in the eye retina during hypercapnia (Gidday and Zhu 1995).

The reason for the variability between the results obtained in various studies may originate from the use of different animals, different protocols, the technology used, the type of inhibitor used or drug dosage. The decrease in the vasodilation obtained by NOS inhibitor during hypercapnia, appeared in the CO_2 range of 40–80 mmHg. The maximal level occurs under exposure to $PaCO_2$ of 54 mmHg and does not occur under $PaCO_2$ of 100 mmHg.

In summary, there is evidence that NOS inhibitors lower the elevation in brain CBF under hypercapnia. Even though NO is not the sole regulator in the vasodilation process.

6.1.4 Effects of Hypoxia

In a state of hypoxia, there is a decrease in the supply of oxygen to the brain, which can result in a disruption of the balance between energy supply and demand, and lead to the development of a pathological condition (Mayevsky 1984). There is also an increase in the level of NADH in the mitochondria, a fact that indicates a decrease in the level of oxygen in the brain (Mayevsky et al. 1986). Recovery from this condition depends on the degree and duration of hypoxia (Hornbein 1991). Tissue hypoxia can result from a decrease in arterial oxygen content, by lowering PO_2 (Hypoxic Hypoxia). Therefore, breathing with a mixture of nitrogen and oxygen, when the percentage of oxygen is lower than the normoxic range (21%), is a way to create experimental hypoxia. If the level of hypoxia is not extreme, oxygen consumption ($CMRO_2$) does not change. A drastic decrease in PO_2 level induces an energy failure accompanied by the loss of high-energy phosphate compounds in brain tissue or peripheral neurons (Siesjo 1978c). The phenomenon of hypoxia in clinical medicine is not so common but appears in industrial accidents, anesthesia and lung diseases. In most of these cases there is a drop in CO_2 pressure, secondary to the decrease in pulmonary ventilation during hypoxia. A more common phenomenon is a combination of low PO_2 and an increase in CO_2 pressure (Asphyxia), when ventilation Pulmonary arrest is prevented by airway injury, respiratory center failure, etc. In humans, general symptoms of hypoxia include mild headaches, restlessness, memory problems, and confusion. Below a PaO_2 of 65 mmHg, ventilation increases, which is reflected in a decrease in $PaCO_2$. Breathing humans at 15% O_2 has been shown to impair judgment and learning ability. Lower concentrations (about 30 mmHg) have been shown to cause loss of consciousness (Siesjo 1978c). Studies have shown that cerebral blood flow increases when PaO_2 falls below 50 mmHg. In addition to the increase in CBF during hypoxia, there is a decrease in extracellular pH in the brain. It is assumed that the decrease in extracellular pH is a stimulus for the increase in cerebral blood flow during hypoxia (Nilsson et al. 1975). Transition from hyperoxia Mild hypoxia (10%–60% O_2 or 40–300 mmHg) in rats caused an increase in blood flow. It is hypothesized that the dilation of blood vessels from hyperoxia to hypoxia is due to tissue acidification (Shinozuka et al. 1989). The initial response to hypoxia is vasodilation, caused by autoregulatory mechanisms, which results in an increase in blood flow. Under hypoxic conditions, mitochondrial NADH increases and there is a decrease in reflectance, suggesting an increase in blood volume. The activity of Na^+/K^+-ATPase pumps is impaired due to the lack of energy, and there is an immediate increase in extracellular potassium. Under severe hypoxia, a massive depolarization develops—Hypoxic depolarization. Another physiological event that occurs during hypoxia is the suppression of electrical activity of the cortex (ECoG) (Mayevsky 1984). During hypoxia, the low frequencies of the EEG become dominant, while the fast frequencies disappear. Finally, the EEG is suppressed in parallel with a severe reduction in cerebral metabolic activity that depends on the degree of hypoxia that occurs (Black and Cucchiara 1990). Changes in PaO_2 from 60 to over 300 mmHg have little effect on CBF. Below 60 mmHg, CBF increases rapidly (Drummond and Shapiro 1990). At a PaO_2 of 30 mmHg, CBF increases almost twofold the normoxic value. In hypoxia, there is an increase in the concentrations of the excitatory amino acids glutamate and aspartate, which act through the NMDA receptor and can contribute to neuronal injury. Very severe hypoxia can cause loss of endothelial integrity, cerebral edema, Increased intracranial pressure (ICP) and infarcted areas (Hornbein 1991). In a study in anesthetized cats, hypoxia (PaO_2 >45 mmHg) increased CBF, heart rate, and ICP, while mean arterial blood pressure decreased (Ishimura et al. 1996). In a study in anesthetized pigs, hypoxia induced by ventilation with 10% O_2 reduced PaO_2 to ~30 mmHg. Arterioles in the pia dilated rapidly during the first 2 min, and moderate further dilation was observed between 2 and 5 min. No further dilation was observed between 5 and 10 min. Five minutes of hypoxia caused minimal change in blood pressure, but with longer periods of hypoxia, blood pressure decreased significantly (Leffler et al. 1997). In a study in Wistar rats, ventilation with a gas mixture containing 4%–7% O_2 reduced arterial blood pressure during 15 min of ventilation. Although cortical blood vessels were maximally dilated, CBF gradually decreased and eventually the cortex became partially ischemic as a result of the gradual decrease in CPP. Extracellular potassium concentrations in the cortex did not change during hypoxia induced by ventilation with 12%–15% O_2, but began to increase after 2 and 3 min of hypoxia induced by ventilation with concentrations of 4%–7% O_2 and 8%–11% O_2 (Dora et al. 1979). Among the mediators proposed for cerebral vasodilation during hypoxia are: Adenosine, potassium channels, perivascular neurons, and NO (Pelligrino et al. 1995). In rats, hypoxia induced by breathing 6%–12% O_2 increased parietal cortex blood flow and decreased systemic blood pressure. This response was prevented after cutting the afferent neurons with peripheral chemoreceptors, suggesting that activation of peripheral chemoreceptors contributes to the increase in CBF during hypoxia (Sato et al. 1992). For the past decade, extracellular adenosine has been thought to play an important role in mediating cerebral vasodilation during hypoxia, but several studies have contradicted this assumption. In these studies, administration of theophylline (an adenosine antagonist) did not block vasodilation during hypoxia (Reid et al. 1995; Haller and Kuschinsky 1987; Dora 1986).

6.1.5 *Role of NO in Hypoxia*

Several studies have examined the role of NO in cerebral vasodilation during hypoxia in various species. Most studies have found that NOS inhibitors do not reduce cerebral vascular responsiveness during hypoxia (Iadecola et al. 1994). In a rat study, topical administration of L-NAME did not alter pia arteriolar diameter during mild hypoxia ($PaO_2 = 45$ mmHg), but under severe hypoxia ($PaO_2 = 33$ mmHg), a vasodilation of pia arteriolar diameter was observed (Pelligrino et al. 1993). In another study in anesthetized rats, IV administration of L-NAME did not alter the increase in CBF achieved during 8% O_2 ventilation for 25 s (Buchanan and Phillis 1993). Administration of L-NMMA to piglets did not alter retinal arteriolar diameter during hypoxia (20–30 mmHg = PaO_2) (Gidday and Zhu 1995). In a study in dogs anesthetized with isoflurane, administration of L-NAME (40 mg/kg IV) did not reduce cerebral vasodilation during hypoxia (McPherson et al. 1994). In contrast, administration of IV L-NA to conscious sheep reduced the increase in cerebral blood flow by 63% during hypoxia (37–38 mmHg = PaO_2) (Iwamoto et al. 1992). In a study in anesthetized cats, IV infusion of L-NAME (30 mg/kg) inhibited vasodilation of arterioles smaller than 200 μm during hypoxia (PaO_2 <45 mmHg) (Ishimura et al. 1996). Administration of L-NMMA (100 μM–5 mM into the cortex of anesthetized rats blocked the vasodilation achieved during hypoxia by 60%–70%. The effect was partially reversed by administration of L-Arginine (Reid et al. 1995). In a study in neonatal pigs, local administration of L-NNA blocked the vasodilation of arterioles in the pia mater during hypoxia ($PaO_2 = 36$ mmHg) (Armstead 1995). According to the above findings, it can be indicated that the contribution of NO to the vasodilation of cerebral blood vessels during hypoxia is relatively small, and there are probably other vasodilator factors that influence the response. It is speculated that in mild hypoxia, adenosine is the main mediator of vasodilation, at least in rats (Iadecola et al. 1994).

In conclusion, it can be seen that in situations where the delicate balance of blood gas concentrations is disturbed, NO may be involved in the autoregulation process, which aims to maintain a regular supply of oxygen to brain cells. In light of the fact that most of the works published to date have mainly examined the mechanism of autoregulation and not the physiological significance of changes in brain activity, it is of great importance to simultaneously and continuously measure several parameters in real time in order to examine the involvement of NO in the regulation of blood flow and brain activity. In this work, the following parameters were measured: blood flow and blood volume in the brain, mitochondrial redox state, extracellular potassium concentration and DC potential, brain temperature and spontaneous electrical activity.

6.1.6 Research Objectives

The general objective of the chapter is to discuss and examine the involvement of the NO molecule in processes related to the regulation of blood flow and whole brain activity in rats, using a multiparametric measurement system. The work steps are as follows:

1. Examination of the hemodynamic, metabolic, ionic and electrical activity of the brain in rats exposed to hypoxia and hypercapnia.
2. Examination of the involvement of NO in hypoxia and hypercapnia by administering an inhibitor of its synthesis (L-NAME) and exposing the animals to the above conditions.
3. Examination of the relationship between mitochondrial function, blood flow regulation mechanisms and other brain activities during exposure to NO synthesis inhibitors.

6.2 Methods

6.2.1 Measurement Methods

6.2.1.1 In Vivo Fluorometric Measurement of Mitochondrial NADH

The method is based on the difference between the absorption spectra of NADH and NAD^+. The exposed brain area is irradiated with light at a wavelength of 366 nm using a mercury lamp. The light is transmitted to the tissue using quartz fibers arranged in a Y shape, with one arm transmitting the excitation light at a wavelength of 366 nm towards the brain and the other arm receiving the light emerging from the brain and transmitting it to a detection, amplification and recording system. The reflected light contains two components: 366 nm—Reflectance (Ref) light, and 450 nm—Fluorescence (Flu) light. Mitochondrial NADH is excited and emits fluorescent light at a wavelength of 450 nm. When NADH decreases, Flu will also decrease (Chance et al. 1962). The intensity of Ref is inversely proportional to the absorption of light by the tissue. The absorption of light by the tissue is affected by hemodynamic changes (vasoconstriction or vasodilation), ion movement, and water movement. These changes will affect both Ref and Flu, and therefore, in order to correct for the hemodynamic artifacts that appear in the measurement, the changes in Ref are subtracted from the changes in Flu in a 1:1 ratio, thus obtaining corrected fluorescence (CF), which is a measure of the NAD^+/NADH ratio in mitochondria only (Mayevsky 1984).

6.2.1.2 Continuous Measurement of Local Blood Flow by (LDF) Laser Doppler Flowmeter

The operating principle of the LDF device we used is—irradiating the tissue with a fixed wavelength of 632.8 nm. The light that hits the static tissue returns at the same wavelength while the light that hits the red blood cells moving in the tissue is randomly shifted so that its frequency and wavelength change (Doppler shift). The more red blood cells move in the tissue, the more rays will be shifted, which means that there is a greater blood flow in the tissue. Since the rays are randomly shifted, the measurement is relative and there is no possibility of calibration (Haberl et al. 1989; Dirnagl et al. 1989).

6.2.1.3 Measurement of Potassium Concentration Using a Specific Mini-Electrode

Potassium is measured using a mini-electrode selective for the measured ion (from WPI Inc., USA). Selectivity is achieved by a specific membrane equipped with the potassium ion carrier (Freiser 1978). The mini-electrode contains the antibiotic Valinomycin, which serves as a potassium ion carrier, and is filled with a solution containing 3 mM KCl, 1.2 mM $CaCl_2$, and 149.8 mM NaCl. The electrode's sensitivity to potassium ions was found to be close to the values of the Nernst equation, i.e., a range of 50–60 mV for a tenfold change in potassium concentration. Additional use has been made of electrodes of this type in previous works (Mayevsky 1983; Friedli et al. 1982; Rogatsky et al. 1996; Mayevsky 1990).

6.2.1.4 Measurement of DC Potential

The DC (Direct Current) potential of the brain is measured as the potential between the brain and a reference point outside the brain, on the back of the animal under the skin. The DC potential of the brain was measured around the electrode specific for the K^+ ion. Each electrode was filled with physiological solution and connected to an Ag/AgCl electrode (WPI Inc.).

6.2.1.5 Reference Electrode

This electrode is identical to the DC electrode, and it is not integrated into the measurement systems but is placed on the animal's back between the skin and muscles.

6.2.1.6 Temperature Measurement

Body and brain temperature were measured using calibrated and linear thermistors in the range of 25–42°C. Brain temperature was measured using a thermistor (Fenwal Electronics, model 120-202EAJ-Q01) located within the measurement system placed on the brain. Body temperature was measured in the rectum of the animal (with a TELE-THERMOMETER from YSI, Model 402).

6.2.1.7 Electrocorticography (ECoG) Measurement

The measurement of spontaneous electrical activity of the brain surface is performed using a bipolar method by measuring the voltage between two points on the surface of the cortex using two electrodes.

6.2.1.8 Arterial Blood Pressure Measurement

Arterial blood pressure is measured in the femoral artery using a Statham P23 pressure transducer or a UTR Disposable Transducer pressure transducer. Additional details on all measurement methods have been published in articles from our laboratory (Mayevsky et al. 1990, Mayevsky 1984).

6.2.2 Monitoring System

6.2.2.1 Brain Function Multiprobe (BFM) Multichannel System

The multichannel system allows for the measurement of metabolic, ionic, hemodynamic and electrical activity in one hemisphere or site of the rat brain. The system allows for the measurement of several parameters simultaneously and includes fibers for measuring NADH, fibers for measuring blood flow and blood volume, an electrode for measuring extracellular potassium, a DC electrode, ECoG electrodes located between the optic fibers and a sensor for measuring brain temperature. In addition, there is a sensor for measuring intracranial pressure, but no data was collected from it during the measurements. All components were assembled in a Plexiglas cannula that during the operation was placed on the surface of the brain in one hemisphere (see Fig. 6.4). All information was stored during the experiment on a computer using National Instruments' LabView software, and polygraph charts were obtained at the same time. To help us, software was developed in the laboratory that allows for qualitative and quantitative analysis of the various parameters. The results underwent statistical calculations in Excel and SAS programs and were presented in the form of graphs or tables.

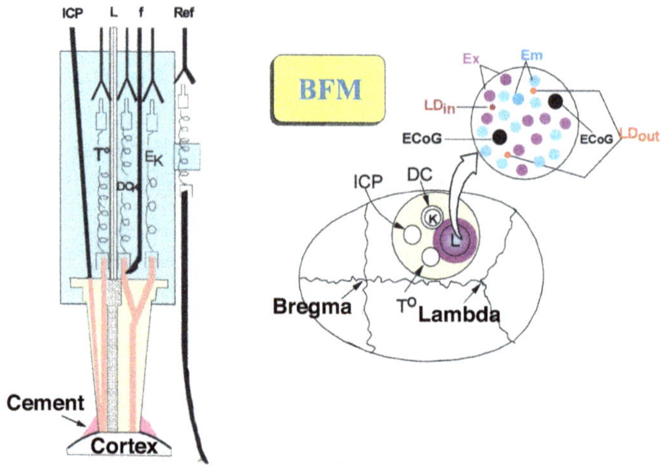

Fig. 6.4 Schematic diagram of the BFM system. On the right is the location of the system on the brain, on the left is a longitudinal section. Right: K potassium-specific electrode, DC surface potential measurement area, Ex and Em excitation and emission fibers for measuring NADH, LDin and LDout optical fibers for measuring blood flow and volume, T° thermistor for measuring local temperature, ECoG electrode for measuring spontaneous electrical activity, ICP sensor for measuring intracranial pressure. Left: ICP intracranial pressure sensor, T° temperature sensor, DCk and Ek potassium and DC specific electrodes, L (light guide) integrated fiber bundle for measuring NADH, blood flow and ECoG, f filler tube for DC electrode, Ref reference electrode, Cement acrylic cement

6.2.3 *Methods for Calculating the Results*

The parameters recorded on the computer and on the polygraph are expressed in mV values. In order to evaluate the ionic changes throughout the experiment, the mV values were converted to mM values, while the reflection, NADH and blood flow values were converted to percentages. The results section presents the means and standard errors in the different experimental groups. A significance level of $p < 0.05$ was used to reject the null hypothesis, in two-tailed calculations. Student *t*-test, paired *t*-test and ANOVA (Analysis of variance) tests were conducted (Figs. 6.5–6.8).

Formulas for calculating the relative change of various parameters throughout the experiment:

6.2.3.1 Calculation of Reflection in Percentages

$Rx(\%) = Rx(mV) - Rbl(mV) \times 100/\text{divided by } 500$
Rx—value measured at a desired point.
R(bl)—Reflection value at the time of control.
500—Measurement range in mV set for all experiments.

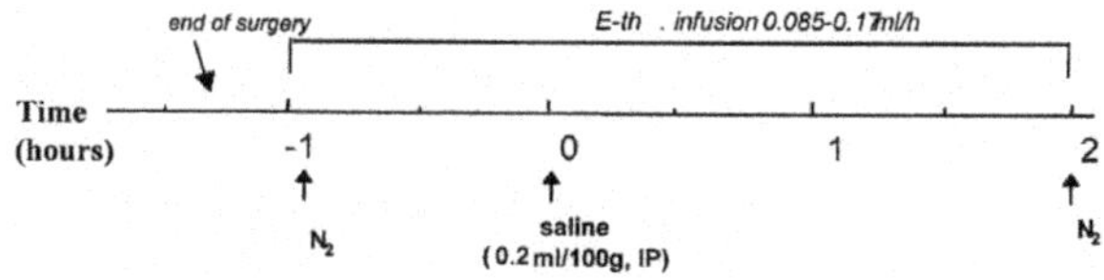

Fig. 6.5 Schematic description of the protocol for the control experiments in which saline was injected into the abdominal cavity

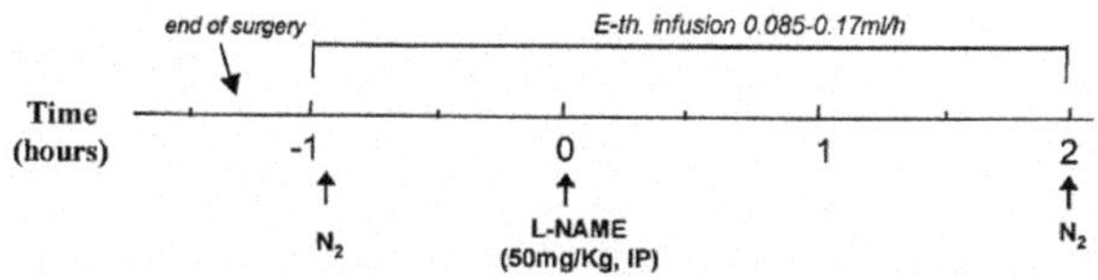

Fig. 6.6 Schematic description of the experimental protocol describing experiments in which L-NAME was injected into the abdominal cavity

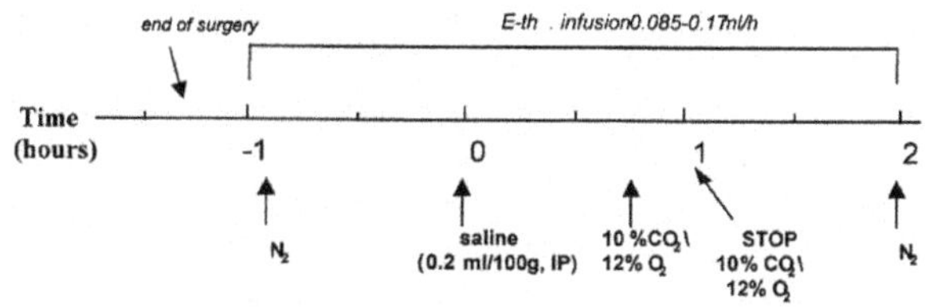

Fig. 6.7 Schematic description of the experimental protocol describing experiments in which hypercapnia or hypoxia were induced 45 min after saline injection into the abdominal cavity

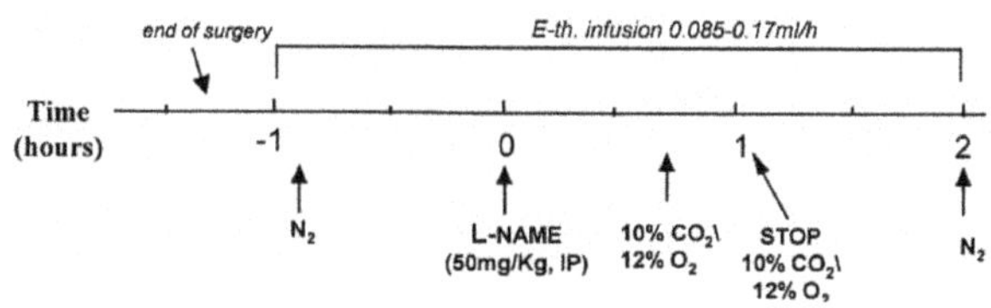

Fig. 6.8 Schematic description of the experimental protocol describing experiments in which hypercapnia or hypoxia were induced 45 min after L-NAME injection into the abdominal cavity

6.2.3.2 Calculation of NADH in Percentages

The change in NADH is measured according to the CF values and the formula is the same as the reflection formula.

6.2.3.3 Calculation of Blood Flow in Percentages

CBFx(%) = CBFx(mV) − CBF death(mV) × 100 divided by CBFbl(mV) − CBF death(mV)

CBFx—Value measured at a desired point relative to the control.
CBF(bl)—Value measured at the control point.
CBF death—Value measured at the death point. This value is considered the physiological zero point.

6.2.3.4 Calculation of Extracellular Potassium Concentration in mM

From the Nernst equation, it can be seen that the specific membrane potential for a particular ion depends on the concentration of the ion on both sides of the membrane. For example, the membrane potential of a potassium electrode, which behaves similarly to a membrane inside a cell, depends on the concentration of potassium inside the electrode and its concentration outside the electrode, on the brain surface, as follows:

$$Em(mV) = k(-\log[K^+]_{in}/[K^+]_{out}$$

k = Nernst equation constant
$[K^+]_{in}$ = potassium concentration inside the electrode.
$[K^+]_{out}$ = potassium concentration outside the electrode.

The sensitivity of the electrodes was determined by two calibrated solutions. The first solution is the solution in which the electrode was filled (3 mM K^+, 1.2 mM Ca^{2+}, 149 mM Na^+) and was therefore set as 0 voltage. In the second solution, the concentration of potassium ions was ten times greater 30 mM K^+).

The formula for converting each point in mV to mM is:

$$K^+\left(mM\right) = 3 \times 10^a \qquad a = \frac{(K(mV) - K(0))}{Range}$$

$K(mV)$ = the value at the measurement point in mV.
$K(0)$ = the value in mV at the 0 point.
$K^+(mM)$ = the ion value after the calculation in mM.
Range = the sensitivity ranges of the electrode in mV which was obtained by subtracting the values measured by the electrode in the transition from one solution (3 mM K^+) to a second solution (30 mM K^+).

6.2.3.5 Calculation of the Changes in DC Potential

DC potential values were sampled in mV. To measure the changes, the control value at time zero was subtracted from the values of each sampled point. In this way, the changes at each sampled point could be compared.

6.2.3.6 Calculation of Mean Arterial Blood Pressure (MAP)

Before each experiment, the pressure transducer and polygraph were calibrated to a measurement range of 200 mmHg. The mean arterial blood pressure was calculated according to the following formula:

$$MAP = (X_{max} - X_{min})/3 + X_{min}$$

X_{max} = Systolic blood pressure at the sampled point
X_{min} = Diastolic blood pressure at the sampled point

6.2.4 *Work Steps*

6.2.4.1 Animal Preparation for the Experiment

The experiments were performed on male rats weighing 180–300 g. The animals were anesthetized by injection of Equithesin (Eth) prepared in the laboratory (each ml contains: Chloral hydrate 42.51 mg; Magnesium sulfate 21.25 mg; Alcohol 11.5%; Propylene glycol 44.34%; Pentobarbital 9.72 mg and supplemented with water) administered IP (Intraperitoneal) at a dose of 0.3 ml/100 g. The femoral artery was cannulated to measure blood pressure during the experiment. The skull was exposed by cutting the skin of the skull at the line connecting the nostrils and the nape. The animal was placed on the operating table on its stomach, and the head was fixed using a head holder. The skull was exposed and a 6 mm diameter hole was drilled in the parietal bone. The dura was gently removed from the exposed area and the BFM system was placed on the brain using a micromanipulator (to prevent pressure on the brain). In addition, four holes for reinforcement screws were drilled in the skull bone: two on the side contralateral to the measurement system and two in the frontal bone (one hole on each side). Upon completion of the surgical phase and placement of the system on the brain, acrylic cement was used to reinforce and fix the system to the skull. During the experiment, the animals remained anesthetized by an IP infusion of Eth at a rate of 0.085–0.17 ml/h.

In the experiments in which the animals were treated with L-NAME at a dose of 50 mg/kg, 50 mg of L-NAME from Sigma was dissolved in 2 ml saline.

6.2.4.2 Experimental Protocols

The final protocols of the experiments were preceded by many preliminary experiments (40 rats) in which an attempt was made to reach the best experimental conditions. First, various treatments were performed on the same rat, which included breathing in different gas mixtures to create hypercapnia, hypoxia, and hyperoxia. The rats were artificially ventilated through a tube connected to the trachea. In addition, the femoral artery was cannulated to measure blood pressure and blood samples were taken to measure the composition of arterial blood gases simultaneously with measurements in the brain. The blood volumes removed were completed by saline injection. In order to eliminate the effects of previous treatments in the same rat, it was decided to separate the different treatments and shorten the protocol times. In the next step, L-NAME was injected into the abdominal cavity at a dose of 50 mg/kg in combination with hypercapnia or hypoxia. Due to the extensive manipulation of the femoral artery during the experiment (blood pressure measurements, blood sample withdrawal, saline injection, and moving the rat for this purpose) in addition to the various treatments in the same animal (creating hypercapnia or hypoxia and administering L-NAME), many rats died during the experiment. Moreover, proper control of the respiratory rate and volume in the ventilator required extensive training and a lot of time. Therefore, it was decided to simplify the complexity of the protocols by separating experiments that included measurements of the various parameters in the brain along with arterial blood pressure in freely breathing rats and experiments that included blood sample withdrawal to measure the composition of arterial blood gases. Without brain surgery, but with the same experimental procedure. The working steps of the different experimental series are described below:

Protocol Number 1: Control Experiments

1. After releasing the animal from the fixation and starting to record the results, wait 10–15 min.
2. Breathe the animal in N_2 for a short time to check the general condition of the brain.
3. Wait about an hour until the animal recovers from the surgery.
4. Inject saline in a volume of 0.2 ml/100 g into the abdominal cavity (IP) Intraperitoneal.
5. Wait about 2 h for recovery.
6. Breathe the animal in nitrogen and end the measurement.

Protocol Number 2: Induction of the NOS Enzyme by Administering L-NAME

Performing a protocol identical to protocol number 1, except for Step 4, in which L-NAME was injected at a concentration of 50 mg/kg and in the same volume (0.2 ml/100 g).

Protocol Number 3: Exposing the Animal to Respiration in Different Gas Mixtures

Performing a protocol identical to protocol Number 1. In addition, approximately 45 min after Step 4, performing respiration for 15 min in one of the following gas mixtures:

A. $69\%N_2 + 21\%O_2 + 10\%CO_2$ to create hypercapnia.
B. $87\%N_2 + 12\%O_2 + 1\%CO_2$ to create hypoxia.

Waiting about an hour for recovery.

The different gas mixtures were infused from balloons gas (each separately), at a pressure of 2.5 atmospheres and at a weak current that did not interfere with the rat's spontaneous breathing. The gases flowed through a polyethylene tube to which a mask adapted to the size of the rat's head was attached. The mask was placed near the rat's nose but not adjacent to it.

Protocol Number 4: Typical Protocol for Experiments in Which Animals Were Exposed to Breathing in Different Gas Mixtures During NOS

Implementation of a protocol combining Protocols Numbers 2 and 3.

Note: According to experiments conducted in our laboratory, it was found that the administration of L-NAME at a concentration of 50 mg/kg and a waiting time of about 45 min are optimal for obtaining a response.

Measurement of $PaCO_2$, PaO_2, pH, and Percentage of Oxygen Saturation in Arterial Blood

Experiments were conducted with protocols identical to those described to take arterial blood samples to measure $PaCO_2$, PaO_2, pH and percentage of oxygen saturation in arterial blood (% Sat). The rats were anesthetized with Eth IP at a dose of 0.3 ml/100 g. The femoral artery was cannulated and a cannula connected to a three-way stopcock was inserted to allow blood samples to be withdrawn from the artery. Saline was injected into the artery to complete the blood volumes sampled. No brain surgery was performed in these experiments. Four blood samples were taken during the experiment:

1. Control sample, before injection of L-NAME or saline.
2. Blood sample at 45 min from the start of the experiment (Constitutes a control for the different respirations in experiments with protocols 1 and 2).
3. Blood sampling at 55 min from the start of the experiment (at 10 min of respiration corresponding to the recording of the steady state in experiments with protocols 1 and 2).
4. Blood sampling at 90 min from the start of the experiment. The blood samples were inserted into an Acid Base Analyzer Radiometer (Radiometer model ABL30) which measured the $PaCO_2$, PaO_2 and pH values and calculated the percentage of oxygen saturation in the arterial blood.

6.3 Results

In this work, the effect of Nitric Oxide (NO) on energy metabolism, electrical and ionic activity, and the mechanism that regulates Cerebral Blood Flow (CBF) was examined. The involvement of NO was examined by blocking its production via the administration of L-NAME.

In light of the recent findings in literature indicating that NO is an important regulator involved in regulating cerebral blood flow in various conditions, in the present work its effect was examined in three conditions: normoxia, exposure to hypercapnia, and exposure to hypoxia.

The results section includes the following topics:

6.3.1 Effect of NO Synthesis Inhibition by L-NAME Administration

1. Effect in the periphery: blood pressure and arterial blood gas levels.
2. Effect in the brain: energy metabolism, hemodynamic activity and ionic and electrical activity.
3. The relationship between changes in CBF and brain temperature.

6.3.2 Examination of NO Involvement in the Response to Hypercapnia

6.3.2.1 Effects of Hypercapnia

1. Effect of hypercapnia in the periphery: blood pressure and arterial blood gas levels.
2. Effect of hypercapnia in the brain: energy metabolism, hemodynamic activity and ionic and electrical activity.

6.3.2.2 Effect of Hypercapnia During NO Synthesis Inhibition

1. Effect in the periphery: blood pressure and arterial blood gas levels.
2. Effect in the brain: energy metabolism, hemodynamic activity and ionic and electrical activity.

6.3.3 Examination of NO Involvement in the Response to Hypoxia

6.3.3.1 Effects of Hypoxia

1. Effect of hypoxia in the periphery: blood pressure and arterial blood gas levels.
2. Effect of hypoxia in the brain: energy metabolism, hemodynamic activity and ionic and electrical activity.

6.3.3.2 Effect of Hypoxia During NO Synthesis Inhibition

1. Effect in the periphery: blood pressure and arterial blood gas levels.
2. Effect in the brain: energy metabolism, hemodynamic activity and ionic and electrical activity.

6.3.4 *Effect of L-NAME on NO Synthesis Inhibition*

6.3.4.1 Effect of L-NAME on the Periphery: Blood Pressure and Arterial Blood Gas Levels

Table 6.1 presents data on levels of pH, $PaCO_2$, PaO_2, and arterial blood oxygen saturation levels in the experimental group treated with L-NAME ($n = 8$) and in the control group ($n = 7$) prior to saline/L-NAME injection and 45 min after injection.

In the control group, no significant differences were found in pH, $PaCO_2$, PaO_2, and arterial blood oxygen saturation levels. Under the influence of L-NAME, a significant decrease in pH was observed ($p < 0.01$) without changes in the other parameters. A comparison between the two groups at the same time points did not reveal a significant difference in any of the parameters (results not shown).

Figure 6.9 shows data on changes in arterial blood pressure in mmHg (absolute values are not shown because the mean initial blood pressure was different in each group). The data were collected over 120 min, with each point representing the mean with a standard error of eight experimental animals or six control animals. At zero-time point, L-NAME was injected into the experimental animals and saline solution into control animals. The significance level between the experimental group and the control group was tested (*t*-test), and an ANOVA test was also conducted to examine the changes within each group (results not shown).

A marked increase in blood pressure could already be observed 5 min after L-NAME injection (21.08 mmHg ± 5.2, $p < 0.05$), while the ANOVA test indicated that the experimental group did not all behave as a straight line ($p = 0.001$). Steady state was already observed after 30 min ($p = 0.1351$) with a mean maximal increase of 59.58 mmHg. In the control group, no changes in arterial blood pressure were observed during the experiment.

Table 6.1 Means ± standard errors of levels of pH, $PaCO_2$, PaO_2, and arterial blood oxygen saturation (%Sat.) in control animals (Control Experiments) and in the experimental group (L-NAME Experiments). Two time points are shown for each group: control (Control)—before saline injection or before L-NAME injection and 45 min after injection. The significance level (*p*) was calculated relative to the control time within each group (ns = nonsignificant)

Control experiments				
Treatment	pH	$PaCO_2$ (mmHg)	PaO_2 (mmHg)	% Sat.
Control	7.25 ± 0.01	40.91 ± 1.58	112.60 ± 6.8	96.29 ± 0.49
45’ Saline	7.24 ± 0.01	38.83 ± 2.25	117.80 ± 5.86	96.70 ± 0.31
Significance	ns	ns	ns	ns

L-NAME experiments				
Treatment	pH	$PaCO_2$ (mmHg)	PaO_2 (mmHg)	% Sat.
Control	7.26 ± 0.02	40.79 ± 3.08	108.30 ± 6.96	95.19 ± 0.94
45’ L-NAME	7.22 ± 0.02	43.38 ± 2.17	104.95 ± 6.15	94.55 ± 1.2
Significance	$p < .01$	ns	ns	ns

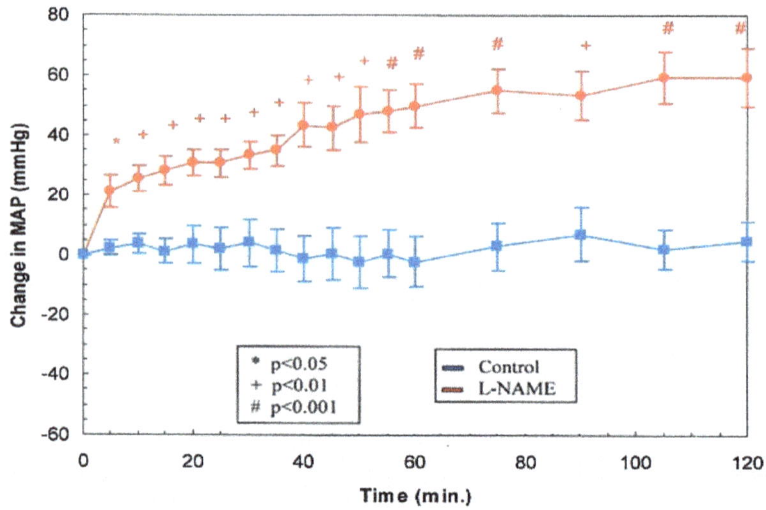

Fig. 6.9 Effect of L-NAME injection on mean arterial blood pressure. The image shows mean values ± standard error in mmHg in experimental rats treated with L-NAME at zero-time point (red) and in control rats injected with saline at zero-time point (blue). The significance level between the experimental group and the control group (*p*) is detailed in the graph's key

The absolute mean value of blood pressure at zero-time point was: 124.21 mmHg ± 4.1 in the experimental group and 127.11 mmHg ± 6.7 in the control group.

6.3.4.2 The Effect of L-NAME in the Brain: Energy Metabolism, Hemodynamic Activity, and Ionic and Electrical Activity

Figure 6.10 illustrates brain responses using six parameters: Reflectance, CF and CBF (expressed in percentages), Brain Temperature (changes in degrees Celsius), K^+ (in mM values, with statistical calculations made according to changes in mM) and DC (changes in mV). The data were collected for 2 h, with each time point representing the mean with SE of the nine experimental animals and eight control animals. At zero-time point, L-NAME was injected into the experimental animals and saline into the control animals. Body temperature was kept constant (around 38.5°C). The significance level between the experimental group and the control group was tested via *t*-test, and an ANOVA test was also conducted to find the time at which the experimental group reached steady state (results not shown).

The ANOVA test showed that no significant change was found in the control group 120 min after saline injection in any of the parameters tested, except for DC which increased throughout the experiment. In the L-NAME treated group, no significant changes were found within the group for 120 min in Reflectance, CF, and K^+. Blood flow and brain temperature showed a decreasing trend, with steady state already recorded at 30 min and beyond. DC potential increased during the experiment.

According to a *t*-test between the experimental group and control group, which compared mean values at parallel time points, it was found that:

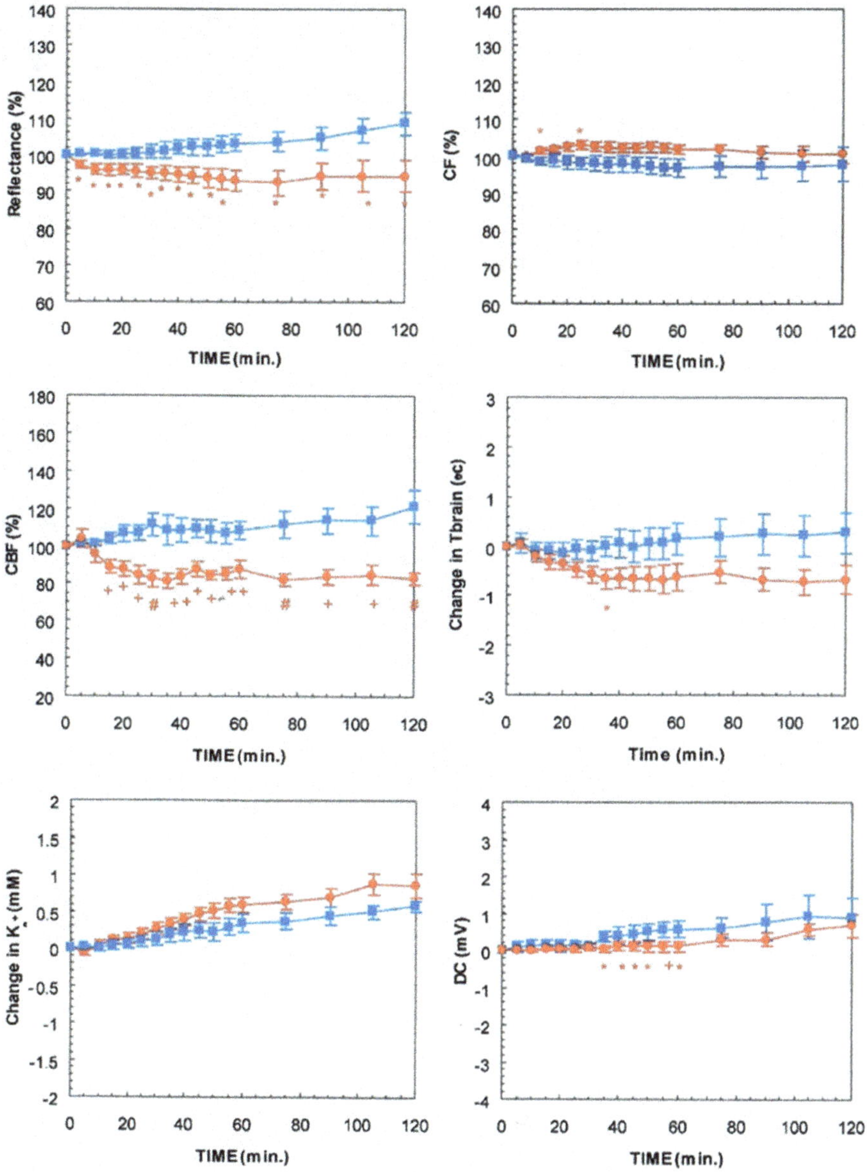

Fig. 6.10 Effects of L-NAME injection on brain parameters: R reflectance, CF mitochondrial NADH level, CBF cerebral blood flow—expressed as percentages, Tbrain brain temperature—changes in degrees Celsius, Ke^{+} extracellular potassium—changes in mM, and DC—changes in mV. The image shows mean values ± standard error in the experimental animals treated with L-NAME at zero-time point (red) and in the control animals injected with saline at zero-time point (blue). The significance level between the experimental group and the control group (p), according to the following key: $^{*}p<0.05$; $+p<0{\cdot}01$; $^{\#}p<0.001$

Reflectance

Reflectance decreased significantly ($p < 0.05$) already 5 min after L-NAME injection and remained low throughout the experiment (minimum mean level was 92.47%).

Mitochondrial NADH Level (CF)

After L-NAME injection, no significant changes were observed in CF for 120 min, except for time points 10 and 25 min (an increase of 1.43% ± 0.6 and 3.01% ± 1.3, $p < 0.05$).

Cerebral Blood Flow (CBF)

A significant decrease was already spotted 15 min after L-NAME injection ($p < 0.01$), with cerebral blood flow remaining low until the end of the experiment at a minimum mean level of 81.07%.

Brain Temperature (Tbrain)

A decrease in brain temperature can be observed after L-NAME injection, but this decrease was not significant compared to the control group, except for the 35-min time point where a decrease of −0.67°C ± 0.2 was observed ($p < 0.05$).

Extracellular K^+ Level

No significant changes were found between the two groups.

DC Potential

An increasing trend was recorded over time in both groups. Significant differences were found between 30 and 60 min ($p < 0.05$, $p < 0.01$), but these changes were very small and did not exceed 1 mV.

6.3.4.3 The Relationship Between Changes in Cerebral Blood Flow and Brain Temperature

Figure 6.11 depicts the relationship between cerebral blood flow and brain temperature in the L-NAME-treated group ($n = 9$). The regression line shows a close relationship ($R^2 = 0.8173$) between the decrease in cerebral blood flow and the decrease in brain temperature following L-NAME injection.

Conclusions
Inhibition of NO synthesis by L-NAME administration caused the following changes:

- Increase in arterial blood pressure.
- Decrease in cerebral blood flow.
- Decrease in reflectance.

A marked correlation was observed between the decrease in cerebral blood flow and the decrease in brain temperature.

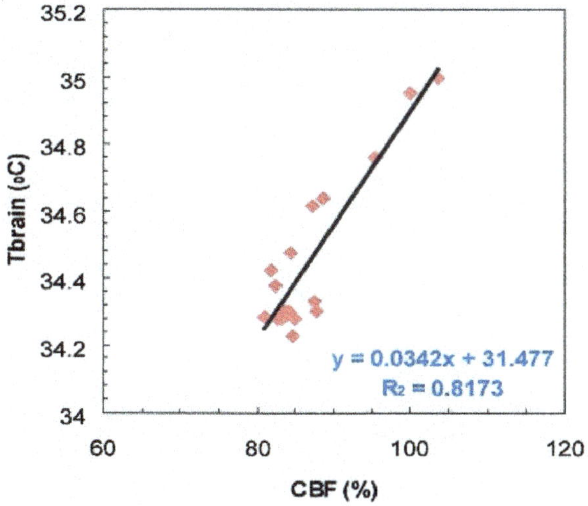

Fig. 6.11 Relationship between CBF and brain temperature in animals treated with L-NAME. Each point represents the mean of CBF and brain temperature at a given time point. R^2 was calculated by running a regression line through the points

6.3.5 Examining the Involvement of NO in the Response to Hypercapnia

Figure 6.12 shows the changes in blood pressure and various brain parameters over a 2-h period. Each point represents the means with standard errors at a given time point in four experimental groups:

Control group (in blue): did not undergo any treatment for two hours from the moment of saline injection at zero-time point ($n = 6$ for blood pressure, $n = 8$ for brain parameters).

CO_2 group (in green): 45 min after saline administration, the following gas mixture was introduced: 69% N_2 + 21% O_2 + 10% CO_2, for 15 min to induce hypercapnia ($n = 6$ for blood pressure, $n = 9$ for brain parameters).

L-NAME group (in red): treated at zero-time point with L-NAME and did not undergo any further treatment for two hours ($n = 8$ for blood pressure, $n = 9$ for brain parameters).

L-NAME + CO_2 group (in black): treat at zero-time point with L-NAME and after 45 min hypercapnia was induced for 15 min ($n = 6$ for blood pressure, $n = 10$ for brain parameters) (Fig. 6.13).

These figures show the general variation of the various parameters during the experiment. A detailed statistical analysis at the 45 and 55 min time points is presented later in this chapter, in Figs. 6.16, 6.17, and 6.18.

The main changes that can be observed are:

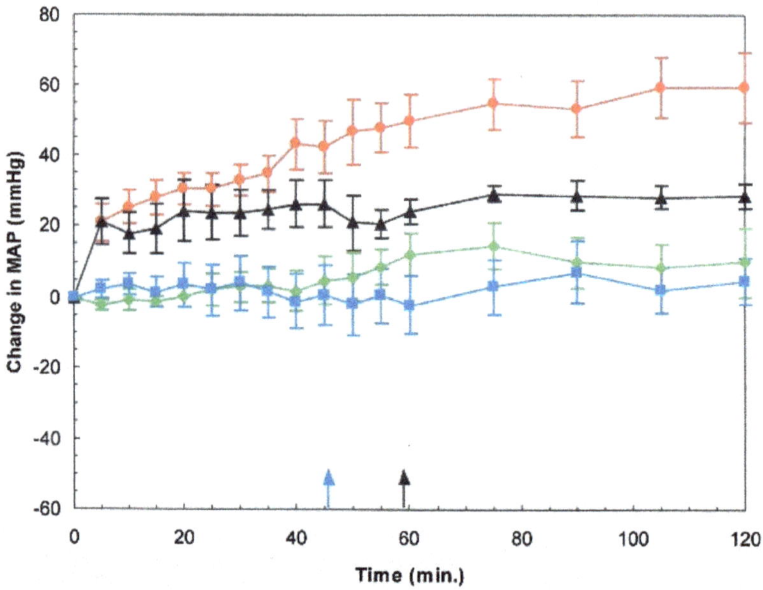

Fig. 6.12 Changes in arterial blood pressure over a 2-h period. Each point represents the mean and standard error values at a given time point in the following experimental groups: control group (blue), CO_2 group (green), L-NAME group (red), and L-NAME+CO_2 group (black). The blue arrow marks the start of hypercapnia induction and the black arrow marks the end of ventilation in the CO_2 and L-NAME+CO_2 groups

Mean Arterial Blood Pressure (MAP)

The trend of change in the control group and the CO_2 group appears to be similar until the end of the experiment. Following treatment with L-NAME, an increase in MAP was observed in the L-NAME and L-NAME+CO_2 groups. Hypercapnia induction caused a small decrease in MAP in the L-NAME+CO_2 group.

Reflectance

The control group showed an increasing trend over time. The other groups showed the opposite trend. Under the influence of hypercapnia, a decrease in reflectance was observed.

Mitochondrial NADH Level (CF)

In general, no changes could be observed over time due to L-NAME injection. Under the influence of hypercapnia, the NADH level decreased slightly in the CO_2 and L-NAME+CO_2 groups. After cessation of ventilation, a trend of returning to the initial state was observed.

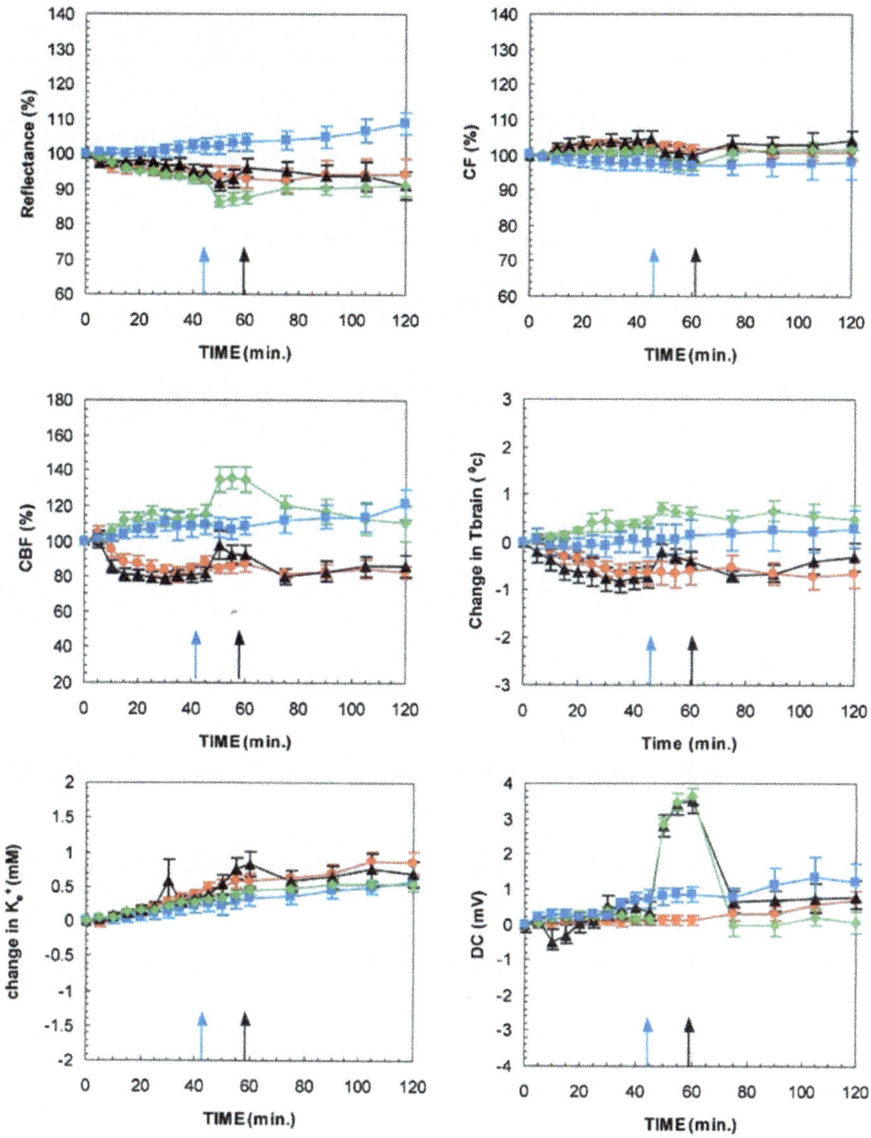

Fig. 6.13 Changes in the various brain parameters over a 2-h period. Each point represents the mean and standard error values at a given time point in the following experimental groups: control group (blue), CO_2 group (green), L-NAME group (red), and L-NAME+CO_2 group (black). The blue arrow marks the start of hypercapnia induction and the black arrow marks the end of ventilation in the CO_2- and L-NAME+CO_2 groups

Cerebral Blood Flow (CBF)

The trend of change in the control group and the CO_2 group was similar up to 45 min, when hypercapnia was induced. At the same time, a trend of decrease in blood flow was observed due to L-NAME injection in the L-NAME and L-NAME+ CO_2 groups. In hypercapnia induction, a trend of increase in CBF was observed in the CO_2 and L-NAME+CO_2 groups (a statistical test comparing the levels of increase is described in Sect. 6.3.2.2).

Brain Temperature (Tbrain)

A similar trend was observed with regard to blood flow.

Extracellular K Level

An increase was recorded with changes in time in all experimental groups.

DC Potential

No effect was seen following L-NAME treatment. Under the influence of hypercapnia induction, an increase of approximately 3.8 mV in DC potential was observed.

6.3.5.1 Effects of Hypercapnia

6.3.5.1.1 Effects of Hypercapnia in the Periphery

In Table 6.2, changes in pH, $PaCO_2$, and PaO_2 levels were recorded in an experimental group exposed to hypercapnia (10% CO_2) for 15 min ($n = 6$), compared to the control value before ventilation. A *t*-test showed that hypercapnia ($PaCO_2 = 69.94$ mmHg) resulted in a significant increase in $PaCO_2$ ($p < 0.01$), a significant decrease in pH ($p < 0.001$) and a significant increase in PaO_2 ($p < 0.001$), without change in arterial blood oxygen saturation percentages.

Hypercapnia induction in experimental rats ($n = 6$) did not cause significant changes in mean arterial blood pressure compared to the value before ventilation and compared to the value at the corresponding time point in the control group (results not shown).

Table 6.2 Effects of hypercapnia on pH, $PaCO_2$, PaO_2, and arterial blood oxygen saturation percentages (% Sat.). Values are presented as means ± standard errors. *P* values indicate the significance level between values during hypercapnia and control values before ventilation (ns = non-significant)

Treatment	pH	$PaCO_2$ (mmHg)	PaO_2 (mmHg)	% Sat.
Control	7.26 ± 0.05	37.47 ± 6.83	113.83 ± 7.65	96.43 ± 0.52
Hypercapnia	7.07 ± 0.05	69.94 ± 5.95	160.49 ± 7.7	97.64 ± 0.13
Significance	$p < 0.001$	$p < 0.01$	$p < 0.001$	n.s

6.3.5.1.2 Effect of Hypercapnia on the Brain: Energy Metabolism, Hemodynamic Activity, and Ionic and Electrical Activity

Figure 6.14 is a typical diagram of controlled experiments in rats that did not undergo treatment at the same time points. Figure 6.15 describes typical responses to exposure of animals to hypercapnia for 15 min (45 min after the start of the experiment).

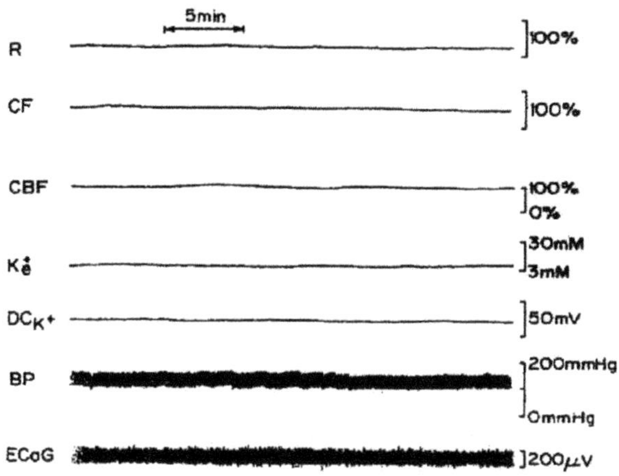

Fig. 6.14 Controlled experiment at parallel time points in an untreated rat. The following parameters were monitored: R reflectance, CF mitochondrial NADH level, CBF cerebral blood flow, Ke$^+$ extracellular potassium, DC_{K^+} potential, BP blood pressure, ECoG electrical activity

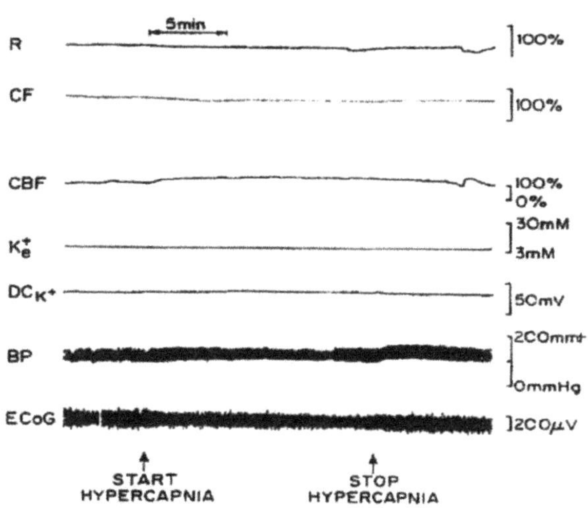

Fig. 6.15 Effects of hypercapnia on the following parameters: R reflectance, CF mitochondrial NADH level, CBF cerebral blood flow , Ke+ extracellular potassium, DC_{K+} potential, BP blood pressure, ECoG electrical activity. The arrows indicate the start and end of ventilation times

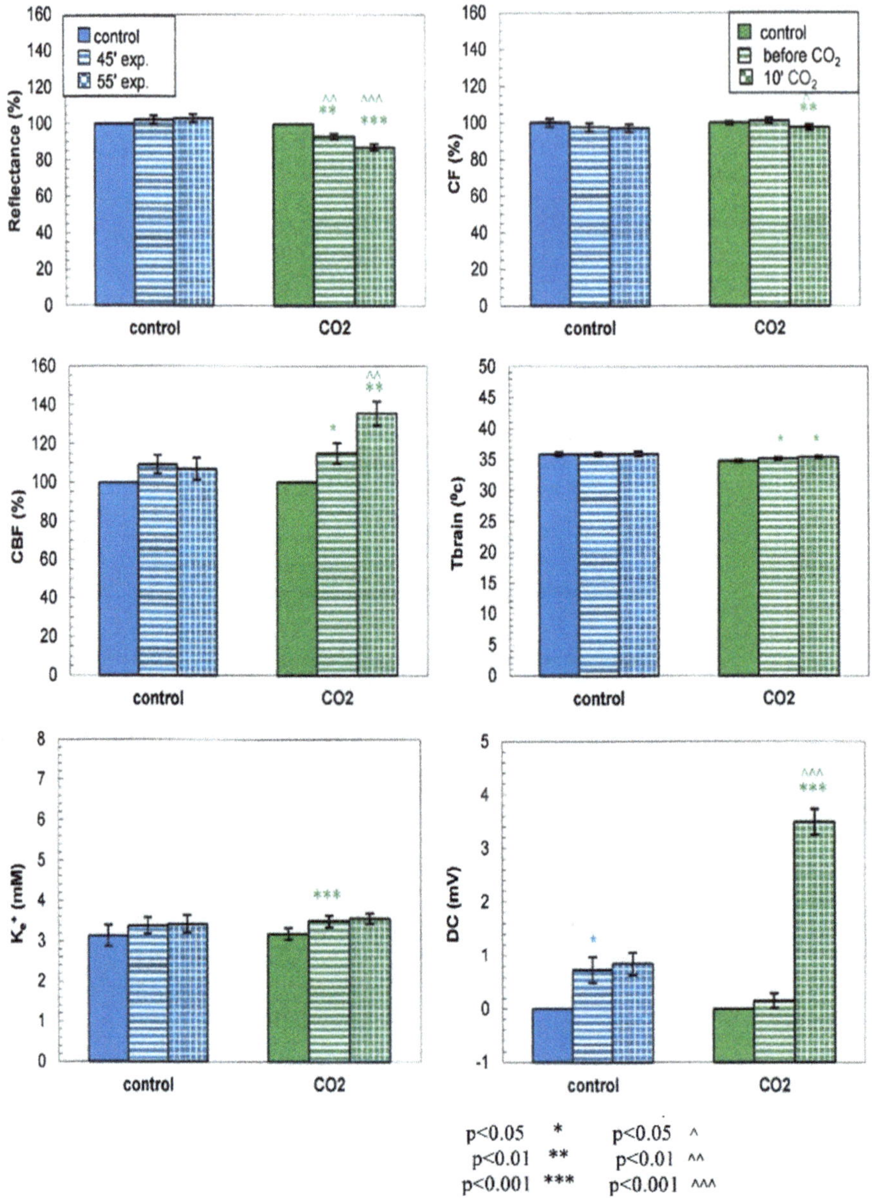

Fig. 6.16 Effects of hypercapnia on the following parameters: R reflectance, CF mitochondrial NADH, CBF cerebral blood flow—expressed in percentages, Tbrain brain temperature—in degrees Celsius, Ke+ extracellular potassium—in mM values, and DC—changes in mV. The figure shows means ± SE in the experimental group (CO_2) and in the control group. The left column shows the mean of the control value at the moment of saline injection at zero-time point in both groups. The middle column shows the mean of the value at the 45-min time point, which serves as a control for hypercapnia in the CO_2 group. The right column shows the mean after 10 min of ventilation in the CO_2

It may be observed that hypercapnia caused a decrease in CF, an increase in CBF, a small increase in DC, and a decrease of about 20% in the ECoG amplitude.

Figure 6.16 shows the mean and standard error values of the following parameters: Reflectance, CF and CBF (percentage changes), Brain Temperature (degrees Celsius changes), K^+ (in mM values) and DC (changes in mV). Two experimental groups are shown: ***CO_2 group*** = rats that were treated with a gas mixture (69% N_2 + 21% O_2 + 10% CO_2) for 45 min from saline injection, for 15 min to create hypercapnia ($n = 9$). ***Control group*** = rats that did not receive any treatment from the time of saline injection at zero-time point ($n = 8$).

In each group, three time points are shown during the experiment (see key). The significance level n the experimental groups was tested via paired t-test and between the two experimental groups in the corresponding values on the time axis via student t-test.

Reflectance

In the CO_2 group: a significant decrease in reflectance (−6.98% ± 1.56, $p < 0.01$) was recorded after 45 min from the start of the experiment. This decrease was also found to be significant in relation to the value at the corresponding time point in the control group ($p < 0.01$). Ventilation with 10% CO_2 caused a further decrease (−5.81% ± 1.04, $p < 0.001$) compared to the value before ventilation (45 min from the start of the experiment). This decrease is also significant compared to the control group at the same time point ($p < 0.001$).
In the control group: No significant changes were observed at the 45 and 55 min time points.

Mitochondrial NADH Level (CF)

In the CO_2 group: During ventilation with 10% CO_2, a significant decrease in CF (−3.64% ± 0.86, $p < 0.01$) was found compared to the value before treatment ($p < 0.01$). This decrease is also significant compared to the control group at the same time point ($p < 0.05$).
In the control group: No changes in CF were observed at 45- and 55-min time points.

Cerebral Blood Flow (CBF)

In the CO_2 group: 45 min after the start of the experiment, a significant increase in cerebral blood flow was recorded (15.04% ± 5.35, $p < 0.05$). This increase was not

Fig. 6.16 (continued) group and the value at the corresponding time point in the control group (55 min from saline injection). The asterisks indicate the significance level within the experimental groups where: The asterisks in the middle column refer to the comparison of the *level of change* between the value at the 45 mi time point and the control value at the zero time point. The asterisks in the right column refer to the comparison of the *level of change* between the value during ventilation (55-min time point) and the control before ventilation (45-min time point). ^= Significant changes between the two groups in the corresponding values on the time axis. * $p<0.05$, ^ $p<0.05$, ** $p<0.01$, ^^ $p<0.01$, *** $p<0.001$, ^^^ $p<0.001$

significant compared to the value at the corresponding time point in the control group. During 10 min of ventilation with 10% CO_2, an additional increase was recorded which was found to be significant (20.62% ± 5.57, $p < 0.01$) compared to the value before treatment. This increase is also significant compared to the value at the corresponding time point in the control group ($p < 0.01$).
In the control group: Small, but not significant increases were recorded at the 45 and 55 min time points.

Brain Temperature (Tbrain)

In the CO_2 group: 45 min after the start of the experiment, a small and significant increase (0.41°C ± 0.12, $p < 0.05$) in brain temperature was observed. This increase was not significant compared to the control group. During 10 min of ventilation with 10% CO_2, an additional increase was recorded (0.24°C ± 0.1, $p < 0.05$), which was found to be significant compared to the value before treatment, but not significant compared to the control animals at the same time point.
In the control group: No changes in brain temperature were observed during the experiment.

Extracellular K^+ Level

In the CO_2 group: after 45 min since the beginning of the experiment, a significant increase in potassium levels was observed relative to the control value (0.3 mM ± 0.05, $p < 0.001$). This increase was not significant compared to the value at the corresponding time point in the control group.
In the control group: No significant changes in potassium levels were observed during the experiment.

DC Potential

In the CO_2 group: 45 min from the beginning of the experiment, no significant change in DC potential was observed. Under the influence of hypercapnia, DC potential increased significantly (3.34 mV ± 0.14, $p < 0.001$) compared to the value before ventilation. This increase was also found to be significant compared to the value at the corresponding time point in the control group.
In the control group: after 45 min from the beginning of the experiment, a significant increase in DC levels was recorded (0.73 mV ± 0.24, $p < 0.05$).

6.3.5.2 Effects of Hypercapnia During NO Synthesis Inhibition

6.3.5.2.1 Effects of Hypercapnia During NO Synthesis Inhibition in the Periphery

Table 6.3 shows changes in pH, $PaCO_2$ and PaO_2 levels in an experimental group exposed to hypercapnia for approximately 45 min after L-NAME administration ($n = 5$). A t-test to test the significance level between the value at the 10-min time point of ventilation and the control value before ventilation showed that hypercapnia ($PaCO_2 =$

Table 6.3 Effects of hypercapnia during NO synthesis inhibition on pH, $PaCO_2$, PaO_2, and arterial blood oxygen saturation (%Sat.). Values are presented as means ± standard errors. *P* values indicate the significance level between values during hypercapnia and control values before ventilation (ns = nonsignificant)

Treatment	pH	$PaCO_2$ (mmHg)	PaO_2 (mmHg)	% Sat.
Control	7.28 ± 0.03	31.78 ± 2.02	106.26 ± 6.27	95.95 ± 0.96
Hypercapnia	7.07 ± 0.03	57.30 ± 4.02	150.00 ± 7.24	97.19 ± 0.29
Significance	$p < 0.001$	$p < 0.01$	$p < 0.01$	ns

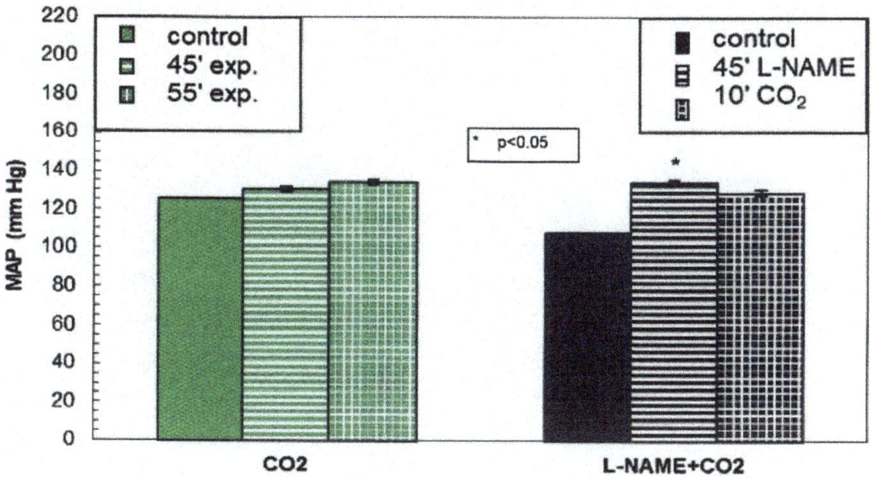

Fig. 6.17 Effects of hypercapnia, during NO synthesis inhibition, on mean arterial blood pressure (MAP). The figure shows mean values ± standard error in mmHg in the experimental group (L-NAME+CO_2) and the CO_2 group. The left column shows the mean and standard error of the control value at the moment of L-NAME or saline injection at zero-time point in both experimental groups. The middle column shows the mean and standard error of the value at the 45-min time point. This value serves as a control for hypercapnia in both experimental groups. The right column shows the mean and standard error of the value at the 10-min time point of ventilation in both experimental groups. The asterisk in the middle column refers to the comparison of the change level between the value at the 45-min time point and the control value at zero-time point

57.3 mmHg) caused a significant increase in $PaCO_2$ ($p < 0.01$), a significant decrease in pH ($p < 0.001$) and a significant increase in PaO_2 ($p < 0.01$). The percentages of arterial blood oxygen saturation did not change. Comparing these parameters with the corresponding parameters in the CO_2 group did not reveal a significant difference between them.

Figure 6.17 shows the mean ± standard errors of arterial blood pressure in mmHg values in two experimental groups:

L-NAME + CO_2 group = rats treated at the beginning of the experiment with L-NAME and after 45 min were ventilated with the following mixture: 69% N_2 + 21% O_2 + 10% CO_2 for 15 min to induce hypercapnia ($n = 6$).

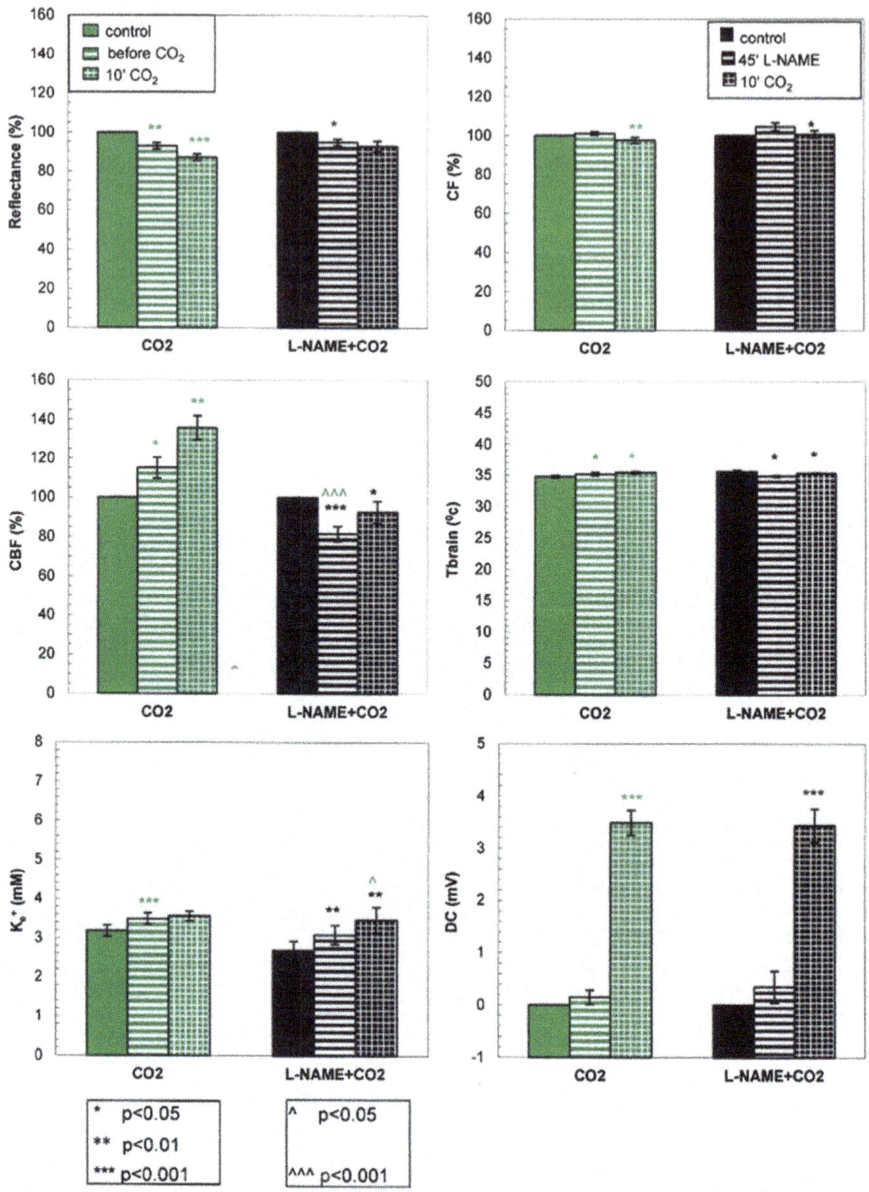

Fig. 6.18 The effects of hypercapnia, during NO synthesis inhibition, on various parameters in the brain (detailed in Fig. 6.16). The figure shows the experimental group (L-NAME+CO_2) and the CO_2 group (details of the histograms appear in Fig. 6.17). The asterisks indicate the significance level within the experimental groups where: The asterisks in the middle column refer to the comparison of the change level between the value at the 45-min time point and the control value at zero-time point. The asterisks in the right column refer to the comparison of the change level between the value during ventilation (55-min time point) and the control before ventilation (45-min time point). ^ = Significant changes between the two groups in the corresponding values on the time axis

CO_2 group = 45 min after saline injection, ventilated with the abovementioned gas mixture for 15 min to induce hypercapnia. This group is a control for the L-NAME + CO_2 group ($n = 6$).

In each group, three time points are shown during the experiment (see key), when the significance level was tested in the experimental groups using a paired *t*-test and between the two experimental groups in the corresponding values on the time axis using a student *t*-test.

It is demonstrable that in the L-NAME + CO_2 group, the mean arterial blood pressure increased significantly after 45 min from L-NAME administration (26.36 mmHg ± 6.87, $p < 0.05$). However, induction of hypercapnia after L-NAME administration did not cause significant changes in blood pressure.

6.3.5.2.2 The Effect of Hypercapnia on the Brain During NOS Inhibition: Energy Metabolism, Hemodynamic Activity, and Ionic and Electrical Activity

Figure 6.18 shows the means and standard errors of the parameters: Reflectance, CF and CBF (changes in percentage), Brain Temperature (changes in degrees Celsius), K^+ (in values of mM) and DC (changes in mV). The figure shows the same experimental groups as in Fig. 6.18 with $n = 10$ in the L-NAME + CO_2 group and $n = 9$ in the CO_2 group. The significance level was tested in the experimental groups (paired *t*-test) and between the two experimental groups in the corresponding values on the time axis (student *t*-test).

Reflectance

In the L-NAME+CO_2 group: 45 min after L-NAME administration, a significant decrease in reflectance was recorded (5.15% ± 1.64, $p < 0.05$). Under the influence of hypercapnia, during NOS inhibition, there was a small decrease in reflectance (−2.02% ± 1.84), which was not found to be significant compared to the pre-ventilation value.
In the CO_2 group: 45 min after the start of the experiment (saline injection), a significant decrease in reflectance (−6.98% ± 1.56, $p < 0.01$) was recorded. Ventilation with 10% CO_2 caused an additional decrease (−5.81% ± 1.04, $p < 0.001$) compared to the pre-ventilation value.

Statistical comparison between the two experimental groups at identical time points revealed no significant change in reflectance following L-NAME injection. During hypercapnia, similar changes were observed in both groups.

Mitochondrial NADH Level (CF)

In the L-NAME+CO_2 group: the CF level decreased significantly during hypercapnia (−3.97% ± 1.68, $p < 0.05$) compared to the value before ventilation.
In the CO_2 group: during ventilation at 10% CO_2, a significant decrease in CF was found (−3.64% ± 0.86, $p < 0.01$), compared to the value before ventilation.

Statistical comparison between the two experimental groups at the same time points did not reveal a significant change in CF after L-NAME injection. The degree of decrease in CF following hypercapnia was similar in both groups.

Cerebral Blood Flow (CBF)

In the L-NAME+CO_2 group: 45 min after L-NAME administration, cerebral blood flow decreased significantly ($-18.48\% \pm 3.85$, $p < 0.001$). Hypercapnia caused a significant increase in CBF after NO synthesis inhibition relative to the value before ventilation ($10.97\% \pm 4.41$, $p < 0.05$)
In the CO_2 group: 45 min after the start of the experiment, a significant increase in cerebral blood flow was recorded ($15.04\% \pm 5.35$, $p < 0.05$). Ventilation with 10% CO_2 provided an additional increase that was found to be significant ($20.62\% \pm 5.57$, $p < 0.01$) compared to the value before ventilation.

Statistical comparison between the two experimental groups at the same time points showed a significant decrease in cerebral blood flow following L-NAME injection ($p < 0.001$). The degree of increase in CBF observed during hypercapnia in both experimental groups was similar.

Brain Temperature (Tbrain)

In the L-NAME+CO_2 group: 45 min after L-NAME administration, a significant decrease in brain temperature was recorded (-0.74°C ± 0.23, $p < 0.05$). Under the influence of hypercapnia, a small and significant increase was observed compared to the value before ventilation (0.44°C ± 0.14, $p < 0.05$).
In the CO_2 group: 45 min from the beginning of the experiment, a small and significant increase (0.41°C ± 0.12, $p < 0.05$) was observed in brain temperature. Ventilation with 10% CO_2 showed an additional increase that was found to be significant (0.24°C ± 0.1, $p < 0.05$) compared to the value before ventilation.

A statistical comparison between the two experimental groups at the same time points revealed no significant change in brain temperature following L-NAME injection. During hypercapnia, similar changes were observed in both groups.

Extracellular K^+ Level

In the L-NAME+CO_2 group: 45 min after L-NAME administration, a significant increase in the extracellular potassium level was found compared to the control value (0.39 mM ± 0.09, $p < 0.01$). Hypercapnia caused a further significant increase in potassium levels (0.38 mM ± 0.11, $p < 0.01$).
In the CO_2 group: 45 min after the start of the experiment, a significant increase in extracellular potassium levels was observed compared to the control value (0.3 mM ± 0.05, $p < 0.001$).

A statistical comparison between the two experimental groups at the same time points revealed no significant change in extracellular potassium levels following L-NAME injection, while significant differences were found in the degree of change in potassium during hypercapnia.

Table 6.4 Effects of hypercapnia (10% CO_2) on measured parameter

Parameter	Hypercapnia normal NO	Hypercapnia inhibition of NO	Statistical significance
$PaCO_2$	↑	↑	ns
pH	↓	↓	ns
PaO_2	↑	↑	ns
% Sat	No change	No change	ns
Mean arterial pressure	No change	No change	ns
Reflectance	↓	No change	ns
NADH level	↓	↓	ns
CBF	↑	↑	ns
Brain temperature	↑	↑	ns
Extracellular potassium	No change	↑	$P < 0.05$
DC potential	↑	↑	ns

ns not significant, *NO* nitric oxide

DC Potential

In the L-NAME+CO_2 group: 45 min after the start of the experiment, no significant increase in DC potential was observed. Under the influence of hypercapnia, DC potential increased significantly (3.09 mV ± 0.25, $p < 0.001$) compared to the value before ventilation.

In the CO_2 group: 45 min from the beginning of the experiment, no significant increase in DC potential was observed. Under the influence of hypercapnia, DC potential increased significantly (3.34 mV ± 0.14, $p < 0.001$) compared to the value before ventilation.

Statistical comparison between the two experimental groups at the same time points revealed no significant change in DC potential following L-NAME injection and during hypercapnia.

Conclusions

Ventilation with 10% CO_2 (hypercapnia) produced the following results seen in Table 6.4.

6.3.6 *Examination of NO Involvement in the Response to Hypoxia*

Figures 6.19 and 6.20 display the changes in blood pressure and various parameters in the brain over a two-hour period. Each point represents the means with standard errors at a given time in four experimental groups:

Control group (in blue): did not undergo any treatment for two hours from the moment of saline injection at zero-time point ($n = 6$ for blood pressure, $n = 8$ for brain parameters).

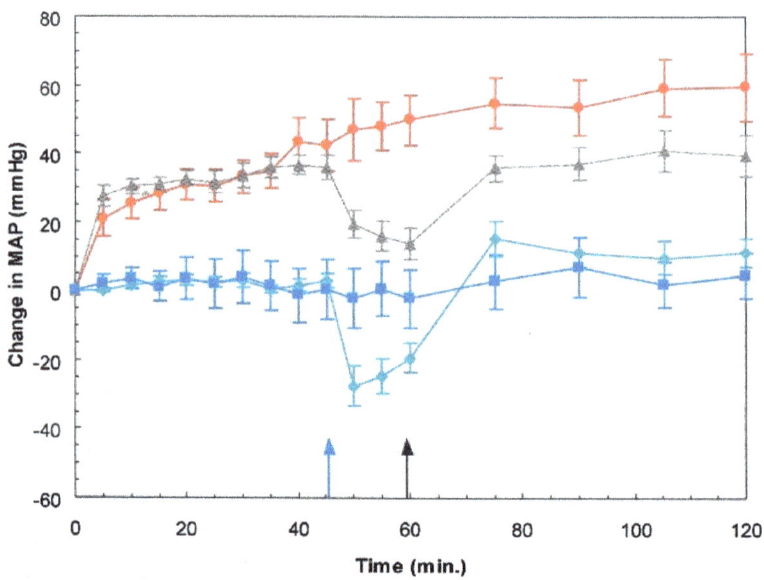

Fig. 6.19 Changes in arterial blood pressure over a 2-h period. Each point represents the mean and standard error at a given time point in the following experimental groups: control group (blue), HPX group (light blue), L-NAME group (red), and L-NAME+HPX group (grey). The blue arrow marks the start of hypoxia induction and the black arrow marks the end of ventilation in the HPX and L-NAME+HPX groups

HPX group (in light blue): 45 min after saline administration, the animals were ventilated with the following gas mixture: 87% N_2 + 1% CO_2 + 12% O_2, for 15 min to induce hypoxia ($n = 7$ for blood pressure, $n = 9$ for brain parameters).
L-NAME group (in red): treated at zero-time point with L-NAME and did not undergo any further treatment for two hours ($n = 8$ for blood pressure, $n = 9$ for brain parameters).
L-NAME+HPX group (in gray): L-NAME was administered at zero-time point and after 45 min hypoxia was induced for 15 min ($n = 8$ for blood pressure, $n = 11$ for brain parameters).

This figure generally shows the trend of change in the parameters measured during the experiment. A detailed statistical analysis at the 45- and 55-min time points is presented later in the chapter in Figs. 6.21 and 6.24–6.26.

The main changes that could be observed are as follows:
Mean Arterial Blood Pressure (MAP): The trend of change in the control group and the HPX group is similar up to the 45 min time point, when hypoxia is induced. Concurrently, a trend of increase in MAP was observed following L-NAME injection in the L-NAME and L-NAME+HPX groups. Following the induction of hypoxia, a trend of decrease in MAP was observed in the HPX and L-NAME+HPX groups. A return to the initial state occurred at the end of ventilation.
Reflectance: The control group and the L-NAME+HPX group showed an increasing trend over time. The other groups showed the opposite trend.

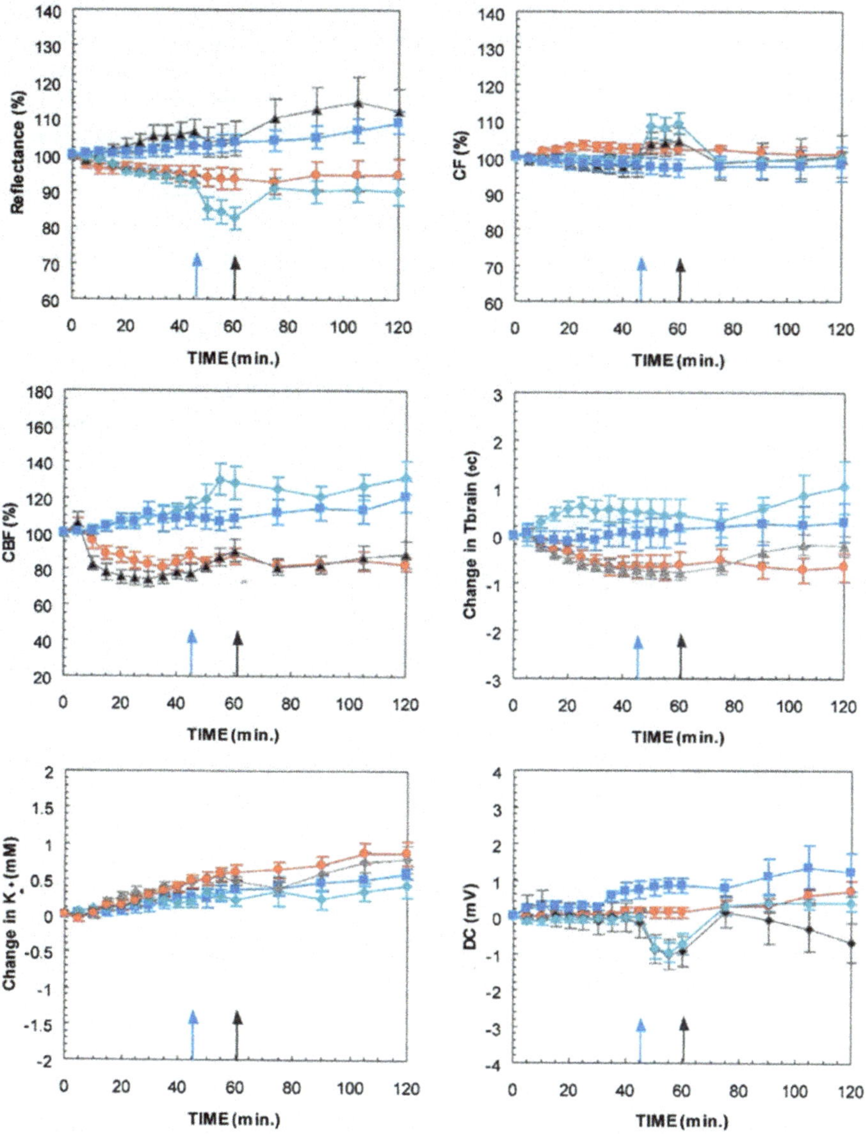

Fig. 6.20 Changes in blood pressure and various brain parameters over a two-hour period. Each point represents the mean and standard error at a given time point in the following experimental groups: Control group (blue), HPX group (light blue), L-NAME group (red), and L-NAME+HPX group (grey). The blue arrow marks the start of hypoxia induction and the black arrow marks the end of ventilation in the HPX and L-NAME+HPX groups

Mitochondrial NADH Level (CF): No changes over time following L-NAME injection could be observed. Following hypoxia induction NADH levels increased in the HPX and L-NAME+HPX groups. After cessation of ventilation, the values returned to baseline.

Cerebral Blood Flow (CBF): The trend of change in the control group and the HPX group is similar up to 45 min, when hypoxia was induced. Concurrently, a decreasing trend in blood flow was observed following L-NAME injection in the L-NAME and L-NAME+HPX groups. Following hypoxia induction, an increasing trend was observed in CBF in the HPX and L-NAME+HPX groups (a statistical test comparing the levels of increase is described in Sect. 6.3.3.2).
Brain Temperature (Tbrain): In the control group, brain temperature did not show a changing trend during the experiment. The HPX group showed an increasing trend over time. Concurrently, a decreasing trend in brain temperature was observed under the influence of L-NAME in the L-NAME and L-NAME+HPX groups.
Extracellular K^+ Level: A small increasing trend was recorded as a function of time in all experimental groups.
DC Potential: No effect was seen following L-NAME treatment. Following the induction of hypoxia, a decrease of approximately −0.9 mV in DC potential was observed.

6.3.6.1 Effects of Hypoxia

6.3.6.1.1 Effect of Hypoxia in the Periphery

Table 6.5 shows changes in levels of pH, $paCO_2$, paO_2, and arterial blood oxygen saturation percents in an experimental group exposed to hypoxia for 15 min ($n = 5$). A t-test showed that hypoxia ($paO_2 = 54.39$ mmHg) caused a marked decrease in paO_2 ($p < 0.001$), a significant increase in pH ($p < 0.001$), a significant decrease in $paCO_2$ ($p < 0.05$) and a significant decrease in percentage of arterial blood oxygen saturation ($p < 0.01$) (Figs. 6.27 and 6.28).

Figure 6.21 shows the mean arterial blood pressure in mmHg values in two experimental groups:
HPX group = rats ventilated 45 min after saline injection with the following gas mixture: 87% N_2 + 12% O_2 + 1% CO_2 for 15 min to create hypoxia ($n = 7$).
Control group = rats that did not receive any treatment from the time of saline injection at zero time points ($n = 6$).

Three time points are shown for each group during the experiment (see key). The significance level was tested within the experimental groups using a paired t-test and between the two experimental groups at the corresponding values on the time axis using a student t-test.

Table 6.5 Effects of hypoxia on pH, $PaCO_2$, PaO_2, and arterial blood oxygen saturation percentages (% Sat.). Values are presented as means ± standard errors. P-values indicate the significance level between values during hypoxia and control values before ventilation

Treatment	pH	$PaCO_2$ (mmHg)	PaO_2 (mmHg)	% Sat.
Control	7.22±0.03	36.42±2.82	108.27±5.59	95.39±0.84
Hypoxia	7.27±0.03	30.04±3.18	54.39±3.88	77.42±4.63
Significance	$p < 0.001$	$p < 0.05$	$p < 0.001$	$p < 0.01$

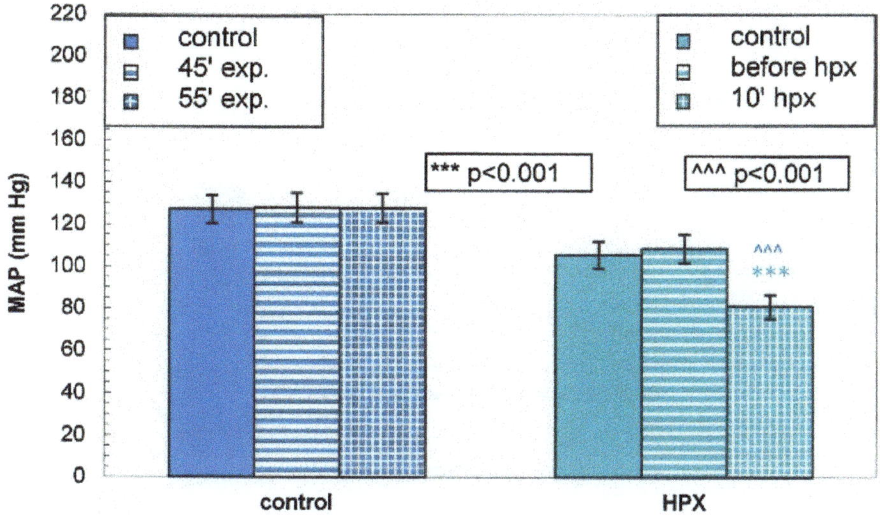

Fig. 6.21 Effects of hypoxia on mean arterial blood pressure (MAP). The figure shows mean values ± standard error in mmHg values in the experimental group (HPX) and the control group. The left column shows the mean control value and standard error at the moment of saline injection at zero time point in both experimental groups. The middle column shows the mean and standard error of the value at the 45-min time point. This value is a control for hypoxia in the HPX group. The right column shows the mean and standard error at 10 min from start of ventilation in the HPX group and the value at the corresponding time point in the control group (55 min from saline injection). The asterisks in the middle column refer to a comparison of the levels of change between the value at the 45-min time point and the control value at zero time point. The asterisks in the right column refer to a comparison of the levels of change between the value during ventilation (at the 55-min time point) and the control before ventilation (45-min time point). ^ = significant changes between the two experimental groups in the corresponding values on the time axis

This figure shows a significant decrease in mean arterial blood pressure following the induction of hypoxia compared to the value before ventilation (−27.64 mmHg ± 3.9, $p < 0.001$). This decrease was also significant when compared to the value at the corresponding time point in the control group ($p < 0.001$).

6.3.6.1.2 Effects of Hypoxia in the Brain: Energy Metabolism, Hemodynamic Activity, and Ionic and Electrical Activity

Figure 6.22 is a typical diagram for control experiments in untreated rats, at the corresponding time points.

Figure 6.23 shows the typical reactions of the animals to hypoxia exposure for 15 min

(45 min after the start of the experiment). It can be readily observed that hypoxia caused a decrease in reflectance, an increase in CF, an increase in CBF and a decrease in arterial blood pressure.

Figure 6.24 shows the mean and standard error of brain parameters. The same experimental groups as in Fig. 6.21 are shown, with $n = 9$ in the HPX group and $n =$

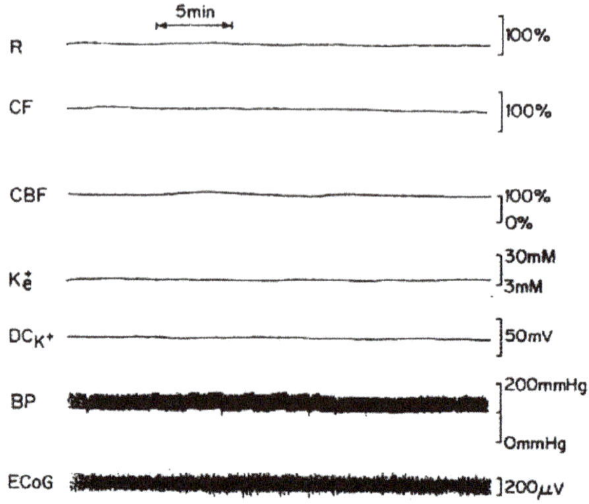

Fig. 6.22 Control experiments at the corresponding time points in untreated rats. R reflectance, CF Mitochondrial NADH levels, CBF cerebral blood flow, Ke+ extracellular potassium, DC potential, BP blood pressure, ECoG electrocorticogram

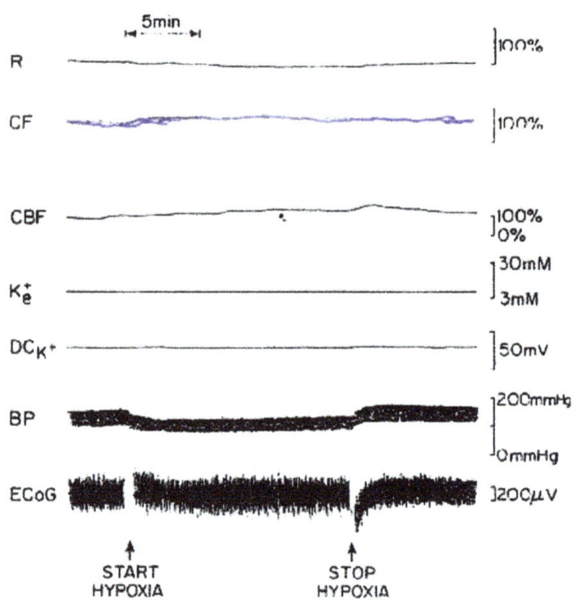

Fig. 6.23 Effects of hypoxia on the following parameters: R reflectance, CF mitochondrial NADH levels, CBF cerebral blood flow, Ke+ extracellular potassium, DC potential, BP blood pressure, ECoG electrocorticogram. The arrows mark the times of the start and end of ventilation

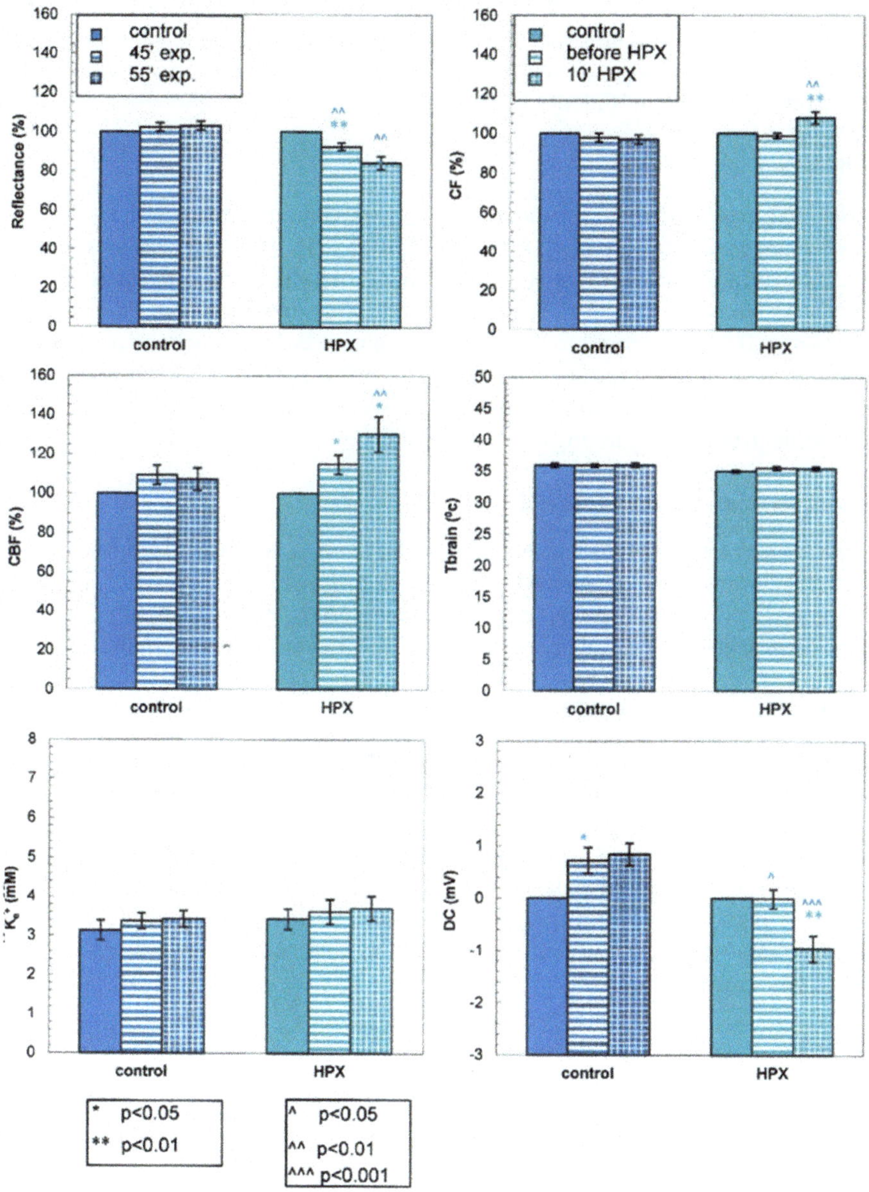

Fig. 6.24 The effects of hypoxia on the various brain parameters (detailed in Fig. 6.16). The figure shows the experimental group (HPX) and the control group (details of the histograms appear in Fig. 6.21). The asterisks in the middle column refer to a comparison of the change levels between the value at the 45-min time point and the control value at zero time point. The asterisks in the right column refer to a comparison of the change levels between the value during ventilation (55-min time point) and the control before ventilation (45-min time point). ^ = significant changes between the two experimental groups in the corresponding values on the time axis

8 in the control group. Three time points are shown for each group during the experiment (see key), with significance levels tested within the experimental groups using a paired *t*-test and between the two experimental groups at the corresponding values on the time axis using a student *t*-test.

Reflectance

In the HPX group: 45 min from the beginning of the experiment, a significant decrease in reflectance was observed (−7.47% ± 1.92, $p < 0.01$). This decrease was also significant in relation to the value at the corresponding time point in the control group ($p < 0.01$). Ventilation with 12% O_2 caused a further significant decrease (−8.23% ± 2.24, $p < 0.05$) compared to the value before ventilation. This decrease was also significant in relation to the control group at the corresponding time point ($p < 0.01$).
In the control group: no significant changes were observed at the 45- and 55-min time points from the beginning of the experiment.

Mitochondrial NADH Level (CF)

In the HPX group: hypoxia caused a significant increase in CF levels (9.07% ± 2.41, $p < 0.01$) compared to the value before ventilation. This increase was also significant compared to the control group at the corresponding time point ($p < 0.01$).
In the control group: no changes were observed in CF at the 45- and 55-min time points.

Cerebral Blood Flow (CBF)

In the HPX group: 45 min from the beginning of the experiment, a significant increase in cerebral blood flow was recorded compared to the control value (14.87% ± 4.94, $p < 0.05$). This increase was not significant compared to the value at the corresponding time point in the control group. During hypoxia, a further significant increase was observed (15.5% ± 4.68, $p < 0.05$) compared to the value before ventilation. This increase was significant compared to the value at the corresponding time point in the control group ($p < 0.01$).
In the control group: small, but non-significant increases were recorded at the 45 and 55 min time points.

Brain Temperature (Tbrain)

No changes in brain temperature were observed in either one of the experimental groups at the examined time points.

Extracellular K^+ Level

No changes in potassium levels were observed in either of the experimental groups at the examined time points.

DC Potential

In the HPX group: there was no change in DC potential after 45 min from the beginning of the experiment. This value was found to be significantly lower than the value at the corresponding time point in the control group ($p < 0.05$). Hypoxia resulted in a significant decrease in DC potential ($-0.95\,mV \pm 0.15$, $p < 0.01$) compared to the value before ventilation. This decrease was also found to be significant when compared to the value at the corresponding time point in the control group ($p < 0.001$). ***In the control group:*** a significant increase in DC potential ($0.73\,mV \pm 0.24$, $p < 0.05$) 45 was recorded 45 min from the beginning of the experiment.

6.3.6.2 Effects of Hypoxia During NO Synthesis Inhibition

6.3.6.2.1 Effects of Hypoxia During NO Synthesis Inhibition in the Periphery

Table 6.6 shows the values of pH, $PaCO_2$, PaO_2 and percentages of oxygen saturation in arterial blood in the experimental group exposed to hypoxia approximately 45 min after L-NAME administration ($n = 6$). A *t*-test was used to test the significance level between the value after 10 min of ventilation and the control value before ventilation showed that under the influence of hypoxia ($PaO_2 = 52.46$ mmHg), PaO_2 decreased significantly ($p < 0.001$). A small increase in pH and a small decrease in $PaCO_2$ were observed, but the changes were not found to be significant. Oxygen saturation percentages decreased significantly ($p < 0.01$). Comparison of these parameters with the corresponding parameters in the HPX group (*t*-test) did not reveal a significant difference between them.

Figure 6.25 shows the means and standard errors of arterial blood pressure in mmHg values in two experimental groups:

L-NAME+HPX group = rats treated at the beginning of the experiment with L-NAME and after 45 min ventilated with the following gas mixture: 87% N_2 + 1% CO_2 + 12% O_2 for 15 min to induce hypoxia ($n = 8$).

HPX group = serves as a control group for the L-NAME+HPX group (see Sect. 6.3.3.1 in results).

Three time points are shown for each group during the experiment (see key). The significance level was tested within the experimental groups using a paired *t*-test and between the two experimental groups in the corresponding values on the time axis using a student *t*-test.

It can be readily observed that in the HPX+L-NAME group the mean arterial blood pressure increased significantly after 45 min from the administration of L-NAME

Table 6.6 Effects of hypoxia during NO synthesis inhibition on pH, $PaCO_2$, PaO_2, and oxygen saturation percentages in arterial blood. The values are presented as means ± standard errors

Treatment	pH	$PaCO_2$ (mmHg)	PaO_2 (mmHg)	% Sat.
Control	7.23 ± 0.01	37.30 ± 2.77	100.82 ± 6.61	94.53 ± 1.01
Hypoxia	7.25 ± 0.02	33.38 ± 1.55	52.47 ± 4.55	74.07 ± 4.26
Significance	ns	ns	$p < 0.001$	$p < 0.01$

[P]-values indicate the significance level between the values during hypercapnia and the control values before ventilation (ns = nonsignificant).

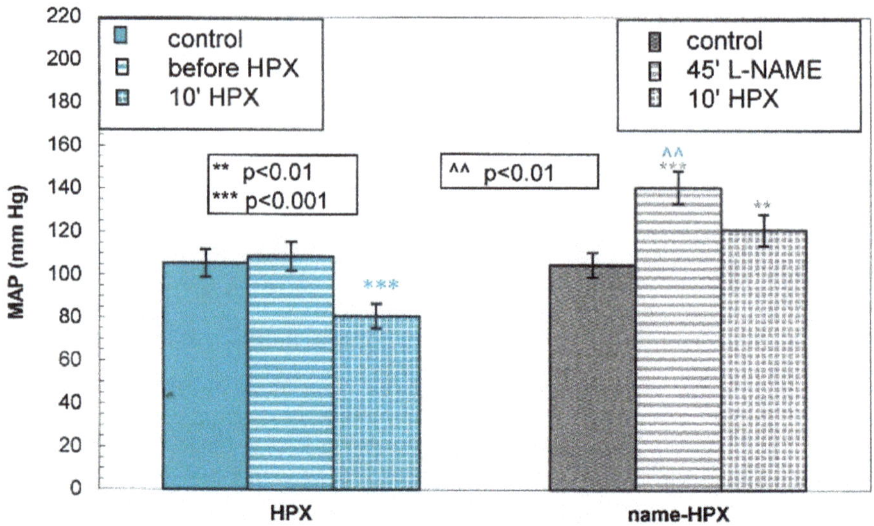

Fig. 6.25 Effects of hypoxia on mean arterial blood pressure (MAP) during NO synthesis inhibition. The figure shows the experimental group (L-NAME+HPX) and HPX group. The left column shows the mean and standard error of the control value at the moment of L-NAME or saline injection at zero-time point in the two experimental groups. The middle column shows the mean and standard error of the value at the 45-min time point. This value serves as a control for hypoxia in both experimental groups. The right column shows the mean and standard error of the value after 10 min of ventilation in both experimental groups. The asterisks in the middle column refer to a comparison of the levels of change between the value at the 45-min time point and the control value at zero-time point. The asterisks in the right column refer to a comparison of the levels of change between the value during ventilation (55-min time point) and the control before ventilation (45-min time point). ^= Significant changes between the two groups in the corresponding values on the time axis

(35.85 mmHg ± 3.58, $p < 0.001$), while in the HPX group the mean arterial blood pressure did not change 45 min after saline administration. Under the influence of hypoxia, there was a significant decrease in blood pressure in both experimental groups (in the HPX+L-NAME group: −19.8 mmHg ± 3.81, $p < 0.01$, in the HPX group: −27.64 mmHg ± 3.9, $p < 0.001$).

6.3.6.2.2 Effects of Hypoxia in the Brain During NOS Inhibition: Energy Metabolism, Hemodynamic Activity, and Ionic and Electrical Activity

Figure 6.26 shows the mean and standard error of the various brain parameters. The figure displays the same experimental groups as in Fig. 6.25, with $n = 11$ in the L-NAME+HPX group and $n = 9$ in the HPX group. Three time points are shown for each group during the experiment (see key). The significance level was tested within the experimental groups via a paired t-test and between the two experimental groups at the corresponding values on the time axis via a student t-test.

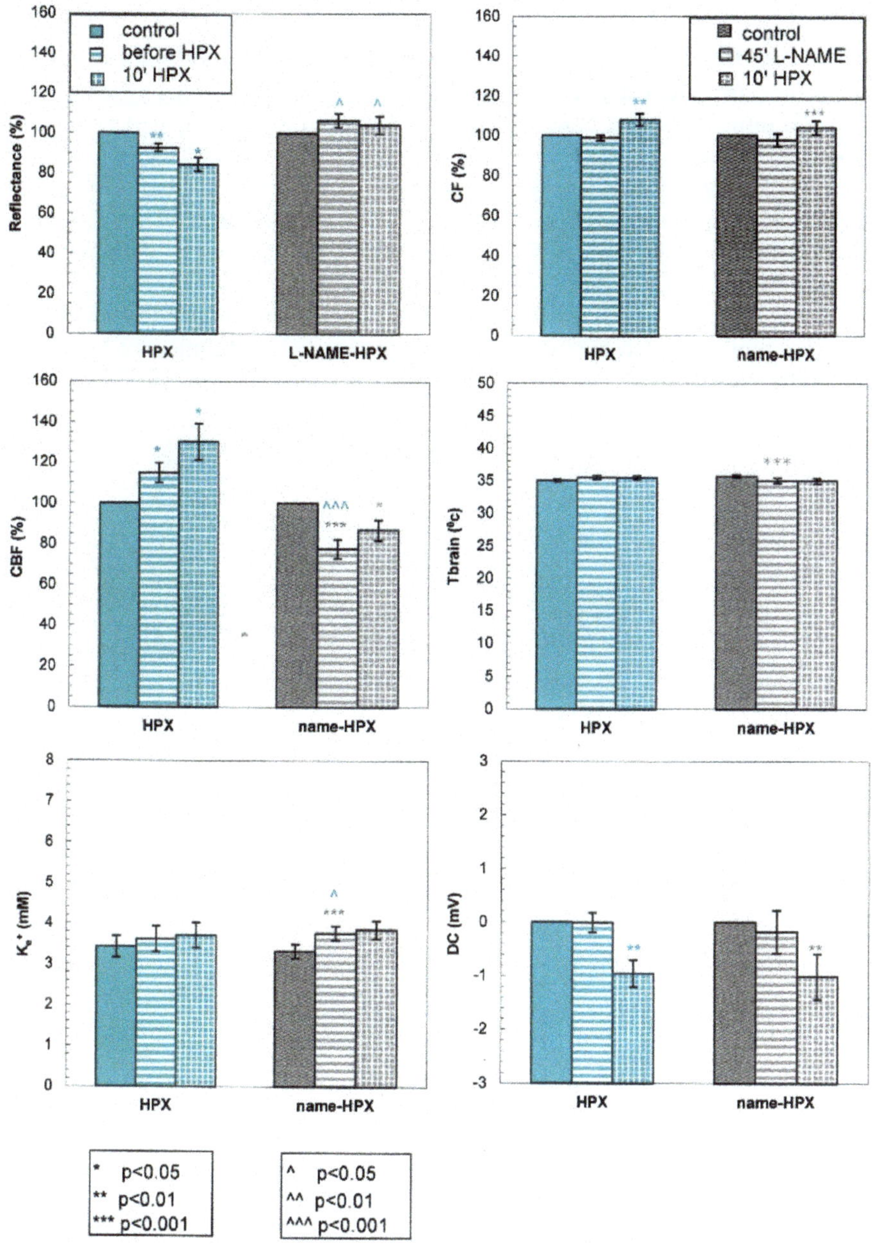

Fig. 6.26 The effects of hypoxia on various brain parameters during NO synthesis inhibition (detailed in Fig. 6.16). The figure shows the experimental group (L-NAME+CO_2) and the CO_2 group. The histograms and details of significance level calculations within the experimental groups (*) and between the groups (^) are detailed in Fig. 6.25

Reflectance

In the L-NAME+HPX group: 45 min after L-NAME administration, a small, non-significant increase in reflectance was observed (6.3% ± 3.51). Following hypoxia induction, no significant change in reflectance was observed when compared to the value before ventilation.

In the HPX group: 45 min from the beginning of the experiment, a significant decrease in reflectance was recorded (−7.47% ± 1.92, $p < 0.01$). Under the influence of hypoxia, a further significant decrease was recorded (−8.23% ± 2.24, $p < 0.05$) compared to the value before ventilation (Figs. 6.27 and 6.28).

Statistical comparison between the two experimental groups at identical time points showed a significant increase in reflectance following L-NAME injection ($p < 0.05$), and a higher level of reflectance during hypoxia + L-NAME, compared to hypoxia alone ($p < 0.05$).

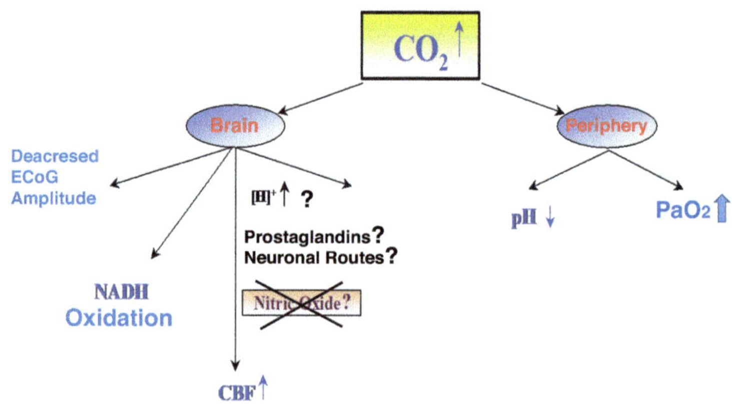

Fig. 6.27 The effect of CO_2 on the various parameters in brain and periphery

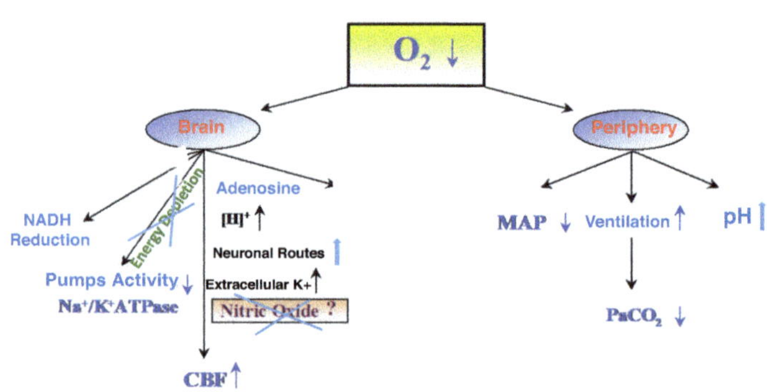

Fig. 6.28 The effect of hypoxia on the various parameters in brain and periphery

Mitochondrial NADH Level (CF)

In the L-NAME+HPX group: Under the influence of hypoxia, a significant increase in the level of CF (6.13% ± 1.16, $p < 0.001$) was found compared to the value before ventilation.
In the HPX group: Under the influence of hypoxia, a significant increase in the level of CF (9.07% ± 2.41, $p < 0.01$) was found when compared to the value before ventilation. Statistical comparison between the two experimental groups at identical time points revealed no significant change in the level of CF following L-NAME injection. Under the influence of hypoxia, a similar increase in CF was observed in both groups.

Cerebral Blood Flow (CBF)

In the L-NAME+HPX group: 45 min after L-NAME administration, a significant decrease in CBF was recorded (−22.44% ± 4.6, $p < 0.001$), which was found to be significantly lower than the value at the corresponding time point in the HPX group ($p < 0.001$). Under the influence of hypoxia, a significant increase in cerebral blood flow was recorded compared to the value before ventilation (10.32% ± 4.53, $p < 0.05$). This increase in CBF is not significantly different from the increase recorded in the HPX group during the same time period.
In the HPX group: 45 min from the start of the experiment, a significant increase in cerebral blood flow was found (14.87% ± 4.94, $p < 0.05$) compared to the control value. Hypoxia resulted in a further significant increase (15.5% ± 4.68, $p < 0.05$) compared to the value before ventilation.
Statistical comparison between the two experimental groups at identical time points revealed a significant decrease in cerebral blood flow following L-NAME injection ($p < 0.001$). The degree of increase in CBF observed during hypoxia in the two experimental groups was similar.

Brain Temperature (Tbrain)

In the L-NAME+HPX group: 45 min after L-NAME administration, brain temperature decreased significantly compared to the control value (−0.71°C ± 0.15, $p < 0.001$). During hypoxia, no further change was observed.
In the HPX group: No changes in brain temperature were observed.
Statistical comparison between the two experimental groups at identical time points did not reveal a significant change in brain temperature following L-NAME injection and during hypoxia.

Extracellular K^+ levels

In the L-NAME+HPX group: 45 min after L-NAME administration, potassium levels increased significantly compared to the control value (0.45 mM ± 0.07, $p < 0.001$). Under the influence of hypoxia, the potassium level did not change.
In the HPX group: no changes in potassium levels were observed.
Statistical comparison between the two experimental groups at identical time points showed a significant increase in potassium levels following L-NAME injection (*p*

Table 6.7 Effects of hypoxia (12% O_2) on measured parameter

Parameter	Hypoxia normal NO	Hypoxia inhibition of NO	Statistical significance
$PaCO_2$	↓	No change	ns
pH	↑	No change	ns
PaO_2	↓	↓	ns
% Sat	↓	↓	ns
Mean arterial pressure	↓	↓	ns
Reflectance	↓	No change	$P < 0.05$
NADH level	↑	↑	ns
CBF	↑	↑	ns
Brain temperature	No change	No change	ns
Extracellular potassium	No change	No change	ns
DC potential	↓	↓	ns

ns not significant, *NO* nitric oxide

< 0.05), with no significant differences found in the degree of change in potassium levels during hypoxia.

DC Potential

In the L-NAME+HPX group: Under the influence of hypoxia, DC potential decreased significantly ($-0.84\,\text{mV} \pm 0.21, p < 0.01$) compared to the value before ventilation.

In the HPX group: Under the influence of hypoxia, DC potential decreased significantly ($-0.95\,\text{mV} \pm 0.15, p < 0.01$) compared to the value before ventilation.

Statistical comparison between the two experimental groups at identical time points revealed no differences in DC potential following L-NAME injection. During hypoxia, the DC potential decreased similarly in both groups.

In Conclusion

Ventilation with 12% O_2 (hypoxia) produced the following results as seen in Table 6.7.

6.4 Discussion

Nitric Oxide (NO) has been discovered in recent years to have many physiological roles, including relaxation of smooth muscle in the vascular system, inhibition of blood platelet function, stimulation and induction of cytotoxicity, modulation of synaptic neurotransmission, and various other functions in the brain.

Different groups around the world have studied the effect of NO on a variety of biochemical and physiological processes in the brain under conditions of hypoxia and hypercapnia. In the current work, the involvement of NO in the regulation of blood flow and brain activity in vivo under conditions of hypercapnia and hypoxia was examined using a multichannel measurement system. The BFM measurement system, developed in our laboratory, allows for the monitoring of hemodynamic (local blood flow), metabolic (mitochondrial NADH levels), ionic (extracellular potassium), and electrical (DC potential and ECoG) processes. All data were collected simultaneously during the experiment and stored on a computer using special software (Labview) that allows for quantitative analysis of the results.

6.4.1 *Effects of NO Synthesis Inhibition Via L-NAME Administration*

In order to examine the involvement of NO in the regulation of various processes in the brain (hypercapnia and hypoxia), it was necessary to use a control group in which the effects of NO inhibition alone were examined. In these experiments, rats were treated with L-NAME at the beginning of the experiment, with measurements being taken for the space of two hours from the time of injection. The controls for these experiments were rats treated with saline solution (Figs. 6.9 and 6.10). It can be readily observed that the most prominent phenomenon following L-NAME treatment is an increase in peripheral blood pressure and a decrease in cerebral blood flow.

Previous work reports that NO plays an important role in maintaining a constant blood pressure. Many studies have shown that NOS inhibition in the periphery increased arterial blood pressure (Benyo et al. 1991; Iadecola and Zhang 1994; Ishimura et al. 1996; Iwamoto et al. 1992; McPherson et al. 1995; Rees et al. 1989; Sandor et al. 1994; Sokoloff et al. 1992; Toda and Okamura 1992). In the present work, arterial blood pressure increased significantly as early as 5 min after L-NAME administration and stabilization was observed at 30 min after injection, with a maximal increase of 59.58 mmHg. This strengthens the assumption that NO plays an important role in maintaining peripheral vascular resistance.

In the literature, NO has been attributed a specific role in maintaining normal blood flow in the brain. Many studies have shown that inhibiting the production of NO by administering NOS enzyme inhibitors resulted in a decrease in cerebral blood flow (Sokoloff et al. 1992; Wang et al. 1995; McPherson et al. 1995; Iadecola and Zhang 1994).

In the current study, a significant decrease in cerebral blood flow was observed 15 min after L-NAME administration. Blood flow remained low until the end of the experiment, at an average minimal level of 81.07%. In addition, gas levels in arterial blood were not altered following NOS inhibition, in line with what has been described in the literature (Macrae et al. 1993; Horvath et al. 1994; Sandor et al. 1994; Sokoloff et al. 1992), except for pH, which decreased significantly but this decrease was not significant when compared to the corresponding time point in the control group. In addition, it has been reported in the literature that NOS inhibitors do not affect cerebral oxygen consumption and do not suppress cerebral energy metabolism (Iadecola and Zhang 1994). Our study revealed that the decrease in CBF was not accompanied by significant changes in NADH levels. The stability of NADH levels can be explained by the fact that there was no oxygen deficiency in the tissue and that normal metabolic activity in the brain did not change, apparently due to an increase in the utilization of oxygen by brain tissue. Experiments conducted in our laboratory have similarly shown that following L-NAME administration, the hemodynamic changes observed in the animals were not accompanied by changes in NADH levels (Avivit 1996; Dror 1994). In addition, the present work has shown that there were no significant changes in extracellular potassium levels in the brain following L-NAME administration, i.e., sufficient ATP was generated for proper activity of sodium-potassium pumps and oxygen balance in the brain did not seem to change. Another possibility that could explain the observed stability in NADH levels is a decrease in cellular energy metabolism due to a decrease in membrane permeability or electrical breakdown. In this case, the demand for ATP by sodium-potassium pumps decreased, and despite the decrease in blood flow and oxygen supply extracellular NADH and potassium levels did not change. Another possibility is that L-NAME administration led to better utilization of oxygen in the respiratory chain and increased the coupling between electron transfer in the respiratory chain to ATP production. Some studies have shown that exogenous NO inhibits cellular respiration. In work with isolated mitochondria, NO was found to inhibit cytochrome oxidase (Brown et al. 1997). In another study, it was found that administration of NO reduced the level of ATP in isolated pneumocytes (Miles et al. 1996). The inhibition of cellular respiration was reversible by administering NO inhibitors. However, it is difficult to draw conclusions from these studies with regards to the activity of endogenous NO.

After L-NAME administration, brain temperature decreased, but not significantly when compared to the control experiments. Examination of the relationship between changes in cerebral blood flow and brain temperature in the context of L-NAME showed a close relation between the two ($R^2 = .8173$), which emphasizes the dependence of the brain's temperature on blood flow into it.

In the present work, an increase in reflectance was observed as time progressed. These results are unexpected, since many previous works in our laboratory have shown an inverse correlation between cerebral blood flow and reflectance, i.e., an increase in reflectance when blood flow decreases. Hence, the changes are unclear and there is a suspected drift in the signal over time due to a technical problem in the measurement system. I will therefore not refer to the results of this parameter in long-term measurements.

6.4.2 *Effects of Hypercapnia*

Inhalation exposure for the rats with a mixture containing 10% CO_2 increased the partial pressure of PCO_2 and PO_2 in arterial blood. In contrast, arterial blood pH decreased.

The increase in PaO_2 could stem from hyperventilation, since the rats breathed freely. In Kallinen et al. (1991), inhalation exposure of freely breathing piglets with a mixture containing 10% CO_2 caused similar results in PaO_2 values (an increase of about 50 mmHg), whereas artificial inhalation exposure with the same mixture with stroke volume control increased PaO_2 by only 20 mmHg (Kallinen et al. 1991). In the present study, an increase in blood flow was observed, which was probably due to the direct effect of the increase in $PaCO_2$ and decrease in arterial blood pH values. It is possible that the observable increase in PaO_2 prevented to some extent a further increase in blood flow. In addition, a decrease in reflectance was observed, which corresponds to an increase in hyperemia.

Many studies have shown that hypercapnia does not change oxygen consumption (Kety and Schmidt 1948; Siesjo 1978a), although some works that reported an increase in oxygen consumption (Yang and Krasney 1995; Horvath et al. 1994) while others showed a decrease in oxygen consumption (Ligeti et al. 1991). In the present work, mitochondrial NADH oxidation occurred, in all likelihood due to an improvement in oxygen balance as a result of increased blood flow and increased PO_2 in arterial blood, while extracellular potassium levels did not change. This suggests that it is possible that oxygen consumption increased due to hypercapnia, although the magnitude of change in NADH was very small (−3.64%). A decrease in NADH levels in hypercapnia has been reported in other studies (Mayevsky 1984; Gyulai et al. 1982). Another possibility is that oxygen consumption did not change during hypercapnia and there was a decrease in the utilization of available oxygen and the efficiency of electron transfer in the respiratory chain for the formation of ATP.

In hypercapnia, an increase in brain temperature was observed in the present study, probably due to the increase in cerebral blood flow. In addition, a decrease in ECoG amplitude and an increase in DC potential were observed. Other studies have shown a decrease in amplitude and activity of EEG (Artru and Michenfelder 1980; Schindler and Betz 1976) and an increase in DC potential (Katsura et al. 1994). According to the latter, the increase in DC potential stems from acidosis in the extracellular space. In the present study, extracellular pH was not measured, but arterial blood pH decreased under the influence of hypercapnia, so it is possible that this fact contributed to the decrease of pH in the extracellular space. In addition, in the work of Katsura et al. (1994), no change in extracellular potassium levels was observed, similar to the present study, so it seems that potassium ions are not involved in the changes in DC potential.

6.4.2.1 Effects of Hypercapnia During the Course of NO Synthesis Inhibition

The involvement of the NO molecule in the vasodilation mechanism in cerebral blood vessels is still unclear. Most studies have found that NOS inhibition reduced the increase in cerebral blood flow that was present in hypercapnia (Wang et al. 1995; Sandor et al. 1994; Horvath et al. 1994; Iadecola 1992; Estevez and Phillis 1997; Fabricius and Lauritzen 1994). Findings suggest that the reduction in vasodilation achieved by NOS inhibitors occurs in the range of $PaCO_2$ between 40 and 80 mmHg

(Iadecola and Zhang 1994). In the present study, the PCO_2 level in arterial blood during hypercapnia was 57.3 mmHg, i.e. within the range of NO's effect on the vasodilation mechanism.

The results of the present work have shown that the response to hypercapnia during NOS inhibition was similar to the response to hypercapnia without L-NAME administration in all parameters measured, both in the periphery and in the brain, except for extracellular potassium levels.

In the present study, 10% CO_2 inhalation exposure increased cerebral blood flow to the same extent in both the L-NAME+CO_2 and the CO_2 groups, i.e., cerebral blood flow responded to hypercapnia similarly under both normoxic conditions and L-NAME (Fig. 6.18). In the L-NAME + CO_2 group, the CBF baseline before hypercapnia induction was lower (81.61% ± 3.9) than in the CO_2 group (115.1% ± 5.3), due to NOS inhibition. During hypercapnia, a smaller increase in cerebral blood flow was observed in this group (10.97% ± 4.4) compared to the CO_2 group (20% ± 6.2), but there was no significant difference in the degree of increase between the two groups. The reasons for the inconsistency between the results of this work and results from other studies may arise from differences in experimental protocols, measurement methods, the inhibitor type used, the dose, the time between administration of the inhibitor and measurement of the response, differences between different species, etc. On the other hand, several studies have reported that NOS inhibitors did not reduce cerebral vasodilation during hypercapnia, similar to the present work (Adachi et al. 1992; McPherson et al. 1995; Sokoloff et al. 1992). According to the results of this work, it can be assumed that NO does not participate in the mechanism of cerebral vasodilation during hypercapnia.

Concurrently with the increase in cerebral blood flow due to hypercapnia, NADH oxidation was observed and in addition the PO_2 levels in arterial blood significantly increased, probably due to hyperventilation, an effect similar to hypercapnia by itself. It can therefore be concluded that there was an increase in oxygen supply to the brain and an improvement in oxygen balance induced by hypercapnia in the context of NOS inhibition.

In the present study, an increase in extracellular potassium concentration was observed, following the induction of hypercapnia during NOS inhibition, in contrast to hypercapnia without NOS inhibition. This result is puzzling, since NADH oxidation was observed, which suggests that the respiratory chain was not disrupted, and apparently enough ATP should have been generated for the operation of Na^+-K^+ ATPase pumps. Several explanations for this phenomenon can be proposed; one hypothesis is the possibility that L-NAME administration combined with hypercapnia directly caused the inhibition of Na^+/K^+-ATPase pumps and an increase in extracellular potassium levels. In this case, NO itself has no role in maintaining proper mitochondrial activity during hypercapnia. Another explanation is that NOS inhibition caused a decoupling in oxidative phosphorylation during hypercapnia. In this case, electron transfer in the respiratory chain is adequate, but not enough ATP is formed for the proper operation of the pumps. As a result, the respiratory chain works faster to provide more ATP and NADH levels decrease. Since not enough ATP is formed for the activity of the Na^+/K^+-ATPase pumps, the extracellular potassium level increases. It can therefore be assumed that NO has some impact on the efficiency of oxidative phosphorylation during hypercapnia by increasing the coupling between the respiratory chain and ATP formation. In addition, NOS inhibition may have affected the Na^+/K^+-ATPase pumps either by increasing their demand for ATP or by increasing the permeability of the membrane to Na^+/K^+ ions during hypercapnia. In such a case, the Na^+/K^+-ATPase pumps required more ATP when compared to hypercapnia alone. The reason for the differences in the results between L-NAME administration combined with hypercapnia and L-NAME administration alone may stem from the decrease in pH induced by hypercapnia. This may have resulted in changes to enzymatic activity in the system. Whatever the case may be, further tests are needed to confirm the above hypotheses.

In addition, an increase in DC potential was recorded, similar to the increase observed in the CO_2 experiments. It appears that potassium ions are not involved in the changes in DC potential, because in experiments where hypercapnia was induced without NOS inhibition, no changes in extracellular potassium ion levels were detected. This result reinforces the hypothesis of Katsura et al. (1994) that potassium ions are not involved in changes in DC potential.

According to the current work, the NO molecule is not involved in the regulation of the various brain parameters during hypercapnia, except for the regulation of potassium ions, which raises the possibility that other factors play a role in this response. Further studies are needed to elucidate the exact mechanism.

6.4.3 *Effects of Hypoxia*

Inhalation exposure of rats with a mixture containing 12% O_2 caused a decrease in the partial pressure of PO_2 and PCO_2 in arterial blood, a decrease in oxygen saturation percentage in arterial blood and an increase in pH. It is possible that the decrease in PaO_2 led to hyperventilation and thus a decrease in $PaCO_2$ levels. The compensatory mechanisms during hypoxia usually precede other changes, and therefore, despite the decrease in arterial blood PCO_2 levels, an increase in cerebral blood flow was observed (Nunn 1987) and a decrease in reflectance (Fig. 6.24). In addition, there was a decrease in arterial blood pressure, as shown in other works (Ishimura et al. 1996; Sato et al. 1992).

The increase observed in cerebral blood flow during hypoxia was probably unable to compensate for the lack of oxygen, which serves as the final electron acceptor in the respiratory chain. As a result, mitochondrial NADH levels increased significantly by 9.1% (Fig. 6.24). An increase in Mitochondrial NADH levels during hypoxia was also observed in another rat study (Mayevsky et al. 1984).

Despite the increase in mitochondrial NADH, extracellular potassium levels remained constant, which suggests that there was no slowing or disruption in the activity of Na^+/K^+-ATPase pumps and in all likelihood no deficiency in ATP. The results of the current study show that in hypoxia (at a PaO_2 level of 54.39 mmHg) the membrane potential and normal activity of Na^+/K^+-ATPase pumps are maintained at the expense of an increase in mitochondrial NADH levels, which may cause mitochondrial damage due to a negative oxygen balance. Similar results were obtained in a study performed on mitochondria isolated from the cortex of mice exposed to hypoxia. There was an increase in NADH levels with no change in ATP levels (Chavez et al. 1995). Since no changes in potassium levels were detected. In the present work, the involvement of potassium ions in the vasodilation that was detected can be ruled out. Similar conclusions have been reached in other works available in the literature (Nilsson et al. 1975; Dora et al. 1979).

In the present work, a decrease in DC potential was recorded. According to Katsura et al. (1994), a decrease in DC potential is consistent with a rise in pH and depolarization of neurons. According to the results of the present work, it does not appear that depolarization of the cells occurred, since no changes in extracellular potassium levels were detected. Results regarding pH levels in the tissues are unclear; on the one hand, pH levels in arterial blood increased, but on the other hand, the metabolic changes observed (an increase in NADH levels) hint that acidosis occurred in the tissue.

6.4.3.1 Effects of Hypoxia During NO Synthesis Inhibition

The present work's results have shown that the reaction to hypoxia during NOS inhibition was similar to the reaction to hypoxia without L-NAME administration in all parameters measured, except for reflectance.

Inhalation exposure with a mixture containing 12% O_2 during NO synthesis inhibition caused a decrease in the partial pressure of PO_2 in arterial blood, a decrease in oxygen saturation percentage in arterial blood, and a decrease in blood pressure. The decrease in PaO_2 resulted subsequently in a decrease in oxygen supply to the brain, which probably served as a direct cause for vasodilation of blood vessels in the brain and an increase in cerebral blood flow. The degree of increase in CBF observed during hypoxia with NOS inhibition was similar to hypoxia experiments without NOS inhibition. This suggests that NO plays no role in the mechanism of vasodilation observed during hypoxia. Other studies have shown similar results (McPherson et al. 1994; Gidday and Zhu 1995; Iadecola and Zhang 1994; Buchanan and Phillis 1993).

Similar to the HPX experiments with no NOS inhibition, the increase in cerebral blood flow probably did not suffice to compensate for the lack of oxygen in the respiratory chain, and therefore an increase in mitochondrial NADH levels was observed. However, extracellular potassium levels remained constant despite the increase in NADH levels, an outcome that suggests that there was probably no slowdown in the activity of Na^+/K^+-ATPase pumps.

According to the findings of the current study, the NO molecule does not appear to play any role in regulating the various brain parameters during hypoxia, and the reaction must therefore be attributed to other factors. Further experiments must be conducted to understand the exact mechanism.

6.4.4 Summary

This work dealt with the evaluation of the involvement of the NO molecule in processes related to the regulation of blood flow and whole brain activity in rats, during exposure to hypercapnia and hypoxia.

According to the findings of the current work, it appears that the main effect of L-NAME (which inhibits NO synthesis) is over arterial blood pressure and cerebral blood flow. Therefore, NO plays a vital role in maintaining blood pressure and normal blood flow into the brain, as posited in many other works. Since no changes in mitochondrial NADH levels were observed after L-NAME administration despite the decrease in cerebral blood flow, it can be assumed that there was no oxygen deficiency, perhaps due to an increase in oxygen extraction by the brain tissue or due to better utilization of oxygen in the respiratory chain.

Exposure of rats to hypercapnia without L-NAME administration and during NOS induction caused similar results in all parameters measured in the periphery and brain, except for extracellular potassium. In both cases, hypercapnia caused an increase in cerebral blood flow and mitochondrial NADH oxidation. It can therefore be assumed that there was an increase in oxygen supply to the brain which improved oxygen balance. NOS inhibition during hypercapnia caused an increase in extracellular potassium. According to this result, it is possible that during hypercapnia NO optimizes the oxygen balance in the cell or reduces the ATP demands of Na^+/K^+-ATPase pumps. Another possibility is that NO affects membrane permeability by preventing the leakage of sodium and potassium ions. On the other hand, NO might not have any involvement in the activity proposed in case L-NAME itself inhibited the Na^+/K^+-ATPase pumps.

According to the findings of the present work, the NO molecule does not play an important role in the vasodilation mechanism observed during hypercapnia, and it may therefore be the case that other vasodilating factors are responsible for this reaction. In any event, further experiments must be performed to understand the exact mechanism. The following diagram describes the reaction to exposure to hypercapnia (10% CO_2) according to the findings of the present work.

Exposure of rats to hypoxia, both without L-NAME and during NOS stimulation, resulted in similar results in all parameters measured in the periphery and brain, except for reflectance. In both cases, exposure to hypoxia was accompanied by an increase in cerebral blood flow and an increase in mitochondrial NADH levels. The observed increase in CBF was probably not sufficient to compensate for the lack of oxygen in the respiratory chain, and therefore an increase in mitochondrial NADH levels was observed. Despite the increase in NADH levels, extracellular potassium levels remained constant, a result indicating that ATP deficiency probably did not occur. Since no changes in extracellular potassium levels were detected, it is unlikely that potassium ions are involved in the vasodilation effect observed.

According to the findings of the present work, the NO molecule does not play an important role in the mechanism of vasodilation observed during hypoxia and other vasodilation factors (such as adenosine, hydrogen ions, neuronal pathways, etc.) in all likelihood play a role in this reaction. The following diagram describes the reaction to exposure to hypoxia (12% O_2) according to the findings of the present work.

References

Adachi, T., O. Inanami, and A. Sato. 1992. Nitric oxide (NO) is involved in increased cerebral cortical blood flow following stimulation of the nucleus basalis of Meynert in anesthetized rats. *Neuroscience Letters* 139:201–204.

Armstead, W. M. 1995. Opioids and nitric oxide contribute to hypoxia-induced pial arterial vasodilation in newborn pigs. *American Journal of Physiology* 268:H226–H232.

Artru, A. A., and J. D. Michenfelder. 1980. Effects of hypercarbia on canine cerebral metabolism and blood flow with simultaneous direct and indirect measurement of blood flow. *Anesthesiology* 52:466–469.

Avivit, M. 1996. *Effect of carbon monoxide (CO) on cerebral function of the brain exposed to various conditions in vivo*. Masters, Bar-Ilan University.

Backman, J. S., K. A. Conger, J. Chen, M. J. Ten, and J. H. Halsey. 1991. Nitric oxide helps regulate cerebral blood flow. *Journal of Cerebral Blood Flow and Metabolism* 11:S629.

Barbiro-Michaely, E., A. Mendelman, and A. Mayevsky. 2011. The evaluation of nitric oxide involvement in metrazol induced status epilepticus using multiparametric monitoring. *Brain Research* 1377:50–59.

Beckman, J. S., T. W. Beckman, J. Chen, P. A. Marshall, and B. A. Freeman. 1990. Apparent hydroxyl radical production by peroxynitrite: Implications for endothelial injury from nitric oxide and superoxide. *Proceedings of the National Academy of Sciences of the United States of America* 87:1620–1624.

Benyo, Z., G. Kiss, C. Szabo, C. Csaki, and A. G. Kovach. 1991. Importance of basal nitric oxide synthesis in regulation of myocardial blood flow. *Cardiovascular Research* 25:700–703.

Betz, E. 1972. Cerebral blood flow: its measurement and regulation. *Physiological Reviews* 52:595–630.

Black, S., and R. F. Cucchiara. 1990. Neurologic monitoring. In *Anesthesia*. Third edition. ed. R. D. Miller. New York: Churchill Livingstone.

Bolanos, J. P., and A. Almeida. 1999. Roles of nitric oxide in brain hypoxia-ischemia. *Biochimica et Biophysica Acta* 1411: 415–436.

Bolanos, J. P., and A. Almeida. 2007. Nitric oxide in regulation 7 o; mitochondrial function, respiration, and glycolysis. In: *Handbook of neurochemistry and molecular neurobiology*, eds. A. Lajtha, G. E. Gibson, and G. A. Dienel. Springer US.

Boughton-Smith, N. K., S. M. Evans, B. J. Whittle, and S. Moncada. 1993. Induction of nitric oxide synthase in rat intestine and its association with tissue injury. *Agents and Actions,* 38:C125–C126.

Brown, G. C. 1995. Nitric oxide regulates mitochondrial respiration and cell functions by inhibiting cytochrome oxidase. *FEBS Letters* 369:136–139.

Brown, G. C. 2003. Cell biology. NO says yes to mitochondria. *Science* 299:838–839.

Brown, G. C., A. G. Mcbride, E. J. Fox, K. S. Mcnaught, and V. Borutaite. 1997. Nitric oxide and oxygen metabolism. *Biochemical Society Transactions* 25:901–904.

Buchanan, J. E., and J. W. Phillis. 1993. The role of nitric oxide in the regulation of cerebral blood flow. *Brain Research* 610: 248–255.

Buerk, D. G. 2007. Nitric oxide regulation of microvascular oxygen. *Antioxidants & Redox Signaling* 9:829–843.

Bures, J., O. Buresova, and J. Krivanek. 1974. *The mechanism and applications of leao's spreading depression of electroencephalographic activity*. New York: Academic.

Buresova, O. 1957. Changes in cerebral circulationin rats during spreading EEG depression. *Physiologia Bohemoslovenica* 6:1–11.

Chance, B., P. Cohen, F. Jobsis, and B. Schoener. 1962. Intracellular oxidation-reduction states *in vivo*. *Science* 137:499–508.

Chavez, J. C., P. Pichiule, J. Boero, and A. Arregui. 1995. Reduced mitochondrial respiration in mouse cerebral cortex during chronic hypoxia. *Neuroscience Letters* 193:169–172.

Cosby, K., K. S. Partovi, J. H. Crawford, R. P. Patel, C. D. Reiter, S. Martyr, B. K. Yang, M. A. Waclawiw, G. Zalos, X. XU, K. T. Huang, H. Shields, D. B. Kim-Shapiro, A. N. Schechter, R. O. Cannon, and M. T. Gladwin. 2003. Nitrite reduction to nitric oxide by deoxyhemoglobin vasodilates the human circulation. *Nature Medicine* 9: 1498–1505.

Crouser, E. D., M. W. Julian, D. V. Blaho, and D. R. Pfeiffer. 2002. Endotoxin-induced mitochondrial damage correlates with impaired respiratory activity. *Critical Care Medicine* 30:276–284.

Culotta, E., and D. E. Koshland Jr. 1992. NO news is good news. *Science* 258:1862–1865.

Danielisova, V., M. Nemethova, and J. Burda. 2004. The protective effect of aminoguanidine on cerebral ischemic damage in the rat brain. *Physiological Research* 53:533–540.

DeSarro, G. B., E. Donato di Paola, A. deSarro, and M. J. Vidal. 1991. Role of nitric oxide in the genesis of excitatory amino acid-induced seizures from the deep prepiriform cortex. *Fundamental & Clinical Pharmacology* 5:503–511.

Dezfulian, C., N. Raat, S. Shiva, and M. T. Gladwin. 2007. Role of the anion nitrite in ischemia-reperfusion cytoprotection and therapeutics. *Cardiovascular Research* 75:327–338.

Dirnagl, U., B. Kaplan, M. Jacewicz, and W. Pulsinelli. 1989. Continuous measurement of cerebral cortical blood flow by laser-Doppler flowmetry in a rat stroke model. *Journal of Cerebral Blood Flow and Metabolism* 9:589–596.

Dora, E. 1986. Effect of theophylline treatment on the functional hyperaemic and hypoxic responses of cerebrocortical microciruclation. *Acta Physiologica Hungarica* 68:183–197.

Dora, E., T. Zeuthen, I. A. Silver, and A. G. B. Kovach. 1979. Effect of arterial hypoxia on the cerebrocortical redox state, vascular volume, oxygen tension, electrical activity and potassium ion concentration. *Acta Physiologica Academiae Scientiarum Hungaricae* 54:319–331.

Dror, S. 1994. *In vivo evaluation of brain function under carbon monoxide exposure*. Masters, Bar-Ilan.

Drummond, J. C., and Shapiro, H. M. 1990. Cerebral physiology. In *Anesthesia*, ed. R. D. Miller. New York: Churchill Livingstone.

Duckrow, R. B. 1993. A brief hypoperfusion precedes spreading depression if nitric oxide synthesis is inhibited. *Brain Research* 618:190–195.

Duelli, R., and W. Kuschinsky. 1993. Changes in brain capillary diameter during hypocapnia and hypercapnia. *Journal of Cerebral Blood Flow and Metabolism* 13:1025–1028.

Ehrenkranz, R. A., and G. Neonatal Inhaled Nitric Oxide Study. 1997. Inhaled nitric oxide in full-term and nearly full-term infants with hypoxic respiratory failure. *The New England Journal of Medicine* 336:597–604.

Eibeshitz, E., E. Barbiro-Michaely, and A. Mayevsky. 2008. The role of nitric oxide in the ischemic brain evaluated by spectroscopic monitoring of mitochondrial NADH, microcirculatory blood flow, and HbO. In *Photonics and optoelectronics meetings*, eds. Q. Luo, L. V. Wang, and V. V. Tuchin, 72801J-72801J-10. International Society for Optics and Photonics.

Erecinska, M., and I. A. Silver. 1989. ATP and brain function. *Journal of Cerebral Blood Flow and Metabolism* 9:2–19.

Estevez, A. Y., and J. W. Phillis. 1997. Hypercapnia-induced increases in cerebral blood flow: Roles of adenosine, nitric oxide and cortical arousal. *Brain Research* 758:1–8.

Fabricius, M., and M. Lauritzen. 1994. Examination of the role of nitric oxide for the hypercapnic rise of cerebral blood flow in rats. *American Journal of Physiology* 266:H1457–H1464.

Faraci, F. M., and D. D. Heistad. 1998. Regulation of the cerebral circulation: Role of endothelium and potassium channels. *Physiological Reviews* 78:53–97.

Fedele, E., and M. Raiteri. 1999. *In vivo* studies of the cerebral glutamate receptor/NO/cGMP pathway. *Progress in Neurobiology* 58:89–120.

Feihl, F., and C. Perret. 1994. Permissive hypercapnia. How permissive should we be?. *American Journal of Respiratory and Critical Care Medicine* 150:1722–1737.

Freiser, H. 1978. Ion-selective electrodes. In *Analytical chemistry*, ed. H. Freiser. New York, London: Plenum Press.

Friedli, C. M., D. S. Sclarsky, and A. Mayevsky. 1982. Multiprobe monitoring of ionic, metabolic, and electrical activities in the awake brain. *American Journal of Physiology* 243:R462–R469.

Garthwaite, J. 1991. Glutamate, nitric oxide and cell-cell signalling in the nervous system. *Trends in Neuroscience (TINS)* 14:60–67.
Gidday, J. M., and Y. Zhu. 1995. Nitric oxide does not mediate autoregulation of retinal blood flow in newborn pig. *American Journal of Physiology* 269:H1065–H1072.
Go, K. G. 1991. Intracranial pressure. In *Cerebral pathophysiology*. Amsterdam: Elsevier.
Goadsby, P. J., H. Kaube, and K. L. Hoskin. 1992. Nitric oxide synthesis couples cerebral blood flow and metabolism. *Brain Research* 595:167–170.
Guevara-Guzman, R., P. C. Emson, and K. M. Kendrick. 1994. Modulation of *in vivo* striatal transmitter release by nitric oxide and cyclic GMP. *Journal of Neurochemistry* 62:807–810.
Gursoy-Ozdemir, Y., H. Bolay, O. Saribas, and T. Dalkara. 2000. Role of endothelial nitric oxide generation and peroxynitrite formation in reperfusion injury after focal cerebral ischemia. *Stroke* 31:1974–1980; discussion 1981.
Gutierrez, G. 1993. Regional blood flow and oxygen transport: implications for the therapy of the septic patient. *Critical Care Medicine* 21:1263–1264.
Gyulai, L., E. Dora, and A. G. B. Kovach. 1982. NAD/NADH: Redox state changes on cat brain cortex during stimulation and hypercapnia. *American Journal of Physiology* 243:H619–H627.
Haberl, R. L., M. L. Heizer, A. Marmarou, and E. F. Ellis. 1989. Laser-Doppler assessment of brain microcirculation: Effect of systemic alterations. *American Journal of Physiology Heart and Circulatory Physiology* 256:H1247–H1254.
Haller, C., and W. Kuschinsky. 1987. Moderate hypoxia: Reactivity of pial arteries and local effect of theophylline. *Journal of Applied Physiology (1985)* 63:2208–2215.
Hansen, A. J. 1985. Effect of anoxia on ion distribution in the brain. *Physiological Reviews* 65:101–148.
Hansen, A. J., B. Quistorff, and A. Gjedde. 1980. Relationship between local changes in cortical blood flow and extracellular K^+ during spreading depression. *Acta Physiologica Scandinavica* 109:1–6.
Harper, A. M., V. D. Deshmukh, J. O. Rowan, and W. B. Jennett. 1972. The influence of sympathetic nervous activity on cerebral blood flow. *Archives of Neurology* 27:1–6.
Hershey, J. C., and R. F. Bond. 1993. Oxygen extraction and vascular dilation are dependently increased in skeletal muscle during canine endotoxemia. *Circulatory Shock* 40:132–138.
Hornbein, T. F. 1991. Hypoxia and the brain. In *The lung*, eds. R. G. C. J. B. West, P. J. Barnes, N. J. Cherniack, E. R. Weibel. New York: Raven Press.
Horvath, I., N. T. Sandor, Z. Ruttner, and A. C. Mclaughlin. 1994. Role of nitric oxide in regulating cerebrocortical oxygen consumption and blood flow during hypercapnia. *Journal of Cerebral Blood Flow and Metabolism* 14:503–509.
Hosseinzadeh, H., G. R. Karimi, and M. Rakhshanizadeh. 2005. Anticonvulsant effect of hypericum perforatum: role of nitric oxide. *Journal of Ethnopharmacology* 98:207–208.
Hudetz, A. G., J. J. Smith, J. G. Lee, Z. J. Bosnjak, and J. P. Kampine. 1995. Modification of cerebral laser-Doppler flow oscillations by halothane, PCO_2, and nitric oxide synthase blockade. *American Journal of Physiology* 269:H114–H120.
Iadecola, C. 1992. Does nitric oxide mediate the increases in cerebral blood flow elicited by hypercapnia?. *Proceedings of the National Academy of Sciences of the United States of America* 89:3913–3916.
Iadecola, C., D. A. Pelligrino, M. A. Moskowitz, and N. A. Lassen. 1994. Nitric oxide synthase inhibition and cerebrovascular regulation. *Journal of Cerebral Blood Flow and Metabolism* 14:175–192.
Iadecola, C., and F. Zhang. 1994. Nitric oxide-dependent and -independent components of cerebrovasodilation elicited by hypercapnia. *American Journal of Physiology* 266:R546–R552.
Ignarro, L. J. 1991. Heme-dependent activation of guanylate cyclase by nitric oxide: A novel signal transduction mechanism. *Blood Vessels* 28:67–73.
Ignarro, L. J., G. Cirino, A. Casini, and C. Napoli. 1999. Nitric oxide as a signaling molecule in the vascular system: An overview. *Journal of Cardiovascular Pharmacology* 34:879–886.

Irikura, K., K. I. Maynard, and M. A. Moskowitz. 1994. Importance of nitric oxide synthase inhibition to the attenuated vascular responses induced by topical L-nitroarginine during vibrissal stimulation. *Journal of Cerebral Blood Flow and Metabolism* 14:45–48.

Ishimura, N., K. Kitaguchi, K. Tatsumi, and H. Furuya. 1996. Nitric oxide involvement in hypoxic dilation of pial arteries in the cat. *Anesthesiology* 85:1350–1356.

Iwamoto, J., S. P. Yang, M. Yoshinaga, E. Krasney, and J. Krasney. 1992. N omega-nitro-L-arginine influences cerebral metabolism in awake sheep. *Journal of Applied Physiology (1985)* 73:2233–2240.

James, I. M., R. A. Millar, and M. J. Purves. 1969. Observations on the extrinsic neural control of cerebral blood flow in the baboon. *Circulation Research* 25:77–93.

Johnson, P. C. 1986. Autoregulation of blood flow. *Circulation Research* 59:483–495.

Jung, K. H., K. Chu, S. Y. KO, S. T. Lee, D. I. Sinn, D. K. Park, J. M. Kim, E. C. Song, M. Kim, and J. K. Roh. 2006. Early intravenous infusion of sodium nitrite protects brain against *in vivo* ischemia-reperfusion injury. *Stroke* 37:2744–2750.

Kader, A., V. I. Frazzini, R. A. Solomon, and R. R. Trifiletti. 1993. Nitric oxide production during focal cerebral ischemia in rats. *Stroke* 24:1709–1716.

Kallinen, J., A. Didier, J. M. Miller, A. Nuttall, and R. Grenman. 1991. The effect of CO_2- and O_2-gas mixtures on laser Doppler measured cochlear and skin blood flow in guinea pigs. *Hearing Research* 55:255–262.

Katsura, K., T. Kristian, R. Nair, and B. K. Siesjo. 1994. Regulation of intra- and extracellular pH in the rat brain in acute hypercapnia: a re-appraisal. *Brain Research* 651:47–56.

Kety, S. S., and C. F. Schmidt. 1948. The effects of altered arterial tensions of carbon dioxide and oxygen on cerebral blood flow and cerebral oxygen consumption of normal young men. *Journal of Clinical Investigation* 27:484–492.

Knowles, R. G., and S. Moncada. 1994. Nitric oxide synthases in mammals. *Biochemical Journal* 298(Pt 2):249–258.

Kogure, K., R. Busto, P. Scheinberg, and O. Reinmuth. 1975. Dynamics of cerebral metabolism during moderate hypercapnia. *Journal of Neurochemistry* 24:471–478.

Lancelot, E., L. Lecanu, M. L. Revaud, R. G. Boulu, M. Plotkine, and J. Callebert. 1998. Glutamate induces hydroxyl radical formation *in vivo* via activation of nitric oxide synthase in Sprague-Dawley rats. *Neuroscience Letters* 242:131–134.

Lauritzen, M. 1984. Long-lasting reduction of cortical blood flow of the brain after spreading depression with preserved autoregulation and impaired CO2 response. *Journal of Cerebral Blood Flow and Metabolism* 4:546–554.

Lauritzen, M. 1987a. Cerebral blood flow in migraine and cortical spreading depression. *Acta Neurologica Scandinavica Supplementum* 113:1–40.

Lauritzen, M. 1987b. Regional cerebral blood flow during cortical spreading depression in rat brain: Increased reactive hyperperfusion in low-flow states. *Acta Neurologica Scandinavica* 75:1–8.

Lauritzen, M., M. B. Jorgensen, N. H. Diemer, A. Gjedde, and A. J. Hansen. 1982. Persistent oligemia of rat cerebral cortex in the wake of spreading depression. *Annals of Neurology* 12:469–474.

Leão, A. A. P. 1944a. Pial circulation and spreading depression of activity in the cerebral cortex. *Journal of Neurophysiology* 7:391–396.

Leão, A. A. P. 1944b. Spreading depression of activity in cerebral cortex. *Journal of Neurophysiology* 7:359–390.

Leffler, C. W., J. S. Smith, J. L. Edrington, S. L. Zuckerman, and H. Parfenova. 1997. Mechanisms of hypoxia-induced cerebrovascular dilation in the newborn pig. *American Journal of Physiology* 272:H1323–H1332.

Lehninger, A. L. 1982. *Principles of biochemistry, Chap. 30, organ interrelationships in the metabolism of mammals*. Worthe.

Lei, S. Z., Z. H. Pan, S. K. Aggarwal, H. S. Chen, J. Hartman, N. J. Sucher, and S. A. Lipton. 1992. Effect of nitric oxide production on the redox modulatory site of the NMDA receptor-channel complex. *Neuron* 8:1087–1099.

Li, H., A. Samouilov, X. Liu, and J. L. Zweier. 2001. Characterization of the magnitude and kinetics of xanthine oxidase-catalyzed nitrite reduction. Evaluation of its role in nitric oxide generation in anoxic tissues. *Journal of Biological Chemistry* 276: 24482–24489.

Ligeti, L., K. Hines, E. Dora, T. Sinnwell, M. T. Huang, and A. C. Mclaughlin. 1991. Cerebral blood flow and metabolic rate in the conscious, freely moving rat: The effects of hypercapnia, and acute ethanol administration. *Alcoholism, Clinical and Experimental Research* 15:766–770.

Lubbe, A. S., R. N. Garrison, H. M. Cryer, N. L. Alsip, and P. D. Harris. 1992. EDRF as a possible mediator of sepsis-induced arteriolar dilation in skeletal muscle. *American Journal of Physiology* 262:H880–H887.

Macrae, I. M., D. A. Dawson, J. D. Norrie, and J. Mcculloch. 1993. Inhibition of nitric oxide synthesis: Effects on cerebral blood flow and glucose utilisation in the rat. *Journal of Cerebral Blood Flow and Metabolism* 13:985–992.

Maginniss, L. A., H. Connolly, R. W. Samsel, and P. T. Schumacker. 1994. Adrenergic vasoconstriction augments tissue O_2 extraction during reductions in O_2 delivery. *Journal of Applied Physiology (1985)* 76:1454–1461.

Manzoni, O., L. Prezeau, P. Marin, S. Desagher, J. Bockaert, and L. Fagni. 1992. Nitric oxide-induced blockade of NMDA receptors. *Neuron* 8:653–662.

Marletta, M. A. 1993. Nitric oxide synthase structure and mechanism. *Journal of Biological Chemistry* 268:12231–12234.

Mayer, B., P. Klatt, E. Bohme, and K. Schmidt. 1992. Regulation of neuronal nitric oxide and cyclic GMP formation by Ca^{2+}. *Journal of Neurochemistry* 59:2024–2029.

Mayevsky, A. 1976. Brain energy metabolism of the conscious rat exposed to various physiological and pathological situations. *Brain Research* 113:327–338.

Mayevsky, A. 1983. Multiparameter monitoring of the awake brain under hyperbaric oxygenation. *Journal of Applied Physiology* 54:740–748.

Mayevsky, A. 1984. Brain NADH redox state monitored *in vivo* by fiber optic surface fluorometry. *Brain Research Reviews* 7:49–68.

Mayevsky, A. 1990. Level of ischemia and brain functions in the Mongolian gerbil *in vivo*. *Brain Research* 524:1–9.

Mayevsky, A., R. B. Duckrow, E. Yoles, N. Zarchin, and D. Kanshansky. 1990. Brain mitochondrial redox state, tissue hemodynamic and extracellular ion responses to four-vessel occlusion and spreading depression in the rat. *Neurological Research* 12:243–248.

Mayevsky, A., K. Frank, M. Muck, S. Nioka, M. Kessler, and B. Chance. 1992. Multiparametric evaluation of brain functions in the Mongolian gerbil *in vivo*. *Journal of Basic and Clinical Physiology and Pharmacology* 3:323–342.

Mayevsky, A., T. Manor, E. Pevzner, A. Deutsch, R. Etziony, and N. Dekel. 2002. Real-time optical monitoring of tissue vitality *in vivo*. *Proceeding of SPIE* 4616:30–39.

Mayevsky, A., and H. R. Weiss. 1991. Cerebral blood flow and oxygen consumption in cortical spreading depression. *Journal of Cerebral Blood Flow and Metabolism* 11:829–836.

Mayevsky, A., E. Yoles, and N. Zarchin. 1986. Metabolic, ionic and electrical responses to oxygen deficiency in the newborn dog *in vivo*. *Advances in Experimental Medicine and Biology* 200:261–269.

Mayevsky, A., N. Zarchin, and B. Tannenbaum. 1984. Brain responses to experimental oxygen deficiency in the Mongolian gerbil. *Advances in Experimental Medicine and Biology* 180:191–202.

Mayevsky, A., T. Zeuthen, and B. Chance. 1974. Measurements of extracellular potassium, ECoG and pyridine nucleotide levels during cortical spreading depression in rats. *Brain Research* 76:347–349.

McPherson, R. W., J. R. Kirsch, R. F. Ghaly, and R. J. Traystman. 1995. Effect of nitric oxide synthase inhibition on the cerebral vascular response to hypercapnia in primates. *Stroke* 26:682–687.

McPherson, R. W., R. C. Koehler, and R. J. Traystman. 1994. Hypoxia, α2-adrenergic, and nitric oxide-dependent interactions on canine cerebral blood flow. *American Journal of Physiology* 266:H476–H482.

Meilin, S., N, Zarchin, A., Mayevsky, and S. Shapira. 1995. Multiparametric responses to cortical spreading depression under nitric oxide synthesis inhibition. In *Biochemical pharmacological and clinical aspects of nitric oxide*, eds. B. A. Weissman, N. Alon, and S. Shapira. New York: Plenum Press.

Mies, G., and W. Paschen. 1984. Regional changes of blood flow, glucose, and ATP content determined on brain sections during a single passage of spreading depression in rat brain cortex. *Experimental Neurology* 84:249–258.

Miles, P. R., L. Bowman, and L. Huffman. 1996. Nitric oxide alters metabolism in isolated alveolar type II cells. *American Journal of Physiology* 271:L23–L30.

Millar, T. M., C. R. Stevens, N. Benjamin, R. Eisenthal, R. Harrison, and D. R. Blake. 1998. Xanthine oxidoreductase catalyses the reduction of nitrates and nitrite to nitric oxide under hypoxic conditions. *FEBS Letters* 427:225–228.

Moncada, S., R. M. Palmer, and E. A. Higgs. 1991. Nitric oxide: Physiology, pathophysiology, and pharmacology. *Pharmacological Reviews* 43:109–142.

Moncada, S., M. W. Radomski, and R. M. Palmer. 1988. Endothelium-derived relaxing factor. Identification as nitric oxide and role in the control of vascular tone and platelet function. *Biochemical Pharmacology* 37:2495–2501.

Muizelaar, J. P., H. G. Van DerPoel, Z. C. Li, H. A. Kontos, and J. E. Levasseur. 1988. Pial arteriolar vessel diameter and CO_2 reactivity during prolonged hyperventilation in the rabbit. *Journal of Neurosurgery* 69:923–927.

Nagababu, E., S. Ramasamy, D. R. Abernethy, and J. M. Rifkind. 2003. Active nitric oxide produced in the red cell under hypoxic conditions by deoxyhemoglobin-mediated nitrite reduction. *Journal of Biological Chemistry* 278:46349–46356.

Nelson, D. P., C. Beyer, R. W. Samsel, L. D. Wood, and P. T. Schumacker. 1987. Pathological supply dependence of O_2 uptake during bacteremia in dogs. *Journal of Applied Physiology (1985)* 63:1487–1492.

Nilsson, B., K. Norberg, C. H. Nordstrom, and B. K. Siesjo. 1975. Influence of hypoxia and hypercapnia on CBF in rats. In *Blood flow and metabolism in the brain,* ed. M. E. A. Harper. London, New York: Edinburg.

Nohl, H., K. Staniek, B. Sobhian, S. Bahrami, H. Redl, and A. V. Kozlov. 2000. Mitochondria recycle nitrite back to the bioregulator nitric monoxide. *Acta Biochimica Polonica* 47: 913–921.

Nunn, J. E. 1987. Hypoxia. In *Applied respiratory physiolog*, ed. J. E. Nunn. London: Butterworths.

Palmer, R. M., D. S. Ashton, and S. Moncada. 1988a. Vascular endothelial cells synthesize nitric oxide from L-arginine. *Nature* 333:664–666.

Palmer, R. M., D. S. Ashton, and S. Moncada. 1988b. Vascular endothelial cells synthesize nitric oxide from L-arginine. *Nature* 333:664–666.

Pelligrino, D. A., H. M. Koenig, and R. F. Albrecht. 1993. The role of nitric oxide (NO) in cerebral arteriolar relaxation during hypoxia in the rat. *Anesthesiology* 79:A757.

Pelligrino, D. A., Q. Wang, H. M. Koenig, and R. F. Albrecht. 1995. Role of nitric oxide, adenosine, N-methyl-D-aspartate receptors, and neuronal activation in hypoxia-induced pial arteriolar dilation in rats. *Brain Research* 704:61–70.

Rampil, I. J., L. Litt, and A. Mayevsky. 1992. Correlated, simultaneous, multiple-wavelength optical monitoring *in vivo* of localized cerebrocortical NADH and brain microvessel hemoglobin oxygen saturation. *Journal of Clinical Monitoring* 8:216–225.

Rao, A. M., A. Dogan, J. F. Hatcher, and R. J. Dempsey. 1998. Fluorometric assay of nitrite and nitrate in brain tissue after traumatic brain injury and cerebral ischemia. *Brain Research* 793:265–270.

Rees, D. D., R. M. Palmer, and S. Moncada. 1989. Role of endothelium-derived nitric oxide in the regulation of blood pressure. *Proceedings of the National Academy of Sciences of the United States of America* 86:3375–3378.

Rees, D. D., R. M. Palmer, R. Schulz, H. F. Hodson, and S. Moncada. 1990. Characterization of three inhibitors of endothelial nitric oxide synthase *in vitro* and *in vivo*. *British Journal of Pharmacology* 101:746–752.

Reid, J. M., A. G. Davies, F. M. Ashcroft, and D. J. Paterson. 1995. Effect of L-NMMA, cromakalim, and glibenclamide on cerebral blood flow in hypercapnia and hypoxia. *American Journal of Physiology* 269:H916–H922.

Reivich, M. 1964. Arterial PCO_2 and cerebral hemodynamics. *American Journal of Physiology* 206:25–35.

Revelly, J. P., T. Ayuse, N. Brienza, H. E. Fessler, and J. L. Robotham. 1996. Endotoxic shock alters distribution of blood flow within the intestinal wall. *Critical Care Medicine* 24:1345–1351.

Rifkind, J. M., E. Nagababu, E. Barbiro-Michaely, S. Ramasamy, R. M. Pluta, and A. Mayevsky. 2007. Nitrite infusion increases cerebral blood flow and decreases mean arterial blood pressure in rats: A role for red cell NO. *Nitric Oxide* 16:448–456.

Rifkind, J. M., E. Nagababu, and S. Ramasamy. 2006. Nitric oxide redox reactions and red cell biology. *Antioxidants & Redox Signaling* 8:1193–1203.

Roberts Jr., J. D., J. R. Fineman, 3rdF. C. Morin, P. W. Shaul, S. Rimar, M. D. Schreiber, R. A. Polin, M. S. Zwass, M. M. Zayek, I. Gross, M. A. Heymann, and W. M. Zapol. 1997. Inhaled nitric oxide and persistent pulmonary hypertension of the newborn. The Inhaled Nitric Oxide Study Group. *The New England Journal of Medicine* 336: 605–610.

Rogatsky, G. G., A. Mayevsky, N. Zarchin, and A. Doron. 1996. Continuous multiparametric monitoring of brain activities following fluid-percussion injury in rats: Preliminary results. *Journal of Basic and Clinical Physiology and Pharmacology* 7:23–43.

Sandor, P., K. Komjati, M. Reivich, and I. Nyary. 1994. Major role of nitric oxide in the mediation of regional CO_2 responsiveness of the cerebral and spinal cord vessels of the cat. *Journal of Cerebral Blood Flow and Metabolism* 14:49–58.

Sato, A., A. Trzebski, and W. Zhou. 1992. Local cerebral blood flow responses in rats to hypercapnia and hypoxia in the rostral ventrolateral medulla and in the cortex. *Journal of the Autonomic Nervous System* 41:79–86.

Schindler, U., and E. Betz. 1976. Influence of severe hypercapnia upon cerebral cortical metabolism, CSF electrolyte concentrations and EEG in the cat. *Bulletin Europeen De Physiopathologie Respiratoire* 12:277–284.

Schmitz, B., B. W. Bottiger, and K. A. Hossmann. 1996. Brief hypercapnia enhances somatosensory activation of blood flow in rat. *Journal of Cerebral Blood Flow and Metabolism* 16:1307–1311.

Shinozuka, T., E. M. Nemoto, and P. M. Winter. 1989. Mechanisms of cerebrovascular O_2 sensitivity from hyperoxia to moderate hypoxia in the rat. *Journal of Cerebral Blood Flow and Metabolism* 9:187–195.

Siesjo, B. K. 1978a. Carbon dioxide and brain energy metabolism. In *Brain energy metabolism*, ed. B. K. Siesjo. Chichester, New York, Brisbone, Toronto: John Wiley & Sons.

Siesjo, B. K. 1978b. Circulation and oxygen consumption in the brain. In *Brain energy metabolism*, ed. B. K. Siesjo. Chichester, New York, Brisbone, Toronto: John Wiley & Sons.

Siesjo, B. K. 1978c. Hypoxia. In *Brain energy metabolism,* ed. B. K. Siesjo. Chichester, New York, Brisbone, Toronto: John Wiley & Sons.

Siesjo, B. K. 1978d. Metabolism, free energy and work – An overview. In *Brain energy metabolism,* ed. B. K. Siesjo. Chichester, New York, Brisbone, Toronto: John Wiley & Sons.

Snyder, S. H. 1992. Nitric oxide: First in a new class of neurotransmitters. *Science* 257:494–496.

Sokoloff, L., C. Kennedy, K. Adachi, F. Wang, S. Takahashi, and P. Melzer. 1992. Effects of inhibition of nitric oxide syntase on resting local cerebral blood flow and on changes induced by hypercapnia or local functional activity. In *Pharmacology of cerebral ischemia,* eds. J. Krieglstein and Oberpichler-Schwenk, H. Marburg.

Spain, D. A., M. A. Wilson, M. F. Bar-Natan, and R. N. Garrison. 1994. Role of nitric oxide in the small intestinal microcirculation during bacteremia. *Shock* 2:41–46.

Stamler, J. S., D. J. Singel, and J. Loscalzo. 1992. Biochemistry of nitric oxide and its redox-activated forms. *Science* 258: 1898–1902.

Szabo, C. 1996. Physiological and pathophysiological roles of nitric oxide in the central nervous system. *Brain Research Bulletin* 41:131–141.

Toda, N., and T. Okamura. 1992. Regulation by nitroxidergic nerve of arterial tone. *NIPS* 7:148–152.

Tolmasov, M., E. Barbiro-Michaely, and A. Mayevsky. 2009. The involvement of nitric oxide in the hemodynamic and metabolic activities of the brain and small intestine. In *SPIE BiOS: Biomedical optics,* eds. A. Mahadevan-Jansen, T. Vo-Dinh and W. S. Grundfest. International Society for Optics and Photonics.

Tonnesen, J., A. Pryds, E. H. Larsen, O. B. Paulson, J. Hauerberg, and G. M. Knudsen. 2005. Laser Doppler flowmetry is valid for measurement of cerebral blood flow autoregulation lower limit in rats. *Experimental Physiology* 90:349–355.

Traystman, R. J. 1997. Regulation of cerebral blood flow by carbon dioxide. In *Primer on cerebrovascular diseases*, ed. K. M. E. A. Welch. New York: Academic Press.

Vallet, B., N. Lund, S. E. Curtis, D. Kelly, and S. M. Cain. 1994. Gut and muscle tissue PO_2 in endotoxemic dogs during shock and resuscitation. *Journal of Applied Physiology (1985)* 76:793–800.

VanHarreveld, A., and S. Ochs. 1957. Electrical and vascular concomitants of spreading depression. *American Journal of Physiology* 189:159–166.

VanHarreveld, A., and J. S. Stamm. 1952. Vascular concomitants of spreading cortical depression. *Journal of Neurophysiology* 15:487–496.

Vyskocil, F., N. Kritz, and J. Bures. 1972. Potassium-selective microelectrodes used for measuring the extracellular brain potassium during spreading depression and anoxic depolarization in rats. *Brain Research* 39:255–259.

Wang, Q., O. B. Paulson, and N. A. Lassen. 1992. Effect of nitric oxide blockade by NG-nitro-L-arginine on cerebral blood flow response to changes in carbon dioxide tension. *Journal of Cerebral Blood Flow and Metabolism* 12:947–953.

Wang, Q., D. A. Pelligrino, V. L. Baughman, H. M. Koenig, and R. F. Albrecht. 1995. The role of neuronal nitric oxide synthase in regulation of cerebral blood flow in normocapnia and hypercapnia in rats. *Journal of Cerebral Blood Flow and Metabolism* 15:774–778.

Weiss, H. R., and J. A. Cohen. 1978. Changes in small vessel blood content of the rat heart induced by hypercapnic, hyperoxic or asphyxic conditions. *Cardiology* 63:199–207.

Yang, S. P., and J. A. Krasney. 1995. Cerebral blood flow and metabolic responses to sustained hypercapnia in awake sheep. *Journal of Cerebral Blood Flow and Metabolism* 15:115–123.

Young, J. D. 1997. Nitric oxide and related vasodilators. *Canadian Journal of Anaesthesia* 44:R23–R33.

Zhang, F., J. G. White, and C. Iadecola. 1994. Nitric oxide donors increase blood flow and reduce brain damage in focal ischemia: Evidence that nitric oxide is beneficial in the early stages of cerebral ischemia. *Journal of Cerebral Blood Flow and Metabolism* 14:217–226.

Zhang, J., H. Benveniste, B. Klitzman, and C. A. Piantadosi. 1995. Nitric oxide synthase inhibition and extracellular glutamate concentration after cerebral ischemia/reperfusion. *Stroke* 26:298–304.

Zhu, X. Z., and L. G. Luo. 1992. Effect of nitroprusside (nitric oxide) on endogenous dopamine release from rat striatal slices. *Journal of Neurochemistry* 59:932–935.

Chapter 7
Responses to Epileptic Activity Developed Under Various Pathophysiological Conditions

Abstract Epilepsy is the most common neurological disorder, affecting approximately 0.5%–2% of the general population. The condition is characterized by bursts of abnormal electrical activity in the brain, which can lead to various types of clinical seizures. A higher prevalence has been recorded among men and children. Thanks to the development of technologies such as EEG, deeper research into the mechanisms of the disorder has become possible. However, existing methods like PET and SPECT are limited in their ability to measure real-time brain changes during seizures.
The current study examined metabolic, electrical, ionic, and hemodynamic changes occurring in the brains of awake rats during epileptic seizures induced by the convulsant agent pentylenetetrazole (PTZ-Metrazol). Six experimental conditions were studied:

1. Exposure to PTZ only (control)
2. Ischemia (reduced blood flow)
3. Inhibition of NO synthesis using a non-selective inhibitor (L-NAME)
4. Inhibition of NO synthesis using a neuron-selective inhibitor (7NI)
5. Hyperoxia (inhalation of 100% oxygen)
6. Anesthesia

To conduct the study, Multiprobe Assembly (MPA) technology was used, which enables continuous measurement of mitochondrial NADH, ion concentrations (Na^+, K^+, Ca^{2+}, H^+), cerebral blood flow (CBF), cerebral blood volume (CBV), electrical activity (ECoG, DC), and temperature.
A statistical comparison was conducted among the first five groups, while the sixth group was presented qualitatively only.
Initially, there was an increase in blood flow and NADH oxidation during seizures. However, in the later seizures, energetic deterioration occurred NADH became reduced, with no change in blood flow between seizures.
There was a decrease in the number of seizures compared to the control. Metabolic function was limited, with a decrease in tissue oxygenation.

A. Mayevsky et al., *Physiological Mapping of the Cerebral Cortex in Neuropathology*,
https://doi.org/10.1007/978-3-032-13019-8_7

Reduced the number of seizures, indicating NO's involvement in seizure initiation. However, blood flow during seizures remained similar to the control group, suggesting the presence of an additional vasodilation mechanism.
Did not prevent seizures and was even accompanied by a decrease in tissue oxygenation in the later seizures. This implies that excess oxygen is not necessarily protective. Anesthesia: Upon administration of Nembutal, seizures ceased completely, and not even isolated spikes were observed on the ECoG. Spreading Depression (SD) waves: Observed in all groups, with no difference in their frequency, indicating a stable mechanism of cortical generation. Potassium ions: No significant differences were recorded between the groups in terms of concentration levels or recovery time following seizures.
This study highlights the advantage of continuous monitoring methods that simultaneously assess a wide range of brain parameters. The results indicate that the NO molecule plays a role in seizure initiation and that its inhibition may serve as an anticonvulsant mechanism. Additionally, ischemia also appears to suppress seizures through a mechanism not solely dependent on NO. Hyperoxia did not show any benefit in seizure reduction or metabolic improvement. Hemodynamic, electrical, and metabolic parameters change over time with seizure progression, indicating a gradual physiological deterioration leading toward death, which is not directly influenced by the treatments administered.

Keywords Mitochondrial activity in vivo · Extracellular potassium · Cortical spreading depression · Cerebral blood flow · Nitric oxide

7.1 Introduction

7.1.1 Energy Metabolism in the Brain

Under normal conditions, the brain consumes about 20% of the total oxygen used by the body, even though it only weighs 2% of the body weight. This oxygen consumption indicates high metabolic activity, which is expressed in the production of ATP to provide metabolic energy. The energy is used to maintain two main activities of the brain (Lehninger 1982):

1. Maintaining the resting potential of nerve cells, which allows the transmission of nerve impulses. Maintaining the resting potential is done by the sodium-potassium pump, which consumes over 50% of the total energy consumed by nerve cells (Erecinska and Silver 1989).
2. Biosynthesis of neurotransmitters and proteins.

ATP is produced by two main processes: glycolysis in the cytosol and the oxidative phosphorylation process that occurs in the mitochondria. In the process of glycolysis, a glucose molecule is converted to pyruvate, yielding two ATP molecules. The products of glycolysis pass into the citric acid cycle, which supplies electrons that recycle the carriers of the respiratory chain. The carriers are oxidized by oxygen, which is the final electron acceptor in the respiratory chain. In the process of oxidative phosphorylation, 36 additional molecules of ATP are obtained from the breakdown of glucose, for which the brain needs the aerobic pathway (Go 1991).

Since the energy reserves in the brain are small (Erecinska and Silver 1989), its normal activity depends on a continuous supply of oxygen by the bloodstream. A decrease in the supply of oxygen to the brain will disrupt the existing balance between oxygen consumption and supply, and will consequently cause a decrease in the function of the intracellular and intercellular systems of the brain tissues (Jobsis 1964; Balazs 1970). A decrease in oxygen supply can result from a decrease in blood flow to the brain (ischemia) or a decrease in oxygen in the inhaled air (hypoxia). As a result of this decrease in oxygen supply, mitochondrial activity will slow down (Mayevsky 1976) and the level of activity of the respiratory chain will slow down so that the chain components will be recycled more (increase in NADH). The decrease in ATP levels will be rapid and depend on the level of oxygen deficiency and will cause a slowdown in pump activity (Mayevsky 1984a) and the result will be the accumulation of potassium in the extracellular space and subsequently the entry of Na^+, and Ca^{2+}.

Excessive activity of the energy utilization system (ion pumping, e.g., will cause activation of mitochondrial activity and the level of NADH will decrease, meaning that NADH will be in a more oxidized state).

The rate of glucose transfer from the blood to the brain is higher than the metabolic requirements of the brain because there is an active blood–brain barrier. It is known that the rate of glucose entry into the brain is two to four times greater than the rate of its consumption by the brain tissue (Betz et al. 1976). However, the brain is an aerobic organ that is so dependent on oxygen supply that anaerobic glycolysis can never satisfy the tissue's demand for ATP (Rosental and Thomas 1989).

There is another mechanism for the formation of ATP, the creatine phosphokinase (CPK) mechanism (Watts 1973). This enzyme catalyzes a reversible transfer of phosphorus between phosphocreatine (PCr) and ATP ($PCr + ADP + H^+ = ATP + Cr$).

The mechanism by which the delicate balance between energy supply and consumption is maintained is the autoregulation of blood flow in the brain. This mechanism regulates blood flow to the brain by dilating blood vessels when cerebral perfusion pressure (CPP) to the brain decreases, and by constricting blood vessels when CPP increases (Heistad and Kontona 1983; Johnson, 1986). It is now known that there are several substances that have a direct effect on the endothelial layer of blood vessels. These substances are called endothelial-derived relaxation factors (EDRF), and they cause blood vessel dilation, including the molecule nitric oxide (NO), which plays an important role in regulating blood vessel diameter (Palmer et al. 1987). The idea of a connection between brain tissue activity and blood supply was raised many years ago (Paulson and Newman 1987). Paulson cites researchers who suggested a link between tissue activity and blood flow as early as 1890.

Since then, many studies have accumulated showing that this coupling between brain cells and blood flow has many aspects. For example, it is known that hydrogen and potassium ions, as well as adenosine and other neurotransmitters secreted by brain cells, affect the diameter of cerebral arterioles. Only potassium ions released from active neurons are transported by astrocytes, which release them into the blood vessels. Cerebral arteries and arterioles are very sensitive to changes in potassium levels and can change their diameter by up to 50% as a result of changes in the concentration of these ions (Paulson and Newman 1987). Acidosis of the CSF fluid is another example of full cooperation between the various brain cells and the blood vessels. A high concentration of hydrogen ions in the CSF fluid causes glial cells to increase their activity in order to restore the pH level to a normal state. In addition, a low pH level causes the blood vessels that supply oxygen to dilate so that increased cellular activity is possible (glial cells, including astrocytes, are drawn in blue in Fig. 7.1).

Figure 7.1 schematically depicts the relationship between brain cells, blood vessels, and intercellular fluid. The image emphasizes the brain's dependence, for its proper functioning, on the interrelationship between oxygen supply, oxygen consumption, and ATP production and consumption.

7.1.2 Scientific Background of Epilepsy

7.1.2.1 Scope of the Problem and Etiological Factors

Epilepsy is the most common neurological disorder. According to various global statistics, it affects approximately 0.5% to 2% of the general population. It is more prevalent in males than in females, and its incidence in children is twice that of adults. Seizures can begin at any age—from birth to old age—with causes varying according to age. In most cases, the disease manifests in childhood. It is found with relatively equal frequency across all races, countries, and social classes.

For many generations, epilepsy was considered a mysterious condition of supernatural origin. Seizures were thought to be the result of the actions of good or evil spirits. Consequently, the disorder was historically referred to as *Morbus demoniacus* (the disease of demons) or *Morbus sacer* (the sacred disease). In Jewish tradition, it was believed that the afflicted individual was seized by a spirit, which led to the Hebrew terms **"nichpeh"** and **"kefiyon."** Only at the end of the nineteenth century did medicine begin to be established as a science, gradually shedding its superstitious interpretations.

The British neurologist John Hughlings Jackson was the first to propose that epileptic seizures originate from electrical discharges in the gray matter—an idea he formulated decades before Hans Berger discovered brain waves and demonstrated these discharges in a measurable form.

The discovery of electroencephalography (EEG), a technique for recording the brain's electrical activity, significantly advanced our understanding of epilepsy and contributed to its demystification. German psychiatrist Hans Berger was the first to record an EEG in humans in 1932 and the first to identify specific changes in the EEG patterns of people with epilepsy. Until that point, the precise brain mechanisms underlying seizures were poorly understood (Fisher 1989; Lennox and Lennox 1960; Prince 1978).

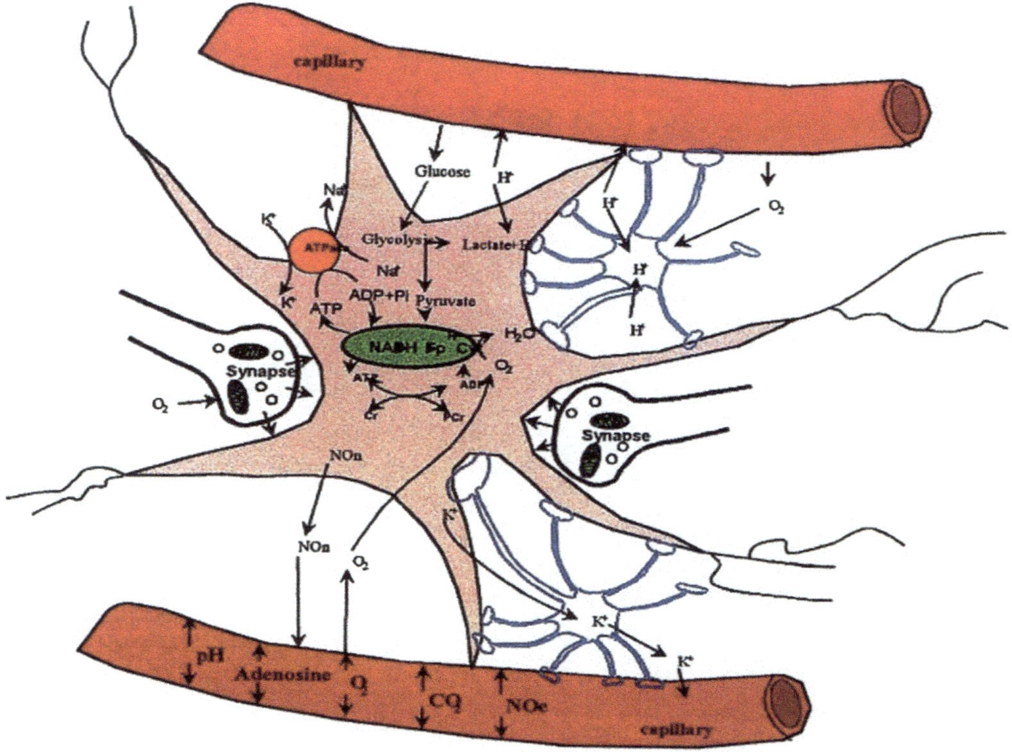

Fig. 7.1 A diagram depicting some of the cells that make up brain tissue and the processes of oxygen and glucose supply by the bloodstream, its consumption by mitochondria for ATP production, and energy consumption by ion pumps

Pathological examination of the epileptic brain may reveal sclerosis, inflammation, calcifications, tumors, abscesses, foreign bodies, and even parasites that have managed to penetrate and inhabit brain tissue. The most common findings are regions of scarring and atrophy, typically resulting from trauma, inflammation, infection, or infarction due to impaired blood supply. However, it is also known that similar brain damage can be found in individuals who do not suffer from epilepsy. In other words, there are individuals with epilepsy both with and without identifiable brain damage, just as there are individuals with brain damage who do or do not have epilepsy. Nevertheless, in many cases, the pathologist finds no pathological abnormalities in the brain—even under meticulous microscopic examination. Such cases are referred to as **primary idiopathic epilepsy**, meaning epilepsy with no detectable brain damage (Annegers et al. 1980).

The age at which epilepsy begins often provides clues regarding its etiology:

- In early childhood: perinatal brain injury.
- In school-age children: primary (idiopathic) epilepsy.
- In the third and fourth decades: trauma and tumors.
- In old age: vascular disorders and tumors (Moses III and Fisher 1988).

7.1.2.2 Terminology and Classification of Epilepsies

There are various forms of epilepsy. The International League Against Epilepsy (ILAE) has classified the different types based on their neurophysiological mechanisms and clinical manifestations (Dreifus 1981).

A. Focal (Partial) Epilepsy, including:

a. *Simple partial seizures*—focal seizures without loss of consciousness.
b. *Complex partial seizures*—also known as psychomotor epilepsy, accompanied by impaired consciousness.
c. *Partial seizures evolving into generalized seizures*—secondarily generalized tonic-clonic seizures (formerly called grand mal).

B. Generalized Epilepsy, including:

a. *Absence seizures* (petit mal)—brief episodes of impaired consciousness without falling.
b. *Generalized convulsive seizures*, including:

   Tonic-clonic seizures (grand mal),
   Tonic seizures (sustained muscle contraction),
   Clonic seizures (short, repetitive muscle contractions),
   Myoclonic-astatic seizures (brief, rapid muscle jerks throughout the body).

c. **Unclassified Seizures**—seizures that cannot be definitively categorized (Dreifus 1981).

Focal epilepsy always originates in the **cerebral cortex**. In these cases, the electrical discharge is restricted to a specific area in one hemisphere and results in seizures affecting a localized region of the body on the contralateral side.

In **generalized epilepsy**, the EEG shows generalized epileptic discharges that are recorded simultaneously and uniformly from all electrodes—i.e., across all brain regions. Research has shown that these discharges originate in the **upper brainstem**, specifically in the **medial thalamic region** bordering the walls of the third ventricle. This area belongs to the **reticular activating system**, which extends along the brain axis from the thalamus to the spinal cord. Because these reticular nuclei are centrally located, the term **centrencephalic discharges** was coined to describe seizures originating from this area. Accordingly, **centrencephalic epilepsy** is now considered synonymous with **primary generalized epilepsy** (Fromm et al. 1987).

7.1.2.3 Electroencephalography (EEG)

EEG is the only laboratory test capable of detecting and demonstrating *epileptic activity*. It is the sole diagnostic tool that can confirm the presence of epilepsy and differentiate between generalized and focal epilepsy. Moreover, EEG can indicate whether the epilepsy is idiopathic—i.e., not associated with brain damage—or secondary to underlying brain pathology. Consequently, EEG is the most critical and meaningful examination in the diagnosis of epilepsy.

Through EEG, researchers have learned that specific alterations in the brain's electrical activity are characteristic of individuals with epilepsy. There is a distinct pathological pattern in the brain waves that is diagnostically indicative of epilepsy (Niedermeyer 1987).

However, it is important to recognize two major limitations of EEG:

a. **Interictal recordings** (i.e., EEG performed between seizures) may appear entirely normal or show nonspecific abnormalities, such as background irregularities or slowing.
b. **Subclinical epilepsy** may be detected via EEG without any corresponding clinical seizures. In other words, the presence of epileptiform activity on EEG is not definitive proof that the individual suffers from epilepsy.

For these reasons, EEG results—whether from clinical practice or research laboratories—must be interpreted with caution (Goodin and Aminoff 1984).

EEG findings in epilepsy: Epileptic activity is typically characterized by the appearance of *sharp waves* and/or *spikes*. A spike is defined as a sharp, rapid waveform lasting 70 ms or less, with a rapid rise to half its peak amplitude.

- In **focal epilepsy**, spikes are observed in a localized area of the brain, while the EEG remains normal in other regions.
- In **generalized epilepsy**, epileptic activity appears simultaneously and symmetrically across all brain regions—right and left, anterior and posterior. Typically, the EEG returns to its baseline pattern within a few minutes after the seizure ends.

- In **petit mal (absence seizures)**, characteristic EEG findings include regular *spike-and-wave discharges* at a frequency of 3 Hz (three cycles per second). Most discharges last between 5 and 15 s, with a maximum duration of approximately 30 s. Only in *status petit mal* do these discharges persist for minutes, hours, or even days.
- In **grand mal (generalized tonic-clonic seizures)**, ictal EEG recordings show repeated generalized discharges of spikes, polyspikes, or spike-and-wave complexes, typically at a frequency of 15–25 per second (Fisch and Pedley 1987).

7.1.2.4 Pathophysiology of the Seizure: The Source of Abnormal EEG Activity

Normal neuronal activity in the brain is regulated through a balance between excitation and inhibition. During an epileptic seizure, cerebral neurons exhibit abnormal behavior and fire in a chaotic and uncontrolled manner, disrupting their typical functional pathways.

Three main conditions can lead to such disorganized neuronal activity:

1. **Excessive excitation**, induced by substances such as Metrazol or cocaine.
2. **Insufficient inhibition**, due to decreased levels of inhibitory neurotransmitters like GABA (e.g., as seen with vitamin B6 deficiency).
3. **Development of intrinsically hyperexcitable neurons**, which fire spontaneously and repetitively.

In each of these states, neurons are so close to their firing threshold that they discharge spontaneously from time to time (Tyner et al. 1989).

7.1.2.5 Changes in Cerebral Blood Flow and Metabolism in Epilepsy

Alterations in cerebral blood flow (CBF) and metabolism in epilepsy are of great interest both in terms of their role in the pathogenesis of neuronal damage during seizures and more recently, for their potential in localizing epileptogenic foci to guide surgical treatment.

7.1.2.5.1 Historical Foundations

John Hughlings Jackson (1835–1911) defined epilepsy as a sudden physiological hyperactivity—an excess of continuous discharges in neural tissue. He hypothesized that the metabolic rate in epileptogenic foci increases significantly due to this overactivity. This concept has been supported over the years (Ingvar 1983).

Gibbs and colleagues (Gibbs et al. 1934) used thermoelectric methods to observe increased cerebral circulation during epileptic seizures. As early as 1836, Cooper demonstrated that seizures could be induced by ligating the carotid arteries, suggesting an ischemic mechanism (Cooper 1836).

Penfield was the first to investigate seizure-related hemodynamic changes directly during surgery. As a pioneer of **respective epilepsy surgery**, he electrically stimulated the cerebral cortex in epileptic patients during craniotomy to induce seizures, enabling real-time visual observation of vascular changes in the pia mater (Penfield 1933; Penfield 1937). Although such intraoperative seizure induction is no longer performed, Penfield documented notable findings. He observed that increased cortical blood flow at the epileptic focus was accompanied by hyperemia in adjacent cortical structures. While Penfield initially attributed these changes to ischemia-induced seizures, Gibbs et al. (1934) later demonstrated that CBF actually increases prior to seizures, leading Penfield to revise his theory.

In subsequent work, Penfield and colleagues (1939) measured blood flow in humans and monkeys using Gibbs' thermocouple technique. They found that:

- Blood flow increased contralaterally to the epileptic focus during focal seizures.
- Generalized seizures led to bilateral increases in CBF.
- Basal ganglia exhibited greater perfusion increases than the cortex.
- Subcortical and white matter changes were smaller and slower.

CBF increases began 4–10 s after seizure onset and persisted beyond seizure termination. These observations established a key principle: seizures are accompanied by blood flow changes—focal in focal epilepsy and global in generalized seizures. These changes are secondary to the seizure and independent of systemic circulation (Mosier et al. 1957, White et al. 1961).

Plum et al. (1968) showed that systemic blood pressure elevation is essential for the substantial rise in CBF during generalized seizures. The conclusion: **blood flow follows the metabolic demands of brain tissue**.

These results were consistent with prior studies:

- Jasper and Erickson (1941): cortical pH decreases during seizures.
- Schmidt et al. (1945): using nitrous oxide, showed increased CBF and metabolism in monkeys.
- Plum et al. (1968) and Posner et al. (1969): cerebral oxygen consumption increases during seizures, with oxygen supply meeting the elevated demand, though some brain regions showed mismatches.

7.1.2.5.2 Contemporary Research in Hemodynamics and Metabolism

In the past two decades, intensive research has focused on cerebral metabolism and blood flow in epilepsy, both in humans and animal models.

- **In humans**: PET and SPECT imaging (Di Chiro and Brooks 1988).
- **In animals**: Autoradiographic techniques using C^{14} iodoantipyrene and C^{14} deoxyglucose.

Two factors motivated these studies:

1. Until recently, no imaging techniques allowed for real-time measurement of blood flow/metabolic changes during or after seizures. Most human recordings were interictal.
2. There was a lack of suitable chronic epilepsy animal models for real-time seizure measurement.

Modern data divides seizure-related measurements into three phases:

- Pre-ictal
- Ictal
- Post-ictal

Measurements also differ between **generalized** and **focal** epilepsy:

7.1.2.5.3 Generalized Epilepsy

- **Pre-ictal**: Normal regional cerebral blood flow (rCBF) and glucose metabolism (CMRglc) (Engel 1984; Sakai et al. 1978; Theodore et al. 1983, 1984; Touchon et al. 1986).
- **Ictal**: Significant increases in blood flow (Grant et al. 1947; Kety and Schmidt 1948) and metabolism (Engel et al. 1982b, 1983b; Mazziotta and Engel 1984), with CBF increasing by 140%–500%.
- **Post-ictal**: Suppression of CBF and metabolism.

7.1.2.5.4 Focal Epilepsy

- **Pre-ictal**: Studies show reduced CBF and metabolism in the epileptogenic focus (Engel et al. 1982a; Bonte et al. 1983; Theodore et al. 1986).
- **Ictal**: Increased CBF and metabolism during the seizure (Engel et al. 1983a).
- **Post-ictal**: Reduced CBF and metabolism within 15 min of seizure termination (Duncan 1992).

Since these hemodynamic and metabolic characteristics often correlate with EEG findings, there's a need for techniques capable of measuring rCBF with spatial and temporal resolution similar to EEG. However, **comparative studies between EEG changes and rCBF dynamics are still lacking** (Ronne-Engstrom et al. 1993).

7.1.2.5.5 Measurement Techniques

Historical techniques include:

- **Thermoelectric probes** during surgery (Penfield 1937).
- **Hydrogen clearance** (Gotoh et al. 1966).

- **Cerebral angiography** (Yarnell et al. 1974).
- **Xe133 clearance** (Hougaard et al. 1976).

Although these methods confirmed rCBF changes, they had limitations in temporal and spatial resolution. Today, PET and SPECT are commonly used, but they do not allow **continuous measurement** or reliable **ictal** recordings (Engel et al. 1983a).

7.1.2.5.6 Laser Doppler Flowmetry (LDF)

LDF is an optical method for measuring **relative microcirculation** in cerebral blood flow. It allows for continuous measurements before and during seizures.

Three key publications using LDF to measure human CBF during seizures:

- Ronne-Engstrom et al. (1993): Measured epileptic foci and showed EEG-correlated blood flow changes—either a 100% increase or a 30% decrease. This contradicted previous PET/SPECT findings, which showed only increases. The study concluded that **blood flow dynamics may vary between epileptogenic foci**.
- Nehlig et al. (1996): Studied absence seizures in GAERS rats and children. They observed a **20%–24% decrease in CBF** during seizures, despite normal or elevated pre-ictal flow. Induced seizures using kainate or picrotoxin resulted in **CBF increases of 175%–664%**.
- Carlson et al. (1991): Presented an abstract on LDF-based human seizure measurements.

7.1.2.5.7 Mitochondrial Redox State and NADH

Few studies have explored how epileptic activity affects the mitochondrial redox state, specifically the NADH/NAD ratio. Most used **fluorometric methods** to track NADH levels.

Findings include:

- Decrease in NADH levels during seizures, indicating **increased mitochondrial respiration** and energy demand (O'Connor et al. 1972b; Van Buren et al. 1978; Mayevsky 1984b).
- Correlation between extracellular potassium rise and NADH reduction (O'Connor et al. 1973).
- Changes in oxygen, carbon dioxide, pH, and NADH levels during seizures (Tenny et al. 1980).

All studies support the idea that **epileptic activity increases energy demands**, activating the sodium-potassium pump, which consumes ATP and leads to enhanced mitochondrial activity and NADH oxidation.

Mayevsky (1984b) also demonstrated that **exposure of rats to hyperbaric oxygen** triggered seizures accompanied by a **decrease in NADH** levels.

7.1.2.6 Epilepsy and the Ionic Environment

The role of changes in the extracellular ionic environment in epilepsy was first proposed by Green (1964) to explain epileptogenesis in the hippocampus. He hypothesized that excessive extracellular K^+, released from neurons, could accumulate to levels that destabilize neuronal activity and precipitate seizures. This idea was supported by findings that extracellular K^+ rises during seizures (Fertziger and Ranck 1970) and that superfusion of High-K^+ solutions can induce seizures in hippocampal preparations (Zuckermann and Glaser 1968). Recording from glial cells in animals with K^+-sensitive electrodes also demonstrated K^+ increases during focal seizures (Kuffler and Nicholls 1966).

The first direct proof of significant ionic shifts during epileptic activity came from cortical microelectrode measurements of extracellular K^+ (Prince et al. 1973). They showed that extracellular K^+ begins to rise minutes before seizure onset—both in neocortex (Prince et al. 1973) and hippocampus (Fisher et al. 1976)—reaching 10–12 mM during seizures. Though applying high-K^+ externally evokes seizures, direct equivalence to endogenous K^+ accumulation is complex. In vitro brain slices exposed to high K^+ (10–12 mM) reliably exhibit epileptiform activity (Ogata 1975).

Green's theory was refined by Dichter et al. (1972), who suggested that sustained extracellular K^+ elevation leads to neuronal depolarization block, thereby limiting the seizure propagation. Focal epilepsy in humans often corresponds with glial scarring (Falconer et al. 1964). Pollen and Trachtenberg (1970) theorized that this scarring impairs glial K^+ buffering. Conversely, Glotzner (1973) found that glial scars may enhance K^+ regulation post aluminum-induced epileptogenesis.

While much emphasis is placed on K^+, other ions shift during seizures. Calcium regulates spike genesis and membrane conductivity postdepolarization (Barrett and Barret 1976). During seizures, extracellular Ca^{2+} falls to 0.5–0.9 mM, likely due to Ca^{2+} influx into neurons. Applying similar Ca^{2+} levels to hippocampal slices triggers spontaneous potentials in CA3/CA1 pyramidal cells (Spencer and Kandel 1961).

These findings show that ion concentration shifts during heightened neural activity can be powerful drivers of epileptic events. Increases in K^+ combined with decreases in Ca^{2+} and Cl^- may facilitate the transition from interictal to ictal states, highlighting the necessity of studying ion changes before and during seizures.

7.1.2.7 Role of Glutamate and Aspartate in Epilepsy

Glutamate is the primary excitatory neurotransmitter in the brain, activating three ionotropic and three metabotropic postsynaptic receptors. Both receptor types are also present on glial cells. Glutamate and aspartate are crucial in seizure initiation, propagation, and maintenance (Meldrum 1992).

Animal experiments and patient studies have shown alterations in glutamate release and receptor function. Local administration of glutamate antagonists can modulate seizure development and spread. Antiepileptic drugs acting on NMDA and other glutamate receptors further support glutamate's pivotal role.

However, the mechanisms driving pre-ictal or ictal increases in extracellular glutamate/aspartate remain unclear. These changes likely enhance excitatory amino acid effects and activate postsynaptic glutamate receptors (Martin et al. 1991).

7.1.2.8 Spreading Depression and Epilepsy

Spreading Depression (SD)—first described by Leão (1944)—consists of:

- Cessation of spontaneous cortical electrical activity,
- A negative shift in local DC potential,
- Increases in blood flow and extracellular K^+,
- Mitochondrial NADH oxidation,
- Increased tissue reflectance.

Rapid electrical bursts can elevate extracellular $[K^+]$ to ~10–12 mM, triggering SD. In contrast, gradual epileptogenic buildup recruits neurons that activate metabolic pumps and buffer K^+ uptake, preventing SD (Koroleva and Bures 1983). Thus, epileptic activity can both induce and inhibit SD.

Prolonged interictal penicillin application elevates K^+ clearance and prevents SD, whereas brief focal penicillin exposure before seizures induces SD (Ueda and Bures 1977). Subconvulsive doses of PTZ (40 mg/kg IP) result in SD, but higher doses that cause generalized seizures do not, suggesting SD disrupts epileptic activity (Koroleva and Bures 1983). Mayevsky (1984b) further observed SD waves in hyperbaric-oxygen-exposed rats, suggesting SD may function to abort seizure activity.

7.1.2.9 Selection of Epilepsy Models

Epilepsy research relies heavily on animal and in vitro models due to ethical constraints on human experimentation. Thousands of compounds cannot be screened clinically; thus, much of our current understanding derives from animal studies (Fisher 1989).

Although many species exhibit spontaneous seizures, they are often sporadic and unsuited to experimental design (Löscher and Schmidt 1988). Therefore, multiple models are utilized because:

1. No single model fully recapitulates human epilepsy; multiple models ensure validity.
2. A single model may represent more than one epilepsy type (Fisher 1988).

Key research questions include:

1. Why do seizures begin, spread, and terminate? Is there a "central pacemaker"?
2. Why do seizures vary in phenotype?
3. What causes the latent period?
4. What are the mechanisms of antiepileptic drugs (Fisher 1989)?

Thus, model selection must align with the specific research question.

7.1.3 Induction of Seizures in Various Pathological and Physiological Conditions

7.1.3.1 Induction of Seizures Using Pentylenetetrazol (PTZ—Metrazol)

Pentylenetetrazol (Metrazol) is considered a prototype compound (Papy and Naquet 1971; Starzl et al. 1953), among systemic convulsant agents. PTZ is a tetrazole derivative (Stone 1970) that exhibits proconvulsant activity in mice, cats, and primates following intravenous administration. Initially, Metrazol induces isolated convulsions that progress into persistent seizures and may culminate in generalized clonic-tonic seizures. EEG recordings show spike or polyspike wave activity (Hout et al., 1973).

The mechanism of action of Metrazol remains only partially understood. Early studies suggested that its effects were primarily cortical (Ajmone-Marsan 1972; Bircher et al. 1962), while more recent findings indicate that nonspecific neurons in the reticular formation are activated by systemic Metrazol prior to cortical neurons. Mirski and Ferrendelli (1986, 1987) demonstrated the critical role of the mammillary bodies, anterior thalamus, and mammillothalamic tracts in mediating Metrazol-induced seizures in mice and guinea pigs. The administration of γ-vinyl-GABA, a GABA degradation inhibitor, into the reticular formation, the anterior medial hypothalamus, and the caudal hypothalamus blocked PTZ-induced seizures in rats (Miller et al., 1987). In contrast, in cats, the clonic phase appears to be mediated by the forebrain (Magistris et al. 1988).

At the synaptic level, Metrazol appears to interact with the GABA receptor-benzodiazepine-chloride ionophore complex (Olsen 1981), likely reducing inhibitory tone and thus promoting seizure activity. A reduction in GABAergic inhibition (Macdonald and Barker 1977), or activation of NMDA receptors (Nevins and Arnolde 1989), have both been implicated in the initiation and general mechanisms of PTZ-induced seizures. Wilson and Escueta (1974) proposed that Metrazol acts by blocking GABA-mediated inhibition. Notably, benzodiazepine antagonists are generally ineffective in preventing Metrazol-induced seizures (Chavoix et al. 1988), suggesting the involvement of additional mechanisms.

7.1.3.2 Ischemia

Cerebral ischemia is defined as a disturbance in normal blood flow to the brain. The literature distinguishes between two types of ischemia: global and focal. Global ischemia refers to a reduction in blood flow to the entire brain, whereas focal ischemia pertains to localized disruptions affecting specific brain regions. In both cases, blood flow to other organs (e.g., the heart) remains unaffected, and the observed responses are attributable solely to cerebral ischemia. In the current study, global ischemia was modeled via occlusion of the carotid arteries in rats. Cerebral blood supply is derived from four major arteries: two carotid arteries located in the neck on either side of the trachea, and two vertebral arteries located in the posterior region of the neck.

These arteries converge in the region known as the **Circle of Willis**, which provides full collateral circulation. In rats, the Circle of Willis is intact, ensuring complete collateral compensation between the carotid and vertebral arteries.

As the brain is entirely dependent on a continuous oxygen supply, cessation of blood flow causes immediate damage to neural tissue, particularly through disruption of mitochondrial function (Chance et al. 1962, 1973; Mayevsky 1984a; Mayevsky and Chance 1982). This energy failure results in widespread ionic disturbances, including increased extracellular potassium and decreased extracellular calcium concentrations. These ionic shifts occur in two phases: an initial slow potassium leakage into the extracellular space, followed by a rapid and global depolarization event (Mayevsky 1990; Mayevsky et al. 1986).

Models investigating the pathogenesis of tissue damage in the ischemic penumbra suggest that maintaining a minimal threshold of cerebral blood flow is essential for preventing secondary injury. These studies highlight correlations among blood flow levels, ion pump activity, and tissue damage (Revett et al. 1998). Sick et al. reported that when cerebral blood flow drops below 15 ml/g/min, extracellular potassium levels rapidly rise from ~3 mmol/L to 50 mmol/L (Sick et al. 1998). Below this threshold, there is also a substantial increase in the release of excitatory amino acids such as glutamate (Hakim 1998). Excess extracellular glutamate leads to uncontrolled activation of NMDA and AMPA receptors, promoting the influx of sodium and calcium ions and resulting in cellular damage.

Over the past decade, substantial evidence has accumulated indicating that elevated intracellular calcium levels during ischemia play a central role in secondary injury (Shuaib and Breker-Klassen 1997; Siesjö and Siesjö 1996; Siesjo and Bengtsson 1989). These studies suggest that increased intracellular calcium leads to mitochondrial overload and ultimately to neuronal death.

Another mechanism implicated in ischemia/reperfusion injury involves the formation of free radicals. These reactive species initiate destructive processes such as phospholipid peroxidation, protein degradation or inactivation, and mitochondrial dysfunction (Hall 1997; Lee 1995; Sims 1992; Sims and Zaidan 1995). Recent studies also suggest that neuronal death following ischemia results not only from necrosis but also from apoptotic mechanisms (Freeman and Crapo 1982; Guglielmo et al. 1998; Li et al. 1998; Maeda et al. 1998), indicating that ischemia may trigger DNA fragmentation and programmed cell death in neurons.

7.1.3.2.1 Ischemia and Epilepsy

In many stroke patients, a chronic epileptic focus develops at the infarct site, typically emerging 1–2 weeks poststroke. These findings have led to the conclusion that ischemic risk factors increase the likelihood of developing epileptic seizures (Engle and Pedley 1998).

Animal studies corroborate these observations. In one study, gerbils developed hippocampal and cortical seizures 10 min after bilateral carotid artery occlusion (Marciani et al. 1993). Another study using the same occlusion protocol reported seizure onset and neuronal loss in the CA1 region of the hippocampus within 10 min (Grazia Marciani et al. 1990). These findings suggest that epileptogenesis may involve synaptic alterations following ischemic injury. Both global and focal cerebral ischemia, as well as traumatic brain injury, are associated with a rapid rise in extracellular levels of glutamate, aspartate, glycine, and other amino acids, beginning 5–20 min before the ischemic or traumatic event. This increase may, in part, be linked to calcium released from nerve terminals. However, most of the elevation results from reversed amino acid transport and is associated with gradual ionic gradient changes (K^+, Na^+) or anoxic depolarization.

The precise role of glutamate and aspartate in ischemia-induced and trauma-induced epileptogenesis remains unclear. It is not yet known whether these amino acids act as risk factors for later-onset epilepsy. What is known, however, is that pharmacological blockade of NMDA, AMPA, and kainate receptors during or after ischemia significantly protects against both focal and global ischemic damage (Meldrum 1990).

7.1.3.3 Nitric Oxide (NO)

Until 1987, nitric oxide (NO) was primarily known as a toxic gas. However, in 1987, it was discovered to play a physiological role in the relaxation of smooth muscle (Maines 1993), functioning via an increase in intracellular cGMP levels. This discovery marked a significant breakthrough, revealing that such a small and radical molecule can act as a neurotransmitter. NO has since been shown to be active in blood vessels, the immune system, the liver, the pancreas, the peripheral nervous system, and in various processes within the brain (Maines 1993).

The NO molecule exists in three forms: the cation (+NO), the anion (−NO), and the radical form (NO*), each exhibiting distinct reactivity with oxygen, metals, and other NO molecules (e.g., forming 2NO). From a biological perspective, the radical form (NO*) is of principal interest.

7.1.3.3.1 Synthesis and Inhibition of NO

L-Arginine ⟶ L-Citrulline + Nitric Oxide (NO) (in the presence of O_2 and NADPH)

This reaction is catalyzed by the enzyme **Nitric Oxide Synthase (NOS)**, with NADPH and oxygen serving as cofactors (Marletta 1993). There are three known NOS isoenzymes: two **constitutively expressed NOS (c-NOS)**, and one **inducible NOS (i-NOS)**, which is upregulated in response to cytokines and/or bacterial lipopolysaccharides.

The three isoenzymes are partially membrane-bound and partially cytosolic proteins:

1. **Endothelial Constitutive NOS (e-NOS)**: Found in vascular endothelial cells and activated by acetylcholine. Inhibition of this enzyme leads to vasoconstriction (Moncada et al. 1989).

2. **Neuronal Constitutive NOS (n-NOS)**: Synthesizes NO in neurons.
3. **Inducible NOS (i-NOS)**: Expressed in activated macrophages (Knowles and Moncada 1994).

c-NOS enzymes are calcium-dependent, while i-NOS is calcium-independent (Iadecola et al. 1994). Importantly, c-NOS produces very small quantities of NO (picomolar range), whereas i-NOS generates much larger amounts (nanomolar range), reflecting their physiological versus pathological roles.

Various inhibitors selectively target these isoenzymes. Their inhibitory potency depends on both the specific enzyme and the inhibitor used. These inhibitors prevent the conversion of L-arginine to L-citrulline by acting as structural analogs that competitively bind to NOS in place of L-arginine (Iadecola et al. 1994).

- **L-NAME** (N^ω-Arginine-Methyl-Ester): Highly effective in inhibiting brain NOS but also strongly inhibits e-NOS (Rees et al. 1990).
- **7NI** (7-nitroindazole): A specific inhibitor of n-NOS in the brain.
- Other effective inhibitors include **L-NA** (Nitro-L-Arginine) and **L-NMMA** (N-monomethyl L-Arginine).

7.1.3.3.2 The Role of NO in the Brain

Since NO is synthesized from L-arginine by NOS, researchers have used NOS inhibitors to study NO's function by blocking its production. Experimental use of such inhibitors led to reduced cerebral blood flow without affecting glucose utilization or oxygen consumption. These findings suggest that while NO maintains basal vascular tone, it does not directly regulate cerebral energy metabolism (Iadecola et al. 1994). Inhibiting NOS was found to reduce cerebral blood flow by 20%–50%, a range that reflects species-specific differences and the type of NOS inhibitor used (Prado et al. 1992; Backman et al. 1991). Administration of L-NAME caused decreased blood flow and increased systemic blood pressure, effects reversed by subsequent L-arginine injection (Rees et al., 1990, el Karib et al., 1993).

L-NAME and L-NMMA both elevate blood pressure by inhibiting e-NOS, which complicates their use in studies of the central nervous system (CNS). In contrast, **7NI** selectively inhibits brain NOS and does not affect acetylcholine-induced endothelial dilation, making it more suitable for studying CNS function. In addition to altered blood flow, inhibitor use led to increased blood viscosity and vascular morphological changes, resulting in localized blockages in both cortical and subcortical regions (Prado et al. 1992). Smooth muscle cells in the vascular walls regulate vessel diameter and are indirectly influenced by NO. NO activates **guanylate cyclase**, increasing cGMP synthesis. Elevated cGMP inhibits smooth muscle contraction by reducing intracellular calcium levels (Galla 1993). Because NO regulates cGMP—a key mediator in intracellular signaling that activates specific protein kinases-NO is crucial for intracellular and intercellular communication in the brain (Maines 1993).

7.1.3.3.3 Regulation of cGMP Synthesis

1. **Brain NOS** is regulated by cofactors NADPH and oxygen, which determine the rate of NO synthesis.
2. **cGMP levels** depend on the activity of guanylate cyclase, itself regulated by intracellular calcium (Mayer et al. 1992).

Furthermore, NO facilitates the release of neurotransmitters such as **acetylcholine** and **glutamate** (Guevara-Guzman et al. 1994; Zhu and Luo 1992).

7.1.3.3.4 The Role of NO in Epileptic Seizures

Research on NO's role in epilepsy has yielded contradictory findings. Some studies suggest that NO acts as an **anticonvulsant** (Buisson et al. 1993; Przegaliński et al. 1994), while others indicate a **pro-convulsant** effect (Mülsch et al. 1994; De Sarro et al. 1991). These discrepancies likely stem from differences in experimental conditions, animal models, epileptogenic agents, and seizure types.

Two key arguments support NO's involvement in seizure generation:

1. **Excitatory amino acids**, such as NMDA and kainate, are potent seizure inducers. NO has been shown to activate NMDA receptors (Garthwaite 1991).
2. **Epileptogenic agents**—e.g., excitatory amino acids, pentylenetetrazol (PTZ), and pilocarpine (PIL)—increase cerebellar cGMP levels, while anticonvulsants reduce them. Since NO raises cGMP levels, it may play a pro-convulsant role (Wood 1991).

Conversely, some studies propose that NO has an **anticonvulsant function** during seizures induced by excitatory amino acids (Buisson et al. 1993; Haberny et al. 1992; Rondouin et al. 1993). This effect may arise from a **negative feedback mechanism**, whereby NO production-triggered by NMDA receptor activation—ultimately inhibits further receptor activity, possibly via competitive antagonism at the NMDA binding site (Manzoni et al. 1992). In the present study, we used the NOS inhibitors **L-NAME** and **7NI** to investigate NO's role in epileptic seizures. L-NAME inhibits both neuronal and endothelial NOS, whereas 7NI selectively inhibits neuronal NOS.

Given that L-NAME-induced inhibition of endothelial NOS elevates blood pressure and thereby activates neurons (Przegalinski et al. 1994), we also employed 7NI in this study due to its lack of cholinergic or cardiovascular effects, making it more appropriate for studying the central nervous system.

7.1.3.4 The Effect of Hyperoxia on Brain Function

Hyperoxia refers to a state in which the tissue oxygen level exceeds the normoxic range. This condition is induced by ventilating the subject with 100% oxygen. Variations in

PaO_2 between 60 and over 300 mmHg exert only a modest influence on cerebral blood flow (CBF). Below a PaO_2 of 60 mmHg, CBF increases rapidly, while at higher PaO_2 values, CBF decreases moderately. At one atmosphere of O_2, cerebral blood flow is reduced by approximately 15% (Drummond and Shapiro 1990).

Under conditions of hyperbaric oxygen (1.5 ATA), oxygen becomes neurotoxic within the central nervous system (CNS). This toxicity is manifested by convulsions, which may lead to death. Toxic hyperoxia induces various biochemical reactions in the CNS, including changes in the concentrations of amino acid neurotransmitters and increased production of reactive oxygen species (ROS) and oxidation of essential enzymes.

Experiments conducted in our laboratory show that the initial response to hyperoxia is vasoconstriction, resulting from autoregulatory mechanisms. This response leads to a reduction in CBF and an increase in tissue reflectance. As expected, tissue pO_2 rises, leading to oxidation of NADH. Under hypertoxic conditions, Na^+/K^+-ATPase pump activity may be either inhibited or activated, and there is a very slow increase in extracellular potassium. In ECoG recordings under hyperbaric hyperoxia, tonic seizures were observed due to the toxic effects of oxygen. Posthypoxic ventilation with 100% O_2 facilitated recovery from hypoxia and significantly increased tissue pO_2 levels (Mayevsky 1984b).

7.1.3.5 The Effects of Barbiturate Anesthesia

Barbiturates are a class of compounds used for anesthesia in both clinical and experimental animal settings. All barbiturates are derivatives of barbituric acid, and their mechanism of action involves binding to GABA receptors—the primary inhibitory neurotransmitter in the mammalian CNS. Binding of GABA to its receptor opens chloride channels, allowing Cl⁻ ions to enter the cell, leading to hyperpolarization and inhibition of postsynaptic neurons (Fragen and Avram 1990).

Barbiturates decrease the dissociation rate of GABA from its receptor, thereby prolonging the opening duration of the chloride channel. At higher concentrations, barbiturates can directly activate the channel even in the absence of GABA (Franks and Lieb 1994).

7.1.3.5.1 Anesthetics

Animal anesthesia is associated with reductions in cerebral blood flow (CBF), as well as in oxygen and glucose consumption in brain tissue (Siesjo 1978; Nilsson and Siesjo 1975). The extent of the effect depends on the anesthetic dose, and under extreme conditions, both CBF and cerebral metabolic rate of oxygen ($CMRO_2$) can decrease by up to 50%. Deep anesthesia also reduces the amplitude of spontaneous EEG activity, sometimes resulting in near-complete signal suppression (Lam et al. 1995).

7.1.3.5.2 Synthesis of Barbiturate Anesthetics

This study employed pentobarbital, an anesthetic belonging to the barbiturate family. Barbituric acid is synthesized by combining urea with malonic acid. Substituting the two hydrogen atoms at carbon-5 of barbituric acid with alkyl or aryl groups yields various barbiturate derivatives with sedative properties. Barbiturates exist in both ionic and nonionic forms in solution. They cross cell membranes in their nonionic state but exert local anesthetic effects in their ionic form.

7.1.3.5.3 Effect of Anesthetics on Brain Energy Metabolism

Barbiturates are general CNS depressants. They inhibit neural activity, skeletal and smooth muscle function, and cardiac muscle, while also reducing oxygen consumption in various mammalian tissues. Additionally, they inhibit mitochondrial activity in preparations from liver and brain cells (Goodman and Gilman 1970).

Barbiturates exert four main effects on the CNS:

1. Hypnosis
2. Reduction in cerebral metabolic rate ($CMRO_2$)
3. Decreased CBF due to increased cerebrovascular resistance
4. Anticonvulsant activity

Due to these properties, barbiturates are among the most commonly used agents in neuroanesthesia. The most effective dose is one that reduces cerebral metabolism and CBF until EEG activity reaches an isoelectric point. At this point, $CMRO_2$ is reduced by approximately 50%. The reduction in metabolism slightly exceeds the decrease in blood flow, thereby ensuring that metabolic supply exceeds demand (Aaslid 1987; Droste et al. 1989; Harders et al. 1989; Ibayashi et al. 1991; Ngai et al. 1988). Since barbiturates induce vasoconstriction only in normal tissue, they help redistribute blood flow from healthy to ischemic regions. Additional barbiturate effects may include suppression of free radicals and attenuation of cerebral edema following ischemic brain injury. These actions support the hypothesis that barbiturates offer neuroprotection. Further evidence suggests that barbiturates reduce maximum transmembrane conductance of potassium and sodium and influence synaptic transmission by rendering postsynaptic membranes unresponsive to acetylcholine-induced depolarization (Morgan and Mikhail 1992).

7.1.4 Conclusion

The EEG is the only laboratory test capable of visualizing and demonstrating epileptic activity. It remains the sole diagnostic tool that can confirm epilepsy and classify its type. Since epileptic seizures influence cerebral blood flow—typically increasing it—it is essential to understand the relationship among EEG activity, blood flow, mitochondrial state (as indicated by NADH levels), and the ionic environment (i.e., pump activity).

To date, most epilepsy studies have assessed CBF using PET and SPECT techniques. These methods, however, do not permit continuous measurement and are less reliable during seizures. Therefore, the use of Laser Doppler Flowmetry (LDF) is crucial, as it allows for real-time measurement of CBF before, during, and after seizures.

Few studies have examined NADH levels during seizures. This study combined continuous, real-time monitoring of electrical, hemodynamic, metabolic, and ionic activity during and after seizures. Given that chronic epileptic foci are often observed in stroke survivors (an ischemic event), examining the relationship between epilepsy and ischemia was warranted. Status epilepticus (SE) is known to be associated with significant increases in CBF and tissue pO_2. Research indicates that neurons are the source of nitric oxide (NO), which contributes to vasodilation during seizures. However, the role of NO remains debated: some argue it provokes seizures, while others view it as anticonvulsant. These contrasting perspectives underscore the need to examine the relationship between NO and epilepsy by pharmacologically inhibiting NO synthesis.

During SE, changes occur in both CBF and tissue oxygenation. Consequently, this study aimed to evaluate whether altering oxygen balance—by ventilating with 100% O_2—might improve oxygenation during seizures and exert a protective effect, or whether hyperoxia would instead exacerbate seizure activity. Since barbiturates are known neural suppressants and serve as effective anticonvulsants, it was also important to assess the effects of anesthetic administration during SE. This study sought to determine whether seizure activity would cease following anesthetic injection and how the monitored physiological parameters would respond.

7.2 Methods

7.2.1 Measurement Techniques

7.2.1.1 In Vivo Mitochondrial NADH Measurement by Fluorometry

This method is based on the spectral absorption difference between NADH and NAD^+. The exposed brain surface is illuminated with light at a wavelength of 366 nm, using an air-cooled 100-watt mercury lamp. Focusing lenses and a band pass filter isolate the desired wavelength (366 nm). Light transmission to and from the brain is performed via quartz optical fibers arranged in a Y-shaped configuration. One arm delivers the excitation light (366 nm) to the brain, while the other collects emitted signals and directs them to a dichroic mirror. Ninety percent of the emitted light passes through a 450 nm band pass filter and reaches a photomultiplier, which detects the fluorescence emitted by excited NADH (Chance et al. 1962). The remaining 10% passes through a 366 nm filter to detect reflected light from the tissue. Both fluorescence and reflectance signals are amplified via a high-voltage power supply and converted to electrical signals by the photomultipliers. These signals are then calibrated and recorded using a polygraph and data acquisition system.

Since tissue absorbance is influenced by hemodynamic changes (vasoconstriction/vasodilation), ion movement, or water shifts—factors affecting both fluorescence and reflectance—changes in fluorescence may not necessarily reflect mitochondrial NADH dynamics. To correct this, reflectance changes are subtracted from fluorescence on a 1:1 ratio, yielding a corrected fluorescence signal (CF) (Mayevsky 1984a): **CF = Flu − R**

7.2.1.2 Continuous Measurement of Local Cerebral Blood Flow Using Laser Doppler Flowmetry (LDF)

Tissue is irradiated with light at a wavelength of 632.8 nm. Light scattered by static tissue maintains its wavelength, while light scattered by moving red blood cells is Doppler-shifted in wavelength and frequency. The extent of the Doppler shift correlates with both red blood cell number and velocity, reflecting changes in blood flow. The backscattered light is collected by optical fibers and directed to a photodetector, which converts photons into electrical pulses. These signals contain frequency and intensity information, though the output is inherently relative and cannot be absolutely calibrated (Haberl et al. 1989; Dirnagl et al. 1989). A Perimed PF2B (Sweden) LDF system was used in this study.

7.2.1.3 Measurement of Extracellular Ion Concentrations Using Ion-Selective Microelectrodes

Extracellular concentrations of potassium (K^+), calcium (Ca^{2+}), and hydrogen (H^+) ions were measured using ion-selective microelectrodes. Ion selectivity is achieved via a membrane incorporating specific ion carriers. Electrodes for K^+ and Ca^{2+} were filled with a solution containing 3 mM KCl, 1.2 mM $CaCl_2$, and 149.8 mM NaCl. The pH electrode was filled with a pH 7 buffer. All electrodes were supplied by WPI (World Precision Instruments), USA.

7.2.1.4 Measurement of Direct Current (DC) Potential

Cortical DC potential was measured between the brain and a reference electrode placed externally. Each ion-selective electrode (K^+, Ca^{2+}, H^+) was filled with a physiological solution and connected to a Ag/AgCl electrode for signal transduction.

7.2.1.5 Reference Electrode

The reference electrode was identical to those used for DC potential measurement but placed subcutaneously in the animal's neck, between the skin and muscle.

7.2.1.6 Temperature Measurement

Body and brain temperatures were measured using calibrated thermistors (YSI, USA). Brain temperature was recorded with a needle thermistor (model 400) integrated into the measurement system and placed on the brain surface. Rectal thermometry was used to monitor core body temperature.

7.2.1.7 Electrocorticography (ECoG)

Spontaneous cortical electrical activity was recorded bipolarly using two stainless steel electrodes embedded in the measurement assembly. The potential difference between the electrodes reflects cortical electrical activity.

7.2.1.8 Intracranial Pressure (ICP) Measurement

ICP was recorded using the Camino V420 ICP Monitoring System (Camino, USA). Pressure changes were detected via a membrane at the tip of an optical fiber, which altered light transmission based on applied pressure. All measurement techniques and technical specifications are described in detail in previous publications (Mayevsky 1984a, 1990a, 1992).

7.2.2 Measurement System

7.2.2.1 Multiprobe Assembly (MPA)

The MPA enables simultaneous recording of metabolic, ionic, and electrical brain activity in one hemisphere of an awake or anesthetized rat. It allows for multiparametric measurements including:

1. **Direct measurement of cerebral blood flow (LDF) and volume (LDV)** via laser Doppler flowmetry.
2. **Assessment of mitochondrial metabolic state** through in vivo NADH fluorescence.
3. **Extracellular ion concentrations** (K^+, Ca^{2+}, H^+) using specific minielectrodes.
4. **DC potential recording**.
5. **ECoG recording** of spontaneous cortical electrical activity.
6. **Body and brain temperature monitoring** using thermistors.

All components were mounted on a custom-fabricated Perspex cannula surgically implanted on the cortical surface of one hemisphere (see Fig. 7.2). All 16 channels of data acquired during experiments were recorded via a **National Instruments A/D interface card**, with **LabVIEW software** for synchronized acquisition and qualitative/

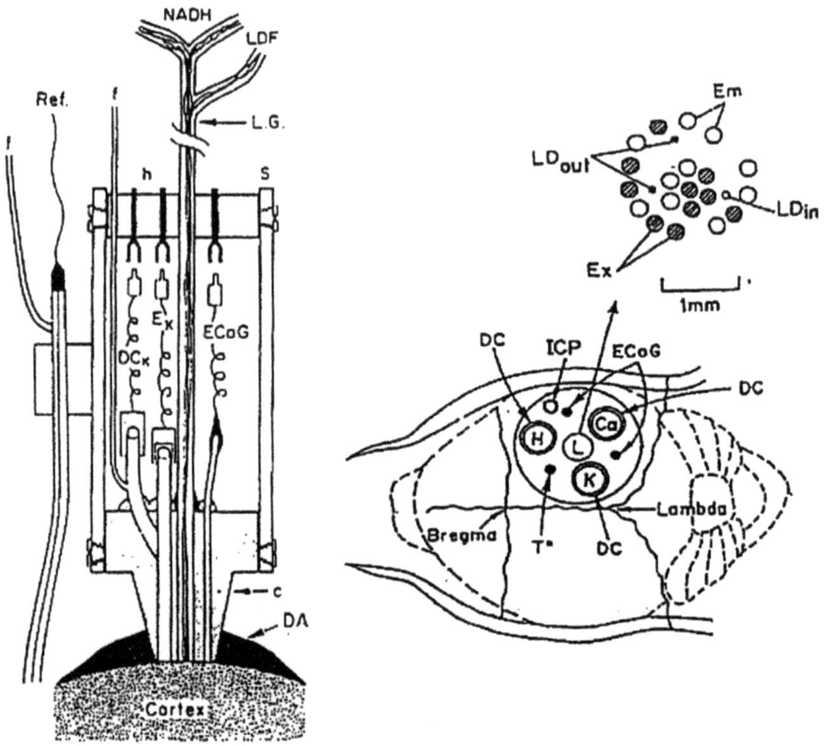

Fig. 7.2 Schematic image of the MPA multiparameter system: NADH, the two NADH fiber bundles; LDF, fibers for measuring blood flow and blood volume; Ref, reference electrode; f, filling tube for the reference electrode or DC electrode; S, aluminum tube; h, connector holder. ECoG electrodes for measuring spontaneous electrical activity on the surface of the cortex; EH, EK, ECa, ion-specific electrodes; DC, area where the DC potential is measured; C, plexiglass cannulas; DA, dental cement; Cortex, brain surface area; T°, thermistor for measuring temperature in the measurement area; Em, Ex, fibers for measuring excitation and emission of NADH; LD_{in}, LD_{out}, optical fibers for measuring blood flow and blood volume

quantitative analysis. Simultaneously, 12 parameters were monitored and documented using a **Grass 12-channel polygraph**.

7.2.3 Methods for Calculating the Results

The parameters recorded on the computer and on the polygraph are expressed in mV values. In order to evaluate the ionic changes throughout the experiment, the mV values were converted to ion concentration values (mM), while the reflectance, NADH and blood flow values were converted to percentages. The values presented in the section discussing the results are the average within the experimental group. The standard error values and the means are found in the tables. A *t*-test was also performed to check for significance between the experimental group and the control group.

Formulas for calculating the relative change of various parameters throughout the experiment:

1. Calculating the Reflectance in percentages:

$$R_x(\%) = \frac{R_x[mV] - R_{bl}[mV]}{500} \times 100$$

R_x—value measured at a desired point.
R_{bl}—reflectance value at the time of the control.
500—the measurement ranges in mV set for all experiments.

2. Calculation of NADH in percent:

The change in NADH is measured according to the CF values and the formula is the same as the reflection formula.

$$CF_x(\%) = \frac{CF_x[mV] - CF_{bl}[mV]}{500} \times 100$$

CFx—value measured at a desired point.
CF_{bl}—corrected fluorescence value at the time of the control.
500—the measurement ranges in mV set for all experiments.

3. Calculation of blood flow in percent:

$$CBF_x(\%) = \frac{CBF_x[mV] - CBF_{death}[mV]}{CBF_{bl}[mV] - CBF_{death}[mV]} \times 100$$

CBFx—value measured at a desired point relative to the control.
CBF_{bl}—value measured at the control point.
CBF_{death}—value measured at the death point. This value is considered the physiological 0-point.

4. Calculation of ion concentration in mM:

The sensitivity of the electrodes to the specific ion was found to be close to the values of the Nernst equation. That is, a range of 50–60 mV for a tenfold change in ion concentration for potassium and hydrogen ions and 25–30 mV for calcium ions.

The sensitivity of the electrodes was determined using four different salt solutions. The first solution is the solution in which the potassium and calcium electrodes were also filled (1.2 mM $CaCl_2$, 3 mM KCl, 149.8 mM NaCl) and was therefore set as 0 voltage. The second solution contains 0.12 mM $CaCl_2$, 30 mM KCl, 123.9 mM NaCl meaning that the potassium ion concentration is ten times greater and the calcium ion concentration is ten times less than their concentration in the electrode. The third solution is 7 pH, which is also the pH electrode, and the fourth is 6 pH, so that the hydrogen ion concentration is ten times higher. The conversion formulas for each point in mV to mM are as follows:

$$a = \frac{(K\,[mV] - K_0\,[mV])/1}{Range}\quad,\quad K^{+}[mM] = 3\times 10^{a}$$

$$z = \frac{(Ca[mV] - Ca_0\,[mV])/2}{Range}\quad,\quad Ca^{2+}[mM] = 1.2\times 10z$$

$$x = \frac{(pH[mV] - pH_0\,[mV])/1}{Range}\quad,\quad pH[mM] = 7 + x$$

$K[mV]$, $Ca[mV]$, $pH[mV]$—the value at the measurement point in mV.

$K_0[mV]$, $Ca_0[mV]$, $pH_0[mV]$—the value in mV at the -0 point.

$K[mM]$, $Ca[mM]$, $pH[mM]$—the ion value after calculation in mM.

Range—The range obtained by the potassium and calcium electrodes when moving from the first solution to the second, and the range obtained by the hydrogen electrode when moving from the third to the fourth solution.

7.2.4 Work Steps

7.2.4.1 Preparing the Animal for the Experiment

The experiments were conducted on male Wistar rats weighing 180–250 g. The animals were anesthetized by injection of Equithesin (each ml contains: Chloral hydrate 42.51 mg; Magnesium sulfate 21.25 mg; Alcohol 11.5%; Propylene glycol 44.34%; Pentobarbital sodium 9.72 mg) injected IP (Intraperitoneal) at a dose of 0.3 ml/100 g. The skull was exposed by cutting the skin of the skull at the line connecting the nostrils and the nape. The animal was placed on the operating table with its belly, and the head was fixed using a head holder. The skull was exposed and a 6 mm diameter hole was drilled in the parietal bone. The dura was gently removed over the exposed area, and the MPA system was placed on the brain using a micromanipulator. In addition, four holes for reinforcement screws were drilled in the skull bone: three on the side contralateral to the measurement system and one in the frontal bone on the same side as the hole for the MPA system was drilled. After the surgical phase and the placement of the system on the brain, acrylic cement was used to strengthen and fix the system.

7.2.4.2 Experimental Protocols

A typical experimental protocol:

1. After surgery, the animal was released from the immobilization and data recording began. After about 10–15 min from the moment the animal was released from the immobilization, the animal was ventilated with 100%-2N in order to determine the maximum NADH level. It is known that breathing the animal with nitrogen for 20–30 s does not cause brain (Mayevsky et al. 1990b; Mayevsky 1984a, 1990c) and therefore we used it as a standard method in all animals.
2. Waiting about an hour and a half until the animal wakes up and recovers from the surgery.
3. Placing the animal in the cage.
4. In order to prevent the disconnection of the measurement system during seizures, the MPA system was connected to the manipulator, which made the system more stable during seizures, and the rat was unable to disconnect the system from the brain.
5. Intravenous injection of Metrazol at a concentration of 10 mg every 5–10 min until continuous seizures developed.
6. Measurement ended when the animal died spontaneously.

Protocol 1: Intravenous injection of Metrazol at a concentration of 10 mg every 5–10 min until continuous seizures developed (Fig. 7.3).

Control experiment: After the animal woke up and recovered from the surgery, the animal was placed in a cage, the system was fixed using a manipulator, and saline was injected instead of Metrazol. The duration of the experiment was the same as the longest survival time of the animals injected with Metrazol.

Protocol 2: Same as Protocol 1 except that 24 h before the start of the experiment, the rats underwent occlusion of two carotid arteries (Fig. 7.4).

Control experiment: Same as the control experiment of Protocol 1, except that 24 h before the start of the experiment, the rats underwent occlusion of two carotid arteries.

Protocol 3: Same as Protocol 1, except that 40 min before the injection of Metrazol, L-NAME was injected at a concentration of 50 mg/kg into the abdominal cavity (IP) (Fig. 7.5).

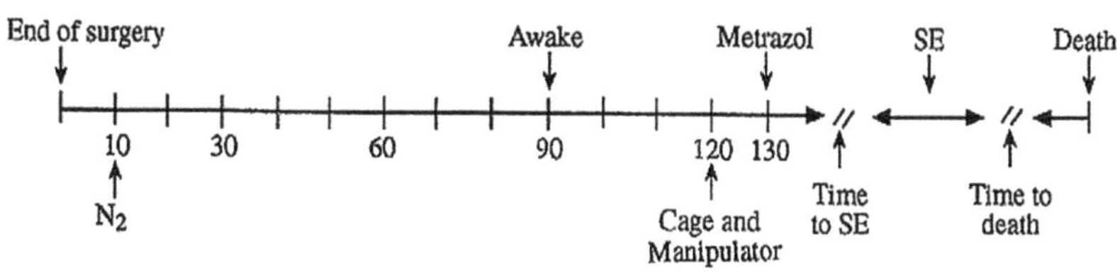

Fig. 7.3 Protocol describing the course of the experiment (in minutes) in awake rats injected with Metrazol

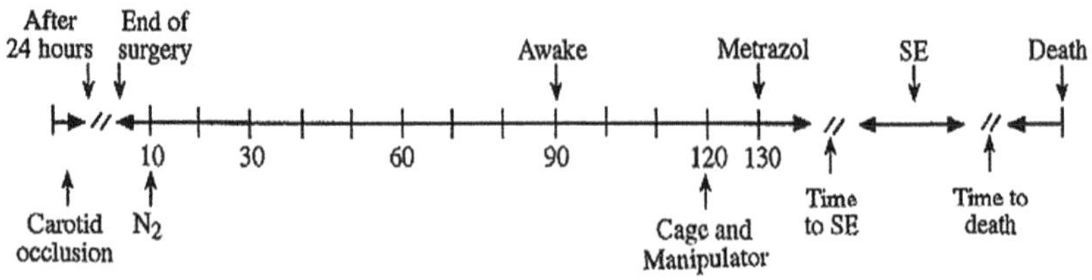

Fig. 7.4 Protocol describing the course of the experiment (in minutes) in awake ischemic rats injected with Metrazol

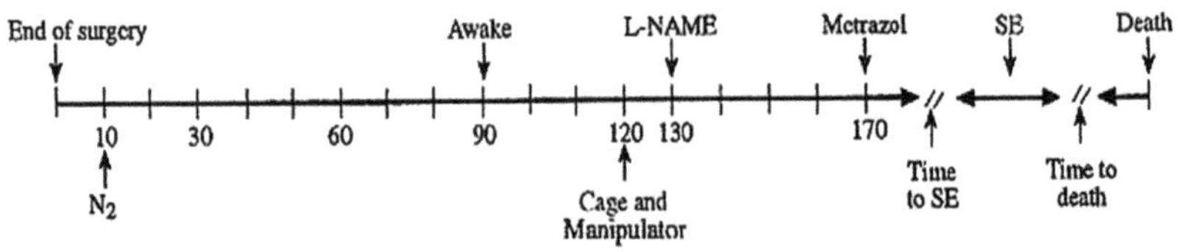

Fig. 7.5 Protocol describing the course of the experiment (in minutes) in rats injected with L-NAME and Metrazol

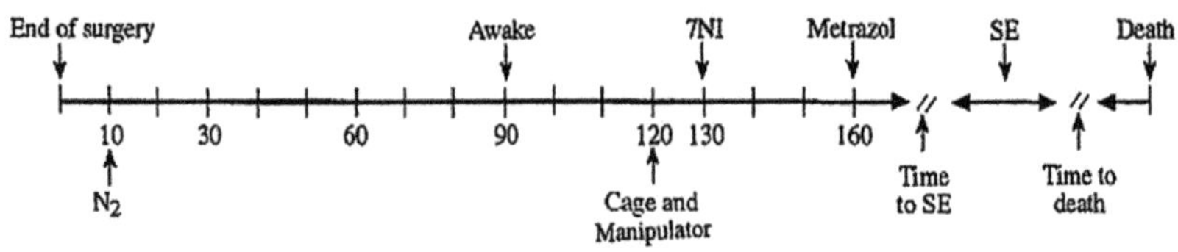

Fig. 7.6 Protocol describing the course of the experiment (in minutes) in rats injected with 7NI and Metrazol

Control experiment: After the animal woke up, L-NAME was injected and 40 min later, saline was injected (instead of Metrazol).

Protocol 4: Same as Protocol 1, except that 30 min before Metrazol injection, 7NI at a concentration of 30 mg/kg into the abdominal cavity (Fig. 7.6).

Two control experiments were performed in this group because 7NI is not soluble in saline. In order to dissolve the substance, two conditions were required: dissolution in peanut oil and sonication. Therefore, the two control experiments were as follows:

Control experiment No. 1: After the animal woke up, 7NI (the substance diluted in oil) was injected into the abdominal cavity (IP) at a concentration of 50 mg/kg and after 30 min, saline was injected instead of Metrazol.

Control experiment No. 2: After the animal woke up, 1 ml of peanut oil was injected into the abdominal cavity and after 30 min, saline was injected instead of Metrazol. (The purpose of the above control was to examine whether peanut oil had any effect.)

Protocol 5: Same as Protocol 1 except that 40 min before Metrazol injection, the animal was placed in another cage and ventilated with 100% oxygen throughout the entire duration of the experiment (Fig. 7.7).

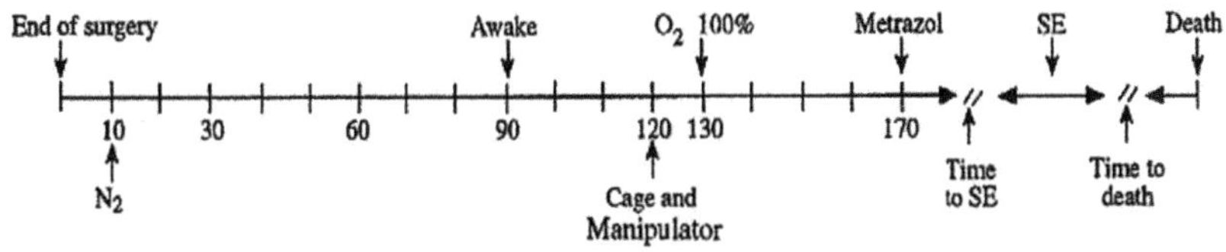

Fig. 7.7 Protocol describing the course of the experiment (in minutes) in rats ventilated with 100% oxygen and injected with Metrazol

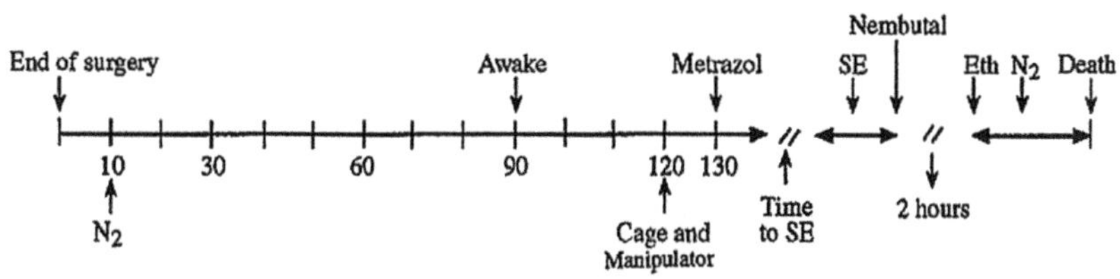

Fig. 7.8 Protocol describing the course of the experiment (in minutes) in rats that were injected with Metrazol, and in the SE condition, they were injected with Nembutal

Control experiment: After the animal woke up, the animal was placed in a cage and ventilated with 100% oxygen, and after 40 min, saline was injected (instead of Metrazol), and the animal was ventilated with 100% oxygen for the same duration of the experiment.

Protocol 6: Same as Protocol 1 except that once the animal reached a state of seizures Continuous (SE) Nembutal was injected intraperitoneally (IP) at a concentration of 40 mg/kg. Since the animal continued to live, the administration of the anesthetic breathed the animal with nitrogen until the end of the measurement (Fig. 7.8).

Control experiment: After the animal woke up, saline was injected (instead of Metrazol and at a time more or less the same as the experiment). Nembutal was injected intraperitoneally at a concentration of 40 mg/kg.

7.3 Results

This study examined the effects of sustained epileptic seizures—status epilepticus (SE)—induced by Metrazol, a prototypical convulsant agent, on cerebral energy metabolism, electrical activity, ionic changes, and hemodynamic responses. These parameters were measured using the MPA (Multiparametric Analysis) system.

We investigated how various pathophysiological conditions (e.g., ischemia, hyperoxia) influence seizure development, particularly focusing on the brain's ability to maintain energetic balance during seizures and whether these conditions exacerbate or attenuate seizure activity. Additionally, we assessed the involvement of nitric oxide (NO) in seizure generation by inhibiting NO synthesis using L-NAME (a general

NOS inhibitor, targeting both neuronal and endothelial isoforms) or 7-NI (a selective neuronal NOS inhibitor).

We also examined whether Nembutal, a known antiepileptic drug, could terminate seizures induced by Metrazol and how it affected the measured cerebral parameters.
The results are presented in three parts:
Part I: Model Development
Part II: Seizure Induction under Various Pathological and Physiological Conditions
Part III: Statistical Comparison across Experimental Groups

7.3.1 Model Development

Approximately 50 preliminary experiments were conducted to establish a stable and reproducible protocol. The objective was to induce sustained epileptic seizures (rather than a single episode) using an appropriate concentration and injection protocol of Metrazol. This would allow reliable comparisons across metabolic, hemodynamic, ionic, and electrical parameters. The selected dose needed to reflect all MPA system variables without compromising any single parameter—a problem encountered in several preliminary trials.

Preliminary experimental groups included:

- **Group 1**: Rats administered Metrazol intraperitoneally (IP) at varying doses (Table 7.1).
- **Group 2**: Rats administered Metrazol intravenously (IV) using various concentrations and injection methods (bolus vs. infusion) (Table 7.2).
- **Group 3**: Rats receiving Metrazol via IV infusion at 70 mg/kg over one hour (Tables 7.3I and 7.3II).

Following analysis, the final protocol involved IV injections of 10 mg Metrazol every 5–10 min in awake rats until the first seizure occurred. If the initial seizure did not lead to additional seizures, another injection was administered.
Group 4 (Table 7.4) represents the finalized protocol and was compared against earlier groups. This injection method and dosage were deemed optimal for the study. To illustrate the characteristics of the seizures, spreading depression (SD), and status epilepticus (SE), a representative recording from the finalized protocol is presented prior to detailing the preliminary tables.

Figures 7.9, 7.10, and 7.11 show a representative recording from one of ten awake rats injected with Metrazol IV. A total of 40 mg was administered in 10 mg increments. The first injection triggered immediate SD, evidenced by suppression of spontaneous ECoG activity, a drop in DC potential, increased cerebral blood flow (CBF), elevated extracellular potassium, mitochondrial NADH oxidation, and increased reflectance. Later, seizures developed, accompanied by 18 additional SD waves. (Note: In later phases of the experiment, some parameters ceased responding to SD events, so SD identification was based on DC signal depression.)

Table 7.1 IP injection of Metrazol at various concentrations

Exp. no.	Dose of Metrazol (mg/kg)	Time (min) to first SD wave after inj.	Number of SD waves	Response of NADH and CBF during SD wave	Time (min) to first Seizure after inj.	Type of ECoG response before death	Symbols explanation
1	3 × 75	3	10	V	55	85	C
2	2 × 70	2	15	V	61	108	B
3	3 × 70	17.5	3	V	20	39	C
4	1 × 70	–	–	–	–	N_2	–
5	1 × 70 and at the end of exp. Administration of Metrazol until death	55	1	–	–	76	D
6	Administration of the substance until seizure	29	3	+	57	67	C
7	210	28	1	+	–	54	D
8	200	26	8	+	61	N_2	–
9	2 × 100	34	15	+	34	316	C

+: Metabolic and ionic responses during SD waves V: Metabolic and ionic responses during SD wave after second injection –: No response B: ECoG with high amplitude C: Strong epileptic attack and death D: ECoG with spikes and not depressed

Both seizures and SD waves displayed similar metabolic, hemodynamic, and ionic responses, although seizure-related changes were milder in amplitude and shorter in duration across all parameters. During SD events, NADH was initially oxidized, with a switch to reduction in later waves (see "flipped NADH" in the figure). From the third injection onward, seizures co-occurred with SD, and following the fourth injection, sustained seizures (SE) emerged. Comparing SD and seizure responses—e.g., SD wave #8—shows that both potassium and CBF waves appear in pairs. The first wave (seizure) is shorter and of lower amplitude, while the second wave (SD) is longer and of higher amplitude. At times, seizure-associated waves are obscured by SD waves (e.g., wave #11) (Table 7.5).

Eight minutes before death, there is a marked increase in NADH, decreased CBF, and elevated potassium, accompanied by large ECoG amplitude. Death ensues shortly thereafter. The experiment is segmented into the following phases, which serve as analytical anchors throughout the study (Table 7.6):

1. **Baseline (Control) Period**: Before Metrazol injection.
2. **Treatment Period**, further divided into:

   (a) Time from injection to first seizure.
   (b) Seizure period—sustained seizure activity (SE).

Table 7.2 Various experimental protocols of IV injection of Metrazol, duration at various concentrations and type of injection

Exp. no.	Route of injection	Dose of Metrazol (mg/Kg)	Time (min) to first SD wave after Inj.	Number of SD waves	Response of NADH and CBF during SD waves	Time (min) to first seizure after inj.	Time to spontaneous death	Type of EcoG response before dath
1	IV—Manual injection, injection duration 10 min	70	14	Many	–	14	160	b
2	IV—Manual injection, in a bolus of nine times	200	23	10	–	70	Given Nembutal	–
3	IV—Manual Injection of 0.2 ml every 10 min for an hour	37	43	5	+	45	106	a
4	IV—Infusion during 1 h	37	28	5	◆	60	70	b
5	IV—Infusion during 68 min	75	21	7	–	48	Min. 128	b
6	IV—Infusion during 85 min	50	10	38	•	55	Given Nembutal	–
7	IV—Infusion during 83 min	85	26	6	–	55	Nembutal given as a result of extremely hard SE	b
8	IV—Infusion during 46 min	70	12	8	–	45	73	d

Symbol 1: In the fourthwave, hemodynamic and ionic responses Symbol 2: In the sixth wave hemodynamic and ionic responses +: Metabolic and ionic responses during SD waves –: No response a: Repeated epileptic attacks b: ECoG with high amplitude d:ECoG with spikes and not depressed.

Table 7.3I One hour IV infusion of Metrazol (70 mg/kg)

Experiment no.	Time (min) to first SD wave after injection	Number of SD waves	Response of NADH and CBF during SD waves	Time (min) to first seizure after injection	Time to spontaneous death	Type of EcoG response before death
1	10	14	+	54	130	a
2	5	9	+	23	60	b
3	25	8	$	45	92	d
4	25	5	$	29	56	c
5	25	6	+	42	52	a
6	23	6	X	23	96	c
7	18	8	X	44	69	b
8	19	4	X	24	116	b
9	18	6	X	40	100b	
10	26	12	X	70	112	b
11	23	7	X	44	87	c
12	21	6	X	43	70	b

a: Type of seizure b: ECoG with wide amplitude c: Strong epileptic attack and death d: ECoG with spikes and not depressed X: Only ion responses during SD waves +: Metabolic and ionic responses during SD waves $: Ionic and CBF responses without NADH change

Table 7.3II Injection of Metrazol into vein (IV) by infusion at 70 mg/kg within 1 h

Experiment no.	Time (min) to first SD wave after injection	Number of SD waves	Response of NADH and CBF during SD waves	Time (min) to first seizure after injection	Time to spontaneous death	Type of EcoG response before death
1	14	7	+	47	72	b

b: ECoG with wide amplitude +: Metabolic and ionic responses during SD waves

(c) Deterioration period—marked by increased NADH, reduced CBF, and elevated potassium.
(d) Death following seizures.

3. **Survival Time**: Entire treatment period from first Metrazol injection to death.

Analysis of both preliminary and finalized protocols revealed four distinct ECoG response patterns preceding death (Fig. 7.12):

(A) Repeated consecutive seizures.
(B) High-amplitude ECoG.
(C) Intense seizure leading to suppressed ECoG with spikes and death.
(D) Unrepressed ECoG with spikes.

Table 7.4 IV injection of 10 mg Metrazol into awake rats until continuous attacks

Exp. no.	Dose of Metrazol (mg/kg)	Time (min) to first SD wave after injection	No. of SD waves	Responses of NADH and CBF during SD waves	The SD wave that NADH response was reversed	Time (min) to first seizure after injection	Exp. in which NADH increased. CBF decreased and K^+ increase before death	Time to spontaneous death	Type of EcoG response before death
1	50	20	14	+	11	27	–	222 ETH given IV	a
2	60	8	3	+	–	32	+	67	b
3	80	6	4	+	4	18	+	46	c
4	30	20	19	+	10	21	+	146	b
5	40	9	4	+	–	27	+	35	c
6	55	68	10	+	5	68	+	112	a
7	45	1	15	+	4	1	+	180	c
8	45	49	12	+	2	49	+	199	c
9	50	21	4	+	4	30	+	74	B
10	50	–	–	+	–	21	–	68	c

a: Continuous attacks b: EcoG with amplitude c: Strong epileptic attack and death +: – –: –

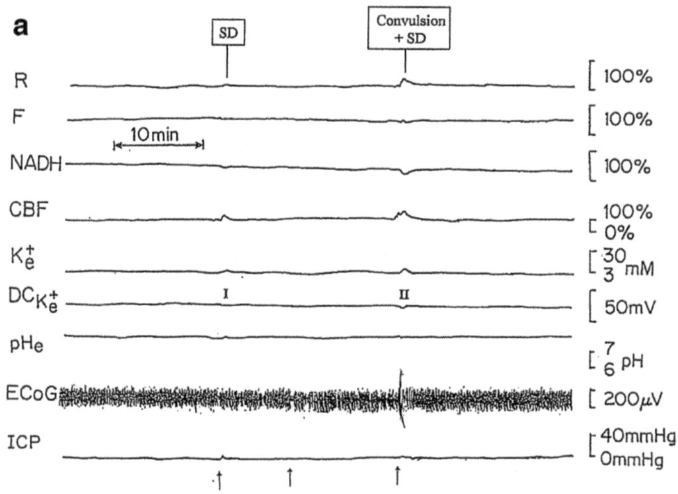

Fig. 7.9 Illustration of metabolic, hemodynamic, ionic, and electrical activity over time in a representative awake rat, injected with Metrazol. The parts a, b and c are in a continuation of the time scale. R: Reflectance at 366 nm; F: Fluorescence; NADH: Mitochondrial NADH levels; CBF: Cerebral blood flow; pHe, K^+_e: Extracellular ion concentrations; DCK^+_e: Surface potential over cortex at potassium electrode ECoG; Spontaneous cortical electrical activity; ICP: Intracranial pressure; ↑: 10 mg Metrazol IV injection

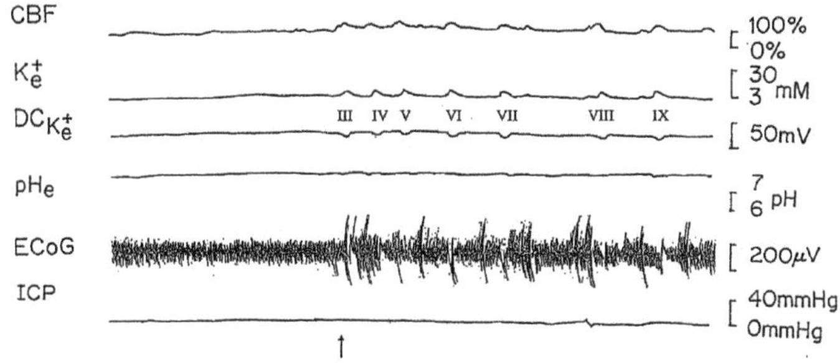

Fig. 7.10 Illustration of metabolic, hemodynamic, ionic, and electrical activity over time in a representative awake rat, injected with Metrazol. The parts a, b and c are in a continuation of the time scale. R: Reflectance at 366 nm; F: Fluorescence; NADH: Mitochondrial NADH levels; CBF: Cerebral blood flow; pHe, K^+_e: Extracellular ion concentrations; DCK^+_e: Surface potential over cortex at potassium electrode ECoG; Spontaneous cortical electrical activity; ICP: Intracranial pressure; ↑: 10 mg Metrazol IV injection

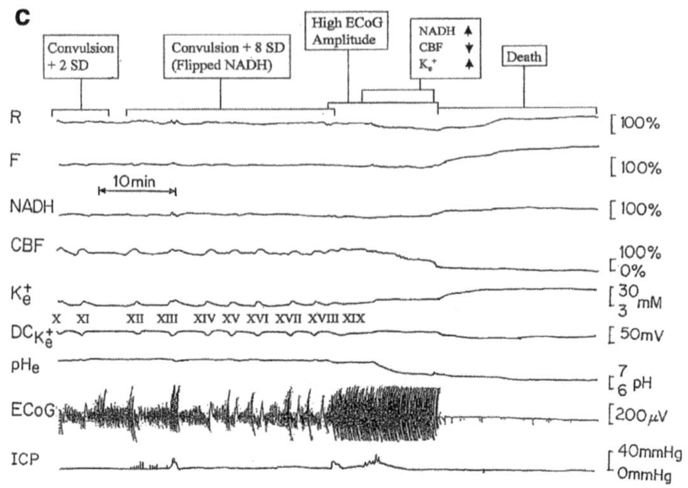

Fig. 7.11 Illustration of metabolic, hemodynamic, ionic, and electrical activity over time in a representative awake rat, injected with Metrazol. The parts a, b and c are in a continuation of the time scale. R: Reflectance at 366 nm; F: Fluorescence; NADH: Mitochondrial NADH levels; CBF: Cerebral blood flow; pHe, K^+_e: Extracellular ion concentrations; DCK^+_e: Surface potential over cortex at potassium electrode ECoG; Spontaneous cortical electrical activity; ICP: Intracranial pressure; ↑: 10 mg Metrazol IV injection

Table 7.5 NADH amplitude (%) in first and last waves

Treatment group	Percentage NADH level	
	First wave	Second wave
Metrazol	−16 ± 1.76	0.5 ± 2.77***
Ischemia	−12.5 ± 6.5	185.13*
L-NAME	−26 ± 8.18	−12 ± 5.46*
7NI	−28.8 ± 2.08	−11.3 ± 3.0***
Hyperoxia	−25.3 ± 5.35	−6.6 ± 2.45*

$*p < 0.05$ $***p < 0.001$

7.3.1.1 Summary of Preliminary Findings

Group 1 (IP injection of Metrazol)

- A single 70 mg/kg IP dose failed to induce seizures or SD. Repeated doses triggered both epileptiform activity and SD.
- At lower doses, SD elicited only ionic changes. At higher doses, metabolic responses emerged.
- The number of SD waves, latency to first seizure, survival time, and ECoG response before death varied widely.

Table 7.6 Duration (min) of first and last wave (according to K^+ wave)

Treatment group	Duration (min) K^+ wave	
	First wave	Last wave
Metrazol	2.16 ± 0.14	3.45 ± 0.76
Ischemia	3.23 ± 0.67	6.98 ± 1.96*
L-NAME	3.41 ± 0.36	5.11 ± 0.64*
7NI	3.44 ± 0.43	4.83 ± 1.08
Hyperoxia	3.00 ± 0.40	5.21 ± 0.61*

*$p < 0.05$ comparison in the same group

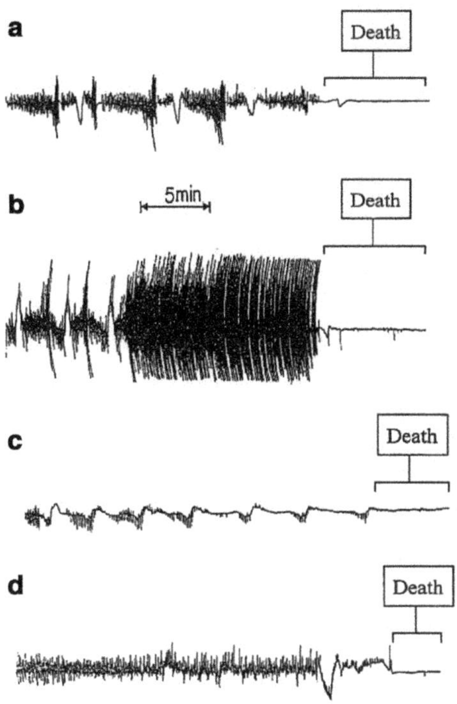

Fig. 7.12 Four typical responses of the ECoG traces recorded before the death of the rats

- No sustained seizures (SE) were achieved, indicating poor reproducibility and control with IP injections.

Group 2 (IV injection—bolus/infusion at varying doses)

- Multiple SD waves were observed, but most showed only ionic—not hemodynamic—responses.

- First seizure latency was fairly consistent; SE and eventual death followed in all cases.
- This supported IV injection as preferable and led to fixed-dose infusion trials.

Group 3 (70 mg/kg IV infusion over 1 h)

- High reproducibility across experiments: similar SD latency, number of SD waves, time to first seizure (~40 min), and SE development.
- However, metabolic responses were largely absent during SD waves.
- Control experiments confirmed intact system function (Table 7.3II).
- Two hypotheses were proposed: (1) dose too low, or (2) slow infusion damages mitochondria, inhibiting response.

Group 4 (10 mg IV injections every 5–10 min)

- This protocol consistently triggered seizures ~40–50 min postinitial injection.
- All trials showed SD with both metabolic and ionic responses.
- Consistent SE development was achieved.
- This method was adopted for the remainder of the study.

7.3.2 Seizure Induction Under Pathological and Physiological Conditions

7.3.2.1 Metabolic, Hemodynamic, Ionic, and Electrical Responses in Awake Rats Injected with Metrazol

In this group, awake rats were injected intravenously (IV) with 10 mg Metrazol every 5–10 min until sustained seizures (status epilepticus) developed. Experiments were conducted on ten experimental animals and six control animals. Control rats, also awake, were injected with saline instead of Metrazol (see Protocol No. 1 in the Materials and Methods section).

7.3.2.1.1 Analog Recordings

The representative analog recording of this experimental group is shown in Fig. 7.9 in Part A of the results, as it was used to illustrate the expected phenomena (epileptic seizures, spreading depression (SD), and the pattern of SE) observed in preliminary experiments. Figure 7.9 displays the response of one of the ten experimental rats receiving Metrazol injections at 5–10 min intervals until the first seizure occurred (see detailed explanation, Figures 7.9, 7.10, and 7.11).

Figure 7.13 presents a representative recording from one of six control animals (awake rats injected with saline). Throughout the entire recording period, all measured parameters remained stable. The ECoG trace is presented separately from the metabolic and physiological parameters.

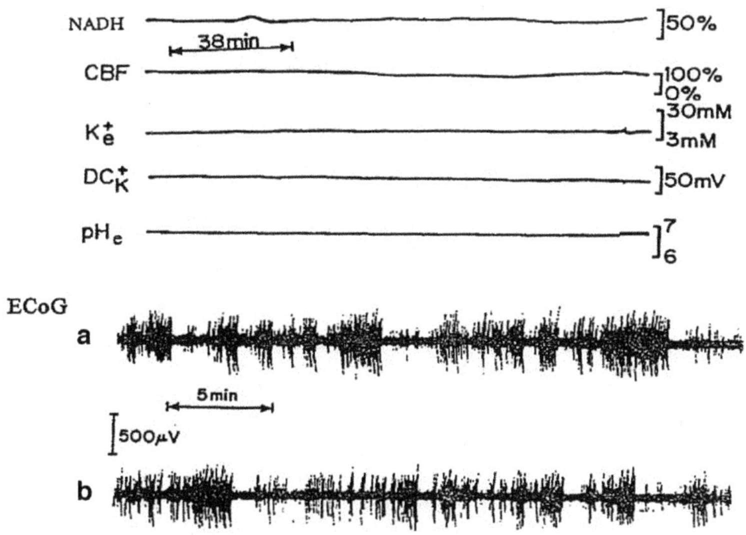

Fig. 7.13 Control recording of metabolic, hemodynamic, ionic, and electrical activity over 3 h in an awake rat. NADH: Mitochondrial NADH levels; CBF: Cerebral blood flow; pHe, K^+_e: Extracellular ion concentrations; DCK^+_e: Surface cortical potential at the potassium electrode; ECoG (a, b): Spontaneous cortical electrical activity recorded at two time points in the same rat

7.3.2.1.2 Quantitative Analysis

All experimental data were quantitatively analyzed, and the experimental group was compared to its corresponding control group. In the experimental group, Metrazol was injected every 5–10 min until SE was achieved (although only the first injection is marked in the figure). The control group consisted of awake rats that received saline instead of Metrazol.

In Fig. 7.14, three parameters are presented: **NADH (%), CBF (%)**, and **Delta ICP (mmHg)**. Data were collected every 10 min throughout the experiment. Each point on the graph represents the mean of ten experimental animals (black filled squares) or six control animals (open circles). Error bars represent standard error (SE). Asterisks indicate statistically significant differences between the groups ($p < 0.05$).

Seventy minutes after the first Metrazol injection, a significant drop in NADH was observed, lasting approximately 60 min. As the animals deteriorated near death, NADH levels began to rise. CBF increased slightly over time in the experimental group but was not statistically significant due to similar changes in control animals. ICP differences were physiologically minimal and statistically insignificant.

Figure 7.15 displays **Delta K^+ (mM), DCK^+ (mV)**, and **Delta pH**. Potassium levels showed no significant differences except for a late rise near death. DCK^+ differed significantly 120 min after Metrazol injection, persisting for 30 min. A nonsignificant decline in pH was observed.

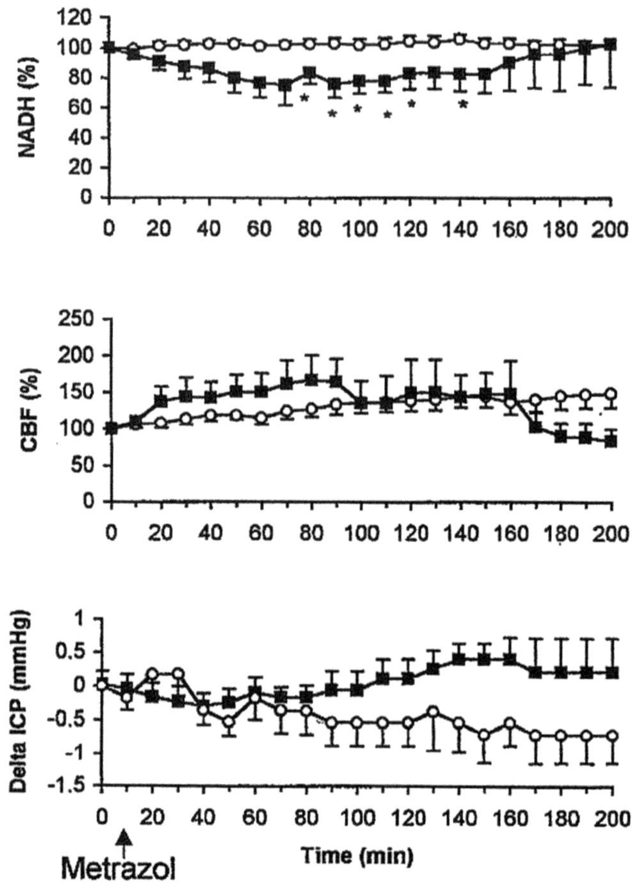

Fig. 7.14 Percent changes in NADH, CBF, and Delta ICP in awake rats injected with Metrazol (black squares) versus controls (open circles). Significant difference at $p < 0.05$ compared to controls

7.3.2.2 Effects of Ischemia on Metabolic, Hemodynamic, Ionic, and Electrical Brain Responses in Awake Rats Injected with Metrazol

In this group, rats underwent bilateral carotid artery occlusion 24 h prior to the experiment. During the experiment, awake rats received 10 mg Metrazol IV every 5–10 min until sustained seizures developed. Seven experimental animals and seven controls were included. The control animals underwent carotid occlusion but were injected with saline (see Protocol No. 2 in Materials and Methods).

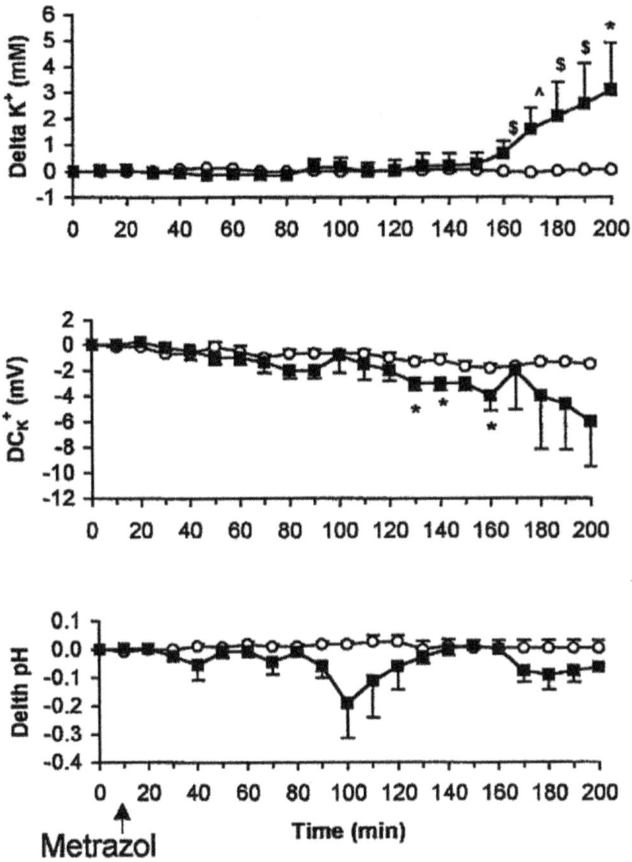

Fig. 7.15 Delta K^+, DCK^+, and Delta pH in awake rats injected with Metrazol (black squares) versus controls (open circles). $p < 0.05$ (*), $p < 0.01$ ($), $p < 0.001$ (^)

7.3.2.2.1 Analog Recordings

Figures 7.16, 7.17, and 7.18a–c present a representative recording from one of the seven ischemic animals. A total of 80 mg Metrazol was administered IV. After seven injections, four SD waves were observed, and only following the eighth injection did a seizure co-occur with SD. SD waves are labeled with boxes, and seizure + SD events are marked accordingly. Roman numerals indicate individual SD waves based on changes in the DC potential.

Spreading depression (SD)—a reversible or transient phenomenon—is characterized by suppression of spontaneous ECoG activity, a drop in DC potential, increased CBF and extracellular potassium, mitochondrial NADH oxidation, and increased reflectance.

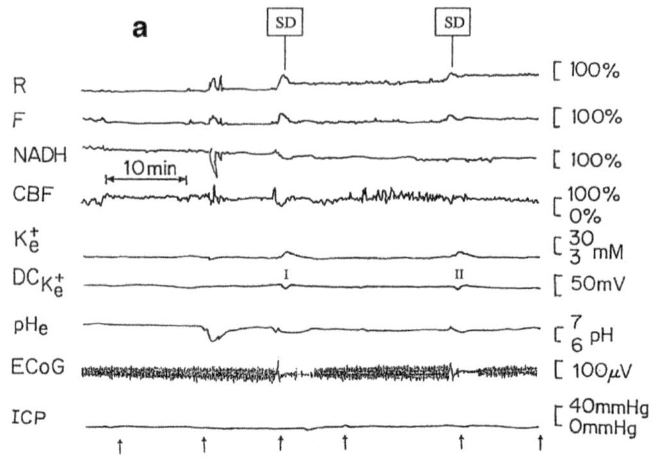

Fig. 7.16 (a–c) Continuous metabolic, hemodynamic, ionic, and electrical recordings from an ischemic awake rat injected with Metrazol. R: Reflectance at 366 nm; F: Fluorescence; NADH: Mitochondrial NADH; CBF: Cerebral blood flow; pH_e, K^+_e: Extracellular ion concentrations; DCK^+_e: Cortical surface potential (potassium electrode); ECoG: Cortical electrical activity; ICP: Intracranial pressure; ↑: Injection of 10 mg Metrazol IV

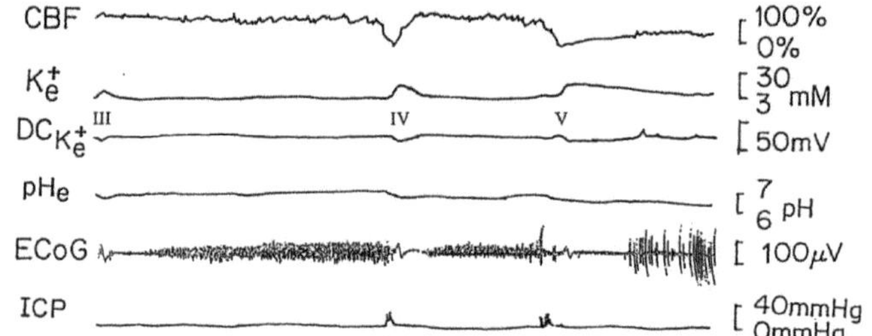

Fig. 7.17 (a–c) Continuous metabolic, hemodynamic, ionic, and electrical recordings from an ischemic awake rat injected with Metrazol. R: Reflectance at 366 nm; F: Fluorescence; NADH: Mitochondrial NADH; CBF: Cerebral blood flow; pH_e, K^+_e: Extracellular ion concentrations; DCK^+_e: Cortical surface potential (potassium electrode); ECoG: Cortical electrical activity; ICP: Intracranial pressure; ↑: Injection of 10 mg Metrazol IV

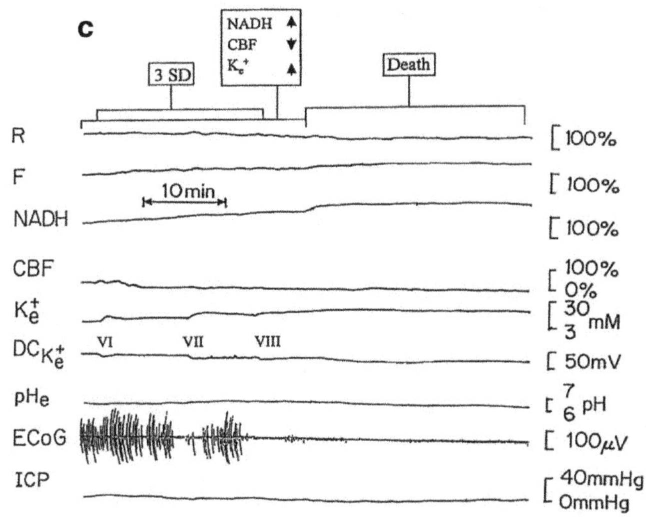

Fig. 7.18 (a–c) Continuous metabolic, hemodynamic, ionic, and electrical recordings from an ischemic awake rat injected with Metrazol. R: Reflectance at 366 nm; F: Fluorescence; NADH: Mitochondrial NADH; CBF: Cerebral blood flow; pH_e, K^+_e: Extracellular ion concentrations; DCK^+_e: Cortical surface potential (potassium electrode); ECoG: Cortical electrical activity; ICP: Intracranial pressure; ↑: Injection of 10 mg Metrazol IV

The first SD wave shows a biphasic response in both CBF and potassium. In the first phase, there is a modest increase in K^+ and CBF, while in the second phase both parameters show larger, prolonged waves. These biphasic patterns are typical when SD is triggered by chemical or epileptogenic stimuli. Non-epileptic SD typically shows only the second phase. Each SD wave is also accompanied by a decrease in DC potential and pH. The second SD wave shows reduced NADH oxidation and no significant CBF increase. The third SD wave is similar to the second in fluorescence and reflectance, and again lacks a CBF rise. The fourth SD wave differs significantly: NADH undergoes **reduction** (rather than oxidation), CBF **decreases sharply**, and the potassium wave has higher amplitude and duration than previous SDs. ICP also increases for the first time.

The **fifth wave** is a **seizure-associated SD**. NADH is reduced, CBF drops with a wave-like profile, potassium amplitude is high, and the duration is long. ECoG amplitude increases during the seizure, followed by suppression during SD. ICP rises significantly during the seizure. Three additional SD waves occur during the SE phase and are associated with a deteriorating physiological state: NADH increases, CBF progressively declines, and potassium levels rise in a single-phase pattern, culminating in death. Figure 7.19 shows a representative 62-min control recording from one of the ischemic rats injected with saline. All parameters remained stable throughout the recording period.

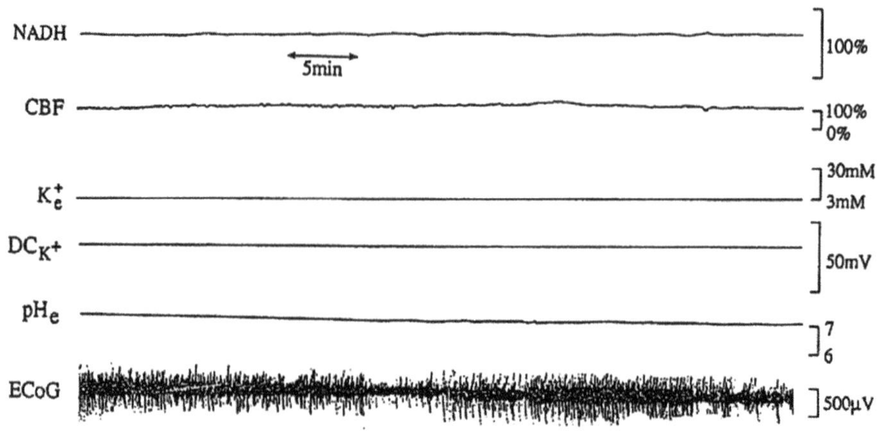

Fig. 7.19 Control recording from an ischemic awake rat (carotid occlusion 24 h prior) over 62 min. NADH, CBF, pH_e, K^+_e, DCK^+_e, ECoG as defined above

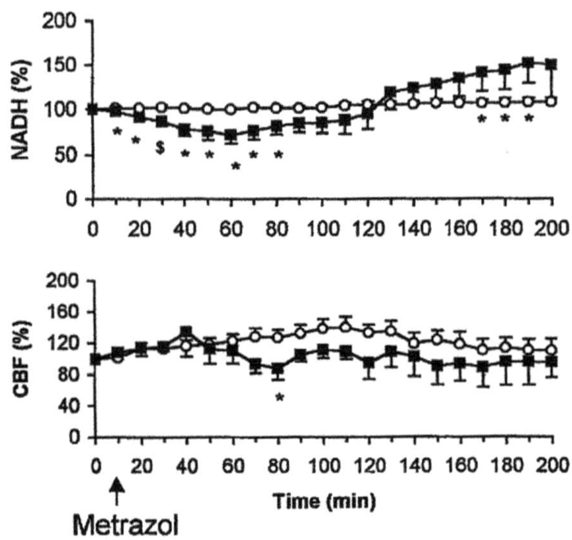

Fig. 7.20 Percent changes in NADH and CBF in ischemic rats injected with Metrazol (black squares) vs. ischemic controls (open circles). $p < 0.05$ (*), $p < 0.01$ ($)

7.3.2.2.2 Quantitative Analysis

Data were collected and analyzed quantitatively. Experimental rats (carotid ligation + Metrazol) were compared to ischemic controls (carotid ligation + saline). Figure 7.20 displays changes in **NADH (%)** and **CBF (%)**. Each point represents the mean of seven animals per group. A significant decrease in NADH is observed immediately

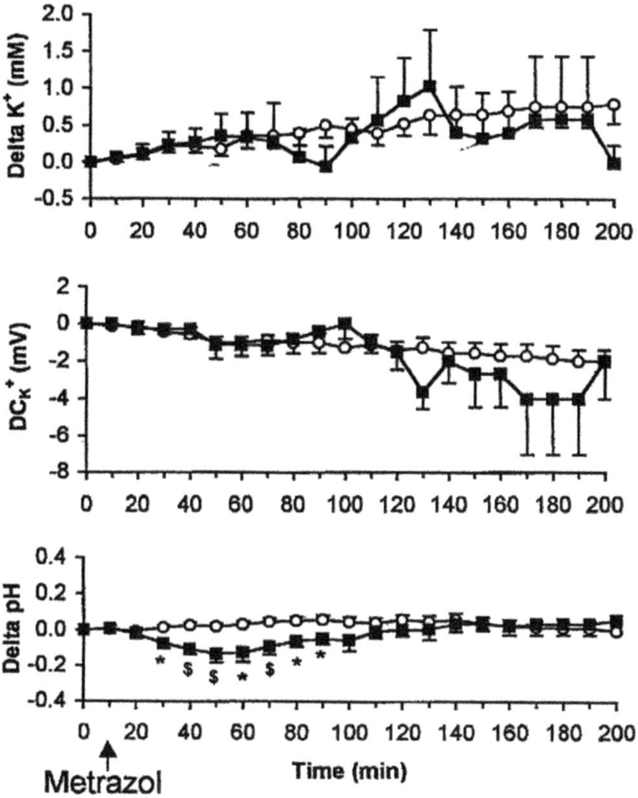

Fig. 7.21 Delta K^+, DCK^+, and Delta pH in ischemic rats injected with Metrazol (black squares) vs. ischemic controls (open circles). $p < 0.05$ (*), $p < 0.01$ ($)

postinjection, persisting for 70 min. After 110 min, NADH begins to increase, and after 150 min—coinciding with deterioration—a significant rise is observed. CBF begins to decline 60 min after injection, becoming significant after 70 min. A minor rebound occurs later, but levels do not return to baseline. Figure 7.21 shows **Delta K^+ (mM), DCK^+ (mV)**, and **Delta pH**. No significant changes were found in K^+ or DCK^+. However, pH significantly decreases 20 min postinjection, remains low for 60 min, then begins to recover.

7.3.2.3 The Effect of Inhibiting NO Synthesis Using the Inhibitor L-NAME on Metabolic, Ionic, and Electrical Brain Responses in Awake Rats Injected with Metrazol

In this experimental group, awake rats were intraperitoneally (IP) injected with L-NAME at a concentration of 50 mg/kg. Forty minutes later, Metrazol was administered intravenously (IV) every 5–10 min until continuous seizures were induced. The experiments were conducted on 11 experimental animals and five control animals. The control animals received an IP injection of L-NAME at the same concentration, followed 40 min later by an IV injection of saline instead of Metrazol (see Protocol 3 in the Materials and Methods Sect. 7.2.4.2, Fig. 7.5).

7.3.2.3.1 Analog Recordings

Figure 7.22 presents a representative brain response of one of the ten experimental animals that received L-NAME followed by intravenous injections of Metrazol until continuous seizures developed (a total of 40 mg Metrazol over six injections). As seen in Fig. 7.23, following the

L-NAME injection, cerebral blood flow (CBF) decreased and blood pressure (BP) increased, while other parameters remained unchanged. Upon subsequent Metrazol injection, a seizure accompanied by a spreading depolarization (SD) was observed (Fig. 7.24). In the absence of spontaneous seizures, injections continued every 15–20 min until continuous spontaneous seizures emerged, each accompanied by SD. During seizures, BP slightly increased and returned to baseline following seizure cessation. Unlike the other animals in the group, this rat was euthanized with an anesthetic followed by 100% N2 ventilation until death; the others died naturally as a result of status epilepticus (SE).

Figure 7.25 shows a 100-min control chart of a representative brain response in one of the five control animals that received an IP injection of L-NAME. Following the injection, CBF decreased and BP increased. Extracellular potassium and pH levels dropped slightly over the experiment, while other parameters remained unchanged.

7.3.2.3.2 Statistical Data

Quantitative analyses were conducted on all experimental results, comparing the experimental and control groups. The experimental group received L-NAME, followed 40 min later by IV Metrazol (10 mg every 5–10 min) until continuous seizures were induced. The control group received saline instead of Metrazol after L-NAME.

Figure 7.26 displays four parameters: (%) NADH, (%) CBF, Delta BP (mmHg), Delta ICP (mmHg). Data were collected every 10 min, and each point represents the average of 11 experimental rats (black squares) or five control rats (empty circles). Error bars represent standard deviation, and asterisks indicate statistically significant differences between the groups at each time point. NADH levels began to decline mildly after 80 min from Metrazol injection, and a significant drop was observed after 130 min, indicating animal deterioration and approaching death. CBF significantly increased upon Metrazol administration and remained elevated for 20 min. BP increased immediately after L-NAME injection in both groups, so no significant difference was observed. ICP in the control group significantly decreased 20 min after L-NAME injection for about 30 min, with no further significant differences thereafter.

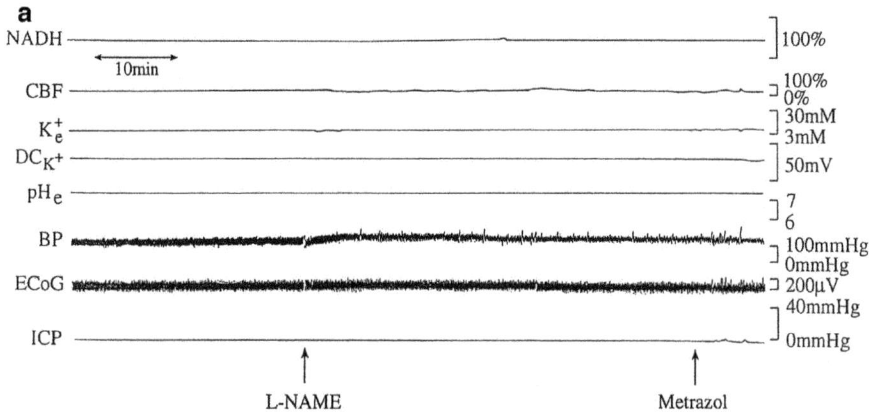

Fig. 7.22 (**a–c**) Recordings of metabolic, hemodynamic, ionic, and electrical activity over time in an awake rat injected with L-NAME followed by Metrazol. NADH mitochondrial NADH levels, CBF cerebral blood flow, pH_e, K^+_e extracellular ion concentrations, DCK^+_e surface potential near the potassium electrode, BP blood pressure, ECoG spontaneous electrical activity on the cortical surface; ICP intracranial pressure

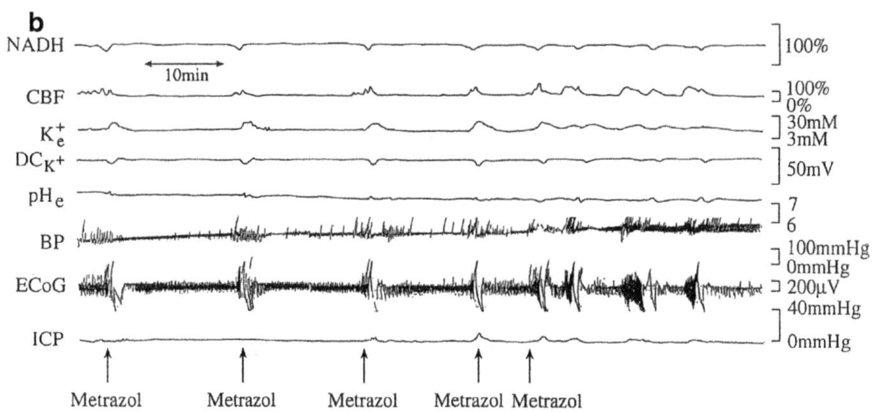

Fig. 7.23 (**a–c**) Recordings of metabolic, hemodynamic, ionic, and electrical activity over time in an awake rat injected with L-NAME followed by Metrazol. NADH mitochondrial NADH levels, CBF cerebral blood flow, pH_e, K^+_e extracellular ion concentrations, DCK^+_e surface potential near the potassium electrode, BP blood pressure, ECoG spontaneous electrical activity on the cortical surface, ICP intracranial pressure

Figure 7.27 shows the parameters Delta K^+ (mM), DCK^+ (mV), and Delta pH. Potassium and DCK^+ levels increased slightly but not significantly. The pH rose significantly 10 min prior to Metrazol administration and remained elevated 10 min after injection, then decreased without statistical significance.

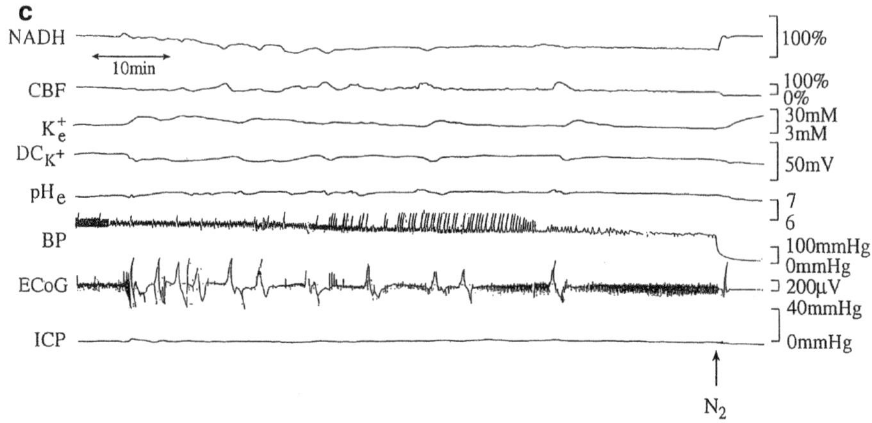

Fig. 7.24 (**a–c**) Recordings of metabolic, hemodynamic, ionic, and electrical activity over time in an awake rat injected with L-NAME followed by Metrazol. NADH mitochondrial NADH levels, CBF cerebral blood flow, pH_e, K^+_e extracellular ion concentrations, DCK^+_e surface potential near the potassium electrode, BP blood pressure, ECoG spontaneous electrical activity on the cortical surface, ICP intracranial pressure

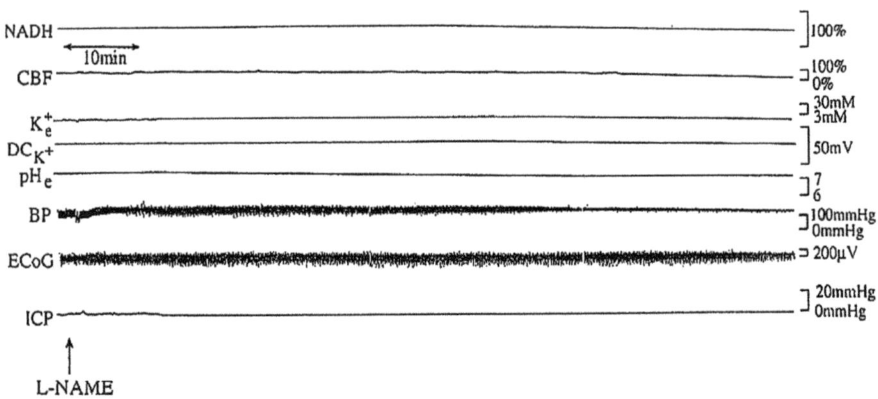

Fig. 7.25 Control chart for 100 min showing metabolic, hemodynamic, ionic, and electrical activity in an awake rat injected with L-NAME only

7.3.2.4 The Effect of Inhibiting NO Synthesis Using the Inhibitor 7NI on Metabolic, Hemodynamic, Ionic, and Electrical Brain Responses in Awake Rats Injected with Metrazol

This section presents one experimental group and two control groups. In the experimental group, awake rats were intraperitoneally injected with 7NI at a concentration of 30 mg/kg. Thirty minutes later, Metrazol was administered intravenously

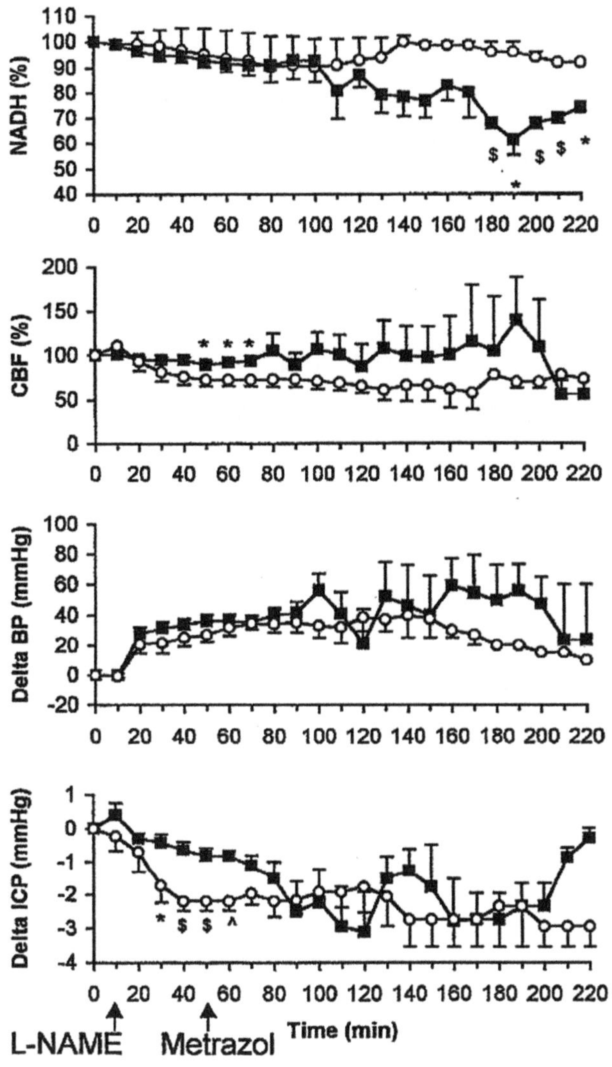

Fig. 7.26 Responses of NADH (%), Flow (%), Delta BP (mmHg), (Delta ICP (mmHg) in awake rats injected with L-NAME followed by Metrazol (black squares) or in control rats injected with L-NAME (open circles). * Difference at the $p < 0.05$ level compared to control animals. $ Differences at the $p < 0.01$ level compared to control animals. ^ Differences at the $p < 0.001$ level compared to control animals

(10 mg every 5–10 min) until continuous seizures developed. The experiments were conducted in eleven experimental rats and nine control rats that received 7NI (same concentration and method) followed 30 min later by saline instead of Metrazol. Since 7NI was dissolved in peanut oil, an additional control group was necessary to evaluate

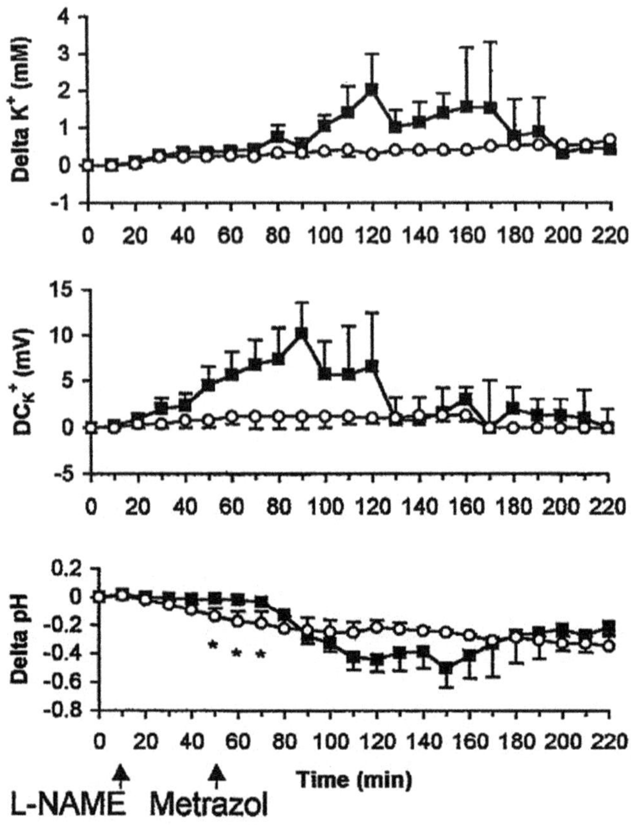

Fig. 7.27 Delta K^+ (mM), DCK^+ (mV), Delta pH responses in alert rats injected with L-NAME and then with Metrazol (black squares) or in control rats injected with L-NAME alone (open circles). * Differences at the $p < 0.05$ level compared to control animals

any physiological effects of the vehicle. This group included six rats that were injected with peanut oil only (the vehicle for 7NI; see Protocol 4 in Materials and Methods).

7.3.2.4.1 Analog Recordings

Figure 7.28 shows a representative 100-minute brain response in one of the ten experimental animals that received 7NI followed 30 min later by Metrazol. After 7NI injection, a slight increase in BP and NADH was observed. A total of 10 mg Metrazol was sufficient to induce spontaneous seizures, which were accompanied by SD waves. During seizures, both BP and ICP increased slightly, returning to baseline postseizure.

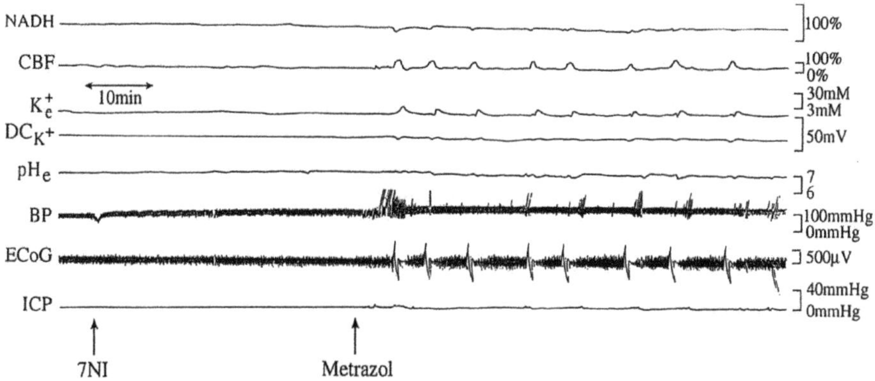

Fig. 7.28 Recording of metabolic, hemodynamic, ionic and electrical activity for 100 representative minutes in an awake rat injected with 7NI and then with Metrazol. NADH mitochondrial NADH level, CBF cerebral blood flow, pH_e, K^+_e extracellular ion concentration, DCK^+_e surface potential on the brain at the potassium electrode, BP blood pressure, ECoG spontaneous electrical activity over the cortex, ICP intracranial pressure

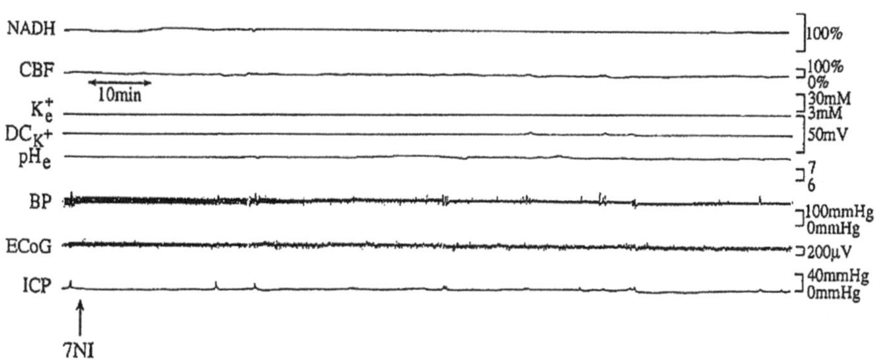

Fig. 7.29 Control chart, recording metabolic, hemodynamic, ionic and electrical activity over time in an awake rat injected with 7NI. NADH mitochondrial NADH level, CBF $C_e$rebral blood flow, pH_e, K^+_e extracellular ion concentration, DCK^+_e surface potential on the brain in the area of the potassium electrode, BP blood pressure, ECoG spontaneous electrical activity over the cortex, ICP intracranial pressure

Figure 7.29 shows a representative response from one of nine control ratsinjected with 7NI. Following the injection, CBF dropped significantly, BP increased slightly, and the other parameters remained stable.

Figure 7.30 depicts a representative response from one of six control rats injected with peanut oil only. All parameters remained stable throughout the 120-min experiment.

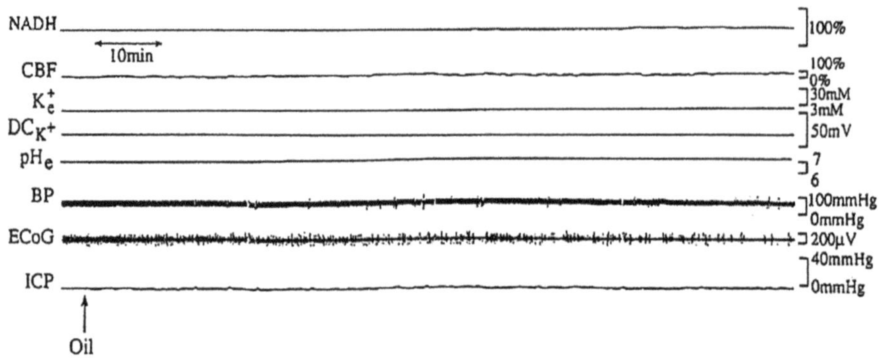

Fig. 7.30 A 120-min control chart representing a recording of metabolic, hemodynamic, ionic, and electrical activity in an awake rat injected with peanut oil. NADH mitochondrial NADH level, CBF C$_e$rebral blood flow, pH$_e$, K^+_e extracellular ion concentration, DCK^+_e brain surface potential at the potassium electrode, BP blood pressure; ECoG spontaneous electrical activity over the cortex, ICP: intracranial pressure

7.3.2.4.2 Statistical Data

Quantitative analysis was performed across all experiments. In addition to the comparison between the experimental group (7NI + Metrazol) and its control (7NI + saline), a comparison was also made between the two control groups (7NI vs. peanut oil). The experimental rats received 7NI followed 30 min later by repeated Metrazol injections until continuous seizures developed.

Figure 7.31 shows: (%) NADH, (%) CBF, Delta BP (mmHg), Delta ICP (mmHg). Data were collected every 10 min. Each data point represents the average of ten experimental rats (black squares), nine 7NI control rats (empty circles), or six peanut oil controls (empty blue triangles). Error bars denote standard deviation. Black markers indicate significant differences between the experimental and 7NI-only control groups. Blue markers denote significant differences between the two control groups.

NADH levels remained largely unchanged in all three groups. CBF in the 7NI control group dropped significantly upon injection and remained low, showing a significant difference compared to the peanut oil group throughout the experiment. When comparing the experimental group (7NI + Metrazol) to the 7NI-only control, both experienced an initial CBF drop from 7NI, but after 80 min, CBF in the experimental group increased significantly compared to the control. No significant differences were found in BP or ICP.

Figure 7.32 presents Delta K^+ (mM), DCK$^+$ (mV), and Delta pH. Potassium levels differed significantly 10 min before death—between both the experimental and control groups and between the two control groups. Comparing the two control groups revealed intermittent significant differences in DCK$^+$. Between the experimental and 7NI-only groups, significant differences were observed at the final two time points. In terms of pH, a significant difference between the experimental and 7NI-only group appeared 20 min after Metrazol injection and persisted for 20 min.

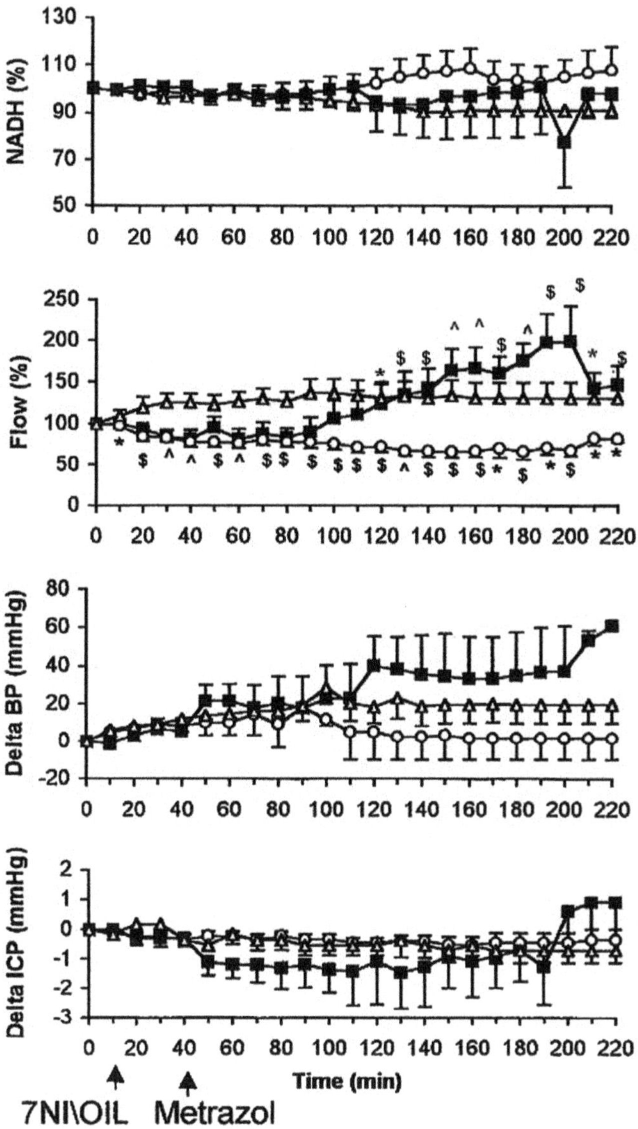

Fig. 7.31 Responses of NADH (%), Flow (%), Delta BP (mmHg), Delta ICP (mmHg) in awake rats injected with 7NI followed by Metrazol (black squares) or in control rats injected with 7NI (open circles) or in control rats injected with peanut oil (blue open triangles). The black markers indicating significant differences indicate differences between the experimental group (group injected with 7NI and Metrazol) and the control group injected with 7NI alone. The blue markers indicate significant differences; – Differences between the two control groups. * Differences at the $p < 0.05$ level; $ Differences at the $p < 0.01$ level; ^ Differences at the $p < 0.001$ level

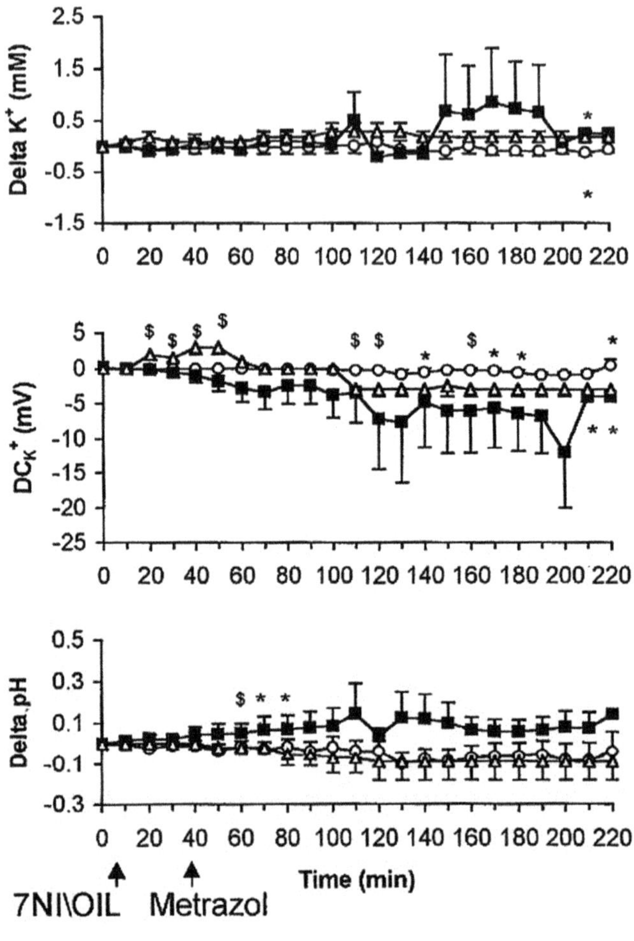

Fig. 7.32 Delta K+ (mM), DCK+ (mV), Delta pH responses in awake rats injected with 7NI followed by Metrazol (black squares) or in control rats injected with 7NI (open circles) or in control rats injected with peanut oil (blue open triangles). * Differences at the $p < 0.05$ level; $ Differences at the $p < 0.001$ level

7.3.2.5 Responses in Awake Rats Injected with Metrazol under Hyperoxia

In this group, awake rats were ventilated with 100% oxygen throughout the entire experiment. Forty minutes after the start of ventilation, they received intravenous (IV) injections of Metrazol at a concentration of 10 mg every 5–10 min until continuous seizures developed. The experiments were conducted on seven experimental animals and eight control animals. The control animals were also ventilated with 100% oxygen and received IV saline instead of Metrazol (see Protocol 5 in the Materials and Methods section).

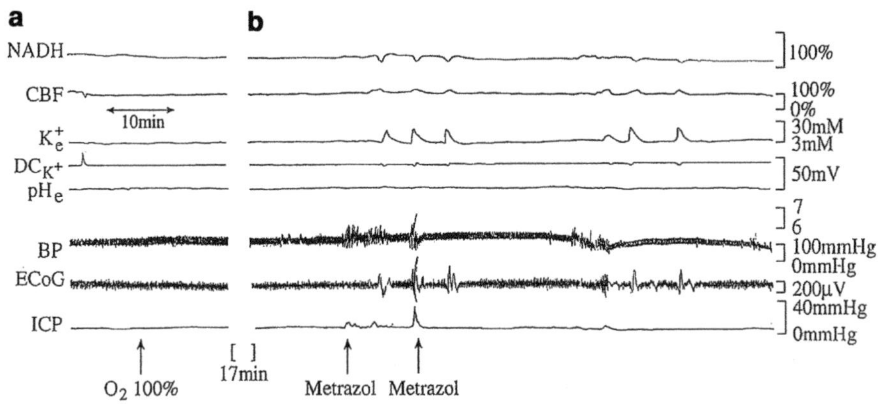

Fig. 7.33 Recording of metabolic, hemodynamic, ionic and electrical activity over time in an awake rat that was ventilated with 100% oxygen throughout the experiment and injected with Metrazol 40 min after the moment of ventilation. NADH mitochondrial NADH level, CBF cerebral blood flow, pH_e, K^+_e extracellular ion concentration, DCK^+_e surface potential on the brain in the area of the potassium electrode, BP blood pressure, ECoG spontaneous electrical activity over the cortex, ICP intracranial pressure, a, b the timeline is the same as in parts a and b

7.3.2.5.1 Analog Recordings

Figure 7.33 presents a representative brain response from one of the seven experimental animals ventilated with 100% oxygen and subsequently injected with Metrazol until repeated seizures were induced (throughout the injection process, the rat continued breathing 100% oxygen). As seen in Fig. 7.33a, following the start of oxygen ventilation, NADH levels decreased, blood pressure (BP) increased, and the other parameters remained at baseline levels. Upon Metrazol administration, seizures accompanied by spreading depolarization (SD) occurred (Fig. 7.33b). It is evident that an SD wave occurred following the first injection. During this initial SD wave, NADH was oxidized, cerebral blood flow (CBF) increased, extracellular potassium (K^+) levels rose, BP increased slightly, ECoG was suppressed, and intracranial pressure (ICP) remained nearly unchanged.

The second injection induced a seizure accompanied by another SD wave. During this episode, NADH oxidation occurred, CBF increased, potassium levels rose, BP increased more than in the first SD, ECoG showed excitation followed by suppression, and ICP increased. Immediately after the first seizure, another SD wave occurred. A seizure-free interval followed, lasting approximately 30 min, after which another seizure appeared, again accompanied by an SD wave. During this later seizure, NADH was oxidized, CBF increased, potassium levels rose, BP dropped significantly and continued to decline until the end of the experiment (indicating vasoconstriction), and ICP increased slightly. In all SD waves observed during the experiment, NADH oxidation and CBF increase were consistently noted.

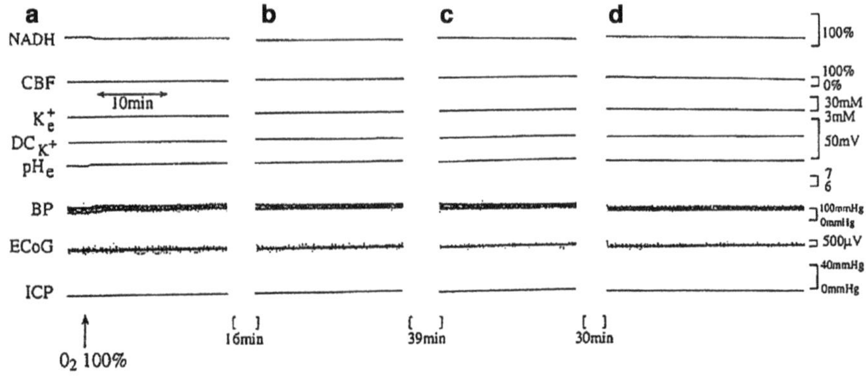

Fig. 7.34 Control chart, recording metabolic, hemodynamic, ionic and electrical activity over time in an awake rat ventilated with 100% oxygen throughout the experiment. NADH mitochondrial NADH level, CBF cerebral blood flow, pH_e, K^+_e extracellular ion concentration, DCK^+_e surface potential on the brain in the area of the potassium electrode, BP blood pressure, ECoG spontaneous electrical activity over the cortex, ICP intracranial pressure, a, b, c, d timeline in all sections is the same as a

Figure 7.34 presents a representative brain response from one of the eight control animals ventilated with 100% oxygen throughout the experiment. It is evident that NADH levels decreased (oxidation) following the onset of oxygen ventilation, CBF also gradually decreased, and BP increased. All other parameters remained stable during the course of the experiment.

7.3.2.5.2 Statistical Data

The results collected from all experiments were subjected to quantitative analysis, comparing the experimental group with its corresponding control group. In the experimental group, rats were ventilated with 100% oxygen throughout the experiment and received IV Metrazol every 5–0 min starting 40 min after ventilation began, until continuous seizures developed (the figure refers only to the first Metrazol injection). The control group consisted of awake rats ventilated with 100% oxygen for the entire experiment and injected with saline instead of Metrazol.

Figure 7.35 presents four parameters: **(%) NADH, (%) CBF, Delta BP (mmHg)**, and **Delta ICP (mmHg)**. Data were recorded every 10 minutes throughout the experiment. Each point represents the average value of seven experimental rats (solid black squares) or eight control rats (empty circles). Error bars indicate the standard error of the mean. Asterisks denote statistically significant differences between the experimental and control groups at each time point.

It can be seen that CBF in the control group declined 120 min after the start of oxygen ventilation. The experimental group began to diverge from the control group

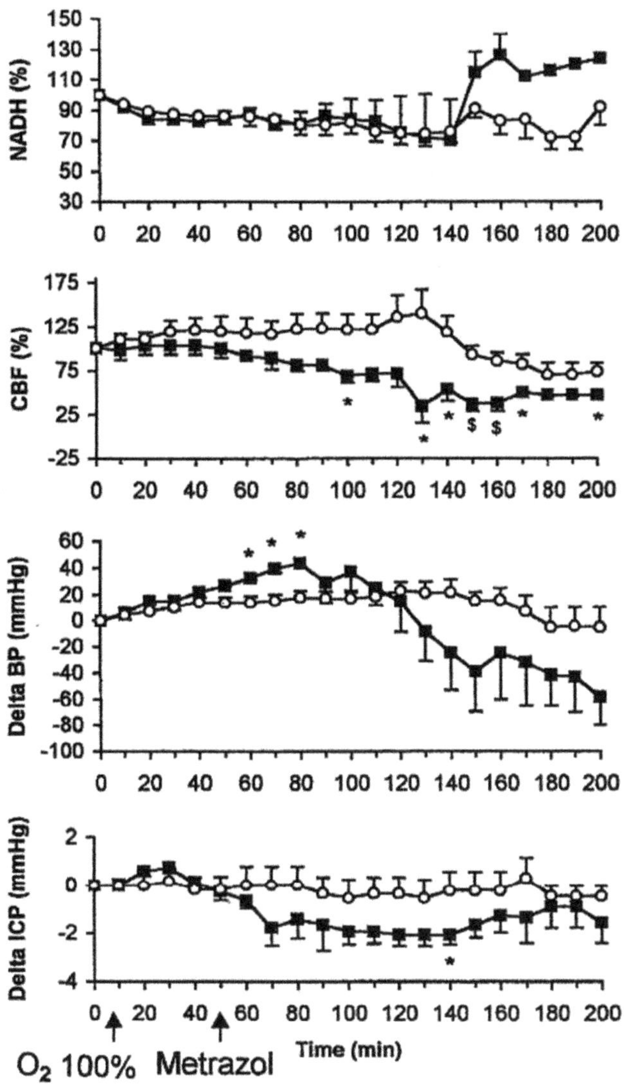

Fig. 7.35 Responses of parameters NADH (%), Flow (%), Delta BP (mmHg), and Delta ICP (mmHg) in awake rats that were ventilated with 100% oxygen and injected with Metrazol (black squares) or in control rats, rats that were ventilated with 100% oxygen throughout the experiment. * Differences at the $p < 0.05$ level compared to control animals. $ Differences at the $p < 0.01$ level compared to control animals

50 min after Metrazol administration. After 80 min, a significant drop in CBF was observed, which persisted for an additional 40 min, up to 150 min post-Metrazol. CBF remained significantly different from the control throughout that period.

BP began to rise significantly 10 min after Metrazol administration, continued rising for 20 min, then reversed direction and gradually declined—although this decline was not statistically significant. NADH levels did not differ significantly from the control group. ICP showed a significant difference at only one-time point—90 min after Metrazol injection—while at all other time points, it remained stable.

Figure 7.36 presents three parameters: **Delta K$^+$ (mM), DCK$^+$ (mV)**, and **Delta pH**. Potassium levels rose modestly and non-significantly starting 60 min after Metrazol administration. pH began to decline at the moment Metrazol was injected. Although some significant differences are indicated by asterisks, they are scattered across time points, showing no clear trend of consistent increase or decrease across any of the three parameters.

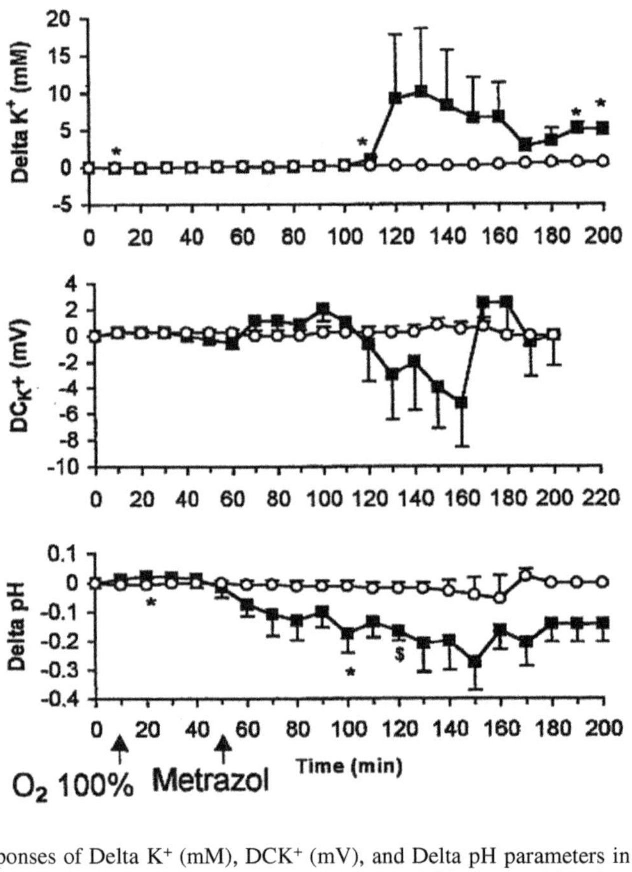

Fig. 7.36 Responses of Delta K$^+$ (mM), DCK$^+$ (mV), and Delta pH parameters in alert rats that were ventilated with 100% oxygen and injected with Metrazol (black squares) or in control rats that were ventilated with 100% oxygen throughout the experiment. * Differences at the $p < 0.05$ level compared to control animals. $ Differences at the $p < 0.01$ level compared to control animals

7.3.2.6 Effects of Anesthesia on the Metabolic, Hemodynamic, Ionic, and Electrical Brain Responses in Rats Injected with Metrazol

In this group, alert rats were injected intravenously (IV) with Metrazol at a concentration of 10 mg every 5–10 min until the development of continuous seizures, and once the animal reached a state of continuous seizures, the anesthetic Nembutal was injected intraperitoneally (IP) at a concentration of 40 mg/kg. The experiments were performed on ten experimental animals and six control animals, alert animals that were injected with saline (instead of Metrazol). At approximately the same time as the experiments, the alert rats were injected intraperitoneally with Nembutal at a concentration of 40 mg/kg.

As already noted at the beginning of the results section, this section will be presented qualitatively only.

7.3.2.6.1 Analogous Diagram

Figure 7.37 depicts a representative brain response of one of the ten experimental animals, a rat that received repeated injections of Metrazol until continuous seizures were formed. As soon as continuous seizures spontaneously occurred without interruption, an anesthetic (Nembutal) was injected to stop the seizures.

In the figure, it can be seen that Metrazol was injected and continuous seizures developed. NADH during the experiment decreases, blood pressure increases, and the other parameters between seizures maintain a relatively constant basic level. As soon as the anesthetic Nembutal is injected, the seizures immediately stopped. It can be

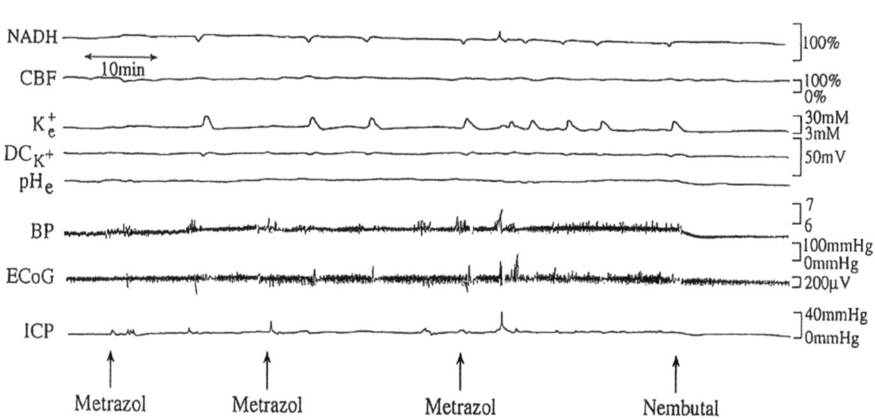

Fig. 7.37 Recording of metabolic, hemodynamic, ionic and electrical activity over time in an awake rat injected with Metrazol and injected with the anesthetic Nembutal to stop the seizures. NADH mitochondrial NADH level, CBF $C_e$rebral blood flow, pH_e, K^+_e extracellular ion concentration, DCK^+_e surface potential on the brain in the area of the potassium electrode, BP blood pressure, ECoG spontaneous electrical activity over the cortex, ICP intracranial pressure

seen that immediately after the injection of Nembutal, blood flow decreases slightly, the potassium level decreases, the pH level decreases, blood pressure decreases very clearly, the ECoG amplitude decreases and ICP decreases slightly.

In the ten experimental rats, as soon as the anesthetic Nembutal was administered, the seizures completely stopped, and not even a single spike was observed in the ECoG.

Figure 7.38 depicts a representative brain response of one of the six control animals that was injected with the anesthetic Nembutal while awake. It can be seen that following anesthesia, blood flow decreases and at the same time, reflexes increase, and

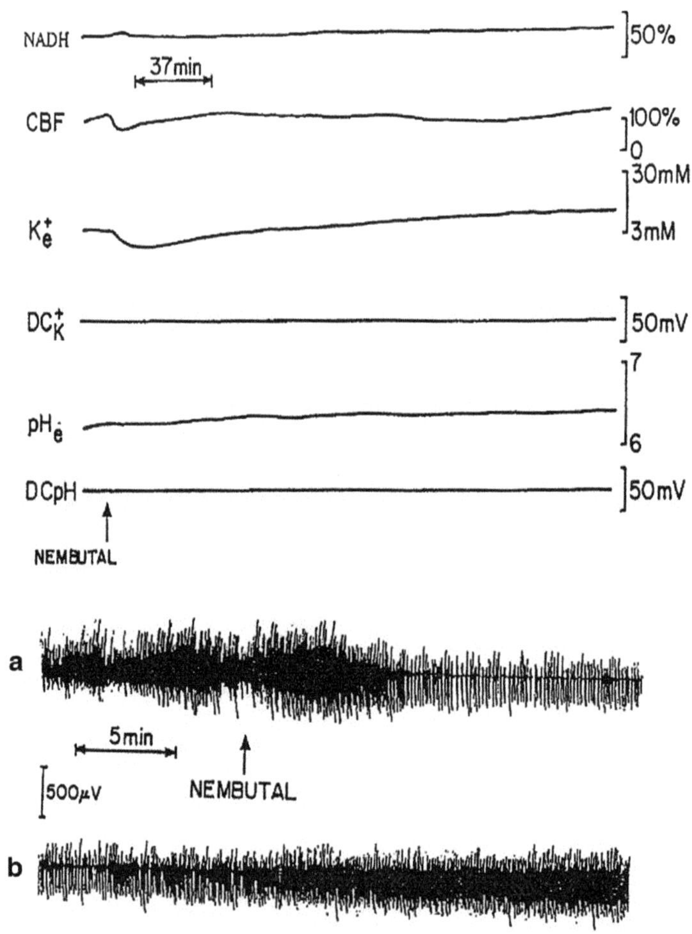

Fig. 7.38 Control chart, recording metabolic, hemodynamic, ionic and electrical activity over time in a rat anesthetized with Nembutal. NADH mitochondrial NADH level, CBF C$_e$rebral blood flow, pH$_e$, K^+_e extracellular ion concentration, DCK^+_e surface potential on the brain in the area of the potassium electrode, ECoG (A, B) spontaneous electrical activity over the cortex

the basal level of extracellular potassium decreases. Toward the end of the experiment, blood flow increases above its basal level. So doreflectance and potassium. The reason for this is probably due to the fact that the animal is in the stages of awakening. The other parameters maintained their basal level.

It can be seen that following anesthesia, there was a drastic decrease in spontaneous electrical activity, but after about half an hour there is a recovery in electrical activity. (The ECoG graph [a, b] appears separately from the parameter graph).

7.3.3 Statistical Comparisons Between the Five Experimental Group

In the third part, which deals with statistical comparisons between the different experimental groups, the results of the five experimental groups (without their control group) were taken and they are:

1. Normoxic group injected with Metrazol (called Metrazol in the graphs)
2. Ischemic group injected with Metrazol (called Ischemia in the graphs).
3. Group treated with L-NAME and injected with Metrazol (called NAME in the graphs).
4. Group treated with 7NI- and injected with Metrazol (called 7NI in the graphs).
5. Group ventilated with 100% oxygen and injected with Metrazol (called Hyperoxia in the graphs).

The images that appear in the above section will be described as follows: Each column belongs to one group and indicates the mean of the parameter under test and the Standard Error (SE). Each column has its own specific color and does not change throughout the presentation of the images. The statistical test conducted is an ANOVA (GLM) test (to find differences between the groups, a Duncan test is performed under ANOVA) and the foreign letters represent the results of the test. The foreign letters that appear above the columns are compared between the five groups. Identical letters indicate that there is no difference between the groups; different letters indicate that there is a difference between the groups.

The following parameters were tested in the five groups:

7.3.3.1 The SD Phenomenon

In each experimental group, the SD waves that appeared from the moment of Metrazol injection were counted (Fig. 7.39I). Each column indicates the average number of SD waves in that group, and the standard error (SE) of that group. It can be seen in the graph that there is no significant difference between the groups according to the ANOVA test.

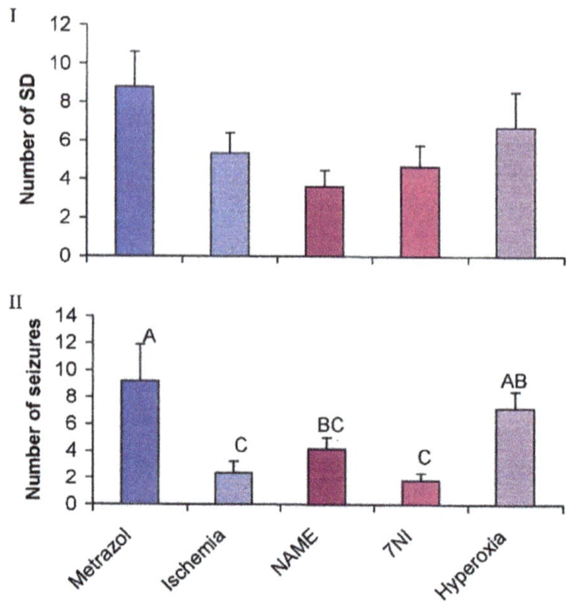

Fig. 7.39 **I**. Average number of SD waves from the moment of Metrazol injection in the five groups. **II**. Average number of seizures from the moment of Metrazol injection in the five groups. Overall significance of the ANOVA test ($F = 4.08$; df = 4.38) $p < 0.0075$

7.3.3.2 Number of Seizures

In each experimental group, the number of seizures that appeared from the moment of Metrazol injection was counted (Fig. 7.39II). When the seizures became continuous, and it was impossible to distinguish between one seizure and another, the sequence was counted as one seizure (see Fig. 7.11, it can be seen that the ECoG is continuous). Each column indicates the average of the seizures in each experimental group, and the standard error (SE) of that group. The letters indicate whether there are differences between the five groups. According to the ANOVA test, it can be seen that the only group that is similar in terms of the number of seizures to the Metrazol group is the hyperoxia group. The groups that are completely different from the Metrazol group are the ischemia and the 7NI-group (the letter A appears in the Metrazol group compared to the letter C that appears in the ischemia and 7NI-groups). The L-NAME group is similar to both the ischemia and 7NI-groups and the hyperoxia group, but is completely different from the Metrazol group.

7.3.3.3 Decay Period

The data extraction method for the above parameter was as follows: In each animal before death, a deterioration of the parameters could be observed. During the

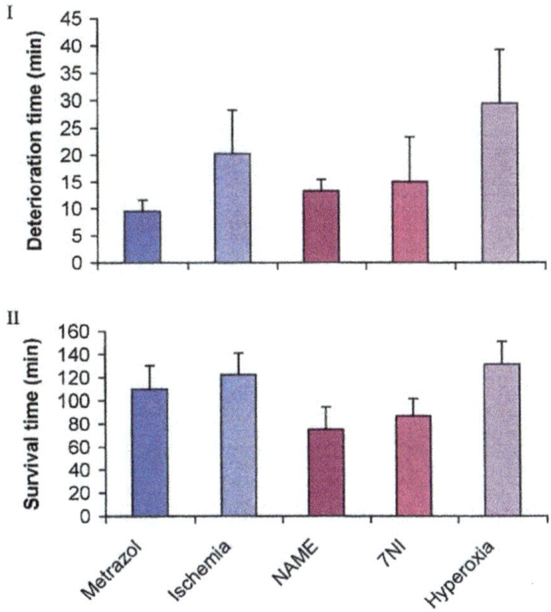

Fig. 7.40 **I**. Decay time (in minutes) in each experimental group from the moment of deterioration in parameters (NADH, blood flow and potassium) to the point at which death begins. **II**. Survival time (in minutes) from the moment of administration of Metrazol to the point at which death begins

deterioration, NADH increases, blood flow decreases, and extracellular potassium increases (see Fig. 7.11). In each experiment, the time in minutes from the moment of deterioration in the parameters NADH), blood flow, and potassium) to the point at which death begins was measured (Fig. 7.40, part I). An average was performed within the same group. Each column represents the average decay time of the experimental group and the standard error of that group. According to the ANOVA test that tested differences between the five experimental groups, it can be seen that there is no significant difference between the groups.

7.3.3.4 Survival Time

The survival time was determined as the time from the moment of administration of Metrazol to death (Fig. 7.40, part II). The survival time was recorded in each experiment and averaged in each experimental group. Each column represents the average survival time of the experimental group and standard error of the same group. An ANOVA test conducted showed that there was no difference between the five groups.

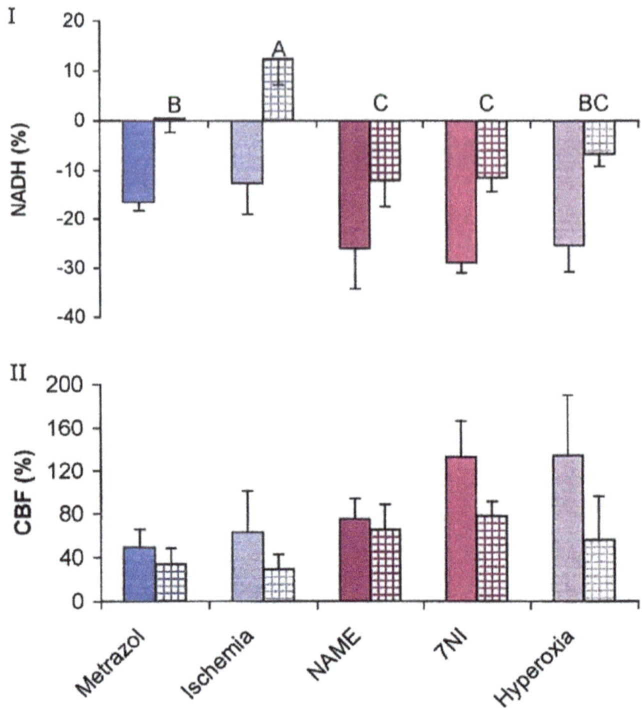

Fig. 7.41 **I**. NADH amplitude in percentages of the first and last wave. The last wave is significantly different in the overall significance ANOVA test of the last wave ($F = 6.91$; df = 4.31) $p < 0.0004$. **II**. Blood flow amplitude in percentages of the first and last wave

7.3.3.5 NADH Level

The data extraction method for the above parameter was as follows: NADH amplitude was taken at the time of the appearance of the first SD wave, and the last SD wave in all experiments (Fig. 7.41I). The amplitude was measured in percent compared to the starting point of the wave. An average for the first wave and an average for the last wave were performed in each group. The graph shows two columns for each group. The first column (solid color) represents the average amplitude of the first wave and the standard error of that group and the second column (checkered) represents the average amplitude of the last wave and the standard error of that group. It can be seen that in the group of animals that received an injection of Metrazol alone, a reversal of the NADH wave occurred. The NADH wave went from an oxidation state to a reduction state, but it can be seen that the reduction level is very low. Also, in the ischemic group of animals, NADH goes from an oxidation state to a high reduction state. In the other three groups, the L-NAME, 7NI, and hyperoxia groups, the second wave does not undergo a reversal from an oxidation state to a reduction state, but there is a transition from a higher oxidation level to a lower oxidation level. A t-test was conducted within the same experimental group, comparing the first wave to the last wave, and the following are the results observed:

It can be seen that there is a statistically significant difference in all groups between the first and second waves. The second wave either undergoes a reversal from oxidation to reduction (in the Metrazol and ischemia groups) or the second wave is less oxidized than the first (in the three groups: L-NAME, 7NI, and hyperoxia).

An ANOVA test that examined differences between the five groups showed that the first wave is not statistically different between the five groups. On the other hand, the last wave is statistically different between the five groups. It can be seen that the last wave in the Metrazol group is similar only to the hyperoxia group and different from the other groups. The last wave of the ischemic group is different from the last wave of the other four groups. The last wave of the L-NAME and 7NI-groups is similar to the last wave of the hyperoxia group but different from the ischemia group and the Metrazol group. The hyperoxia group is similar to the Metrazol, L-NAME and 7NI-groups but statistically different from the ischemic group.

7.3.3.6 Blood Flow

The data extraction method for the above parameter was as follows: Blood flow amplitude was taken at the time of the first and last SD waves in all experiments (Fig. 7.41, part II). The amplitude was measured in percent compared to the starting point of the wave. In each group, an average for the first wave and an average for the last wave were calculated. The graph shows two columns for each group, the first column (solid color) represents the average amplitude of the first wave and the standard error of that group and the second column (checked) represents the average amplitude of the last wave and the standard error of that group. A *t*-test that compared within the same experimental group, compared the first wave with the last wave. According to the *t*-test, there seems to be no significant difference between the first and last waves in all five groups, so no table was presented. An ANOVA test that examined differences between the five groups showed that there was no statistical difference in either the first or last wave in the five groups.

7.3.3.7 Potassium Amplitude

The data extraction method for the above parameter was as follows: Potassium amplitude was taken at the time of the first SD wave and the last SD wave in all experiments (Fig. 7.42, part I). The amplitude was measured in mM compared to the starting point of the wave. In each group, an average for the first wave and an average for the last wave were performed. Two columns appear in the graph for each group. The first column (solid color) represents the average amplitude of the first wave and the standard error of that group, and the second column (checkered) represents the average amplitude of the last wave and the standard error of that group. A *t*-test that compared within the same experimental group compared the amplitude of the first wave and

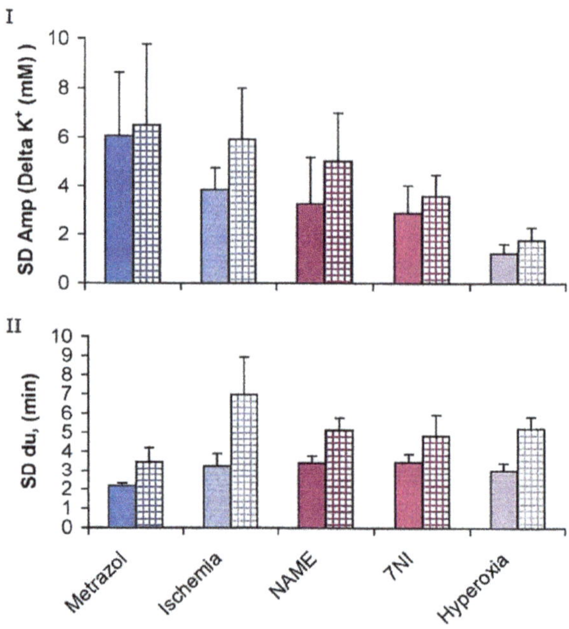

Fig. 7.42 **I**. NADH amplitude in percentages of the first and last wave. The last wave is significantly different in the overall significance ANOVA test of the last wave ($F = 6.91$; df = 4.31) $p < 0.0004$. **II**. Blood flow amplitude in percentages of the first and last wave

the amplitude of the last wave. The results according to the *t*-test showed that there were no significant differences within the same group in all five groups, therefore a table was not presented.

An ANOVA test examining differences between the five groups showed that the first wave was not significantly different in the five experimental groups.

7.3.3.8 Duration of the SD Wave According to the Potassium Parameter

The data extraction method for the above parameter was as follows: At the time of the first SD wave, the duration in minutes of the first potassium wave was measured. Also, at the time of the last SD wave, the duration in minutes of the last potassium wave was measured. In each group, an average was calculated for the first wave and an average for the last wave (Fig. 7.42, part II). Two columns appear in the graph for each group. The first column (solid color) represents the average duration of the first wave, and the second column (checkered) represents the average duration of the last wave. In a *t*-test conducted within the same experimental group and comparing the duration in minutes of the first potassium wave to the duration in minutes of the last potassium wave, the following results were observed:

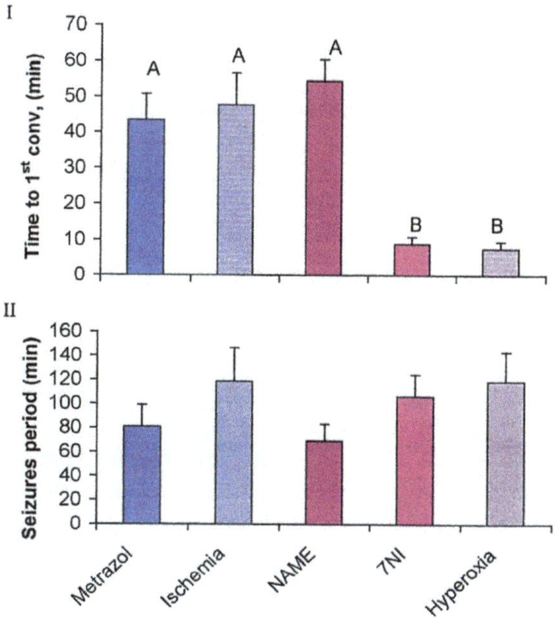

Fig. 7.43 **I**. Time to first seizure (in minutes) from the moment of Metrazol injection. Overall significance in the ANOVA test ($F = 13.63$; df = 4.31). $p < 0.001$. **II**. Duration of seizure occurrence (in minutes)

It can be seen that the duration of the second wave in the ischemic group, in the L-NAME group and in the hyperoxia group was significantly different from the duration of the first wave of the same group. In the Metrazol group and in the 7NI-No significant difference was observed between the duration of the last wave and the duration of the first wave. An ANOVA test that examined differences between the five groups showed that the duration of the first potassium wave was not significantly different in the five groups, and likewise the duration of the last wave was not significantly different in the five groups.

7.3.3.9 Time to the Appearance of the First Seizure

The data extraction method for the above parameter was as follows: The time to the appearance of the first seizure was measured (in minutes) from the moment of Metrazol injection (Fig. 7.43, part I). The time to the appearance of the first seizure was recorded in each experiment and averaged in each experimental group. Each column indicates the average time to the appearance of the first seizure in each experimental group and the standard error (SE) of that group. The letters indicate whether there are differences between the five groups.

In the ANOVA test comparing the five experimental groups, it can be seen that the Metrazol, ischemia, and L-NAME groups are statistically similar (the letter A appears in all). The 7NI- and hyperoxia groups are also similar to each other but different from the other three groups (the letter B appears in both).

7.3.3.10 Duration of Seizures

The data extraction method for the above parameter was as follows: The time from the moment of the first seizure to the moment of death was measured (Fig. 7.43, part II). The moment of death was determined as the moment when the seizures stopped, although sometimes it seemed that the seizures stopped before death. The moment of death was taken as the last point because sometimes there are mild seizures that are not expressed in the electrical activity, the ECoG or that appear as short spikes. Therefore, it is not possible to determine exactly when the seizures stop, and it was decided that the moment of death would be the moment when the seizures stopped. The duration of the seizures was recorded in each experiment and an average was calculated for each experimental group. Each column indicates the average duration of the seizures in each experimental group, and the standard error of that group.

According to the ANOVA test, a test that compared the five experimental groups, it was found that the groups did not differ significantly from each other.

7.4 Discussion

Epilepsy is a complex disease in its various forms, with increasing numbers in the population. A large proportion of patients are treated with medications that prevent seizures to one degree or another. The scientific basis for explaining the development of seizures, as well as the physiological and biochemical changes that accompany them, is not yet sufficiently clear. Most of the work described so far has dealt with anesthetized or paralyzed and artificially ventilated animals in which a limited number of brain parameters were measured simultaneously. No detailed measurements have been made to date that would indicate the relationship between ionic and electrical metabolic activity of the brain in an awake animal. In the present work, we used awake animals exposed to epileptic seizures, which were induced with Metrazol, and in which various parameters from the same brain region were examined. These parameters reflect the relationship between the ionic and electrical energetic activity of the brain and were measured continuously starting from the control phase (before the first seizure), during the seizures, and after the end of this period. Since most researchers did not conduct continuous measurements, their results focus on the seizures themselves and the phenomena that occur around the seizures (such as SD waves, survival time, time until the first seizure occurs, and more), and less on what occurs between seizures. Since Sect. 7.4.1 focuses on measurements between seizures, this chapter will provide a few quotes from other studies (the chapter refers to the second part of the results chapter). On the other hand, in Sect. 7.4.5, which deals with the phenomena that occur around the seizure and the seizure itself, a comparison will be made between the current study and various studies (the chapter refers to the third part of the results chapter).

7.4.1 The Effect of Metrazol on Brain Function

In these experiments, alert animals were injected with Metrazol until consecutive seizures occurred, and the control for the experiment was alert rats that were injected with saline instead of Metrazol for the same experimental period.

It can be seen that the most prominent phenomenon following Metrazol injection is a decrease Significant increase in the level of NADH, i.e. oxidation, during the experiment. However, toward the end of the experiment, an insignificant increase in the level of NADH begins (see Fig. 7.14). A decrease in the level of NADH can be seen, but the blood flow does not change.

A partially similar result was observed by the following researchers: Van Burn et al. (1978), O'Connor et al. (1972a, 1973), Mayevsky (1984b), Tenny et al. (1980). The researchers' observation showed a decrease in the level of NADH and an increase in the blood flow during the seizures. In all the above works, it was proven that epileptic activity increases the energetic activity due to an increase in extracellular potassium, which leads to the activation of the sodium-potassium pump, which requires a large amount of energy.

The researchers' result is similar only in terms of the level of NADH and different in terms of the blood flow, and this is because all the parameters shown in Fig. 7.14 represent points taken between The seizures were not during the seizure, while the researchers measured during the seizures and not between them. It is important to note that the points were taken between the seizures and not during them because when you look at the experimental charts, you can see that during the seizures the NADH level does indeed decrease, the blood flow increases and the potassium level increases (similar to the researchers mentioned above). However, when you measure between the seizures, you can see that NADH undergoes oxidation during the experiment (not until the end) while the blood flow does not change, nor does the potassium level change. Only at the end of the experiment, toward death, does deterioration occur and the potassium level increases. In other words, between the seizures the blood flow and potassium return to their baseline level before the seizure, while the NADH between the seizures undergoes oxidation. This means that the epileptic activity increases the energetic activity so that the NADH undergoes oxidation, but the system, despite the repeated seizures, is in a state where there is no lack of oxygen, and therefore the blood flow between the seizures returns to its baseline level.

In a state of oxygen deficiency (hypoxia), we would expect blood flow to increase (Mayevsky 1992). In this case, there is no oxygen deficiency in the system, and the blood flow that does exist manages to provide enough oxygen, so that despite the change in oxygen consumption (oxygen consumption is increased following the seizures), the system is not required to increase blood flow.

It can be seen that the NADH level decreases significantly. This means that the respiratory chain increases its activity even between seizures. This may be due to the fact that energy is required to get out of the seizures. During the experiment, the NADH level decreases, but toward the end of the experiment, there is a deterioration in the animal's condition. It can be seen that NADH begins to increase (not significantly).

When looking at Fig. 7.41I, it can also be seen that measuring NADH during the first SD, which appears following the Metrazol injection, shows oxidation (in the first column), while measuring NADH in the last SD (in the second column) The second column shows a state of reduction (in the Metrazol group and the ischemic group). In the other groups, the second column is not in a state of reduction, but it can still be seen that the NADH in the last attack is less oxidized than the first attack (in all groups, a significant difference was found between the first and last waves). According to the results, it can be concluded that the respiratory chain is able to work less because the blood flow does not increase, and there is less oxygen in the system. There is less oxygen in the system but there is no problem of lack of oxygen, because in a state of lack of oxygen (hypoxia), we would expect the blood flow to increase (Mayevsky 1992).

7.4.2 Effect of Ischemia on Brain Function During Metrazol Exposure

In these experiments, Metrazol was injected into awake ischemic animals until consecutive seizures were induced, and the control for the experiment was ischemic rats that were injected with saline instead of Metrazol for the same experimental period. Several changes can be seen in this group.

At the beginning of the experiment, NADH undergoes oxidation, and toward the end of the experiment, when the animal's condition deteriorated, NADH undergoes reduction (Fig. 7.20). The decrease in NADH levels is significant, as is the increase in NADH levels. Blood flow following Metrazol injection is in a downward trend. It can be seen that blood flow decreases and reaches an even lower level than the control group. Seventy minutes after Metrazol injection, it can be seen that there is a significant difference between the experimental group and the control group (Fig. 7.20). (Since the Metrazol solution is injected intravenously, it can be seen that the changes in the parameters are immediate).

When comparing the group of ischemic animals that were injected with Metrazol with the group of animals that did not undergo any treatment other than Metrazol injection (Discussion—Sect. 7.4.2), it can be seen that NADH in the ischemic group begins to decrease immediately after Metrazol administration (Fig. 7.20), in contrast to Group 1, where NADH begins to decrease 70 min after Metrazol administration (Fig. 7.14). From this, it can be concluded that in the ischemic group of animals, already in the first attacks, the metabolic demand is high and in order to get out of the attacks, the respiratory chain works harder and therefore the NADH level decreases. This is due to the fact that the ischemic group of animals begins the experiment in a lower energetic state, since both carotids are tied and blood flow to the brain is relatively low compared to rats that did not undergo carotid ligation. Hence, the oxygen supply to the brain of ischemic rats is lower than the oxygen supply to the group that did not undergo carotid ligation. On the other hand, in the group of animals that did not undergo ischemia, the NADH level between the first seizures is less oxidized, meaning that less energy is required to get out of the first seizures.

In the ischemic group of rats, NADH reaches levels higher than the control level 120 min after the Metrazol injection, still not significantly. And 160 min after the injection, the NADH level is significantly higher than the control. On the other hand, in the group of animals that did not undergo ischemia, toward the end of the experiment, NADH is in an upward direction, but the increase is not significant.

If we look at the blood flow in both groups, we can see that in animals that did not undergo ischemic treatment, the blood flow trended downward relative to the control, while in the ischemic rats, the blood flow trended downward relative to the control rats (except for one significant point 70 min after the Metrazol was administered). In other words, it is possible that in the ischemic animals, the oxygen supply is lower because the blood flow is unable to increase between the attacks, because the oxygen supply is lower, and therefore NADH toward the end of the experiment in the ischemic animals goes from a level of oxidation to a level of reduction.

This means that it can be concluded that in the ischemic animals, an energy problem occurs earlier than in animals that did not undergo ischemia, the blood flow is lower, there is less oxygen in the system and therefore toward the end of the experiment NADH increases.

7.4.3 Involvement of NO in Brain Responses to Metrazol

Some argue that the NO molecule functions as an anticonvulsant (anti-seizure) (Buisson et al. 1993; Przegaliński et al. 1994) and some argue that the NO molecule causes convulsions (seizures) (Mülsch et al. 1994; De Sarro et al. 1991).

In order to examine the role of the NO molecule in epileptic seizures, we used in the present study the inhibitors L-NAME and 7NI- that inhibit the enzyme (NOS) NO synthase and thus prevent the formation of NO. L-NAME inhibits neuronal NOS and endothelial NOS and 7NI- is a specific inhibitor of neuronal NOS.

Since inhibition of endothelial NOS by the inhibitor L-NAME increases blood pressure, which causes neuronal activation (Przegalinski et al. 1996), in the present study, we also used the inhibitor 7NI, which does not have cholinergic or cardiovascular effects, and is therefore more suitable for studying the central nervous system.

In the experimental part in which we used the inhibitor L-NAME, L-NAME was injected into rats while they were awake and after 40 min they were injected with Metrazol until consecutive seizures were induced. The control rats for this group were awake rats that were injected with L-NAME and after 40 min saline instead of Metrazol for the same experimental period.

Since it is known that epileptic seizures increase energetic activity so that blood flow increases during seizures, and it is known that the NO molecule participates in the regulation of cerebral blood flow, it was decided to examine the involvement of NO in the hemodynamic processes that occur following Metrazol injection that induces consecutive seizures. The inhibitor L-NAME was used, which is a nonspecific inhibitor of the NOS enzyme, which inhibits both the neuronal and endothelial NOS enzymes.

It is known that neuronal NO inhibition using L-NAME increases blood pressure, which causes excitability and excitation of neurons (Fewell and Johnson 1984). Blood pressure increases because the substance has cholinergic effects (Moore et al. 1993a, 1993b). Indeed, it can be seen in this work that the fastest response following the injection of L-NAME is an increase in blood pressure (Fig. 7.26).

It can be seen that following the injection of L-NAME, blood flow decreases in the control rats (although blood pressure increased) and in the experimental rats, but 10 min after the injection of Metrazol, the blood flow in the experimental rats is significantly higher than in the control rats (Fig. 7.26). Although the blood flow in the experimental rats is higher than the blood flow in the control rats, it can be seen that the blood flow in the experimental rats maintains a more or less constant level of blood flow. The blood flow following the Metrazol injection reaches a maximum of $107.5 \pm 31.9\%$ 70 min from the time of the Metrazol injection, meaning an increase of only 18% in blood flow relative to the basic level reached by the blood flow following the injection of L-NAME (Fig. 7.26). In comparison, Group no. 1 (Fig. 7.14) rats that underwent an injection of Metrazol alone (Chapter A) in which the blood flow reaches a maximum of $167.5 \pm 34\%$ 70 min from the time of the Metrazol injection, meaning an increase of 67.5%.

Although blood flow following Metrazol increases slightly compared to the control, it is seen that, unlike the two previous groups (rats that did not undergo treatment except for Metrazol injection—Fig. 7.14, and ischemic rats that underwent Metrazol injection—Fig. 7.18), the NADH level does not change throughout the experiment, only toward the end of the experiments, 130 min after Metrazol injection. There is a significant decrease in the NADH level and the pH in the experiment becomes slightly more basic.

In another episode, experiments were conducted in which awake rats were injected with 7NI and after 30 min, Metrazol was injected until consecutive seizures were induced. The control rats for this group were awake rats that were injected with 7NI and after 30 min, saline instead of Metrazol for the same experimental period.

There is morphological evidence in the literature that supports the assumption that neurons are the source of NO responsible for vasodilation during epileptic seizures (Bredt et al. 1990). In Chapter 3, we used the inhibitor L-NAME. The inhibitor L-NAME inhibits the activity of the enzyme NOS that acts on endothelial blood vessels, and as a result, its use showed an increase in blood pressure (Fig. 7.22). The effect of increasing blood pressure complicates the possibility of examining how NOS inhibitors affect central nervous system activity. It was found that 7NI- is a specific inhibitor of brain NOS, and does not affect the endothelium that undergoes dilation by acetylcholine (Moore et al. 1993b). Therefore, using the 7NI inhibitor is more suitable for studying the general nervous system, since the aforementioned inhibitor does not have cholinergic or cardiovascular effects (Moore et al. 1993a, 1993b). Indeed, it can be seen that the injection of 7NI caused a very slight increase in blood pressure (Fig. 7.27), so by injecting 7NI instead of L-NAME we neutralized the cholinergic and cardiovascular effects and thereby enabled us to examine how 7NI (a neuronal NOS inhibitor) affects the activity of the central nervous system, without any associated effects. It can be seen that in the first control group, which was injected with 7NI mixed with oil (Fig. 7.27, circles), blood flow decreases significantly compared to the second control group, which was injected with oil alone (triangles). Blood flow also decreases in the experimental group (group injected with 7NI and later with Metrazol); however, blood flow in the experimental group begins to increase significantly. Significant 80 min after Metrazol injection. (The delayed increase in blood flow is due to the injection method of 7NI—see explanation below). It has been reported that intravenous administration of 30 mg/kg 7NI produces an 80%–85% inhibition of the NOS enzyme activity within 30 min of injection. Four hours after injection, the enzyme inhibition is around 36%–54%, after 24 h there is no inhibition at all and the enzyme is fully active.

According to articles (Przegaliński et al. 1994), the 7NI inhibitor has a transient effect, so it is necessary to inject the substance every half hour or hour. Since the solubility of the substance is very low, it is necessary to dissolve it in 1 ml of oil, because the amount of liquid that needs to be injected is relatively large and the frequency of injection is high (every half hour) it was decided to inject 7NI only once One. And be content with the fact that after 4 h the enzyme is still inhibited at a level of 36%–54%. However, the transient effect of 7NI was reflected in the results (Figs. 7.27 and 7.28), as it can be seen that 90 min after the injection of the 7NI inhibitor, its effect decreased and blood flow began to increase (still insignificantly), and 110 min after the injection of the 7NI inhibitor, blood flow managed to increase significantly relative to the control. In contrast, NADH maintained its baseline level almost throughout the experiment.

When comparing this group to the group injected with the L-NAME inhibitor, it can be seen that the L-NAME inhibitor inhibited the increase in blood flow almost throughout the experiment (Fig. 7.22). Blood flow increased significantly only in the first attacks and later more or less maintained a baseline level. In contrast, NADH maintains its basal level almost throughout the experiment, in contrast to rats that did not undergo any additional treatments except for the injection of Metrazol (Fig. 7.14) and in contrast to ischemic rats that were injected with Metrazol (Fig. 7.18).

This finding is consistent with the findings of the researchers (Montécot et al. 1998), who examined the effect of 7NI injection and the creation of seizures in the hippocampus (our measurement, in contrast to theirs, was made in the cortex). They claim that 7NI- in addition to reducing blood flow in the hippocampus, reduces the level of pO_2 in the hippocampus, which prevents hyperoxia in the tissue during SE. They claim that the observed changes in pO_2 are probably a direct result of the local changes in blood flow. In their results, no changes in arterial pO2- and glucose metabolism were observed, but only a tissue change in pO_2.

These results are consistent with the results observed in this work, although our measurement was made on the surface of the cortex, and instead of measuring the tissue pO_2 level, we measured the NADH level. The NADH level represents for us the oxygen level inside the cells. When the oxygen level decreases, the NADH level increases, and in contrast, when the oxygen level in the tissue increases, the NADH level decreases if there is no effect of enzymatic factors such as ATPase activity. It can be seen that the blood flow following the injection of 7NI initially decreases, and there are no significant changes in the NADH level (Fig. 7.27). When compared to the other groups, it can be seen that in the group of animals that did not undergo additional treatments except for the injection of Metrazol (Fig. 7.14) and in the group of ischemic animals (Fig. 7.18), NADH decreased. In other words, it can be concluded that in the other groups, because the blood flow increased following the seizures, oxygen reached the system, the respiratory chain worked in an increased manner, and therefore NADH decreased. Later, toward the end of the experiment, when the animal's condition deteriorates due to the SE, NADH goes from an oxidation state to a reduction state because there is a lack of oxygen in the system, a lack of oxygen in the system because blood flow decreased toward the end of the experiment. In contrast, in the group of animals injected with 7NI, blood flow does not initially increase, so there is no excess oxygen in the system and NADH remains at its baseline level. However, toward the end of the experiment, blood flow increases, so the late increase supplies the oxygen so that the NADH level does not change.

Likewise, in the group injected with the inhibitor L-NAME and Metrazol (Fig. 7.22), a significant decrease in NADH levels can be seen toward the end of the experiment. In contrast, in the group of rats that did not undergo additional treatments except for the Metrazol injection, in which NADH increased toward the end of the experiment (Fig. 7.14). And in contrast to the group of ischemic animals that were injected with Metrazol, NADH increased significantly (shifted from oxidation to reduction) toward the end of the experiment (Fig. 7.18).

7.4.4 Effect of Hyperoxia on Brain Responses to Metrazol

This chapter describes the brain's response to exposure to high oxygen levels (hyperoxia) (Mayevsky 1984b; Mayevsky et al. 1986), a process accompanied by a decrease in blood flow, due to the increased presence of oxygen. And in the oxidation of mitochondrial NADH. In this chapter, the goal was to test whether improving the oxygen level would change the animal's condition during SE.

In these experiments, alert animals were ventilated with 100% oxygen and later injected with Metrazol until they had consecutive seizures. The control for the experiment was a group of alert rats that were ventilated with 100% oxygen for the same experimental period.

When exposed to hyperoxia, several changes were observed (Figs. 7.31 and 7.32). Fifty minutes after the Metrazol injection, blood flow decreased significantly relative to the control. NADH decreased in parallel with blood flow, both in the treated group and in the control group. Toward the end of the experiment, an increase in the NADH level could be seen in the experimental group relative to the control, but this increase was insignificant. Blood pressure in the first attacks increased slightly and then it is seen that the blood pressure began to decrease. According to these images (Figs. 7.31 and 7.32), it can be seen that as a result of excess oxygen in the system, the blood vessels narrow (vasoconstriction) and the blood flow decreases. The control group in this group maintains a relatively constant blood flow at the beginning of the experiments, after 120 min of 100% oxygen flow, there is a decrease in blood flow. In the experimental group, an insignificant decrease in blood flow can be seen that occurs following the Metrazol injection, and the decrease becomes significant relative to the control 50 min after the Metrazol injection.

Although blood flow decreases as a result of vasoconstriction, the NADH level does not change. Only toward the end of the experiment can a nonsignificant increase be seen in the experimental group compared to the control group. The NADH level does not change despite the decrease in blood flow, and this is because the rats are ventilated with 100% oxygen (there is no lack of oxygen in the system).

In the experimental group of rats exposed to 100% oxygen and epileptic seizures, blood flow is lower than in the control rats exposed to 100% oxygen only. That is, in the rats that have had epileptic seizures, stronger vasoconstriction occurs. It is possible that the reason is that, in addition to vasoconstriction that occurs from excess oxygen, vasoconstriction also occurs as a result of spasm that occurs in the muscles following the seizures, leading to further narrowing of the blood vessels.

When this group is compared to the group of ischemic animals that were injected with Metrazol (Figs. 7.18 and 7.19), although a decrease in blood flow is observed in both groups, the groups are different. In ischemic conditions, it is known that the oxygen level in the brain is lower; the decrease in blood flow in the ischemic condition is accompanied by a decrease in the rate of energy metabolism, which is expressed in an increase in the level of NADH (Mayevsky 1984a; Chance et al. 1973; Sick et al. 1998).

The diagnosis between the two conditions (hyperoxia and ischemia) is made by monitoring the two indicators blood flow and NADH level, and indeed it can be seen that when the blood flow in the ischemic experimental group decreased relative to the control, the level of NADH and potassium began to increase (80 min from the moment of Metrazol injection) and 160 min from the moment of injection the difference is significant. This is because there is an energy failure in the system, and the lack of oxygen led to an increase in the level of NADH and potassium. In contrast, when looking at the group of animals that underwent hyperoxia (Figs. 7.31 and 7.32), although blood flow decreased significantly, the NADH level did not increase but rather decreased because the system did not suffer from a lack of oxygen. Only toward the end of the experiment does a slight increase in the NADH level appear.

When comparing the hyperoxia condition (Figs. 7.31 and 7.32) to rats injected with the L-NAME inhibitor (a general inhibitor of the NOS enzyme of neuronal and endothelial origin) or the 7NI inhibitor (a specific inhibitor of the NOS enzyme of neuronal origin), it can be seen that the aforementioned inhibitors reduce blood flow. The situation is somewhat similar to the hyperoxia condition, since both conditions act on the blood vessels. The inhibitors cause the blood vessels not to dilate (vasodilation) because they inhibit the production of the NO molecule, which is the molecule that causes the expansion of blood vessels, and the hyperoxia condition causes a similar situation because the blood vessels, as a result of excess oxygen, undergo narrowing (vasoconstriction). However, it is seen that the inhibitors create a different situation from the ischemic condition because although the blood vessels do not undergo dilation and blood flow decreases. Nevertheless, there is no energy problem in the system until the end of the experiments, and NADH undergoes oxidation. (in the group injected with L-NAME, Fig. 7.22) or maintains its basal level (in the group injected with 7NI, Fig. 7.27), a situation similar to the NADH level in the group that underwent hyperoxia (Fig. 7.31). On the other hand, when looking at the blood flow, it can be seen that the inhibitors do reduce blood flow but no vasoconstriction occurs, similar to the group that was ventilated with 100% oxygen (in the 7NI group, it can be seen that when the effect of the inhibitor wears off, the blood flow increases, as already noted earlier).

When comparing the hyperoxia group to the group of animals that did not undergo any additional treatments except for the Metrazol injection (Figs. 7.14 and 7.15), it is seen that the blood flow in this group increases relative to the control and reaches a maximum value of 167.5 ± 34 after 70 min from the time of Metrazol injection. In this group, NADH also decreases, that is, it undergoes oxidation. In other words, the increase in blood flow is sufficient oxygen enters the system and NADH undergoes oxidation, but toward the end of the experiment NADH is on an increasing trend. In this group, in contrast to the hyperoxia group, there was an increase in blood flow during the experiments and not a decrease.

Under hyperoxia conditions, the activity of Na^+/K^+-ATPase pumps can be inhibited or attenuated, and a very slow increase in extracellular potassium occurs (Mayevsky 1984b). Indeed, it can be seen in Fig.7.36 that there was a significant increase in the extracellular potassium level at several points and at some time points an increase is observed but it is not significant due to the large variation between the experimental animals.

7.4.5 *Effect of Metrazol in Various Hemodynamic and Metabolic States*

Sect. 7.4 in this chapter will deal with the comparison between the five experimental groups in terms of several parameters (the parameters examined were statistically examined in Part 3 of the Results chapter). At the end of the five experimental groups, several questions were asked and this chapter will attempt to answer them.

7.4.5.1 Is there a Difference Between the Groups in the Number of SD Waves That Appeared Following the Metrazol Injection?

The phenomenon of the spreading depression wave is a suppression of the spontaneous electrical activity ECoG (due to tissue depolarization), a decrease in the local electrical potential (DC), an increase in blood flow, an increase in extracellular potassium, mitochondrial NADH oxidation and an increase in reflectance.

The SD wave propagates at a slow rate of 3–5 mm/min on the outer surface of the cortex from one hemisphere to the other and does not pass to the other hemisphere (Krivanek 1962).

In these experiments, we wanted to test the relationship between SD and epilepsy. The results of the present study showed that intravenous injection of PTZ 10 mg every 5–10 min is characterized by three phases: in the first phase, only SD waves appear. Later, short seizures appear that are accompanied by SD waves (see Fig. 7.10) and in the last phase, toward the end of the experiment, when the seizures are consecutive without a break between them, it can be seen that the ECoG amplitude increases and SD waves do not appear between the seizures, and one long seizure is obtained (see Fig. 7.11).

Researchers Koroleva and Bures (1983) claim that a burst of fast electrical frequency increases $[K^+]_e$ to a level of 10–12 mM. At this threshold level, an SD wave is formed. On the other hand, gradual epileptic development produces a gradual recruitment of the population of neurons responsible for the activation of metabolic pumps and they increase the transport of potassium ions and neurons that undergo depolarization. The transport is carried out at a level that neutralizes the accumulation of potassium in the extracellular space. This situation prevents the SD phase. It follows that epileptic activity induces SD or prevents SD.

Other researchers (Ueda and Bures 1977), who studied epileptic foci produced by penicillin, claim that prolonged intra-ictal activity (activity that precedes the epileptic seizure) of penicillin increases the clearance of extracellular potassium so that its level decreases to a level that prevents the development of SD. But brief local intervention with penicillin produces an SD wave.

A group of researchers using PTZ at an anticonvulsant concentration (40 mg/kg/IP) observed SD in the thalamus and cortex when they administered repeated daily injections of PTZ 40 mg/kg, and when generalized seizures developed, the number of SDs decreased. They claim that the early appearance of SD waves in the thalamus and neocortex after a single injection of 40 mg/kg PTZ indicates that the "naive" brain that first encounters PTZ responds to the initial application of PTZ by rapidly increasing potassium levels. The increase in potassium levels triggers the formation of SD, and this SD disrupts epileptic activity and returns activity to normal.

Epileptic activity produces epileptic seizures that generate high rate, repetitive firings that penetrate the entire brain, which disrupts the balance mechanism and skeletal muscle activation. SD blocks the spread of the seizures in the brain and creates a wave of return to normal activity (Koroleva and Bures 1983). Indeed, in the present study, it can also be seen that SD waves appear at the end of seizures.

Mayevsky exposed rats to hyperbaric oxygen and observed SD waves accompanying the seizures. His conclusion was that the phenomenon the SD observed after the epileptic activity may serve as a mechanism for the cessation of the electrical activity that occurs during the seizure (Mayevsky 1984b).

When the five groups are analyzed using an ANOVA test for the number of SDs counted in each experimental group, it can be seen that the groups do not differ from each other. This means that the different treatments did not affect the mechanism of SD generation in the cortex (Fig. 7.39I). It should be noted that all measurements were performed in the cortex, and therefore it is difficult to answer the mechanism of SD generation because it is possible that if the measurement had included different areas we would have seen differences between the groups.

7.4.5.2 Did the Number of Seizures Change Follow a Particular Treatment?

It can be seen from the results (Fig. 7.39II) that the number of seizures in the three groups: the ischemic group, the group treated with L-NAME and the 7NI group decreased relative to the group of animals treated with Metrazol alone. The group of animals that were ventilated with 100% oxygen was similar in terms of the number of seizures. The group that received only Metrazol.

The results of the L-NAME group and the group that was injected with 7NI are consistent with some researchers. Some claim that NO is an anticonvulsant (Buisson et al. 1993; Przegaliński et al. 1994) and some claim that NO causes tremors (convulsions) (Mülsch et al. 1994; De Sarro et al. 1991). The above debate is explained by different experimental conditions, different species, different epileptic agents and the type of seizures.

Those who support the assumption that NO is responsible for convulsions make two claims:

1. Excitatory amino acids such as NMDA and kinin are known to be potent in causing seizures, and since the activation of the NMDA receptor is done by NO.
2. Several tremor-inducing substances, such as amino acids Excitatory agents, PTZ, and pilocarpine (a substance that induces tonic-clonic seizures) increase cGMP in the cerebellum. Anticonvulsants lower their level. And the NO molecule also causes an increase in cGMP levels.

The relationship between the NO molecule and PTZ is explained as follows: lack of inhibition of the inhibitory neurotransmitter GABA (Macdonald and Barker 1977) or activation of the NMDA receptor (Nevins and Arnolde 1989; Velísek et al. 1990) are both factors involved in the initiation of seizures, and in general in the induction of seizures by PTZ. And since the main action of NMDA and GABA is to increase the level of intracellular Ca^{2+}, and since it is known that the enzyme NOS depends on Ca^{2+} calmodulin, it is possible that PTZ activates the enzyme NOS so that the use of L-NAME and 7NI- (which are NOS inhibitors) causes a reduction in the number of seizures.

Since L-NAME is a neuronal and endothelial inhibitor, inhibition by L-NAME causes an increase in blood pressure. The increase in blood pressure results from inhibition of the activity of the NOS enzyme that acts in endothelial blood vessels. The above effect complicates the possibility of testing the activity of NOS inhibitors on the activity of the central nervous system, since the increase in blood pressure causes the agitation and excitation of the neurons (Fewell and Johnson 1984). It was found that 7NI- inhibits the NOS enzyme in the brain and does not affect the endothelium, which is dilated by acetylcholine. The disadvantage of the increase in blood pressure shows that 7NI- is more suitable for studying the general nervous system. Therefore, it was decided in this study to inhibit the brain NOS enzyme using 7NI. It can be seen from the present results that both inhibitors L-NAME and 7NI, similar to other studies (Mülsch et al. 1994; De Sarro et al. 1991), acted as anticonvulsants, since they reduced the number of seizures. It can be concluded from the results that the NO molecule participates in the induction of epileptic seizures, since it can be seen that inhibition of the NO molecule by the inhibitors reduced the number of seizures. The current results show that there is no difference between the inhibitors in terms of chemical or electronic effect, and the cardiovascular effect is not involved in the induction of seizures, similar to other studies (Przegaliński et al. 1994).

However, according to the results, it can be seen that the group injected with 7NI and the ischemic group are completely different from the Metrazole group and the group of animals that were ventilated with 100% oxygen (the hypercapnia group), and in contrast, the L-NAME group is different from the Metrazol group but similar to the hyperoxia group.

The answer to this lies in the fact that, in addition to the fact that L-NAME and 7NI inhibit the activation of the NMDA receptor and thus inhibit seizures, another mechanism is probably involved in the generation of seizures.

Since the ischemic group also showed a decrease in the number The seizures and also the groups injected with an inhibitor (7NI and L-NAME), the factor that plays a role in the decrease in the number of seizures is probably tissue oxygen. It has been proven that during hyperoxia, a group with a high tissue oxygen level had multiple seizures. In an ischemic state, the tissue oxygen level is lower. Also, findings from various studies show that NO inhibitors, in addition to affecting blood flow, reduce tissue oxygen levels (Montecot et al. 1997). There is morphological evidence that supports the assumption that neurons are responsible for the NO that increases blood flow during epileptic seizures (Montecot et al. 1997). This finding may explain the differences observed between the 7NI and L-NAME groups.

The research group (Montécot et al. 1998) claims that during SE there is a large increase in blood flow and tissue pO_2. More oxygen comes to the tissue than it needs. Excess oxygen endangers neurons because it oxidizes free fatty acids. The above situation apparently leads to oxidative stress, which leads to damage to brain tissue. The claim of the aforementioned group of researchers may explain the current results by stating that the damage caused to the neurons probably causes them to become activated, and they begin to fire, causing repeated seizures and an increase in the number of seizures.

However, it should be noted again that all the measurements conducted in this study were made in the cortex, and therefore it is difficult to answer the question of what the mechanism for creating the seizure is, since the measurement was made in the cortex. It is also possible that the number of seizures decreased in the cortex, and did not change in the other areas or was increased in the other areas.

7.4.5.3 Does the Decay Time Before Death Differ Between the Groups? Does It Lengthen Or Shorten?

In each animal before death, it was possible to observe a deterioration in the physiological state of the brain, as expressed in changes in the parameters. During the deterioration, NADH increased, blood flow decreased, and extracellular potassium increased. From Fig. 7.40I, it can be seen that there is no difference between the groups regarding the decay time. That is, the different treatments did not accelerate or prolong the deterioration before death.

7.4.5.4 Is Survival Time Prolonged Or Shortened?

Survival time was determined as the time in minutes from the moment of Metrazol administration until death, and the aim was to test whether survival time was prolonged or shortened in the five experimental groups. It can be seen that survival time does not differ between the five experimental groups (Fig. 7.40II). In other words, the different treatments did not prolong or shorten the life span of the rats, and the factor that affected the duration of life was only the epileptic seizures, which were created by Metrazol.

7.4.5.5 Is the Level of NADH in the First and Last Seizures Different Within the Group and Between the Groups?

In order to test the energetic state during the seizures, the level of NADH was measured in the first SD waves that appear after the Metrazol injection, and in the last SD waves. The first and last SD waves were chosen as representative in the five experimental groups. The first SD waves were chosen because during the SD wave The first wave is the animal in the best energetic state relative to the other waves that appear later. The last SD wave was chosen because the last wave appears after a sequence of seizures, and the aim of the study was to examine whether, following the seizures, the energetic state will deteriorate and NADH will move from an oxidation level to a reduction level.

Based on the results of the ANOVA test (Fig. 7.41I), it can be seen that the first NADH wave does not differ between the five groups. That is, in the above situation, the energetic state of all groups is similar. However, in the last wave, energetic changes occur and it can be seen that the Metrazol group is similar to the hyperoxia group, but different from the ischemic group, the 7NI- group, and the L-NAME group. The ischemic group is different from all groups and the L-NAME group is similar to the 7NI- group and the hyperoxia group.

It can be seen that the Metrazol group and the ischemic group switched from an oxidation state to a reduction state, while the L-NAME, 7NI, and hyperoxia groups did not switch to a reduction state, but even in the above groups the NADH wave was less oxidized. The above results are partially consistent with the results of other researchers who measured the NADH level using the fluorometric method (this study also measured using the above method) and examined the relationship between the oxidation-reduction state of NADH and epileptic seizures (Van Buren et al. 1978; O'Connor et al. 1972a; Mayevsky 1984b). Most studies have shown a decrease in NADH levels during an attack. In all of the above works, it was proven that epileptic activity increases energetic activity due to an increase in extracellular potassium, which leads to the activation of sodium-potassium pumps that require a large amount of energy.

In this work, unlike the other works, continuous measurement was made so that it was possible to see in each seizure how the NADH level changes. The current results support the other researchers because a decrease in NADH levels can be seen in the first seizures, meaning that NADH undergoes oxidation. However, as the number of seizures increases, the energetic state in the animal's brain changes and NADH goes from an oxidation state to a reduction state (such as the Metrazol group and the ischemic group) or the oxidation level is lower (such as the L-NAME, 7NI and hyperoxia groups). It can be seen that the group with the greatest energetic problem is the ischemic group, because from oxidation there is a transition to a high reduction state. Also, the Metrazol group goes from an oxidation state to a reduction state. The other groups do not go into a reduction state, but the oxidation level in the last wave is lower.

The ischemic group probably goes into a high reduction state because this group underwent closure of two carotid arteries 24 h before the experiment. Apparently, this group starts from a lower oxidation level at the beginning of the experiment. Indeed, it can be seen that already in the first SD wave, the oxidation in the ischemic group is the lowest, but not significantly.

A very interesting phenomenon is observed when looking at Fig. 7.41I. It can be seen that NADH in the L-NAME, 7NI and hyperoxia groups is less oxidized in the last wave but does not go into a reduction state. We would expect that in the 7NI and L-NAME groups, blood flow during seizures will decrease because NO is inhibited, less oxygen will be found in the system, so that the NADH level will be higher than in the group of animals that received only Metrazol. But this is not the case. While looking at Fig. 7.41II, which measures the increase in blood flow during the first and last seizures, it can be seen that during the first and last seizures, the blood flow in the L-NAME and 7NI- groups does not decrease relative to the other groups. This means that following treatment with the inhibitors L-NAME and 7NI-, the blood flow between seizures decreased, as noted in the previous chapter (Fig. 7.41I), but during the seizure the inhibitors do not inhibit the increase in blood flow and apparently another mechanism is involved in the dilation of blood vessels during seizures besides NO. It can be seen that the 7NI- and L-NAME groups are in a better energetic state than the Metrazol group. This situation may be due to several reasons:

1. L-NAME and 7NI- showed a reduction in the number of seizures, so it is possible that the energetic state of the animals is better in these groups because they appeared less seizures.

2. The group of researchers who measured the hippocampus showed that during SE there is a large increase in blood flow and tissue pO_2. As a result, more oxygen reaches the tissue, more than it needs. During SE, NO molecules are involved in the destruction of neurons not only because they participate in the toxicity of glutamate on the NMDA receptor and the kinetics. But also because they increase local blood flow and cause a greater supply of oxygen to the tissue, the high level of oxygen in the tissue causes oxidative stress (Montecot et al. 1997). Therefore, when we neutralized the NO molecule using the inhibitors, the increased oxygen supply to the tissue was prevented, and therefore the 7NI- and L-NAME groups are in a better energetic state.

It can be seen that the hyperoxia group is similar in terms of the level of oxidation in the last waves to the L-NAME group and the 7NI- group. When we think about what the three groups have in common The above conclusion is that in the three groups, the blood vessels are undergoing narrowing. In a state of hyperoxia, the blood vessels are vasoconstricted due to excess oxygen, and in the L-NAME and 7NI groups, the blood vessels are less dilated because the inhibitors impair the activity of the NOS enzyme. (In the ischemia group, blood flow is lower, but the decrease in blood flow is not due to vasoconstriction, but rather to the closure of the carotid arteries.) Although the three groups are in a state where their blood vessels are less dilated, it can be seen in Fig. 7.41II that the increase in blood flow in the above groups in both the first and last SD waves is no different from the Metrazol group and the ischemic group. Moreover, it can be seen that the blood flow in the above groups is insignificantly higher than the ischemic group and the Metrazol group. So it is possible that when the blood vessels are in a state of vasoconstriction, an additional mechanism is at work that causes an increase in blood flow during the attacks.

7.4.5.6 Blood Flow

Since the creation of ischemia, inhibition of the NOS enzyme, and the creation of hyperoxia reduce blood flow, the goal was to test whether blood flow would also decrease during seizures. As can be seen (Fig. 7.41II), there are no significant differences between the first and last SD waves between the groups, and there are no significant differences within the group. However, it can be seen that the blood flow in the last waves in the ischemic group of animals is lower than in the other groups, but not significantly. This means that it is more difficult for the ischemic animals to increase blood flow as a result of the carotid arteries being closed. The blood flow in the last waves in the L-NAME and 7NI- groups is insignificantly higher than in the other groups, contrary to expectations, since the substances L-NAME and 7NI- reduce blood flow. We expected to see a decrease in blood flow also in SD and concluded that, apart from the NO molecule that causes vasodilation, another mechanism is probably involved in vasodilation during epileptic seizures.

7.4.5.7 Potassium Amplitude

The potassium amplitude was measured in the first and last SD waves. The meaning of the amplitude is the ratio between the leakage of potassium ions and their pumping in. Potassium amplitude = leakage/pumping in It can be seen from the image (Fig. 7.42II) that there is no difference between the groups in terms of potassium amplitude and also there are no significant differences within the group between the first and last waves. In other words, it can be concluded that in all experiments, there is no damage to the sodium and potassium pumps.

7.4.5.8 The Duration of Time It Takes for the Potassium Wave During the SD To Drop to Its Baseline Level

Duration means how long it takes for the system to return the potassium to its baseline level, and duration depends on the energy available in the system. It can be seen that according to the ANOVA test (Fig. 7.42II), there are no differences between the groups, neither in the first nor in the last waves. This means that there is no energy problem in any of the five groups.

7.4.5.9 Time to First Seizure

It can be seen from the ANOVA test (Fig. 7.43I) that the time to first seizure does not differ between the Metrazol, ischemic and L-NAME groups. In contrast, the 7NI- and hyperoxia groups differ in that the first seizure occurs earlier in them. As is known, breathing 100% oxygen (hyperbaric hyperoxia) causes vasoconstriction and the formation of tonic-clonic seizures (Bitterman and Bitterman 1998). In the present study, hyperoxia was created using 100% oxygen and at atmospheric pressure. Despite this, it was found that the first seizure following Metrazol injection appeared faster. This means that the tissue is probably more sensitive to the formation of seizures as a result of the high oxygen concentration. Similar to the hyperoxia group, the inhibitor 7NI also caused the first seizure to occur more rapidly, similar to the study by (Montecot et al. 1997) and different from the study by (Bitterman and Bitterman 1998). Apparently, in addition to the fact that 7NI competes with the NOS enzyme, the substance 7NI probably works through a neuronal mechanism that increases the sensitivity to the onset of the first seizure (Montecot et al. 1997). However, although the first seizure appeared earlier, the substance 7NI reduced the total number of seizures during the experiment (Fig. 7.39II).

7.4.5.10 Duration Of Seizures

The above parameter was very difficult to determine, because after the Metrazol injection, the first seizure appears at some time. After which, continuous seizures develop.

Sometimes, the seizures seem to stop electrically (no expression of the seizures is seen in the ECoG parameter or in the other parameters) but there are still mild seizures that are not expressed in the ECoG but in the form of spikes. Therefore, it was decided in the current study not to determine the time at which the seizures stop, since it is possible to make a mistake based on the polygraph chart and the computer program. Rather, to take the time from the moment the first seizure appears until death and to call this time—the seizure time. According to the ANOVA test (Fig. 7.43II) it can be seen that the time from the moment of the first seizure to death is not different in the five groups. In other words, it can be concluded that the different treatments did not affect the duration of the seizures.

7.4.5.10.1 Summary of the Discussion

This work deals with the assessment of brain function during epileptic seizures induced by Metrazol, when additional factors were combined: ischemia (closure of two carotid arteries), inhibition of the NO molecule using the inhibitors L-NAME and 7NI-, and hyperoxia. In the present work, we used alert animals exposed to epileptic seizures induced by PTZ, in which various parameters from the same brain region were examined. These parameters indicate the relationship between the hemodynamic, energetic, ionic, and electrical activity of the brain and were measured continuously starting from the control phase (before the first seizure), during the seizures, and after the end of this period. The results were examined on two levels. One, focusing on measurements between seizures, and the other focusing on the seizures themselves and the phenomena that occur around the seizures (such as SD waves, survival time, time until the appearance of the first seizure, etc.). Experimental models (external factors) are known in the literature, the use of which creates epileptic seizures in animals. Internal factors are also known in humans and animals that cause the occurrence of epileptic seizures. The external factors that can be used to create epileptic seizures are: (1) chemical stimulation, use of a variety of different compounds, local use (drip over the cortex or systemically), (2) electrical stimulation, and (3) use of hyperbaric oxygen. The internal factors through which spontaneous seizures are created are: (1) ischemia, (2) head trauma, and (3) genetic epilepsy. As is known, none of the models is a fully reliable imitation of clinical epilepsy. Second, a dual type of epilepsy appears for the same model (Fisher 1988). Therefore, in order to validate important findings, it is necessary to use several models. Many animal species develop spontaneous seizures. However, the above cases are random and not always in animals that are suitable for the experiment (Löscher and Schmidt 1988). All the factors I mentioned above lead to epileptic seizures that are sometimes accompanied by the SD phenomenon. The SD phenomenon is manifested by spontaneous suppression of EEG activity in contrast to epileptic seizures that are accompanied by electrical discharges or spikes. The SD phenomenon observed after the epileptic activity serves as a mechanism for the cessation of electrical activity that occurs during the seizure (Koroleva and Bures 1983; Mayevsky 1984b). It is known that gradual epileptic development produces a gradual recruitment of the population of neurons that are responsible for the activation of metabolic pumps and they increase the transport of potassium ions to neurons that undergo depolarization. The transport is carried out at a level that neutralizes the accumulation of potassium In the extracellular space. This condition prevents the SD phase (Koroleva and Bures 1983), but rapid epileptic development produces an SD wave (Ueda and Bures 1977). According to the model we used in the present study (intravenous injection of 10 mg of Metrazol every 5–10 min), it can be seen that three phases appear in the experiment: in the first phase, only SD waves appear. Later, short seizures appear that are accompanied by SD waves, and in the last phase, toward the end of the experiment, the seizures are continuous without a break between them. It can be seen that the ECoG amplitude increases and SD waves do not appear between the seizures, and one long seizure is obtained. During the seizure, electrical discharges or spikes appear on the EEG, and then when the SD phenomenon appears, the brain's electrical activity is suppressed. The trigger for the successful appearance of an SD wave depends on the concentration of extracellular potassium reaching a level of 10–12 mM (Lothman et al. 1975). In studies where extracellular ion concentrations were measured using ion-selective mini electrodes, it was found that, in parallel with the EEG suppression and changes in the DC potential during SD, large but reversible changes in the extracellular ion concentration were obtained (Mayevsky et al. 1982; Hansen and Zeuthen 1981). An increase in the level of extracellular potassium ions was obtained, and the extracellular pH decreased. According to the scheme (Fig. 7.44), it can be seen that Metrazol leads to the formation of SD and the formation of epileptic seizures. Sometimes the SD waves appear alone (usually at the beginning of the injection) and later the SD waves appear at the end of the seizure. The arrows indicating a relationship between the SD waves and seizures indicate a positive arrow from the direction of the seizures to the SD waves. That is, seizures lead to the development of SD waves, while the SD in the direction of the seizures appears with a negative arrow. That is, the SD phenomenon observed after the epileptic activity serves as a mechanism for the cessation of electrical activity that occurs during the seizure. It can be seen that both the seizures and the SD waves are characterized by similar metabolic, hemodynamic and ionic changes in contrast to the electrical activity. However, the metabolic, hemodynamic and ionic changes that appear during the seizure are expressed at a lower intensity. The seizure is characterized by a wave whose duration is short and whose amplitude is low in all parameters. In order to understand the changes that occur during the seizure compared to the changes that occur during SD, you can look at Fig. 7.10. In Fig. 7.10, when looking at the eighth SD wave, one can see that the potassium level increases in two waves. The first wave is the wave that appears as a result of the seizure. One can see that the duration of the aforementioned wave is short, as well as its amplitude. In contrast, the second potassium wave that characterizes the SD phenomenon is characterized by a longer duration and higher amplitude. It can also be seen that blood flow is also characterized by two waves. In the first wave that appears as a result of the seizure, blood flow increases but its duration is short and the amplitude is low. In contrast, the second wave of blood flow that characterizes the SD phenomenon is characterized by a longer duration and high

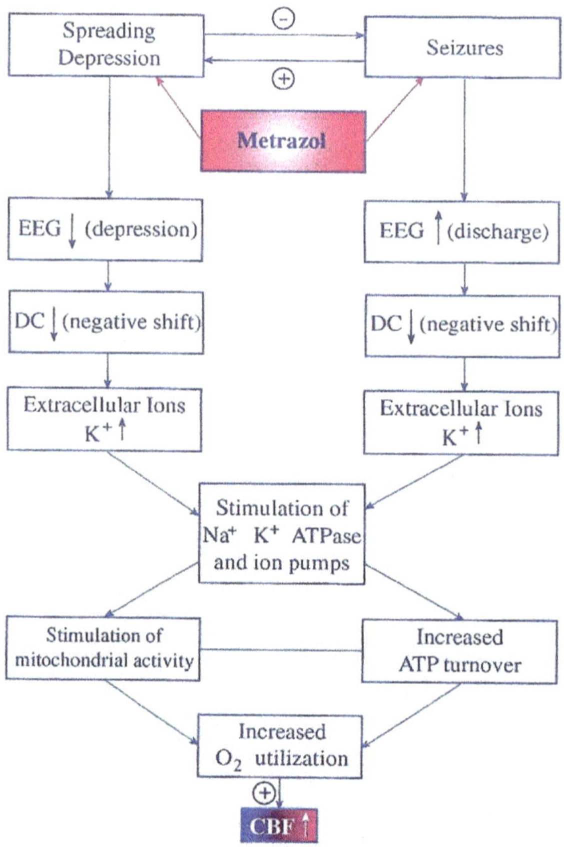

Fig. 7.44 Schematic description of electrical, ionic, metabolic, and hemodynamic processes that occur following seizures and SD waves

amplitude. Sometimes the wave that characterizes the seizure is invisible because it is swallowed up in the SD wave. It can be seen by agreement (Fig. 7.44) that returning the ions to their normal level following the seizure and the SD wave is a process that requires energy. Therefore, an increase in energy metabolism, activation of the respiratory chain and the formation of more ATP are expected. This will allow increased activation of Na+/K+-ATPase and the return of the ions to their normal level on both sides of the membrane. An increase in oxygen consumption and an increase in blood flow are also observed (Mayevsky and Weiss 1991).

The mechanism of action of Metrazol is only partially understood. According to the literature, it is known so far that at the synaptic level, lack of inhibition of the inhibitory neurotransmitter GABA (Macdonald and Barker 1977) or activation of the NMDA receptor (Nevins and Arnolde 1989) are both factors that are initially and generally involved in the induction of seizures by Metrazol. Since the main action of NMDA and GABA is to increase the intracellular Ca^{2+} level and since the enzyme NOS is a

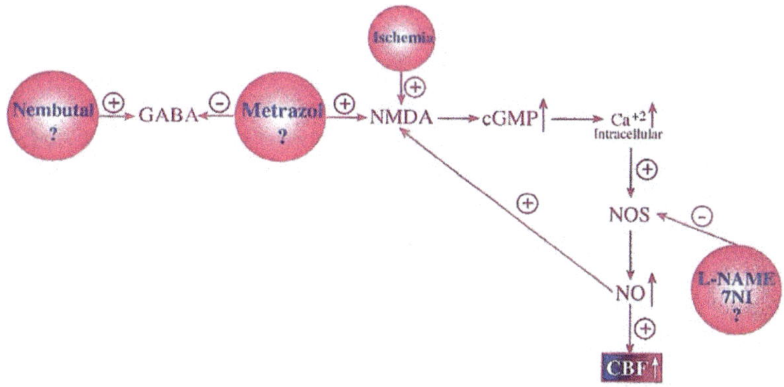

Fig. 7.45 Schematic description of processes that occur in the synapse during Metrazol injection and in combination with various conditions: ischemia, injection of NOS inhibitors, and injection of Nembutal

Ca^{2+} calmodulin-dependent enzymes (Knowles et al. 1989), it is possible that Metrazol activates the enzyme NOS (Osonoe et al. 1994), see scheme.

This is consistent with the results we obtained, since the use of the inhibitors L-NAME and 7NI- showed a reduction in the number of seizures. L-NAME and 7NI-inhibit the increase in the enzyme NOS and thus reduce the level of NO molecules, thereby leading to the suppression of the seizures induced by PTZ. Although the number of seizures decreased, no differences were found in the mortality rate in animals treated with the inhibitors, and while observing the time of the experiments, it was not possible to see that L-NAME or 7NI moderated the intensity and parameters of the seizure in terms of the electrical and motor aspects of the rats. As is known, during ischemia, extracellular glutamate causes uncontrolled activation of the NMDA and AMPA receptors, which allow sodium and calcium to enter the cell (Hakim 1998)—see diagram (Fig. 7.45). When looking at the results of the current study, it can be seen that between seizures, blood flow does not increase and even decreases relative to the control (Fig. 7.20 in the Results section). However, when looking at blood flow during seizures (Fig. 7.41II in the Results section), it can be seen that blood flow manages to increase even in the last seizures. These results indicate that during ischemia and epileptic seizures, when the system is required to increase blood flow, another factor participates that helps to dilate blood vessels and probably during ischemia and epileptic seizures, an increase in NO synthesis occurs, which causes an increase in blood flow. At the synaptic level, the Metrazol probably interacts with the GABA receptor—benzodiazepine chloride ionophore complex and apparently in some way reduces the inhibitory potency of the inhibitory neurotransmitter GABA (Macdonald and Barker 1977). Since barbiturates are known to reduce the rate of dissociation of GABA from its receptor (Franks and Lieb 1994), it was important to examine how barbiturates would affect seizures induced by Metrazol. Indeed, it can be seen in the present study that administration of Nembutal in a state of status epilepticus completely stopped the development of seizures (Fig. 7.37 in the Results section).

References

Aaslid, R. 1987. Visually evoked dynamic blood flow response of the human cerebral circulation. *Stroke* 18: 771–775.

Ajmone-Marsan, C. 1972. Focal electrical stimulation. In *Experimental models of epilepsy: A manual for the laboratory worker*, eds. D. P. Purpura, J. K. Penry, D. M. Woodbury, D. B. Tower, and R. D. Walter. New York: Raven.

Annegers, J. F., J. D. Grabow, R. V. Groover, E. R. Laws, Jr., L. R. Elveback, and L. T. Kurland. 1980. Seizures after head trauma: a population study. *Neurology* 30:683–689.

Backman, J. S., K. A. Conger, J. Chen, M. J. Ten, and J. H. Halsey. 1991. Nitric oxide helps regulate cerebral blood flow. *Journal of Cerebral Blood Flow and Metabolism* 11:S629.

Balazs, R. 1970. Carbohydrate metabolism. In *Handbook of neurochemistry*, ed. E. Lajtha. New York: Plenum Press.

Barrett, E. F., and J. N. Barret. 1976. Separation of two voltage-sensitive potassium currents, and demonstration of a tetrodotoxin-resistant calcium current in frog motoneurones. *The Journal of Physiology* 255:737–774.

Betz, A. L., D. D. Gilboe, and L. R. Drewes. 1976. The characteristics of glucose transport across the blood–brain barrier and its relation to glucose metabolism. *Advances in Experimental Medicine and Biology*. New York: Plenum Press.

Bircher, R. P., T. Kanai, and S. C. Wang. 1962. Intravenous, cortical and intraventricular dose-effect relationship of pentylenetetrazol, picrotoxin and deslanoside in dogs. *Electroencephalography and Clinical Neurophysiology* 14:256–267.

Bitterman, N., and H. Bitterman. 1998. L-arginine-NO pathway and CNS oxygen toxicity. *Journal of Applied Physiology (1985)* 84: 1633–1638.

Bonte, F. J., M. D. Devous, Sr., E. M. Stokely, and R. W. Homan. 1983. Single-photon tomographic determination of regional cerebral blood flow in epilepsy. *American Journal of Neuroradiology* 4:544–546.

Bredt, D. S., P. M. Hwang, and S. H. Snyder. 1990. Localization of nitric oxide synthase indicating a neural role for nitric oxide. *Nature* 347:768–770.

Buisson, A., N. Lakhmeche, C. Verrecchia, M. Plotkine, and R. G. Boulu. 1993. Nitric oxide: An endogenous anticonvulsant substance. *Neuroreport* 4:444–446.

Carlson, H., E. Ronne-Engstrom, S. Blom, R. Flink, B. Gazelius, and L. Hillered. 1991. Electrocorticography and simultaneous recording of regional cerebral blood flow with laser Doppler technique from the human epileptic focus. *Abstracts Social neuroscience* 17:1255.

Chance, B., P. Cohen, F. Jobsis, and B. Schoener. 1962. Intracellular oxidation-reduction states *in vivo*. *Science* 137:499–508.

Chance, B., N. Oshino, T. Sugano, and A. Mayevsky. 1973. Basic principles of tissue oxygen determination from mitochondrial signals. *Advances in Experimental Medicine and Biology* 37A:277–292.

Chavoix, C., P. Hantraye, E. Brouillet, B. Guibert, H. Fukuda, V.de LaSayette, D. Fournier, R. Naquet, and Mazière, M.. 1988. Status epilepticus induced by pentylenetetrazole modulates in vivo [11C]Ro 15-1788 binding to benzodiazepine receptors. Effects of ligands acting at the supramolecular receptor complex. *European Journal of Pharmacology* 146:207–214.

Cooper, A.. 1836. Some experiments and observations on tying the carotid and vertebral arteries, and the pneumogastric, phrenic and sympathetic nerves. *Guy's Hospital Reports* 1:457–459.

De Sarro, G. B., E. Donato Di Paola, A. De Sarro, and M. J. Vidal. 1991. Role of nitric oxide in the genesis of excitatory amino acid-induced seizures from the deep prepiriform cortex. *Fundamental & Clinical Pharmacology* 5:503–511.

DiChiro, G., and R. A. Brooks. 1988. PET quantitation: Blessing and curse. *Journal of Nuclear Medicine: Official Publication, Society of Nuclear Medicine* 29:1603–1604.

Dichter, M. A., C. J. Herman, and M. Selzer. 1972. Silent cells during interictal discharges and seizures in hippocampal penicillin foci. Evidence for the role of extracellular K+ in the transition from the interictal state to seizures. *Brain Research* 48: 173–183.

Dirnagl, U., B. Kaplan, M. Jacewicz, and W. Pulsinelli. 1989. Continuous measurement of cerebral cortical blood flow by laser-Doppler flowmetry in a rat stroke model. *Journal of Cerebral Blood Flow & Metabolism* 9:589–596.

Dreifus, F. E.. 1981. Proposal for revised clinical and electroencephalographic classification of epileptic seizures. From the Commission on Classification and Terminology of the International League Against Epilepsy. *Epilepsia* 22:489–501.

Droste, D. W., A. G. Harders, and E. Rastogi. 1989. A transcranial Doppler study of blood flow velocity in the middle cerebral arteries performed at rest and during mental activities. *Stroke* 20:1005–1011.

Drummond, J. C., and H. M. Shapiro. 1990. Cerebral physiology. In *Anesthesia*, ed. R. D. Miller. New York: Churchill Livingstone.

Duncan, R.. 1992. Epilepsy, cerebral blood flow, and cerebral metabolic rate. *Cerebrovascular and Brain Metabolism Reviews* 4:105–121.

el Karib, A. O., J. Sheng, A. L. Betz, and R. L. Malvin. 1993. The central effects of a nitric oxide synthase inhibitor (N omega-nitro-L-arginine) on blood pressure and plasma renin. *Clinical and Experimental Hypertension* 15:819–832.

Engel, Jr., J.. 1984. The use of positron emission tomographic scanning in epilepsy. *Annals of Neurology* 15(Suppl):S180–S191.

Engel, Jr., J., W. J. Brown, D. E. Kuhl, M. E. Phelps, J. C. Mazziotta, and P. H. Crandall. 1982a. Pathological findings underlying focal temporal lobe hypometabolism in partial epilepsy. *Annals of Neurology* 12:518–528.

Engel, Jr., J., D. E. Kuhl, and M. E. Phelps. 1982b. Patterns of human local cerebral glucose metabolism during epileptic seizures. *Science* 218:64–66.

Engel, Jr., J., D. E. Kuhl, M. E. Helps, R. Rausch, and M. Nuwer. 1983a. Local cerebral metabolism during partial seizures. *Neurology* 33:400–413.

Engel, J., D. E. Kuhl, and M. E. Phelps. 1983b. Regional brain metabolism during seizures in humans. In *Advances in neurology*, eds. A. V. Delgado-Escueta, C. G. Wasterlain, E. M. Traiman, and R. J. Porter. New York: Raven Press.

Engle, J. and T. A. Pedley. 1998. *Epilepsy: A comprehensive textbook.* Lippincott-Raven Pub. Philadelphia, Pennsylvania USA.

Erecinska, M., and I. A. Silver. 1989. ATP and brain function. *Journal of Cerebral Blood Flow and Metabolism* 9:2–19.

Falconer, M. A., E. A. Serafetinides, and J. A. Corsellis. 1964. Etiology and pathogenesis of temporal lobe epilepsy. *Archives of Neurology* 10:233–248.

Fertziger, A. P., and J. B. Ranck, Jr. 1970. Potassium accumulation in interstitial space during epileptiform seizures. *Experimental Neurology* 26:571–585.

Fewell, J. E., and P. Johnson. 1984. Acute increases in blood pressure cause arousal from sleep in lambs. *Brain Research* 311:259–265.

Fisch, B. J., and T. A. Pedley. 1987. Generalized tonic-clonic epilepsies. In *Epilepsy: Electroclinical syndromes*, eds. H. Lüders and R. P. Lesser. New York: Springer.

Fisher, R. S. 1988. Epilepsy. In *Experimental surgery and physiology: Induced animal models of disease*, eds. M. M. Swindle and R. J. Adams. Baltimore: Williams and Wilkins.

Fisher, R. S. 1989. Animal models of the epilepsies. *Brain Research. Brain Research Reviews* 14:245–278.

Fisher, R. S., T. A. Pedley, W. J. Moody, Jr., and D. A. Prince. 1976. The role of extracellular potassium in hippocampal epilepsy. *Archives of Neurology* 33:76–83.

Fragen, R. J., and M. J. Avram. 1990. Barbiturates. In *Anaesthesia*, ed. R. D. Miller. New York: Churchill Livingstone.

Franks, N. P., and W. R. Lieb. 1994. Molecular and cellular mechanisms of general anaesthesia. *Nature* 367:607–614.

Freeman, B. A., and J. D. Crapo. 1982. Biology of disease: Free radicals and tissue injury. *Laboratory Investigation* 47:412–426.

Fromm, G. H., C. L. Faingold, R. A. Browning, and W. M. Burnham. 1987. *Epilepsy and the reticular formation: The role of reticular core in convulsive seizures*. New York: Liss.

Galla, H. J. 1993. Nitric oxide, an intracellular messenger. *Angewandte Chemie International Edition in English* 32:378–380.

Garthwaite, J. 1991. Glutamate, nitric oxide and cell-cell signalling in the nervous system. *Trends in Neurosciences* 14:60–67.

Gibbs, F. A., W. G. Lennox, and E. L. Gibbs. 1934. Cerebral blood flow preceding and accompanying seizures in man. *Archives of Neurology and Psychiatry* 32:257–272.

Glötzner, F. L. 1973. Membrane properties of neuroglia in epileptogenic gliosis. *Brain Research* 55:159–171.

Go, K. G. 1991. *Intracranial pressure: Cerebral pathophysiology*. Amsterdam: Elsevier.

Goodin, D. S., and M. J. Aminoff. 1984. Does the interictal EEG have a role in the diagnosis of epilepsy?. *Lancet* 1: 837–839.

Goodman, L. S., and A. Gilman. 1970. *The pharmacological basis of therapeutics*. London: Macmillan.

Gotoh, F., J. S. Meyer, and M. Tomita. 1966. Hydrogen method for determining cerebral blood flow in man. *Archives of Neurology* 15:549–559.

Grant, F. C., E. B. Spitz, H. A. Shenkin, C. F. Schmidt, and S. S. Kety. 1947. The cerebral blood flow and metabolism in idiopathic epilepsy. *Transactions of the American Neurological Association* 72:82–87.

Grazia Marciani, M., G. Santone, G. Sancesario, R. Massa, P. Stanzione, and G. Bernardi. 1990. Epileptic activity following 10 minute cerebral ischemia in Mongolian gerbils: An electrophysiological study. *Neuroscience Letters* 112:48–53.

Green, J. D. 1964. The hippocampus. *Physiological Reviews* 44:561–608.

Guevara-Guzman, R., P. C. Emson, and K. M. Kendrick. 1994. Modulation of *in vivo* striatal transmitter release by nitric oxide and cyclic GMP. *Journal of Neurochemistry* 62:807–810.

Guglielmo, M. A., P. T. Chan, S. Cortez, E. G. Stopa, P. Mcmillan, C. E. Johanson, M. Epstein, and C. E. Doberstein. 1998. The temporal profile and morphologic features of neuronal death in human stroke resemble those observed in experimental forebrain ischemia: The potential role of apoptosis. *Neurological Research* 20:283–296.

Haberl, R. L., M. L. Heizer, A. Marmarou, and E. F. Ellis. 1989. Laser-Doppler assessment of brain microcirculation: Effect of systemic alterations. *American Journal of Physiology. Heart and Circulatory Physiology* 256:H1247–H1254.

Haberny, K. A., S. Pou, and C. U. Eccles. 1992. Potentiation of quinolinate-induced hippocampal lesions by inhibition of NO synthesis. *Neuroscience Letters* 146:187–190.

Hakim, A. M. 1998. Ischemic penumbra: The therapeutic window. *Neurology* 51:S44–S46.

Hall, E. D. 1997. Brain attack. Acute therapeutic interventions. Free radical scavengers and antioxidants. *Neurosurgery Clinics of North America* 8:195–206.

Hansen, A. J., and T. Zeuthen. 1981. Extracellular ion concentrations during spreading depression and ischemia in the rat brain cortex. *Acta Physiologica Scandinavica* 113:437–445.

Harders, A. G., G. Laborde, D. W. Droste, and E. Rastogi. 1989. Brain activity and blood flow velocity changes: A transcranial Doppler study. *The International Journal of Neuroscience* 47:91–102.

Heistad, D. D., and H. A. Kontona. 1983. Cerebral circulation. In *Handbook of physiology. Section 2: The cardiovascular system*, eds. J. T. Shepherd and F. M. Abboud. Bethesda: American Physiological Society.

Hougaard, K., T. Oikawa, E. Sveinsdottir, E. Skinoj, D. H. Ingvar, and N. A. Lassen. 1976. Regional cerebral blood flow in focal cortical epilepsy. *Archives of Neurology* 33:527–535.

Hout, J., S. Radouco-Thomas, and C. Radouco-Thomas. 1973. Qualitative and quantitative evaluation of experimentally induced seizures. In *Anticonvulsant drugs*, ed. Mercier. Oxford: Pergamon.

Iadecola, C., D. A. Pelligrino, M. A. Moskowitz, and N. A. Lassen. 1994. Nitric oxide synthase inhibition and cerebrovascular regulation. *Journal of Cerebral Blood Flow and Metabolism* 14:175–192.

Ibayashi, S., A. C. Ngai, M. A. Howard, 3rd, J. R. Meno, M. R. Mayberg, and H. R. Winn. 1991. Lack of sympathetic and cholinergic influences on cerebral vasodilation caused by sciatic nerve stimulation in the rat. *Journal of Cerebral Blood Flow and Metabolism* 11:678–683.

Ingvar, D. H. 1983. Current problems in epilepsy: 1. Cerebral blood flow, metabolism and epilepsy. In *Proceedings of the International Symposium, Montpellier, France*, eds. M. B. Moulinier, D. H. Ingvar, and B. S. Meldrum, 2–5.

Jasper, H., and T. C. Erickson. 1941. Cerebral blood flow and pH in excessive cortical discharge induced by metrazol and electrical stimulation. *Journal of Neurophysiology* 4:333–347.

Jobsis, F. F. 1964. Basic processes in cellular respiration. In *Handbook of physiology: Respiration*, eds. W. O. Fenn and H. Rahn. 1st ed.

Johnson, P. C. 1986. Autoregulation of blood flow. *Circulation Research* 59: 483–495.

Kety, S. S., and C. F. Schmidt. 1948. The effects of altered arterial tensions of carbon dioxide and oxygen on cerebral blood flow and cerebral oxygen consumption of normal young men. *The Journal of Clinical Investigation* 27:484–492.

Knowles, R. G., and S. Moncada. 1994. Nitric oxide synthases in mammals. *The Biochemical journal* 298(Pt 2):249–258.

Knowles, R. G., M. Palacios, R. M. Palmer, and S. Moncada. 1989. Formation of nitric oxide from L-arginine in the central nervous system: A transduction mechanism for stimulation of the soluble guanylate cyclase. *Proceedings of the National Academy of Sciences of the United States of America* 86:5159–5162.

Koroleva, V. I., and J. Bures. 1983. Cortical penicillin focus as a generator of repetitive spike-triggered waves of spreading depression in rats. *Experimental Brain Research* 51:291–297.

Krivanek, J. 1962. Concerning the dynamics of the metabolic changes accompanying cortical spreading depression. *Physiologia Bohemoslovenica* 11:383–391.

Kuffler, S. W., and J. G. Nicholls. 1966. The physiology of neuroglial cells. *Ergebnisse der Physiologie, biologischen Chemie und experimentellen Pharmakologie* 57:1–90.

Lam, A. M., B. F. Matta, T. S. Mayberg, and S. Strebel. 1995. Change in cerebral blood flow velocity with onset of EEG silence during inhalation anesthesia in humans: Evidence of flow-metabolism coupling?. *Journal of Cerebral Blood Flow and Metabolism* 15:714–717.

Leão, A. A. P. 1944. Spreading depression of activity in cerebral cortex. *Journal of Neurophysiology* 7:359–390.

Lee, C. P. 1995. Biochemical studies of isolated mitochondria from normal and diseased tissues. *Biochimica et Biophysica Acta* 1271:21–28.

Lehninger, A. L. 1982. Chapter 30: Organ interrelationships in the metabolism of mammals. In *Principles of biochemistry*. New York: Worth.

Lennox, W. G., and M. A. Lennox. 1960. *Epilepsy and related disorders*. Boston and Toronto: Little, Brown and Co.

Li, Y., C. Powers, N. Jiang, and M. Chopp. 1998. Intact, injured, necrotic and apoptotic cells after focal cerebral ischemia in the rat. *Journal of the Neurological Sciences* 156:119–132.

Löscher, W., and D. Schmidt. 1988. Which animal models should be used in the search for new antiepileptic drugs? A proposal based on experimental and clinical considerations. *Epilepsy Research* 2:145–181.

Lothman, E., J. Lamanna, G. Cordingley, M. Rosenthal, and G. Somjen. 1975. Responses of electrical potential potassium levels, and oxidative metabolic activity of the cerebral neocortex of cats. *Brain Research* 88:15–36.

Macdonald, R. L., and J. L. Barker. 1977. Pentylenetetrazol and penicillin are selective antagonists of GABA-mediated post-synaptic inhibition in cultured mammalian neurones. *Nature* 267:720–721.

Maeda, M., T. Sugiyama, F. Akai, I. Jikihara, Y. Hayashi, and H. Takagi. 1998. Single stranded DNA as an immunocytochemical marker for apoptotic change of ischemia in the gerbil hippocampus. *Neuroscience Letters* 240:69–72.

Magistris, M. R., M. S. Mouradian, and P. Gloor. 1988. Generalized convulsions induced by pentylenetetrazol in the cat: participation of forebrain, brainstem, and spinal cord. *Epilepsia* 29:379–388.

Maines, M. D.. 1993. Carbon Monoxide: An emerging regulator of cGMP in the brain. *Molecular and Cellular Neurosciences* 4:389–397.

Manzoni, O., L. Prezeau, P. Marin, S. Desagher, J. Bockaert, and L. Fagni. 1992. Nitric oxide-induced blockade of NMDA receptors. *Neuron* 8:653–662.

Marciani, M. G., G. Santone, G. Sancesario, R. Massa, P. Stanzione, and G. Bernardi. 1993. Protective effect of clonazepam on ischemic brain damage induced by 10-minute bilateral carotid occlusion in Mongolian gerbils. *Functional Neurology* 8:115–120.

Marletta, M. A.. 1993. Nitric oxide synthase structure and mechanism. *The Journal of Biological Chemistry* 268:12231–12234.

Martin, D., G. A. Bustos, M. A. Bowe, S. D. Bray, and J. V. Nadler. 1991. Autoreceptor regulation of glutamate and aspartate release from slices of the hippocampal CA1 area. *Journal of Neurochemistry* 56:1647–1655.

Mayer, B., P. Klatt, E. Bohme, and K. Schmidt. 1992. Regulation of neuronal nitric oxide and cyclic GMP formation by Ca^{2+}. *Journal of Neurochemistry* 59:2024–2029.

Mayevsky, A. 1976. Brain energy metabolism of the conscious rat exposed to various physiological and pathological situations. *Brain Research* 113:327–338.

Mayevsky, A. 1984a. Brain NADH redox state monitored *in vivo* by fiber optic surface fluorometry. *Brain Research Reviews* 7:49–68.

Mayevsky, A. 1948b. Brain oxygen toxicity. Invited review. In *Proceedings of the 8th Symposium on Underwater Physiology, Bethesda, Maryland*, eds. A. J. Bachrach and M. M. Matzen, 69–89. Bethesda, MD: Undersea Medical Society Inc.

Mayevsky, A. 1990. Level of ischemia and brain functions in the Mongolian gerbil *in vivo*. *Brain Research* 524:1–9.

Mayevsky, A. 1992. Cerebral blood flow and brain mitochondrial redox state responses to various perturbations in gerbils. *Advances in Experimental Medicine and Biology* 317:707–716.

Mayevsky, A., and B. Chance. 1982. Intracellular oxidation-reduction state measured *in situ* by a multicannel fiber-optic surface fluorometer. *Science* 217: 537–540.

Mayevsky, A., R. B. Duckrow, E. Yoles, N. Zarchin, and D. Kanshansky. 1990a. Brain mitochondrial redox state, tissue hemodynamic and extracellular ion responses to four-vessel occlusion and spreading depression in the rat. *Neurological Research* 12:243–248.

Mayevsky, A., K. H. Frank, S. Nioka, M. Kessler, and B. Chance. 1990b. Oxygen supply and brain function in vivo: A multiparametric monitoring approach in the Mongolian gerbil. In *Oxygen transport to tissue XII*, eds. J. Piper, T. K. Goldstick, and M. Meyer. New York: Plenum Press.

Mayevsky, A., and H. R. Weiss. 1991. Cerebral blood flow and oxygen consumption in cortical spreading depression. *Journal of Cerebral Blood Flow & Metabolism* 11:829–836.

Mayevsky, A., E. Yoles, and N. Zarchin. 1986. Metabolic, ionic and electrical responses to oxygen deficiency in the newborn dog *in vivo*. *Advances in Experimental Medicine and Biology* 200:261–269.

Mayevsky, A., E. Yoles, N. Zarchin, and D. Kaushansky. 1990c. Brain vascular ionic and metabolic responses to ischemia in the Mongolian gerbil. *Journal of Basic and Clinical Physiology and Pharmacology* 1:207–220.

Mayevsky, A., N. Zarchin, and C. M. Friedli. 1982. Factors affecting the oxygen balance in the awake cerebral cortex exposed to spreading depression. *Brain Research* 236:93–105.

Mazziotta, J. C., and J. Engel, Jr. 1984. The use and impact of positron computed tomography scanning in epilepsy. *Epilepsia* 25(Suppl 2):S86–S104.

Meldrum, B. 1990. Protection against ischaemic neuronal damage by drugs acting on excitatory neurotransmission. *Cerebrovascular and Brain Metabolism Reviews* 2:27–57.

Meldrum, B. S. 1992. Excitatory amino acids in epilepsy and potential novel therapies. *Epilepsy Research* 12:189–196.

Miller, J. W., A. C. Mckeon, and J. A. Ferrendelli. 1987. Functional anatomy of pentylenetetrazol and electroshock seizures in the rat brainstem. *Annals of Neurology* 22:615–621.

Mirski, M. A., and J. A. Ferrendelli. 1986. Anterior thalamic mediation of generalized pentylene-tetrazol seizures. *Brain Research* 399:212–223.

Mirski, M. A., and J. A. Ferrendelli. 1987. Interruption of the connections of the mammillary bodies protects against generalized pentylenetetrazol seizures in guinea pigs. *The Journal of Neuroscience* 7:662–670.

Moncada, S., R. M. Palmer, and E. A. Higgs. 1989. Biosynthesis of nitric oxide from L-arginine. A pathway for the regulation of cell function and communication. *Biochemical Pharmacology* 38:1709–1715.

Montecot, C., J. Borredon, J. Seylaz, and E. Pinard. 1997. Nitric oxide of neuronal origin is involved in cerebral blood flow increase during seizures induced by kainate. *Journal of Cerebral Blood Flow and Metabolism* 17:94–99.

Montécot, C., L. Rondi-Reig, V. Springhetti, J. Seylaz, and E. Pinard. 1998. Inhibition of neuronal (type 1) nitric oxide synthase prevents hyperaemia and hippocampal lesions resulting from kainate-induced seizures. *Neuroscience* 84:791–800.

Moore, P. K., R. C. Babbedge, P. Wallace, Z. A. Gaffen, and S. L. Hart. 1993a. 7-Nitro indazole, an inhibitor of nitric oxide synthase, exhibits anti-nociceptive activity in the mouse without increasing blood pressure. *British Journal of Pharmacology* 108:296–297.

Moore, P. K., P. Wallace, Z. Gaffen, S. L. Hart, and R. C. Babbedge. 1993b. Characterization of the novel nitric oxide synthase inhibitor 7-nitro indazole and related indazoles: Antinociceptive and cardiovascular effects. *British Journal of Pharmacology* 110:219–224.

Morgan, G. E., and M. S. Mikhail. 1992. Neurophysiology and anesthesia. In *Clinical anesthesiology*, eds. G. E. Morgan and M. S. Mikhail. Norwalk, CT: Appleton and Lange.

Moses III, H. M., and R. S. Fisher. 1988. Syncope, seizures, and other episodic disorders. In *The principles and practice of medicine*, 22nd ed., eds. A. M. Harvey, R. T. Johns, V. A. McKusick, A. H. Owens Jr., and R. S. Ross. Norwalk, CT: Appleton and Lange.

Mosier, J. M., P. White, P. Grant, J. E. Fisher, and R. Taylor. 1957. Cerebroautonomic and myographic changes accompanying induced seizures. *Neurology* 7:204–210.

Mülsch, A., R. Busse, P. I. Mordvintcev, A. F. Vanin, E. O. Nielsen, J. Scheel-Krüger, and S. P. Olesen. 1994. Nitric oxide promotes seizure activity in kainate-treated rats. *Neuroreport* 5:2325–2328.

Nehlig, A., M. Vergnes, R. Waydelich, E. Hirsch, R. Charbonne, C. Marescaux, and J. Seylaz. 1996. Absence seizures induce a decrease in cerebral blood flow: Human and animal data. *Journal of Cerebral Blood Flow and Metabolism* 16:147–155.

Nevins, M. E., and S. M. Arnolde. 1989. A comparison of the anticonvulsant effects of competitive and non-competitive antagonists of the N-methyl-D-aspartate receptor. *Brain Research* 503:1–4.

Ngai, A. C., K. R. KO, S. Morii, and H. R. Winn. 1988. Effect of sciatic nerve stimulation on pial arterioles in rats. *The American Journal of Physiology* 254:H133–H139.

Niedermeyer, E. 1987. Epileptic seizure disorder. In *Electroencephalography: Basic principles, clinical applications and related fields*, eds. E. Niedermeyer and F. Lopes da Silva. Baltimore: Urban and Schwarzenberg.

Nilsson, L., and B. K. Siesjo. 1975. The effect of phenobarbitone anaesthesia on blood flow and oxygen consumption in the rat brain. *Acta Anaesthesiologica Scandinavica Supplementum* 57:18–24.

O'Connor, M. J., C. J. Herman, M. Rosenthal, and F. Jobsis. 1972a. Intracellular redox changes preceding onset of epileptiform activity in intact cat hippocampus. *Journal of Neurophysiology* 35:471–483.

O'Connor, M. J., C. J. Herman, M. Rosenthal, and F. F. Jobsis. 1972b. Intracellular redox changes preceding onset of epileptiform activity in intact cat hippocampus. *Journal of Neurophysiology* 35:471–483.

O'Connor, M. J., D. V. Lewis, and C. J. Herman. 1973. Effects of potassium on oxidative metabolism and seizures. *Electroencephalography and Clinical Neurophysiology* 35:205–208.

Ogata, N.. 1975. Ionic mechanisms of the depolarization shift in thin hippocampal slices. *Experimental Neurology* 46:147–155.

Olsen, R. W. 1981. The GABA postsynaptic membrane receptor-ionophore complex. Site of action of convulsant and anticonvulsant drugs. *Molecular and Cellular Biochemistry* 39:261–279.

Osonoe, K., N. Mori, K. Suzuki, and M. Osonoe. 1994. Antiepileptic effects of inhibitors of nitric oxide synthase examined in pentylenetetrazol-induced seizures in rats. *Brain Research* 663:338–340.

Palmer, R. M., A. G. Ferrige, and S. Moncada. 1987. Nitric oxide release accounts for the biological activity of endothelium-derived relaxing factor. *Nature* 327:524–526.

Papy, J. J., and R. Naquet. 1971. Cerebral blood flow and seizures induced by MetrazolR in the baboon. *Clinical & Developmental Medicine* 39:283–296.

Paulson, O. B., and E. A. Newman. 1987. Does the release of potassium from astrocyte endfeet regulate cerebral blood flow?. *Science* 237:896–898.

Penfield, W. G.. 1933. The evidence for a cerebral vascular mechanism in epilepsy. *Annals of Internal Medicine* 7:303–310.

Penfield, W. G. 1937. The circulation of the epileptic brain. *Research publications - Association for Research in Nervous and Mental Disease* 18:605–737.

Penfield, W. G., K. Von Santha, and A. Cipriani 1939. Cerebral blood flow during induced epileptiform seizures in animals and man. *Journal of Neurophysiology* 2:257–267.

Plum, F., J. B. Posner, and B. Troy. 1968. Cerebral metabolic and circulatory responses to induced convulsions in animals. *Archives of Neurology* 18:1–13.

Pollen, D. A., and M. C. Trachtenberg. 1970. Neuroglia: Gliosis and focal epilepsy. *Science* 167:1252–1253.

Posner, J. B., F. Plum, and van A. Poznak. 1969. Cerebral metabolism during electrically induced seizures in man. *Archives of Neurology* 20:388–395.

Prado, R., B. D. Watson, J. Kuluz, and W. D. Dietrich. 1992. Endothelium-derived nitric oxide synthase inhibition. Effects on cerebral blood flow, pial artery diameter, and vascular morphology in rats. *Stroke* 23:1118–1123; discussion 1124.

Prince, D. A.. 1978. Neurophysiology of epilepsy. *Annual Review of Neuroscience* 1:395–415.

Prince, D. A., H. D. Lux, and E. Neher. 1973. Measurement of extracellular potassium activity in cat cortex. *Brain Research* 50:489–495.

Przegalinski, E., L. Baran, and J. Siwanowicz. 1996. The role of nitric oxide in chemically- and electrically-induced seizures in mice. *Neuroscience Letters* 217:145–148.

Przegaliński, E., L. Baran, and J. Siwanowicz. 1994. The role of nitric oxide in the kainate-induced seizures in mice. *Neuroscience Letters* 170:74–76.

Rees, D. D., R. M. Palmer, R. Schulz, H. F. Hodson, and S. Moncada. 1990. Characterization of three inhibitors of endothelial nitric oxide synthase *in vitro* and *in vivo*. *British Journal of Pharmacology* 101:746–752.

Revett, K., E. Ruppin, S. Goodall, and J. A. Reggia. 1998. Spreading depression in focal ischemia: A computational study. *Journal of Cerebral Blood Flow and Metabolism* 18:998–1007.

Rondouin, G., J. Bockaert, and M. Lerner-Natoli. 1993. L-nitroarginine, an inhibitor of NO synthase, dramatically worsens limbic epilepsy in rats. *Neuroreport* 4:1187–1190.

Ronne-Engstrom, E., H. K. Carlson, S. Blom, R. Flink, B. Gazelius, B. Spannare, and L. Hillered. 1993. Monitoring of cortical blood flow in human epileptic foci using laser Doppler flowmetry. *Journal of Epilepsy* 6:145–151.

Rosental, M., and J. S. Thomas. 1989. Brain oxygenation: Monitoring technique and insights into brain function and survival. *Bailliere's Clinics Anesthesiologists* 3:647–674.

Sakai, F., J. S. Meyer, H. Naritomi, and M. C. Hsu. 1978. Regional cerebral blood flow and EEG in patients with epilepsy. *Archives of Neurology* 35:648–657.
Schmidt, C. F., S. S. Kety, and H. H. Pennes. 1945. The gaseous metabolism of the brain of the monkey. *The American Journal of Physiology* 143:33–52.
Shuaib, A., and M. M. Breker-Klassen. 1997. Inhibitory mechanisms in cerebral ischemia: A brief review. *Neuroscience and Biobehavioral Reviews* 21:219–226.
Sick, T. J., Z. C. Feng, and M. Rosenthal. 1998. Spatial stability of extracellular potassium ion and blood flow distribution in rat cerebral cortex after permanent middle cerebral artery occlusion. *Journal of Cerebral Blood Flow and Metabolism* 18:1114–1120.
Siesjo, B. K., and F. Bengtsson. 1989. Calcium fluxes, calcium antagonists, and calcium-related pathology in brain ischemia, hypoglycemia, and spreading depression: A unifying hypothesis. *Journal of Cerebral Blood Flow and Metabolism* 9: 127–140.
Siesjo, B. K. 1978. Anaesthesia, analgesia, and sedation. In *Brain energy metabolism*, ed. B. K. Siesjo. Chichester; New York; Brisbane; Toronto: John Wiley & Sons.
Siesjö, B. K., and P. Siesjö. 1996. Mechanisms of secondary brain injury. *European Journal of Anaesthesiology* 13:247–268.
Sims, N. R.. 1992. Energy metabolism and selective neuronal vulnerability following global cerebral ischemia. *Neurochemical Research* 17:923–931.
Sims, N. R., and E. Zaidan. 1995. Biochemical changes associated with selective neuronal death following short-term cerebral ischaemia. *The International Journal of Biochemistry & Cell Biology* 27:531–550.
Spencer, W. A., and E. R. Kandel. 1961. Electrophysiology of hippocampal neurons. IV. Fast prepotentials. *Journal of Neurophysiology* 24:272–285.
Starzl, T. E., W. T. Niemer, M. Dell, and P. R. Forgrave. 1953. Cortical and subcortical electrical activity in experimental seizures induced by metrazol. *Journal of Neuropathology and Experimental Neurology* 12:262–276.
Stone, W. E. 1970. Convulsant actions of tetrazole derivatives. *Pharmacology* 3:367–370.
Tenny, R. T., F. W. Sharbrough, and R. E. Anderson. 1980. Correlation of intracellular redox states and pH with blood flow in primary and secondary seizure foci. *Annals of Neurology* 8:564–573.
Theodore, W. H., R. Brooks, S. Sato, N. Patronas, R. Margolin, G.DiChiro, and R. J. Porter. 1984. The role of positron emission tomography in the evaluation of seizure disorders. *Annals of Neurology* 15:S176–S179.
Theodore, W. H., R. Dorwart, M. Holmes, R. J. Porter, and G. Dichiro. 1986. Neuroimaging in refractory partial seizures: Comparison of PET, CT, and MRI. *Neurology* 36:750–759.
Theodore, W. H., M. E. Newmark, S. Sato, R. Brooks, N. Patronas, R.de laPaz, G. Dichiro, R. M. Kessler, R. Margolin, R. G. Manninget al. 1983. [^{18}F] fluorodeoxyglucose positron emission tomography in refractory complex partial seizures. *Annals of Neurology* 14:429–437.
Touchon, J., J. Valmier, M. Baldy-Moulinier, and J. Cadilhac. 1986. Regional cerebral blood flow during interictal state: differences between temporal lobe epilepsy and primary generalized epilepsy. *European Neurology* 25:43–52.
Tyner, F. S., J. R. Knott, and W. B. Mayer Jr. 1989. *Fundamentals of EEG technology*. New York: Raven Press.
Ueda, M., and J. Bures. 1977. Differential effects of cortical speading depression on epileptic foci induced by various convulsants. *Electroencephalography and Clinical Neurophysiology* 43:666–674.
Van Buren, J. M., M. D. Lewis, W. H. Schuette, W. C. Whitehouse, and C. A. Marsan. 1978. Fluorometric monitoring of NADH levels in cerebral cortex: Preliminary observations in human epilepsy. *Neurosurgery* 2:114–121.
Velísek, L., R. Kusá, M. Kulovaná, and P. Mares. 1990. Excitatory amino acid antagonists and pentylenetetrazol-induced seizures during ontogenesis. I. The effects of 2-amino-7-phosphonoheptanoate. *Life Sciences* 46:1349–1357.
Watts, D. C.. 1973. Creatine kinase (adenosine 5'-triphosphate-creatine phosphokinase). *Enzymes* 8A:384–455.

White, P. T., P. Grant, J. Mosier, and A. Craig. 1961. Changes in cerebral dynamics associated with seizures. *Neurology* 11(4):354–361.

Wilson, W. A., and A. V. Escueta. 1974. Common synaptic effects of pentylenetetrazol and penicillin. *Brain Research* 72:168–171.

Wood, P. L.. 1991. Pharmacology of the second messenger, cyclic guanosine 3',5'-monophosphate, in the cerebellum. *Pharmacological Reviews* 43:1–25.

Yarnell, P. R., D. Burdick, B. Sanders, and J. Stears. 1974. Focal seizures, early veins, and increased flow. A clinical, angiographic, and radioisotopic correlation. *Neurology* 24:512–516.

Zhu, X. Z., and L. G. Luo. 1992. Effect of nitroprusside (nitric oxide) on endogenous dopamine release from rat striatal slices. *Journal of Neurochemistry* 59:932–935.

Zuckermann, E. C., and G. H. Glaser. 1968. Hippocampal epileptic activity induced by localized ventricular perfusion with high-potassium cerebrospinal fluid. *Experimental Neurology* 20:87–110.

Chapter 8
The Effects of Aging and TBI on Brain Physiological Functions

Abstract The effect of age on the responses of the brain to decapitation (complete ischemia) was tested in the awake and anesthetized rat. The metabolic and reflected light activities were evaluated by the use of a DC fluorometer/reflectometer (one or two hemispheres), and the electrical activity was measured by electrocorticogram. In all 81 animals used in these experiments, the above-mentioned parameters were measured to enable comparison between the age groups. The rats were divided into six age groups ranging from 2 to 120 weeks, and each age group was subdivided into awake and anesthetized animals. The results can be summarized as follows. (1) The rate of energy consumption was lower in the young (2 weeks) and the old rats (120 weeks) compared to the adult rats. (2) The effect of anesthetic was more pronounced in the young and the old rats than in the adult rats. (3) The electrical activity was in good correlation with the energy availability in all animals of various ages as well as in the awake and anesthetized rats. Severe head injury can result in a high mortality rate or irreversible brain damage. One technique used to induce traumatic brain injury (TBI) is exposure of the brain to fluid percussion pressure while monitoring the increase in intracranial pressure (ICP). Since brain injury is a multifactorial, pathological, time-dependent state, the multiparametric monitoring approach was adopted for studying fluid percussion effects on the rat brain. A multiprobe assembly (MPA) connected to the brain in vivo (right hemisphere) enabled the simultaneous monitoring of CBF, NADH redox state, extracellular K^+, Ca^{2+}, H^+ levels as well as DC potential, ECoG and ICP. The animal was connected to the monitoring system and exposed to TBI after a recuperation period of at least 3 h after the end of the operation. Two typical responses to TBI were recorded in our preliminary experiments. When severe injury was induced, ischemic depolarization (ID) developed, whereas mild or moderate injury led to repetitive spreading depression (SD) cycles. The relationship between the ID and SD observed under TBI is important to the understanding of the mechanism of brain injury. ICP before injury was between 2 and 6 mm Hg and increased to 20–22 mm Hg 2–3 min after the ID. After severe head injury, ICP remained high and in some cases increased to critical values causing death of these animals. Some

A. Mayevsky et al., *Physiological Mapping of the Cerebral Cortex in Neuropathology*,
https://doi.org/10.1007/978-3-032-13019-8_8

animals developed seizures at various stages after the TBI. Hyperbaric oxygenation was used as a therapeutic tool to treat severely injured animals.

Keywords Mitochondrial redox state · Cerebral blood flow · Multiparametric monitoring · Head injury · Intracranial pressure

8.1 The Effects of Age on Brain Responses to Complete Ischemia

8.1.1 Introduction

The effects of age on the metabolic activity of the brain have been tested on various animal models. Most of the studies described the differences either between the newborn and the adult (Hernández et al. 1978; Thurston and McDougal 1969; Gregoire et al. 1978) or between the adult and the old animal (Patel 1977; Peng et al., 1977; Pettegrew and Ewing 1971). In our present study, we compared rats of six age groups from 14 days to 2.5 years. Comparison of the tolerance to hypoxia between young and adult animals has been described by various investigators (Duffy et al. 1975; Mares et al. 1976; Stafford and Weatherall 1960; Berlet, 1976). It was found that the young are more tolerant to hypoxia. This resistance cannot be due to the fact that glycolysis is the main source of energy in the young brain (Samson and Dahl 1975), nor to more energy reserves such as ATP and phosphocreatine (Thurston and McDougal 1969; Berlet 1976). It was suggested that the resistance was due to low oxygen consumption (Patel 1977; Mayevsky 1976).

In the brains of old animals, some events relevant to energy metabolism have been described. A decrease in blood flow was reported (Fazekas et al. 1955; Meier-Ruge 1975), as well as a decrease in oxygen consumption (Patel 1977; Peng et al. 1977; Pettegrew and Ewing 1971). It is not clear whether those changes are primary or secondary factors in the aging process (Ferrendelli et al. 1971; Shenkin et al. 1953; Simard et al. 1971). Studies of the effects of age on the brain energy metabolism have been made using various techniques. Lowry and his collaborators (King et al. 1967) described the effects of decapitation, which transferred the brain into a closed system, on the energy metabolism by measuring the levels of substrates and metabolites. The same approach was described by other investigators (King et al. 1967; Swaab and Boer 1972; Maker and Lehrer 1971). In 1976, Mayevsky and Chance (Mayevsky and Chance 1976) described a new approach by which the metabolic and electrical responses were correlated in the decapitated awake brain. Later, they showed the effects of anesthesia on the response to decapitation (Mayevsky 1978a). In the present communication we will describe the effects of age (2 weeks to 2.5 years) on the metabolic, hemodynamic and electrical responses of the awake and anesthetized rat brain to decapitation. The aims of the present study were to correlate in detail the metabolic (NADH fluorescence) and electrical (electrocorticogram) responses to decapitation with the age of the animal tested. A further point was to study the effect of anesthetics on the mentioned correlation. By using the same technique for all ages, we assumed that variability between various ages due to technical reasons would be diminished. Preliminary results of this study were previously described (Zarchin and Mayevsky 1979).

8.1.2 Methods

Male Wistar rats were used in all experiments. The rats were divided into six age groups and each group contained awake and anesthetized animals (for detailed numbers, see Table 8.1). The young rats were kept with their mother until the time of the operation. Two types of DC fluorometer/reflectometers were used: one for unilateral measurement, and the other for bilateral records. The details of the instruments were recently described (Mayevsky 1978b). The basic approach used to study the changes in metabolic activity was to monitor continuously the oxidation–reduction state of pyridine-nucleotide (NADH) from the cortical surface. By illuminating the brain with 366 nm light, the intramitochondrial NADH is excited and fluoresces at 450 nm which can be measured. Another parameter measured is the intensity of the reflected light at the excitation wavelength (366 nm), which is believed to be correlated to blood volume in the tissue monitored (for references, see Mayevsky and Chance 1975; Mayevsky, 1976, 1978a). The reflected light was used as a means of correcting the NADH fluorescence measurements. Animals were anesthetized with Equithesin (ET—a mixture of chloral hydrate, pentobarbital, magnesium sulfate and propylene glycol) by intraperitoneal injection (0.3 ml per 100 g body weight). This dose was used for all age groups except the 2-week-old group, for which diluted ET was used and the dose calculated accordingly. The injection kept the animal anesthetized for the duration of the operation and its connection to the measuring system. The skull was exposed by a midline incision of the skin, and one or two 4 mm diameter holes were drilled in the parietal bone for observation of metabolic changes. Two stainless steel screws were used as electrocorticogram (ECoG) electrodes on each hemisphere. A cannula for application of KC1 solution was used in most experiments in order to test the response of the brain to spreading depression (Mayevsky 1976). The cannula and the ECoG electrodes were cemented to the skull using dental acrylic cement. Figure 8.1a shows schematically the localization of the various cannulae and ECoG electrodes on the skull. The animal was connected to the fluorometer/reflectometer and placed in the specially designed plexiglass guillotine-cage for decapitation (Fig. 8.1b). After connecting the animal to the fluorometer, two different procedures were used to test the metabolic state of the cortical tissue monitored. The brain was exposed to anoxia by flushing nitrogen through a nasal mask. Only animals that showed a "normal" increase of NADH during the anoxia were used. If the response was much smaller (below the average of the group), the animal was discarded. The same test was done by exposing the brain to cortical spreading depression. which

Table 8.1 The effects of age on the hemodynamic, metabolic and electrical responses to decapitation of the awake and anesthetized rat brain

Body weight (g)	Age (weekly)	Treatment[b] and no. of rats	T_0 (s)	T_1 (s)	T_2 (s)	CF_{max} (%)	CF_0/CF_{max} (%)	T_3 (s)	T_4 (s)	CF_{N2} (%)
20	2	Aw (10)	21 ± 1	1.9 ± 0.2	38 ± 3	90 ± 3	88 ± 1	7.0 ± 0.3	110 ± 15	68 ± 3
		An (9)	63 ± 5***	4.3 ± 0.3***	68 ± 7***	95 ± 51 36 ± 5***	95 ± 2**	15.1 ± 1.7***	514 ± 66***	66 ± 4
60	4	Aw (9)	15 ± 1	1.1 ± 0.1	29 ± 3	99 ± 8	90 ± 1	4.6 ± 0.4	74 ± 10	82 ± 7
		An (7)	36 ± 5***	3.2 ± 0.4***	43 ± 3***	11.2 ± 5	94 ± 3	11.2 ± 0.9***	250 ± 26***	77 ± 6
100	6	Aw (7)	16	0.8 ± 0.1	34 ± 2	73 ± 6	91 ± 2	4.3 ± 0.4	66 ± 6	69 ± 4
		An (7)	29 ± 3***	1.4 ± 0.2**	39 ± 5	100 ± 5***	96 ± 1*	5.4 ± 0.6	122 ± 9***	71 ± 6
200	10	Aw (10)	16 ± 1	1.1 ± 0.2	18 ± 2	70 ± 4	96 ± 1	3.3 ± 0.2	82 ± 9	52 ± 2
		An (5)	38 ± 2***	1.3 ± 0.1	45 ± 4***	93 ± 5***	96 ± 1	5.5 ± 0.5***	144 ± 12***	56 ± 6
400	48	Aw (5)	21 ± 2	0.5 ± 0.1	48 ± 5	68 ± 7	82 ± 4	6.1 ± 0.6	73 ± 7	38 ± 4
		An (4)	37 ± 4***	1.5 ± 0.4*	48 ± 4	68 ± 8	92 ± 4	7.1 ± 0.4	135 ± 36	50 ± 3*
300	120	Aw (4)	34 ± 2	1.1 ± 0.3	38 ± 6	65 ± 5	99 ± 1	5.1 ± 0.7	96 ± 29	48 ± 6
		An (4)	60 ± 9*	2.6 ± 0.4**	66 ± 5**	61 ± 9	95 ± 2	13.5 ± 1.5***	202 ± 26*	43 ± 6

Values are expressed as mean ± SEM [a]See text for explanation of parameters and abbreviations shown in the upper line of the table(T_0(s) [b]Aw – awake; An – anesthetized $*p < 0.05$; $**p < 0.01$; $***p < 0.001$; differences between awake and anesthetized were evaluated by *t*-test

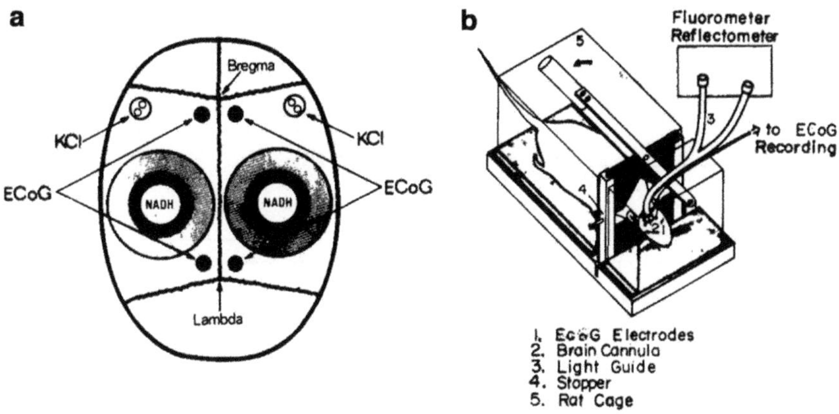

Fig. 8.1 Schematic representations of the location of various cannulas and electrodes on the skull of a rat (**a**) and the special cage used for decapitation (**b**). NADH was measured from one or two sites using a DC fluorometer/reflectometer. KCI (0.5 M) was applied topically in order to initiate spreading depression. The electrocorticogram (ECoG) was recorded using a bipolar technique

should show a mitochondrial State 4 to 3 transition (oxidation or decrease in NADH levels). Decapitation was done in the fully awake rat, as indicated by the ECoG and motor activity. In the anesthetized groups the rats were allowed to awaken and then were anesthetized again by intraperitoneal injection of ET.

8.1.3 Results

A typical metabolic, hemodynamic and electrical response of the brain to decapitation is shown in Fig. 8.2. For each age group bilateral measurements were performed in some rats to test the variability between the brain's two hemispheres. After analyzing the results, we decided to use only unilateral measurements due to the very small variability between the two hemispheres. The four upper traces in Fig. 8.2 were measured from the right hemisphere and the lower four from the left hemisphere. It can be seen that the differences between the two hemispheres are insignificant in most cases. The 366 nm reflected light (R) shows a very small initial response to the decapitation. However, a very large secondary reflectance increase was recorded 1–2 min after ECoG = 0 when NADH reached its maximal level. The uncorrected (F) and corrected (CF) 450 nm fluorescence signals are similar to those described previously (Mayevsky 1978a; Mayevsky and Chance 1976). In order to analyze the effects of age on the responses to decapitation, various parameters were measured and calculated from the analog signals, as shown in Fig. 8.2. The definitions for the various parameters are as follows:
T_0 = time (s) when electrical activity was very low and close to 0.

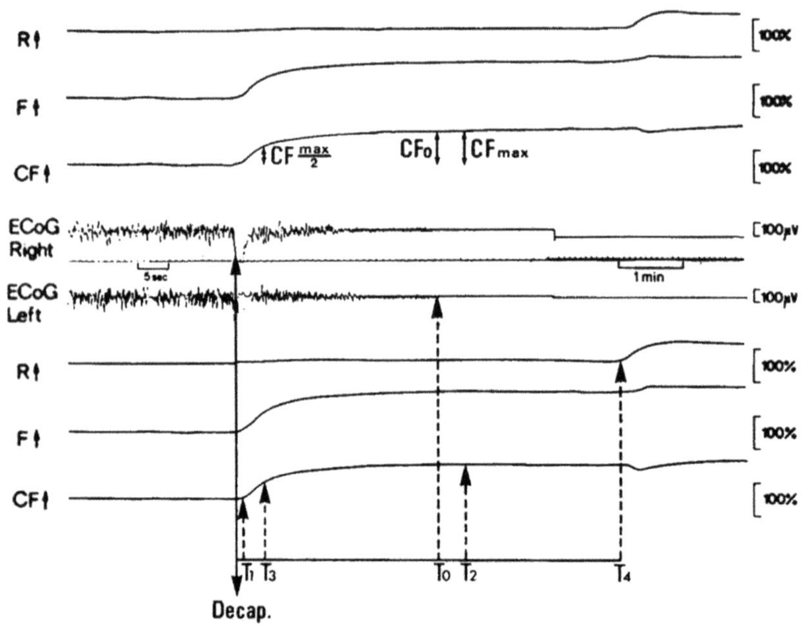

Fig. 8.2 Metabolic, reflected-light and electrical responses to decapitation (measured bilaterally). The upper four traces were measured from the right hemisphere and the lower four from the left hemisphere. R reflectance, F fluorescence, CF corrected fluorescence (F - R), ECoG electrocorticogram. See text for definitions of the measured parameters

$T_1 =$ interval between decapitation and the point when corrected fluorescence started increasing.
$T_2 =$ time when the maximum level of NADH was reached after decapitation.
$T_3 =$ time when NADH reached a level which is half of its maximum increase ($CF_{max/2}$).
$T_4 =$ time when a large increase of the reflectance was measured (SRI = secondary reflectance increase).
$CF_{max} =$ maximum percentage increase of NADH above baseline after decapitation.
$CF_0 =$ percentage increase of NADH above baseline when ECoG = 0.
$CF_0 =$ percentage of NADH increase when ECoG = 0 in proportion to the maximum increase measured in the CF_{max} same rat.
$CF_{N2} =$ percentage increase of NADH above baseline in nitrogen environment (anoxia).

All correlation coefficients between T_0, T_1, T_2, T_3, and T_4 were significant ($p < 0.01$).

Table 8.1 summarizes the results obtained from 81 awake and anesthetized rats of various ages. The differences between the awake and anesthetized rats of each group

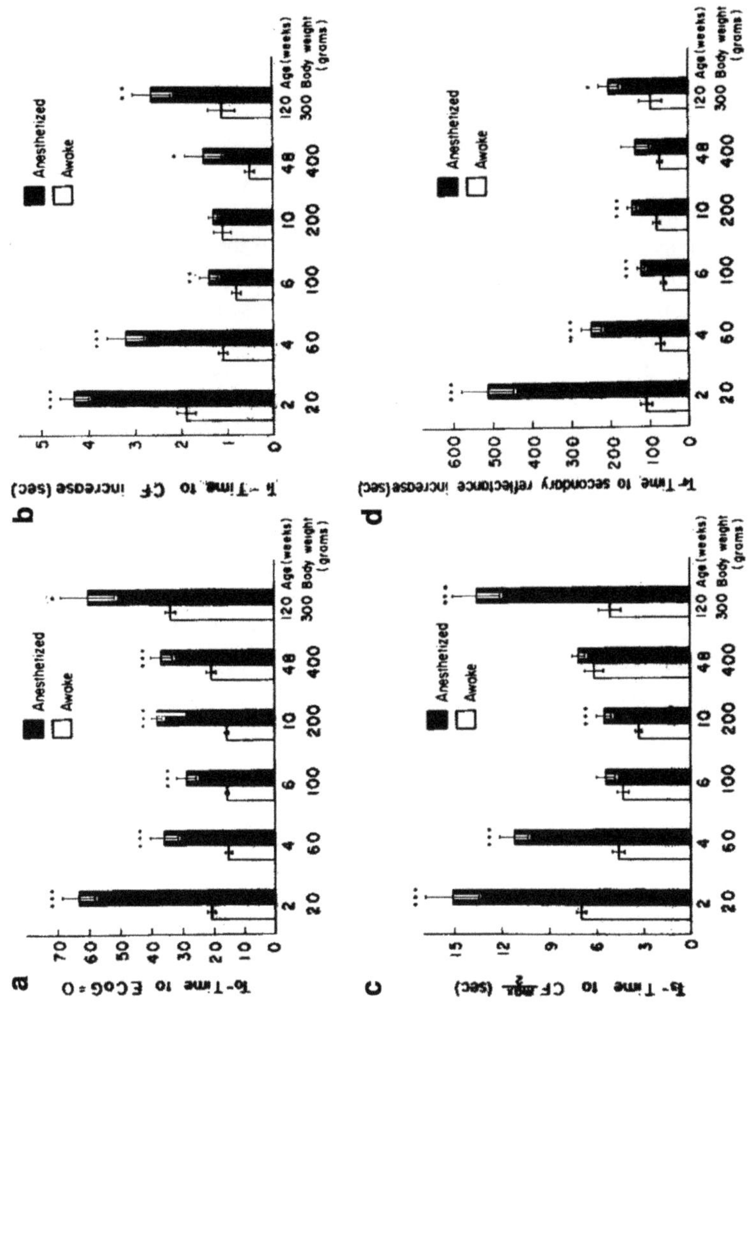

Fig. 8.3 The effects of age on the various responses of the awake and anesthetized rats to decapitation. Significance levels between awake and anesthetized rats $*p < 0.05$; $**p < 0.01$; $***p < 0.001$

Table 8.2 Analysis of variance for differences in parameter T_3 obtained from awake and anesthetized animals of various ages

Source of variation	SS	d.f.	MS	*F*	*p*
Total	1499.1	80	18.7		
Age	397.9	5	79.6	16.5	<0.01
Anesthesia	378.2	1	378.2	78.5	<0.01
Age—anesthesia interaction	184.2	5	36.8	7.6	<0.01
Error	332.2	69	4.8		

were analyzed by Student's *t*-test and are shown in the table. For most of the parameters representing time, the anesthetic had a significant effect. The responses to anoxia and spreading depression (KCl 0.5 M) were measured in each animal after its connection to the measuring system as a standard test of the state of the brain. The response to anoxia (last column in Table 8.1) showed no difference between the rats of each group, with one exception (the 48-week-old group). The response to spreading depression recorded on the NADH trace showed a typical oxidation–reduction cycle as described previously (Mayevsky 1976). Figure 8.3 presents graphically the differences in four temporal parameters between awake and anesthetized rats at various ages. In all four parameters presented in Fig. 8.3, the effect of age on the response to decapitation is pronounced, and a "bell" shape is seen in the anesthetized groups. This shape indicates that in the young and old rats the response times are longer than in the intermediate ages. The same phenomenon was seen, though not as clearly in the awake animals. We analyzed the differences between the various age groups in the awake and anesthetized animals. Table 8.2 shows the results obtained from a two-way analysis of variance for parameter T_a. Both age and anesthetic effects were significant ($p < 0.01$). The interaction between age and treatment (awake-anesthetized) was also significant ($p < 0.01$). Analysis of variance was also used in testing the other parameters, namely, T_0, T_1, and T_4. The interaction between the age of the animal and the treatment was found to be significant for the three parameters ($p < 0.01$).

8.1.4 Discussion

The advantages of our fluorometric technique for studying brain energy metabolism by monitoring the NADH oxidation–reduction state have been described previously (Mayevsky 1976, 1977, 1978a, 1978b; Mayevsky and Chance 1975, 1976). In the present study, we tested the effects of age (2 weeks to 2.5 years) on the awake and anesthetized rat brain during decapitation (complete ischemia).

The metabolic (NADH fluorescence), hemodynamic (reflectance) and electrical (ECoG) responses were obtained. Two groups of parameters on variables were calculated from the analog signals. The first group are temporal variables such as T_0–T_4 and the other one described the magnitude of redox changes during the decapitation.

Longer times to obtain isoelectric recording (T_0) indicate that the energy needed to maintain electrical activity was available due to the slow consumption or utilization (production stopped due to the lack of O_2). The longer times regarding NADH changes also indicate a slower O_2 utilization and expressed resistance to the ischemia induced by the decapitation. This can be demonstrated in parameter T_1; namely, that when oxygen utilization is slow, the CF started to increase later suggesting a lower metabolic rate as well as resistance of the brain. Two of the parameters calculated from the analog signal measured after decapitation did not show any significant relation to age or treatment of the animal. CF_{max} was not affected by the anesthetic, except in two age groups (6 and 10 weeks) as shown in Table 8.1. This finding can be explained by the fact that the resting oxygen supply to the brain is not affected dramatically by the anesthetic, so that the oxidation–reduction state of NADH remained the same. The parameter CF_0/CF_{max} had a very high value (90–100) in all animals, and was also not dependent on age or anesthesia.

The electrical activity disappeared at a certain level of oxygen availability as was found previously (Mayevsky and Chance, 1975). From the present results and together with what was published in the past (Mayevsky and Chance 1975) we can conclude that in order to maintain electrical activity the brain has to have certain amounts of available energy not depending upon the age or the treatment given to the animal. Three main parameters measured were affected by the state of the animal. Two are metabolic—T_1 and T_3—and the third represents electrical activity—T_0. All three parameters were related to the age of the animal and to the existence of anesthetics. Differences between the various ages were more pronounced in the anesthetized state, as can be seen from Fig. 8.3. In all parameters described in Fig. 8.3, the difference between the awake and anesthetized state was very noticeable in the young and the old animals, so that a “bell” shape was formed. The effects of anesthesia were previously described by others who showed that the rate of oxygen consumption was lower in the anesthetized brain (Fink and Haschke 1973; Lowry et al. 1964), and that cerebral blood flow was also lower under anesthetics (Cohen and Mayevsky 1989).

The direct interrelation between energy availability and electrical activity indicates that the cortical tissue needs a constant supply of oxygen for normal electrical activity (Swaab and Boer 1972). The late disappearance of the ECoG after decapitation in the young (2 weeks) and the old (120 weeks) animals corresponds to the slower rate of utilization of high-energy phosphate compounds (Simard et al. 1971). Under anesthesia the time to disappearance of the ECoG was longer, due to the slower rate of energy utilization (Fink and Haschke 1973).

The results presented in this study are very similar to those described previously on the effect of anesthetics on the hemodynamic, metabolic and electrical responses to decapitation (Mayevsky 1978a). In two parameters measured (T_0 and T_4) the effects of anesthetics were smaller in our previous study (Mayevsky 1978a). Two reasons may explain the differences: (1) in the present study, Equithesin was used as an anesthetic in contrast to Pentobarbital or Urethane used previously; (2) the difference in time between the two studies (2 years) may account for small variations in the strain of the rats. For this last reason, we always compared awake to anesthetized animals in the same set of experiments.

Another parameter which was sensitive to the age and treatment of the animal was the secondary increase in reflectance (SRI). This SRI was found in all animals tested and described a unique response of light absorption change after the death of the brain. This phenomenon was found by us previously (Mayevsky 1978a; Mayevsky and Chance, 1976), and we assume that it originates from the large decrease in blood content of the vessels, or a change in the size of the extracellular space. As was shown previously by various investigators (Hansen 1978; Van Harreveld 1972; Dora and Zeuthen 1976), complete depolarization induced a shift of water and electrolytes between the intra- and extracellular spaces. If such a change occurs after decapitation, then an increase in reflectance is possible.

In the early 1980s, we started the measurements of changes in NADH redox state, reflected-light, extracellular K^+, as well as the DC potential from the cortex of an awake animal (Friedli et al. 1982; Mayevsky et al. 1982). It was found that the large increase in reflectance (SRI) appeared simultaneously with a large negative shift in the DC potential and a large increase in extracellular K^+. Therefore, we can interpret the SRI event as a response to the complete depolarization occurring after complete ischemia (decapitation).

8.2 Physiological Responses to Cortical Spreading Depression in the Aging Rats

8.2.1 Introduction

Spreading depression (SD) is a phenomenon that has been described in the literature as depression of spontaneous electrical activity (Leão 1944). It is accompanied by increased metabolic activity (Hansen et al. 1980; Rosenthal and Somjen 1973; Mayevsky and Chance 1974), changes in cerebral blood flow (Piper et al. 1991; Shibata et al. 1990) and fluctuations in ionic homeostasis (Mayevsky et al. 1982; Hansen and Zeuthen 1981). Previous results from our laboratory indicate that the nature of the response to spreading depression is dependent on the ability of the brain to compensate for extra oxygen needed by increasing blood flow to the activated site (Mayevsky and Weiss 1991; Mayevsky et al. 1982). It was also found that this ability of the brain is age-dependent, namely, adult brain showed "oxidation cycles" of the NADH in most conditions, while the young rat and the anesthetized gerbil brains exhibited "reduction cycle" of the NADH. On the other hand, other studies demonstrated that aging acts intracellularly to limit the brain's capacity to increase oxidative phosphorylation to meet heightened ATP damage (Benzi et al. 1992). Furthermore, it is also already known that the metabolic functions of the brain as well as its blood supply inevitably decline with age (Meyer et al. 1993).

The aim of this study was to evaluate the function during cortical spreading depression of the aging brain using the multiparametric monitoring system (Mayevsky et al. 1995; Mayevsky 1992).

8.2.2 Methods

We used the multiparametric assembly (MPA) for the continuous simultaneous online measurement of hemodynamic, metabolic, ionic and electrical activities of the rat cerebral cortex (Mayevsky et al. 1995). The measurement included NADH redox state (using surface fluorometry), CBF (laser Doppler flowmetry), extracellular K^+, Ca^{2+}, and H^+ concentrations and DC steady potential as well as ECoG (Fig. 8.4).

The experiments were performed in 10 aging (25–30 months) Wistar rats and in 7 adult rats (2–3 months). The rats were anesthetized by an intraperitoneal injection (0.3 ml/100 g) of Equithesin. The skull was exposed by a midline incision in the skin,

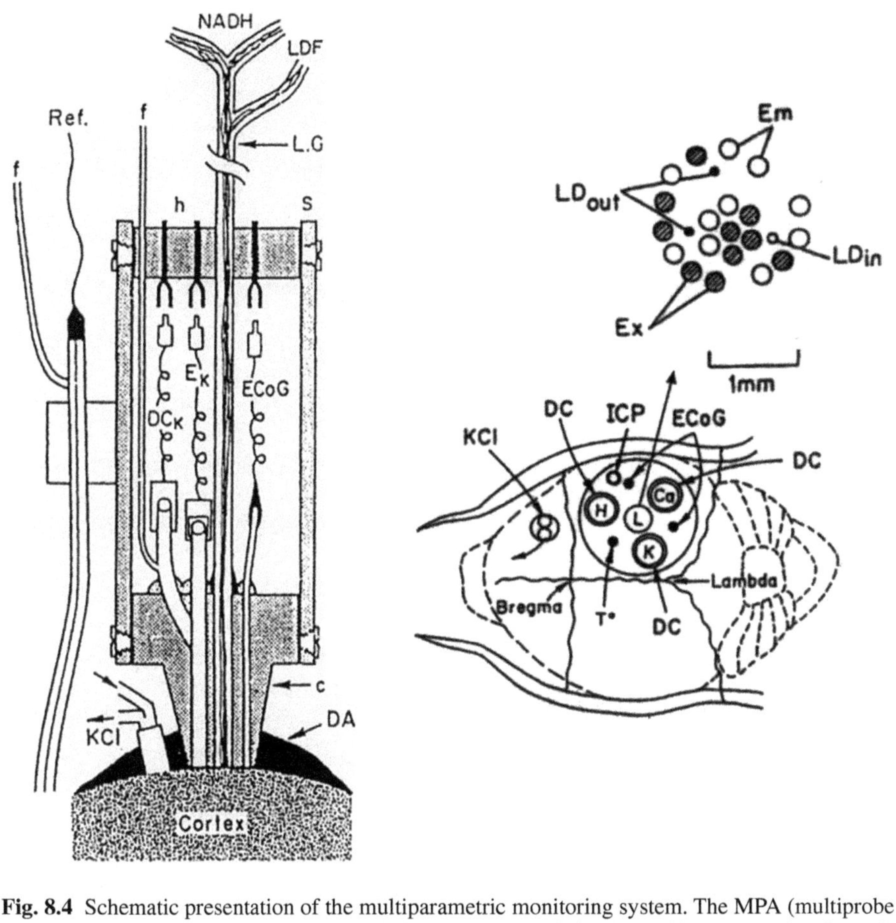

Fig. 8.4 Schematic presentation of the multiparametric monitoring system. The MPA (multiprobe assembly) presented here contains three minielectrodes (K^+, Ca^{2+}, H^+) and is connected to the various amplifiers. The light guide (L) contains two sets of optical fibers, one for NADH measurement and the other for laser Doppler flowmetry measurement of the cerebral blood flow (CBF). A thermistor for measuring brain temperature (T^0) as well as two electrodes for ECoG measurements (ECoG) are located between the other probes

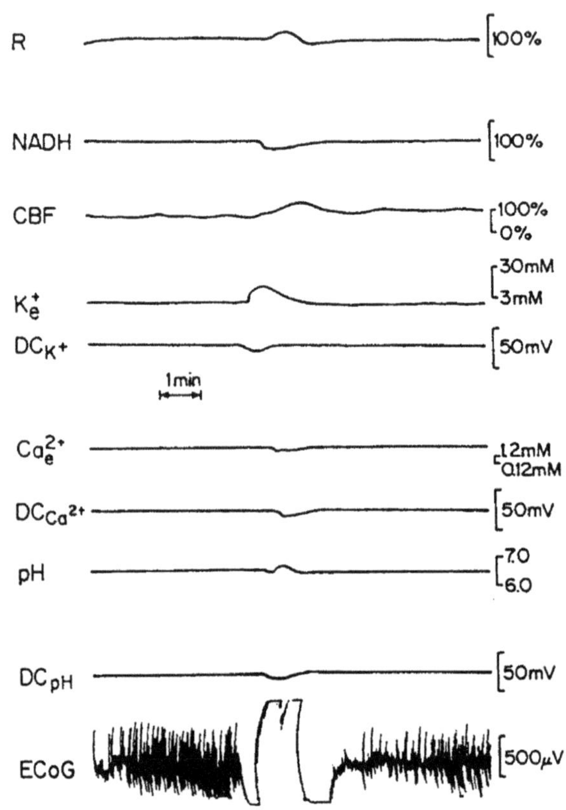

Fig. 8.5 Polygraph tracing of the responses of the adult rat to cortical stimulation by KCI. R reflectance, NADH intramitochondrial NADH levels, CBF cerebral blood flow, K^+ Ca^{2+} - extracellular potassium and calcium ions, pH extracellular pH, DC_{K+}, DC_{Ca2+} DC_{pH} direct current concentric the potassium ion, calcium ion and pH electrodes respectively, ECoG electrocorticogram

and a 6 mm hole was drilled in the parietal bone. The dura mater was gently removed, and the MPA was located on the parietal cortex with a micromanipulator so that undue pressure on the brain was avoided. A 2 mm hole was drilled in the frontal bone ipsilateral to the MPA for a push-pull cannula enabling flushing the brain with KCL solution in order to initiate SD. The MPA cannula as well as four stainless steel screws and the push-pull cannula were cemented to the skull with dental acrylic cement.

8.2.3 Results

Spreading depression (SD) was induced in seven adults and ten old rats. A typical response to cortical SD is demonstrated in Fig. 8.5. This figure is a polygraph tracing taken from an adult animal after induction of SD by KCl solution. The typical

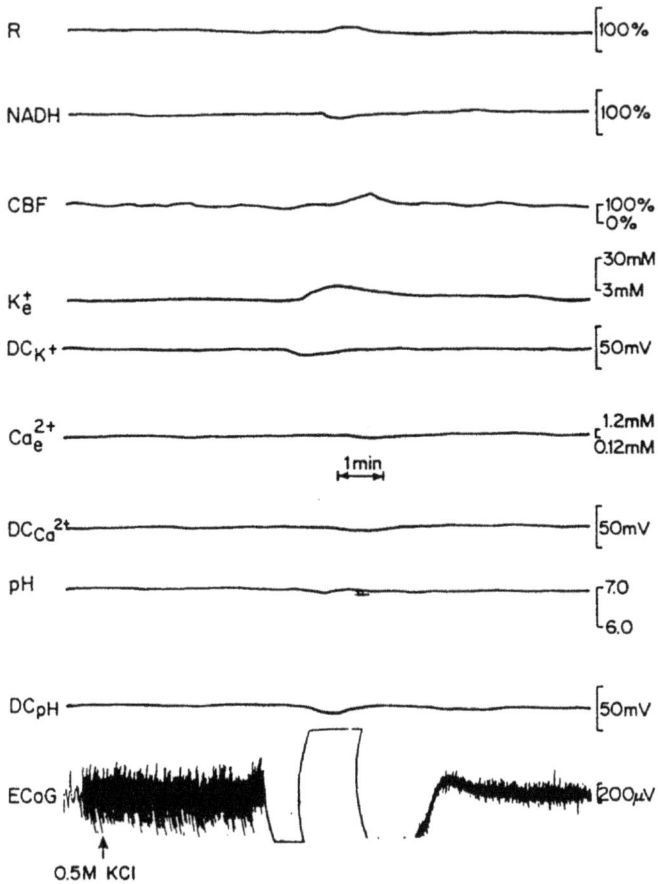

Fig. 8.6 Polygraph tracing of an old rat cortex demonstrating a typical normoxic response to KCl stimulation

SD is characterized by a transient increase in CBF followed by a decrease of NADH level (oxidation), electrical changes, namely: ECoG suppression and DC negative shift and ion leak from the cells to the extracellular space (elevation in extracellular K^+ concentration, a decrease in extracellular Ca^{2+} concentration and bi-phasic change pH).

Figures 8.6–8.9 are polygraph tracings of old rats' cortex demonstrating the various responses to cortical SD induction. The general response to SD induction demonstrated in Fig. 8.6 is similar to the responses obtained in adult animal cortex (as was demonstrated previously in Fig. 8.5). Although CBF and NADH responses are smaller than those monitored from the adult rat's cortex and the SD wave is longer, this SD response can be categorized as a typical normoxic response.

In one old animal we observed "flipped SD" namely, as a result of flushing the cortex with KCl the CBF increased but immediately decreased and NADH increased. These changes were followed by a negative shift of the DC steady potential and by

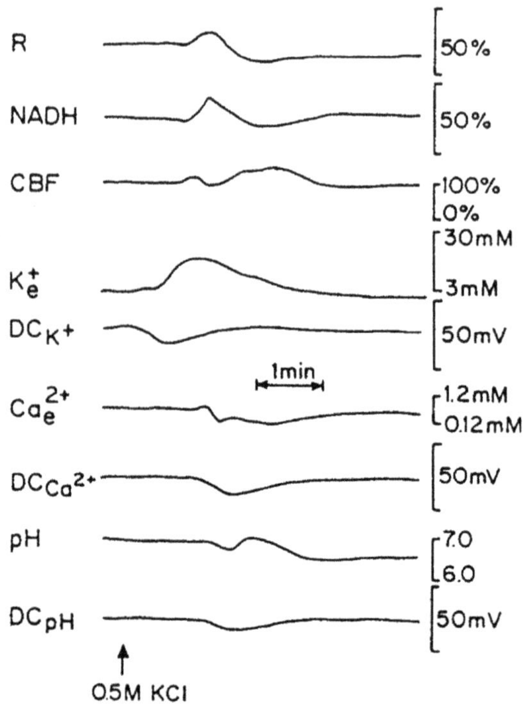

Fig. 8.7 Polygraph tracing of an old rat cortex demonstrating "flipped" response to SD stimulation by KCl

changes in the extracellular ions (elevation in K^+ level, reduction in Ca^{2+} level and biphasic change in pH) (Fig. 8.7). Unfortunately, the ECoG was not monitored in this experiment due to technical problems.

Another response of the old rat's cortex to KCl flush is demonstrated in Fig. 8.8. In this animal, the CBF increased but no metabolic change was observed as was recorded from the NADH signal. All other parameters responded normally, namely, ions leak, DC negative shift and ECoG suppression (extracellular Ca^{2+} concentration was not monitored in this experiment due to technical difficulties).

Figure 8.9 is a polygraph tracing of an old animal representing five out of ten animals used in this study, which responded in the same way to SD inuction by KCl. Those animals did not exhibit any hemodynamic, metabolic, or ionic responses. The only recorded change was the ECoG suppression.

8.2.4 *Discussion*

In this study, we observed various responses of the aged rat cortex during SD. Typical SD is characterized by a large efflux of potassium to the extracellular space leading to elevation in Na^+/K^+-ATPase pump activity. In order to supply the elevation in ATP

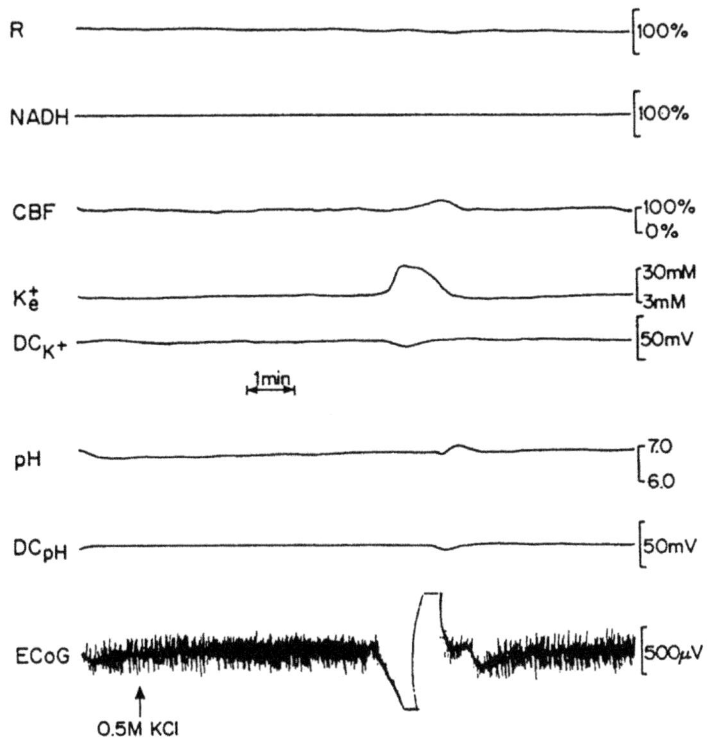

Fig. 8.8 Polygraph tracing of an old rat cortex demonstrating cortical response to KCI stimulation. No changes were recorded in the NADH

demand, the mitochondria are more activated and NADH oxidizes. As a result of these events, CBF increases and this facilitates oxygen supply. Negative shift of DC steady potential and a depression in ECoG occur (Mayevsky 1992).

One out of ten old rats exhibited such a typical response, suggesting that this animal's brain cortex was not influenced by age in terms of tissue adaptation to an increase in energy demand.

However, the remaining nine animals exhibited different levels of brain tissue dysfunction. The first no-typical response to KCl flush was "flipped SD" namely, reduction of NADH and a decrease in CBF. This kind of response is well known as typical response of SD induced under limitation of oxygen delivery (such as ischemia) in adult rats (Mayevsky 1993; Mayevsky et al. 1982, 1996). All the remaining animals exhibited different levels of brain tissue dysfunction under brain stimulation by SD. Only the monitoring of the ECoG was the same. These results are similar to a large number of aging studies describing a decline in mitochondrial number and function (Ames et al. 1995; Takai et al. 1995; Fattoretti et al. 1996). A possible explanation for the reduction in mitochondrial activity is an appreciable loss of cytochrome oxidase with age, such as reported in heart, skeletal and liver tissue (Wallace 1992; Buerk and Nair 1993).

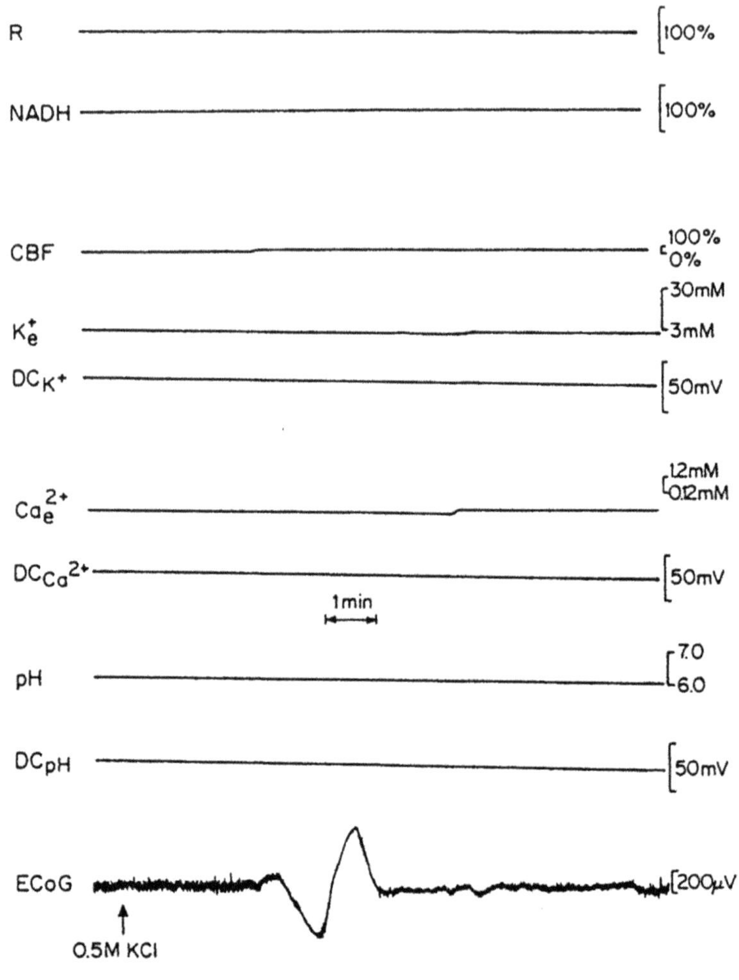

Fig. 8.9 Polygraph tracing of an old rat cortex demonstrating cortical response to KCI stimulation. The only responding parameter was the ECoG

Furthermore, documentation of morphological changes in the cerebral vasculature of the aging rats included thinning of the endothelium, an increase in collagen and impaired reactivity to vasodilator stimuli, which is thought to result from a decline in endothelial function (Davis et al. 1995). In addition, NO is thought to have a vasodilatation function during SD (Meilin et al. 1995). However, later studies measuring NO synthesis found an elevation in NO synthesis probably to compensate for the reduced sensitivity of the NO system in the aging process (Dorheim et al. 1994; Inada et al. 1996).

In five out of ten old animals, no ion (potassium, calcium, and hydrogen) flux was observed. Roberts et al. found age-related alteration in K^+ homeostasis in anoxic rats (Roberts et al. 1997). Early studies carried out by us demonstrated different kinds of age-dependent responses to SD in young animals (Mayevsky et al. 1982). This study underlines the findings suggesting that SD is an age-dependent phenomenon.

In conclusion, our findings show that in the old rats' brain the response to SD is in one of three conditions:

1. Normal brain-exhibiting typical response for SD stimulation.
2. Pathological condition type a—"flipped SD" as occurred in ischemic adult brain.
3. Pathological condition type b—metabolic, hemodynamic and ionic stability—suggesting dysfunction of the mitochondria with inability of blood vessels responding to the increasing oxygen demand.

8.3 Brain Multiparametric Responses to Carbon Monoxide Exposure in the Aging Rat

8.3.1 Introduction

Carbon monoxide (CO) has been shown to cause impairment of the normal binding mechanisms of hemoglobin and oxygen. It has also been implicated in various physiological processes (Coburn and Forman 1987; Thom and Keim 1989). The various effects of exposure to carbon monoxide (CO) were summarized 10 years ago in a review (Penney 1990) and also in a book (Hilfaer and Traystman 1996). Several lines of evidence have recently led to the suggestion that CO may act as a neuronal messenger in a manner similar to nitric oxide (NO) (Verma et al. 1993; Marks et al. 1991; Barinaga 1993; Maines 1993).

The physiological versus the toxic effects of CO would be expected to depend on the pharmacodynamics of this molecule in the various tissues. The central nervous system is the most sensitive tissue to CO and usually suffers the greatest lasting damage from CO poisoning (Penney 1990). We have, therefore, initiated a study of the effects of CO on the central nervous system. Using the recently developed multiparametric monitoring system, we were able to delineate various concentration-dependent effects of CO (Mayevsky et al. 1993a). It was shown that 1000 ppm CO increases cerebral blood flow (CBF), without any sign of tissue hypoxia as evaluated by the mitochondrial redox state (Mayevsky et al. 1995; Meilin et al. 1996). In experiments where rats were exposed to 3000 ppm CO, we found simultaneous effects on CBF (an increase) and a state of minor hypoxia as indicated by an increase in the NADH redox state (Mayevsky et al. 1993a). In our study of adult rats, exposure to 2000 ppm CO had a more complicated mode of action. Under these conditions (2000 ppm CO), the CBF, both in awake and anesthetized rats, increased significantly while NADH showed a mild nonsignificant decrease compared to the control group. In those experiments, the extracellular levels of K^+ did increase significantly in the awake unanesthetized rats apparently due to an inhibition of the Na^+/K^+-ATPase. We concluded that under 2000 ppm CO, the mitochondria were in an uncoupling state as indicated by the apparent oxidation of NADH.

The clinical effects of toxic exposure to CO are dependent on the age of the patient (Kindwall 1996). In this study, we therefore have investigated the toxic effects of CO on the central nervous system using our animal model. Our hypothesis is that blood flow levels as well as its regulation vary in the aging rat. In order to test CBF regulation, the responses to low levels of CO were tested. We have limited this initial study to the effect of 2000 ppm CO, because it shows the most complex effect, including an increase in CBF as well as inhibition of the Na^+/K^+-ATPase via possible uncoupling of the mitochondria. To accomplish this, we performed a study in which aging rats were exposed to 2000 ppm CO while monitoring brain functions using the multiparametric monitoring system (Mayevsky et al. 1995).

8.3.2 Methods

8.3.2.1 Multiprobe Assembly (MPA)

The multiparametric assembly enables the continuous and simultaneous measurement of hemodynamic, metabolic, ionic and electrical activity in one hemisphere of an awake or anesthetized rat brain. Details of the measuring techniques and the technical details have been described previously (Mayevsky et al. 1990, 1995; Mayevsky, 1984a, 1992), therefore, only a short description will be given here. This approach was developed several years ago by Mayevsky et al. (1992) and has been used successfully in rats and gerbils. The MPA used in this study contained a fiber-optic (1.8 mm in diameter) probe connected to two instruments: a fluorometer/reflectometer for monitoring NADH and a Laser-Doppler flowmeter (LDF) for monitoring CBF. For the LDF, one 50-μm input fiber and two 100-μm output fibers were glued in a triangular shape (0.7 mm separation). For measuring NADH redox state, ten excitation and ten emission 200-μm fibers (General Fiber, Cedar Grove, NY) were randomly mixed between and around the fiber used for the LDF (i.e., each excitation fiber was adjacent to an emission fiber) (Mayevsky et al. 1995). This fiber configuration enabled the monitoring of CBF and NADH redox state from nearly the same tissue volume. The MPA also contained three ion-selective electrodes capable of detecting millimolar changes in extracellular ion concentrations of K^+, Ca^{2+}, and H^+. Details of the preparation of these electrodes have been published (Mayevsky et al. 1992, 1995; Meilin et al. 1996). Direct current (DC) potential was measured by a highly stable Ag-AgCl electrode located around the K^+, Ca^{2+}, and H^+ electrodes. A reference electrode (also Ag-AgCl) was located below the skin in the animal's neck. Two polished stainless steel rods or platinum wires were inserted into the MPA for electrocorticography (ECoG) measurement. Brain temperature was measured with a thermistor probe (Yellow Springs Instruments) located inside the MPA (made of Plexiglass) and connected to the thermometer. The MPA was constructed to hold the electrodes to the cannula (with epoxy glue) and was therefore suitable for use in the awake state.

8.3.2.2 Animal Preparation

All experimental protocols were approved by the institutional Animal Care and Use Committee of Bar-Ilan University, which restricted the study to the minimal number of rats necessary for the designed study.

Twelve aged (24 months of age) male Wistar and 13 normal adults (3 months of age) rats (Biodynamics Department, Weizmann Institute of Science, Rehovot, Israel) weighing 500–650 g and 200–220 g, respectively, were used in this study. The rats were anesthetized by a 0.3 ml/100 gr IP injection of Equithesin (each ml contains 9.72 mg pentobarbital sodium, 42.51 mg chloral hydrate, 21.25 mg magnesium sulfate, 44.34% (w/v) propylene glycol, 11.5% alcohol, and water). The skull was exposed and a 6 mm hole was drilled in the parietal bone (right or left). The dura was gently removed and the MPA was located on the brain with a micromanipulator to avoid undue pressure on the brain. The MPA and screws were cemented to the skull with dental acrylic cement. The monitoring of the various parameters was started immediately after the MPA was placed on the brain. The exposure of the rat to CO started at least 1 h after the end of the surgical procedure.

8.3.2.3 CO Exposure Procedure

The rat was placed in an airtight Plexiglas cage and CO at a concentration of 2000 ppm was flushedW through the chamber at room temperature, which was maintained at 23°C (Mayevsky et al. 1995).

The first group of seven rats was exposed to 2000 ppm CO for 1 h while awake. The control experiments for the first group were performed on five awake animals allowed to breathe air without added CO for the entire experimental period. Since the aim of the study was to compare the aging rat to the adult one, we used only two age groups. The effect of various ages was, therefore, not within the scope of this study. Blood gas analysis was not performed during the experiment to eliminate the possibility of bleeding resulting from the use of heparin, required for blood gas analysis.

8.3.2.4 Statistical Analysis

The student's *t*-test was used to examine the effects of 2000 ppm CO on the various parameters measured. A value of $p < 0.05$ was considered to be significant. All values are presented as means ± SEM.

8.3.3 Results

The responses of adult rats (3 months of age) to 2000 ppm CO are presented in Figs. 8.10 and 8.11. Figure 8.10 shows the responses, with respect to all of the

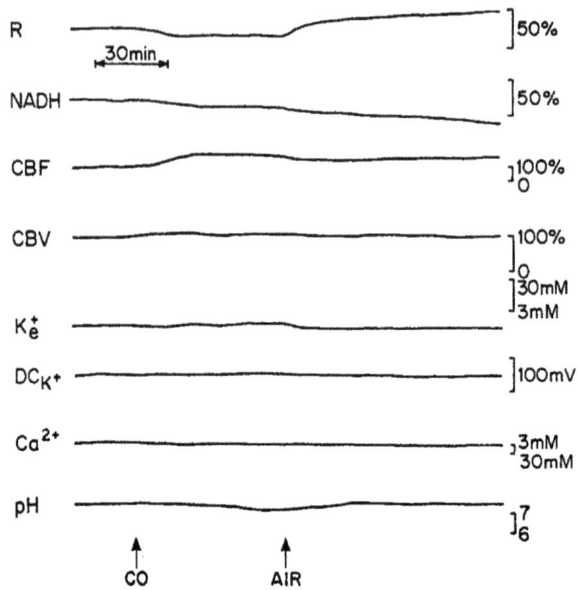

Fig. 8.10 Tracing of the metabolic (NADH), hemodynamic (CBF) and ionic responses of an adult awake rat exposed to 2000 ppm CO for 1 h and recovery upon breathing air for another 2 h. *R* Reflectance at 366 nm; NADH corrected NADH fluorescence, CBF, CBV relative cerebral blood flow and volume, DC_{K+} steady direct current potential measured concentric to the K^+ electrode, K_e^+ Ca^{2+}, pH extracellular ion concentration

parameters monitored for a single rat exposed to CO and Fig. 8.11 shows the average results for both groups of adult animals. Figure 8.11 only shows the data for NADH, CBF, and potassium. These results show that the significant increase in CBF recorded previously under 1000 or 3000 ppm CO (Mayevsky et al. 1995) is also observed under 2000 ppm. Thus, in Fig. 8.10, a clear increase in CBF was recorded immediately after the exposure to CO with a partial recovery after re-breathing air. The decrease in the signal of the mitochondrial NADH indicates oxidation of NADH. There is a clear correlation between the increased CBF and the oxidation of NADH under CO environment. Although during the recovery phase, the CBF decreased, the NADH remained oxidized. During the CO exposure period, a small, clear and reversible increase in extracellular K^+ was found together with a decrease in brain pH (Fig. 8.10).

The detailed results and statistical analysis of the effects of 2000 ppm CO in the normal adult (3 months of age) rat are shown in Fig. 8.11. As can be seen, the NADH levels in the control group (air) were quite stable during the entire recording period (six rats). In the CO exposed group (seven rats) a small (not significant) decrease in the NADH levels without a recovery upon breathing normal air mixture was recorded. As can be seen in both groups, a general trend (not significant) of increased CBF was recorded. Superimposed on this gradual increase, the CBF was elevated after CO

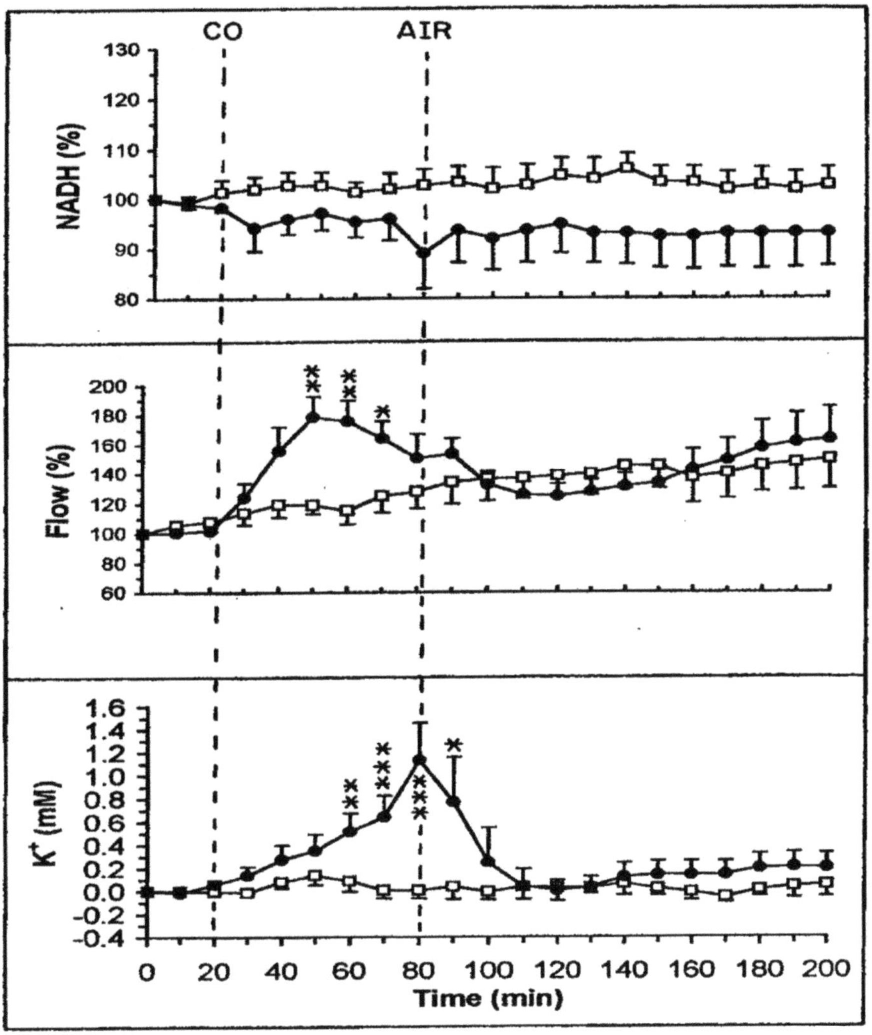

Fig. 8.11 Responses of the adult rat brain to 2000 ppm (●) or air (□). The main three parameters measure by the MPA, namely NADH, CBF and extracellular ΔK^+ (mM) are presented. Note that extracellular K^+ is expressed as ΔK^+ values and not in absolute values. $*p < 0.01$, $**p < 0.05$, $***p < 0.001$ between the experimental and the control group.

breathing, and the difference became significant at three time intervals 30–50 min after the exposure to CO. The effect of 2000 ppm CO on extracellular K^+ was significant at four time points. Also, a significant decrease in pH was found between the CO exposed and control adult rats as seen in Fig. 8.10. The detailed data will be published elsewhere together with more relevant data. After re-breathing air, the recovery of the ionic homeostasis to the baseline was observed.

The results for the old (24 months of age) rats are presented in Figs. 8.12 and 8.13. In these figures, a comparison is made between the responses of rats exposed to 2000 ppm CO (seven rats) as compared to control animals of the same age (five rats) allowed to breath air for the entire experimental period.

In the old rat, the CBF response was clearly different from that of the adult rat as shown in Fig. 8.12. During the 60 min exposure time to CO the CBF showed an insignificant decrease followed by an insignificant increase, meaning no significant difference in CBF was recorded during the entire period of exposure to CO.

The effects of 2000 ppm CO on the reflectance trace were insignificant as shown in Fig. 8.12. The NADH response to CO shown in Fig. 8.12 is somewhat unique. Thus, though the NADH signal increased during the entire course of the experiment for the CO group, during the 60 min of CO exposure, no significant change was recorded.

The effects of 2000 ppm CO on the extracellular K^+ levels for the aged rats (Fig. 8.13) were similar to those found for the adult rats (Fig. 8.11). Thus, the exposure to CO led to a significant increase and complete recovery was recorded when air was returned.

The effects of 2000 ppm CO on the pH values in the old rats (Fig. 8.13) show the same trend found for the adult rat (Fig. 8.10). Thus, the control air exposed old group was very stable during the monitoring period while the CO exposed group showed a significant decrease in pH and recovery when re-breathing air.

The results presented in Fig. 8.11 were selected in order to emphasize the main difference between the adult and old age rats. The main significant difference was in the CBF response without any difference in the K^+ or H^+ responses, therefore, only the K^+ is presented in Fig. 8.11.

8.3.4 Discussion

The exact nature of the physiological and pathological effects of CO on the brain of adult animals or humans, at the cellular level, is still unclear. The influence of age on these physiological and pathological effects of CO not previously been investigated even in animal models. Therefore, in the present study, we investigated the effect of aging on the response to 2000 ppm CO, which has previously been investigated in adult rats (Mendelman et al. 1996).

It may be suggested that this increase in blood flow (in adult rats) is part of the regulatory response following oxygen deficiency (MacMillan 1975a, 1975b, 1978). However, this increase in blood flow is not accompanied by an elevation of basal NADH levels, which are an indirect measure of the oxidative state of the cells. Thus, the effect of exposure to 2000 ppm CO is distinct from the effect of hypoxia. During hypoxia, a correlation has been found between oxyhemoglobin, tissue oxygen and NADH levels (Mayevsky et al. 1992). Hypoxia resulted in increased blood flow, while NADH levels also increased (Mayevsky 1992). This phenomenon is explained in terms of increased blood flow, which is not sufficient to compensate for the oxygen deficiency, causing impairment of the respiratory chain. Several theoretical explanations

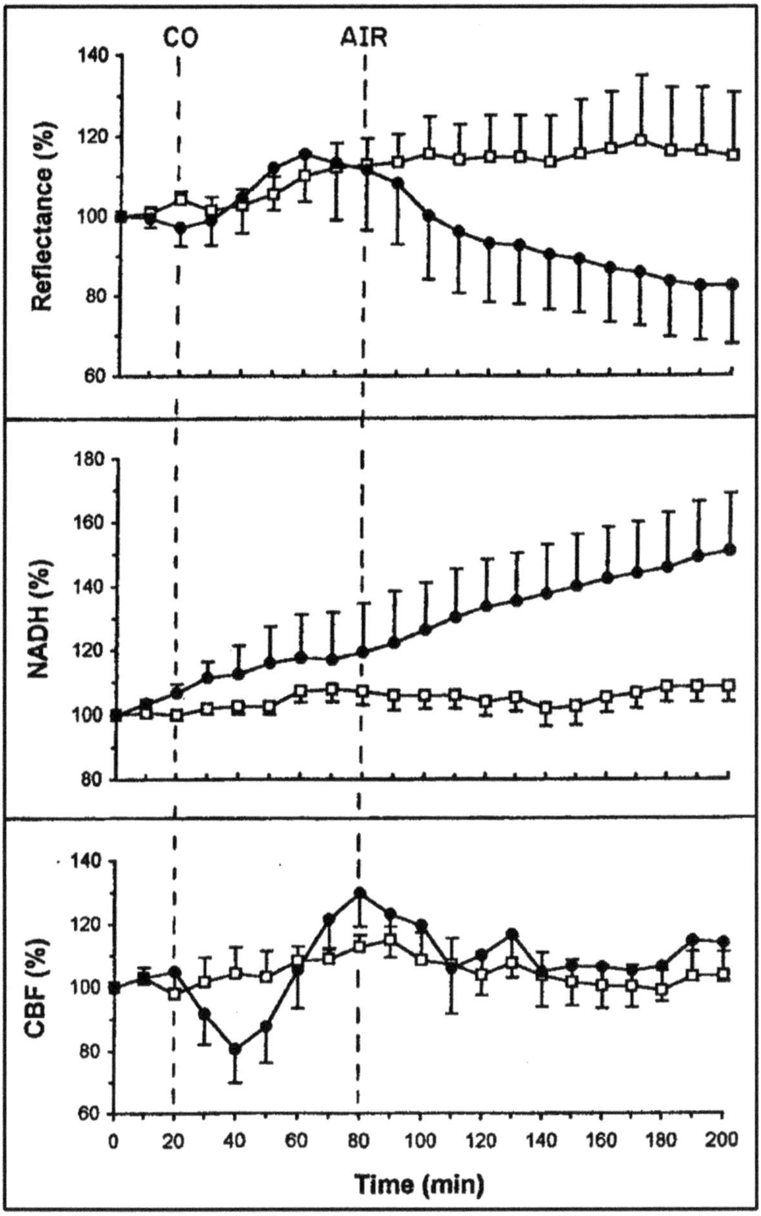

Fig. 8.12 Graph presenting the responses (in percentages) of the reflectance, NADH and cerebral blood flow in aging rats exposed to 2000 ppm CO (●) or air (□)

can be proposed to explain the different relationships between the changes in blood flow and NADH. The increase in blood flow may be able to supply enough oxygen (no hypoxia) such that there is no change in the oxygen consumption or its supply

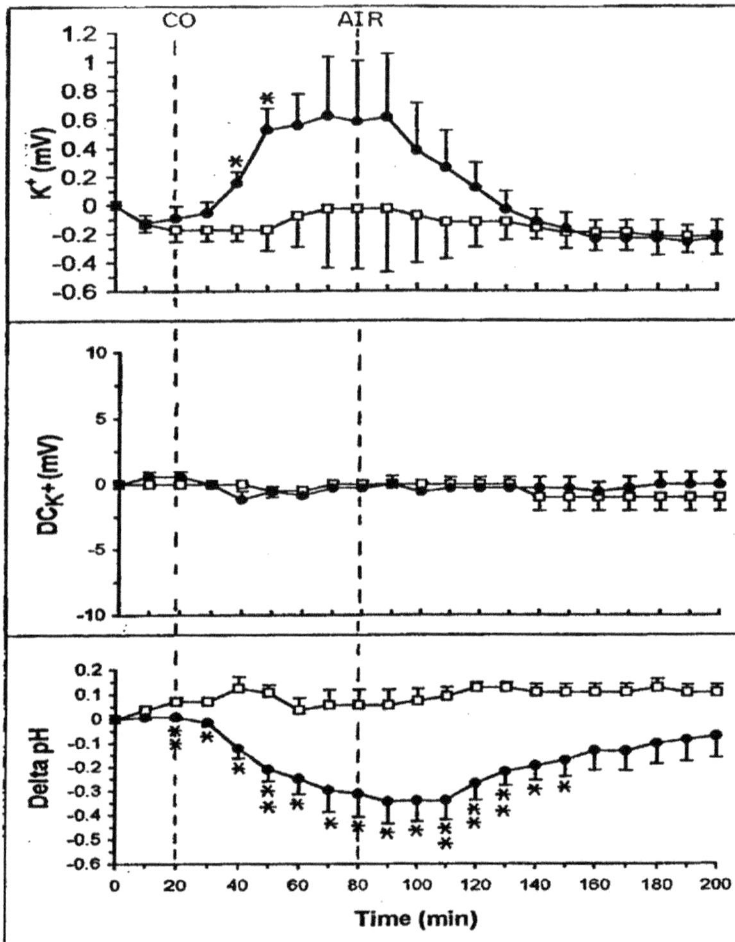

Fig. 8.13 Graph presenting the reaction of the ~K^+ (mM), DC and pH in aging rats exposed to 2000 ppm CO (●) or air (□). Note that extracellular K^+ is expressed as ~K^+ values and not in absolute values. *$p < 0.05$, **$p < 0.01$ between the experimental and the control group

to the tissue, and thus no change in NADH is observed. Furthermore, the increase in blood flow may be due to a dilatory agent (such as NO), which causes vasodilation independently of oxygen levels (Thom and Keim 1989). Another explanation may be that oxygen extraction is increased, and this can maintain adequate tissue oxygen levels (Mayevsky and Weiss 1991). In fact, improved oxygen delivery can result in a decreased demand on the respiratory chain resulting in a decrease in NADH.

Koehler et al. suggest that an increase in the affinity of hemoglobin for oxygen, which takes place when CO binds to hemoglobin, impairs the delivery of oxygen and, thereby, plays a major role in blood flow regulation (Koehler et al. 1982, 1983). The findings in this study do not support this theory since the increase in blood flow was not accompanied by changes in NADH in the adult rats. A significant change was observed in either CBF or NADH. month rats.

In the present study, we found that in the aging rats no significant change was observed in either CBF or NADH. These results indicate that aging influences the response to CO of CBF. We have already suggested that low levels of CO (1000 or 2000 ppm) affect the CBF in the adult rats via a non-hypoxic mechanism but rather through an NO-dependent or NO-involving process (Meilin et al., 1996). One of the possible explanations for the lack of a clear CBF response to 2000 ppm CO in the aging rat is a weaker NO vasodilatory response in the old brain than in the adult brain. This conclusion is consistent with other reports in the literature.

Toyoda et al. (1997) showed, using laser Doppler flowmetry, that aging diminished the compensatory dilatation of brain stem arterioles during hypotension. The changes in autoregulation in aging rats were also studied in spontaneously hypertensive animals. Fujishima et al. (1984) found a diminished vasodilatory response of the brain to hypotension in 20-month-old rats.

The extracellular ionic environment was responsive to CO in both the aging rat and the adult rat (Figs. 8.11 and 8.13). These findings demonstrate the strength of the multiparametric monitoring approach developed by us and applied to various pathological and physiological conditions (Mayevsky et al. 1992, 1993b, 1995, 1996; Mayevsky 1993). The differentiation between the responses of the aging brain to CO seen in the energy metabolism and ionic homeostasis is very challenging. The fact that both extracellular K and H levels were elevated without a clear corresponding change in energy metabolism indicates that two different mechanisms or effects of CO may exist. The presence of more than one CO-related process has previously been suggested for the adult rat (Mendelman et al. 1996).

The increase in extracellular potassium ion concentration during exposure to CO without a change in NADH (Fig. 8.12) indicates that the respiratory chain was not impaired and, therefore, ATP formation was not affected. Why then did the extracellular potassium ion concentration increase? Several explanations for this can be offered. As a result of the exposure to CO, free radicals are formed, changing the membrane permeability to ions (Zhang and Piantadosi 1992) and a general depolarization of the cells occurs. This depolarization is expressed in a massive efflux of potassium ions out of the cells accompanied by a decrease in pH, suggesting increased glycolysis and accumulation of lactic acid. Another possible explanation is that exposure to 2000 ppm CO directly activates an outward potassium channel, as shown for smooth muscle (Wang and Wu 1997).

The effect of aging on brain energy metabolism has been studied. The vascular architecture of the brain shows a severe decrease in the vascularity of 124-week-old rats (Wilkinson et al. 1981). Using the gerbil brain exposed to complete ischemia, Buerk and Nair (1993) found a significant decrease in O_2 consumption in the old animals (18–21 months). In another study, it was found that the infarct size in old rats (28–36 months) was larger as compared to the normal adult controls (Davis et al., 1995). Ohata et al. (1981) found that CBF was not significantly different between 24-month and 34-month rats.

Parmacek et al. (1979) found in 34-month-old mice a decrease in O_2 consumption as compared to normal adults (6 months). Sylvia and Rosenthal (1979), using dual wavelength spectrophotometry, found that in rats, the resting mitochondrial redox state was not affected by age.

8.4 Brain Energy Metabolism in the Aging Rat

8.4.1 Introduction

A great deal of knowledge has been accumulated during the last decade concerning the aging brain in health and disease. It is well documented that functional metabolism of the aging brain and its blood supply inevitably declined (Meyer et al. 1993). Old rats are more sensitive to pathophysiological manipulations such as ischemia (Funahashi et al. 1994), anoxia, hypoxia and hypercapnia. However, the mechanism determining recovery from these pathological conditions is unknown.

During anoxia, the fast decrease in blood oxygen concentration will eventually lead to a decrease in the metabolic-energy state. The aging process is accompanied by morphological and functional changes that occur in the blood vessels and by a decline in the number and function of the mitochondria (Ames et al. 1995; Takai et al. 1995; Fattoretti et al. 1996). Furthermore, slow oscillations of cerebral blood flow and oxidative metabolism have been previously reported in animals under different pathological conditions (concerning lack of oxygen), such as ischemia (Meilin et al. 1999). Whether the occurrence of oscillations is a pathological condition or a normal phenomenon of the CNS is not yet known.

The aim of the present study was to evaluate the effect of aging on the energy metabolism of the brain during anoxia and normoxia and to further investigate the occurrence of oscillations under normoxia in the aging rat brain.

8.4.2 Methods

We used the multiparametric assembly (MPA) for the continuous, simultaneous online measurement of hemodynamic, metabolic, ionic and electrical activities of the rat cerebral cortex (Mayevsky et al. 1995). Measurements included NADH redox state using surface fluorimetry (Mayevsky and Chance 1982), CBF by using laser Doppler flowmetry (Mayevsky et al. 1996), extracellular K^+, Ca^{2+}, and H^+ concentrations, and DC steady potential as well as ECoG.

The experiments were performed on ten aging (25–30 months) Wistar male rats and ten adult male rats (2–3 months). The rats were anesthetized by an intraperitoneal injection of Equithesin (0.3 ml/100 g). The skull was exposed by a midline incision in the skin, and a 6 mm hole was drilled in the parietal bone. The dura mater was gently removed, and the MPA was located on the parietal cortex with a micromanipulator so that undue pressure on the brain was avoided. The MPA cannula and four stainless steel screws were cemented to the skull with dental acrylic cement. The monitoring of the various parameters started immediately after the MPA was placed on the brain. The exposure of the rat to 100% N_2 started at least 30 min after the end of the surgical procedure. Data was analyzed using a Student *t*-test. A value of $p < 0.05$ was considered as significant.

8.4.3 Results

Each experiment started with a 30-s exposure of the animal to 100% N_2 (transient anoxia) for the measurement of maximal accumulation of NADH (Fig. 8.14). Exposure of the adult rats (Part a) to anoxia resulted in significantly higher elevation of the NADH level compared to the aged rat (Part b). Following exposure of the adult rats to anoxia, a 36 ± 2% elevation of NADH was observed (Fig. 8.15). Exposure of the aged group to anoxia resulted in significantly lower changes of the NADH level (9.6 ± 4%, $p < 0.05$). These

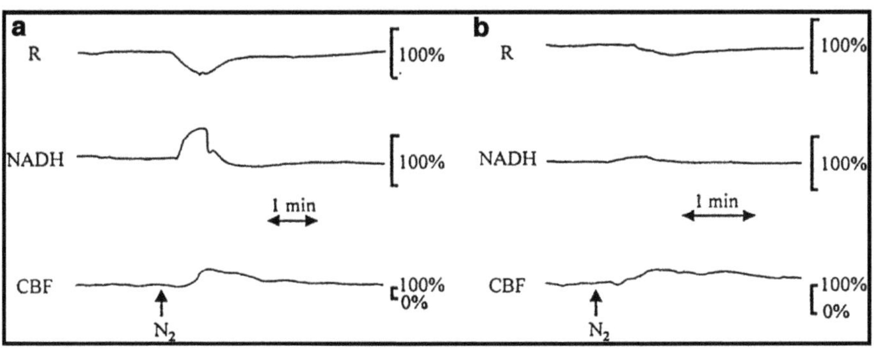

Fig. 8.14 Typical responses to anoxia (40–50 s, 100% N_2) in the adult (**a**) and aging (**b**) rats. R 366 nm reflectance measured by the total back scattered light at the excitation wavelength. NADH corrected 450 nm NADH fluorescence. CBF cerebral blood flow

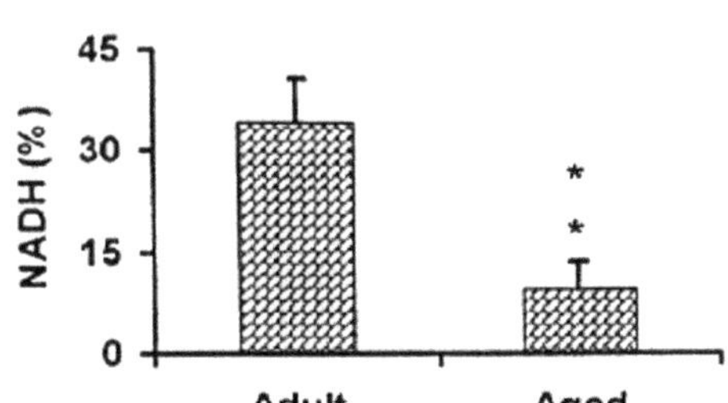

Fig. 8.15 Average responses to anoxia in the adult and aging rats. The NADH (%) represents the increase due to N_2 as compared to the normoxic levels. $^{**}p < 0.01$ significant difference between the two groups of rats

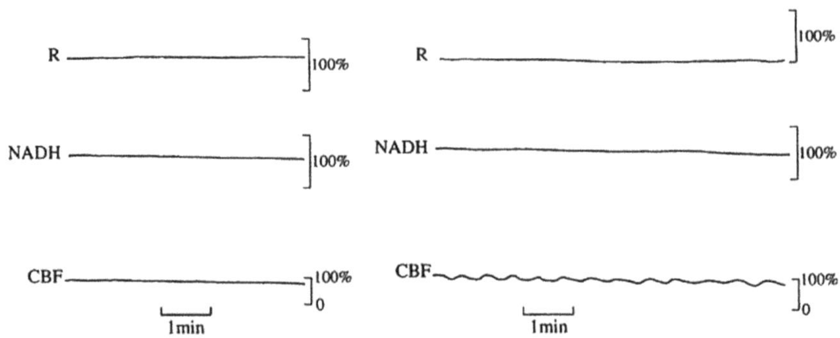

Fig. 8.16 Typical analog recording of the three parameters monitored in the adult (left) and aging (right) rats demonstrating the oscillations in the CBF

changes in the NADH level were followed by CBF increase. A high, negative correlation was found between the age of the animal and the NADH maximum accumulation potential of the animal. After the anoxia ceased and the animals were allowed to breathe air (normoxia), the NADH level immediately returned to baseline level in all animals. However, in 95% of the aged animals, oscillations of the CBF (three waves/min) were recorded during the normoxic period. The hemodynamic oscillations were not accompanied by any changes in the mitochondrial NADH level. The CBF of all adult animals remained unchanged (Fig. 8.16).

8.4.4 Discussion

Exposure rats to 100% N_2 (anoxia) caused complete cessation of mitochondrial activity due to lack of oxygen delivery. Thus, during the anoxic period, NADH accumulated and its maximal level could be monitored. Our study shows that the maximal accumulation of NADH recorded during anoxia of the adult rats was significantly higher compared to the maximal NADH level recorded from the aged rat. This may indicate that the NADH redox range shrinks with age. A previous study carried out in our laboratory demonstrates that newborn rats (2 weeks) exposed to anoxia show an even higher peak of NADH level than adult rats (Zarchin and Mayevsky 1981). This may indicate that the shrinking of the NADH redox range with age is probably a progressive rather than an acute event. This study is in agreement with other studies reporting deterioration of mitochondrial number and function with age (Ames et al. 1995). Furthermore, our data show that only the aged rats but none of the adult rats underwent hemodynamic oscillation under normoxic conditions. A comparative study of oscillation carried out in our laboratory using adult animals suggests that the occurrence of hemodynamic-metabolic oscillations is a pathological phenomenon and apparently not a normal physiological phenomenon (Meilin et al. 1999; Mayevsky and Ziv 1991). Furthermore, we have shown that the oscillation phenomena are related to lack of oxygen in brain tissue. The typical oscillations monitored from the adult rats under various oxygen stress conditions were hemodynamic and metabolic oscillations, meaning that the wave-like changes of the CBF were followed by matching wave-like changes of the NADH. However, the aged rats in this study performed only hemodynamic oscillations and without any changes of the NADH, were undetectable, maybe because the NADH redox range became so narrow that the oscillations were too small to notice.

8.5 Age-Related Alteration of Brain Function During Cerebral Ischemia

8.5.1 Introduction

The aging process involves, among other things, morphological changes. A great deal of knowledge has been accumulated during the last decade concerning the aging brain in health and disease. One of the major diseases causing death and disability in the elderly is ischemic stroke. However, in vivo studies, including the evaluation of brain function under ischemic conditions, have relied on models of focal cerebral ischemia in young brains. It is well documented that even in normal aging, the functional metabolism of the brain and its blood supply inevitably declines in cerebral vasculature, including thinning of the endothelium, thickening of the basal lamina in hippocampal vessels, and increases in collagen and elastin within vessel walls with arteriosclerosis (Davis et al. 1995; Wolinsky 1972; Marín 1995). These morphological changes result in a decline of endothelial function to dilate and contract. Recently, it was reported that the activity of NO synthase declines with age as well (Mollace et al. 1995; Inada et al. 1996) enhancing the morphological studies. The structural and functional changes occurring in the aging brain impair the ability of cerebro-microvessels to optimally deliver nutrients and oxygen to the brain (Meyer et al. 1993; de la Torre 1997), thus affecting mitochondrial function. Brain tissue function deterioration with age is related to the decline of mitochondria number (Fattoretti et al. 1996) and function (Takai et al. 1995) occurring in the aging brain.

The aim of this study was to evaluate aging brain function under ischemic conditions and to estimate the effect of aging on recovery from ischemic events.

8.5.2 Methods

8.5.2.1 Animal Preparation

All experimental protocols were approved by the Animal Care and Use Committee of Bar-Ilan University. Seven adult (2–3 months) and 10 aged (25–30 months) male rats were anesthetized by an intraperitoneal (IP) injection (0.3 ml/100 g) of Equithesin (each ml contains: 9.72 mg pentobarbital; 42.51 mg chloral hydrate; 21.25 mg magnesium sulfate; 44.34% w/v propylene glycol; 11.5% alcohol; and water). Additional IP injections of equithesin (0.03–0.05 ml) were administered every 20–30 min to keep the rats in a slightly anesthetized state during the entire experimental period. The common carotid arteries were isolated just prior to brain surgery and 4-0 silk thread ligatures were placed around them. After a midline incision of the skin of the skull, a 6 mm hole was drilled in the parietal bone, and the dura mater was gently removed. The MPA was held by a micromanipulator and lowered, allowing gentle contact with the brain. A reference electrode was introduced into the neck area between the skin and the muscles. Two stainless steel screws located in the skull were used to fixate the Plexiglas holder of the MPA to the skull using dental acrylic cement.

8.5.2.2 The Multiparametric Assembly (MPA) and Monitoring System

This system enables the simultaneous measurement of physiological, biochemical and electrical activity from the same relatively small area of the cerebral cortex. Hemodynamic changes were measured using Doppler flowmeter. The metabolic state of the intracellular space was monitored using the light guide fluorometer-reflectometer measuring changes in the intramitochondrial NADH redox state. Changes in the extracellular concentration of potassium ion were measured with the aid of specific minielectrodes. This system also included electrodes suitable for measuring spontaneous electrical activity of the brain (ECoG) and electrodes for measuring DC potential. All these sensors were mounted in a Plexiglas cannula, which was positioned on the cerebral cortex as described previously (further details on the MPA can be found in Meilin and Mayevsky (Mayevsky et al. 1995; Meilin et al. 1995, 1996). Body temperature was monitored in all animals using a rectal thermistor and was regulated to 37°C. A schematic presentation of the MPA located on the cerebral cortex of a rat is shown in Fig. 8.17.

8.5.2.3 Experimental Protocol

Ninety minutes after surgery, unilateral occlusion was performed for 10–12 min and then released in order to test the integrity of the brain before starting the protocol. All parameters were both monitored online and recorded on a Grass polygraph and stored in a computer using and A/D converter (Codas system, DATAQ Inc., MN, USA).

8.5.3 Results

The responses of the aged brain to bilateral carotid occlusion were different in nature compared to the responses of the adult brain. Figure 8.18 is a polygraph tracing demonstrating the changes of the various parameters following bilateral carotid occlusion in

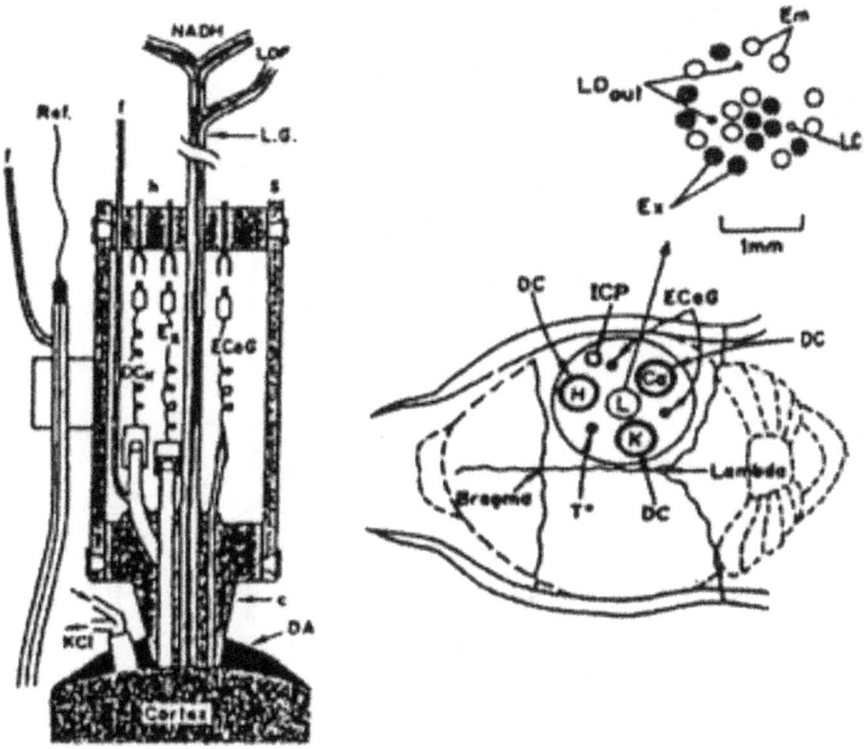

Fig. 8.17 Multiprobe assembly used in the awake rat to monitor hemodynamic, metabolic, ionic and electrical activities of the brain. The left side shows a longitudinal section, while the right side shows the location of the MPA and skull. C Plexiglas probe holder, DA dental acrylic cement, ECoG electrocortical electrodes, K, Ca, H ion specific electrodes, DC area of DC steady potential monitoring, NADH 2 arms of NADH monitoring light guide, h connector holder, s aluminum sleeve, Ref reference electrode, f feeling tube of the reference or the other DC electrodes, T^O thermistor for local temperature measurement, Ex, Em excitation and emission fibers for NADH monitoring, LD_{in}, LD_{out} optical fibers for monitoring blood flow and volume

an adult (A) and an aged (B) rat. Carotid occlusion for 12 min in the adult rat resulted in a transient decrease in CBF followed by a short episode of NADH elevation and reflectance (R) changes. A slight depression of the spontaneous electrical activity was recorded from the ECoG as well. All parameters returned to their initial levels during the ischemic episode while the carotids were still occluded. However, bilateral carotid occlusion of the aged brain resulted in far more radical changes in the various parameters. Immediately after occlusion CBF decreased, followed by an increase in NADH level. Unlike the responses of the adult brain, these changes remained permanent during the entire ischemic episode. As ischemia continued, a secondary elevation of reflectance was observed, followed by major changes of the extracellular ions (i.e., increased potassium level, decreased calcium level and decrease of pH). The electrical activity was also depressed as recorded from the ECoG and DC potential. After the

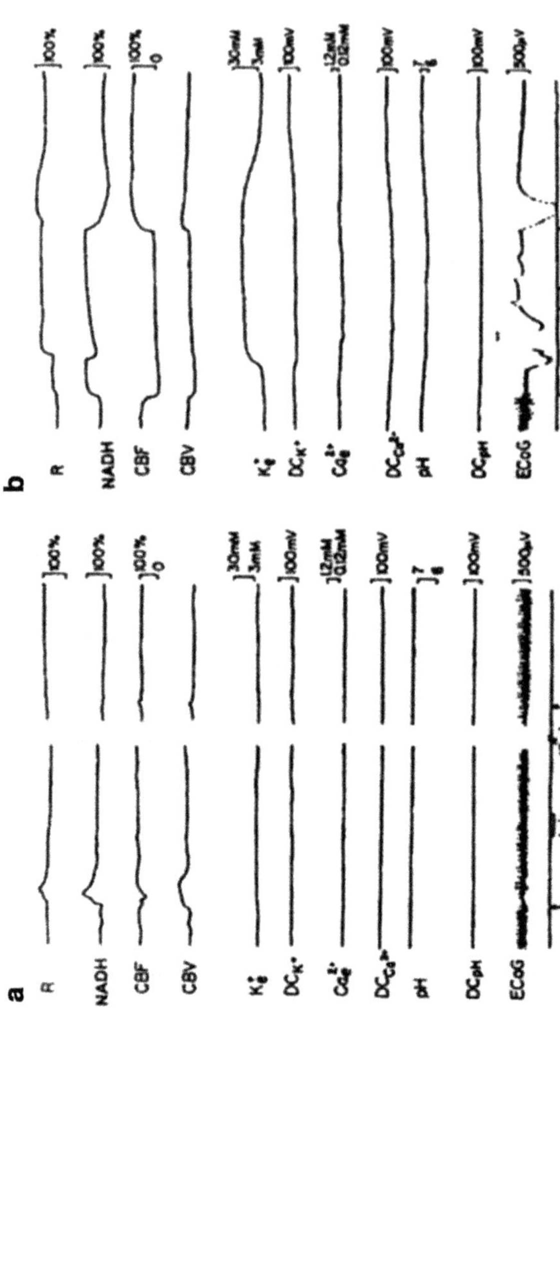

Fig. 8.18 Polygraph tracing demonstrating the changes of various parameters following bilateral carotid occlusion in an adult (**a**) and aged (**b**) rat

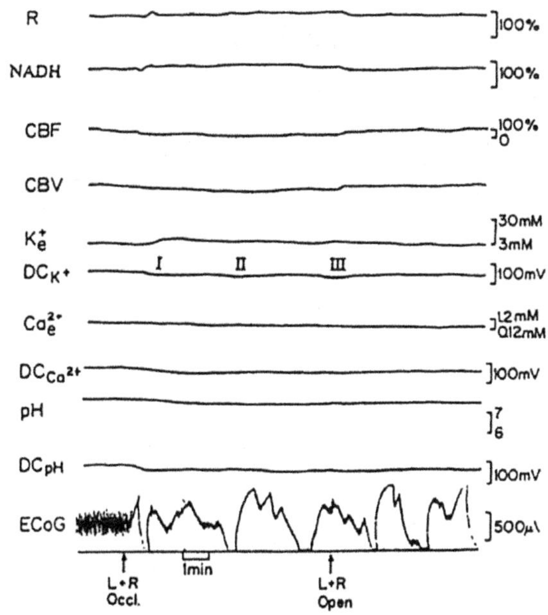

Fig. 8.19 Polygraph tracing demonstrating SD following ischemia in an aged rat. I, II, and III are three waves of SD. R, CF reflectance and corrected NADH fluorescence, CBF, CBV cerebral blood flow and volume, K^+, Ca^{2+}, pH extracellular levels of potassium, calcium and hydrogen, DC_{K^+}, $DC_{Ca^{2+}}$, DC_{pH} direct current steady potential measured concentric to each of the ion-specific electrodes

carotids were opened and the blood supply renewed, all parameters returned to their initial levels within minutes.

Two minutes after the carotids were occluded in seven out of ten aged rats, ischemic depolarization was observed. Namely, after the decrease in CBF and the increase in NADH level, secondary increase of reflectance, a major increase of the extracellular potassium followed by a decrease of the DC potential, occurred. The rest of the 3-day-old rats demonstrated spreading depression-like responses (SD). That is, while CBF decreased and NADH increased, the potassium level and DC potential were transiently changed in three waves (Fig. 8.19). The activation of the brain (SD-like response) was developed spontaneously in the ischemic brain.

For statistical analysis, NADH and CBF levels from the adult rats as well as from the aged rats were sampled at four main points: (1) Before the ischemic episode; (2) 30 s after the two carotid arteries were occluded; (3) Before the carotid arteries were opened; and (4) 30 s after the carotid arteries were reopened. Figure 8.20 is a histogram presentation describing the CBF and NADH levels at these four time points. Immediately after carotid artery occlusion, CBF in the adult rats decreased. However, these changes in CBF recovered during the ischemic period, so that when the carotids were reopened, the level of CBF was similar to the initial level. In contradistinction,

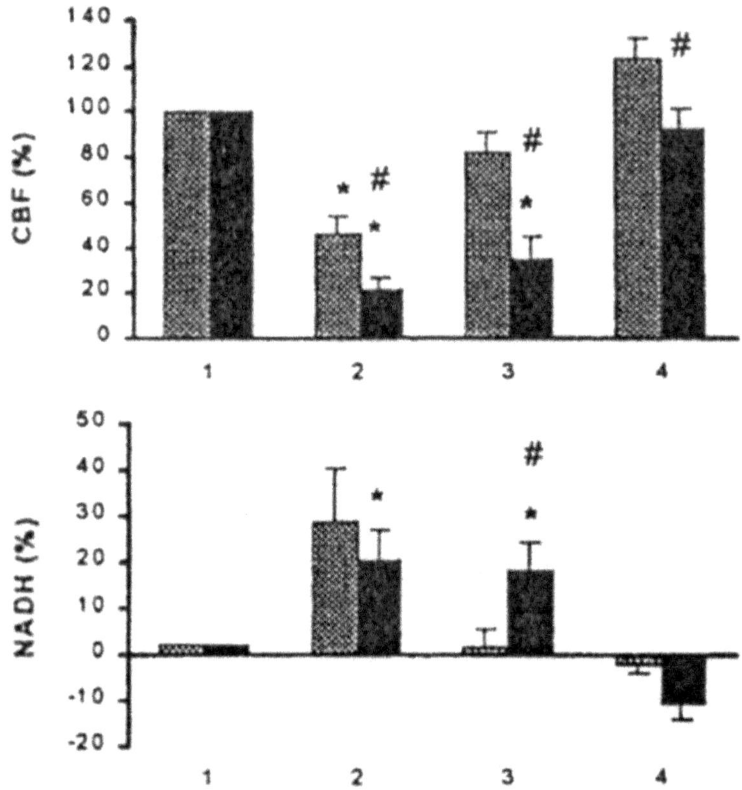

Fig. 8.20 CBF and NADH levels after ischemia and reperfusion in adult rats (gray) and old rats (black). 1. CBF and NADH levels before occlusion. 2. CBF and NADH levels 30 s after carotid artery occlusion. 3. CBF and NADH levels at the end of the ischemic period (just before carotid arteries were reopened). 4. CBF and NADH levels 30 s after reperfusion. *$p < 0.05$ within the group. #$p < 0.05$ between adult and aged groups

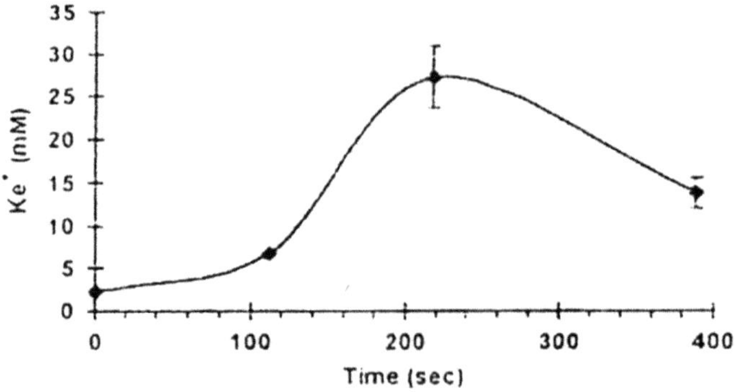

Fig. 8.21 Changes of extracellular potassium (K^+) during ischemia in the aged rat

the decrease in CBF after occlusion recovered only after the carotid arteries were reopened. These changes in CBF were accompanied by changes in NADH level. The elevation of the NADH level in the adult animal was not significant because of the great difference between the animals.

In addition to the hemodynamic and metabolic changes, extracellular potassium level was elevated after carotid artery occlusion only in the old rats (Fig. 8.21). The maximum level of extracellular potassium (23 ± 3.61 mM) was recorded 217 s (216.86 ± 16.27) after the beginning of the ischemia. Furthermore, extracellular pH decreased significantly ($p < 0.01$) (results not shown) during the ischemic period as well. The pH did not fully recover even after the opening of the carotid arteries. Control animals that did not undergo carotid artery occlusion managed to keep their pH levels unchanged (7.16 ± 0.5 and 7.18 ± 0.5 in adult rats and aged rats, respectively).

8.5.4 Discussion

A normal compensation mechanism in the adult rat enables quick recovery from ischemia due to the existence of the “circle of Willis”. Bilateral carotid artery occlusion in the adult rat caused transient changes in CBF and energy metabolism, as recorded from the NADH. These changes recovered during the ischemia period (i.e., while the carotid arteries were still occluded). No other changes were monitored during the ischemia or reperfusion period, suggesting that bilateral occlusion of the carotid arteries in the adult rats did not lead to radical tissue damage.

However, the aging process is accompanied by well-known morphological as well as functional changes in the arteries. During aging, the endothelium becomes thinner, the basal lamina becomes thicker and accumulation of collagen and elastin within vessel walls occurs. These morphological changes result in a decline of endothelial ability to dilate and contract. The structural and functional changes that occur in the aging brain impair the ability of cerebral microvessels to optimally deliver nutrients and oxygen to the brain, thus affecting mitochondrial function. Also, it is possible that in the aging brain the blood flow compensation via the vertebral arteries and circle of Willis is limited after bilateral occlusion of the carotid arteries. Furthermore, deterioration with age is related to the decline of mitochondria number and function occurring in the aging brain. The combination of blood vessels changes and mitochondrial function has a major impact on the responses of the aged brain to ischemic condition. Indeed, bilateral carotid arteries occlusion in the aged rats resulted in a major decrease in CBF followed by impairment of energy metabolism, as recorded from the NADH elevation. However, unlike the adult rat, these changes recovered only when the carotid arteries were opened and blood supply was renewed. Furthermore, the changes in the CBF led to general cortical depolarization expressed by increase of extracellular potassium level, electrical activity suppression, as well as secondary elevation of the reflectance. These responses are very similar to the responses of the adult gerbil to general ischemia. As reported by Breuer and Mayevsky (1992) and by Mayevsky and Breuer (1992), the gerbil suffers from a structural genetic defect of its artery. Therefore, these results support other studies demonstrating vascular disabilities of the aged brain and suggest that these disabilities have an impact on the severity of ischemia.

It is important to note that extracellular pH declines during ischemia only in the aged rats. However, unlike the other parameters, the extracellular pH did not recover even after reopening of the carotid arteries. These findings support other studies suggesting that regarding pH level, recovery of the aged brain from ischemia is much slower than recovery of the adult brain (Funahashi et al. 1994). We suggest that under ischemia, there is partial damage to the mitochondria, which makes them unable to produce the necessary amount of ATP required for maintaining the critical brain activities. Therefore, glycolytic activity increase resulted in long-term acidosis.

In conclusion, the two major processes accompanied "normal aging," namely, vascular changes together with mitochondrial damage, deteriorate brain tissue energy metabolism ability. As a result, the aged brain is less capable of handling reversible ischemia and reperfusion.

8.6 Continuous Multiparametric Monitoring of Brain Activities Following Fluid-Percussion Injury in Rats: Preliminary Results

8.6.1 Introduction

Brain trauma is one of the most complicated and real problems in today's neurosurgical intensive care unit. Approximately 20% of severe head traumas end in death or irreversible brain damage. Data accumulated to date indicate that brain trauma is a multifactorial pathophysiological process (Mattson and Scheff 1994). In order to understand this complicated process, numerous basic parameters must be measured continuously and simultaneously. Therefore, monitoring of the metabolic, hemodynamic, ionic and electrical brain activities, combined with intracranial pressure (ICP), may reflect the functional deviation of the brain resulting from primary events developed during brain injury. Most investigators studying various animal models for brain injury only monitored one to three parameters simultaneously (Meyer et al. 1970; Duckrow et al. 1981; Lewelt et al. 1982; DeWitt et al. 1986; Muir et al. 1992b; Shackford et al. 1992; Ishige et al. 1987b; Kubota et al. 1989; Katayama et al., 1990).

We adopted the multiparametric monitoring approach (Friedli et al. 1982; Mayevsky 1983a) to the fluid percussion traumatic brain injury (TBI) model in the rat. Brain hemodynamics were measured by monitoring microcirculatory blood flow using Laser Doppler flowmetry (Wadhwani et al. 1990). Metabolic activity was evaluated by the intramitochondrial redox state as measured by NADH fluorescence (Mayevsky 1984a). Ionic homeostasis was measured by ion-sensitive K^+, Ca^{2+}, and H^+ mini-electrodes (Mayevsky et al. 1991, 1995; Moravec et al. 1972; Crowe et al. 1977). Electrical activity was measured by the spontaneous ECoG and steady DC potential. Intracranial pressure was continuously measured using a fiber optic probe (Crutchfield et al. 1990).

The aims of the present study were:

1. To develop the rat TBI model for recording typical responses using the multi-probe assembly (MPA);
2. To test the effect of cortical spreading depression (brain activation) induced after TBI on the metabolic responses of the brain;
3. To evaluate qualitatively the potential use of hyperbaric oxygen therapy in the TBI model using these monitoring approaches. It is assumed that our approach would only detect changes developed during the acute phase after the TBI (up to 12 h), although it is clear that secondary damage may also develop after this monitoring period.

8.6.2 Methods

8.6.2.1 NADH Fluorometry

The principle of NADH monitoring from the surface of the brain is that excitation light (366 nm) is passed from the fluorometer to the brain via a bundle of quartz optical fibers. The emitted light (450 nm), together with the reflected light at the excitation wavelength, is transferred to the fluorometer via another bundle of fibers. The changes in the reflected light correlate to changes in tissue blood volume and also serve to correct for hemodynamic artifacts appearing in the NADH measurement. The changes in fluorescence and reflectance signals are calculated relative to the calibrated signals under normoxic conditions. This type of calibration is not absolute, but provides reliable and reproducible results from different animals and between different laboratories. Details on this technique have been published previously (Mayevsky 1978b; Mayevsky and Chance 1982, 1983).

8.6.2.2 Local Cerebral Blood Flow

In order to measure CBF in real-time from the same cortical area as the MPA, we used the Laser Doppler flowmeter (LDF) technique (Stern et al. 1977; Haberl et al. 1989; Dirnagl et al. 1989). The principle of Laser Doppler flowmetry is to utilize the Doppler shift, namely, the frequency change that light undergoes when reflected by moving red blood cells. A low-power laser beam or diode laser light is transmitted by an optical fiber to the tissue. After multiple scattering of the light another optical fiber picks up the reflected light and enters it into the photodetector. The run signal is analyzed by a complicated algorithm developed by the manufacturers, and the results are presented as percent of a full scale (0%–100%) providing relative flow values. The change in the total back-scattered light is an indirect indicator for the blood volume in the tested tissue volume. During the past few years, this technique has been calibrated against two well-established methods for quantitative monitoring of CBF, i.e., the H_2 clearance probe and [^{14}C]-iodoantipyrine autoradiography. LDF apparently measures relative changes which are significantly correlated to the relative changes in CBF measured by the two other quantified approaches (Wadhwani et al. 1990).

8.6.2.3 Ion Selective Electrodes

Mini-electrodes made by WPI (Sarasota, Florida, USA) were used for monitoring extracellular levels of K^+, Ca^{2+}, and H^+. Flexible PVC tubing was sealed at one end with a membrane sensitive to a specific ion. The tube was filled with the appropriate solution and connected to an electrode holder (WPI) with a saline bridge between the membrane and the Ag/AgCl pellet located inside the holder. The interface between the PVC tubing and the holder was glued using 5-min epoxy. Such an electrode was usable for several weeks. The electrode sensitivity for the specific ion was near its Nernstian value, namely, 50–60 mv/decade for K^+ and H^+ or 25–30 mv/decade for Ca^{2+}.

8.6.2.4 DC Steady Potential Electrode

DC potential was measured concentrically around the ion electrodes (K^+, Ca^{2+}, and H^+). Each electrode had a saline bath around its perimeter and an Ag/AgCl electrode (WPI) was connected to it.

8.6.2.5 Reference Electrode

An Ag/AgCl electrode connected via a saline bridge to the neck area of the animal was used. Polyethylene tubing, stuffed with a cotton string that expanded when wet, was inserted into the electrode holder (WPI) and glued with 5-min epoxy.

8.6.2.6 Electrocorticography

Spontaneous electrical activity of the brain surface was measured by two polished stainless steel rods or silver wires inserted into the MPA.

8.6.2.7 Intracranial Pressure Monitoring

In order to monitor parenchymal ICP, a fiber optic ICP probe was used (Camino Lab., USA) as detailed in Fig. 8.22a. This probe has recently been introduced into the MPA, such that it has the same contact area with the brain surface as the other probes.

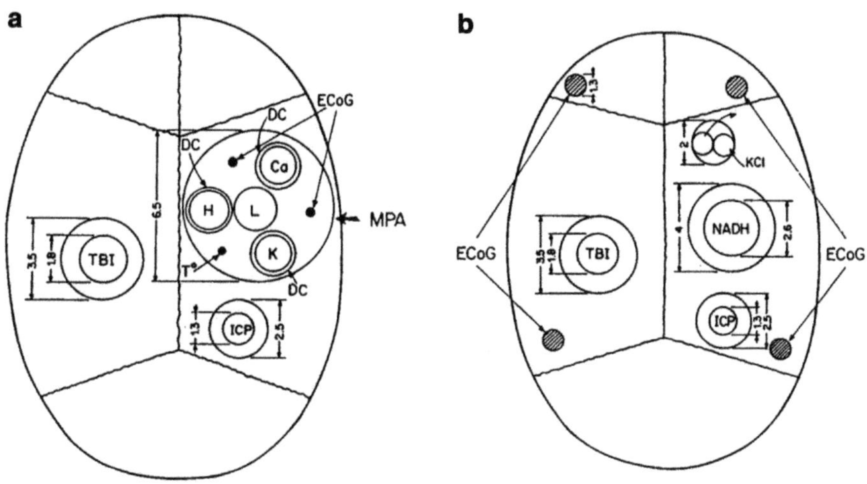

Fig. 8.22 The relative location of various sensing elements on the rat skull. (**a**) Using the MPA. L light guide monitoring NADH redox state and CBF, ECoG electrocorticography electrodes, T° temperature thermistor probe, K, Ca, H mini-electrodes for monitoring extracellular potassium, calcium and hydrogen ion concentrations, DC DC steady potential electrodes. (**b**) Using MSA. NADH cannula for light guide monitoring the NADH redox state, KCl push-pull cannula for inducing cortical spreading depression, ECoG electrocorticography electrodes, ICP cannula for intracranial pressure sensor, TBI cannula for inducing brain injury (dimensions, external and internal diameters for cannula, in mm)

The values measured by the optical probe were found to be well correlated with the standard ICP monitoring device (Crutchfield et al. 1990).

8.6.2.8 Multiparametric Monitoring System

This approach was developed over 30 years ago (Friedli et al. 1982; Mayevsky 1983a) and has been used successfully in rats and gerbils. The multiprobe assembly (MPA) used in the present study had the following features:

A. All parameters had similar noninvasive contact with the brain.
B. All signals monitored by the MPA exhibited only very slight sensitivity to animal movements so that awake animals could be used.
C. The Laser Doppler flowmeter probe was combined with the NADH optical probe. Details regarding the preparation of the electrodes have been described previously (Friedli et al. 1982; Mayevsky 1983a). For the present studies, a modified assembly with the following differences was developed:

1. Ca^{2+} and H^+ electrodes were added to the original MPA (Mayevsky et al. 1980).

2. The DC steady potential was measured by a highly stable Ag/AgCl electrode located around the K^+, H^+, and Ca^{2+} electrodes.
3. The reference electrode was an Ag/AgCl electrode (via a saline bridge), located under the skin of the animal's neck.

8.6.2.9 Construction of the MPA

The solution adopted for the electrode holder was basically a modification of the Lucite cannula described by Mayevsky et al. (1980) for the light guide and the potassium-sensitive electrode. To offer space for more probes, the new cannula was shaped as a truncated cone instead of a cylinder. The holes accommodating short electrode probes (K^+, Ca^{2+}, and H^+) were convergent toward the lower surface so as to occupy less space on the brain and divergent at the top to facilitate handling and sealing of the probes (Friedli et al. 1982; Mayevsky 1983a). An additional hole was drilled obliquely from the upper surface to merge with each sensor channel at approximately mid-distance from the lower surface. This hole accommodated polyethylene tubing used for recording the local DC potential concentrically to the sensor.

8.6.2.10 Multisite Monitoring (MSA)

In experiments in which the effects of spreading depression after TBI were tested or in which the animal was exposed to hyperbaric oxygenation (HBO) another monitoring setup was used. NADH was monitored either from one hemisphere (Fig. 8.22b) or bilaterally (not shown). A push-pull cannula (for KCl application) was located anteriorly to the NADH cannula, while the ICP probe was located posteriorly. TBI was induced in the contralateral hemisphere. One pair of ECoG electrodes was implanted in each hemisphere.

8.6.2.11 Fluid-Percussion Injury

Traumatic brain injury (TBI) was induced by the lateral fluid percussion method (Dixon et al. 1987; Sullivan et al. 1976; McIntosh et al. 1989). Hydraulic impact of moderate or severe magnitude (3.1 to 5.0 atm) was applied to the left parietal cortex (monitoring of all parameters was performed in the right hemisphere).

8.6.2.12 Surgical Preparation

The experimental protocols were approved by the University Institutional Animal Care and Use Committee. All animals were anesthetized by an intraperitoneal injection (0.3 ml/100 g) of Equithesin and the skull was exposed. In experiments using the MPA, three holes were drilled: a 6.0 mm hole in the right parietal bone, a 2.5 mm hole posterior to the MPA for the ICP sensor and a 3.5 mm diameter hole in the left parietal bone for TBI induction (Fig. 8.22a). The dura mater was gently removed, and the MPA was located on the brain such that pressure on the brain was avoided. The cannula used for fixing the sensor for ICP and the Leur-Lock syringe hub (1.8 mm inner diameter) were inserted in the appropriate holes. In experiments in which MSA was used, appropriate holes were drilled for the NADH fiber optic (4 mm), the ECoG (1.3 mm each, right and left), the ICP (2.5 mm) and for TBI induction (3.5 mm). A special push-pull cannula (2 mm) was implanted for inducing cortical spreading depression (Fig. 8.22b).

All implantable parts were fixed to the skull by dental acrylic cement after implantation of two to three stainless steel screws.

8.6.2.13 Hyperbaric Oxygenation (HBO)

Rats were monitored inside an HBO chamber (Bethlehem Steel). The fiber optic probes and the ECoG cables were inserted through the chamber wall as described previously (Mayevsky 1983b, 1984b).

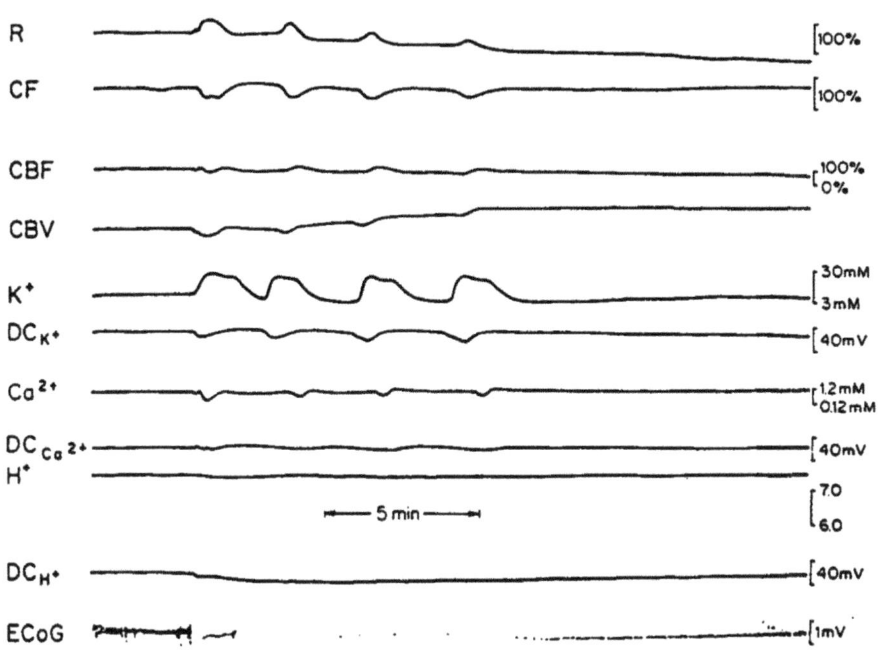

Fig. 8.23 Effect of traumatic brain injury (3.0 atm) on the metabolic, hemodynamic, ionic and electrical activities of the rat brain. R 366 nm reflectance, CF NADH corrected fluorescence, CBF, CBV - cerebral blood flow flow and volume monitored by a Laser Doppler flowmeter, K^+, Ca^{2+}, H^+ extracellular concentrations of potassium, calcium and hydrogen ions, DCk_{+}, DC_{Ca2+}, DC_{H+} DC steady potential measured concentric to the three electrodes, ECoG electrocorticogram, ICP intracranial pressure

8.6.3 Results

The results presented in this study are only preliminary (24 experiments), and their purpose is to illustrate the possibilities of modern monitoring in research on the pathophysiological post-concussion brain.

Spreading depression (SD) is one of the brain's reactions to mechanical trauma. A typical SD response is presented in Fig. 8.23, which shows that directly after fluid percussion impact cyclic changes occur on a background of a severe decrease in brain electrical activity (ECoG), in parameters representing brain blood supply (R, CBF and CBV), oxidative metabolism (CF) and ion transport (K^+, Ca^{2+}, and H^+). After four waves of post-traumatic SD, gradual normalization of the measured parameters was observed.

The ICP level in this animal remained unchanged during the posttraumatic period (not shown in the figure). The correlation between the changes in extracellular K^+ level and the metabolic response was very clear, and the increase in K^+ correlated well with NADH oxidation (decreased CF) and increases in CBF. TBI led to severe depression of ECoG activity, which returned to normal levels after 20–30 min.

Spreading depression following brain injury was observed in 12 experiments. The duration of SD periods in the various experiments varied from 3 to 42 min. In three rats, SD episodes recurred for more than 10 h.

Ischemic depolarization (ID) developed when severe injury was induced (Fig. 8.24). The ICP level before injury was between 2 and 6 mm Hg and increased to 20–22 mm Hg 2–3 min after the ID (Fig. 8.24a). The injury led to an immediate ischemic depolarization, a large decrease in CBF (LDF), while 1 h after the injury (Fig. 8.24b) a small decrease in extracellular K^+ (recovery) was recorded simultaneously with an increase in Ca^{2+}. The DC and ECoG exhibited partial recovery, while ICP remained elevated. Recovery of the brain improved within 3 h (Fig. 8.24c). This rat was monitored for several more hours without any further visible improvement. The development of post-concussion status epilepticus was recorded in five experiments. The dynamics of parameter changes in one of the experiments is illustrated in Fig. 8.25. Immediately after trauma, four typical SD waves appeared on a background of decreased CBF (not shown in the figure). Two hours after the trauma the first epileptic episode appeared in the ECoG, followed by severe depression similar to spreading depression. This was accompanied by typical ion transport fluxes (increase in concentration of extracellular K^+ and H^+ and decrease in extracellular Ca^{2+}) as well as a transient two-fold increase in CBF (Fig. 8.25a). Another seizure event is shown in Fig. 8.25b. Termination of this seizure was not followed by an SD wave. The increase in CBF was due to the seizure alone. In this case, the large increase in K^+ was not found and the ECoG depression was less pronounced and of shorter duration. The absence of a DC negative shift was another indicator that SD did not occur after this seizure. Figure 8.25c (recorded 260 min after TBI) shows events similar to those observed in A, i.e., a combination of seizure and SD waves (note the change in time scale). Furthermore, the same episodes (with or without SD) recurred many times either cyclically or at 8–12 min intervals. Fifteen episodes of this kind were recorded during a 1.5-h period. Most episodes were

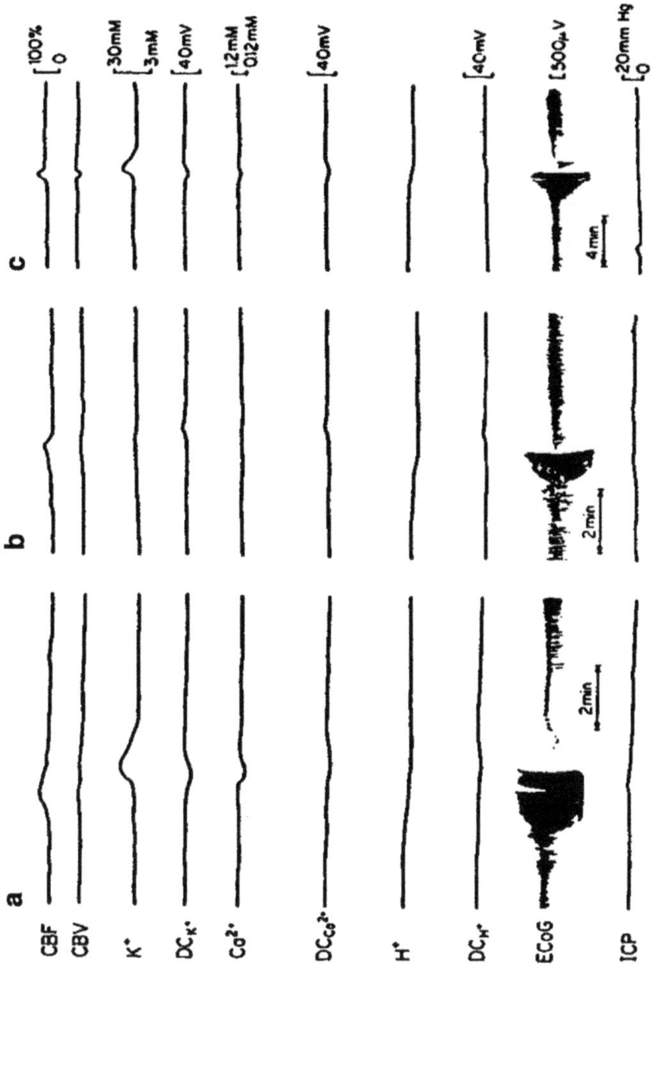

Fig. 8.24 Hemodynamic, ionic and electrical responses to lateral fluid percussion (4.5–5 atm). The distinct epileptiform episodes as observed in the electrical activity (ECoG). (**a**), (**b**), and (**c**) were recorded 120, 160, and 260 min after induction to TBI, respectively. Abbreviations as in Fig. 8.23

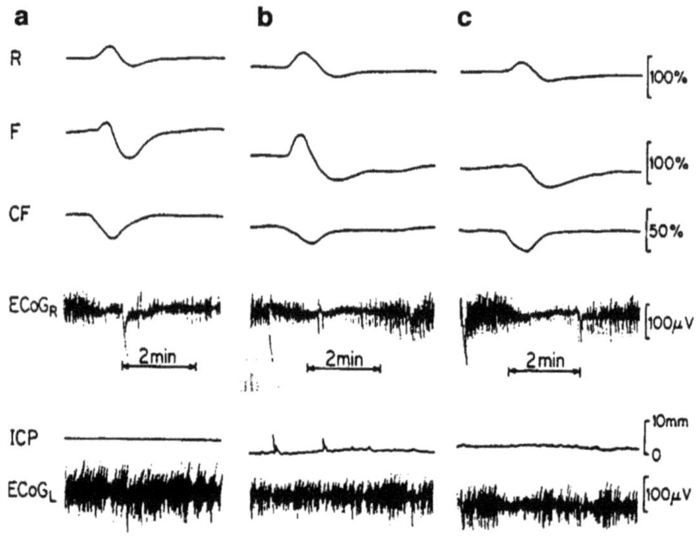

Fig. 8.25 Metabolic and electrical responses to right-side cortical application of a 0.5 M solution of KCI. R 366 nm reflectance, F 450 nm fluorescence, CF NADH corrected fluorescence, ECoG R, L electrocorticogram measured from the right and left hemispheres, ICP intracranial pressure. In (**a**), the ICP was not measured due to technical difficulties. (**a**) Measured before the TBI, and (**b**) and (**c**), 30 and 60 min after TBI (4.5–5 atm)

accompanied by changes in ion balance (spreading depression type) and a transient rise in the CBF level, which did not, however, return to initial levels. The amplitude of the ICP changes was small during these cycles of seizure activity. Immediately after TBI, ICP reached the 20–21 mm Hg level and then decreased to 10 mm or lower. The data presented above demonstrate that during the early posttraumatic period changes in electrical activity and ion balance correlated with changes in CBF and in oxidative metabolism adjustments. A second experimental setup was used to examine the functional state of the brain exposed to TBI. NADH and ECoG were monitored from one or both hemispheres as well as from one ICP probe. The brain was exposed to spreading depression (epidural application of KCl) before and after the TBI event.

Typical responses measured in 12 experiments are shown in Fig. 8.25. The typical control response to SD, which indicates an oxidation wave (decrease in CF) as well as ECoG depression in the ipsilateral hemisphere are presented in Fig. 8.25a. Responses measured 30 min after TBI (Fig. 8.25b) show that the amplitude of the NADH wave decreased. Sixty minutes after TBI the response to SD was similar to the control response (Fig. 8.25c). The ICP level was very low after TBI and was not affected by the SD waves.

In experiments with a more severe initial postconcussion period, the metabolic response was recorded as a progressive increase in NADH (Fig. 8.26). One minute after brain injury, a sharp decrease in ECoG amplitude, concomitant with an increase in intramitochondrial NADH, was observed. Three minutes later, under increased

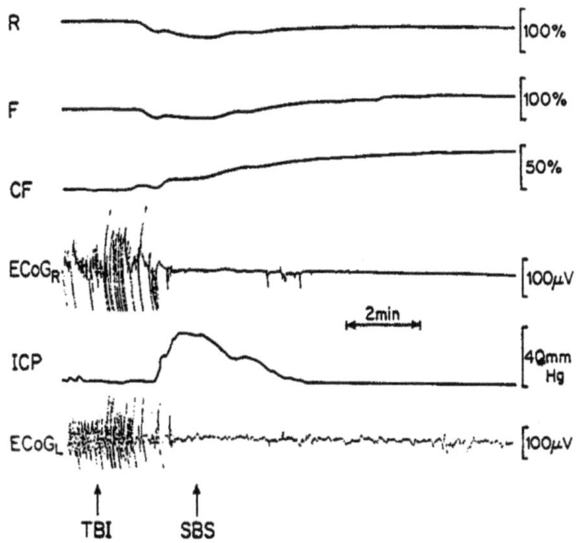

Fig. 8.26 Metabolic and electrical activities were measured during severe lateral fluid percussion. Death was caused by TBI as indicated by changes in the various parameters. Abbreviations as in Fig. 8.25

NADH levels and zero brain electrical activity, spontaneous breathing stopped (SBS) and the animal died. ICP reached a maximal level of 35 mm Hg at the moment of cessation of spontaneous breathing. Exposure to hyperbaric oxygenation (HBO) after severe TBI, when the mitochondria may not be able to supply sufficient energy, is a possible potential therapeutic approach. The brain of a control animal exposed to HBO (Fig. 8.27) exhibited increased oxidation levels (decreased NADH signal). After decompression, the brain redox state returned to control levels (not shown in the figure). The reflectance changes, indicating changes in blood volume, demonstrate a decrease in blood volume (increased R signal). Effects of HBO treatment after the exposure to severe TBI are presented in Fig. 8.28a. In this animal TBI led to cessation of spontaneous respiration. After 2 min of artificial respiration the chamber was washed with 100% 0_2 and compressed to 50 psi. NADH oxidation occurred during the compression period. NADH remained oxidized for the duration of the decompression period and recovery of ECoG was observed. When the rat was exposed to normal air, the NADH reduction occurred concomitant to an increase in blood volume as reflected by a decrease in the reflectance signal.

This same rat was exposed to spreading depression (while breathing normal air) before TBI as well as after exposure to HBO. A typical normal response to SD, namely, an NADH oxidation cycle and a biphasic response in reflectance, is presented in Fig. 8.28b. Thirty minutes after the induction of TBI and exposure to HBO, the amplitude of the oxidation cycle was only 50% of that occurring under control conditions (Fig. 8.28c). Furthermore, the changes in the reflectance trace were much smaller. At

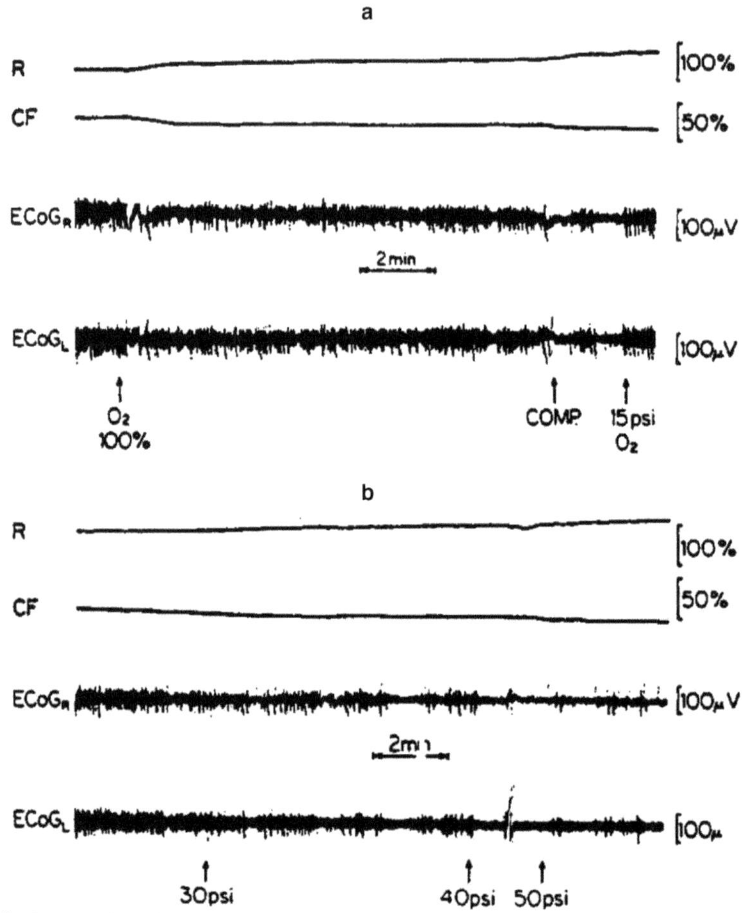

Fig. 8.27 Effects of hyperbaric hyperoxia on the metabolic and electrical activities measured from an awake brain. (**a**) Initial step; (**b**) A continuation in real time of (**a**). The pressure level reached 50 psi (4.5 atmospheres absolute); Comp compression. Other abbreviations as in Fig. 8.25

200 min after TBI, the response to SD demonstrated further recovery, although the duration of the cycles was longer (Fig. 8.28d).

8.6.4 Discussion

The efficacy of the treatment given to a head injury patient is dependent upon the effective and early diagnosis of the pathological state of the brain. If treatment is immediate, irreversible damage, which is a critical factor in the prognosis, will be

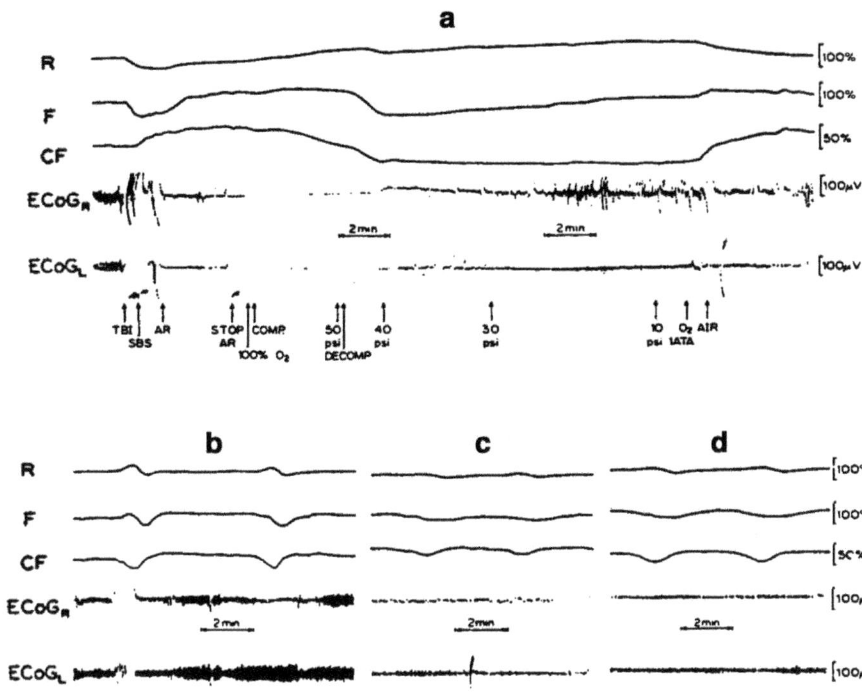

Fig. 8.28 Metabolic and electrical responses to spreading depression in the rat brain before and after severe lateral fluid percussion treated by hyperbaric oxygenation. (**a**) The effects of TBI (4.5–5 atm) and recovery by hyperbaric oxygenation: SBS spontaneous breathing stopped, AR artificial ventilation, stop AR stop artificial ventilation, Comp compression with 100% O_2, Decomp initial decompression, 50, 40, 10 psi the level of compression O_2 in chamber, Air breathing of room air. (**b, c** and **d**) Metabolic and electrical responses to cortical spreading depression during control (**b**); (**c** and **d**) 30 and 200 min after TBI. Other abbreviations as in Fig. 8.25

minimal or prevented. In order to achieve significant progress in the treatment of TBI patients, it is necessary to establish a representative animal model in which as many physiological and biochemical activities as possible can be monitored simultaneously in real time. In order to achieve this goal, we adopted the multiparametric monitoring approach, developed by our group over 10 years ago (Friedli et al. 1982; Mayevsky 1983a), and applied it to the rat fluid percussion model.

The main advantage of the MPA approach is its easy application to patients after severe head injury (Mayevsky et al., in preparation). Our development is a response to the challenge made by Wiebers et al. (1990) who declared that: "Ultimately, however, the answers to many of our questions regarding the underlying pathophysiology and treatment of stroke do not lie with continued attempts to model the human situation perfectly in animals, but rather with development of techniques to enable the study of more basic metabolism, pathophysiology and anatomical imaging detail in living humans". This statement was used by Wiebers for stroke and it can well be applied to TBI. The MBA sensors touch the surface of the brain after dura mater removal, without penetrating the brain tissue. Under such conditions, the brain itself remains very close to normal, as opposed to other methods in which penetration of the tissue by a microdialysis probe or K^+ electrode is necessary (Katayama et al. 1990; Hubschmann and Kornhauser 1983; Nilsson et al. 1990; Persson and Hillered 1992). Since most real-time probes sample a very small tissue volume, this brain region should be maintained in as normal a condition as possible. Indeed, the MPA has this advantage over other techniques. When TBI was induced in an animal model or developed in head injured patients, the nature of the pathological state could be defined better using the multiparametric monitoring approach. We added two important parameters to the MPA, ICP and tissue temperature, which can identify other aspects of TBI. The extensive publications, as well as our preliminary experiments regarding TBI, led us to conclude that TBI induced by fluid percussion is a gradual dynamic process which can be divided into two intensity levels, severe and mild or moderate. The initial effect of TBI affects CBF autoregulation (severe) or induces disturbances in ionic and electrical homeostasis. These events may lead to the development of the following pathophysiological states: ischemia, spreading depression or seizures. Each of these states is a very well defined condition and can be described in terms of changes in various physiological and biochemical parameters. Ischemia, for example, affects brain activity primarily by changing the energy availability measured as a decrease in CBF (O_2 supply), an increase in the intramitochondrial redox state (NADH accumulation) and an inhibition of ion pumps. These changes in energy supply affect ionic homeostasis mechanisms resulting in an increase in extracellular K^+ levels while Ca^{2+} and Na^+ will enter the cellular compartments (Hansen 1985). pH decreases due to the stimulation of glycolysis. The brain enters a phase of ischemic depolarization (ID) characterized by a negative shift in the DC potential as well as depression of spontaneous electrical activity (ECoG). Spreading depression developing spontaneously by TBI (or induced experimentally by KCl application) is primarily a result of changes in the ion homeostasis of the brain monitored as changes in the extracellular levels of various ions. These changes stimulate the activity of various ion pump s (in order to restore ion homeostasis) and as a result, energy metabolism is enhanced (Mayevsky et al. 1982). This is characterized by the oxidation of mitochondrial NADH as well as by a large increase in CBF and O_2 consumption (Mayevsky and Weiss 1991). Seizures, induced by epileptogenic drug injection (Mayevsky 1983b) or by exposure to hyperbaric oxygenation (Mayevsky 1984b), will usually affect electrical activity as well as ion homeostasis. Similar responses were measured when seizures developed after TBI. The metabolic responses were similar to those measured under SD, although in some cases, ischemic effects overlap the seizure episode and the responses are different.

SD developed after TBI without any significant change in ICP (Fig. 8.23). Coupling between the ionic and metabolic events is similar to that observed in the response of the normal brain to SD (Mayevsky and Weiss 1991; Mayevsky 1992). In some animals, SD develops sometime after TBI and the nature of the response to SD may serve as an indicator for the integrity of the microcirculation, mitochondrial activity and ionic homeostasis. Therefore, we used the rat TBI model in which SD was later induced (Fig. 8.25), and indeed, in this specific animal, the response was very similar to that monitored before TBI. The same SD test was applied to the rat shown in Fig. 8.28, which was exposed to severe TBI and treated by HBO. In some cases, the response to SD is abnormal, as identified by an increase instead of a decrease in NADH (Mayevsky et al. 1982; Mayevsky 1978b). Under these conditions, the CBF response will exhibit a large, typical, hyperemic response to SD (Mayevsky and Breuer 1990; Mayevsky 1992). The seizure activity shown in Fig. 8.24, developed more than an hour after TBI, led to the typical response published by us and other groups (Jobsis et al. 1971; Mayevsky 1983a, 1983b; Kreisman et al. 1987). The responses to TBI shown in Fig. 8.23 are typical of those recorded under ischemic depolarization, as seen by the decrease in energy supply as well as by the accumulation of K^+ in the extracellular space. The slow recovery of K^+ demonstrates the importance of continuous monitoring with the MPA.

The combination of the MPA and the fiber optic ICP sensor enables the study of the correlation between recorded values. This is important for understanding the mechanisms causing increased ICP after trauma as well as for elucidating the clinical relevance of ICP as a universal and integral parameter indicating the status of the injured brain (Bouma et al. 1991). The facts presented above together with information in the literature show that an immediate and severe energy deficiency resulting from CBF disorders may occur during the posttraumatic period. Therefore, attempts must be made to counteract the rapid reduction of intramitochondrial NADH by using the powerful modem means of improved tissue oxidation via hyperbaric oxygenation (HBO). Recently, clinical and experimental data have accumulated on the application of HBO in the treatment of acute functional disturbances caused by brain injury (Sukoff et al. 1968; Sukoff and Ragatz 1982; Brown et al. 1988). Whether HBO treatment is beneficial for these conditions is still not known. However, several studies reported the normalization of ICP levels following several HBO exposures (Elitorai and Montroy 1991; Rockswold et al. 1992). The mechanisms of this effect are not yet clear, but can be clarified by studying ICP changes with the concomitant recording of ionic balance parameters and redox state. Since HBO is known to be a powerful means for tissue oxidation, such information might be used as a basis for the elaboration of pathogenesis therapy methods for acute disturbances caused by brain injury.

8.7 Relationship Between ICP and CSD Following Brain Injury in Rats

8.7.1 Introduction

THE mechanism underlying the pathophysiology of cerebral contusion is complex and currently not well understood (Graham et al. 2000; Marmarou et al. 2000; McIntosh et al. 1998). However, in recent years, cortical spreading depression (CSD) cycles were found during clinical investigations following traumatic brain injury (TBI) (Mayevsky et al. 1996, 1998; Strong et al. 2002) as well as in models of traumatic injury in the rat cortex (Rogatsky et al. 1996; Kubota et al. 1989; Nilsson et al. 1993, 1996; Ozawa et al. 1991; Sunami et al. 1989).

CSD causes a disruption in ionic homeostasis (specifically, an increase in extracellular K^+ levels, intracellular Ca^{2+} and Na^+ levels and acidosis), which augments ion pump activities, such as Na^+/K^+-ATPase, consequently increasing energy metabolism and oxygen consumption (Sonn and Mayevsky 2000; Lauritzen 1987; Somjen 2001; Mayevsky et al. 1993b). Thus, in the normal brain, where delivery of oxygen to the tissue is not disturbed, an increase in oxygen supply will compensate for the energetic demands, thus inducing ionic homeostasis. In contrast, under conditions where the tissue suffers from a lack of oxygen (such as ischemia and hypoxia), an increase in oxygen demand by CSD waves may cause decreased CBF and tissue pO_2, delaying the recovery phase of ionic homeostasis (Mayevsky et al. 1993b; Back et al., 1994; Sonn and Mayevsky 2000). Further, the disruption in ionic homeostasis during ischemia, hypoxia or head trauma, can stimulate the development of continuous CSD cycles, thereby inducing acidosis, and the release of noxious substances such as glutamate and other excitatory amino acids (EAAs), thus enhancing tissue damage (Koura et al. 1998; Mutch and Hansen 1984; Otsuka et al. 2000; Benveniste et al. 1988; Bullock et al. 1998; Dietrich et al. 1994; Gido et al. 1994a, 1994b; Somjen 2001; Koroleva and Bures 1996). When occurring after traumatic brain injury, this cascade of events can adversely affect brain tissue, resulting in varied damage, linked to the degree of tissue oxygenation (Koroleva and Bures 1996; Somjen 2001; Nedergaard and Hansen 1993; Sonn and Mayevsky 2000). In a recent study, we found that the number of CSD cycles, together with the mean changes in ICP levels after severe FPBI, were the most significant parameters linked to morbidity when compared to the others measured by a multiparametric assembly (Manor et al. 2003). However, the relationship of CSD cycles to the severity of TBI was not clear. Therefore, the relationship between the level of post-traumatic intracranial hypertension and the number of CSD cycles generated, as well as the linkage between CSD and the severity of brain injury, is yet unknown.

To clarify these issues, the present work was carried out in rats in order to study the dynamic relationship between the level of intracranial pressure (ICP) and the number of CSD cycles during the early post-traumatic brain injury period under various levels of TBI severity.

8.7.2 Methods

A multiparametric assembly (MPA) was developed in our laboratory and used to study several pathophysiological states in experimental animals (Mayevsky et al. 1995; Mayevsky 1992) and in clinical research (Mayevsky et al. 1996, 1999b, 1999a). We used the MPA for continuous, simultaneous online measurement of hemodynamic, metabolic, ionic, and electrical activity of the rat cerebral cortex. The measurements included NADH redox state (using surface fluorometry), CBF (using laser Doppler flowmetry), extracellular K^+, Ca^{2+}, and H^+ concentrations, DC steady potential, as well as ECoG and ICP (Fig. 8.29b, c). The characteristic simultaneous increase in

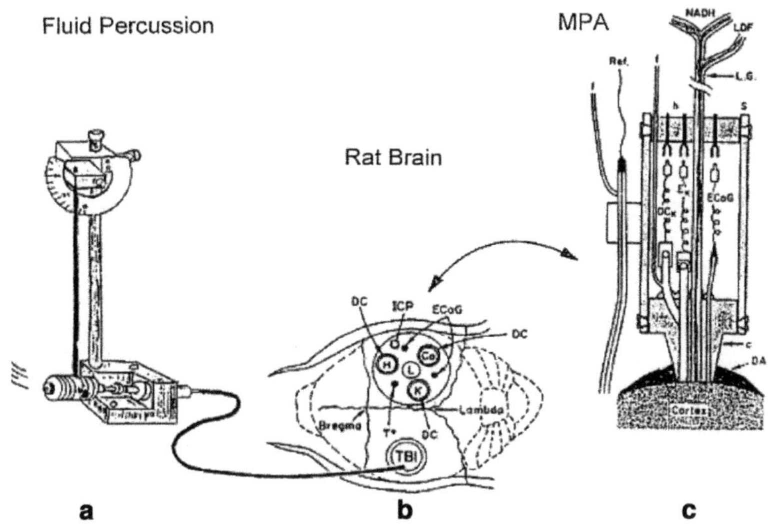

Fig. 8.29 Schematic presentation of the multiprobe assembly (MPA) used for monitoring the brain of head-injured rats. (**a**) Schematic presentation of the fluid percussion impactor. (**b**) Location of the MPA in the skull. (**c**) Longitudinal section of the MPA. DC steady DC potential, ECoG electrocorticography, ICP intracranial pressure probe, K, Ca, H potassium, calcium and hydrogen electrodes, L fiber optic light guide probe to measure local cerebral blood flow (CBF) and mitochondrial redox state (NADH), TBI cannula for inducing traumatic brain injury

extracelullar K^+ levels, the decrease in DC potential as well as the depression in ECoG (if ECoG was not previously depressed by the FPBI) were used to define CSD initiation. Normalization of these parameters to the baseline levels showed that CSD phenomenon stopped. From these data, the changes in ICP levels and the number of CSD cycles generated every 30 min under various severity levels of fluid-percussion brain injury (FPBI), were analyzed and calculated.

8.7.2.1 Animal Preparation

The experimental protocol was approved by the institutional Animal Care and Use Committee of Bar-Ilan University. All experiments were performed with 27 adult male rats (180–220 g). Each rat was anesthetized by an intraperitoneal injection (IP) of Equithesin 0.3 ml/100 g (each ml contains 42.51 mg chloral hydrate, 9.72 mg pentobarbital, 21.25 mg magnesium sulfate, 44.43% w/v propylene glycol, 11.5% alcohol and distilled water). First, the rat skull was exposed, and two holes were drilled: a 6 mm diameter hole in the right parietal bone for the MPA, and a 3.5-mm diameter hole in the left parietal bone for inducing TBI (Fig. 8.29b). Next, the dura mater was removed and the MPA was located on the brain, avoiding any pressure on it. A special cannula was used for fixing the Leur-Lock syringe hub (1.8 mm inner diameter) and inserted into the second hole (3.5 mm diameter). Screws were inserted into the skull to secure it mechanically. Finally, the MPA, the cannula, and the screws were cemented to the skull with dental acrylic.

8.7.2.2 Fluid-Percussion Brain Injury

TBI was induced by the lateral fluid-percussion brain injury (FPBI) technique (Dixon et al. 1987; Rogatsky et al. 1996; Sullivan et al. 1976; McIntosh et al. 1989). The magnitudes of the hydraulic impact with the Peak Pressure Pulses (PPP) were 2.9, 3.3, 4.1, and 5.0 atmospheres (ATM)—groups A, B, C, and D, respectively, which were applied to the left parietal cortex (Fig. 8.29a). More details concerning this technique can be found in our previous publication (Rogatsky et al. 1996). The monitoring session following FPBI for the four experimental groups of animals was 5 h. Each experimental group consisted of four rats in group A, seven in B, seven in C, and nine in D. Observation of the highest trauma level group (D) continued for more than 5 h (up to 18 h only for determining survival). Group D was divided into two subgroups: subgroup D1—animals that died during monitoring (four animals), and subgroup D2—survivors (five animals).

8.7.2.3 Statistical Analysis

A Pearson correlation analysis was performed initially to determine the most pronounced relationships between the investigated variables. Then one-way analysis of variance (ANOVA) was used to test possible differences between the trauma level groups. Null hypotheses were rejected at the 5% level of significance. Linear regression analysis was applied to obtain quantitative evaluations of the most interesting trends. The computer program SPSS 10.0 for Windows (Norusis 1999) was used for all statistical analyses. Data are presented in mean ± SE.

8.7.3 Results

The mean ICP levels of the four experimental groups, as a function of time, are presented in Fig. 8.30. FPBI caused a marked increase in ICP (240 ± 2.52, 250 ± 1.5, 250 ± 2.23, and 312 ± 0.95 mm Hg in groups A, B, C, and D, respectively). However, immediately afterwards, a rapid decline was observed. Then, a slow secondary increase in ICP was seen (30–90 min after the traumatic impact). This was especially pronounced in groups C and D. Next, a trend of increased ICP was noted (the most striking trend was found in group D). At the end of the 5 h experiment, the mean ICP levels were 6.75 ± 2.87, 8.40 ± 2.70, 12.75 ± 4.03, 29.56 ± 9.25 for groups A, B, C, and D, respectively.

Using one-way ANOVA with Bonferroni correction for multiple comparisons, we found significant differences in the ICP levels ($p < 0.05$) affected by the PPP levels

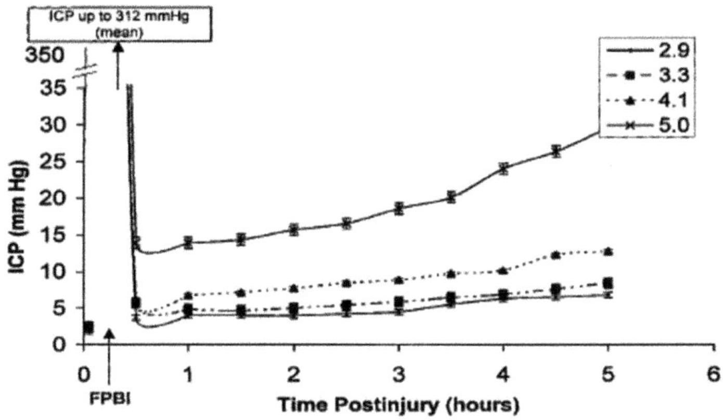

Fig. 8.30 The effect of FPBI levels (2.9, 3.3, 4.1, and 5.0 ATM, for groups A, B, C, and D, respectively) on ICP levels (mean ± SE mm Hg) in four experimental groups of animals [A ($n = 4$), B $n =$ 7), C ($n = 7$), and D ($n = 9$)]. The first mean ICP values in all experimental groups were under control conditions before FPBI. Lower arrow FPBI induction, upper arrow maximal ICP levels immediately after FPBI

as a function of time, starting 30 min after trauma. These differences were conserved and tended to increase with time.

Significant differences in ICP were found between the experimental groups. ICP in group D was significantly higher than in all other groups during hours 1–5 post injury and was higher than group A at 30 min post injury. These differences were more significant at later periods. Thus, the statistical results support the graphical presentation in Fig. 8.30 and show that group D differed from the other three groups during the monitored time after FPBI.

CSD cycles were observed in all experimental groups. Usually, one to four cycles were observed for each animal during the first 10–15 min immediately after FPBI (Fig. 8.31). Then, the total number of CSD cycles increased in all groups, except in group A. During the 300 min of monitoring, the mean total number of CSD cycles increased with the impact magnitude. The maximum number was found in group D and the minimum in group A. The mean cumulative number of CSD cycles found was 2.00 ± 1.41 in group A, 4.29 6 ± 4.23 in group B, 11.71 6 ± 13.29 in group C, and 20.11 ± 19.26 in group D. Regarding the CSD total count, the differences increased with time, but even for the latest moments, they were not significant. The highest number of CSD cycles was observed during a specific mode that we term "CSD status"—when CSD cycles repeated almost continuously, until the end of the monitoring period (Fig. 8.32). Here, the total number of CSD cycles increased to 53 during 300 min of observation. The "CSD status" was noted in 50% of the animals in group D and only in 14% of the animals in group C. In groups A and B, the "CSD status" was not found. The intensity of CSD cycles generated was expressed by the number of CSD cycles observed every 30 min after brain trauma. Figure 8.33 shows the intensity of CSD cycles during the postimpact period. After an initial high level

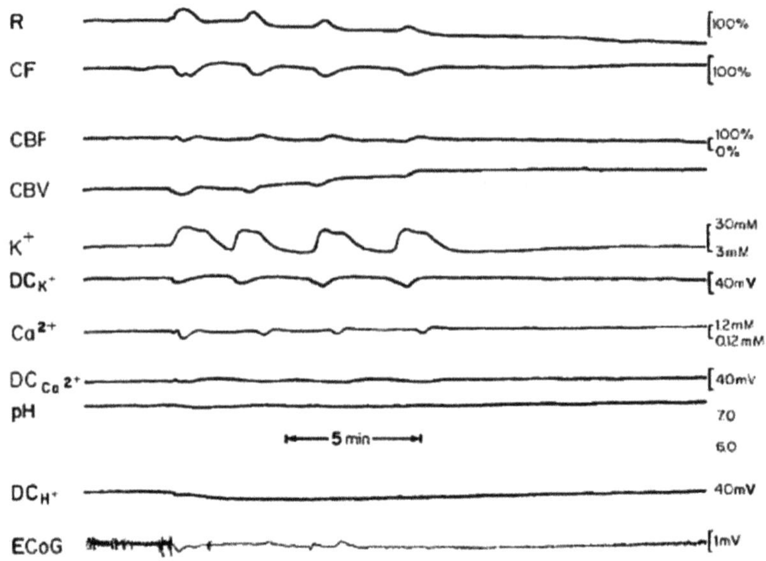

Fig. 8.31 The effect of traumatic brain injury (TBI) on metabolic, hemodynamic, ionic and electrical activities of the rat brain from a rat of group A. Note that TBI led to cortical spreading depression (CSD)-like cycles. R 366 nm reflectance, CF NADH corrected fluorescence,CBF, CBV cerebral blood flow and volume, K^+, Ca^{2+}, H^+ extracellular concentrations of potassium, calcium and hydrogen ions, $DC_K{}^+$, $DC_{Ca}{}^{2+}$, $DCH_H{}^+$ DC steady potential measured concentric to the three electrodes, ECoG electrocorticogram

(during the first 30 min following FPBI), the number of CSD cycles declined in all four groups for the next 60–90 min. Then, a new stage began, characterized by an increase in CSD cycle intensity (most pronounced in groups C and D). After this stage, the second CSD intensity decline took place in group C, with no clear tendency in group B. Only group D, with almost constant activity (up to two CSD cycles per 30 min), was maintained until the end of the experiment. To estimate the ICP level (ICP_{xx}), the intensity of CSD cycle generation (CSD_{xx}), and the total number of CSD cycles (CSD_{tot}), approximation regression equations were obtained (Fig. 8.34a, b; Table 8.3). For all time points measured, the ICP regression equations were significant ($p < 0.001$; $n = 27$), while equations for CSD had p-level, 0.05 only beginning 180 min (t180) after traumatic impact. The coefficient of determination (r^2) and significance level for CSD equations were lower than the corresponding values for ICP equations. Using the mean levels for each PPP-level ($n = 4$), a highly significant linear equation was obtained for the total number of CSD cycles: $CSD_{mean} = a + bX$, where X is the PPP-level, $a = -23.35 \pm 2.05$, $b = 8.62 \pm 0.52$, $r^2 = 0.992$, $p = 0.004$. These results indicate that ICP level is a much more reliable parameter for evaluating the trend of head trauma than the number of CSD cycles.

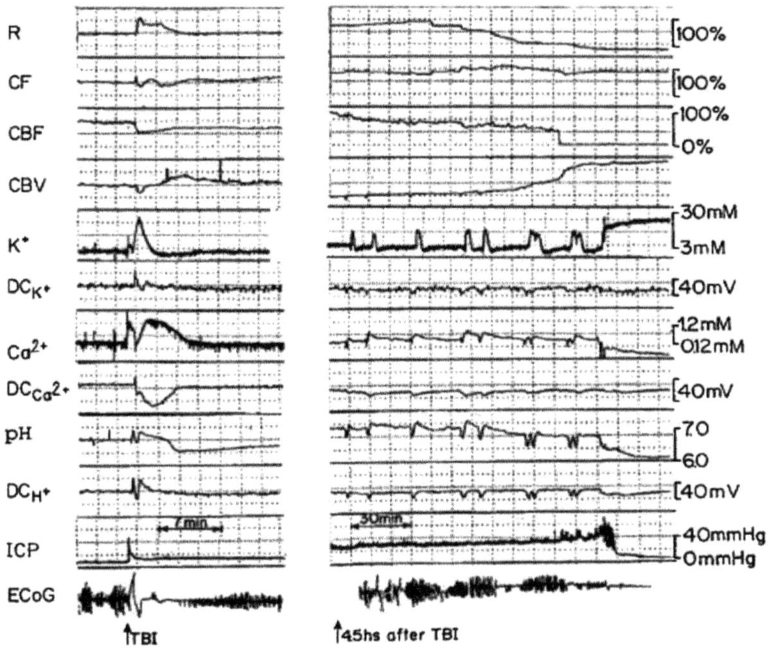

Fig. 8.32 Multiparametric responses to severe head injury induced in a rat from group D. The TBI led to a cortical spreading depression (CSD)-like response. The CBF decreased and reached a new steady state. CSD cycles and epileptic activity were recorded 4.5 h after TBI induction. The ICP level was elevated gradually until the final point, when the rat died and ICP dropped to 0. ICP intracranial pressure. For other abbreviations, see Fig. 8.31

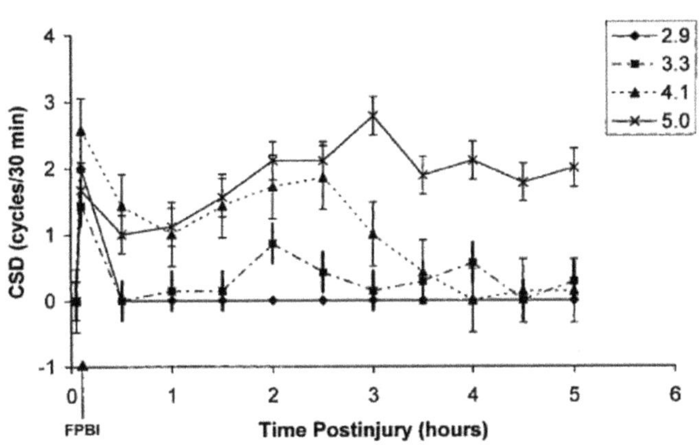

Fig. 8.33 Changes in the mean number of CSD cycles generated (per 30 min mean ± SE) during 5 h of monitoring after induction of different levels of FPBI: 2.9, 3.3, 4.1, and 5.0 ATM, respectively. CSD cycles started immediately after FPBI (lower arrow). Note that before FPBI, no CSD cycles were observed (0 cycles)

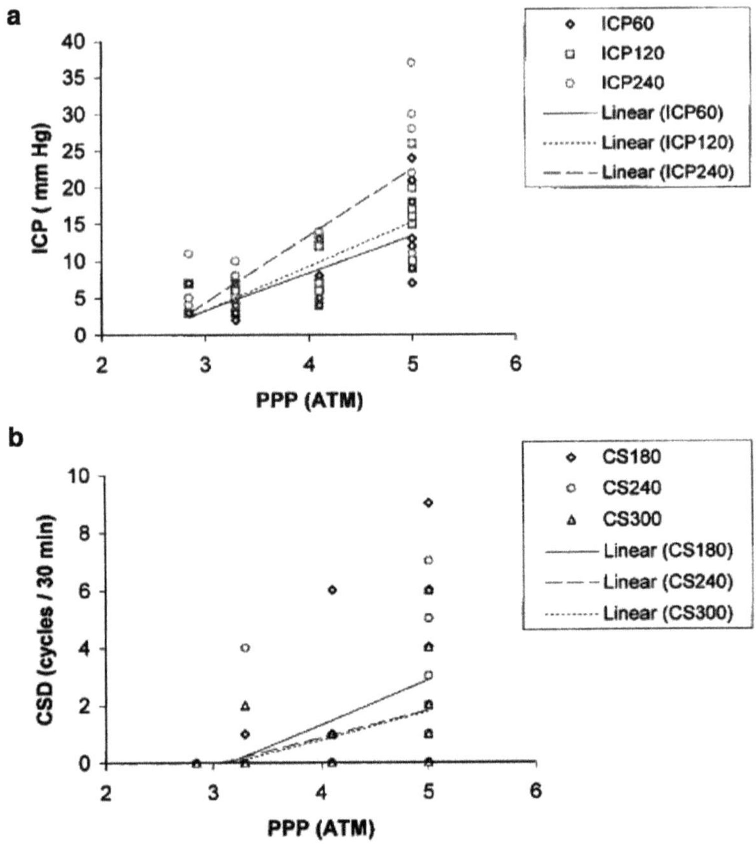

Fig. 8.34 Dependence of the ICP level (**a**) and number of CSD cycles generated per 30 min (**b**) on the peak pressure pulse (PPP). Parameters of the regression equations are given in Table 8.1. ICP_{xx}, CSD_{xx}, CSD_{xx} mean estimates for time xx. Lines CS240 amd CS300 overlap since they have very close parameters (Table 8.3)

8.7.3.1 Correlation between Parameters

The Pearson coefficient (r) of paired correlation between PPP and ICP levels was calculated for all time points measured. A significant, strong correlation was found 30 min after trauma (t30) ($r = 0.644, p < 0.001, n = 27$), then the correlation coefficient increased to 0.706 (t60) and $0.777 - 0.841$ (t90–t300). The coefficient of correlation between PPP level and the intensity of CSD cycle generation was also calculated and became significant starting from t180. For time periods t180–t300, $r = 0.411–0.496$ ($p < 0.05, n = 27$). Thus, the growing effect of the impact factor on the level of ICP was observed during the whole experimental period, whereas the analogical effect on the intensity of CSD cycle generation was noted only during the second half of the experimental period (from 180 min after FPBI)

Table 8.3 Values of linear regression parameters (±SEM), the coefficient of determination (r^2), and significance levels for the model. Y = a + bX (Rogatsky et al. 2003)

Dependent variable (Y)	Time (xx)	Intercept a (SEM)	Slope b (SEM)	r^2	*p*-level
ICP_{xx}	60	−11.12 (3.95)	4.82 (0.97)	0.499	<0.001
	120	−13.80 (3.79)	5.72 (0.92)	0.604	<0.001
	240	−22.28 (5.35)	8.89 (1.30)	0.649	<0.001
CSD_{xx}	180	−4.35 (2.09)	1.39 (0.51)	0.229	0.012
	240	−2.76 (1.63)	0.90 (0.40)	0.169	0.033
	300	−2.95 (1.33)	0.93 (0.32)	0.246	0.009
CSD_{tot}	300	−24.17 (12.49)	8.81 (3.05)	0.250	0.008

Where x is the PPP level (ATM); ICP_{xx} is the intercranial pressure at time xx (min); CSD_{xx} is the intensity of CSD cycle at time xx (min); and CSD_{tot} is the number of CSD cycles generated during the experimental period. $n = 27$ for each regression

A significant coefficient of correlation between ICP level and the total number of CSD cycles was found for ICP measurements ($r = 0.47$–0.63, $p < 0.05$, $n = 27$) for each time point. For time periods t120–210, $p < 0.001$. Therefore, from the results of the correlation analysis, we can select t120–t210 as the best period to analyze the relation between intracranial pressure and cortical spreading depression.

8.7.4 Discussion

We found a high correlation between the magnitude of PPP and intracranial hypertension level in our experiments, showing that ICP levels can provide an estimate of the trauma severity. These results support those of Engelborghs et al. (1997), who concluded earlier that a significant correlation exists between the severity of the traumatic challenge and the resultant pressure rise. This dependence, and the fact that the maximal correlation coefficients previously listed between PPP and ICP were noted in group D (i.e., at the maximal level of FPBI), demonstrate the diagnostic importance of ICP as a parameter used to estimate the risk level of the unfavorable development of head trauma (Stocchetti et al. 1991; Lang and Chesnut 1995; Czosnyka et al. 1996; Bardt et al., 1998).

Recent clinical observations have shown that TBI outcome was often unfavorable when an ICP level >20 mm Hg was reached (Miller et al. 1977; Pfenninger 1984; Piper et al. 1990; Signorini et al. 1999). In our experiments, such a level of ICP was reached during 5 h of monitoring in seven of nine rats (almost 80%) in group D, and only in this group. More than half (four of seven) of those animals died spontaneously within 2–3 h after the first 5 h of monitoring. The ICP increase that persisted in group D may have been caused by the appearance of brain contusion in the contralateral hemisphere (Kochanek et al. 1995), with the development of significant brain edema (Kochanek et al. 1995; Marmarou and Shima 1990) and traumatic subarachnoid hemorrhage ICP (Thomas et al. 2000). Results of linear regression analysis (Table 8.3) reveal that the slope coefficient (b) was positive and almost constant (0.9–1.39) for t>180 for CSD; b was positive and grew with time (4.82–8.89) for ICP. Therefore, mean CSD generation intensity grew with the level of traumatic impact (PPP) and did not change with time. The ICP level also grew with PPP level, and these differences are more and more pronounced with time (Table 8.3; Fig. 8.30).

A positive correlation between the total number of CSD cycles and the PPP levels ($r^2 = .250$, $n = 27$, $p < 0.01$) emphasizes the correlation between the intensity and duration of the CSD wave generation period and the severity level of TBI. It has been noted in clinical studies evaluating this problem that the CSD phenomenon was only noted in patients with well-developed forms of TBI (Strong et al. 2002; Mayevsky et al. 1996). As CSD data are heteroscedastic, separate *t*-tests were used to find significant differences between the groups, rather than the use of ANOVA for ICP differences. Significant differences ($p < 0.05$) were found between groups of 5.0 and 4.1 ATM for t240 and t300, between 5.0 and 3.3 ATM for t180, between 5.0 and 2.9 ATM for t120–t240, and t300.

Our results showed that in the groups with maximal injury, an increase in the mean number of CSD cycles was found together with the highest levels of ICP (subgroups D1 and D2). The growing number of CSD cycles found under these circumstances of steadily increasing ICP could be related to secondary hypoxic insult, which may result from pronounced metabolic disorders ongoing in the severely traumatized brain (Bergsneider et al. 2000; Ishige et al. 1987b; Signoretti et al. 2001). In such cases, progressing posttraumatic intracranial hypertension is a factor of major importance in influencing ischemic brain damage (Graham et al. 1978, 1989; Miller et al. 1992; Obrist et al. 1984). Our results of high ICP levels accompanied by CSD cycles in the contralateral hemisphere may indicate the global and severe development of the secondary brain ischemic/hypoxic insult.

It is also known that tissue damage caused by brain trauma can induce marked disorders in ionic homeostasis characterized by increasing extracellular K^+ levels, which can in turn initiate CSD cycles (Somjen 2001; Mayevsky et al. 1996). The lack of an adequate oxygen supply under these circumstances can induce continuous CSD cycles without progressing to sustained depolarization. Furthermore, the high ICP levels that develop after brain trauma may further disturb the availability of oxygen to the tissue, partially compromising ionic homeostasis, thus causing a continuation of the progression of CSD cycles and a further increase in tissue damage.

An indication of this hypoxic insult can be found in our results 60–90 min after FPBI in the groups of maximal impact level (5.0 ATM), where a tight coupling can be seen between the total number of CSD cycles, the increase in ICP levels, and the high mortality of rats in group D.

It can be presumed that in such cases, the cumulative damaging influences of many spontaneous and continuous CSD cycles can, at a specific stage of development of severe brain pathology, promote an increase in ICP levels. This raises the possibility that a vicious cycle can exist between increased ICP levels and the generation of numerous CSD cycles, especially during a period of CSD status. Such a relationship between these pathophysiological processes can take place under conditions induced by secondary hypoxic insult after severe TBI. Some additional support for this hypothesis can be found, as noted above, in the strengthening correlation between increased ICP levels and the generation of a vicious cycle of CSD intensity, beginning 2–3 h after the trauma and continuing until the end of the experiment. It, therefore, seems reasonable that during the secondary hypoxic insult, under conditions of increasing ICP, the increased number of CSD cycles becomes a more important factor in brain trauma pathogenesis.

As shown in Fig. 8.33, the diminishing intensity of CSD cycles after 2–3 h in group C (especially in comparison to group D), paralleled favorable posttraumatic development. It can be assumed that the appearance and growth of the number of CSD cycles (even multiple cycles) in group C, nonconfluent only with growing ICP, could be initiated by different mechanisms. For example, such phenomena, associated with only acute ischemic lesion, are described in studies dealing with CSD (Dietrich et al. 1994; Otsuka et al. 2000; Takano et al. 1996). Among the most probable causes of CSD cycle induction are those originating from local and regional CBF disorders, provoked by TBI. Theoretically, some such CSD generation zones can appear, and such CSD waves can be induced without any considerable elevation of ICP, as can be seen for group C. With low-level trauma, the number of CSD cycles can be minimal, or the CSD cycles can cease (Groups B and A).

From our results, we conclude that repeating numerous CSD cycles are not rare but rather a typical phenomenon for all (including the moderate and heavy) forms of TBI. The possibility of CSD cycles appearing and the augmentation of their number grows with increased intracranial hypertension. The growing number of CSD cycles under conditions of ICP level—>20 mm Hg, demonstrates with high probability the unfavorable development of posttraumatic syndrome. This increasing number of spontaneous repeating CSD cycles under conditions of growing intracranial hypertension has been interpreted as manifestation of progressive brain mitochondrial dysfunction. Probably, this paroxysmal character of generation of repeated CSD cycles, characterized by us "CSD status," may be a compensatory mechanism required for elimination of local ionic disbalance and reactivation of transmembrane ionic transport. It is possible that this mechanism of adaptation toward mitochondrial insufficiency under conditions of increased ischemic-hypoxic insult was not effective in animals in group D, in comparison to animals in group C, or A and B. This supposition is consistent with the recent results on the development of brain mitochondrial insufficiency after severe traumatic brain injury (Zauner et al. 2002). In contrast, a discontinuation or a slowdown in the generation of CSD cycles, even under persisting hypertension, can be considered a sign of recovery.

As was pointed out previously, the classical detection of the CSD phenomenon requires special multi-parametric monitoring. Nevertheless, an analysis of the clinical and experimental materials that we have collected makes it reasonable to assume that repeating numerous CSD cycles may be possibly identified using EEG monitoring alone. In combination with the monitoring of ICP and other usually measured parameters, the monitoring of the number of CSD cycles can provide important information for monitoring the progression of trauma and its overall implications for mortality and/or morbidity.

8.8 Multisite Multiparametric Responses to Traumatic Brain Injury

8.8.1 Introduction

Severe head injury is one of the major causes of high mortality or irreversible brain damage. One of the techniques used to induce traumatic brain injury (TBI) in experimental studies is the exposure of the brain to fluid percussion damage. The lateral fluid percussion has become, since 1989, the most extensively used model for experimental TBI (Laurer et al. 2002), since it reproduces the histopathology associated with multiple types of human TBI, including focal cortical contusion with parenchymal hemorrhage (Cortez et al. 1989; Thompson et al. 2005; McIntosh et al. 1989). The fluid percussion injury is characterized by rapid radial movement within the epidural space that induces pathology and dysfunction of the brain surface in a diffuse mechanical loading manner (Dixon et al. 1988).

Recently, the processes following TBI were characterized by four overlapping phases, which include primary injury, evolution of the primary injury, secondary or additional injury and regeneration (Cernak 2005). The primary damage develops immediately after the injury and persists during the first few hours' postinjury, while the secondary damage develops during a period of hours to days (Sullivan et al. 2005; McIntosh et al. 1996). In the present study, we describe the damage that develops during the initial two stages following the injury up to the first 4 h after the injury. As was described previously, the primary injury consists of rapid deformation of the brain tissue due to rupture of cell membranes and blood vessels (Finnie and Blumbergs 2002), leakage of intracellular content, a massive increase in extracellular level of K^+ (Manor et al. 2003; Kubota et al. 1989; Katayama et al. 1990), a massive Ca^{2+} entry into the intracellular compartment (Nilsson et al. 1993) and disruption of blood flow (McIntosh 1994) eventually leading to necrotic cell death. Although these phenomena exist in the core of the contusion site, they also appear, in different levels, near the contusion site, namely, at the penumbra zone of the injury.

Even though a considerable number of studies were published on secondary brain injury, which is often involved with ischemia/hypoxia, the pathophysiologic mechanisms of the early stage of head trauma have not been elucidated sufficiently. One of the promising approaches to study this problem may be the simultaneous multi-focal monitoring of the dynamics of physiologic and metabolic parameters of the brain, which would allow examining in detail the functional map of a traumatized brain. Such an approach was successfully applied in experimental studies, where a number of modern methods for the observation and evaluation of brain blood supply and metabolic variability were used (Schneider et al. 2002; Van Putten et al. 2005; Walter et al. 2004; Albensi et al. 2000; Ishige et al. 1987a; Ito et al. 1996; Ginsberg et al. 2001). However, such techniques are difficult to conduct and generally provide discontinuous information.

Muir et al. (1992a) and other authors (DeWitt et al. 1997; Prat et al. 1997) demonstrated a methodologic approach using cortical laser Doppler flowmetry. Although Wang used this technology for bilateral brain monitoring (Wang et al. 1998; Nilsson et al. 1996) most of the cited studies used laser Doppler flowmetry from a single location on the cerebral cortex. This fact limits the possibilities to characterize the functional and metabolic changes in focal and perifocal brain areas of the contusion.

In 1983, we described, for the first time, the method for multi-site measurements of NADH redox state from the cerebral cortex of awake animals (Mayevsky and Chance 1983). During the following years, this approach was used by us in various experimental models, such as under CO poisoning (Meilin et al. 1998), cortical spreading depression (Sonn and Mayevsky 1995) and following ethanol infusion (Sonn and Mayevsky 2001), where multi-site monitoring of the brain was performed.

In the present study, we present the accumulated results of the use of the multi-site monitoring approach in the early period (4 h) following TBI in rats. The aim of this study was to find the best location for real-time monitoring of the cerebral cortex at the time of injury, as well as several hours after the induction of the injury. To achieve this, we placed four probes at four different locations on the cerebral cortex. Two were placed on the ipsilateral hemisphere, one on the parietal lobe and one on the frontal lobe and two on the contralateral hemisphere at the same locations.

8.8.2 *Methods*

The multi-site multi-parametric monitoring system. The monitoring of the cerebral cortex in four locations on the cerebral cortex was performed using the multi-site multi-parametric monitoring system (MSMP) (Fig. 8.35). Each probe of the MSMP (1–4), 3 mm in diameter, contains optical fibers for NADH redox state fluorometry, a laser Doppler probe and a dc potential electrode. The position of each probe on the cerebral cortex was similar in all the experiments (Fig. 8.35b).

The principle of NADH monitoring from the surface of the tissue (1 mm depth) is that excitation light (366 nm) passes from the fluorometer, which contains appropriate filters, to the brain via a bundle of optical fibers made of quartz. The emitted light (450 nm), together with the reflected light at the excitation wavelength (366 nm), is transferred to the fluorometer via another bundle of optical fibers (Mayevsky 1993; Mayevsky and Rogatsky 2007). Changes in the reflected light are correlated to changes in tissue blood volume and therefore serve to correct for hemodynamic artifacts in NADH monitoring (Mayevsky 1984a). The corrected fluorescence (NADH) is obtained by subtracting the reflectance from the fluorescence signal at a 1:1 ratio (Mayevsky 1984a). The fluorometer is calibrated in such a way that a range of 500 mV is equal to a change of 100% in the monitored signal.

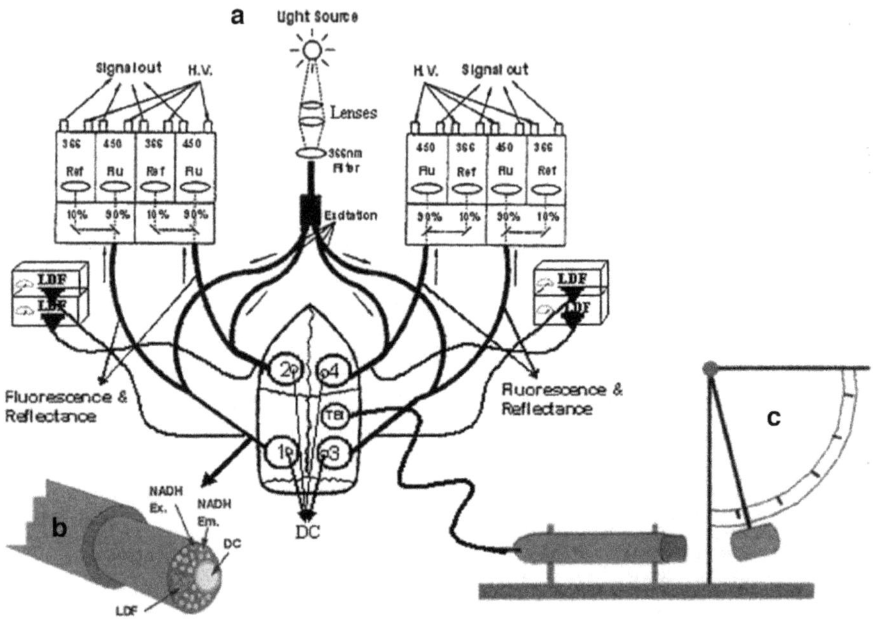

Fig. 8.35 (**a**) Schematic representation of the MSMP and the location of the various probes on the cerebral cortex. The MSMP includes four fluorometer units, optical fibers to conduct the excitation light from the light source to the cerebral tissue, and the emission light (fluorescence and reflectance) from the tissue to the photo-detectors. For CBF monitoring, four laser Doppler flowmeters (LDF) were used. (**b**) The tip of each probe includes optical fibers for NADH monitoring (Ex.—excitation: Em.—emission), for blood flow monitoring (LDF) and a dc electrode for dc steady potential monitoring. (**c**) The fluid percussion injury device for the induction of traumatic brain injury (TBI)

Real-time monitoring of CBF from the same cortical area was achieved using the laser Doppler flowmetry technique, which is based on the Doppler shift reflecting the flow of red blood cells in the microcirculation at a depth of 1–2 mm^3. For CBF, the flow at death was considered as 0% flow, and the range between the basal level and the flow in death was considered as 100%.

The DC steady potential was measured by highly stable Ag/AgCl electrodes connected via a salt bridge. A reference electrode was located below the skin of the animal's neck. All of the signals monitored during the experiment were digitized and transmitted to a multi-channeled computerized data acquisition and recording system (Labview A/D software, National Instruments Co., USA) for further analysis.

8.8.2.1 Statistical Analysis

Monitoring was performed at a rate of one sample per second, and the results were presented at intervals of 5 min for 4 h. Statistical analysis was performed using the

SAS and SPSS software. One-way ANOVA test was used for the evaluation of the differences between the various TBI groups versus the control group.

8.8.2.2 Animal preparation

All experiments were conducted in accordance with the Guidelines of the Animal Care Committee of Bar-Ilan University. Wistar male rats (250–300 g) were anesthetized with an i.p. injection (0.3 ml/100 g) of equithesin (each ml contains: pentobarbital 9.72 mg, chloral hydrate 42.51 mg, magnesium sulfate 21.25 mg, propylene glycol 44.34% w/v, alcohol 11.5% and distilled water).

The rat was fixed in a head holder and a midline incision in the skin was made exposing the parietal and frontal bones of the rat's skull. Two holes (5 mm) were drilled in the right and left parietal bones in such a way that their center was 3 mm posterior to the bregma and another two holes (5 mm) in the right and left frontal bones in such a way that their center was 3 mm anterior to the lambda. Four cannulae made of stainless steel (probe holders) were placed in these holes and fixated to the brain surface using acrylic cement. Later on, the four probes were placed in these holders for the monitoring of the cerebral cortex. Another two holes were drilled in the parietal bones (one on each side). In the right hemisphere, a special connector was fixated for the connection of the brain injury device (Fig. 8.35c). Additionally, two stainless steel screws were inserted at the parietal bones to ensure better fixation of the monitoring probes to the cortical surface by acrylic cement.

In the group of rats in which spreading depression (SD) was induced, all four probes were located on the parietal lobes (two in each hemisphere), and the holes for the induction of SD were drilled at the frontal lobes (Fig. 8.36). SD was induced using a special push–pull cannula located epidurally that was washed with KCl solution (0.5 M) until SD waves appeared. In order to stop the waves, the cerebral cortex was washed with saline.

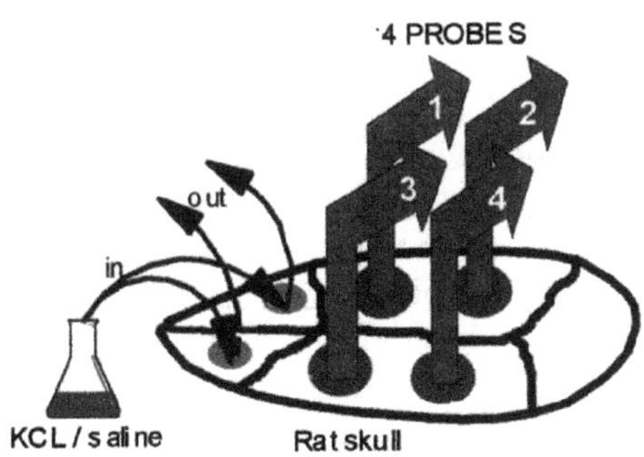

Fig. 8.36 The location of the MSMP probes on the cerebral cortex in experiments on cortical spreading depression (SD) induction

The rats were left to recover from the operation for 2 h and thereafter, TBI was induced in the awake animals. Monitoring proceeded before, during and for a period of 4 h after TBI and then the rats were anesthetized by equithesin and then killed using pure N_2.

8.8.2.3 Fluid Percussion Injury

TBI was induced according to the model of fluid percussion injury (Dixon and Hayes 1996; Dixon et al. 1987; Sullivan et al. 1976) first applied by Sullivan et al. (1976) using a special device (VCU Biomedical ENG. Facility, Richmond, VA, USA). The device was calibrated using nitrogen gas (Fig. 8.36) in such a manner that an angle of 15° yielded an injury of 1.15 ± 0.14 ATA (mild), $20° = 1.78 \pm 0.28$ ATA, $25° = 2.13 \pm 0.34$ ATA (moderate) and $30° = 4.34 \pm 1.08$ ATA. TBI was induced in awake rats. In the present study, we focus on two levels of injury: the mild and the moderate.

8.8.3 Results

Before the induction of percussion injury, we had to confirm that the model of multi-site (four locations) monitoring of the rat cerebral cortex is feasible, yielding reliable and comparable results. Moreover, since the multi-site model involves complicated craniotomy, in which five holes (5 mm diameter) are drilled in the rat skull, leaving only a small skull portion intact for a duration of at least 5 h, we had to confirm that the combination of all these factors does not cause any deterioration of the parameters monitored from the cerebral cortex, or does not lead to early death. To achieve this, we run several protocols, frequently used in our laboratory. The first model was the control group in which the probes were placed on the cerebral cortex and the monitoring proceeded for 5 h without any type of perturbation. The results confirmed stable monitoring in all of the locations and parameters during the entire experimental period (results not shown). To ensure proper location of the probes on the brain surface and normal responses of the cerebral tissue to complete oxygen deficit, a short anoxia (30 s) was induced. As seen in Fig. 8.37, at the start of anoxia, the reflectance decreased gradually by, 25% in all of the locations. CBF showed a minor decrease followed by an increase to hyperemic levels, which returned to basal levels later on (not shown). Mitochondrial NADH levels rapidly increased after the initiation of anoxia by, 50% and recovered immediately after the rat resumed breathing air. The steady DC potential was the first parameter to show a biphasic response: a minor decrease followed by an increase of 3 mV, which gradually returned to the basal level. It was also clearly demonstrated that all four locations responded similarly to anoxia.

To test the ability of the monitoring system to reliably represent the hemodynamic and metabolic state of the brain, not only in worsened cerebral conditions such as anoxia, but also when the brain is highly activated, we used the SD model. As seen in Fig. 8.38, in probes 1 and 2, three SD waves were developed; whereas in probes

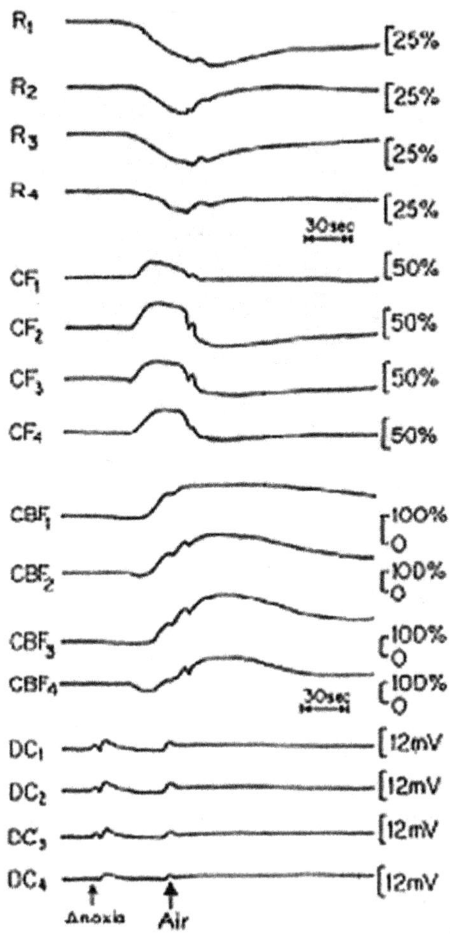

Fig. 8.37 A typical response to short anoxia (30 s) monitored by the MSMP device in the cerebral cortex. R reflectance, CF corrected NADH fluorescence, CBF cerebral blood flow, DC DC steady potential

3 and 4, only two waves were generated. The waves in each hemisphere resemble one another, clearly showing the spreading character of the SD waves, as probes 1 and 3 show the waves before probes 2 and 4. The reflectance showed a biphasic response with an increase of 25%–50% followed by a decrease back to the basal levels. CBF increased by 100%, mitochondrial NADH was oxidized by 25%–50% and DC potential decreased by 6–10 mV.

When the TBI model was induced, a large variance in the results was observed, although a strict protocol was maintained. Therefore, the results were summarized in two tables (Tables 8.4 and 8.5) in such a way that the main trend of change observed immediately after the impact and through most of the monitoring phase is presented by

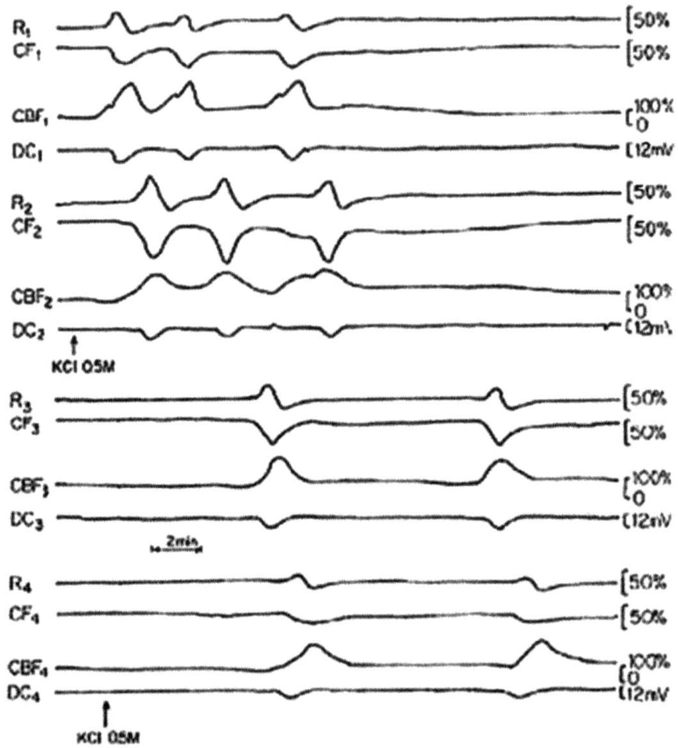

Fig. 8.38 Typical responses to cortical spreading depression (SD) induced by 0.5 M KCl solution. R reflectance, CF corrected NADH fluorescence, CBF cerebral blood flow, DC DC steady potential

Table 8.4 The changes in the different parameters during the recovery phase following mild percussion injury ($n = 3$).

Probe location (probe number		CBF	NADH	Reflectance	DC potential
Ipsilateral	Parietal (3)	↓, ↓, ↑	↑, ↑	↓, ↓, –	↓, ↓, –
	Frontal (4)	↓, ↑, ↑	↑, –	↑, ↑, –	↓, ↓, ↓
Contralateral	Parietal (1)	↓, ↑, ↑	↓, –	↓, ↓, –	↓, ↓, –
	Frontal (2)	↓, ↓, ↑	↓, ↑, ↑	↑, ↑, –	↓, ↓, –

CBF cerebral blood flow, *NADH* redox stte. (↑) increase, (↓) decrease, (↓↑) biphasic response, (–) no change (stable)

arrows as will be later explained. Nevertheless, we also present an example of changes in all parameters and locations recorded from one of the rats in the moderate TBI group (Fig. 8.39). As it is seen, immediately after the impact, CBF dramatically decreased, reaching the level of 0%–25%, thereafter, CBF in probes 2–4 (frontal lobes) increased, while in probe 3, CBF remained very low (0%). From 30 min on, CBF remained in

Table 8.5 The changes in the different parameters during the recovery phase following moderate percussion injury ($n = 4$)

Probe location (probe number)		CBF	NADH	Reflectance	DC potential
Ipsilateral	Parietal (3)	↓, ↓, ↓	↑, ↑, –, –	↑, ↑, ↓↑	↓, ↓,↓, ↓
	Frontal (4)	↑, ↑,↓, ↓	↑, ↑,↑, ↑	↑, ↑, ↓↑	↓, ↓,↓, ↓
Contralateral	Parietal (1)	↓, ↓,↓, ↓	↓↑, ↓↑, ↑	↑, ↑, ↓↑	↓, ↓,↓, ↓
	Frontal (2)	↑, ↑,↓, ↓	↑, ↑, ↑, –	↑, ↑, ↓↑	↓, ↓,↓, ↓

CBF cerebral blood flow, *NADH* mitochondrial NADH redox state. (↑) increase. (↓) decrease, (↓↑) biphasic response, (–) no change (stable)

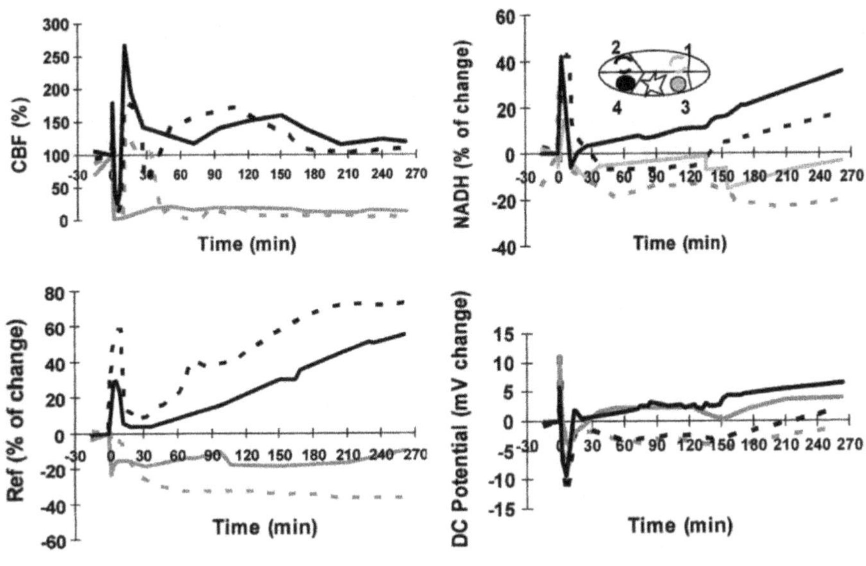

Fig. 8.39 Typical responses of CBF, NADH, reflectance (Ref) and DC steady potential (DC), in all four locations on the cerebral cortex following moderate brain injury

the range of 100%–170% in probes 2 and 4, whereas in probe 1, 35 min after the impact, CBF completely declined reaching 0% flow. The reflectance increased in all locations immediately after the impact. Fifteen minutes after the impact, the reflectance decreased below the basal levels in probes 1 and 3 (parietal), whereas in probes 2 and 4 (frontal), the reflectance increased by 60%–70%. Mitochondrial NADH also increased immediately after the impact and then decreased in all probes. Then, NADH gradually increased in probes 2 and 4 (frontal) by 30%–35%, while in positions 1 and 3 (parietal), NADH remained low (below the basal levels). Immediately following the impact, the DC potential increased and then decreased by ~12 mV. Later on (15 min later), at the contralateral side (probes 1 and 2), the DC potential was around −3 mV.

As formerly demonstrated, the impact caused an immediate increase or decrease in the first 15 min in all of the parameters, followed by different continuous changes from this point further on until the end of the experiment. Tables 8.4 and 8.5 present the direction of change of each of the parameters in the various locations following the two levels of impact (mild and moderate) as were detected by the MSMP during the recovery period. There were three rats that underwent mild percussion injury and four rats that underwent moderate injury. Since there were cases in which monitoring was impossible probably due to massive intracranial bleeding, there are cases in which only partial results are presented (namely, there are less arrows in the table than shown). In cases where the parameter returned to its basal level, the '-' was used.

Under mild TBI (Table 8.4) in awake rats, the parietal ipsilateral probe (probe 3), located nearest to the injury, showed a decrease in CBF (in two rats) and an increase in NADH in two rats (the tired rat could not be monitored). The reflectance (Ref) and the DC potential mostly decreased (in two out of three rats). However, in the ipsilateral frontal location (probe 4), CBF increased in two rats and NADH increased or remained unchanged. The reflectance increased and DC potential decreased. In the contralateral parietal position (probe 1), CBF increased in two out of three rats and NADH remained stable or was oxidized. The reflectance and the DC potential decreased in two rats and remained stable in one rat. The measurements of the contralateral frontal probe (probe 2) showed a decrease in CBF in two rats and an increase in the third rat. NADH level and the reflectance were mainly increased, and DC potential decreased in two rats and was stable in the third.

Following moderate TBI (Table 8.5), CBF decreased in three of the rats in the parietal ipsilateral location, probe 3, and NADH increased in two rats and remained stable in the other two. The reflectance increased in two rats and showed a biphasic response (an increase followed by a decrease) in one rat. The DC potential decreased in all four rats in all of the locations. In both frontal probes (probes 2 and 4), CBF increased in two rats and decreased in two rats, NADH level mostly increased, and the reflectance increased in two rats and showed a biphasic change in one rat. At the parietal contralateral location (probe 1), CBF decreased in all rats, NADH showed a biphasic response in two rats and increased in the third one, while the reflectance increased in two rats and showed a biphasic response in the third one.

8.8.4 Discussion

Real-time evaluation of the brain metabolic state after TBI is extremely important, especially in the initial period after contusion, before secondary damage to the tissue develops. In the present study, we demonstrate the use of the MSMP monitoring approach for the simultaneous monitoring of blood flow and mitochondrial function in the cerebral cortex after mild or moderate TBI, using fluid percussion model (FPM).

Before conducting the FPM, we had to ensure that the use of the MSMP device supplies reliable information on the hemodynamic and metabolic state of the cerebral cortex. Hence, we run several control experiments in which prolonged monitoring

(5 h) was conducted, and we stimulated the cerebral cortex with KCl solution to evoke SD waves. Since SD waves do not pass from one hemisphere to another (Bures et al. 1974), we used two KCl cannulas, one for each hemisphere. Our results showed that the multi-site multi-parametric monitoring approach enables reliable monitoring of the cerebral cortex in the control experiments, since the monitoring of the various parameters was stable throughout the entire monitoring period. The SD waves had a typical shape, including oxidation of mitochondrial NADH, CBF hyperemia, a negative shift in DC potential and a biphasic response in the reflectance parameter (Sonn and Mayevsky 1995, 2000).

Before discussing the results of the FPM, the following clarification should be mode. In the present brain injury model, four additional holes were drilled in the skull in addition to the hole through which the impact was induced. However, all of the holes were closed by the fiber optic probes. Furthermore, the cannulae in which the probes were placed were also cemented to the skull inducing a closed microenvironment during the impact. Moreover, in a similar model in which a large hole of 6 mm in diameter was drilled in the skull as well as a 3 mm hole for the fluid percussion injury, the ICP that was monitored showed a large increase immediately after the impact. This data clearly indicates the validity of the model, pointing out for the nonsignificant effect of the area of the brain that is exposed during the operation under the condition that the entire monitoring system is fixed to the brain, inducing a close environment.

The responses of the brain to TBI in our multimonitoring sites model showed a large that may be partly due to the fact that TBI was induced in awake rats, and the results were presented in two steps. First of all, we presented a typical example of the changes in CBF, mitochondrial NADH, reflectance and DC steady potential recorded in one rat following moderate TBI. These results clearly show a basal control level of each parameter before the induction of the injury, as well as major changes immediately after the onset of TBI and throughout the 4 h of monitoring. However, due to a large variability in the degree of change in each parameter as well as the kinetics of changes, we decided to present the major trends of changes (relative to the control level) in each parameter as represented in Tables 8.4 and 8.5. As it is seen, the parameter that showed the most uniform responses to TBI, in all locations, was the DC steady potential, which mostly decreased, although not at the same level and not with the same kinetics. Fabricius et al. (2006) recently emphasized the importance of changes in DC steady potential following brain injury in humans. These authors measured the slow potential changes (SPC) on the cerebral cortex and were able to record preinfarct depolarization in the electrically 'silent' cortical tissue. Alterations in the SPC indicate depolarization of the brain tissue and according to the authors, may serve as an indicator for the presence of a penumbra area in the brain after head injury. In our study, a negative shift in DC potential probably originates from the mechanical injury, which includes the rupture of the cell membranes and massive leakage of the cell content to the extracellular space yielding with tissue depolarization.

In contrast to cortical SD, the depolarization augments neuronal damage in the penumbra may be considered as a principal mechanism of progressive deterioration of the brain after head injury (Strong and Dardis 2005). In light of this, we may assume that our results also confirm the development of widely spread penumbra in different areas of ipsilateral and contralateral brain cortex within the first hours after the impact. Moreover, the increase in NADH levels in most of the animals in both severities of injury also strengths this point, namely, that both hemispheres are injured following the induction of fluid percussion injury in one hemisphere. Similarly, to recent studies dealing with this issue (Strong and Dardis 2005; Fabricius et al. 2006), we believe that real-time monitoring of the penumbra can be considered a promising approach for the development of adequate methods of diagnosis and therapy of the outcomes of brain contusion.

The decrease in steady DC potential was previously reported in several studies in which this phenomenon was related not only to mechanical tissue damage but also to the development of spreading depression at the site of injury and at its borders (Manor et al. 2003; Rogatsky et al. 2003). For example, Sunami et al. (1989) found an increase in local cerebral glucose utilization, a depression of EEG and DC potential negative shift at the injured cortex 2 h postinjury, presumably due to cortical SD. Nilsson et al. (1993) also showed a negative shift in DC potential following TBI induced by the compression contusion model of weight dropping. Even though SD is known to be associated with an increase in CBF, in the present study, CBF was mostly decreased after TBI. This is probably because the vascular response to spreading depression was abolished within the first 4 h postinjury (Ozawa et al. 1991) or because of impaired autoregulation mechanisms (Prat et al. 1997).

In our study and according to other authors (Prat et al. 1997; Schuhmann et al. 2002; Ahn et al. 2004; DeWitt et al. 1997) in the first few hours after the trauma, there is a clear tendency of brain hypoperfusion (reaching 50% and below of the basal level). It is important to emphasize that the observed decrease in CBF occurred not only in the ipsilateral but also in the contralateral brain hemispheres, the same as in the studies by Muir and Yunjie (Muir et al. 1992a; Wang et al. 1998). The increase in NADH levels that we observed in all cases of CBF decline may additionally indicate the development of acute cerebral cortex hypoxia accompanying its hypoperfusion. As it is known, the emergence in the early head injury period of the circulatory brain hypoxia (often accompanied by respiratory hypoxia) can initiate the development of one of the severest complications of brain trauma, the so-called secondary damage (Pilitsis and Rengachary 2001).

The CBF increase observed in the less injured tissue, namely, the frontal positions, may be due to the development of the "steal" phenomenon in which blood is transferred from the more injured tissue, at the site nearest to the impact, to the less injured tissue or as a response to focal cerebral ischemia (Hudetz et al. 1993). The intracerebral "steal" phenomenon was also observed in a group of patients with significant unilateral carotid artery stenosis during pharmacologic provocation, when blood was transferred from the contralateral to the ipsilateral hemisphere (Hartl et al. 1994).

Nilsson et al. (1996), who measured rCBF in the ipsilateral and contralateral hemispheres using laser Doppler flowmetry in a rat model following contusion trauma, showed a decrease in rCBF in the ipsilateral hemisphere, while in the contralateral hemisphere, CBF increased and thereafter decreased. Considering the effect of the reverse "stealing" phenomenon on the injured brain tissue, mostly seen during the first 3 days post-head injury (Cold et al. 1977), showed that this phenomenon was associated with poor outcome, namely, dementia, vegetative survival or death.

As for changes in the reflectance, its decrease resulted probably from the development of local hemorrhagic insults, which were the major factor preventing successful monitoring in many of the rats, as was also reported by others (Dixon et al. 1988; Wang et al. 1997). However, the increase in the reflectance at the ipsilateral and contralateral lobes in both severities of injury may be due to the development of edema at those locations. This was previously reported in several studies where edema appeared 1–6 h after mild fluid percussion injury in the ipsilateral lobe (Van Putten et al. 2005; Albensi et al. 2000) and in the contralateral hemisphere (Bareyre et al. 1997). While under severe injury, marked edema in the white matter was observed in both ipsi- and contralateral cerebral hemispheres (Wang et al. 1997).

In our experiments, the reflectance is also used for the correction of the mitochondrial NADH fluorescence, yielding the corrected NADH value in such a way that NADH oxidation appears as a decrease in NADH level, whereas an increase in NADH indicates a reduction of mitochondrial NADH in the tissue. The effect of TBI induced by the fluid percussion model on energy metabolism was previously tested in brain tissue or acid extracts (Tavazzi et al. 2005; Vagnozzi et al. 1999), where the involvement of mitochondrial dysfunction was suggested (Jiang et al. 2000; Lifshitz et al. 2003), yet without using real-time technology such as the present fluorometric technique. Some of our results showed no correlation between CBF and NADH, namely, when CBF decreased, NADH did not rise but rather declined. Although a decrease in NADH level means that NADH is oxidized, the changes in the reflectance at that moment must also be considered. In other words, when there are large changes in the reflectance, the calculated NADH is affected and the results may be misinterpreted (Mayevsky and Rogatsky 2007). Therefore, it should be emphasized that in order to correctly evaluate the hemodynamic and energetic state of the injured tissue by our monitoring method, all four parameters should be taken into account.

Regarding the location of monitoring, clearly, "in vivo" measurements of the impact "core" are very problematic due to the development of hemorrhage insults, edematous tissue, contusion injury, etc. Therefore, in the past decade, the study of TBI has included the measurements of biochemical, histological and immunologic parameters in the surrounding penumbra (Bell et al. 1998; Bellander et al. 2001; Schuhmann et al. 2002; Vlodavsky et al. 2005). The present study suggests monitoring the distant location on the cerebral cortex, in the current case, the contralateral hemisphere. It appears that tissue monitoring using the MSMP in the contralateral hemisphere is less affected by factors that may interfere with the hemodynamic and metabolic characteristics of the tissue and can supply with reliable information considering the impact on the tissue.

References

Ahn, M. J., E. R. Sherwood, D. S. Prough, C. Y. Lin, and D. S. DeWitt. 2004. The effects of traumatic brain injury on cerebral blood flow and brain tissue nitric oxide levels and cytokine expression. *Journal of Neurotrauma* 21:1431–1442.

Albensi, B. C., S. M. Knoblach, B. G. Chew, M. P. O'Reilly, A. I. Faden, and J. J. Pekar. 2000. Diffusion and high resolution MRI of traumatic brain injury in rats: Time course and correlation with histology. *Experimental Neurology* 162:61–72.

Ames, B. N., M. K. Shigenaga, and T. M. Hagen. 1995. Mitochondrial decay in aging. *Biochimica Et Biophysica Acta (BBA) - Molecular Basis of Disease* 1271:165.

Back, T., K. Kohno, and K. A. Hossmann. 1994. Cortical negative DC deflections following middle cerebral artery occlusion and KCl-induced spreading depression: Effect on blood flow, tissue oxygenation, and electroencephalogram. *Journal of Cerebral Blood Flow and Metabolism: Official Journal of the International Society of Cerebral Blood Flow and Metabolism* 14:12–19.

Bardt, T. F., A. W. Unterberg, R. Hartl, K. L. Kiening, G. H. Schneider, and W. R. Lanksch. 1998. Monitoring of brain tissue PO_2 in traumatic brain injury: Effect of cerebral hypoxia on outcome. *Acta neurochirurgica. Supplement* 71:153–156.

Bareyre, F., F. Wahl, T. K. Mcintosh, and J. M. Stutzmann. 1997. Time course of cerebral edema after traumatic brain injury in rats: Effects of riluzole and mannitol. *Journal of Neurotrauma* 14:839–849.

Barinaga, M.. 1993. Carbon monoxide: Killer to brain messenger in one step. *Science* 259:309.

Bell, M. J., P. M. Kochanek, J. A. Carcillo, Z. MI, J. K. Schiding, S. R. Wisniewski, R. S. Clark, C. E. Dixon, D. W. Marion, and E. Jackson. 1998. Interstitial adenosine, inosine, and hypoxanthine are increased after experimental traumatic brain injury in the rat. *Journal of Neurotrauma* 15:163–170.

Bellander, B. M., S. K. Singhrao, M. Ohlsson, P. Mattsson, and M. Svensson. 2001. Complement activation in the human brain after traumatic head injury. *Journal of Neurotrauma* 18:1295–1311.

Benveniste, H., M. B. Jorgensen, N. H. Diemer, and A. J. Hansen. 1988. Calcium accumulation by glutamate receptor activation is involved in hippocampal cell damage after ischemia. *Acta Neurologica Scandinavica* 78:529–536.

Benzi, G., O. Pastoris, F. Marzatico, R. F. Villa, F. Dagani, and D. Curti. 1992. The mitochondrial electron transfer alteration as a factor involved in the brain aging. *Neurobiology of Aging* 13:361–368.

Bergsneider, M., D. A. Hovda, S. M. Lee, D. F. Kelly, D. L. Mcarthur, P. M. Vespa, J. H. Lee, S. C. Huang, N. A. Martin, M. E. Phelps, and D. P. Becker. 2000. Dissociation of cerebral glucose metabolism and level of consciousness during the period of metabolic depression following human traumatic brain injury. *Journal of Neurotrauma* 17:389–401.

Berlet, H. H. 1976. Hypoxic survival of normoglycaemic young adult and adult mice in relation to cerebral metabolic rates. *Journal of Neurochemistry* 26:1267–1274.

Bouma, G. J., J. P. Muizelaar, S. C. Choi, P. G. Newlon, and H. F. Young. 1991. Cerebral circulation and metabolism after severe traumatic brain injury: The elusive role of ischemia. *Journal of Neurosurgery* 75:685–693.

Breuer, Z., and A. Mayevsky. 1992. Brain vasculature and mitochondrial responses to ischemia in Gerbils: II. Strain differences and statistical evaluation. *Brain Research* 598:251–256.

Brown, J. A., M. C. Preul, and A. Taha. 1988. Hyperbaric oxygen in the treatment of elevated intracranial pressure after head injury. *Pediatric Neuroscience* 14:286–290.

Buerk, D. G., and P. Nair. 1993. $PtiO_2$ and $CMRO_2$ changes in cortex and hippocampus of aging gerbil brain. *Journal of Applied Physiology* 74:1723–1728.

Bullock, R., A. Zauner, J. J. Woodward, J. Myseros, S. C. Choi, J. D. Ward, A. Marmarou, and H. F. Young. 1998. Factors affecting excitatory amino acid release following severe human head injury. *Journal of Neurosurgery* 89:507–518.

Bures, J., O. Buresova and J. Krivanek. 1974. Initiation of spreading depression. In *The mechanism and application of Leao's spreading-depression of electroencephalographic activity*, eds. J. Bures, O. Buresova and J. Krivanek. New York and London: Academic Press.

Cernak, I.. 2005. Animal models of head trauma. *NeuroRx: The Journal of the American Society for Experimental NeuroTherapeutics* 2:410–422.

Coburn, R. F. and H. J. Forman. 1987. Carbon monoxide toxicity. In *Handbook of physiology. The respiratory system, gas exchange*, eds. I. E. Farhi and S. M. Tenney. Bethesda, Maryland: American Physiology Society.

Cohen, S. and A. Mayevsky. 1989. Oxygen transport to tissue XI. In *Oxygen transport to tissue XI*, eds. K. Rakusan, G. P. Biro, T. K. Goldstick and Z. Turek. Plenum Press.

Cold, G. E., F. T. Jensen, and R. Malmros. 1977. The effects of $PaCO_2$ reduction on regional cerebral blood flow in the acute phase of brain injury. *Acta Anaesthesiologica Scandinavica* 21:359–367.

Cortez, S. C., T. K. Mcintosh, and L. J. Noble. 1989. Experimental fluid percussion brain injury: Vascular disruption and neuronal and glial alterations. *Brain Research* 482:271–282.

Crowe, W., A. Mayevsky, and L. Mela. 1977. Application of a solid membrane ion selective electrode to *in vivo* measurements. *American Journal of Physiology* 233:C56–C60.

Crutchfield, J. S., R. K. Narayan, C. S. Robertson, and L. H. Michael. 1990. Evaluation of a fiber-optic intracranial pressure monitor. *Journal of Neurosurgery* 72:482–487.

Czosnyka, M., E. Guazzo, M. Whitehouse, P. Smielewski, Z. Czosnyka, P. Kirkpatrick, S. Piechnik, and J. D. Pickard. 1996. Significance of intracranial pressure waveform analysis after head injury. *Acta Neurochirurgica* 138:531–541. discussion 541–542.

Davis, M., A. D. Mendelow, R. H. Perry, I. R. Chambers, and O. F. James. 1995. Experimental stroke and neuroprotection in the aging rat brain. *Stroke* 26:1072–1078.

De LaTorre, J. C. 1997. Cerebromicrovascular pathology in Alzheimer's disease compared to normal aging. *Gerontology* 43: 26–43.

DeWitt, D. S., L. W. Jenkins, E. P. Wei, H. Lutz, D. P. Becker, and H. A. Kontos. 1986. Effects of fluid-percussion brain injury on regional cerebral blood flow and pial arteriolar diameter. *Journal of Neurosurgery* 64:787–794.

DeWitt, D. S., T. G. Smith, D. J. Deyo, K. R. Miller, T. Uchida, and D. S. Prough. 1997. L-arginine and superoxide dismutase prevent or reverse cerebral hypoperfusion after fluid-percussion traumatic brain injury. *Journal of Neurotrauma* 14:223–233.

Dietrich, W. D., Z. C. Feng, H. Leistra, B. D. Watson, and M. Rosenthal. 1994. Photothrombotic infarction triggers multiple episodes of cortical spreading depression in distant brain regions. *Journal of Cerebral Blood Flow and Metabolism: Official Journal of the International Society of Cerebral Blood Flow and Metabolism* 14:20–28.

Dirnagl, U., B. Kaplan, M. Jacewicz, and W. Pulsinelli. 1989. Continuous measurement of cerebral cortical blood flow by laser-Doppler flowmetry in a rat stroke model. *Journal of Cerebral Blood Flow and Metabolism* 9:589–596.

Dixon, C. E. and R. L. Hayes. 1996. Fluid percussion and cortical impact models of brain injury. In *Neurotrauma*, eds. R. K. Narayan, J. E. Wilberger and J. T. Povlishock. New York: McGraw-Hill.

Dixon, C. E., J. W. Lighthall, and T. E. Anderson. 1988. Physiologic, histopathologic, and cineradiographic characterization of a new fluid-percussion model of experimental brain injury in the rat. *Journal of Neurotrauma* 5:91–104.

Dixon, C. E., B. G. Lyeth, J. T. Povlishock, R. L. Findling, R. J. Hamm, A. Marmarou, H. F. Young, and R. L. Hayes. 1987. A fluid percussion model of experimental brain injury in the rat. *Journal of Neurosurgery* 67:110–119.

Dora, E. and T. Zeuthen. 1976. Brain metabolism and ion movements in the brain cortex of the rat during anoxia. In *Ion and enzyme electrodes in biology and medicine*, ed. M. Kessler. Baltimore: University Park Press.

Dorheim, M. A., W. R. Tracey, J. S. Pollock, and P. Grammas. 1994. Nitric oxide synthase activity is elevated in brain microvessels in Alzheimer's disease. *Biochemical and Biophysical Research Communications* 205:659–665.

Duckrow, R. B., J. C. Lamanna, M. Rosenthal, J. E. Levasseur, and J. L. Patterson, Jr. 1981. Oxidative metabolic activity of cerebral cortex after fluid-percussion head injury in the cat. *Journal of Neurosurgery* 54: 607–614.

Duffy, T. E., S. J. Kohle, and R. C. Vannucci. 1975. Carbohydrate and energy metabolism in perinatal rat brain: Relation to survival in anoxia. *Journal of Neurochemistry* 24: 271–276.

Elitorai, I., and R. Montroy. 1991. Hyperbaric oxygen therapy leading to recovery of a 6-week comatose patient afflicted by anoxic encephalopathy and posttruamatic edema. *Journal of Hyperbaric Medicine* 6:189–198.

Engelborghs, K., J. Verlooy, B.VanDeuren, J.VanReempts, and M. Borgers. 1997. Intracranial pressure in a modified experimental model of closed head injury. *Acta neurochirurgica. Supplement* 70:123–125.

Fabricius, M., S. Fuhr, R. Bhatia, M. Boutelle, P. Hashemi, A. J. Strong, and M. Lauritzen. 2006. Cortical spreading depression and peri-infarct depolarization in acutely injured human cerebral cortex. *Brain* 129:778–790.

Fattoretti, P., C. Bertoni-Freddari, U. Caselli, R. Paoloni, and W. Meier-Ruge. 1996. Morphologic changes in cerebellar mitochondria during aging. *Analytical and Quantitative Cytology and Histology* 18:205–208.

Fazekas, J. F., J. Kleh, and F. A. Finnerty. 1955. Influence of age and vascular disease on cerebral hemodynamics and metabolism. *The American Journal of Medicine* 18:477–485.

Ferrendelli, J. A., W. G. Sedgwick, and V. Suntzeff. 1971. Regional energy metabolism and lipofuscin accumulation in mouse brain during aging. *Journal of Neuropathology and Experimental Neurology* 30:638–649.

Fink, B. R., and R. H. Haschke. 1973. Anesthetic effects on cerebral metabolism. *Anesthesiology* 39:199–215.

Finnie, J. W., and P. C. Blumbergs. 2002. Traumatic brain injury. *Veterinary Pathology* 39:679–689.

Friedli, C. M., D. S. Sclarsky, and A. Mayevsky. 1982. Multiprobe monitoring of ionic, metabolic, and electrical activities in the awake brain. *American Journal of Physiology* 243:R462–R469.

Fujishima, M., S. Sadoshima, J. Ogata, F. Yoshida, O. Shiokawa, S. Ibayashi, and T. Omae. 1984. Autoregulation of cerebral blood flow in young and aged spontaneously hypertensive rats (SHR). *Gerontology* 30:30–36.

Funahashi, T., R. A. Floyd, and J. M. Carney. 1994. Age effect on brain pH during ischemia/reperfusion and pH influence on peroxidation. *Neurobiol Aging* 15:161–167.

Gido, G., T. Kristian, K. Katsura, and B. K. Siesjo. 1994a. The influence of repeated spreading depression-induced calcium transients on neuronal viability in moderately hypoglycemic rats. *Experimental Brain Research* 97:397–403.

Gido, G., T. Kristian, and B. K. Siesjo. 1994b. Induced spreading depressions in energy-compromised neocortical tissue: Calcium transients and histopathological correlates. *Neurobiology of Disease* 1:31–41.

Ginsberg, M. D., W. Zhao, L. Belayev, O. F. Alonso, Y. Liu, J. Y. Loor, and R. Busto. 2001. Diminution of metabolism/blood flow uncoupling following traumatic brain injury in rats in response to high-dose human albumin treatment. *Journal of Neurosurgery* 94:499–509.

Graham, D. I., J. H. Adams, and D. Doyle. 1978. Ischaemic brain damage in fatal non-missile head injuries. *Journal of the Neurological Sciences* 39:213–234.

Graham, D. I., I. Ford, J. H. Adams, D. Doyle, G. M. Teasdale, A. E. Lawrence, and D. R. Mclellan. 1989. Ischaemic brain damage is still common in fatal non-missile head injury. *Journal of Neurology, Neurosurgery & Psychiatry* 52:346–350.

Graham, D. I., T. K. Mcintosh, W. L. Maxwell, and J. A. Nicoll. 2000. Recent advances in neurotrauma. *Journal of Neuropathology and Experimental Neurology* 59:641–651.

Gregoire, N. M., A. Gjedde, F. Plum, and T. E. Duffy. 1978. Cerebral blood flow and cerebral metabolic rates for oxygen, glucose, and ketone bodies in newborn dogs. *Journal of Neurochemistry* 30:63–69.

Haberl, R. L., M. L. Heizer, A. Marmarou, and E. F. Ellis. 1989. Laser-Doppler assessment of brain microcirculation: Effect of systemic alterations. *American Journal of Physiology-Heart and Circulatory Physiology* 256:H1247–H1254.

Hansen, A. J.. 1978. The extracellular potassium concentration in brain cortex following ischemia in hypo- and hyperglycemic rats. *Acta Physiologica Scandinavica* 102:324–329.

Hansen, A. J. 1985. Effect of anoxia on ion distribution in the brain. *Physiological Reviews* 65:101–148.

Hansen, A. J., B. Quistorff, and A. Gjedde. 1980. Relationship between local changes in cortical blood flow and extracellular K^+ during spreading depression. *Acta Physiologica Scandinavica* 109:1–6.

Hansen, A. J., and T. Zeuthen. 1981. Extracellular ion concentrations during spreading depression and ischemia in the rat brain cortex. *Acta Physiologica Scandinavica* 113:437–445.

Hartl, W. H., I. Janssen, and H. Fürst. 1994. Effect of carotid endarterectomy on patterns of cerebrovascular reactivity in patients with unilateral carotid artery stenosis. *Stroke* 25:1952–1957.

Hernández, M. J., R. W. Brennan, R. C. Vannucci, and G. S. Bowman. 1978. Cerebral blood flow and oxygen consumption in the newborn dog. *American Journal of Physiology* 234:R209–R215.

Hilfaer, M. A. and R. J. Traystman. 1996. Cerebrovascular effects of carbon monoxide. In *Carbon monoxide*, ed. D. G. Penney. New York: CRC Press.

Hubschmann, O. R., and D. Kornhauser. 1983. Effects of intraparenchymal hemorrhage on extracellular cortical potassium in experimental head trauma. *Journal of Neurosurgery* 59:289–293.

Hudetz, A. G., K. A. Conger, A. Eke, and J. H. Halsey, Jr. 1993. A mathematical model of the intracerebral steal phenomenon in regional and focal ischaemia. *Neurological Research* 15:117–127.

Inada, K., I. Yokoi, H. Kabuto, H. Habu, A. Mori, and N. Ogawa. 1996. Age-related increase in nitric oxide synthase activity in senescence accelerated mouse brain and the effect of long-term administration of superoxide radical scavenger. *Mechanisms of ageing and Development* 89:95–102.

Ishige, N., L. H. Pitts, I. Berry, S. G. Carlson, M. C. Nishimura, M. E. Moseley, and P. R. Weinstein. 1987a. The effect of hypoxia on traumatic head injury in rats: alterations in neurologic function, brain edema, and cerebral blood flow. *Journal of Cerebral Blood Flow and Metabolism: Official Journal of the International Society of Cerebral Blood Flow and Metabolism* 7:759–767.

Ishige, N., L. H. Pitts, L. Pogliani, T. Hashimoto, M. C. Nishimura, H. M. Bartkowski, and T. L. James. 1987b. Effect of hypoxia on traumatic brain injury in rats: Part 2. Changes in high energy phosphate metabolism. *Neurosurgery* 20:854–858.

Ito, J., A. Marmarou, P. Barzó, P. Fatouros, and F. Corwin. 1996. Characterization of edema by diffusion-weighted imaging in experimental traumatic brain injury. *Journal of Neurosurgery* 84:97–103.

Jiang, X. B., K. Ohno, L. Qian, B. Tominaga, T. Kuroiwa, T. Nariai, and K. Hirakawa. 2000. Changes in local cerebral blood flow, glucose utilization, and mitochondrial function following traumatic brain injury in rats. *Neurologia Medico-chirurgica* 40:16–28. ; discussion 28–29.

Jobsis, F. F., M. O'Connor, A. Vitale, and H. Vreman. 1971. Intracellular redox changes in functioning cerebral cortex. I. Metabolic effects of epileptiform activity. *Journal of Neurophysiology* 3465:735–749.

Katayama, Y., D. P. Becker, T. Tamura, and D. A. Hovda. 1990. Massive increases in extracellular potassium and the indiscriminate release of glutamate following concussive brain injury. *Journal of Neurosurgery* 73:889–900.

Kindwall, E. P. 1996. Delayed sequence in carbon monoxide poisoning and the possible mechanisms. In *Carbon monoxide*, ed. D. G. Penney. New York: CRC Press.

King, L. J., G. M. Schoepfle, O. H. Lowry, J. V. Passonneau, and S. Wilson. 1967. Effects of electrical stimulation on metabolites in brain of decapitated mice. *Journal of Neurochemistry* 14:613–618.

Kochanek, P. M., D. W. Marion, W. Zhang, J. K. Schiding, M. White, A. M. Palmer, R. S. Clark, M. E. O'Malley, S. D. Styren, C. HOet al. 1995. Severe controlled cortical impact in rats: Assessment of cerebral edema, blood flow, and contusion volume. *Journal of Neurotrauma* 12:1015–1025.

Koehler, R. C., M. D. Jones, Jr., and R. J. Traystman. 1982. Cerebral circulatory response to carbon monoxide and hypoxic hypoxia in the lamb. *American Journal of Physiology* 243:H27–H32.

Koehler, R. C., R. J. Traystman, A. A. Rosenberg, M. L. Hudak, and M. D. Jones, Jr.. 1983. Role of O_2-hemoglobin affinity on cerebrovascular response to carbon monoxide hypoxia. *American Journal of Physiology* 245:H1019–H1023.

Koroleva, V. I., and J. Bures. 1996. The use of spreading depression waves for acute and long-term monitoring of the penumbra zone of focal ischemic damage in rats. *Proceedings of the National Academy of Sciences of the USA* 93:3710–3714.

Koura, S. S., E. M. Doppenberg, A. Marmarou, S. Choi, H. F. Young, and R. Bullock. 1998. Relationship between excitatory amino acid release and outcome after severe human head injury. *Acta neurochirurgica. Supplement* 71:244–246.

Kreisman, N. R., R. A. Hodin, B. L. Brizzee, M. Rosenthal, T. J. Sick, R. Busto, and M. D. Ginsberg. 1987. Seizure-associated pulmonary edema and cerebral oxygenation in the rat. *Journal of Applied Physiology* 62:658–667.

Kubota, M., T. Nakamura, K. Sunami, Y. Ozawa, H. Namba, A. Yamaura, and H. Makino. 1989. Changes of local cerebral glucose utilization, DC potential and extracellular potassium concentration in experimental head injury of varying severity. *Neurosurgical Review* 12(Suppl 1):393–399.

Lang, E. W., and R. M. Chesnut. 1995. Intracranial pressure and cerebral perfusion pressure in severe head injury. *New Horizons (Baltimore, Md.)* 3:400–409.

Laurer, H. L., D. F. Meaney, S. S. Margulies et al. 2002. Modeling brain injury/trauma. In *Encyclopedia of the human brain*, ed. V. S. Ramachandran. San Diego, CA: Academic Press.

Lauritzen, M. 1987. Cerebral blood flow in migraine and cortical spreading depression. *Acta neurologica Scandinavica. Supplementum* 113:1–40.

Leão, A. A. P.. 1944. Spreading depression of activity in cerebral cortex. *Journal of Neurophysiology* 7:359–390.

Lewelt, W., L. W. Jenkins, and J. D. Miller. 1982. Effects of experimental fluid-percussion injury of the brain on cerebrovascular reactivity of hypoxia and to hypercapnia. *Journal of Neurosurgery* 56:332–338.

Lifshitz, J., H. Friberg, R. W. Neumar, R. Raghupathi, F. A. Welsh, P. Janmey, K. E. Saatman, T. Wieloch, M. S. Grady, and T. K. Mcintosh. 2003. Structural and functional damage sustained by mitochondria after traumatic brain injury in the rat: Evidence for differentially sensitive populations in the cortex and hippocampus. *Journal of Cerebral Blood Flow and Metabolism: Official Journal of the International Society of Cerebral Blood Flow and Metabolism* 23:219–231.

Lowry, O. H., J. V. Passonneau, F. X. Hasselberger, and D. W. Schulz. 1964. Effect of ischemia on known substrates and cofactors of the glycolytic pathway in brain. *Journal of Biological Chemistry* 239:18–30.

MacMillan, V. 1975a. The effects of acute carbon monoxide intoxication on the cerebral energy metabolism of the rat. *Canadian Journal of Physiology and Pharmacology* 53:354–362.

MacMillan, V. 1975b. Regional cerebral blood flow of the rat in acute carbon monoxide intoxication. *Canadian Journal of Physiology and Pharmacology* 53:644–650.

MacMillan, V. 1978. Cerebral energy metabolism during recovery from carbon monoxide hypoxia-oligemia. *Brain Research* 151:353–368.

Maines, M. D. 1993. Carbon Monoxide: An emerging regulator of cGMP in the brain. *Molecular and Cellular Neurosciences* 4:389–397.

Maker, H. S. and G. M. Lehrer. 1971. Effect of ischemia. In *Handbook of neurochemistry*, ed. A. Lajtha. New York: Plenum Press.

Manor, T., G. Rogatsky, N. Zarchin, and A. Mayevsky. 2003. Effects of fluid percussion injury on rat brain hemodynamics, ionic, electrical activity, and energy metabolism *in vivo*. *Advances in Experimental Medicine and Biology* 510:409–414.

Mares, P., N. Kríz, G. Brozek, and J. Bures. 1976. Anoxic changes of extracellular potassium concentration in the cerebral cortex of young rats. *Experimental Neurology* 53:12–20.

Marín, J.. 1995. Age-related changes in vascular responses: a review. *Mechanisms of ageing and Development* 79:71–114.

Marks, G. S., J. F. Brien, K. Nakatsu, and B. E. Mclaughlin. 1991. Does carbon monoxide have a physiological function?. *Trends in Pharmacological Sciences* 12:185–188.

Marmarou, A., P. P. Fatouros, P. Barzo, G. Portella, M. Yoshihara, O. Tsuji, T. Yamamoto, F. Laine, S. Signoretti, J. D. Ward, M. R. Bullock, and H. F. Young. 2000. Contribution of edema and cerebral blood volume to traumatic brain swelling in head-injured patients. *Journal of Neurosurgery* 93:183–193.

Marmarou, A., and K. Shima. 1990. Comparative studies of edema produced by fluid percussion injury with lateral and central modes of injury in cats. *Advances in Neurology* 52:233–236.

Mattson, M. P., and S. W. Scheff. 1994. Endogenous neuroprotection factors and traumatic brain injury: Mechanisms of action and implications for therapy. *Journal of Neurotrauma* 11:3–33.

Mayevsky, A. 1976. Brain energy metabolism of the conscious rat exposed to various physiological and pathological situations. *Brain Research* 113:327–338.

Mayevsky, A. 1977. Brain energy metabolism of the conscious rat exposed to ischemic conditions. *Bibliotheca anatomica* 15 Pt 1:361–364.

Mayevsky, A.. 1978a. Ischemia in the brain: The effects of carotid artery ligation and decapitation on the energy state of the awake and anesthetized rat. *Brain Research* 140:217–230.

Mayevsky, A. 1978b. Shedding light on the awake brain. In *Frontiers in bioenergetics: From electrons to tissues*, vol. II, eds. P. L. Dutton, J. Leigh and A. Scarpa. New York: Academic Press.

Mayevsky, A. 1983a. Metabolic, ionic and electrical responses to experimental epilepsy in the awake rat. In *Proc. first internat. cong. cerebral blood flow, metabolism and epilepsy*, eds. M. Baldy, D. H. Moulinier, D. H. Ingvar and B. S. Meldrum. John Libbey.

Mayevsky, A.. 1983b. Multiparameter monitoring of the awake brain under hyperbaric oxygenation. *Journal of Applied Physiology* 54:740–748.

Mayevsky, A. 1984a. Brain NADH redox state monitored *in vivo* by fiber optic surface fluorometry. *Brain Research Reviews* 7:49–68.

Mayevsky, A. 1984b. Brain oxygen toxicity. Invited review. In *8th symp. underwater physiol.*, eds. A. J. Bachrach and M. M. Matzen, 69–89. Bethesda, Maryland: Undersea Medical Society Inc.

Mayevsky, A.. 1992. Cerebral blood flow and brain mitochondrial redox state responses to various perturbations in gerbils. *Advances in Experimental Medicine and Biology* 317:707–716.

Mayevsky, A. 1993. Biochemical and physiological activities of the brain as in vivo markers of brain pathology. In *Cerebral revascularization*, eds. E. F. Bernstein, A. D. Callow, A. N. Nicolaides and E. G. Shifrin. Med-Orion Pub.

Mayevsky, A. and Z. Breuer. 1990. The Mongolian gerbil as a model for cerebral ischemia. In *Cerebral ischemia and cerebral resuscitation*, eds. A. Schurr and B. M. Rigor, 27–46. CRC Press.

Mayevsky, A. and Z. Breuer. 1992. Brain vasculature and mitochondrial responses to ischemia in Gerbils: I. Basic anatomical patterns and biochemical correlates. *Brain Research* 598:242–250.

Mayevsky, A., and B. Chance. 1974. Repetitive patterns of metabolic changes during cortical spreading depression of the awake rat. *Brain Research* 65:529–533.

Mayevsky, A., and B. Chance. 1975. Metabolic responses of the awake cerebral cortex to anoxia hypoxia spreading depression and epileptiform activity. *Brain Research* 98:149–165.

Mayevsky, A., and B. Chance. 1976. The effect of decapitation on the oxidation-reduction state of NADH and ECoG in the brain of the awake rat. *Advances in Experimental Medicine and Biology* 75:307–312.

Mayevsky, A., and B. Chance. 1982. Intracellular oxidation-reduction state measured in situ by a multicannel fiber-optic surface fluorometer. *Science* 217:537–540.

Mayevsky, A., and B. Chance. 1983. Multisite measurements of NADH redox state from cerebral cortex of the awake animal. *Advances in Experimental Medicine and Biology* 159:143–155.
Mayevsky, A., W. Crowe, and L. Mela. 1980. The interrelation between brain oxidative metabolism and extracellular potassium in the unanesthetized gerbil. *Neurological Research* 1:213–226.
Mayevsky, A., A. Doron, T. Manor, S. Meilin, N. Zarchin, and G. E. Ouaknine. 1996. Cortical spreading depression recorded from the human brain using a multiparametric monitoring system. *Brain Research* 740:268–274.
Mayevsky, A., A. Doron, S. Meilin, T. Manor, E. Ornstein, and G. E. Ouaknine. 1999a. Brain viability and function analyzer: Multiparametric real-time monitoring in neurosurgical patients. *Acta Neurochirurgica. Supplementum* 75:63–66.
Mayevsky, A., R. B. Duckrow, E. Yoles, N. Zarchin, and D. Kanshansky. 1990. Brain mitochondrial redox state, tissue hemodynamic and extracellular ion responses to four-vessel occlusion and spreading depression in the rat. *Neurological Research* 12:243–248.
Mayevsky, A., E. S. Flamm, W. Pennie, and B. Chance. 1991. A fiber optic based multiprobes system for intraoperative monitoring of brain functions. *SPIE* 1431:303–313.
Mayevsky, A., K. Frank, M. Muck, S. Nioka, M. Kessler, and B. Chance. 1992. Multiparametric evaluation of brain functions in the Mongolian gerbil *in vivo*. *Journal of Basic and Clinical Physiology and Pharmacology* 3:323–342.
Mayevsky, A., T. Manor, S. Meilin, A. Doron and G. E. Ouaknine. 1998. Real-time multiparametric monitoring of the injured human cerebral cortex – a new approach. In *Acta neurochirurgica. Supplementum*, eds. R. Marmarou and R. Bullock. ICPX.
Mayevsky, A., S. Meilin, T. Manor, N. Zarchin, and J. Sonn. 1999b. Optical monitoring of NADH redox state and blood flow as indicators of brain energy balance. *Advances in Experimental Medicine and Biology* 471:133–140.
Mayevsky, A., S. Meilin, G. G. Rogatsky, N. Zarchin, and J. Sonn. 1995. Multiparametric monitoring of the awake brain exposed to carbon monoxide. *Journal of Applied Physiology* 78:1188–1196.
Mayevsky, A., and G. Rogatsky. 2007. Mitochondrial function *in vivo* evaluated by NADH fluorescence: From animal models to human studies. *American Journal of Physiology. Cell Physiology* 292:C615–C640.
Mayevsky, A., G. G. Rogatsky, N. Zarchin and S. R. Thom. 1993a. Interrelation between hyperbaric oxygenation and carbon monoxide intoxication in the rat brain in vivo. In *Basic and applied high pressure biology*, eds. P. B. Bennett and R. E. Marquis. University of Rochester Press.
Mayevsky, A., and H. R. Weiss. 1991. Cerebral blood flow and oxygen consumption in cortical spreading depression. *Journal of Cerebral Blood Flow and Metabolism* 11:829–836.
Mayevsky, A., N. Zarchin, and C. M. Friedli. 1982. Factors affecting the oxygen balance in the awake cerebral cortex exposed to spreading depression. *Brain Research* 236:93–105.
Mayevsky, A., N. Zarchin and J. Sonn. 1993b. Brain redox state and O_2 balance in experimental spreading depression and ischemia. In *Migraine – basic mechanisms and treatment*, eds. A. Lehmenkuhler, K.-H. Grotemeyer and F. Tegtmeyer. Munchen-Wien, Baltimore: Urban & Schwarzenberg.
Mayevsky, A., and I. Ziv. 1991. Oscillations of cortical oxidative metabolism and microcirculation in the ischaemic brain. *Neurological Research* 13:39–47.
McIntosh, T. K. 1994. Neurochemical sequelae of traumatic brain injury: Therapeutic implications. *Cerebrovascular and Brain Metabolism Reviews* 6:109–162.
McIntosh, T. K., M. Juhler, and T. Wieloch. 1998. Novel pharmacologic strategies in the treatment of experimental traumatic brain injury: 1998. *Journal of Neurotrauma* 15:731–769.
McIntosh, T. K., D. H. Smith, D. F. Meaney, M. J. Kotapka, T. A. Gennarelli, and D. I. Graham. 1996. Neuropathological sequelae of traumatic brain injury: Relationship to neurochemical and biomechanical mechanisms. *Laboratory Investigation; A Journal of Technical Methods and Pathology* 74:315–342.

McIntosh, T. K., R. Vink, L. Noble, I. Yamakami, S. Fernyak, H. Soares, and A. L. Faden. 1989. Traumatic brain injury in the rat: Characterization of a lateral fluid-percussion model. *Neuroscience* 28:233–244.

Meier-Ruge, W.. 1975. Experimental pathology and pharmacology in brain research and aging. *Life Sciences* 17:1627–1635.

Meilin, S., G. G. Rogatsky, S. R. Thom, N. Zarchin, E. Guggenheimer-Furman, and A. Mayevsky. 1996. Effects of carbon monoxide exposure on the brain may be mediated by nitric oxide. *Journal of Applied Physiology* 81:1078–1083.

Meilin, S., J. Sonn, N. Zarchin, G. Rogatsky, E. Guggenheimer-Furman, and A. Mayevsky. 1998. Responses of rat brain to induced spreading depression following exposure to carbon monoxide. *Brain Research* 780:323–328.

Meilin, S., N. Zarchin, and A. Mayevsky. 1999. Inter-relation between hemodynamic, metabolic, ionic and electrical activities during ischemia and reperfusion in the gerbil brain. *Neurological Research* 21:699–704.

Meilin, S., N. Zarchin, A. Mayevsky and S. Shapira. 1995. Multiparametric responses to cortical spreading depression under nitric oxide synthesis inhibition. In *Biochemical pharmacological and clinical aspects of nitric oxide*, eds. B. A. Weissman, N. Alon and S. Shapira. New York: Plenum Press.

Mendelman, A., E. Guggenheimer-Furman, N. Zarchin, A. Meilin, and A. Mayevsky. 1996. Protection of rat brain against exposure to CO afforded by barbiturate anesthesia. (Abstract). *Israel Journal of Medical Sciences* 32:832.

Meyer, J. S., A. Kondo, F. Nomura, K. Sakamoto, and T. Teraura. 1970. Cerebral hemodynamics and metabolism following experimental head injury. *Journal of Neurosurgery* 32:304–319.

Meyer, J. S., Y. Terayama, and S. Takashima. 1993. Cerebral circulation in the elderly. *Cerebrovascular and Brain Metabolism Reviews* 5:122–146.

Miller, J. D., D. P. Becker, J. D. Ward, H. G. Sullivan, W. E. Adams, and M. J. Rosner. 1977. Significance of intracranial hypertension in severe head injury. *Journal of Neurosurgery* 47:503–516.

Miller, J. D., N. M. Dearden, I. R. Piper, and K. H. Chan. 1992. Control of intracranial pressure in patients with severe head injury. *Journal of Neurotrauma* 9(Suppl 1):S317–S326.

Mollace, V., P. Rodino, R. Massoud, D. Rotiroti, and G. Nistico. 1995. Age-dependent changes of NO synthase activity in the rat brain. *Biochemical and Biophysical Research Communications* 215:822–827.

Moravec, J., P. Y. Hatt, L. H. Opie, and F. D. Rost. 1972. The application of the cytophotometer to the study of metabolic transitions of isolated rat heart. *Cardiology* 57:61–66.

Muir, J. K., M. Boerschel, and E. F. Ellis. 1992a. Continuous monitoring of posttraumatic cerebral blood flow using laser-Doppler flowmetry. *Journal of Neurotrauma* 9:355–362.

Muir, J. K., M. Boerschel, and E. F. Ellis. 1992b. Continuous monitoring of posttraumatic cerebral blood flow using laser-Doppler flowmetry. *Journal of Neurotrauma* 9:355–362.

Mutch, W. A., and A. J. Hansen. 1984. Extracellular pH changes during spreading depression and cerebral ischemia: Mechanisms of brain pH regulation. *Journal of Cerebral Blood Flow and Metabolism: Official Journal of the International Society of Cerebral Blood Flow and Metabolism* 4:17–27.

Nedergaard, M., and A. J. Hansen. 1993. Characterization of cortical depolarizations evoked in focal cerebral ischemia. *Journal of Cerebral Blood Flow and Metabolism: Official Journal of the International Society of Cerebral Blood Flow and Metabolism* 13:568–574.

Nilsson, P., B. Gazelius, H. Carlson, and L. Hillered. 1996. Continuous measurement of changes in regional cerebral blood flow following cortical compression contusion trauma in the rat. *Journal of Neurotrauma* 13:201–207.

Nilsson, P., L. Hillered, Y. Olsson, M. J. Sheardown, and A. J. Hansen. 1993. Regional changes in interstitial K^+ and Ca^{2+} levels following cortical compression contusion trauma in rats. *Journal of Cerebral Blood Flow and Metabolism: Official Journal of the International Society of Cerebral Blood Flow and Metabolism* 13:183–192.

Nilsson, P., L. Hillered, U. Ponten, and U. Ungerstedt. 1990. Changes in cortical extracellular levels of energy-related metabolites and amino acids following concussive brain injury in rats. *Journal of Cerebral Blood Flow and Metabolism: Official Journal of the International Society of Cerebral Blood Flow and Metabolism* 10:631–637.
Norusis, M. J. 1999. *Guide of data analysis*. Upper Saddle River, NJ: Prentice-Hall.
Obrist, W. D., T. W. Langfitt, J. L. Jaggi, J. Cruz, and T. A. Gennarelli. 1984. Cerebral blood flow and metabolism in comatose patients with acute head injury. Relationship to intracranial hypertension. *Journal of Neurosurgery* 61:241–253.
Ohata, M., U. Sundaram, W. R. Fredericks, E. D. London, and S. I. Rapoport. 1981. Regional cerebral blood flow during develoent and ageing of the rat brain. *Brain* 104:319–332.
Otsuka, H., K. Ueda, A. Heimann, and O. Kempski. 2000. Effects of cortical spreading depression on cortical blood flow, impedance, DC potential, and infarct size in a rat venous infarct model. *Experimental Neurology* 162:201–214.
Ozawa, Y., T. Nakamura, K. Sunami, M. Kubota, C. Ito, H. Murai, A. Yamaura, and H. Makino. 1991. Study of regional cerebral blood flow in experimental head injury: Changes following cerebral contusion and during spreading depression. *Neurologia Medico-chirurgica* 31:685–690.
Parmacek, M. S., J. H. Fox, W. H. Harrison, D. C. Garron, and D. Swenie. 1979. Effect of aging on brain respiration and carbohydrate metabolism of CBF1 mice. *Gerontology* 25:185–191.
Patel, M. S. 1977. Age-dependent changes in the oxidative metabolism in rat brain. *Journal of Gerontology* 32:643–646.
Peng, M. T., Y. I. Peng, and F. N. Chen. 1977. Age-dependent changes in the oxygen consumption of the cerebral cortex, hypothalamus, hippocampus, and amygdaloid in rats. *Journal of Gerontology* 32:517–522.
Penney, D. G. 1990. Acute carbon monoxide poisoning: Animal models: A review. *Toxicology* 62:123–160.
Persson, L., and L. Hillered. 1992. Chemical monitoring of neurosurgical intensive care patients using intracerebral microdialysis. *Journal of Neurosurgery* 76:72–80.
Pettegrew, R. K., and K. L. Ewing. 1971. Life history study of oxygen utilization in the C57BL/6 mouse. *Journal of Gerontology* 26:381–385.
Pfenninger, J. 1984. Early prediction of outcome after severe head injury in children. *Zeitschrift Fur Kinderchirurgie : Organ der Deutschen, der Schweizerischen und der Osterreichischen Gesellschaft Fur Kinderchirurgie = Surgery in Infancy and childhood* 39:223–228.
Pilitsis, J. G., and S. S. Rengachary. 2001. Complications of head injury. *Neurological Research* 23:227–236.
Piper, I. R., J. D. Miller, N. M. Dearden, J. R. Leggate, and I. Robertson. 1990. Systems analysis of cerebrovascular pressure transmission: an observational study in head-injured patients. *Journal of Neurosurgery* 73:871–880.
Piper, R. D., G. A. Lambert, and J. W. Duckworth. 1991. Cortical blood flow changes during spreading depression in cats. *American Journal of Physiology* 261:H96–H102.
Prat, R., V. Markiv, M. Dujovny, and M. Misra. 1997. Evaluation of cerebral autoregulation following diffuse brain injury in rats. *Neurological Research* 19:393–402.
Roberts, Jr., E. L., C. P. Chih, and M. Rosenthal. 1997. Age-related changes in brain metabolism and vulnerability to anoxia. *Advances in Experimental Medicine and Biology* 411:83–89.
Rockswold, G. L., S. E. Ford, D. C. Anderson, I. A. Bergman, and R. E. Sherman. 1992. Results of prospective randomized trail for treatment of severly brain-injured patients with hyperbaric oxygen. *Journal of Neurosurgery* 76:929–934.
Rogatsky, G. G., A. Mayevsky, N. Zarchin, and A. Doron. 1996. Continuous multiparametric monitoring of brain activities following fluid-percussion injury in rats: Preliminary results. *Journal of Basic and Clinical Physiology and Pharmacology* 7:23–43.
Rogatsky, G. G., J. Sonn, Y. Kamenir, N. Zarchin, and A. Mayevsky. 2003. Relationship between intracranial pressure and cortical spreading depression following fluid percussion brain injury in rats. *Journal of Neurotrauma* 20:1315–1325.

Rosenthal, M., and G. Somjen. 1973. Spreading depression, sustained potential shifts, and metabolic activity of cerebral cortex of cats. *Journal of Neurophysiology* 36:739–749.
Samson, F. E., and N. A. Dahl. 1975. Cerebral energy requirement of neonatal rats. *American Journal of Physiology* 188: 277–280.
Schneider, G., P. Fries, D. Wagner-Jochem, D. Thome, H. Laurer, B. Kramann, A. Mautes, and T. Hagen. 2002. Pathophysiological changes after traumatic brain injury: Comparison of two experimental animal models by means of MRI. *Magma (New York, N.Y.)* 14:233–241.
Schuhmann, M. U., D. Stiller, M. Skardelly, M. Mokktarzadeh, S. Thomas, T. Brinker, and M. Samii. 2002. Determination of contusion and oedema volume by MRI corresponds to changes of brain water content following controlled cortical impact injury. *Acta Neurochirurgica. Supplementum* 81:213–215.
Shackford, S. R., J. Zhuang, and J. Schmoker. 1992. Intravenous fluid tonicity: Effect on intracranial pressure, cerebral blood flow, and cerebral oxygen delivery in focal brain injury. *Journal of Neurosurgery* 76:91–98.
Shenkin, H. A., P. Novak, B. Goluboff, A. M. Soffe, and L. Bortin. 1953. The effects of aging, arteriosclerosis, and hypertension upon the cerebral circulation. *Journal of Clinical Investigation* 32:459–465.
Shibata, M., C. W. Leffler, and D. W. Busija. 1990. Cerebral hemodynamics during cortical spreading depression in rabbits. *Brain Research* 530:267–274.
Signoretti, S., A. Marmarou, B. Tavazzi, G. Lazzarino, A. Beaumont, and R. Vagnozzi. 2001. N-Acetylaspartate reduction as a measure of injury severity and mitochondrial dysfunction following diffuse traumatic brain injury. *Journal of Neurotrauma* 18: 977–991.
Signorini, D. F., P. J. Andrews, P. A. Jones, J. M. Wardlaw, and J. D. Miller. 1999. Adding insult to injury: The prognostic value of early secondary insults for survival after traumatic brain injury. *Journal of Neurology, Neurosurgery & Psychiatry* 66:26–31.
Simard, D., J. Olesen, O. B. Paulson, N. A. Lassen, and E. Skinhoj. 1971. Regional cerebral blood flow and its regulation in dementia. *Brain* 94:273–288.
Somjen, G. G.. 2001. Mechanisms of spreading depression and hypoxic spreading depression-like depolarization. *Physiological Reviews* 81:1065–1096.
Sonn, J., and A. Mayevsky. 1995. A new multiprobe assembly for studying the propagating cortical spreading depression. *Journal of Cerebral Blood Flow & Metabolism* 15:S318.
Sonn, J., and A. Mayevsky. 2000. Effects of brain oxygenation on metabolic, hemodynamic, ionic and electrical responses to spreading depression in the rat. *Brain Research* 882:212–216.
Sonn, J., and A. Mayevsky. 2001. The effect of ethanol on metabolic, hemodynamic and electrical responses to cortical spreading depression. *Brain Research* 908:174–186.
Stafford, A., and J. A. Weatherall. 1960. The survival of young rats in nitrogen. *Journal of Physiology* 153:457–472.
Stern, M. D., D. L. Lappe, P. D. Bowen, J. E. Chimosky, G. A. Holloway, Jr., H. R. Keiser, and R. L. Bowman. 1977. Continuous measurement of tissue blood flow by laser-Doppler spectroscopy. *American Journal of Physiology* 232:H441–H448.
Stocchetti, N., M. Barbagallo, G. C. Bellini, A. Furlan, A. Vezzani, and V. Nizzoli. 1991. Arteriojugular difference of oxygen and intracranial pressure in comatose, head injured patients. II. Clinical correlations. *Minerva Anestesiologica* 57:327–334.
Strong, A. J., and R. Dardis. 2005. Depolarisation phenomena in traumatic and ischaemic brain injury. *Advances and Technical Standards in Neurosurgery* 30:3–49.
Strong, A. J., M. Fabricius, M. G. Boutelle, S. J. Hibbins, S. E. Hopwood, R. Jones, M. C. Parkin, and M. Lauritzen. 2002. Spreading and synchronous depressions of cortical activity in acutely injured human brain. *Stroke* 33:2738–2743.
Sukoff, M. H., S. A. Hollin, O. E. Espinosa, and J. H. Jacobson, 2nd. 1968. The protective effect of hyperbaric oxygenation in experimental cerebral edema. *Journal of Neurosurgery* 29:236–241.
Sukoff, M. H., and R. E. Ragatz. 1982. Hyperbaric oxygenation for the treatment of acute cerebral edema. *Neurosurgery* 10: 29–38.

Sullivan, H. G., J. Martinez, D. P. Becker, J. D. Miller, R. Griffith, and A. O. Wist. 1976. Fluid-percussion model of mechanical brain injury in the cat. *Journal of Neurosurgery* 45:521–534.

Sullivan, P. G., A. G. Rabchevsky, P. C. Waldmeier, and J. E. Springer. 2005. Mitochondrial permeability transition in CNS trauma: Cause or effect of neuronal cell death?. *Journal of Neuroscience Research* 79:231–239.

Sunami, K., T. Nakamura, Y. Ozawa, M. Kubota, H. Namba, and A. Yamaura. 1989. Hypermetabolic state following experimental head injury. *Neurosurgical Review* 12(Suppl 1):400–411.

Swaab, D. F., and K. Boer. 1972. The presence of biologically labile compounds during ischemia and their relationship to the EEG in rat cerebral cortex and hypothalamus. *Journal of Neurochemistry* 19:2843–2853.

Sylvia, A. L., and M. Rosenthal. 1979. Effects of age on brain oxidative metabolism *in vivo*. *Brain Research* 165: 235–248.

Takai, D., K. Inoue, H. Shisa, Y. Kagawa, and J. Hayashi. 1995. Age-associated changes of mitochondrial translation and respiratory function in mouse brain. *Biochemical and Biophysical Research Communications* 217:668–674.

Takano, K., L. L. Latour, J. E. Formato, R. A. Carano, K. G. Helmer, Y. Hasegawa, C. H. Sotak, and M. Fisher. 1996. The role of spreading depression in focal ischemia evaluated by diffusion mapping. *Annals of Neurology* 39:308–318.

Tavazzi, B., S. Signoretti, G. Lazzarino, A. M. Amorini, R. Delfini, M. Cimatti, A. Marmarou, and R. Vagnozzi. 2005. Cerebral oxidative stress and depression of energy metabolism correlate with severity of diffuse brain injury in rats. *Neurosurgery* 56: 582–589. discussion 582–589.

Thom, S. R., and L. W. Keim. 1989. Carbon monoxide poisoning: A review epidemiology, pathophysiology, clinical findings, and treatment options including hyperbaric oxygen therapy. *Journal of Toxicology, Clinical Toxicology* 27:141–156.

Thomas, S., F. Tabibnia, M. U. Schuhmann, T. Brinker, and M. Samii. 2000. ICP and MABP following traumatic subarachnoid hemorrhage in the rat. *Acta neurochirurgica. Supplement* 76:203–205.

Thompson, H. J., J. Lifshitz, N. Marklund, M. S. Grady, D. I. Graham, D. A. Hovda, and T. K. Mcintosh. 2005. Lateral fluid percussion brain injury: A 15-year review and evaluation. *Journal of Neurotrauma* 22:42–75.

Thurston, J. H., and D. B. Mcdougal, Jr. 1969. Effect of ischemia on metabolism of the brain of the newborn mouse. *American Journal of Physiology* 216:348–352.

Toyoda, K., K. Fujii, Y. Takata, S. Ibayashi, M. Fujikawa, and M. Fujishima. 1997. Effect of aging on regulation of brain stem circulation during hypotension. *Journal of Cerebral Blood Flow and Metabolism: Official Journal of the International Society of Cerebral Blood Flow and Metabolism* 17:680–685.

Vagnozzi, R., A. Marmarou, B. Tavazzi, S. Signoretti, D.DiPierro, F.DelBolgia, A. M. Amorini, G. Fazzina, S. Sherkat, and G. Lazzarino. 1999. Changes of cerebral energy metabolism and lipid peroxidation in rats leading to mitochondrial dysfunction after diffuse brain injury. *Journal of Neurotrauma* 16:903–913.

Van Harreveld, A. 1972. The extracellular space in the vertebrate central nervous system. In *The structure and function of nervous tissue*, ed. G. H. Bourne. New York: Academic Press.

VanPutten, H. P., M. G. Bouwhuis, J. P. Muizelaar, B. G. Lyeth, and R. F. Berman. 2005. Diffusion-weighted imaging of edema following traumatic brain injury in rats: Effects of secondary hypoxia. *Journal of Neurotrauma* 22:857–872.

Verma, A., D. J. Hirsch, C. E. Glatt, G. V. Ronnett, and S. H. Snyder. 1993. Carbon monoxide: a putative neural messenger. *Science* 259:381–384.

Vlodavsky, E., E. Palzur, M. Feinsod, and J. F. Soustiel. 2005. Evaluation of the apoptosis-related proteins of the BCL-2 family in the traumatic penumbra area of the rat model of cerebral contusion, treated by hyperbaric oxygen therapy: A quantitative immunohistochemical study. *Acta Neuropathologica* 110:120–126.

Wadhwani, K. C., S. I. Rapoport, A. P. Shepherd and P. A. Oberg. 1990. Blood flow in the central and peripheral nervous systems. In *Laser Doppler blood flowmetry*, eds. A. P. Shepherd and P. A. Oberg. Boston: Kluwer Academic Pub.

Wallace, D. C. 1992. Mitochondrial genetics: A paradigm for aging and degenerative diseases?. *Science* 256:628–632.

Walter, B., P. Brust, F. Füchtner, M. Müller, R. Hinz, H. Kuwabara, H. Fritz, U. Zwiener, and R. Bauer. 2004. Age-dependent effects of severe traumatic brain injury on cerebral dopaminergic activity in newborn and juvenile pigs. *Journal of Neurotrauma* 21:1076–1089.

Wang, R., and L. Wu. 1997. The chemical modification of KCa channels by carbon monoxide in vascular smooth muscle cells. *Journal of Biological Chemistry* 272:8222–8226.

Wang, Y. J., T. Shimura, S. Kobayashi, A. Teramoto, and S. Nakazawa. 1997. A lateral fluid percussion model for the experimental severe brain injury and a morphological study in the rats. *Nihon Ika Daigaku Zasshi* 64:172–175.

Wang, Y. J., A. H. WU, L. Qian, and E. AL. 1998. Changes in cerebral blood flow and glucose utilization following fluid percussion brain Injury in rats. *Zhonghua Shenjingwaike Zazhi* 14:61–64.

Wiebers, D. O., H. P. Adams, Jr., and J. P. Whisnant. 1990. Animal models of stroke: Are they relevant to human disease?. *Stroke* 21:1–3.

Wilkinson, J. H., J. W. Hopewell, and H. S. Reinhold. 1981. A quantitative study of age-related changes in the vascular architecture of the rat cerebral cortex. *Neuropathology and Applied Neurobiology* 7:451–462.

Wolinsky, H.. 1972. Long-term effects of hypertension on the rat aortic wall and their relation to concurrent aging changes. Morphological and chemical studies. *Circulation Research* 30:301–309.

Zarchin, N., and A. Mayevsky. 1979. The effects of age on cerebral metabolic and electrical responses to complete ischemia in the awake and anesthetized rat. *Israel Journal of Medical Science* 15:17.

Zarchin, N., and A. Mayevsky. 1981. The effects of age on the metabolic and electrical responses to decapitation in the awake and anesthetized rat brain. *Mechanisms of Ageing and Development* 16:285–294.

Zauner, A., W. P. Daugherty, M. R. Bullock, and D. S. Warner. 2002. Brain oxygenation and energy metabolism: Part I-biological function and pathophysiology. *Neurosurgery* 51: 289–301. discussion 302.

Zhang, J., and C. A. Piantadosi. 1992. Mitochondrial oxidative stress after carbon monoxide hypoxia in the rat brain. *Journal of Clinical Investigation* 90:1193–1199.

Chapter 9
General Discussion and Conclusions

Abstract This book is devoted to the introduction of a new concept named—**Cerebral Cortex Physiological Mapping**. The meaning of this new term is to monitor various physiological parameters from a relatively small brain cortical tissue in real time. This approach enables a better understanding of the physiological activities taking place in a typical brain tissue and that are recorded in real time for a long time (many hours). The meaning of the word "Mapping" is to record the metabolic, hemodynamic, ionic and electrical activities in a typical cortical tissue (up to eight parameters) and in real time. This new experimental approach provides results that enable the understanding of the real-time relationship between various physiological activities under induced Neuropathological states. In addition to better understanding the physiological activities, this approach could be used in the development of new drug development as well as other treatments. Our new approach is a unique one and may contribute to the development of new drug or treatments to patients.

Keywords Mitochondrial NADH · Real time brain monitoring · Brain oxygen balance · Multiparametric brain monitoring · Brain metabolism

9.1 Introduction

9.1.1 From Isolated Mitochondria to Intact Cerebral Cortex Physiological Mapping

In 1973, Chance, Oshino, Sugano, and Mayevsky published a review paper (Chance et al. 1973a) entitled "Basic Principles of Tissue Oxygen Determination from Mitochondrial Signals."

The first subject that was discussed in this review was the following:

A. Mayevsky et al., *Physiological Mapping of the Cerebral Cortex in Neuropathology*,
https://doi.org/10.1007/978-3-032-13019-8_9

9.1.2 Mitochondrial Indication of Intracellular pO_2

"However, the organelles that surely exhibit the intracellular oxygen tension are the ones that are using the oxygen. These are the mitochondria. They do not need to be reached by electrodes; they emit spectroscopically detectable signals themselves. It is the purpose of this paper to show how these signals may be read. The accumulation of evidence since the pioneer work of Otto Warburg on 'Atmungsferment' (Warburg 1949) and David Keilin on cytochromes (Keilin 1966) as the keystones of cellular oxygen utilization led us to a study of redox states of electron carriers in isolated mitochondria as function of oxygen concentration, and to develop techniques to measure the states of anoxia and normoxia in living tissues."

"The morphology of the system is a primary consideration, and we shall begin by showing where the mitochondria are situated and how they may reveal on account of this location. The ubiquity of mitochondria in eukaryotic cells is due to their superlatively efficient energy conservation system. Often, one-third or more of the volume of an active tissue, such as mammalian cardiac tissue, may be filled with mitochondria. Similarly, the oxygen utilization of the brain as compared with that of the whole body stands as testimony to mitochondrial activity in that organ. Thus, there is no doubt that the readout of mitochondrial redox states takes us to the very source and furthermore, the cell membrane does not need to be penetrated in order to do so."

The second relevant subject discussed was:

9.1.3 Sensitive Fluorometry of Tissue NADH

"Having identified that NADH, being the most sensitive component of the respiratory chain to oxygen, is most competent to serve as an indicator of intracellular oxygen concentrations, and having determined the degree of NADH oxidation required for energy-linked function, we may consider how the oxidation-reduction state of NADH may be measured. This component is, in fact, a natural fluorochrome; excitation in the range 340–370 nm gives rise to fluorescence emission of the reduced form with a peak centered around 450 nm. Excitation of NADH fluorescence at 340 nm gives an approximately fourfold increase of fluorescence intensity in the transition from the aerobic, substrate-free state to the anaerobic state of the mitochondria."

In the same paper Chance et al discussed the following subject:

9.1.4 Simultaneous Recordings of Reflectance and Fluorescence

"In order, however, to estimate the reflectance change in a continuous manner during normoxic-anoxic-normoxic cycles, we record a reflectance signal together with the fluorescence emission excited at the same wavelength, as was done some time ago in electric stimulation of slices of the organ of the electric fish in experiments together with Aubert et al. (1964). Jobsis et al. (1971) used 366 nm as a wavelength for monitoring reflectance for some time while Kobayashi et al. (1971) used 620 nm; in the latter case, a computed function of the reflectance was used to correct the fluorescence trace. In this laboratory, Harbig et al. (1976) employed combined reflectance at 366 and 450 nm, based upon a modification of the Ultropak apparatus employed by Chance et al. (1963); these results will appear elsewhere (Harbig et al. 1976). The apparatus described in this accompanying paper by Mayevsky and Chance (1973a) employs reflectance at 366 nm with a linear correction term at 1:1 gain."

"The charts illustrate the current state of technology available with a time-sharing fluorometer based upon an air turbine containing the excitation and emission filters. The turbine employed in the experiments of Fig. 9.1 contains filters appropriate to 366 nm excitation and to reflectance measurements at 366 and 577 nm. The latter wavelength is also employed for the excitation of electrochromic probes, the results of which are reported elsewhere. The fluorescence is picked up by a separate photomultiplier containing a narrow band secondary filter at 445 nm (as close to the 448 nm wavelength indicated in Fig. 9.1 as feasible (Chance et al. 1973a)) of 5 nm half-bandwidth in Fig. 9.1a, and in Fig. 9.1b through a broad-band filter combination comprised of Wratten 47 and a Wratten 2C, which has half-bandwidth points at 420 and 480 nm, with a peak near 450 nm."

During the same year (Chance et al. 1973a), Mayevsky and Chance published the following paper:

A New Long-Term Method for The Measurement of NADH Fluorescence in Intact Rat Brain with Chronically Implanted Cannula.

9.1.5 *Introduction*

The method developed by Chance et al. (1962a) which measures the oxidation-reduction state of NADH in vivo by surface fluorometry is in use in many laboratories. In all cases, acute preparations are used.

The purpose of the present study has been to develop a method by which the NADH fluorescence could be measured on the brain surface of chronically operated rats, and of unanesthetized rats as well. Using this method, one can measure the NADH fluorescence in the same animal over an interval of several days and compare the effect of various physiological and behavioral situations on the oxidation-reduction state of NADH in the brain.

9.1.6 *Methods*

The air-turbine-driven time-sharing fluorometer designed by Chance et al. (1973a) was employed here. This fluorometer permits the measurement of NADH fluorescence and light scattering changes at the excitation wavelength (366 nm) on the surface of

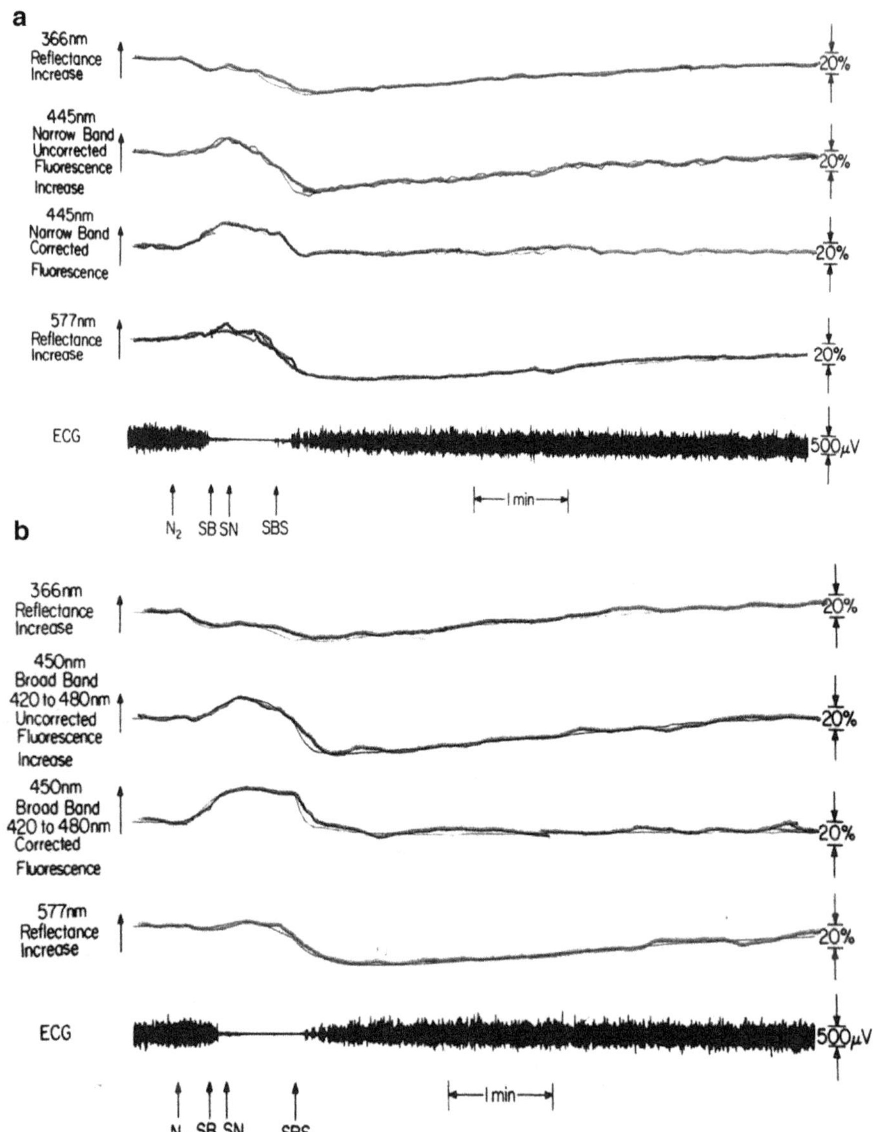

Fig. 9.1 Combined reflectance and fluorescence measurements of the rat brain cortex in cycles of normoxia-anoxia-normoxia. Part (**a**) shows results obtained with the narrow-band filter (5 nm half- bandwidth) centered at 445 nm and taken in the early interval of experimentation, i.e., with a fresh animal. Part (**b**) represents similar data obtained later—in the course of experimentation on the same animal, but employing a wide-band emission filter of half-bandwidth extending from 420 to 480 nm, and peaking at 450 nm. The calibrations appropriate to the eight traces are similar, as are the calibrations for the ECG's (Electrocorticography). Note that increasing intensity is indicated as an upward deflection in this figure, as contrasted with Figs. 9.3–9.5 (Mayevsky and Chance 1973a) of the accompanying paper, where increases are indicated as downward deflections. Symbols: N_2 inspiration of 100% nitrogen, SB stop breathing, SN stop nitrogen, SBS start breathing simultaneously

the brain cortex. In this mode of operation, the light is guided by a Y-shaped flexible (quartz fiber) light pipe with a total length of 20 cm and a common part of 5 cm. The outer quartz fiber diameter of the common part is 4 mm and the inner quartz fiber diameter is 2 mm for each limb, one limb carries the 366 nm excitation light to the cannula and the other carries the fluorescence emission (450 nm) and reflectance (366 nm) back to the air-driven disk where time-sharing wavelength resolution is accomplished by appropriate filters (Chance et al. 1970). The Plexiglas cannula which fits the common part of the light pipe is shown in Fig. 9.2. The thread at the bottom of the cannula enables the cannula to be screwed into the skull and to make a better connection with the acrylic cement. A 0.2 mm step holds the light pipe in the cannula at a constant distance from the brain.

Male Wistar rats were used (200–250 g). The animal was anesthetized (Equithesin, 0.75 ml) and mounted to a head-holder. The skull was exposed and a hole of 6 mm diameter was drilled in the parietal bone on the right side of the brain. Only animals that showed a clear, hemoglobin-free brain surface were used. The cannula was screwed into

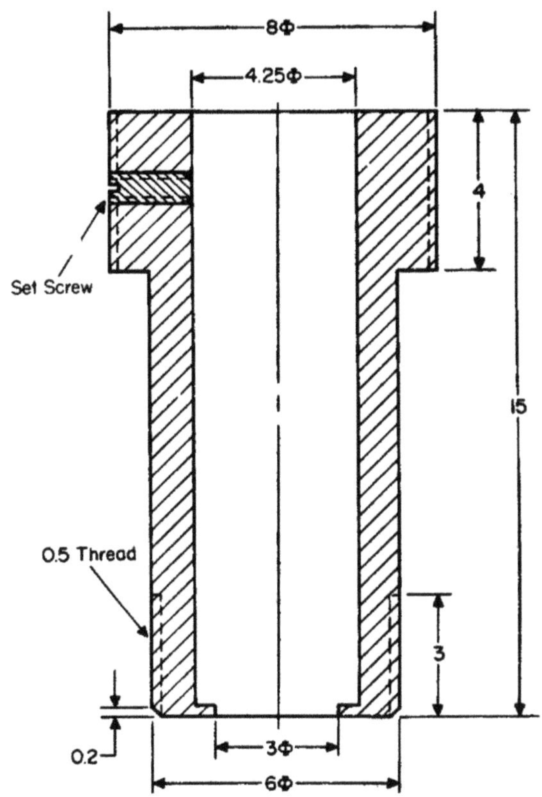

Fig. 9.2 The construction of the implanted cannula. All dimensions in millimeters. The set-screw is used to fix the light pipe to the brain surface

the skull (1 mm deep) above the dura. Two metal screws were used for EEG recording from the contralateral as well as from the ipsilateral hemisphere. Acrylic dental cement fixed the cannula and the screws to the skull. A suitable Plexiglas plug closed the exposed area until measurements were made. Blood pressure was measured by cannulation of the femoral artery. In the chronic preparation, blood pressure was not measured.

During measurements, the common part of the light pipe was plugged into the cannula in contact with the dura to minimize brain movement or fluid accumulation, which would interfere with the measurement. A nitrogen cycle was produced by exposing the animal to 100% N_2 for 60 s. During the N_2 flushing, the animal stopped breathing, and 20–30 s later the animal started breathing spontaneously.

A spreading depression was obtained by flushing a solution of 0.2–0.4 M KCl through a small reservoir in the bottom of the cannula so that the solution was localized externally.

9.1.7 Results and Discussion

Two distinctive conditions of the brain cortex were observed and interpreted to differ with respect to the microcirculatory response to anoxia. The first and simpler case occurs in the later phases of repetitive anoxia where, as shown in Fig. 9.3, there are scarcely any systematic changes in the reflectance trace (top) as compared with the uncorrected fluorescence trace (second from top) during the cycle of anoxia, both traces

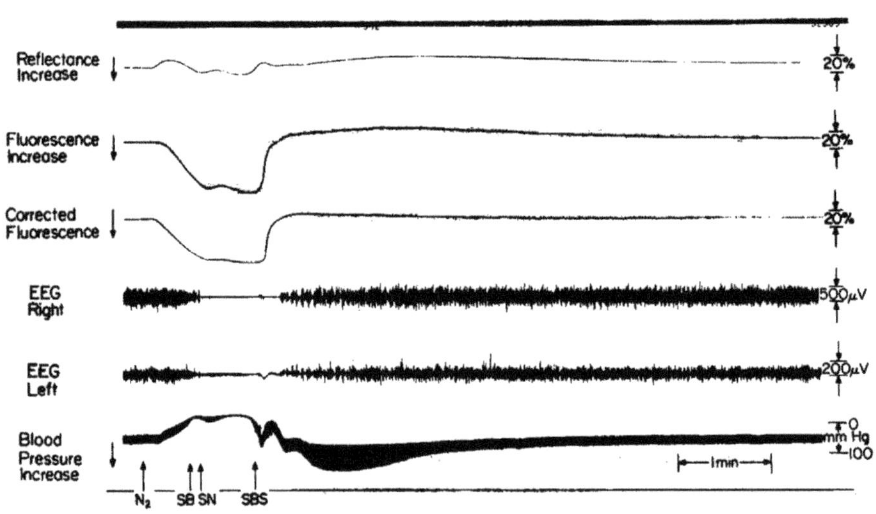

Fig. 9.3 A typical response of the rat brain and the systemic blood pressure to a nitrogen cycle (1/12). The corrected fluorescence trace is obtained by subtracting the reflectance change from the fluorescence change on a 1:1 basis. SB stop breathing, SN stop nitrogen, SBS start breathing spontaneously

being recorded on the same sensitivity (percent of maximum signal). Therefore, the corrected fluorescence trace (third from top) appears nearly identical to the uncorrected trace, the reflectance being subtracted from the uncorrected fluorescence at a 1:1 ratio (Jobsis et al. 1971; Harbig and Rievich 1973).

The characteristic transition from normoxia to anoxia is slow and extends over an interval of about 30 s. The EEG maintains a constant level until the fluorescence increase reaches approximately 80% of maximum, and by the time 95% of maximal fluorescence is reached, no further EEG can be observed. The transition from normoxia to anoxia is slow because the oxygen tension is reduced from the normoxic value to a very low value by a combination of washout of the oxygen and tissue oxygen consumption. The return from anoxia to normoxia is very rapid, since the "front" of oxygen concentration in the freshly oxygenated hemoglobin is high, and the very small span of mitochondrial oxidation-reduction states (from 5×10^{-8} to 5×10^{-7} M oxygen) is bridged in a short time. As indicated in an accompanying paper (Chance et al. 1973a), calibration of the NADH fluorescence increase in isolated mitochondria identifies the point of 50% reduction of NAD with an oxygen concentration of approximately 10^{-7} M (0.07 mm) and with virtual cessation of energy coupling. Thus, only about 10 s elapse between the cessation of energy coupling, the complete reduction of NADH, and the cessation of the EEG signal. The conditions of this experiment are very similar to those employed by Chance and Schoener (1963a, 1963b) whose uncorrected fluorescence traces were also taken in later phases of the anoxic response where reflectance artifacts are negligible.

The response to an early anoxia (#1 for this particular animal) is indicated in Fig. 9.4. Here, the blood pressure pattern is generally similar but the microcirculatory response is such that shortly after the beginning of the pyridine nucleotide reduction,

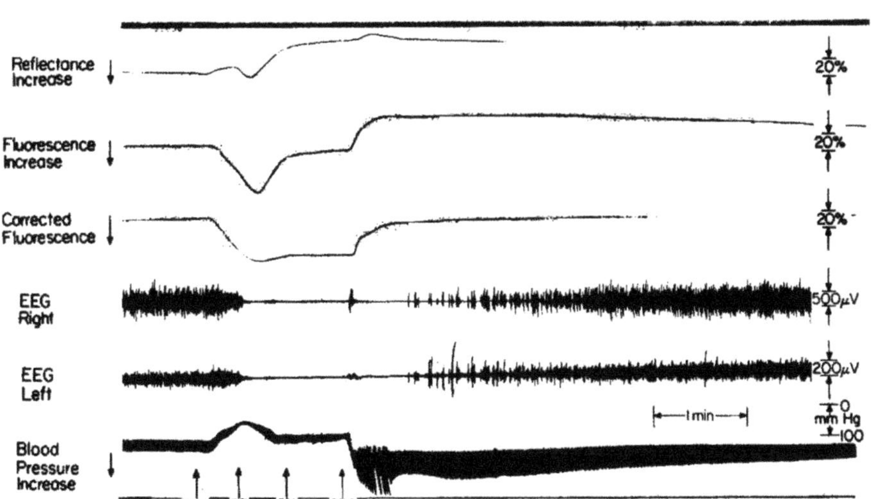

Fig. 9.4 The response of the rat brain to the first nitrogen cycle applied to the animal is shown in Fig. 9.3

a large reflectance decrease causes a fluorescence decrease of comparable magnitude. These effects very nearly counter-balance each other on a 1:1 basis on the corrected trace, with similar changes of reflectance also appropriately cancelled out in the recovery from anoxia. Other features of the recovery from anoxia are the same as in Fig. 9.3, except it should be noted that the recovery of the EEG is considerably delayed.

This change in the reflectance from cycle #1 to #12 may be caused by some damage to the auto-regulatory mechanism controlling the blood flow to the brain. The arterial blood pressure of this animal is not involved, since quite stable values are observed from cycle #1 to cycle #12 during this interval, 110–80 mmHg in the first nitrogen cycle of Fig. 9.4, as compared with 110–90 mmHg in the 12th cycle shown in Fig. 9.3. The amplitude of the EEG cycle decreases with time.

Preliminary experiments have been done in order to correlate the redox state of NADH with the level of electrical activity of the brain. In these experiments, we perfused a small area of the dura (2 mm^2) with 0.4 KCl solution. The distance from the perfused area to the NADH fluorescence measurement was 2 mm. A typical response to spreading depression is shown in Fig. 9.5.

The animal was breathing pure oxygen or normal air during this part of the experiment, so the oxygen supply was in the normoxic range. The change in the reflectance and the NADH fluorescence appears 30–40 s after the application of the KCl solution. The changes in the right EEG (upper) begin even earlier since the recording was made from a point near the KCl application site. From the changes in the reflected light, we assume that the response to spreading depression is biphasic; in the beginning there is a short increase in reflectance, followed by a longer decrease. The large reflectance changes introduce an artifact into the fluorescence measurement, and the corrected trace shows a small NADH oxidation. In some animals where the reflectance changes were smaller, we saw an NADH oxidation even in the uncorrected fluorescence trace.

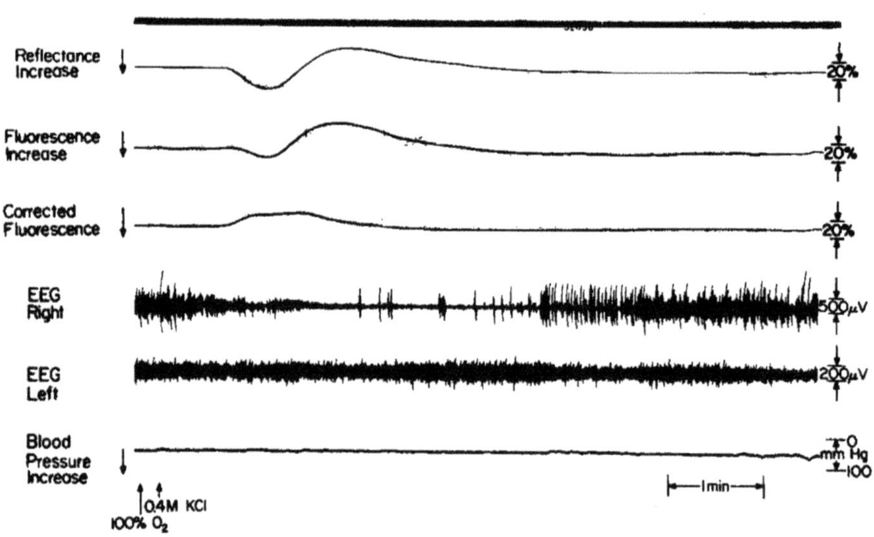

Fig. 9.5 The response of the brain to the application of KCl (0.4 M) on the dura surface

The recovery of the corrected fluorescence trace from spreading depression is much earlier than that of the EEG. The recovery of the reflectance and of the uncorrected fluorescence trace is very close to that of the EEG. It seems that the response to spreading depression begins in a short vasoconstriction response, followed by a longer vasodilation, as suggested also by van Harreveld and Ochs (1957). It is also apparent that EEG activity in the contralateral side of the brain (left side) was unaffected by the spreading depression on the right side. The blood pressure was stable during this interval. We assume that the response to spreading depression contains changes in blood flow or volume as well as changes in oxidation–reduction state of NADH in the brain. That the oxidation of NADH is due to an increase in oxygen consumption is in agreement with other findings in relation to the energy change during spreading depression. Lukyanova and Bures (Lukyanov and Bures 1967), found an increase in oxygen consumption during spreading depression. Davies and Remond (1946), using the oxygen electrode, showed an increase in oxygen consumption during convulsions. Rosenthal and Jobsis reported that direct electrical stimulation of the cortex led to NADH oxidation (Rosenthal and Jobsis 1971) and Brauser et al. (1970) showed that stimulating the isolated rat ganglion led to oxidation of NADH.

In summary, we have presented here a new method which enables us to measure the oxidation-reduction state of the brain. One can use this method in chronic operated rats as well as in acute ones. It also permits measurements on an awake animal.

9.2 From a Single Parameter to Various Multiparametric Monitoring Systems

The development of the optical fibers, that transmit effectively the UV light, open up a new era in which the measurement of NADH fluorescence together with other brain physiological parameter became practical.

Table 9.1 presents the milestones in the development of various brain multiparametric monitoring systems in a relatively small tissue area. As can be seen in Fig. 1.1 (in Chapter 1 of this book), our approach aims to monitor, in real-time, a small volume of the cerebral cortex containing all the tissue elements that are parts of a typical functioning brain. We are interested in the microenvironment of the brain containing neurons, glia, synapses, and the microcirculatory elements (small arterioles and capillaries). The various parameters and the technology developed are presented in Fig. 1.1.

During the development process, we pursued the goal of being minimally invasive in terms of the penetration to the cortical tissue itself. It was obvious that the various probes could not monitor the same volume of tissue due to the size of each probe used. Therefore, we attempted to minimize the diameter of the various probes located in the MPA that had a 5–6 mm contact area with the cerebral cortex. In most of the perturbations used, such as global ischemia, anoxia, hypoxia or hemorrhage, most of the areas in the cortex will respond in the same way.

The list in Table 9.1 is organized in chronological order. All details of the technological aspects and animal preparation appear in the original relevant publications; therefore, a short description of technology relevant to each parameter will appear in the initial part of the methods section. In the results section, typical responses to various types of perturbations will be presented together with the technology used in the specific study. Our approach was to develop a new upgraded version of the monitoring system and presenting initial preliminary results. The next step was to run a large well-designed study on a few groups of animals, and the data were quantitated and analyzed for its statistical significance.

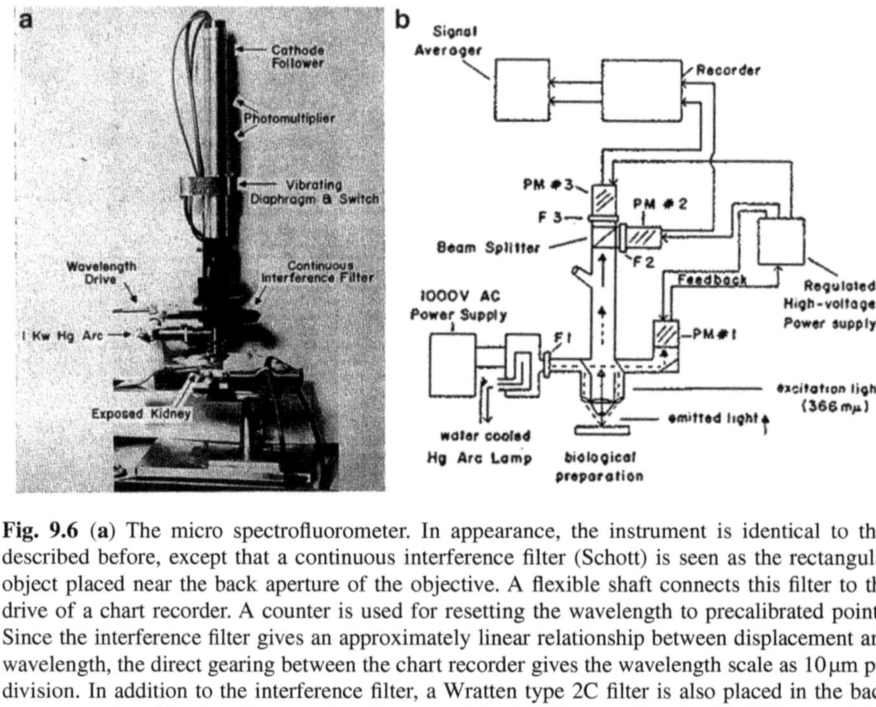

Fig. 9.6 (**a**) The micro spectrofluorometer. In appearance, the instrument is identical to that described before, except that a continuous interference filter (Schott) is seen as the rectangular object placed near the back aperture of the objective. A flexible shaft connects this filter to the drive of a chart recorder. A counter is used for resetting the wavelength to precalibrated points. Since the interference filter gives an approximately linear relationship between displacement and wavelength, the direct gearing between the chart recorder gives the wavelength scale as 10 μm per division. In addition to the interference filter, a Wratten type 2C filter is also placed in the back aperture of the objective. The wavelength range of the interference filter is 400–700 μm, and the specification on its spectral interval is 30–40 μm. In order, however, to compensate somewhat for the loss of light intensity, the measuring aperture has been increased from 15 to 30 mμ. Other features of the high-pressure mercury arc excite the fluorescence of the specimen at 366 mμ by means of an 'Eppendorf' primary filter. Fluorescence excitation and emission pass through the Leitz Ultrapak objective (×11) and a ×3 ocular. Olson, Loeser, and West have described rapid scanning spectrofluorometers which could be suitably modified for these studies (Chance et al. 1962b). (**b**) Final form of the regulated, differential fluorometer. The 366 nm line from the water-cooled AH-6 mercury arc lamp is transmitted by the primary filter (F1) after previous removal of IR radiation by a heat filter within the arc lamp housing (not shown). A small fraction of this light passes to the regulating photomultiplier (PM #1). The excitation light is focused as a cone on the preparation. Diffusely reflected excitation light and emitted fluorescence, enter the microscope optics, are divided by the beam splitter, and measured after passage through the secondary filter, F2 or F3, which respectively, allow passage of the 366 nm and the 465 regions (Jobsis et al. 1971)

Table 9.1 Milestones in the development of brain multiparametric monitoring of NADH fluorescence and other physiological parameters in vivo by Mayevsky et al.

Protocol	Year	Discovery/activity	Author(s)
1	1973	The first fiber optic-based fluorometer-reflectometer was used in the brain of an unanesthetized animal. Monitoring of NADH and electrocortical activity (ECoG)	Chance et al. (1973a) and Mayevsky and Chance (1973a)
2	1974, 1977	Simultaneous monitoring of NADH in vivo together with extracellular K^+ (microelectrode and surface electrode) and ECoG	Mayevsky et al. (1974) and Crowe et al. (1977)
3	1980	Monitoring of brain NADH together with tissue pO_2 and ECoG	Mayevsky et al. (1980)
4	1982	The first multiparametric assembly for NADH, extracellular K^+, H^+, DC steady potential, and ECoG	Friedli et al. (1982)
5	1983	Monitoring of NADH, pO_2, extracellular K^+, DC and ECoG inside hyperbaric chamber	Mayevsky (1983)
6	1990, 1992	Simultaneous real time monitoring of brain NADH, HbO_2, ECoG, DC potential, extracellular K^+, and Ca	Mayevsky et al. (1990, 1992)
7	1995, 1996	Simultaneous monitoring of brain NADH, CBF, ECoG, DC potential, extracellular K^+, Ca^{2+}, H^+	Mayevsky et al. (1995) and Meilin et al. (1996)
8	1996	Multiparametric monitoring of neurosurgical patients	Mayevsky et al. (1996)
9	1997	Monitoring of brain NADH, CBF, DC potential, extracellular K^+, Ca^{2+} together with high energy phosphates by ^{31}P NMR spectroscopy	Mayevsky et al. (1997)
10	2000	Monitoring of the mechanism of CSD propagation	Sonn and Mayevsky (2000)
11	2001	Multiparametric monitoring under ICP elevation	Barbiro-Michaely and Mayevsky (2001)
12	2003	Multiparametric monitoring of rats under traumatic brain injury (TBI)	Rogatsky et al. (2003)
13	2011	Multiparametric monitoring under epileptic activity	Barbiro-Michaely et al. (2011) and Barbiro-Michaely and Mayevsky (2001)

The first in vivo NADH monitoring device was presented in the late 1950s and early 1960s. Figure 9.6a shows an initially developed fluorometer (Chance et al. 1962b). At that stage, the effects of scattered light and tissue absorption due to blood were not taken into consideration when monitoring NADH fluorescence. The first detailed results of in vivo NADH fluorescence measurements were published in 1962 (Chance et al. 1962a, 1962b). Later on, the monitoring of reflectance was added as seen in Fig. 9.6b (Jobsis et al. 1971). Figures 9.6–9.15 illustrate the various monitoring systems presented in Table 9.1.

9.2.1 *Monitoring NADH Fluorescence and Reflectance*

In order to enable the monitoring of NADH fluorescence in unanesthetized animals or other in vivo preparations, a flexible means was needed to connect the fluorometer with the tested organ, for example, the brain. This was achieved in 1972, when UV-transmitting quartz fibers became available (Schott Jena Glass, Germany). We have used the light-guide-based fluorometer for in vivo monitoring of the brain (Chance et al. 1973b; Mayevsky and Chance 1973b) subjected to anoxia or cortical spreading depression. The original device functioned on the time-sharing principle (Fig. 9.7a), where four filters were placed in front of a two arms light guide. Filters two and three enabled the measurement of NADH fluorescence, while filters two and four were used to measure tissue reflectance at the excitation wavelength. The reflectance trace was used to correct the NADH signal for hemodynamic artifacts and to indicate changes in the blood volume of the sampled tissue.

In this original system, only one photomultiplier tube was used for the detection of the two signals. In order to simplify the monitoring system, the time-sharing approach (AC mode) was replaced by splitting the light emitted from the tissue into two unequal fractions for the measurement of fluorescence and reflectance signals. This model, named the DC type fluorometer, had originally a three-way light guide which was later replaced by a two arms light guide probe (Fig. 9.7b). In all configurations, the reflectance signal was used for the correction of the fluorescence signal.

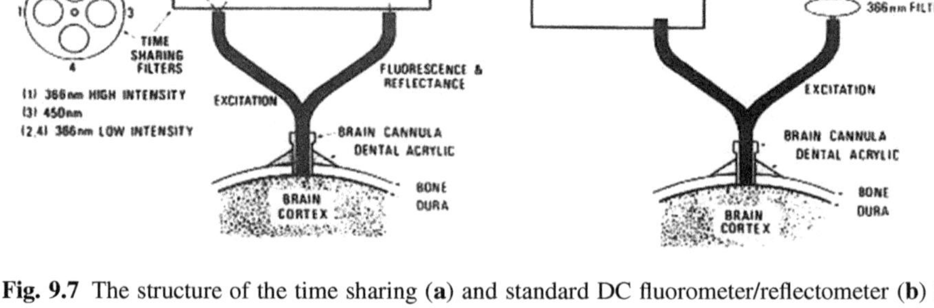

Fig. 9.7 The structure of the time sharing (**a**) and standard DC fluorometer/reflectometer (**b**) Ex excitation, Em emission optical fibers for the monitoring of NADH redox state. H.V. high voltage, PM photomultiplier (Mayevsky 1984)

9.2.2 NADH and pO_2 Measurements

Since Mitochondrial NADH is sensitive to intracellular levels of oxygen, we had developed an MPA that contained a surface oxygen electrode in combination with NADH and ECoG (Fig. 9.8a). It is important to note that oxygen electrode readings average the oxygen level in the vascular, extracellular, and intracellular compartments (Mayevsky et al. 1980).

9.2.3 The First Multiparametric Monitoring System

Our aim was to provide a new holding system allowing an easy implantation of surface probes in various combinations on the brain of small mammals and to reevaluate the potential of such surface electrode assemblies in tracing physiological and pathological events (Friedli et al. 1982). In this stage, we added to the basic assembly system the measurements of extracellular tissue pH and K^+, DC potential, and local temperature, in addition to the other parameters, including NADH fluorescence, reflected light, and electrocorticogram (ECoG). Ideally, the multiprobe assembly designed (Fig. 9.8b) for our project had to be rugged enough to withstand routine

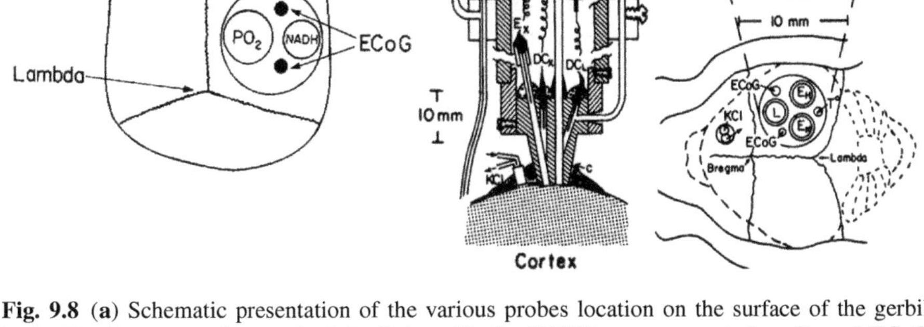

Fig. 9.8 (**a**) Schematic presentation of the various probes location on the surface of the gerbil brain. The large cannula contained the light guide for NADH measurement, the pO_2, and ECoG electrodes. The cannula for KCl application was located 2–3 mm anterior to the large cannula (Mayevsky et al. 1980). (**b**) Presentation of the first developed multiprobe assembly. Left: longitudinal section of the cannula and assembly located above the cortex. A small push-pull cannula was placed near the assembly to initiate spreading depression. Right: topical view of the gerbil skull together with the location of the various electrodes with a cross-section of the wide end of the cannula (Friedli et al. 1982)

use by non-specialists and yet be miniaturized enough to fit the limited surface available for implantation on rat and gerbil skulls (about 6–7 mm in diameter on each hemisphere). For ischemic studies, our interest was focused on monitoring mean values over large areas, suggesting the use of larger electrodes. In contrast, following propagated events such as spreading depression would have required the tip of the multiprobe assembly to be concentrated in a tiny space, perceiving the same phase of the wave, unless each sensor could be located with respect to the wave front by an auxiliary signal. The second option was retained by recording the wave of the DC potential concentrically to each sensor. In addition, the complete assembly had to provide adequate protection and shielding of the high impedance ion-sensitive electrodes (K^+ and pH) and stable enough reference junctions for use in unanesthetized animals. With K^+ surface electrodes, concentric DC potential has been proposed as the best approximation for the DC potential component to be subtracted from the sensor signal (Crowe et al. 1977). In this study, we also attempted to check the accuracy of the method by recording DC potential differentially between a central barrel and the peripheral slit and to reevaluate its usefulness in correcting surface K^+ and pH measurements under various conditions.

9.2.4 An Upgraded Multiparametric Monitoring System

To elucidate the mechanism of brain O_2 toxicity, it is necessary to measure as many physiological parameters as possible from the same site. The present study aimed to develop a new multiparametric monitoring system located in a hyperbaric chamber, using an awake animal. The basic features of the multiparametric assembly used under normal conditions were described previously (Friedli et al. 1982), and in the present study, we extended and adapted the system for the hyperbaric chamber (Mayevsky 1983).

The upgraded multiparametric assembly can measure up to eight parameters representing the metabolic, ionic, and electrical activities on the cerebral cortex. Metabolic activity is evaluated by measuring tissue O_2 tension and by monitoring the intramitochondrial NADH oxidation-reduction state. The ionic state of the tissue is evaluated by monitoring extracellular potassium (K^+_e) activity. Electrical activity was monitored by bipolar electrocorticography (ECoG) and the DC steady potential. In addition, we monitored the surface temperature with a miniature thermistor.

The rat was placed in the hyperbaric chamber (Bethlehem model 1836-HP) 10 min after the end of the operation and connected to the measuring system (located outside the chamber) at the connection points (C in Fig. 9.9). During the CSD episode the NADH showed a clear oxidation cycle (CF trace).

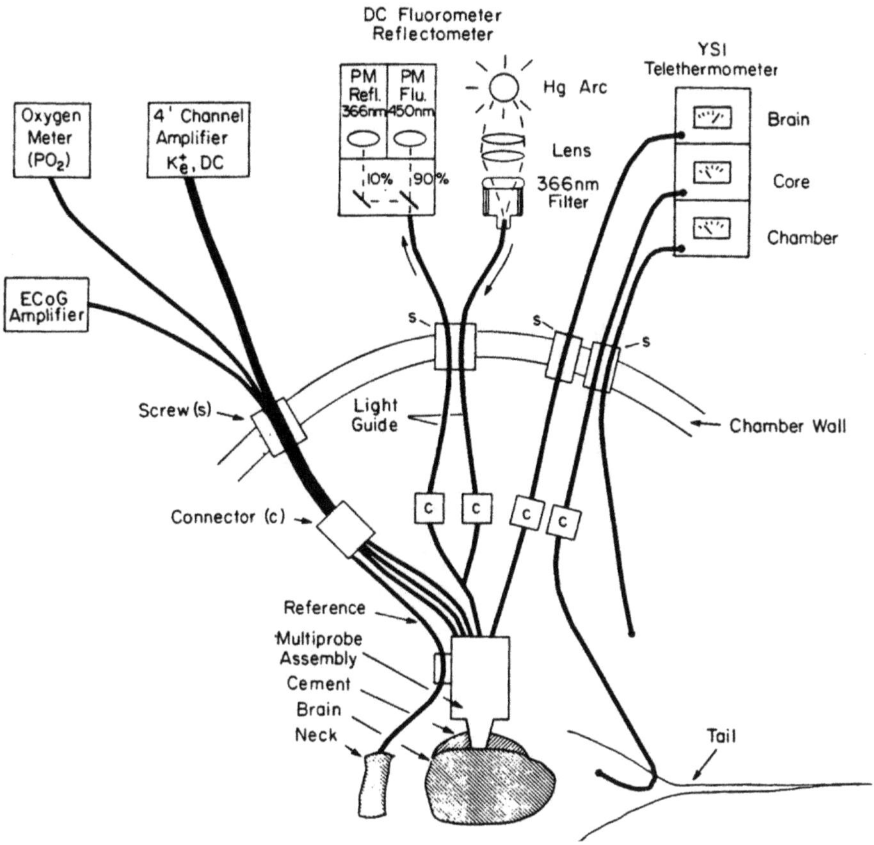

Fig. 9.9 The multiprobe assembly is connected to the rat brain surface while the rat is placed inside a hyperbaric chamber. All the signals were transferred to the monitoring system via the hyperbaric chamber wall (Mayevsky 1983)

9.2.5 The Addition of Hb Saturation to the MPA

The basic concepts of the MPA as previously described (Mayevsky 1983; Friedli et al. 1982) were adopted and modified to accommodate more probes in the same surface area of the brain (Mayevsky et al. 1992; Mayevsky 1988). As shown in Fig. 9.10, the center hole was used for the NADH light guide, which was developed very recently to include fibers for relative CBF measurement by laser Doppler flowmeter (Mayevsky et al. 1991) in addition to NADH fluorescence. In the present study, two light guides were used, namely the central one (L) for NADH and CBF measurements and a microlight guide used for HbO_2 monitoring (Frank et al. 1989). In order to prevent interference between the measurements of HbO_2 and CBF we used them alternatively with the NADH measurements. The longitudinal section of the MPA, shown on the

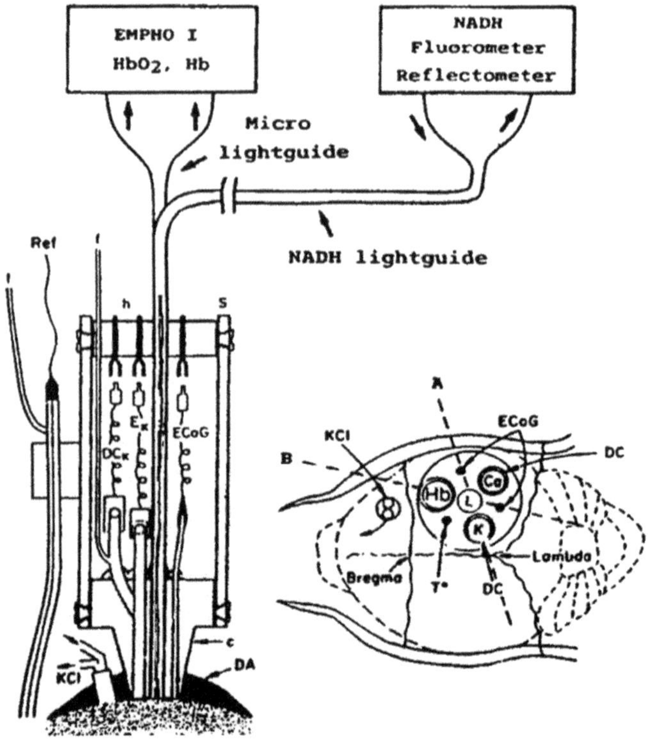

Fig. 9.10 The multiprobe assembly (MPA) that enabled the monitoring of hemoglobin oxygenation together with other parameters. The longitudinal section, at the left, shows the two light guides connected to the EMPHO-I (microlight guide) and the NADH fluorometer and reflectometer (NADH light guide). At the right side of the figure, the various probes are shown as located on the gerbil cerebral hemisphere. L NADH light guide, Hb EMPHO-light guide, K, Ca extracellular potassium and calcium electrodes, DC steady potential electrodes, T temperature probe, ECoG electro-corticography electrodes, KCI push-pull cannula for topical application of KCl solution, Ref reference electrode, DCk, Ek, ECoG cables connecting the various electrodes to the cable holder, f filling tubes connected to the reference and the DC electrodes, C Plexiglas cannula, DA dental acrylic cement, h, s Plexiglas cable holder and sleeve (Mayevsky et al. 1992)

left side of the figure, corresponds to the cross-section line A shown on the right side. In order to see the two light guides in the same longitudinal section, a different sectioning line would need to be used (B). Extracellular K^+ and Ca^{2+} electrodes were made by WPI Inc. (Sarasota, FL, USA) using 1.2 mm diameter PVC tubing (Cole Flex, W. Babylon, NY, USA).

The electrodes were assembled in the holder by the same basic approach described previously for the connection of the DC steady potential electrode (Chance and Schoener 1963a). For the DC electrode, we used the same electrode holder (containing Ag/AgCl pellet) provided by WPI Inc. and polyethylene tubing was connected to it and inserted into the holder. The center light guide (NADH) was covered by a rigid brass tube and was glued by a 5-min epoxy so that after cementation to the cannula (screwed and glued) it served as an axis to hold the cannula (c) and the connector holder (h) at a fixed convenient distance.

9.2.5.1 Hemoglobin Spectrophotometry

In order to evaluate the level of hemoglobin oxygenation, we used the Erlangen microlight guide spectrophotometer EMPHO-1 (Frank et al. 1989). The light source was a water-cooled xenon high-pressure arc lamp (XBO 75W/2 Osram) operated by a starting unit and a power supply (Muller, Moosinning, Germany). The light (400–1200 nm) passing through a heat absorption filter (KGI, Schott), located between two focusing lenses, entered a central 250 μ diameter micro light guide fiber closely surrounded by a hexagon of six detecting fibers (70 μ diameter) and was transmitted to the brain surface. The emitted light from the brain was passed through a rotating interference band pass filter disc (Anders, Nabburg), which works in the spectral domain from 502 to 650 run. All other details regarding data collection, calibration, and calculations were published recently (Frank et al. 1989).

9.2.6 A New Modification of the MPA

The goal of this investigation was to evaluate the responses of brain tissue to CO and to attempt to elucidate differences between the physiological responses and toxic effects (Mayevsky et al. 1995; Meilin et al. 1996). The affinity of CO to hemoglobin is 200 times greater than that of O_2; hence, an obvious aspect of exposure to CO is the possibility of impaired O_2 delivery and cellular hypoxia, depending on the concentration of CO. Because >50% of the energy consumed by the brain is used for the normal functions of active ion transport mechanisms, it is important to evaluate O_2 or energy balance under CO exposure. Oxygen balance can be evaluated by monitoring the level of O_2 supply compared with the O_2 demand. To monitor brain O_2 balance in vivo and in a continuous mode, we adopted a multiparametric monitoring approach (Mayevsky 1975, 1983; Mayevsky and Rogatsky 2007) by which the hemodynamic, metabolic, ionic, and electrical activities could be examined (Fig. 9.11). Our approach was to monitor the brain on the "tissue level" rather than the "cellular level." We believe that the various component elements of the brain, namely, neurons, glia cells, and capillaries, are acting as an integrated system. Therefore, we monitored the functions of the brain by using "mini probes" rather than "microprobes" that are normally used to monitor ionic homeostasis (Mayevsky et al. 1995). Also, the evaluation of other parameters, such as cerebral blood flow (CBF) and NADH redox state, was adapted to the tissue level rather than to the cellular level. Hence, our strategy was to develop a multiparametric monitoring assembly in which all the probes had the same type of contact with a small area of the brain. None of the probes penetrated the brain itself, thus avoiding severe damage to the

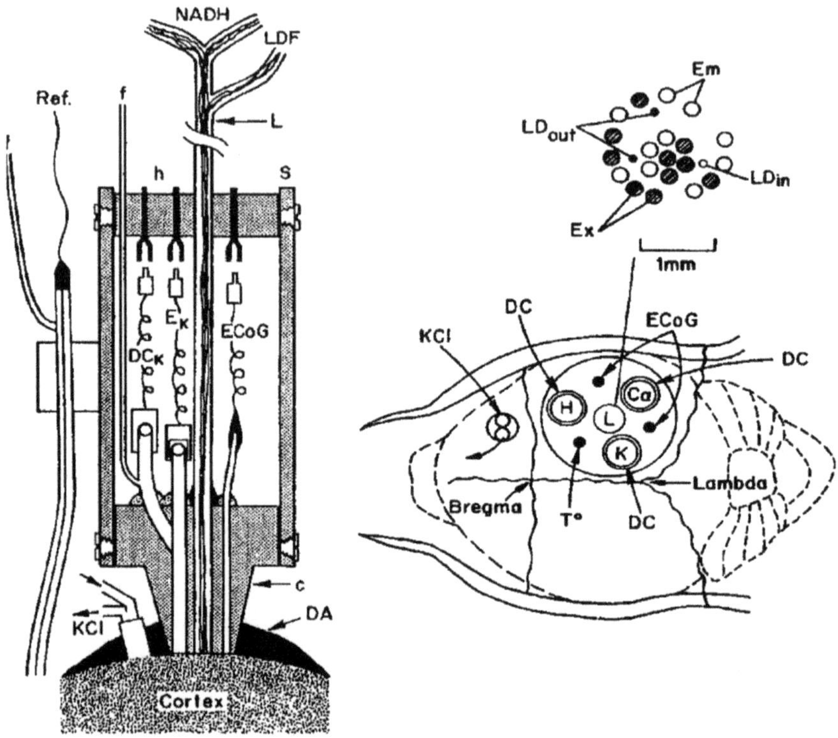

Fig. 9.11 Multiprobe assembly used in awake rat to monitor hemodynamic, metabolic, ionic, and electrical activities from the brain. Left: longitudinal section. Right: location of multiprobe assembly on the skull. Ref reference electrode, f feeling tube of reference or other direct current electrodes, h connectors holder, NADH two arms of NADH monitoring light guide, LDF laser Doppler flowmeter, L light guide, s aluminum sleeve, DCK and EK direct current and ion specific electrodes, respectively, for K, ECoG electrocortical electrodes, c Plexiglas probe holder, DA dental acrylic cement, KCl push-pull cannula for KCl, K, Ca, and H ion specific electrodes, DC area of direct current steady potential monitoring, Ex and Em excitation and emission fibers, respectively, for NADH monitoring, LD_{in} and LD_{out} optical fibers for monitoring blood flow and volume, respectively, To thermistor for local temperature measurement (Mayevsky et al. 1995; Meilin et al. 1996)

brain or the formation of an artificial environment around the sensor that can be created around a penetrating microelectrode.

9.2.7 *Multiparametric Monitoring of Neurosurgical Patients*

The new era in monitoring of mitochondrial function in patients started in 1990, when our team developed a unique multiparametric monitoring system that included the measurement of NADH fluorescence, using a light-guide-based device (Mayevsky et al. 1991).

This system was initially applied to monitor neurosurgical patients undergoing brain surgery. In the second half of the 1990s (Mayevsky et al. 1996, 1998b; Meilin et al. 1999a), we were able to monitor the brain of comatose patients in the ICU and described, for the first time, the development of cortical spreading depression-like responses, including marked NADH responses that were correlated to the micro-circulatory blood flow. A modified multiparametric monitoring system was applied for NADH measurement during neurosurgical procedures in the operating room (Mayevsky et al. 1998a, 1999, 2001, 2002).

Based on the well-developed MultiProbe Assembly (MPA) used for animal experiments (Mayevsky et al. 1995), a new MPA was developed and applied to patients monitored in the neurosurgical operation rooms and ICU. This device allowed us to monitor, in real-time, the hemodynamic, metabolic ionic and electrical activities in the brain of comatose patients. All details regarding the technology and the clinical setup appear in our published paper (Mayevsky et al. 1996). Fig. 9.12a shows the longitudinal section of the MPA (measuring eight parameters) used in the neurosurgical operating room and ICU. In Fig. 9.6a a simplified multiprobe that monitors up to five parameters is shown. The MPA was connected to the multiparametric monitoring system shown in Fig. 9.12b.

9.2.8 The Use of MPA Inside NMR Magnet

We developed a new multi-probe assembly, which can be used inside the magnet, such that the hemodynamic, metabolic, electrical and ionic activities could be monitored continuously and simultaneously with 31P-MRS (Mayevsky et al. 1997). Relative CBF was measured by a Laser Doppler flowmeter using a fiber optic probe (Haberl et al. 1989). Mitochondrial activity was evaluated by monitoring the NADH redox state using surface fluorometry (Mayevsky and Chance 1982; Mayevsky 1984; Chance et al. 1962a). Ionic homeostasis was evaluated by the extracellular levels of K+ and Ca^{2+} measured by surface ion-selective mini-electrodes (Mayevsky et al. 1997; Crowe et al. 1977). The electrical activities were assessed by the DC steady potential as well as by the electro-corticogram (ECoG). Schematic presentation of the monitoring system is shown in Fig. 9.13.

This study was carried out in order to determine whether it is possible to measure, simultaneously and in real time, parameters that are measurable by the multiprobe assembly with parameters that are measurable using NMR spectroscopy so as to more precisely determine injury during pathological interventions or conditions.

The K^+ and Ca^{2+} electrodes (WPI Inc., USA), DC electrodes, ECoG electrodes as well as the light guide were held in a multiprobe assembly as described previously in detail (Friedli et al. 1982; Mayevsky 1983). The probes were mounted in a lucite holder (cannula), which can be cemented directly to the skull and can be removed very easily from the skull at the end of the experiment without any damage, so that the same electrodes can be used for performing repetitive experiments within a short period of time.

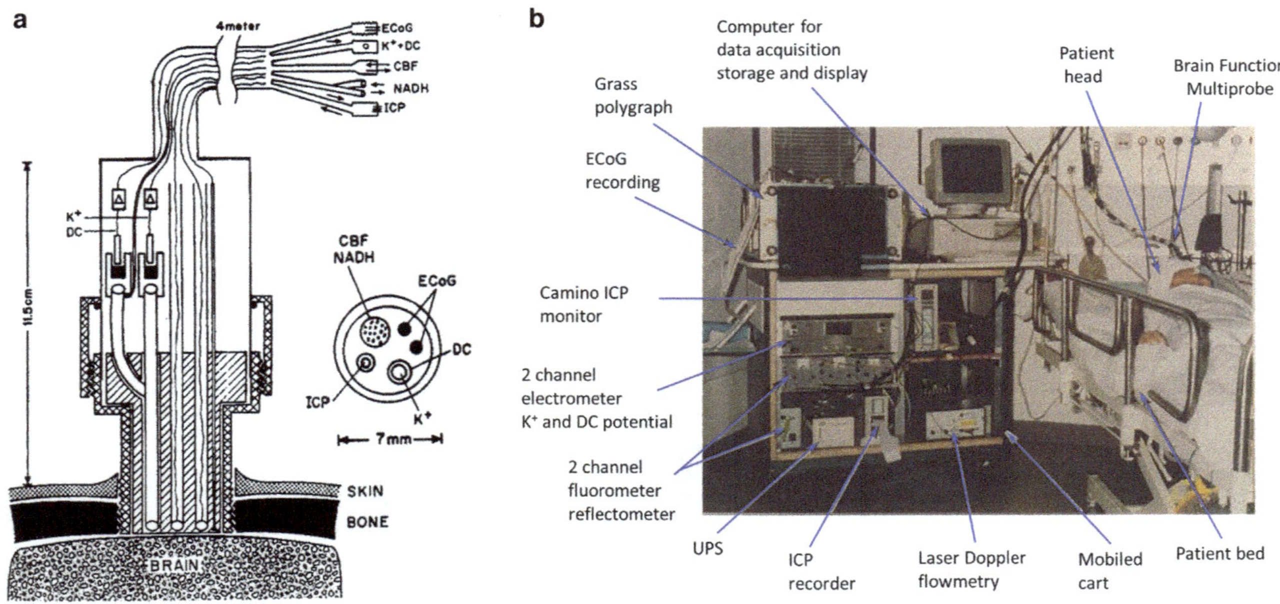

Fig. 9.12 (**a**) Schematic representation of the multiprobe assembly (MPA) used for monitoring the brain of head-injured patients in the ICU. The MPA is connected to the brain via a special holder screwed into the skull. ICP intracranial pressure probe, CBF NADH-fiber optic light guide probe to measure local cerebral blood flow (CBF) and mitochondrial redox state (NADH), K^+, DC extracellular K^+ mini surface-electrode surrounded by a DC steady potential monitoring space, ECoG bipolar electrocortical electrodes (Mayevsky et al. 1996). (**b**) The first clinical monitoring setup ("Brain Monitor") was used in the neurosurgical intensive care unit. The multiparametric monitoring system consists of various devices installed in the same cart. The MPA shown in Part A is connecting the brain of the patient to the Brain monitor (Mayevsky and Barbiro-Michaely 2013)

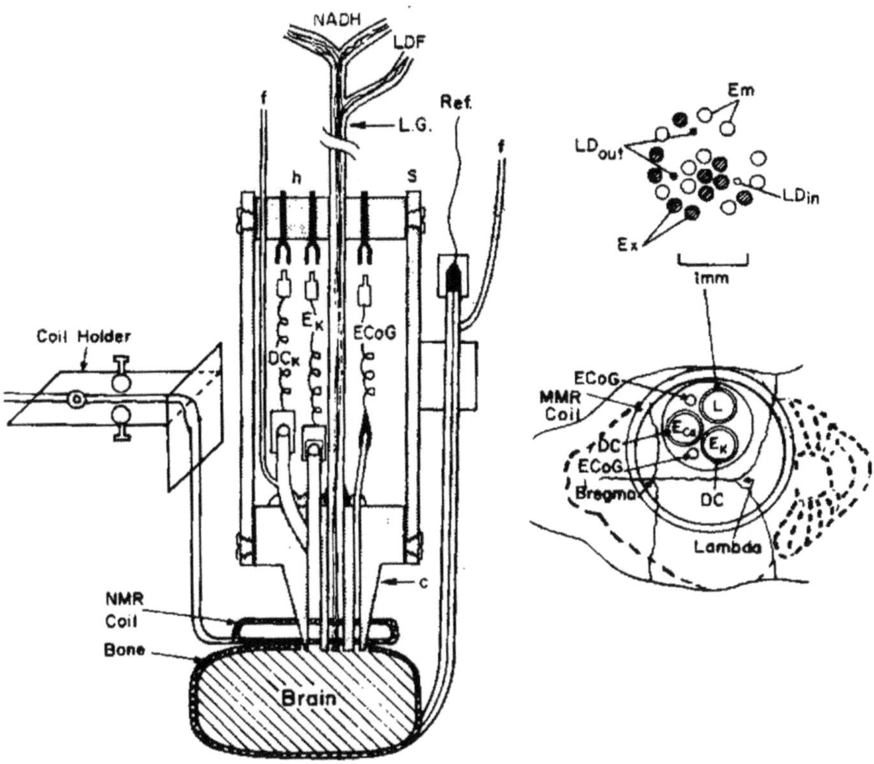

Fig. 9.13 Schematic presentation of the experimental setup used inside the phosphorous NMR magnet. A longitudinal section of the multiprobe assembly connected to the NMR coil and the gerbil brain is shown on the left side of Part (**a**). The relative locations of the various probes on the brain are shown on the right side. ECoG electrocorticographic electrodes, EK^+, ECa^{2+} DCK^+ potassium, calcium, and DC electrode, c Plexiglas multiprobe holder', h connection holder, s Plexiglas sleeve, Ref reference electrode, f filling tube of DC or reference electrode; L.G. light guide, Ex, Em excitation and emission fibers for NADH monitoring, LD_{in} LD_{out} fibers for CBF monitoring by Laser Doppler flow-meter (Mayevsky et al. 1997)

9.2.9 *Propagation of CSD Wave*

The aim of this study was to identify the sequence of responses developed under the CSD phenomenon, initiated during various levels of brain oxygenation: normoxia, hypoxia, and partial ischemia (Sonn and Mayevsky 2000).

The probes were embedded in a Delrin rectangular cannula (6.4 × 3.3 mm) (Fig. 9.14). It was found that the probes should be arranged on a circumference where the KCl cannula (CSD initiation area) is located in the center of the circle. This enabled the probes to be arranged according to the shape of the CSD wave determined and allowed the CSD wave to be monitored from its front line (by all the detected

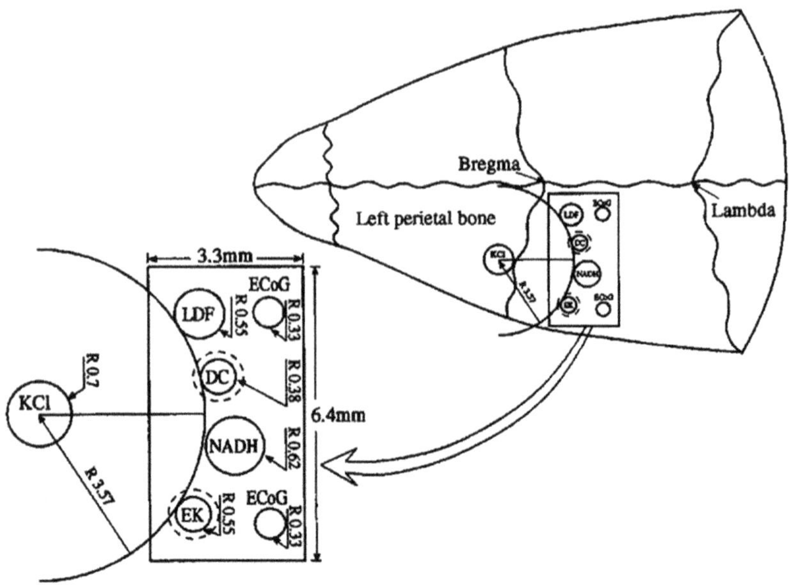

Fig. 9.14 Schematic presentation of the special multiprobe assembly (MPA) developed in order to study the timing of the responses to cortical spreading depression. The right side shows the MPA located on the cerebral cortex. The localization of the various sensors is shown on the left side of the figure (**a**). LDF light guide connected to the laser Doppler flowmeter, NADH light guide connected to the fluorometer/reflectometer for NADH fluorescence and 366 nm reflectance, EK selective K^+ electrode, DC direct current, KCl push-pull cannula, ECoG electrocorticography electrodes. The R values represent the radius of the particular probes (Sonn and Mayevsky 2000)

parameters) until complete recovery. The MPA contained: surface mini-electrodes for measuring $[K^+]e$ and DC steady potential, fiber optic probes that were used to monitor intramitochondrial NADH redox state (NADH fluorescence) and reflected light, as well as relative CBF and CBV by laser Doppler flowmetry and spontaneous bipolar ECoG via two platinum electrodes. All details regarding the various probe construction and calibration were published previously (Mayevsky 1992, 1993).

9.2.10 Addition of ICP Probe to the MPA

Monitoring of brain functions during ICP (intracranial pressure) elevation (Barbiro-Michaely and Mayevsky 2001) was performed by the MPA developed in our laboratory (Meilin et al. 1996, 1999b; Friedli et al. 1982; Mayevsky et al. 1995; Mayevsky 1993). The MPA (Fig. 9.15a) includes a bundle of optical fibers for NADH redox state measurement by surface fluorometry as well as for CBF using laser Doppler flowmetry and Camino transducer for ICP monitoring. Three mini-electrodes were used

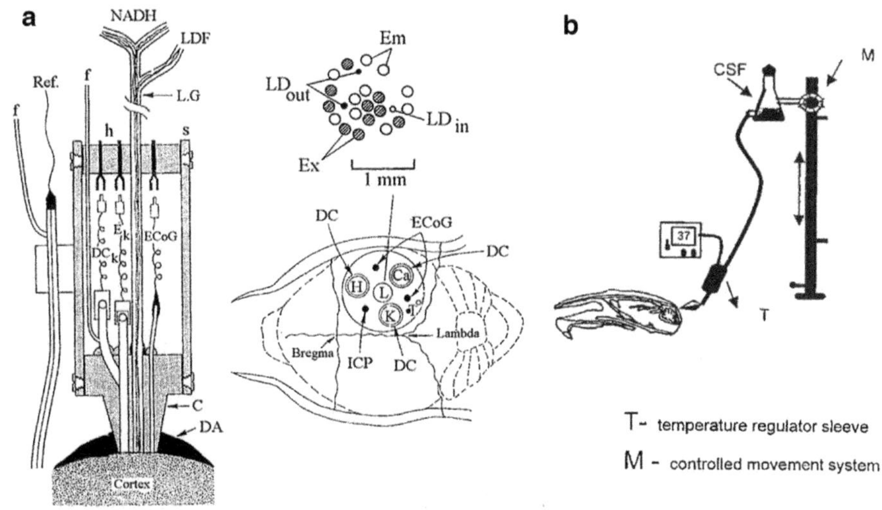

Fig. 9.15 The Multiprobe Assembly (MPA) used in order to study the effects elevated Intracranial pressure (ICP) in the rat brain. (**a**) The MPA includes optic fibers for NADH redox state fluorometry (Em Emission; Ex Excitation), laser Doppler flowmetry ($LD_{out}/_{in}$), and ICP monitoring. The mini-electrodes for the measurement of the extracellular levels of K^+, Ca^{2+}, and H^+ were surrounded by DC electrode, brain temperature thermistor (T), and two ECoG electrodes. (**b**) The model for intracranial pressure elevation: a controlled movement system (M) enables the elevation and reduction of a container full with mock CSF, which is connected to the cisterna magna space. The mock CSF is being warmed to a temperature of 37°C by a temperature regulator sleeve (T) (Qian et al. 1996)

for the measurement of the extracellular levels of K^+, Ca^{2+}, and H^+, each surrounded by a DC electrode. Brain temperature thermistor and two ECoG electrodes were also located inside the MPA. None of the probes, including the ICP probe, penetrates the brain tissue, but rather they are located on the surface of the brain after the removal of the dura mater.

9.2.11 The Use of MPA in Traumatic Brain Injury

To clarify these issues, the work was carried out in rats in order to study the dynamic relationship between the level of intracranial pressure (ICP) and the number of CSD cycles during the early post-traumatic brain injury period under various levels of TBI severity (Rogatsky et al. 2003).

TBI was induced by the lateral fluid-percussion brain injury (FPBI) technique (McIntosh et al. 1989; Sullivan et al. 1976; Dixon et al. 1987; Rogatsky et al. 1996). The magnitudes of the hydraulic impact with the Peak Pressure Pulses (PPP) were 2.9, 3.3, 4.1, and 5.0 atmospheres (ATM)–groups A, B, C, and D, respectively, which were applied to the left parietal cortex (Fig. 9.16). More details concerning this technique can be found in our previous publication (Rogatsky et al. 1996).

9.3 From Brain Physiological Mapping to Brain Metabolic Score (BMS)

The need for new approaches in evaluating the clinical status of patients in real time is very wide. For example, the interpretation of the physiological status of anesthetized patients is very subjective and therefore may be evaluated differently by various anesthesiologists. The same problem is demonstrated in patients hospitalized in the various Operating Rooms or Intensive Care Units that need additional monitoring approaches to evaluate objectively their condition. As of today, there is a discrepancy between the large amount of data collected by various techniques and the ability to integrate it into a practical objective tool to be used in clinical practice. Our effort to apply our developed multiparametric system to clinical daily use required the development of a new indexing technique for data interpretation. After the accumulation of huge amounts of experimental results, we decided to develop a preliminary tool named Tissue Metabolic Score–TMS. The TMS was calculated by using two parameters. In the present study, we are presenting the preliminary developed Brain Metabolic Score BMS that could be used in evaluating the metabolic state of the cerebral cortex in real time. We used the brain as a typical organ for the calculation of the Score, but the same approach could be used for other organs or tissues. The ultimate goal of our study is to develop the Tissue Metabolic Score that could be monitored in the intensive care units or in the operating rooms. In those medical environments the need

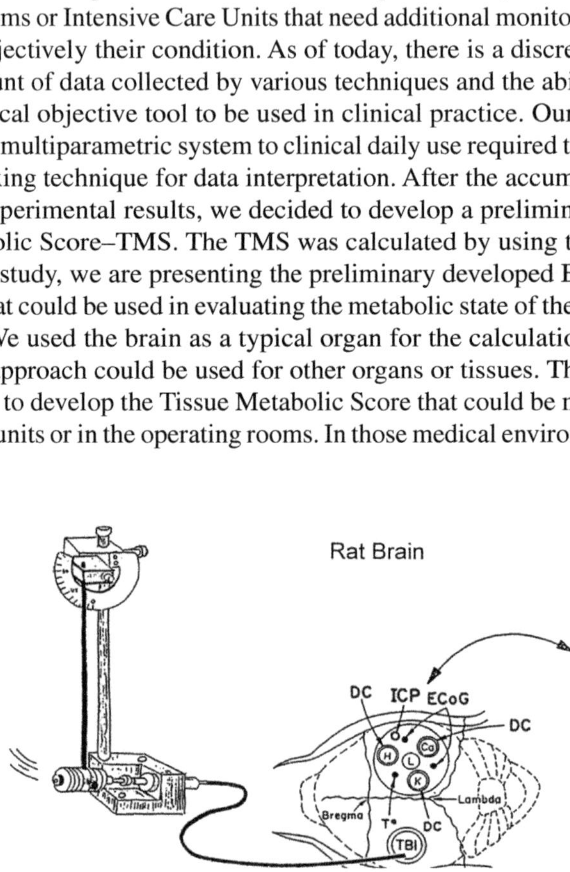

Fig. 9.16 Schematic presentation of the fluid percussion impactor connected to the rat brain together with the multiprobe assembly (MPA) used for monitoring the brain. DC steady DC potential, ECoG electrocorticography, ICP intracranial pressure probe, K, Ca, H potassium, calcium, and hydrogen electrodes, L fiber optic light guide probe to measure local cerebral blood flow (CBF) and mitochondrial redox state (NADH), TBI cannula for inducing traumatic brain injury (Rogatsky et al. 2003)

for objective approach decision making processes is very important as discussed by Hanson and Marshall (2001) and by Bates and Young (2003). In order to develop the BMS, we used the multiparametric monitoring system and the development of the score appeared in our publication (Kutai-Asis et al. 2006).

In the present study the developed new Brain Metabolic Score–BMS based on the monitoring of three physiological parameters is presented. The multiparametric monitoring system provides simultaneously information on microcirculatory blood flow and hemoglobin oxygenation as well as mitochondrial function. Under various pathophysiological conditions, these three parameters change continuously and the expert is able to evaluate the vitality level of the tissue. In order to provide an objective evaluation on the state of the tissue, it was important to develop the BMS. Few techniques are available in order to calculate and evaluate the changes in various physiological and biochemical parameters related to the metabolic activity of the tissue. In most cases, the monitoring device provides information on a single parameter.

Our results demonstrated that the most accurate BMS is calculated when NADH is represented with the highest percentage–80%, and CBF and HbO_2 and reflectance are presented together with about 20%. Under ischemia, the calculated BMS decreases in two Steps (1 and 2 in the lowest trace) and recover to preischemic level at 3. As seen under anoxia (right side), the hyperemia which probably developed in order to compensate for the deficit in oxygen is not affecting the BMS. Moreover, mitochondrial NADH levels fully recovered as the rat started to breathe air, a few seconds before CBF reached its maximum hyperemia. In other words, the hyperemia seen following anoxia is actually not essential for adequate O_2 supply to the mitochondria in the brain cortex.

It should be clarified that this hyperemia does not improve tissue vitality state but rather is a result of microvessels contractility state. During ischemia, cortical arteries are vasoligated in order to improve blood supply to the impacted cerebral tissue. In order to restore vasculature tone to its pre-occlusion state, several seconds are needed. Meanwhile, vessels are relatively vasodilated compared to their normal state. Moreover, here also the hyperemia is needless, since full recovery of tissue energetic state (NADH) was accomplished immediately as arterial obstructions were removed, namely before full hyperemia was achieved. Since the main purpose of a BMS is to give an indication for the state in which a tissue or an organ is in, and since oxygen is a major factor in tissue viability, it is obvious that this Score should include the combination between tissue oxygen supply and demand. The mitochondrion is the cell energy factory (Waltemath 1970), and mitochondrial NADH is the most sensitive indicator for oxygen balance (Mayevsky et al. 1980, 1992, 1998b). Thus, we believe NADH should get the highest importance in a formula for calculating the BMS.

Figure 9.17 presents the responses to SD in a rat model at the upper part and in patients at the lower part. When we induced SD in the animal study, we were sure about the oxygen supply to the animal. The blood supply to the brain decreased between parts a and b. The shift in response to CSD from the normoxic (a) to the ischemic (b) state is clear as discussed before (Mayevsky 1978). When we move to patients' monitoring, we want to use the NADH trace as an indicator to the physiopathological state of the patient or the brain in this example. As seen in the lower part, the two responses developed in the neurosurgical patients (c and d) are not the same. There

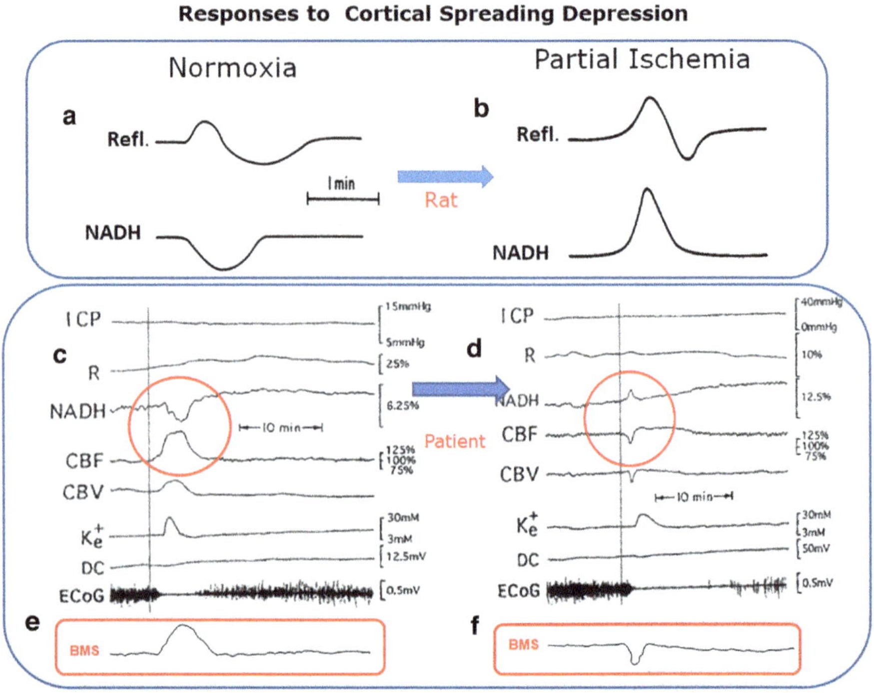

Fig. 9.17 The responses to cortical spreading depression (CSD) measured in normoxic (**a**) and ischemic brain (**b**) in the rat model. Parts (**c**) and (d) were measured in head injured patient. (**e** and **f**) Brain metabolic score (BMS) calculated for the normoxic (**e**) and partial Ischemic (**f**) human brain responses to CSD. TBF cerebral blood flow, R376 Reflectance, NADH mitochondrial NADH redox state, HbO_2 oxygenation of hemoglobin

are similarities between a and c as well as between b and d. Therefore, we concluded that the brain of the patient in state c was probably normoxic while in d the brain was partially ischemic. Indeed, this patient was deteriorating in time, and finally, brain death was determined (Mayevsky et al. 1996). This example demonstrates that the basic pathophysiological events developed in patients present very similar results as compared to those monitored in animal models. The calculated BMS traces, measured from the patient shown in Part c and d are presented in Part e and f and the difference in the responses to CSD is very obvious.

References

Aubert, X., B. Chance, and R. D. Keynes. 1964. Optical studies of biochemical events in the electric organ of Electrophorus. *Proceedings of the Royal Society* 160: 211–245.

Barbiro-Michaely, E., and A. Mayevsky. 2001. Multiparametric monitoring of brain under elevated intracranial pressure in a rat model. *Journal of Neurotrauma* 18: 711–725.

Barbiro-Michaely, E., A. Mendelman, and A. Mayevsky. 2011. The evaluation of nitric oxide involvement in metrazol induced status epilepticus using multiparametric monitoring. *Brain Research* 1377: 50–59.

Bates, J. H., and M. P. Young. 2003. Applying fuzzy logic to medical decision making in the intensive care unit. *American Journal of Respiratory and Critical Care Medicine* 167: 948–952.

Brauser, B., T. Bucher, and M. Dolivo. 1970. Redox transitions of cytochromes and pyridine nucleotides upon stimulation of an isolated rat ganglion. *FEBS Letters* 8: 297–300.

Chance, B., P. Cohen, F. Jobsis, and B. Schoener. 1962a. Intracellular oxidation-reduction states *in vivo*. *Science* 137: 499–508.

Chance, B., V. Legallias, and B. Schoener. 1962b. Metabolically linked changes in fluorescence emission spectra of cortex of rat brain, kidney and adrenal gland. *Nature* 195: 1073–1075.

Chance, B., D. Mayer, and L. Rossini. 1970. A time-sharing instrument for direct readout of oxidation- reduction states in intracellular compartments of cardiac tissue. *IEEE Transactions on Bio-Medical Engineering, BME* 17: 118–121.

Chance, B., N. Oshino, T. Sugano, and A. Mayevsky. 1973a. Basic principles of tissue oxygen determination from mitochondrial signals. *Advances in Experimental Medicine and Biology* 37A: 277–292.

Chance, B., N. Oshino, T. Sugano, and A. Mayevsky. 1973b. Basic principles of tissue oxygen determination from mitochondrial signals. *International Symposium on Oxygen Transport to Tissue, Advances in Experimental Medicine and Biology* 37A: 277–292.

Chance, B., and B. Schoener. 1963a. *Symp. on oxygen in animal organism, abstract.* 1972/02/01 ed.

Chance, B., and B. Schoener. 1963b. Control of oxidation-reduction state of NADH in the liver of anesthetized rats. *Symphony Regul Enzymes Acts Synthesizer Norman Neoplastic Tissue Procedure*, 169–185.

Chance, B., B. Schoener, and F. Schindler F. 1963. Tracellular oxidation-reduction state. In: Dickens F and Neil E (eds). Oxygen In The Animal Organism. London: Pergamon Press.

Crowe, W., A. Mayevsky, and L. Mela. 1977. Application of a solid membrane ion selective electrode to *in vivo* measurements. *The American Journal of Physiology* 233: C56–C60.

Davies, P. W., and A. Remond. 1946. Oxygen consumption of the cerebral cortex of the cat during metrazole convulsions. *Research publications - Association for Research in Nervous and Mental Disease* 26: 205–217.

Dixon, C. E., B. G. Lyeth, J. T. Povlishock, R. L. Findling, R. J. Hamm, A. Marmarou, H. F. Young, and R. L. Hayes. 1987. A fluid percussion model of experimental brain injury in the rat. *Journal of Neurosurgery* 67: 110–119.

Frank, K. H., M. Kessler, K. Appelbaum, and W. Dummler. 1989. The Erlangen micro-lightguide spectrophotometer EMPHO I. *Physics in Medicine and Biology* 34: 1883–1900.

Friedli, C. M., D. S. Sclarsky, and A. Mayevsky. 1982. Multiprobe monitoring of ionic, metabolic, and electrical activities in the awake brain. *The American Journal of Physiology* 243: R462–R469.

Haberl, R. L., M. L. Heizer, A. Marmarou, and E. F. Ellis. 1989. Laser-Doppler assessment of brain microcirculation: Effect of systemic alterations. *American Journal of Physiology - Heart and Circulatory Physiology* 256: H1247–H1254.

Hanson 3rd, C. W., and B. E. Marshall. 2001. Artificial intelligence applications in the intensive care unit. *Critical Care Medicine* 29: 427–435.

Harbig, K., B. Chance, A. G. B. Kovach, and M. Reivich. 1976. *In vivo* measurement of pyridine nucleotide fluorescence from cat brain cortex. *Journal of Applied Physiology: Respiratory, Environmental and Exercise Physiology* 41: 480–488.

Harbig, K., and M. Rievich. 1973. *Stroke* 4: 341. (Abstract VI-43).

Jobsis, F. F., M. O'Connor, A. Vitale, and H. Vreman. 1971. Intracellular redox changes in functioning cerebral cortex. I. Metabolic effects of epileptiform activity. *Journal of Neurophysiology* 3465: 735–749.

Keilin, D. 1966. *The history of cell respiration and cytochrome*. Cambridge, UK: Cambridge University Press.

Kobayashi, S., K. Kaede, K. Nishiki, and E. Ogata. 1971. Microfluorometry of oxidation-reduction state of the rat kidney *in situ*. *Journal of Applied Physiology: Respiratory, Environmental and Exercise Physiology* 31: 693–696.

Kutai-Asis, H., E. Barbiro-Michaely, A. Deutsch, and A. Mayevsky. 2006. Fiber optic based multiparametric spectroscopy in vivo: Toward a new quantitative tissue vitality index. *SPIE Proceedings*, 6083.

Lukyanov, L. D., and J. Bures. 1967. Changes in pO_2 due to spreading depression in cortex and nucleus caudatus of rat. *Physiologia Bohemoslovenica* 16: 449–455.

Mayevsky, A. 1975. The effect of trimethadione on brain energy metabolism and EEG activity of the conscious rat exposed to HPO. *Journal of Neuroscience Research* 1: 131–142.

Mayevsky, A. 1978. Ischemia in the brain: The effects of carotid artery ligation and decapitation on the energy state of the awake and anesthetized rat. *Brain Research* 140: 217–230.

Mayevsky, A. 1983. Multiparameter monitoring of the awake brain under hyperbaric oxygenation. *Journal of Applied Physiology: Respiratory, Environmental and Exercise Physiology* 54: 740–748.

Mayevsky, A. 1984. Brain NADH redox state monitored *in vivo* by fiber optic surface fluorometry. *Brain Research Reviews* 7: 49–68.

Mayevsky, A. 1988. Microcirculatory and ionic responses to ischemia in the Mongolian gerbil. In MANABE, H., ZWEIFACH, B. W. & MASSMER, K. (eds.). Microcirculation in Circulatory Disorders. Springer-Verlag.

Mayevsky, A. 1992. Cerebral blood flow and brain mitochondrial redox state responses to various perturbations in gerbils. *Advances in Experimental Medicine and Biology* 317: 707–716.

Mayevsky, A. 1993. Biochemical and physiological activities of the brain as *in vivo* markers of brain pathology. *In* BERNSTEIN, E. F., CALLOW, A. D., NICOLAIDES, A. N. & SHIFRIN, E. G. (eds.) *Cerebral revascularization* Med-Orion Pub. London: Los Angeles, Nicosia

Mayevsky, A., and E. Barbiro-Michaely. 2013. Shedding light on mitochondrial function by real time monitoring of NADH fluorescence: II: Human studies. *Journal of Clinical Monitoring and Computing* 27: 125–145.

Mayevsky, A., and B. Chance. 1973a. A new long-term method for the measurement of NADH fluorescence in intact rat brain with implanted cannula. *Advances in Experimental Medicine and Biology* 37A: 239–244.

Mayevsky, A., and B. Chance. 1973b. A new long-term method for the measurement of NADH fluorescence in intact rat brain with implanted cannula. *In* BICHER, H. I. & BRULEY, D. F., eds, 239–244. Adv. Exp. Med. Biol. New York: Plenum Press.

Mayevsky, A., and B. Chance. 1982. Intracellular oxidation-reduction state measured *in situ* by a multicannel fiber-optic surface fluorometer. *Science* 217: 537–540.

Mayevsky, A., A. Doron, T. Manor, S. Meilin, N. Zarchin, and G. E. Ouaknine. 1996. Cortical spreading depression recorded from the human brain using a multiparametric monitoring system. *Brain Research* 740: 268–274.

Mayevsky, A., A. Doron, S. Meilin, T. Manor, E. Ornstein, and G. E. Ouaknine. 1999. Brain viability and function analyzer: Multiparametric real-time monitoring in neurosurgical patients. *Acta Neurochirurgica. Supplement (Wien.)* 75: 63–66.

Mayevsky, A., E. S. Flamm, W. Pennie, and B. Chance. 1991. A fiber optic based multiprobes system for intraoperative monitoring of brain functions. *SPIE* 1431: 303–313.

Mayevsky, A., K. Frank, M. Muck, S. Nioka, M. Kessler, and B. Chance. 1992. Multiparametric evaluation of brain functions in the Mongolian gerbil *in vivo*. *Journal of Basic and Clinical Physiology and Pharmacology* 3: 323–342.

Mayevsky, A., K. H. Frank, S. Nioka, M. Kessler, and B. Chance. 1990. Oxygen supply and brain function *in vivo*: a multiparametric monitoring approach in the Mongolian gerbil. In *Oxygen transport to tissue XII* , PIPER, J., GOLDSTICK, T. K and MEYER, M. (eds) Plenum Pub New York.

Mayevsky, A., S. Lebourdais, and B. Chance. 1980. The interrelation between brain PO_2 and NADH oxidation-reduction state in the gerbil. *Journal of Neuroscience Research* 5: 173–182.

Mayevsky, A., T. Manor, S. Meilin, A. Doron, and G. E. Ouaknine. 1998a. Real-time multiparametric monitoring of the injured human cerebral cortex – A new approach. In *Acta neurochirurgica. Supplement (Wien.)*, MARMAROU, R and BULLOCK, R. (eds) ICPX. Wien

Mayevsky, A., T. Manor, S. Meilin, N. Razon, and G. E. Ouaknine. 2001. Multiparametric monitoring of tissue vitality in clinical situations. *SPIE* 4255: 33–39.

Mayevsky, A., S. Meilin, T. Manor, E. Ornstein, N. Zarchin, and J. Sonn. 1998b. Multiparametric monitoring of brain oxygen balance under experimental and clinical conditions. *Neurological Research* 20: S76–S80.

Mayevsky, A., S. Meilin, G. G. Rogatsky, N. Zarchin, and J. Sonn. 1995. Multiparametric monitoring of the awake brain exposed to carbon monoxide. *Journal of Applied Physiology: Respiratory, Environmental and Exercise Physiology* 78: 1188–1196.

Mayevsky, A., S. Nioka, D. J. Wang, and B. Chance. 1997. The functioning gerbil brain *in vivo*. Correlation between 31P NMR spectroscopy and the multiparametric monitoring approach. *Advances in Experimental Medicine and Biology* 411: 41–53.

Mayevsky, A., E. Ornstein, S. Meilin, N. Razon, and G. E. Ouaknine. 2002. The evaluation of brain CBF and mitochondrial function by a fiber optic tissue spectroscope in neurosurgical patients. *Acta Neurochirurgica. Supplement* 81: 367–371.

Mayevsky, A., and G. Rogatsky. 2007. Mitochondrial function *in vivo* evaluated by NADH fluorescence: From animal models to human studies. *American Journal of Physiology - Cell Physiology* 292: C615–C640.

Mayevsky, A., T. Zeuthen, and B. Chance. 1974. Measurements of extracellular potassium, ECoG and pyridine nucleotide levels during cortical spreading depression in rats. *Brain Research* 76: 347–349.

McIntosh, T. K., R. Vink, L. Noble, I. Yamakami, S. Fernyak, H. Soares, and A. L. Faden. 1989. Traumatic brain injury in the rat: Characterization of a lateral fluid-percussion model. *Neuroscience* 28: 233–244.

Meilin, S., A. Mendelman, J. Sonn, T. Manor, N. Zarchin, and A. Mayevsky. 1999a. Metabolic and hemodynamic oscillations monitored optically in the brain exposed to various pathological states. *Advances in Experimental Medicine and Biology* 471: 141–146.

Meilin, S., G. G. Rogatsky, S. R. Thom, N. Zarchin, E. Guggenheimer-Furman, and A. Mayevsky. 1996. Effects of carbon monoxide exposure on the brain may be mediated by nitric oxide. *Journal of Applied Physiology: Respiratory, Environmental and Exercise Physiology* 81: 1078–1083.

Meilin, S., N. Zarchin, and A. Mayevsky. 1999b. Inter-relation between hemodynamic, metabolic, ionic and electrical activities during ischemia and reperfusion in the gerbil brain. *Neurological Research* 21: 699–704.

Qian, L., K. Ohno, T. Maehara, B. Tominaga, K. Hirakawa, T. Kuroiwa, K. Takakuda, and H. Miyairi. 1996. Changes in lCBF, morphology and related parameters by fluid percussion injury. *Acta Neurochirurgica* 138: 90–98.

Rogatsky, G. G., A. Mayevsky, N. Zarchin, and A. Doron. 1996. Continuous multiparametric monitoring of brain activities following fluid-percussion injury in rats: Preliminary results. *Journal of Basic and Clinical Physiology and Pharmacology* 7: 23–43.

Rogatsky, G. G., J. Sonn, Y. Kamenir, N. Zarchin, and A. Mayevsky. 2003. Relationship between intracranial pressure and cortical spreading depression following fluid percussion brain injury in rats. *Journal of Neurotrauma* 20: 1315–1325.

Rosenthal, M., and F. F. Jobsis. 1971. Intracellular redox changes in functioning cerebral cortex. II. Effects of direct cortical stimulation. *Journal of Neurophysiology* 34: 750–762.

Sonn, J., and A. Mayevsky. 2000. Effects of brain oxygenation on metabolic, hemodynamic, ionic and electrical responses to spreading depression in the rat. *Brain Research* 882: 212–216.

Sullivan, H. G., J. Martinez, D. P. Becker, J. D. Miller, R. Griffith, and A. O. Wist. 1976. Fluid-percussion model of mechanical brain injury in the cat. *Journal of Neurosurgery* 45: 521–534.

Van Harreveld, A., and S. Ochs. 1957. Electrical and vascular concomitants of spreading depression. *The American Journal of Physiology* 189: 159–166.

Waltemath, C. L. 1970. Oxygen, uptake, transport, and tissue utilization. *Anesthesia and Analgesia* 49: 184–203.

Warburg, O. 1949. Heavy metal prosthetic groups and enzyme action. In: Lawson TA (ed). Heavy metal prosthetic groups and enzyme is the name of the book written by otto Warburg. Oxford: Clarendon Press.

Index

A. Mayevsky et al., *Physiological Mapping of the Cerebral Cortex in Neuropathology*,
https://doi.org/10.1007/978-3-032-13019-8

GPSR Compliance
The European Union's (EU) General Product Safety Regulation (GPSR) is a set of rules that requires consumer products to be safe and our obligations to ensure this.

If you have any concerns about our products, you can contact us on

ProductSafety@springernature.com

In case Publisher is established outside the EU, the EU authorized representative is:

Springer Nature Customer Service Center GmbH
Europaplatz 3
69115 Heidelberg, Germany

www.ingramcontent.com/pod-product-compliance
Ingram Content Group UK Ltd.
Pitfield, Milton Keynes, MK11 3LW, UK
UKHW020933180926
475692UK00003B/14
* 9 7 8 3 0 3 2 1 3 0 1 8 1 *